Heavy Flavours and
High-Energy Collisions
in the 1-100 TeV Range

ETTORE MAJORANA
INTERNATIONAL SCIENCE SERIES
Series Editor:
Antonino Zichichi
European Physical Society
Geneva, Switzerland

(PHYSICAL SCIENCES)

Recent volumes in the series:

A Continuation Order Plan is available for this series. A continuation order will bring delivery of
each new volume immediately upon publication. Volumes are billed only upon actual shipment.
For further information please contact the publisher.

Heavy Flavours and High-Energy Collisions in the 1-100 TeV Range

Edited by

A. Ali

DESY
Hamburg, Federal Republic of Germany

and

L. Cifarelli

CERN
Geneva, Switzerland
and University of Naples
Naples, Italy

Springer Science+Business Media, LLC

Library of Congress Cataloging in Publication Data

Heavy flavours and high-energy collisions in the 1–100 TeV range / edited by A. Ali and L. Cifarelli.
    p.   cm.—(Ettore Majorana international science series. Physical sciences; v. 44)
    Based on the proceedings of the 6th and 7th INFN ELOISATRON Project workshops held at the Centro di Cultra Scientifica Ettore Majorana, Erice, Italy.
    Includes bibliographical references.
    ISBN 978-1-4684-9983-4   978-1-4684-9981-0 (eBook)
    DOI 10.1007/978-1-4684-9981-0
    1. Heavy particles (Nuclear physics)—Congresses. 2. Quantum flavor dynamics—Congresses. 3. Collisions (Nuclear physics)—Congresses. 4. Proton beams—Congresses. 5. INFN ELOISATRON Project—Congresses. I. Ali, A. (Ahmed) II. Cifarelli, L. (Luisa) III. INFN ELOISATRON Project. IV. Series.
QC793.H43   1989   89-22891
539.7'2167—dc20   CIP

Based on the proceedings of the Sixth and Seventh INFN
ELOISATRON Project Workshops on
Heavy Flavours: Status and Perspectives; and
Novel Features of High-Energy Collisions in the 1–100 TeV Region,
respectively, held June 10–27, 1988, in Erice, Sicily, Italy

© 1989 Springer Science+Business Media New York
Originally published by Plenum Press, New York in 1989.
Softcover reprint of the hardcover 1st edition 1989
A Division of Plenum Publishing Corporation
233 Spring Street, New York, N.Y. 10013

Preface

The present volume is based on the proceedings of the 6*th* and 7*th* INFN ELOISATRON project workshops, held at the Centro di Cultura Scientifica "Ettore Majorana" CCSEM, Erice–Trapani, Sicily, Italy, in the period June 10–27, 1988. The topics of the two workshops were, respectively:

- Heavy Flavours: Status and Perspectives, and

- Novel Features of High Energy Collisions in 1–100 TeV Region.

They were attended by sixty-three physicists. The two workshops were followed by a meeting of the INFN ELOISATRON working group, also held at the CCSEM in the period October 7–15, 1988 in which twenty-five physicists participated. Since there was quite a bit of overlap among speakers, participants and the topics covered at the three meetings, we have decided to issue a joint proceeding, with the first part entitled: Heavy Flavour Physics, and the second: High Energy Physics with 1–100 TeV Proton Beams. Some of the reports included in this volume have been contributed by the INFN ELOISATRON working group members.

The first part of these proceedings deals mostly with the presentation and interpretation of results in the so-called flavour physics sector. New results, which have become available in the last three years from experiments involving kaons, charmed and beauty hadrons, and searches for the still missing top quark at the present and fothcoming colliders are topics of major interest here. The contributions in this part are organized in three categories: Experimental Results, Theoretical Interpretation, and Future Directions.

The field of flavour physics has made rapid progress in the last several years. Though a lot of consolidation work is continuing in the charmed hadron sector providing a wealth of data and their theoretical interpretation, qualitatively new results have been obtained in the kaon and the B-meson sector. In particular, measurements and implications of the particle-antiparticle mixing in the neutral B-meson systems, observed by the UA1 and ARGUS collaborations and subsequenly also confirmed by the CLEO collaboration, and the fundamental measurement at CERN of the CP violating ratio ε'/ε in the neutral kaon sector are presently at the centre of attention in particle physics. These experiments have confronted the standard theory of particle physics with very stringent tests and though the theory is in broad agreement at present with experiments but the experimental constraints will become very precise with the measurements of the top quark mass and the weak mixing angles. Hence, in addition to improving the present experimental measurements the two outstanding goals in flavour physics are the detection of the top quark and determination of the weak mixing angles. Likewise, a new class of CP violating phenomana is expected to become amenable to experimental measurements with high statistics beauty hadron data. The exciting results involving the B-hadrons have spurred numerous proposals for the construction of intense sources of beauty hadrons—the so-called beauty factories. The e^+e^- collider LEP and the ep collider HERA, though not exactly beauty factories, could provide $\mathcal{O}(10^6)$ beauty hadrons and hence may lead to new results in B-physics. In the kaon sector, several ongoing and planned experiments at CERN and Fermilab will attempt at measuring the mentioned CP-violating ratio to a higher precision. These experiments will provide an accurate determination of the phases ϕ_{+-} and ϕ_{00} leading to a precision test of the CPT symmetry. These topics are discussed at considerable length in these proceedings.

A subject of considerable topical interest is the issue of a quantitative understanding of heavy quark production rates in high energy collisions with photon-, lepton-, and hadron-beams. There is complete agreement between theory (QCD) and e^+e^- data on this point. The situation with the photon-hadron and lepton-hadron collisions producing charmed hadrons was somewhat ambivalent in the past. Recent data from CERN and Fermilab and new improved calculation have brought theoretical expectations in the framework of QCD and experiments closer in these sectors. Concerning hadroproduction of charmed hadrons, the disagreement between theory and experiment (as well as among various experiments) at lower energies is still not resolved. Hence hadroproduction of charm at low p_T is not yet understood. On the other hand, large p_T production of charmed and beauty hadrons in the $p\bar{p}$ collisions measured by the UA1 collaboration is in agreement with perturbative QCD calculations. These topics were discussed at great length at the workshop and the contributed reports are included in these proceedings.

The continued evasion of the top quark from showing up in ongoing experiments is somewhat frustrating though understandable due to its large mass. Lower bounds on the top quark mass have steadily moved upwards and there now exists an increasing suspicion that the LEP1/SLC and also HERA may remain topless machines. Probably, $p\bar{p}$ collisions at the Tevatron and CERN provide the only realistic hope of producing top hadrons in near future, apart from e^+e^- collisions at LEP-200. It is now grudgingly admitted that the detection of top quarks in hadron colliders is far from being trivial, and hence new strategies for their search are called for. The production and detection of the top quark at various operating and forthcoming experimental facilities is addressed in a number of reports published here.

The second part of these proceedings is concerned with the physics and experimental aspects of very energetic colliders involving 1–100 TeV proton beams. Experimental high energy physics at present is already in the TeV region. We recall that the Tevatron collider at Fermilab already has and the electron-proton collider HERA at DESY will have $\mathcal{O}(1\ \text{TeV})$ proton beams. Higher energy proton-proton collisions will become available if the proposed colliders LHC (Large Hadron Collider) with $(8+8)$ TeV proton beams, SSC (Superconducting Super Collider) with $(20+20)$ TeV proton beams, and the ELOISATRON (Eurasiatic Long Intersecting Storage Accelerator) with $(100+100)$ TeV proton beams are approved and become operational. Since the scientific and financial committments in muti-TeV particle colliders are very high, it is imperative to conduct independent and dedicated studies to examine their scientific justification. This is one of the aims of the INFN ELOISATRON workshops, as already stated on various occasions.

There are three crucial aspects that have to be carefully studied in the present context:

a. Physics goals: This means identifying the crucial questions that have to be answered to bring about a fundamental progress in the understanding of elementary particle physics, and the best means in terms of the nature, energy and luminosity of particle colliders that are likely to have the potential to address these questions.

b. Technological profile of high energy colliders: Here one has to assess the technology of building and successfully operating a collider with the desired specifications of the machine parameters. This involves extrapolation of prevailing techniques and the evaluation of scientific R & D work required to reach the desired specifications.

c. Technological profile of the detecting devices: This involves assessing the known techniques and the fundamental technological breakthroughs that have to be made in detector building and operation such that the physics potential of the proposed machines could actually be exploited.

The choice advocated here is that of a proton-proton collider with very energetic and luminous beams. Instead of attempting the ambitious, and somewhat premature, task of aiming at a faithful description of all final states at future colliders, we have concentrated on some test case studies in selected areas. The contributed papers in this part are organized in five topics: Experimental Results, Techniques and Detector R & D, Electroweak Symmetry Breaking, Vector Boson Physics, Substructure of Quarks and Leptons, and QCD. Except for one, the reports in this part are all concerned with the physics, experimental and technical issues at future colliders. The reports published here expand and in many cases complement the existing studies on proton-proton colliders with multi-TeV beams, conducted elsewhere. In most cases of interest, the reach of the LHC, SSC and the ELOISATRON colliders are estimated to provide a relative figure of merit for these machines. Where necessary, comparisons are made with high energy e^+e^- collisions.

Clearly, there are many intermediate steps before one could rush to the $(100 + 100)$ TeV proton proton colliders. Fortunately, some of the earlier stages in this long development process are already (or will soon be) available. For example, the superconducting magnets being built and tested for HERA and elsewhere give reasons to be optimistic for the attainment of yet higher magnetic fields needed for the future proton colliders. In the same vein, first encounters with proton beams having a crossing time of 100 ns will be made at HERA. However, going from the HERA/Tevatron proton energies to the full scale ELOISATRON collider requires an extrapolation of two orders of magnitude in proton energy. The experimental problems related with collisions of 10–100 TeV proton beams having luminosities in excess of 10^{33} protons/second, yielding an event rate of 10^8 Hz and a beam crossing of the order of a few ns are formidable. It will be essential to build an intermediate energy proton machine to be confident about the success of the full scale ELOISATRON project. The technological experience and physics input of a reduced scale proton-proton collider (for example, LHC, SSC or a 10% ELOISATRON machine) are simply indispensable.

The case for the full scale ELOISATRON project, of course, depends on the development of our field. High energy physics finds itself at present at cross-roads. It is not entirely clear which of the various options in the choice of beams and their energy provides the best investment of money and human endeavour. However, we would like to stress that in due course of time there may emerge a strong physics case for going to energies much higher than the SSC. For example, this may be necessitated by the scale associated with the physics of the phenomenon of spontaneous symmetry breaking and/or supersymmetry, both of which may lie in excess of 1 TeV, or perhaps it could be something entirely unexpected. Judging from the past, energy has remained the most important parameter. In our opinion, if the LHC and/or SSC machines do find some evidence for the Higgs or Higgs–like particle(s) or some other phenomenon, then the case for going to higher energies such as the ELOISATRON will become compelling. A handful of events which may be discovered at the LHC or SSC, while exciting in their own right, would hardly be enough for a quantitaive understanding of the underlying physics.

The INFN workshops have been supported by the funds provided by INFN, Italy, the Italian Ministry of Education, the Italian Ministry of Scientific and Technological Research and the Sicilian Regional Government. We are thankful to the staff of the Ettore Majorana Centre for their efficient administrative support. In particular, we would like to thank Pinola Savalli, Alberto Gabriele, Jerry Pilarsky and Guido Torelli for their help. Giacomo D'Ali made possible the use of the Erice Vax computer and we would like to thank him for his support. Finally, we thank Marianne Hausser and Zbigniew Jakubowski for their help with the secretarial work and text-editing, respectively.

Ahmed Ali, DESY, Hamburg, FRG
Luisa Cifarelli, University of Naples and INFN, Bologna, Italy
Editors

Hamburg, May 12, 1989

Contents

Heavy Flavour Physics

Experimental Results

Theoretical Interpretations

13. A Theoretical Determination of the CP Violating Phase ϕ_{00}
M. Gourdin

14. Possible Searches for CP Non-Conservation in Z Boson Decays
W. Benreuther, U. Löw, J.P. Ma and O. Nachtmann

15. Heavy Flavour Production in High Energy Electron-Proton Collisions
G.A. Schuler

16. The Lepton-Quark Mass Spectrum—a Guide to the Physics Beyond Standard Model
H. Fritzsch

Future Directions

High Energy Physics with 1–100 TeV Proton Beams

Experimental Results, Techniques and Detector R&D

Electroweak Symmetry Breaking

Vector Boson Physics

Substructure of Quarks and Leptons

QCD

PROPERTIES OF B MESONS - RESULTS FROM ARGUS

Henning Schröder

Deutsches Elektronen-Synchrotron DESY, Hamburg, Germany

1 Introduction

The study of B-mesons provides information on the fundamental parameters of the standard model, namely the masses of the b quark and the t quark, m_b and m_t, and two of the three angles , namely β and γ, which together with the phase δ' describe the Kobayashi-Maskawa-matrix. These angles can in principle be determined by three precise measurements of

- the b-lifetime which yields γ

- the rate of $b \to u$ transitions which yields β

- $B^0\overline{B}^0$ mixing which yields δ'.

The connection between these parameters can be easily read off from the Kobayashi-Maskawa-matrix in the Chau parametrization [1]

$$V = \begin{pmatrix} V_{ud} & V_{us} & V_{ub} \\ V_{cd} & V_{cs} & V_{cb} \\ V_{td} & V_{ts} & V_{tb} \end{pmatrix} = \begin{pmatrix} c_\beta c_\theta & c_\beta s_\theta & s_\beta e^{i\delta} \\ -s_\gamma c_\theta s_\beta e^{i\delta} - s_\theta c_\gamma & c_\gamma c_\theta - s_\gamma s_\theta s_\beta e^{i\delta} & s_\gamma c_\beta \\ s_\gamma s_\theta - c_\gamma c_\theta s_\beta e^{-i\delta} & -s_\gamma c_\theta - c_\gamma s_\theta s_\beta e^{-i\delta} & c_\gamma c_\beta \end{pmatrix}$$

which leads to the unitarity triangle (figure 13)

$$V_{td} + c_\gamma c_\theta V_{ub}^* = s_\gamma s_\theta.$$

To determine this relation, however, it is necessary to establish the properties of B mesons.

2 Reconstruction of B mesons

The reconstruction of B mesons has been performed succesfully by ARGUS in 16 decay channels (table 1) [2]. The B mesons originating from $\Upsilon(4S)$ decays where the $\Upsilon(4S)$ mesons are produced at rest in e^+e^- annihilations have to fulfill the condition that the beam energy E_{Beam} coincides with the energy of the B candidate, and its mass is derived by

$$m_B^2 = E_{Beam}^2 - |\sum_i \vec{p}_i|^2$$

Table 1. Branching ratios of B mesons. The ratio of $\Upsilon(4S)$ decays into charged and neutral B mesons is taken to be 55:45 [2]

Decay	ARGUS Branching Ratio [%]	Theory[3] BR [%]
$\overline{B}^0 \to D^{*+}\pi^-$	$0.35 \pm 0.18 \pm 0.13$	0.45
$\overline{B}^0 \to D^{*+}\pi^-\pi^0$	$2.0 \pm 1.0 \pm 1.0$	
$\overline{B}^0 \to D^{*+}\pi^-\pi^-\pi^+$	$4.3 \pm 1.2 \pm 2.0$	2.0
$\overline{B}^0 \to D^+\pi^-$	$0.33 \pm 0.12 \pm 0.10$	
$\overline{B}^0 \to D^+\rho^-$		1.5
$\overline{B}^0 \to J/\psi K^{*0}$	0.33 ± 0.18	0.15
$B^- \to D^{*+}\pi^-\pi^-$	$0.6 \pm 0.3 \pm 0.4$	
$B^- \to D^{*+}\pi^-\pi^-\pi^0$	$5.6 \pm 1.7 \pm 3.4$	
$B^- \to D^0\pi^-$	$0.21 \pm 0.10 \pm 0.06$	0.41
$B^- \to D^+\pi^-\pi^-$	$0.29 \pm 0.19 \pm 0.11$	
$B^- \to J/\psi K^-$	0.07 ± 0.04	0.06
$B^- \to J/\psi K^-\pi^+\pi^-$	0.11 ± 0.07	
$B^- \to \psi' K^-$	0.22 ± 0.17	
$B^- \to D^0\rho^-$	$2.1 \pm 0.8 \pm 0.8$	1.3
$\overline{B}^0 \to D^{*+}\ell^-\nu$	$7.0 \pm 1.2 \pm 1.9$	6.6

where $\vec{p}_i$ are the momenta of the decay products. This energy constraint improves the B mass resolution by an order of magnitude. The reconstruction of B mesons in low multiplicity channels involving $\overline{D}^0$ or D^- mesons suffers from background, which arises mainly from continuum events and can be efficiently reduced by demanding that the thrust axis of the B candidate does not coincide with the thrust axis of the rest of the event. Specifically the angle α between these two thrust axes must satisfy the requirement $|\cos\alpha| < 0.8$, and furthermore the energy of the B candidate must lie within $\pm 60 MeV$ of the beam energy. ARGUS is then able to observe the decays

$$
\begin{aligned}
B^- &\to D^0\pi^-, \\
B^- &\to D^0\pi^-\pi^0, \\
B^- &\to D^+\pi^-\pi^-, \\
\overline{B}^0 &\to D^+\pi^-, \\
\overline{B}^0 &\to D^+\pi^-\pi^0.
\end{aligned}
$$

The sharp cut on $|E_B - E_{beam}| \leq 60 MeV$ eliminates feed-down from B decays involving D^{*-} mesons.

The observed signals in the decay channels $B^- \to D^0\pi^-\pi^0$ and $\overline{B}^0 \to D^+\pi^-\pi^0$ can be interpreted as completely due to $B^- \to D^0\rho^-$ and $\overline{B}^0 \to D^+\rho^-$ and represent the first observations of these decays. Figure 1 (b) shows the $\pi^-\pi^0$ mass distribution if the mass of the B candidate lies in the signal region: $5.27 \leq M \leq 5.29 GeV/c^2$. A clear ρ^- signal of 16 events is observed on 11 ± 4 events background. Similarly, by requiring a $\pi^-\pi^0$ mass between $0.5 \leq m_{\pi^-\pi^0} \leq 1.04\ GeV/c^2$ a B signal of 17 events is observed over 10 ± 3 background events (Figure 1 a). As a further check, for the decay $B^- \to D^0\rho^-$, and $\overline{B}^0 \to D^+\rho^-$, the helicity angle θ should exhibit a $\cos^2\theta$ distribution, where θ is the angle between the ρ^- helicity axis and the π^0 in the rest frame of the ρ^- meson. Figure 2 shows the measured angular distribution, together with the expected curve

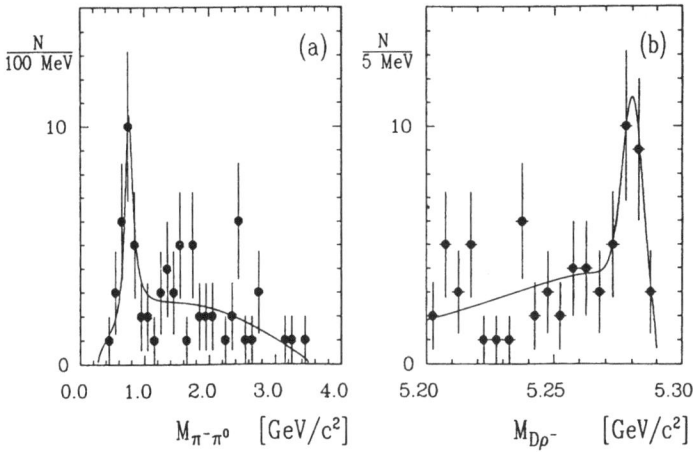

Figure 1. (a) $M_{\pi^-\pi^0}$ distribution in $\overline{B} \rightarrow D\pi^-\pi^0$ (b) $D\pi^-\pi^0$ mass for $0.5 \leq m_{\pi^-\pi^0} \leq 1.04 GeV/c^2$. decays

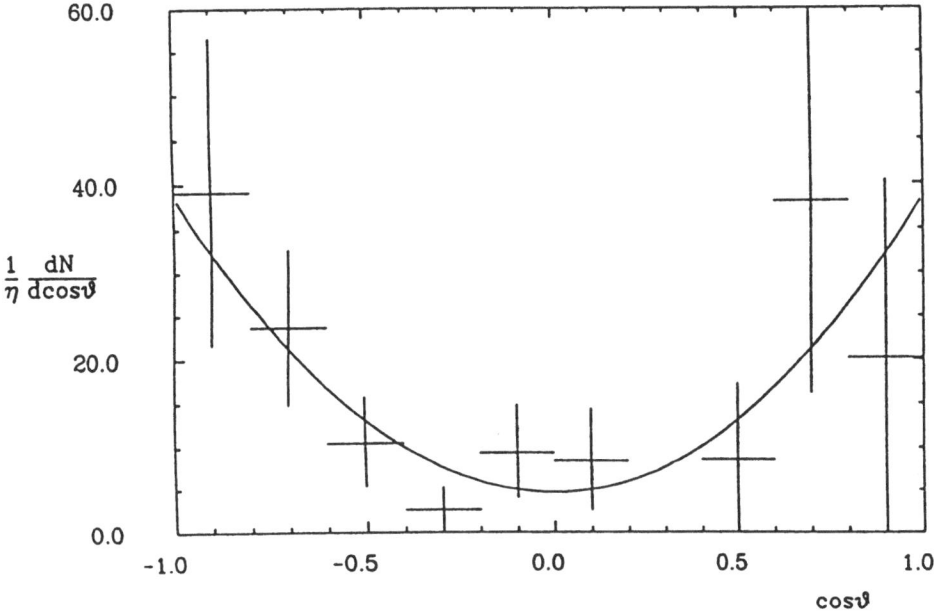

Figure 2. Acceptance corrected angular distribution of π^0 mesons in $B \rightarrow D\rho^-$ decays, $\rho^- \rightarrow \pi^-\pi^0$

for a signal of 17 events in the decays $B^- \rightarrow D^0\rho^-$, and $\overline{B^0} \rightarrow D^+\rho^-$. A background of 10 events, which are expected to have a flat distribution, is also included in the predicted curve . Reasonable agreement between data and expectation is observed.

The masses of the B mesons, $m_{B^-} = (5278.6 \pm 1.5 \pm 3.0)MeV/c^2$ and $m_{B^0} = (5280.9 \pm 1.5 \pm 3.0)MeV/c^2$, are close to half of the mass of the $\Upsilon(4S)$ which implies that the B mesons have small momenta of $p_B \approx 0.3GeV/c$. For two body decays the observed branching ratios agree reasonably well with theoretical predictions [3].

3   The Decay $\overline{B}^0 \to D^{*+}\ell^-\nu$

The decay $\overline{B}^0 \to D^{*+}\ell^-\nu$ has been reconstructed with a sizable rate by ARGUS which has high efficiencies for the electrons and muons as well as for the reconstruction of D^{*+} mesons [4]. The neutrino in the decay is inferred from its mass:

$$M_\nu^2 = [E_B - (E_{D^{*+}} + E_{\ell^-})]^2 - [\vec{p}_B - \vec{p}_{D^{*+}} - \vec{p}_{\ell^-}]^2 = 0$$

The B^0 mesons as decay products of a resting $\Upsilon(4S)$ have $E_B = E_{Beam}$ and $\vec{p}_B \approx 0$. The decay $\overline{B}^0 \to D^{*+}\ell^-\nu$ is therefore seen as a peak in the recoil mass squared spectrum M_{rec}^2 at $M_{rec}^2 = 0$ (figure 3 (a))

$$M_{rec}^2 = [E_{Beam} - (E_{D^{*+}} + E_{\ell^-})]^2 - [\vec{p}_{D^{*+}} + \vec{p}_{\ell^-}]^2.$$

The prominent peak at $M_{rec}^2 = 0$ with 75 ± 11 events on a small background of 23 ± 5 events which is determined from data in size and shape allow the study of the decay $\overline{B}^0 \to D^{*+}\ell^-\nu$ in detail.

First, the branching ratio is determined to be $BR(\overline{B}^0 \to D^{*+}\ell^-\nu) = (7.0 \pm 1.2 \pm 1.9)\%$, thus representing about $(60 - 70)\%$ of the total semileptonic branching ratio. Therefore this decay dominates the inclusive lepton spectrum in B decays, and it is of vital interest to determine the helicity structure of the decay. The total rate for the decay $\overline{B}^0 \to D^{*+}\ell^-\nu$ is given by [5]

$$\Gamma_{\overline{B}^0 \to D^{*+}\ell^-\nu} = \Gamma_L + \Gamma_{T_+} + \Gamma_{T_-}$$

where the 3 rates Γ_L, Γ_{T_+} and Γ_{T_-} correspond to the longitudinally and transversely polarized D^{*+} mesons. The inclusive lepton spectrum is dominated at high momenta by the T_- component. The polarisation of D^{*+} mesons can be determined by using the strong decay $D^{*+} \to \pi^+ D^0$ as an analyser. The distribution in the angle θ^*, where θ^* is the decay angle of the π^+ in the D^{*+} rest frame with respect to the D^{*+} direction is given by

$$\frac{dN}{d\cos\theta^*} = 1 + \alpha_P \cos^2\theta^*$$

where α_P measures the ratio of longitudinal to transverse polarisation:

$$\alpha_P = 2 \cdot \frac{\Gamma_L}{\Gamma_{T_-} + \Gamma_{T_+}} - 1.$$

The resulting $\cos\theta^*$ distribution gives for $p_\ell > 1.0 GeV/c$ a value of $\alpha_P = 0.5^{+0.8}_{-0.6}$ (figure 3(b)) and for $p_\ell > 1.2 GeV/c$ a value of $\alpha_P = 0.8 \pm 0.8$ which fits well to the theoretical models [5].

These results together with the lifetime of the B^0 mesons, which is taken to be the average b lifetime $\tau_b = (1.14 \pm 0.14)ps$ [6] , allows the determination of the KM matrix element $|V_{cb}|$ which is given by

$$\Gamma(\overline{B}^0 \to D^{*+}\ell^-\nu) = \frac{BR(\overline{B}^0 \to D^{*+}\ell^-\nu)}{\tau_{B^0}} = |V_{cb}|^2 \tilde{\Gamma}_T \cdot (1 + \frac{\Gamma_L}{\Gamma_{T_+} + \Gamma_{T_-}})$$

The parameter $\tilde{\Gamma}_T$ is expected to be computed reliably in the form factor approach: $\tilde{\Gamma}_T = 1.21 \cdot 10^{13} s^{-1}$ [7]. Inserting the observed numbers one obtains

$$|V_{cb}| = 0.051 \pm 0.010.$$

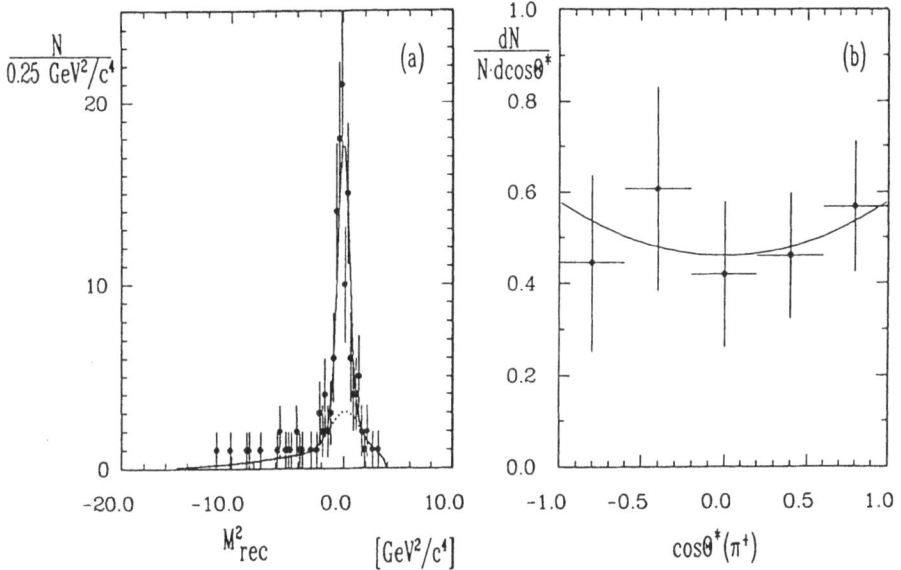

Figure 3. (a) M^2_{rec} distribution in the decay $\overline{B}^0 \to D^{*+}\ell^-\nu$. (b) $\cos\theta^*$ distribution in the decay $\overline{B}^0 \to D^{*+}\ell^-\nu$

4 Semileptonic B Decays

An alternative approach to measure $\mid V_{cb} \mid$ is provided by the determination of the inclusive semileptonic branching ratio $B \to \ell\nu X$ [2]:

$$\Gamma(B \to \ell\nu X) = \frac{BR(B \to \ell\nu X)}{\tau_B} = \frac{G_F m_b^5}{192\pi^2}(0.86 \mid V_{ub} \mid^2 + 0.48 \mid V_{cb} \mid^2).$$

A new measurements by ARGUS on the semileptonic B decays is shown in (figure 4). The determination of the branching ratio is model dependent. Using specific models , ARGUS obtains $BR(B \to \ell^+\nu X) = 10.0 \pm 0.8)\%$ [8] or $BR(B \to \ell^+\nu X) = 9.4 \pm 0.7)\%$ [10]. These values are smaller than the ones previously obtained and are even harder to understand in spectator models. Assuming that all B decays proceed via spectator diagrams one gets the smallest semileptonic branching ratio of $BR(B \to \ell\nu X) \approx 12\%$ if current quark masses and next-to-leading order QCD corrections are used [9].

Contributions of $b \to u$ transitions to the semileptonic B decays can contribute to the lepton momentum spectrum above the kinematical limit for $b \to c$ transitions of $p_\ell = 2.4 GeV/c$. No $b \to u$ contributions are observed in this momentum range by ARGUS, which results in

$$\mid \frac{V_{ub}}{V_{cb}} \mid < 0.16 \text{ at } 90 \%C.L.$$

using a specific model [10] which gives the largest upper limit. Concerning the validity of such a limit severe reservations still remain due to its model dependance.

With the obtained upper limit the above relation allows the determination of $\mid V_{cb} \mid$ and $\mid V_{ub} \mid$: $\mid V_{cb} \mid = 0.045 \pm 0.007$ and $\mid V_{ub} \mid < 0.008$.

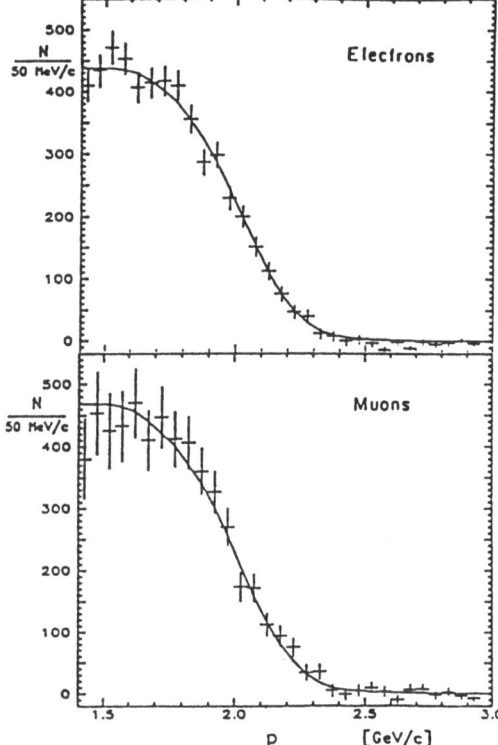

Figure 4. Lepton spectrum in $\Upsilon(4S)$ decays

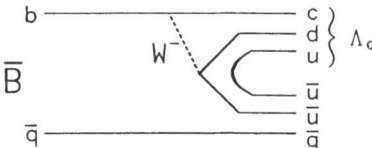

Figure 5. Baryon-antibaryon production in B decays

5   Baryon production in B decays

B mesons can decay into final states containing baryon-antibaryon pairs by a simple mechanism. In the decay $b \rightarrow cd\bar{u}$ for example, a cd diquark can be formed (Figure 5). Only *one* $q\bar{q}$ loop has to be pulled out of the vacuum so that the cd diquark can hadronize into a charmed baryon and likewise the $\overline{uq}$ diquark, together with the spectator anti-quark, forms an ordinary anti-baryon.

For $(b \rightarrow c)$ transitions one expects to have predominantly Λ_c baryons in the final state. Λ_c production in B decays is directly observed by ARGUS [11]. Figure 6 (a) shows the invariant $pK^-\pi^+$ mass from $\Upsilon(4S)$ data, and (b) the same distribution for continuum data. For both spectra a momentum cut of $p(pK^-\pi^+) \leq 2.3 GeV/c$ was applied. A Λ_c signal of 398± 60 events is observed in the $\Upsilon(4S)$ data while in the continuum one finds only 89± 31 Λ_c candidates. Subtracting the continuum contribution results in   208±89 Λ_c baryons which are attributed to genuine $\Upsilon(4S)$ decays. This number yields, after acceptance corrections,

6

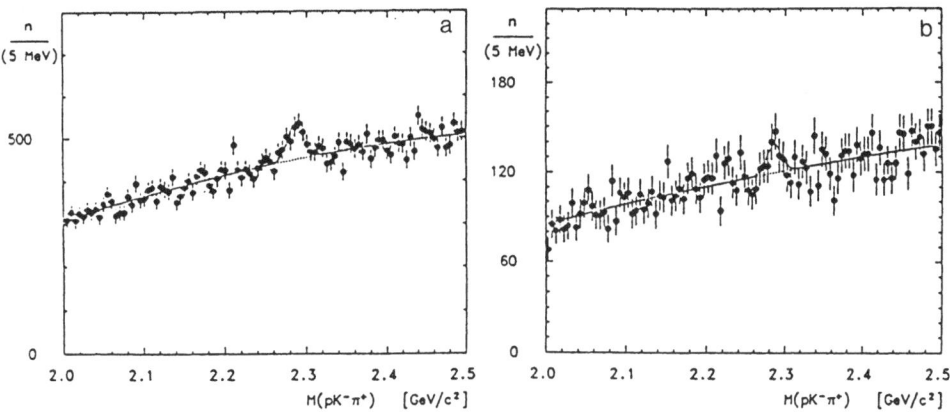

Figure 6. $pK^+\pi^-$ mass distribution with $p \leq 2.3 GeV/c$ in (a) events taken at the $\Upsilon(4S)$ mass, (b) in the continuum below the $\Upsilon(4S)$ mass [11]

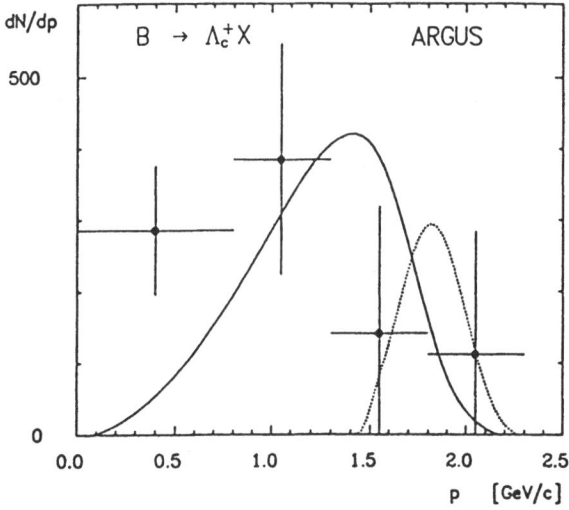

Figure 7. Momentum distribution of Λ_c baryons from B decays. The expected shapes of the contributions from two-body (dotted line) and three-body (solid line) B decays are shown [11]

$$BR(B \rightarrow \Lambda_c^+ X)BR(\Lambda_c^+ \rightarrow pK^-\pi^+) = 0.0030 \pm 0.0012 \pm 0.0006$$

The momentum spectrum of the Λ_c baryons from B decays is shown in Figure 7. It exhibits a soft momentum distribution with no pronounced structure which might be due to two body decays like $\overline{B} \rightarrow \Lambda_c^+ \overline{N}$ (dotted curve in Figure 7). Three body phase-space decays like $\overline{B} \rightarrow \Lambda_c^+ \overline{N}\pi$ gives a qualitativly better description (solid curve in Figure 7).

B decays into protons and Λ's are well established [12] [13]. The proton and Λ momentum spectra are very soft (Figure 8) which reflects the softness of the Λ_c spectrum. The ARGUS results on the inclusive branching ratios of $B \rightarrow \Lambda X = (4.2 \pm 0.5 \pm 0.6)\%$ and $B \rightarrow p X = (5.5 \pm 0.6 \pm 1.5)\%$, as well as on the momentum spectra are in good agreement with the CLEO results [12].

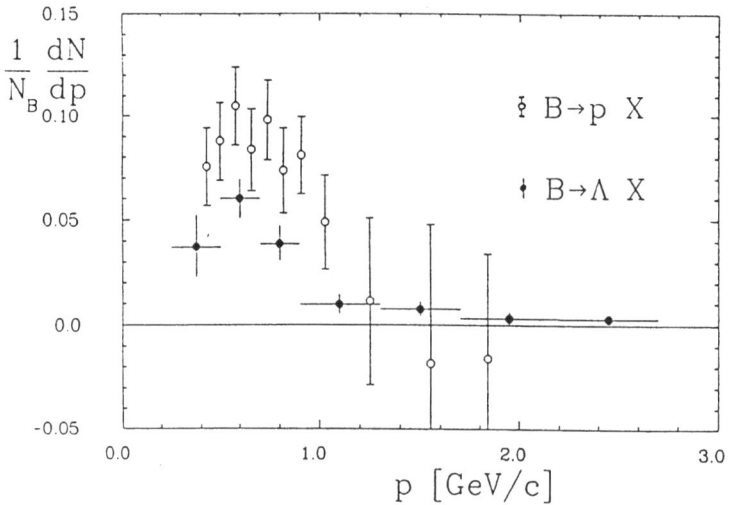

Figure 8. Momentum spectra of protons and Λ's in $\Upsilon(4S)$ decays [13]

The inclusive proton and Λ rates can be translated into a rate for $\overline{B} \rightarrow \Lambda_c X$, assuming that all baryons in B decays are from a decay $\overline{B} \rightarrow \Lambda_c \overline{N} X$ and that $BR(B \rightarrow p\ X) = BR(B \rightarrow n\ X)$. ARGUS obtains then $BR(B \rightarrow \Lambda_c^- X) = (7.6 \pm 1.4 \pm 1.8)\%$. This result can be used to extract a measurement of the branching ratio for the decay $\Lambda_c \rightarrow pK^-\pi^+$ by using the observed result on the Λ_c production in B decays. From the product branching ratio given above one obtains:

$$BR(\Lambda_c \rightarrow pK^-\pi^+) = (3.9 \pm 2.2)\%.$$

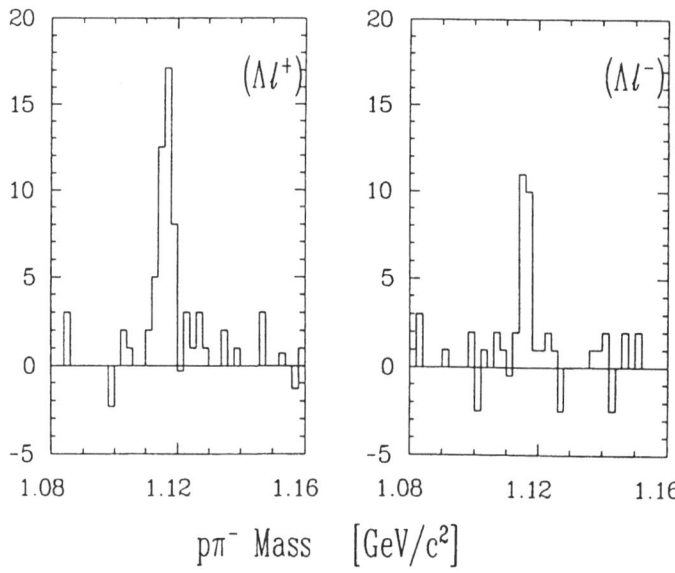

Figure 9. $p\pi^-$ invariant mass for the $\Upsilon(4S)$ events with Λl^+ and Λl^- pairs after subtraction of continuum contribution (ARGUS)

A study of lepton-baryon charge correlation provides additional information on the mechanism of baryon production in B meson decays, which is not yet well understood. Recently CLEO observed a strong lepton-Λ charge correlation in B decays [12] by observing only events with Λl^+ or $\overline{\Lambda} \ell^-$ pairs. This would imply that Λ particles in B decays should come predominantly from the decay of the charmed baryon and not from the hadronization of the spectator quark.

This result is **not** confirmed by ARGUS which performed a similar analysis. In $\Upsilon(4S)$ events with exactly one fast lepton with momentum between 1.5 GeV/c and 2.8 GeV/c and at least one Λ, clear signals for both Λl^- and Λl^+ combinations are observed (Figure 5). The fit results in signals of 40.7 and 25.3 events with Λl^+ and Λl^- pairs respectively. This observation shows that Λ particles in B decays arise in the decay of charmed particles as well as in the fragmentation of the spectator quark. Therefore, Λ particles cannot be used to tag the b-quark content of a B meson.

6 Search for Exclusive $b \rightarrow u$ Transitions

Exclusive charmless B decays have been searched for in channels involving only low multiplicities of pions where no significant signals have been observed [2]. The motivation to search for non-charm final states containing baryons derives from the higher masses of the decay particles absorbing more of the available phase-space. Therefore it is possible that low multiplicity channels involving baryons could have larger branching ratios.

Following this line of thought, ARGUS has investigated the channels [14]:

$$B^- \rightarrow p\bar{p}\pi^-$$
$$B^0 \rightarrow p\bar{p}\pi^+\pi^-$$

Suppression of the continuum component of combinatorial background is similar to that made for the reconstruction of B mesons in the decay $B \rightarrow DX$ (chapter 2) with some additional requirements:

- $| E(p\bar{p}\pi^+(\pi^-)) - E_{Beam} | < 2\sigma_E$ and $\sigma_E < 60 MeV/c^2$

- the angle α between the two thrust axes must satisfy $|\cos\alpha| < 0.8$

- the multiplicity of charged particles in the rest of the event, excluding the particles contributing to the B candidate, must be larger than 3

- the Fox - Wolfram moment H_2 [15], calculated for the remaining charged and neutral particles in the event, must be smaller than 0.3

- the opening angle $\Theta_{p\bar{p}}$ between the proton and the antiproton must be such that $\cos\Theta_{p\bar{p}} \leq -0.98$.

The mass spectrum for candidates which pass these cuts is shown in figure 10 for $\Upsilon(4S)$ data. A pronounced peak in the B mass region on a small background (Figure 10) is observed. From a fit with a Gaussian plus background (solid curve) one observes a signal of 27 events at a mass of $(5278.3 \pm 1.1 \pm 3.0)$ MeV/c^2 with a width, consistent with expectation, of $\sigma_M = (4.2 \pm 1.0)$ MeV/c^2. The probability that the observed excess of events is a statistical fluctuation of the background corresponds to more than five standard deviations.

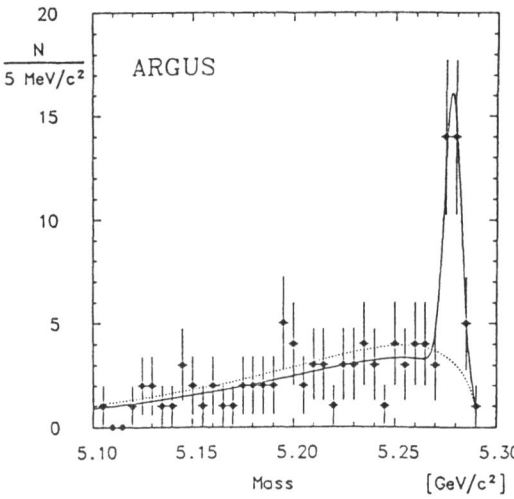

Figure 10. Mass distribution of B meson candidates for $\Upsilon(4S)$ data (points with error bars). The dashed line corresponds to continuum data, fitted and normalized

Protons and antiprotons cannot be identified unambiguously, but all detector information is consistent with the particle assignment. The sample can be divided into neutral and charged B mesons. The obtained masses for the charged and neutral B mesons agree nicely with previous measurements in the decay channels involving charmed mesons. For the branching ratios one finds

$$\text{Br}(B^- \to p\bar{p}\pi^-) = (5.2 \pm 1.4 \pm 1.9) \times 10^{-4}$$
$$\text{Br}(B^0 \to p\bar{p}\pi^+\pi^-) = (6.0 \pm 2.0 \pm 2.2) \times 10^{-4}$$

where the first is the statistical and the second the systematic error, including a contribution introduced by the assumed background function.

From the theoretical side it is not trivial to relate any number for decays like $B \to p\bar{p}\pi^+(\pi^-)$ to a value of $|V_{ub}|$. Using the ARGUS result, estimates range from $|\frac{V_{ub}}{V_{cb}}| > 0.05$ [16] to values which are unacceptable in the standard model [17].

7 $B^0\overline{B}^0$ Mixing

ARGUS observed in 1987 a surprisingly large $B^0\overline{B}^0$ mixing: [18]

$$r = \frac{Prob(B^0 \to \overline{B}^0)}{Prob(B^0 \to B^0)} = 0.21 \pm 0.08.$$

This observation of $B^0\overline{B}^0$ mixing was established through three methods of which one was the observation of one event where the $\Upsilon(4S)$ apparently decays into two B^0 mesons (figure 11) which in turn both decay like $B^0 \to D^{*-}\mu^+\nu$. With the above value for r one expects to find 0.3 events of this type in the ARGUS data sample.

The amount of mixing was measured in a sample of 96 000 $B\overline{B}$ events through the observation of like-sign and unlike-sign lepton events at the $\Upsilon(4S)$ (table 2)

$$r = \frac{[N(\ell^+\ell^+) + N(\ell^-\ell^-)](1+\lambda)}{N(\ell^+\ell^-) - [N(\ell^+\ell^+) + N(\ell^-\ell^-)]\lambda}$$

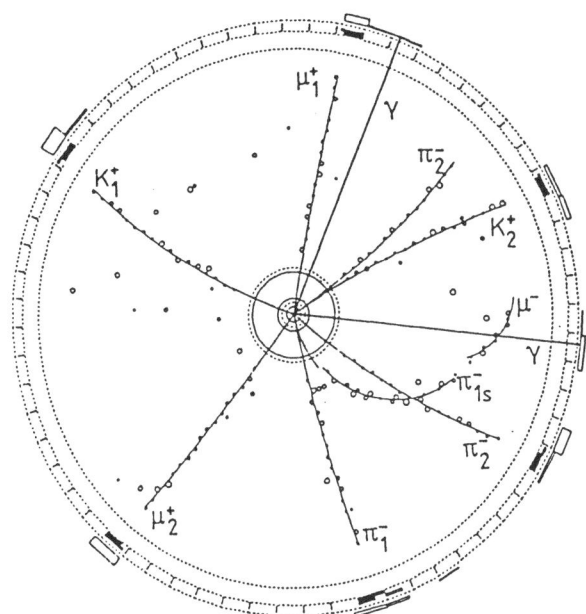

Figure 11. Completely reconstructed event consisting of the decay $\Upsilon(4S) \to B^0\overline{B}^0$

where the leptons originate both from primary B decays. The parameter λ accounts for the fact that the $\Upsilon(4S)$ does not solely decay into a $B^0\overline{B}^0$ pair but more than 50% of the time also into a B^+B^- pair where no mixing can occur and is taken to be $\lambda = 1.2$ [18].

Table 2. $B^0\overline{B}^0$ Mixing results from ARGUS

	ARGUS
N($\Upsilon(4S)$)	96000
Like − sign dilepton candidates	
Total	50
Fakes	12.3
Secondaries	12.9
Signal	$24.8 \pm 7.6 \pm 3.8$
Unlike − sign dilepton candidates	
Total	302
Fakes	24.7
Secondaries	7.2
Signal	$270 \pm 19.4 \pm 5.0$
r	0.22 ± 0.10

That the measurement of like-sign lepton pairs is really due to $B^0\overline{B}^0$ mixing could be checked by investigating events containing one reconstructed B^0 and a fast lepton where ARGUS obtains

$$r = \frac{N_{B^0\ell^+} + N_{\overline{B}^0\ell^-}}{N_{B^0\ell^-} + N_{\overline{B}^0\ell^+}} = 0.20 \pm 0.12.$$

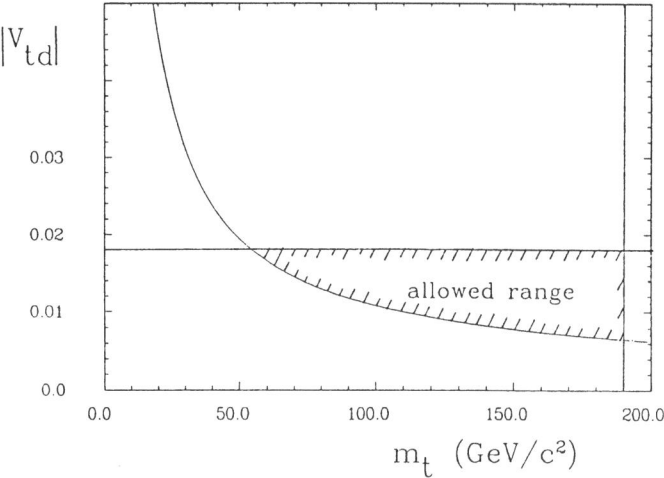

Figure 12. Allowed range for $|V_{td}|$ and m_t

which can be combined to give the final ARGUS result of $r = 0.21 \pm 0.08$.

This result implies a value of

$$x = \frac{\Delta M}{\Gamma} = 0.73 \pm 0.18$$

where

$$r = \frac{x^2}{2 + x^2}.$$

The mass difference ΔM between the CP eigenstates in the neutral B system then amounts to

$$\Delta M = (4.1 \pm 1.1) \cdot 10^{-4} \text{ eV}$$

which is about 100 times larger than the corresponding one between the K_S^0 and K_L^0 masses in the neutral K system where the large $K^0 - \overline{K}^0$ mixing is related to the large lifetime differences.

In the standard model with 3 families the mixing parameter x is given by [19]

$$x \approx \frac{G_F^2}{6\pi^2} B_B f_B^2 m_b \tau_b |V_{td}|^2 m_t^2 F(\frac{m_t^2}{M_W^2}) \eta_{QCD}$$

with $| V_{td} | < 0.018$ from the unitarity of the KM matrix [20], $m_t < 190 GeV/c^2$ from the determination of the electroweak radiation corrections [21]. For $B_B^{1/2} f_B$ only estimates exist based on theoretical computations or experimental extrapolations. A reasonable guess seems to be $B_B^{1/2} f_B \approx (140 \pm 40) MeV$ [22].

Figure 12 sketches the allowed range for $| V_{td} |$ and m_t where the lower curve is calculated for $x = 0.44$ and $B_B^{1/2} f_B = 100 MeV$. The upper bounds on $| V_{td} |$ and m_t then also allow the derivation of the lower bounds:

$$m_t > 50 GeV/c^2 \text{and} | V_{td} | > 0.006.$$

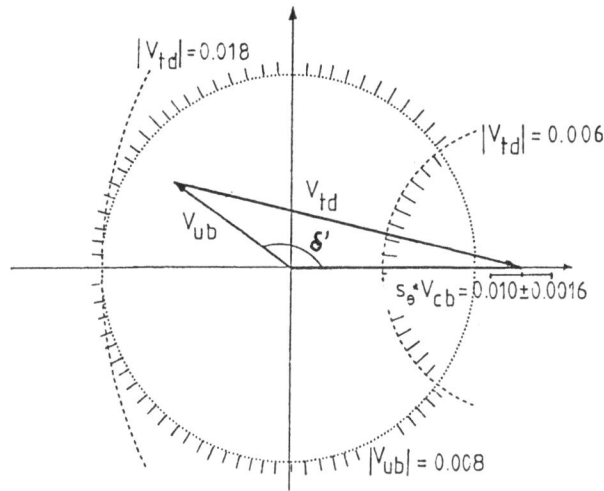

Figure 13. Unitarity triangle

8   Conclusions

Figure 13 shows our knowledge on the standard model parameters γ, β, δ' where we have a measurement only on γ:

$$\sin \gamma = 0.045 \pm 0.007.$$

For β we have an upper limit of $\sin \beta < 0.008$ which comes from the non-observation of $b \rightarrow u$ contributions at the endpoint of the momentum spectrum of leptons from B decays. From the observation of exclusive charmless B decays (section 6) a non-zero value for V_{ub} can be inferred, implying $\beta \neq 0, \pi$. Though the phase δ' is still completely unconstrained, a determination of δ' would be possible from the observed $B^0\overline{B}^0$ mixing result if m_t and especially $B^{1/2}f_B$ would be known. Ongoing experiments should be able to determine these values and also give a definite value for $\mid V_{ub} \mid$.

In summary, the recent period has been especially fruitful in B physics where the breakthrough towards a better understanding of the standard model is achieved by the observation of $B^0\overline{B}^0$ mixing by ARGUS.

References

[1] L.L.Chau and W.Y.Keung, Phys.Rev.Lett. **53** 1802 (1984)

[2] H.Schröder (ARGUS collaboration) DESY preprint DESY 88-101 (1988)

[3] M.Bauer et al, Z.Phys. **C34** 103 (1987)

[4] H.Albrecht et al (ARGUS collaboration), Phys.Lett. **197B** 452 (1987)

[5] J.G.Körner and G.A.Schuler, Z.Phys. **C38** 511 (1988)

[6] S.L.Wu, DESY preprint DESY 87-164 (1987)

[7] T.Altomari and L.Wolfenstein, CMU-HEP87-20 (1987)

[8] G.Altarelli et al Nucl.Phys. **B208** 365 (1982)

[9] R.Rückl, Habilitationsschrift, Universität München, 1984

[10] N.Isgur et al, priv. communication

[11] H.Albrecht et al. (ARGUS collaboration), DESY preprint DESY 88-012 (1988)

[12] M.S.Alam et al. (CLEO collaboration), Phys.Rev.Lett. **59** 22 (1987)

[13] H.Albrecht et al. (ARGUS collaboration), DESY preprint DESY 88-145 (1988)

[14] H.Albrecht et al (ARGUS collaboration), Phys.Lett. **209B** 119 (1987)

[15] G.C.Fox and S.Wolfram, Phys.Lett. **82B** 134 (1979)

[16] N.G.Deshpande et al,University of Oregon preprint OITS 370 (1987)

[17] V.L.Chernyak et al, Novosibirsk preprint Institute of Nuclear Physics 88-65 (1988)

[18] H.Albrecht et al (ARGUS collaboration), Phys.Lett. **192B** 245 (1987)

[19] J.S.Hagelin, Phys.Rev. **D20** 2893 (1979)

[20] Particle Data Group, Phys.Lett. **170B** 1 (1986)

[21] G.Costa et al, Nucl.Phys. **B297** 244 (1988)

[22] G.Altarelli, in Proceedings of the International Europhysics Conference on High Energy Physics, Uppsala (1987), p.1002

2

RECENT RESULTS FROM CLEO ON THE DECAY OF B MESONS

Daniel M. Coffman

Laboratory of Nuclear Studies
Wilson Laboratory
Cornell University
Ithaca, NY 14850
USA

Abstract

The CLEO collaboration has investigated Kobayashi-Moskawa favored and disfavored decays of B mesons. Upper limits on the matrix element V_{bu} have been obtained through searches for $b \rightarrow u$ transitions. There is no evidence for the decay modes $B^- \rightarrow p\bar{p}\pi^-$ or $B^0 \rightarrow p\bar{p}\pi^+\pi^-$ reported by the ARGUS collaboration.

I. The Theory and the Experiment

In the Standard Model of electroweak interactions, the mass and weak eigenstates of quarks are not identical. They are related to one another by a unitary transformation; this is usually represented by a family mixing matrix, the Kobayashi-Moskawa (KM) matrix. If a quark q decays weakly to a different quark q', the effective coupling is determined by the KM matrix element $V_{qq'}$.

Several of these matrix elements have been measured directly. These values and the unitarity of the KM matrix constrain the value of $|V_{bc}|$ to be less than 0.46 and that of $|V_{bu}|$ to be less than 0.084 at the 90% confidence level.[1] The smallness of these elements is in agreement with the lifetime of the b quark which is observed to be quite long. However, the magnitude of V_{bu} is expected to be greater than zero;

quark mixing is capable of explaining the violation of CP symmetry but only if V_{bu} is non-zero. It is very important that the values of $|V_{bc}|$ and $|V_{bu}|$ be measured directly. These parameters are fundamental to the Standard Model, and are as important as say the quark and lepton masses.

The CLEO collaboration is actively engaged in studying the decays of B mesons. The CLEO detector has been described extensively elsewhere.[2] Recently a new main drift chamber with 51 sensitive layers replaced the older 17 layer device, improving the momentum resolution to $(\delta p/p)^2 = (0.007)^2 + (0.0023p)^2$. This chamber affords a measurement of specific ionization (dE/dx) with a resolution of 6.5%. The upgraded detector has been exposed with the Cornell Electron Storage Ring (CESR) to an integrated luminosity of 212 pb^{-1} at the $\Upsilon(4S)$ resonance. Since this resonance is quite small, many studies require non-resonant — or continuum — data for the estimation of backgrounds. For this reason, an additional integrated luminosity of 102 pb^{-1} was obtained at an energy just below the threshold for producing pairs of B mesons.

II. The KM favored* transition $b \to c$

The products of a KM favored decay of a B meson will usually contain a charmed meson or baryon if the intermediary W boson materializes as a light quark pair. If it materializes as a $c\bar{s}$ or $c\bar{d}$ pair, the decay products may contain a J/ψ boson. In either case, these products may be sought inclusively or exclusively.

An inclusive study relies on the fact that a particle of mass m produced by the decay of a particle of mass M may have momentum of at most $(M^2 - m^2)/2M$ in the center-of-mass (cm) frame of the decaying particle. At the $\Upsilon(4S)$, B mesons are formed with small momenta, so the products from their decay have momenta of at most about $\frac{1}{2}M_B$. Particles produced through continuum annihilation may, on the other hand, have momenta up to the much larger value $p_{\max} \equiv \sqrt{E_{\text{beam}}^2 - m^2}$. An excess of events with $x \equiv p/p_{\max} \le 0.5$ from data take at the $\Upsilon(4S)$ *vis-à-vis* the spectrum obtained at energies just below $B\overline{B}$ threshold is considered *prima face* evidence for the decays of B mesons.

$\star$ The bottom quark is known to decay most often to a charmed quark,[3] in spite of the up quark being much less massive. This may be intrepreted as implying that $|V_{bc}| \gg |V_{bu}|$. Thus transitions of the $b \to c$ may be called 'Kobayashi-Moskawa favored' in analogy with the term 'Cabibbo favored'.

$\underline{B \to D^{*+}X}$

The D^{*+} meson is detected through its decay $D^{*+} \to D^0\pi^+$. For inclusive studies, the D^0 is observed in the decay mode with the least combinatoric background $D^0 \to K^-\pi^+$. The mass difference of the $K^-\pi^+\pi^+$ and $K^-\pi^+$ combinations is required to be within 2.0 MeV/c^2 of the known mass difference between the D^{*+} and D^0, 145.4 MeV/c^2. An additional requirement reduces combinatoric backgrounds from continuum and $B\overline{B}$ events. The 'helicity angle' θ_K^* — the angle between the momentum vector of the K in the D^0 rest frame and the momentum vector of the D^0 in the laboratory frame — should be isotropically distributed for a pseudoscalar meson such as the D^0. The situation is different for background events for which θ_K^* peaks strongly in the forward and backward directions. The requirement imposed is that $|\cos(\theta_K^*)| < 0.8$.

The D^{*+} momentum spectra are measured at the $\Upsilon(4S)$ and at an energy 30 MeV below the $\Upsilon(4S)$. The spectra corrected for detection efficiency and relative integrated luminosity are shown in Fig. 1. The yield of D^{*+} is obtained by subtracting these two spectra. This procedure does not distinguish between charged and neutral B mesons, so the resulting product branching ratio[4]

$$B(B \to D^{*+}X) \cdot B(D^{*+} \to D^0\pi^+) \cdot B(D^0 \to K^-\pi^+) = (7.1 \pm 0.7 \pm 0.6) \times 10^{-3} \quad (1)$$

is valid for an unknown mix of charged and neutral B mesons. The first error reported is statistical and the second systematic. This result may be converted into an absolute branching ratio by making use of recent measurements of the branching ratios[5] [6] of D^{*+} and D^0 mesons.

$$B(D^0 \to K^-\pi^+) = (4.2 \pm 0.4 \pm 0.4)\% \quad (2)$$

$$B(D^{*+} \to D^0\pi^+) = (55 \pm 2 \pm 6)\% \quad (3)$$

These yield an inclusive branching ratio of

$$B(B \to D^{*+}X) = (32 \pm 3 \pm 6)\% \ .$$

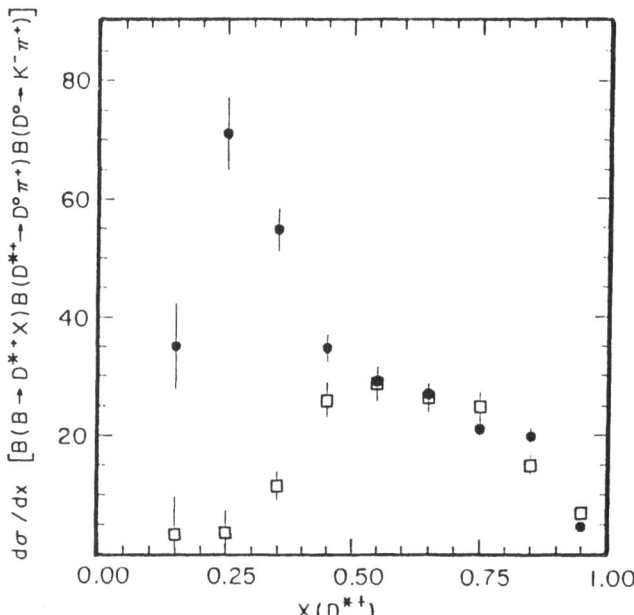

Figure 1. The D^{*+} yield (number of events corrected for detection efficiency) as a function of $x \equiv p/p_{\max}$. The filled circles are from the data taken at the $\Upsilon(4S)$, and the open circles from continuum data.

<u>$B \to D_s X$</u>

The charmed-strange meson D_s^+ is detected through its decay to $\varphi\pi^+$ or $\overline{K}^{*0}K^+$. As the combinatoric background for the $\overline{K}^{*0}K^+$ mode is much worse than that for the $\varphi\pi^+$ mode, the latter is used for inclusive studies. Pairs of oppositely charged kaons are combined to form φ candidates. The candidates are combined with charged pions to form D_s candidates.

Two cuts similar to the 'helicity angle' cut described above are used to limit combinatoric backgrounds. The angular distribution of the single pion should be isotropic in the D_s cm frame, whereas the angular distribution of the kaons is given by $\cos^2(\theta_K^*)$ in the φ cm frame.

The D_s momentum spectra from the $\Upsilon(4S)$ and the continuum are corrected and then subtracted just as in the case of the D^{*+}. The resulting product branching ratio[7] is

$$B(B \to D_s^+ X) \cdot B(D_s^+ \to \varphi\pi^+) = (3.8 \pm 0.58) \times 10^{-3} \qquad (4)$$

<u>$B \to J/\psi$</u>

The J/ψ is detected through its electromagnetic decay to pairs of electrons or muons. These leptons are required to have momenta between 0.8 and 2.7 GeV/c corresponding to the allowed kinematic range. Electrons are identified using measurements of their specific ionization and shower energy. Muons are identified using proportional counters located outside of the magnetic flux return.

There is no evidence of J/ψ production in the continuum data, therefore no continuum background subtraction is necessary. The inclusive branching ratios[8] derived from the $\Upsilon(4S)$ data are

$$B(B \to J/\psi\, X) = (0.96 \pm 0.10 \pm 0.19)\% \quad \text{(muon pairs)} \qquad (5)$$

$$B(B \to J/\psi\, X) = (0.71 \pm 0.09 \pm 0.20)\% \quad \text{(electron pairs)} \qquad (6)$$

These may be combined taking account of common systematic uncertainties yielding

$$B(B \to J/\psi) = (0.86 \pm 0.07 \pm 0.16)\% \qquad (7)$$

19

$\underline{B \rightarrow \Lambda_c X}$

The charmed baryon Λ_c^+ is detected through its decays to $pK^-\pi^+$ and $p\overline{K}^0$. The various charged particles are identified using measurements of their specific ionization. The Λ_c momentum spectra for data at the $\Upsilon(4S)$ and on the continuum are subtracted as before, yielding product branching ratios

$$B(B \rightarrow \Lambda_c^+ X) \cdot B(\Lambda_c^+ \rightarrow pK^-\pi^+) = (0.31 \pm 0.07 \pm 0.05)\% \tag{8}$$

$$B(B \rightarrow \Lambda_c^+ X) \cdot B(\Lambda_c^+ \rightarrow p\overline{K}^0) = (0.10 \pm 0.07 \pm 0.01)\% \tag{9}$$

These cannot be converted into absolute branching ratios as the branching ratios of the Λ_c^+ are at best poorly known.

Exclusive Modes

It has proven possible in a relatively small number of cases to identify all of the final state particles produced in the decay of a B meson. The number of such exclusive measurements is limited by the fact that a number of intermediate particles are produced before only stable particles remain. For example, in the decay chain $\overline{B}^0 \rightarrow D^{*+}\pi^-$; $D^{*+} \rightarrow D^0\pi^+$; $D^0 \rightarrow K^-\pi^+$, two intermediate particles are produced. In addition, the decay products of many of these intermediate states contain neutral pions, a particle CLEO does not detect.

It is interesting to note that the branching ratios of the $\Upsilon(4S)$ itself have not been measured. This is similar to the case of charmed mesons where the branching ratios of the $\psi(3770)$ to D mesons have not been directly measured. The value of the $\Upsilon(4S)$ branching ratio[9] to neutral B mesons used here is

$$B(\Upsilon(4S) \rightarrow B^0\overline{B}^0) = (43 \pm 4)\% \tag{10}$$

Further, the $\Upsilon(4S)$ is assumed only to decay to $B\overline{B}$ pairs.

The charmed particles in exclusive studies are reconstructed largely as described above. In certain cases, decay modes with more background may be used. The results of several exclusive studies are summarized in Table I; the specific analysis techniques are described fully in the indicated references. Common to all of these studies are cuts on the total measured energy and on the 'beam-constrained' mass

Table I. Branching Ratios for KM Favored Decays

Mode	Branching Ratio (%)	Reference
$\overline{B}^0 \rightarrow D^{*+}\pi^-$	$0.46 \pm 0.12 \pm 0.10$	4
$\overline{B}^0 \rightarrow D^{*+}\pi^-\pi^+\pi^-$	$2.6 \pm 0.9 \pm 1.1$	4
$\overline{B}^0 \rightarrow D_s^+ D^-$	1.1 ± 1.1	7
$\overline{B}^0 \rightarrow D_s^+ D^{*-} \oplus D_s^{*+} D^{*-}$	3.7 ± 1.1	7
$\overline{B}^0 \rightarrow \psi \overline{K}^0$	0.04 ± 0.03	8
$\overline{B}^0 \rightarrow \psi \overline{K}^{*0}$	0.06 ± 0.03	8
$B^- \rightarrow \psi K^-$	0.04 ± 0.03	8

$M_{BC} \equiv \sqrt{E_{\text{beam}}^2 - p_{\text{measured}}^2}$. The measured energy is typically required to be within two standard deviations of the the beam energy. This is intended to lessen the likelihood that a pion could have escaped detection. The resolution in M_{BC} is much better than for the usual invariant mass as the beam energy is large and well-known whereas the momenta of the B is much smaller, about 300 MeV/c. Further, the measured momentum does not depend on any identification of particles. Figure 2 summarizes the beam-constrained mass distributions from two exclusive studies.

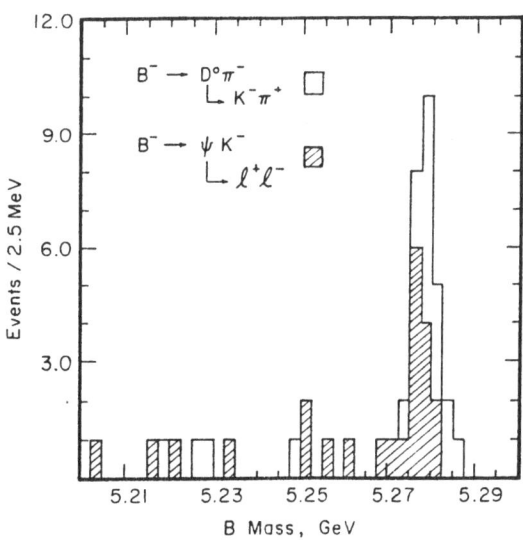

Figure 2. The Beam-Constrained mass distribution for two exclusive modes.

III. The KM disfavored transition $b \to u$

The apparently rare transitions of the bottom quark to the up quark, for which the decay rate is goverened by $|V_{bu}|$, may be sought in exclusive hadronic and semileptonic decay modes, and also in inclusive lepton momentum spectra from semileptonic decays. This latter procedure will not be described here.

The results[10] of a number of exclusive studies are given in Table II. No evidence is seen for any of the indicated channels. The upper limits given for the branching ratios are calculated at the 90% confidence level. For the decay $\overline{B}^0 \to K^{*0}\gamma$, the shower detector is used only to determine the direction of the photon not its energy; the angular resolution of this device is much better than its energy resolution. The quantity $E_\gamma \equiv E_{\text{beam}} - E_{K^*}$ is used instead of the measured energy as an estimate of the correct photon energy.

Table II. Upper Limits on Branching Ratios of Rare Decays

B^0 decays		B^+ decays	
Mode	Upper Limit $(\times 10^{-4})$	Mode	Upper Limit $(\times 10^{-4})$
$\pi^+\pi^-$	0.9	$\rho^0\pi^+$	1.5
$p\overline{p}$	0.4	$\overline{\Delta}^0 p$	3.3
$\Delta^{++}\overline{\Delta}^{--}$	1.3	$\Delta^{++}\overline{p}$	1.3
$\Delta^0\overline{\Delta}^0$	17.6	$K^{*0}\pi^+$	1.3
$K^+\pi^-$	0.9	$\rho^0 K^+$	0.7
$K^{*0}\gamma$	2.8	φK^+	0.8
e^+e^-	0.3	$K^{*+}\gamma$	9.2
$\mu^+\mu^-$	0.4	$K^+\mu^+\mu^-$	1.2
$\mu^\pm e^\mp$	0.3	$K^+e^+e^-$	0.5

The transition $b \to u$ may be sought in the decays $B^- \to \rho^0 l^- \overline{\nu}_l$ and $\overline{B}^0 \to \pi^+ l^- \overline{\nu}_l$. The neutrino, of course, is undetected, so a method is required for distinguishing between these final states and final states actually derived from $b \to c$ transitions in which one of the final state particles escapes detection. The lightest charmed particle is the D meson; the lepton momentum spectrum for $B \to Dl\nu$ falls to zero at 2.4 GeV/c. The ρ and π mesons are much lighter than any charmed particles, so the leptonic momentum spectrum for $b \to u$ transitions extends up to 2.7 GeV/c.

Leptons are required to have momenta* between 2.3 and 2.7 GeV/c. The hadron and lepton momenta are required to be consistent with the three-body decay of a B meson. Since the decay is mediated by a 'V-A' interaction, near the endpoint of the lepton momentum spectrum, the hadron and lepton should be produced nearly back-to-back.

No evidence of any signals is found. In order to calculate upper limits on the branching ratios, it is necessary to know the fraction of the lepton spectrum in the range 2.3 to 2.7 GeV/c, that is, it is necessary to know the spectral shapes. These shapes and corresponding decay rates are provided by a number of models. The upper limits obtained on the basis of three models — those of Grinstein, Isgur and Wise (GIW) [11] ; Wirbel, Bauer, and Stech (WBS) [12] ; and Körner annd Schuler (KS) [13] — are given in Table III. Note that the presence of the undetected neutrino removes a number of kinematic constraints, so it is impossible for CLEO to distinguish between a ρ^0 and an ω^0.

Table III. Upper Limits on $|V_{bu}/V_{bc}|$ from $B^- \to \rho^0 l^- \overline{\nu}_l$ and $\overline{B}^0 \to \pi^+ l^- \overline{\nu}_l$

| | $B^- \to \rho^0 l^- \overline{\nu}_l$ | | $\overline{B}^0 \to \pi^+ l^- \overline{\nu}_l$ | |
| | Upper Limit | | Upper Limit | |
Model	Branching Ratio ($\times 10^{-3}$)	V_{bu}/V_{bc}	Branching Ratio ($\times 10^{-3}$)	V_{bu}/V_{bc}
GIW	1.5	0.25	1.0	0.42
WBS	1.6	0.19	1.5	0.25
KS	0.8	0.11	1.4	0.23

Of particular interest are the decays $B^+ \to p\overline{p}\pi^+$ and $B^0 \to p\overline{p}\pi^+\pi^-$. These have previously been reported by the ARGUS collaboration, [14] and constitute the only positive signals for $b \to u$ transitions yet reported. CLEO finds no evidence for these decays and the upper limits obtained [15] — $B(B^- \to p\overline{p}\pi^-) < 1.4 \times 10^{-4}$ and $B(\overline{B}^0 \to p\overline{p}\pi^+\pi^-) < 2.9 \times 10^{-4}$ — in fact contradict the ARGUS measurements of $B(B^+ \to p\overline{p}\pi^+) = (5.2 \pm 1.4 \pm 1.9) \times 10^{-4}$ and $B(B^0 \to p\overline{p}\pi^+\pi^-) = (6.0 \pm 2.0 \pm$

$\star$ The value of 2.3 GeV/c is used as a low cut-off as the contribution from charm should already be negligible.

2.2) $\times 10^{-4}$. CLEO has found no single background source which could explain the signals seen by ARGUS.

IV. Conclusions and Prospects

The data presented in the preceeding sections may be analyzed on the basis of several models for B meson decay. For example, in the model of Bauer and Stech,[16] the branching ratios

$$B(B^0 \to D^{*-}\pi^+) = 5.0 \times 10^{-3}(a_1/1.1)^2|V_{bc}/0.05|^2 \qquad (11)$$

$$B(B^0 \to \pi^-\pi^+) = 2.1 \times 10^{-3}(a_1/1.1)^2|V_{bu}/0.05|^2 \qquad (12)$$

may be combined with the measurements given above to obtain the limit $|V_{bu}|/|V_{bc}| <$ 0.2 at the 90% confidence level. Different models, not surprisingly produce differing limits. The limits[17] from the semileptonic decays given in Table III range from 0.23 to 0.42 for the former final state and from 0.11 to 0.25 for the latter. The model dependence is apparent. The various models do agree, however, that the magnitude of V_{bu} appears to be much smaller than that of V_{bc}. The precise ratio of these is sadly still to be determined.

The CLEO collaboration hopes to help clarify this situation when data from the new CLEO II detector becomes available in late 1989. This detector combines the excellent charged particle detection capabilities of the original CLEO detector with state-of-the-art photon calorimetry.

Acknowledgements

This work was supported by the National Science Foundation and the US Department of Energy under contracts DE-AC02-76ER01428, DE-AC02-76ER03066, DE-AC02-76ER03064, DE-AC02-76ER01545, DE-AC02-76ER05001, and FG05-86-ER40272. The Cornell National Supercomputer Facility funded in part by the NSF, the State of New York, and IBM was used in this research.

REFERENCES

1) S. Stone, "The Quark Mixing Matrix, Charm Decays and B Decays", in *CP Violation*, ed. C. Jarlskog, World Scientific, Singapore, 1988.

2) D. Andrews *et al.*, *Nucl. Inst. Meth.* **211**, 47(1983).

3) A. Brody *et al.*, *Phys. Rev. Lett.* **48**, 1070(1982);
   M.S. Alam *et al.*, *Phys. Rev. Lett.* **58**, 1814(1987);
   D. Bortoletto *et al.*, *Phys. Rev.* **D35**, 19(1987);
   P. Haas *et al.*, *Phys. Rev. Lett.* **56**, 2781(1986);
   H. Albrecht *et al.*, *Phys. Lett.* **187B**, 425(1987);
   M.S. Alam *et al.*, *Phys. Rev.* **D34**, 3279(1986);
   H. Albrecht *et al.*, *Phys. Lett.* **162B**, 395(1985);
   M.S. Alam *et al.*, *Phys. Rev. Lett.* **59**, 22(1987).

4) R. Poling *et al.*, "Inclusive and Exclusive Decays of B Mesons to D^{*+} Mesons", submitted to XXIV Int. Conf. on High Energy Physics, Munich (1988).

5) D. Hitlin, "Weak Decays of Charmed Particles and τ Leptons", in *Proc. of 1987 Int. Symp. on Lepton and Photon Interactions at High Energies*, ed. W. Bartel and R. Rückl, North Holland, Amsterdam (1987).

6) J. Adler *et al.*, *Phys. Rev. Lett* **60**, 89(1988).

7) A. Jawahery *et al.*, "A Study of Inclusive and Exclusive Decays of B Mesons to D_s Mesons", submitted to XXIV Int. Conf. on High Energy Physics, Munich (1988).

8) C. Bebek *et al.*, "The Decay of B mesons to ψ", submitted to XXIV Int. Conf. on High Energy Physics, Munich (1988).

9) C. Bebek *et al.*, *Phys. Rev.* **D36**, 1289(1987).

10) P. Baringer *et al.*, "Limits on Rare Exclusive Decays of B Mesons", submitted to XXIV Int. Conf. on High Energy Physics, Munich (1988).

11) B. Grinstein, N. Isgur, and M. Wise, *Phys. Rev. Lett.* **56**, 298(1986);
    B. Grinstein and M. Wise, *Phys. Lett.* **197B**, 249(1987).

12) M. Wirbel, M. Bauer, and B. Stech, *Z. Phys.* C29, 637(1985).

13) J.G. Körner and G.A. Schuler, *Z. Phys.* **C38**, 511(1988).

14) H. Albrecht *et al.*, "Observation of Charmless B Meson Decays", DESY 88-056(1988).

15) C. Bebek *et al.*, "Search for the Charmless B Decays $B^- \rightarrow p\bar{p}\pi^-$ and $\overline{B}^0 \rightarrow p\bar{p}\pi^+\pi^-$, submitted to the XXIV Int. Conf. on High Energy Physics, Munich (1988).

16) M. Bauer and B. Stech, *Z. Phys.* **C34**, 103(1987).

17) T. Bowcock *et al.*, "Investigation of the Charmless Decays $\overline{B}^0 \rightarrow \pi^+ l^- \bar{\nu}_l$ and $B^- \rightarrow \rho^0 l^- \bar{\nu}_l$", submitted to the XXIV Int. Conf. on High Energy Physics, Munich (1988).

3

B-PHYSICS POSSIBILITIES AT D0

David Hedin

Northern Illinois University, DeKalb IL 60115

This report presents a study of the capabilities of the D0 detector to do B-related physics by utilizing the presence of a muon both as a trigger and as a tag.[1] The almost complete muon coverage plus the fine calorimeter segmentation will allow D0 to expand upon the B-physics which have been done at UA1 with muons.[2] This report will briefly discuss the physics topics and then will give rate calculations from different muon sources along with studies on how to separate them. Finally, studies of $\psi \rightarrow \mu\mu$ and $B \rightarrow \psi K_S$ will be presented.

I. Physics Issues

The D0 detector has been described elsewhere[3] and its main feature that will allow it to study B-related items is its muon detector. This consists of iron toroids from 1-1.5 m thick located outside the uranium liquid argon calorimeter. Drift chambers are interspersed between the calorimeter and the iron and then outside the iron with coverage down to a minimum angle of 4 degrees. The total thickness varies from 14 to 20 absorption lengths with about half this due to the calorimeter and the rest from the iron. For angles greater then 11 degrees, there are 10 planes of 10 cm drift cells (all with the drift direction in the bend view) and the small angle region has 18 planes (X,Y,U) of 2.5 cm cells. With this, the D0 detector should be able to study a wide range of B-physics items both at high and low p_t. Among them will be:

A. B-production cross sections versus x and p_t including separating direct B-jet production from gluon splitting ($g \rightarrow BB$).

B. Searches for heavy states decaying into B's such as $Z^0 \rightarrow b\bar{b}$ and $H^0 \rightarrow b\bar{b}$.

C. $B^0 - \overline{B}^0$ oscillations through the like-to-unlike sign mu-pair ratio.

D. CP-violation through charge asymmetries.

E. Searches for reconstructible rare decays such as $B_d \rightarrow \psi K_S$ or $B_c \rightarrow \psi\mu\nu$.

The first two items will depend upon being able to separate different sources of B-jets (see reference 1). Though a complete study has not been done, by requiring both B's to decay leptonically, kinematic cuts may reduce the background from $g \rightarrow b\bar{b}$ and allow $Z^0 \rightarrow b\bar{b}$ to be observed (albeit at low rates). Items C-E are essentially exploring low-p_t events and here the strategy is to begin with a muon trigger, identify a sample rich in $B \rightarrow \mu X$ decays, and then look at the rest of the event for either CP-related charge asymmetries or rare decays.

The most accessible CP-violating channel will be the like sign dimuon charge asymmetry. While this asymmetry is expected to be small, statistical limits should be close to current theoretical expectations. For example, an integrated luminosity of 10 pb^{-1} (500 pb^{-1}) will give about 10^5 (5 10^6) like-sign dimuon events with a 1 sigma statistical error of .002 (.0003). Ignoring backgrounds from non-B decays (which should be correct to within 1.4), this should be compared to the expected asymmetry from $B^{\pm}B_d^0$ decays estimated to be a "few" times 10^{-3}. As we cannot separate $B^{\pm}B_d^0$ from B^0B^0 decays, a dilution factor of about 3 will reduce our sensitivity. The systematic error is hard to estimate at the current time though events such as $\Upsilon \to \mu^+\mu^-$ and $\psi \to \mu^+\mu^-$ will help in understanding both μ^+/μ^- and $p/\bar{p}$ differences.

As emphasized by Bigi and Sanda,[4] it will probably also be possible to look for differences between states with K_S's in them. For a tagged CP-eigenstate such as ψK_S, charge asymmetries in the range from 0.05 to 0.30 are predicted. This will be experimentally diluted by 2.5 if charge and neutral $B \to \mu$ decays aren't separated. As we will show below, the rate of ψK_S events is about a picobarn including acceptances and efficiencies. So for 500 pb^{-1}, the 500 expected tagged events will give a statistical error of 0.04. If instead of the exclusive channel we sum over all $B_d \to \psi + K_S + X$ events, then we can gain events but the asymmetry will be diluted. The semi-exclusive branching ratio from B_d is 10 times larger and most of this seems to be CP = 1 three body decays like $\psi K_S \pi^0$. Events from $B_u \to \psi + K_S + X$ will increase the dilution due to not being able to separate charged and neutral B decays to 6. So the numerology will be 20 times more events with the expected asymmetry being (.05-.30)/6/1.5 = .005-.03 (the last factor of 1.5 is for the average CP being reduced). So again, many years at high luminosity will be needed unless new physics gives larger asymmetries.

One can also look for asymmetries in exclusive K_S events. There are two dilutions which enter in. The first is the factor of 6 due to the inability of separating charged and neutral B's (as above). The second is the fraction of $B_d \to K_S + X$ states which are CP-eigenstates and the resultant 'average' CP of inclusive K_S events. This has been estimated as 3 so that the final charge asymmetry is (.05-.30)/3/6 = .003-.02. About 20% of B's will decay into K_S and so, with the $K \to \pi^+\pi^-$ branching ratio, the inclusive K_S rate is comparable to the muon rate. If acceptances/efficiencies are the same then the statistical limits for the dimuon rates given above will apply to K_S events with the same conclusions regarding systematic limits and D0 expectations. There are, of course, experimental difficulties in associating a K_S with a non-leptonically B-decay and only preliminary studies have been done (see Section IV below).

One should note that the dilution of any asymmetry of from 3-6 due to mixing neutral and charge B-decays can, in principle, be eliminated with the proper vertex detector. Rejection versus efficiency studies for different technologies have not been studied but as the dilution factor is quite large, if a sample of predominantly charged or neutral decays could be obtained with modest efficiency (say 30%), this could increase the ability to measure any asymmetry.

II. Muon Rates and Sources

A set of ISAJET events (minimum bias plus unbiased two-jet events) and a D0 trigger simulator were used to estimate muon rates and the effects of kinematic cuts on separating different muon sources (see reference 1 for details). The acceptance versus muon p_t, integrated over angles for QCD events, and versus angle are given

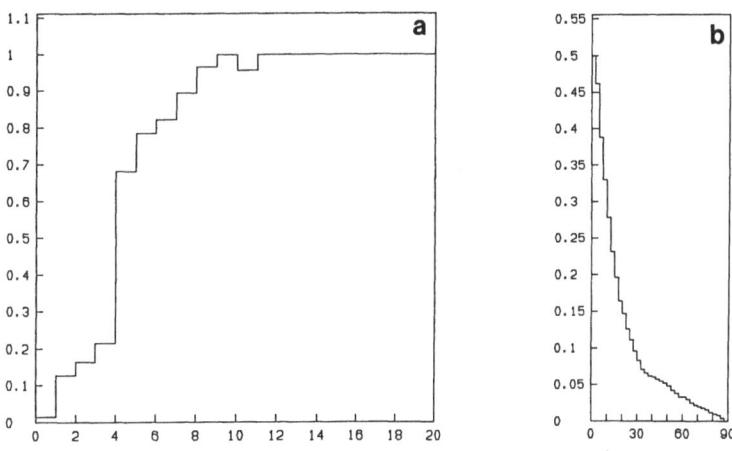

Fig. 1. Single muon trigger acceptance versus a) $p_{t\mu}$ and b) θ_{μ}.

in Figure 1 with the p_t distribution and the pseudorapidity (η) distributions shown in Figure 2. A 5 degree minimum angle is assumed.

The muon rate at low p_t (and angles) will be dominated by π and K decays in flight. Compared to UA1, the B-production cross section is increased by about three and the decay path length in the central region is reduced by two so the relative decay rates in D0 should be appreciably less allowing $B \to \mu$ tagging at lower p_t. For now we will use ISAJET to determine the decay rates while recognizing that the low-p_t region is the place that is hardest to calculate. Again, a 5° minimum angle will be assumed, though the minimum angle at which D0 will be able to trigger can only be determined using the actual detector. Figure 3 gives the fraction of decays as a function of $p_{t\mu}$ while Table 1 summarizes the single muon rates (note that at $L = 10^{30} \ cm^{-2}sec^{-1}$ a 1 μb cross section corresponds to 1 Hz) and the dimuon rates assuming at least one of the muons passed the trigger with the other at least making it through the toroid.

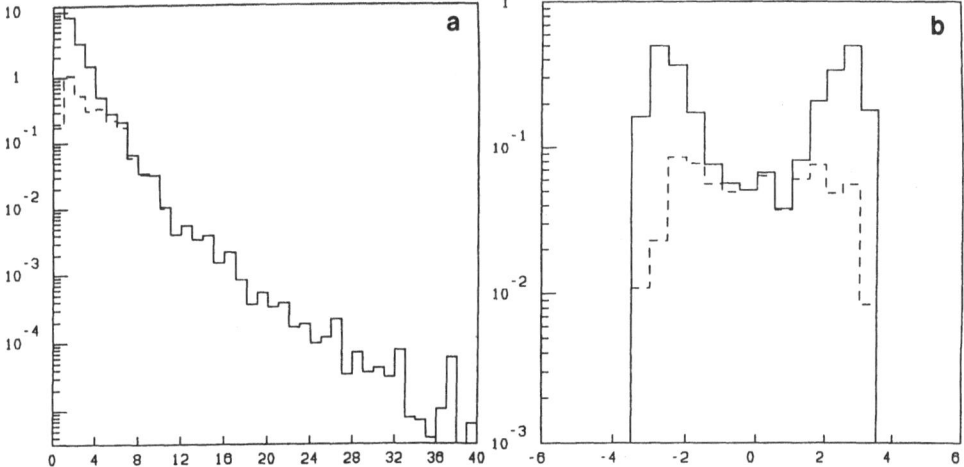

Fig. 2. The a) p_t distributions from QCD sources before (solid) and after (dashed) applying the muon trigger (in μb/GeV/c) and the b) η distributions for all (solid) muons passing the trigger and those with p_t greater than 4 GeV/c (dashed).

Table 1. Muon Trigger Rates

	Source	Raw	Pass Trigger
1 μ	π/K Decays	1000	30
	$B, C \to \mu$	15	3
2 μ	$B, C \to \mu\mu$		.17
	$B, C \to \mu + \pi/K$ decay		.08

$\theta_\mu > 5^o$; rates in μb; no additional cuts

From the above calculation, a $p_{t\mu}$ cut at about 6 GeV/c will give a sample which is more than half from B-decays. We have looked at enhancing this by decay and kinematic cuts applied to single muon events. The average angle between the π/K-direction and its decay muon is 6 mrad for those decay muons which pass the trigger. This obviously depends upon the momentum; for example it is 2 mrad for $p_{t\mu} > 5$ GeV/c or still equal to 6 mrad for $\theta_\mu > 20^o$ (same average momentum). The resolutions of the three inner tracking chambers are 0.3 mrad, 0.7 mrad and 0.7 mrad for the micovertex, central and end chamber systems. These resolutions do not include any vertex or matching enhancements and for the microvertex and central are 2D (r-ϕ) only. We have not yet done a study of how well D0 will reject kink events and for now assign a kink "cut" of 2 mrad. This will tend to eliminate low-p_t decay events.

The second series of cuts relies upon the energy deposition in the calorimeter "near" the muon. B-decays should produce both wider hadronic activity and more isolated muons compared to C-decays especially at lower-p_t. In both cases, some hadronic activity should be related with the muon as opposed to decays which often come from an isolated π or K. We have looked at this using a perfect detector (no segmentation, resolution or shower size) in order to get a first level feeling for the sensitivity we will have. We have defined a few cuts but have made no attempt to optimize the analysis.

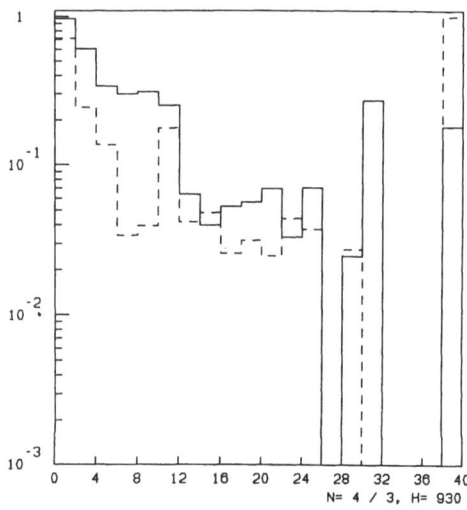

Fig. 3. The fraction of muon events passing the trigger from π/K decays versus $p_{t\mu}$ for all (solid) events and those with kink and b-cuts (dashed).

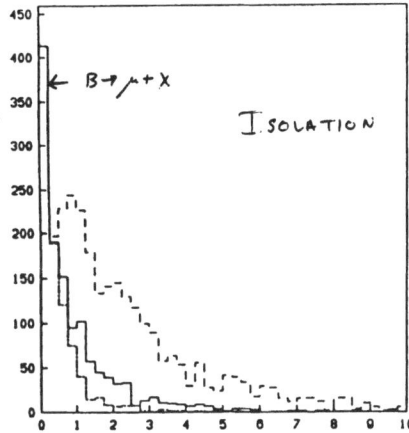

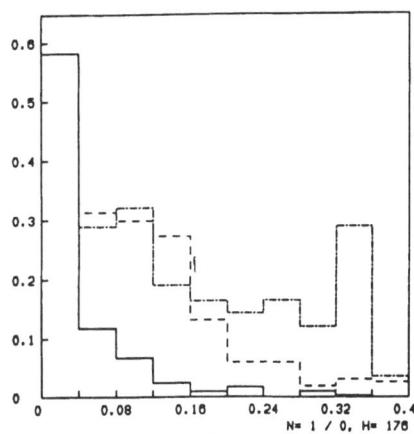

Fig. 4. The $E_T(\Delta R < .3)/p_{t\mu}$ distributions for muons from B (solid), C (dashed) and π/K (dash-dot) decays.

Fig. 5. The $E_T(\Delta R < .3)/E_T(\Delta R < 1)/p_{t\mu}$ distributions for muons from B (solid), C (dashed) and π/K (dash-dot) decays.

The three requirements we have tried are ("b-cuts"):

A. $E_T(\Delta R < .3)/p_{t\mu} \quad < \quad 1 + .15 p_{t\mu} \qquad$: Isolation

B. $E_T(\Delta R < .3)/E_T(\Delta R < 1)/p_{t\mu} \quad < \quad .05 \quad$: Jet Width

C. $p_{t\mu}$ relative to hadron axis > 0.5 GeV/c

where the hadron axis in C is just the sum of all calorimeter energy in a cone of ΔR less than 1 about the muon (i.e. no clustering). Distributions for these items are given in Figures 4-6. The relative number of $B \to \mu$ events to the sum of B and C with these cuts is shown in Figure 7 and the fraction of decay muons to all muon sources with both these cuts and the 2 mrad kink cut is given in Figure 3. Taking these results at face value, for $p_{t\mu} > 2$ GeV/c plus b-cuts and the kink cut gives a .5 μb $b \to \mu$ cross section and about 60% of the muon events coming from B-decays. A matrix of our current best guesses for muon rates is given in Table 2.

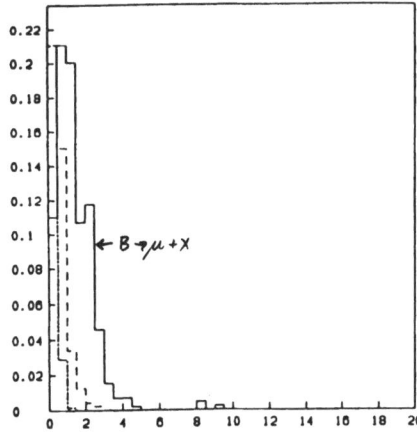

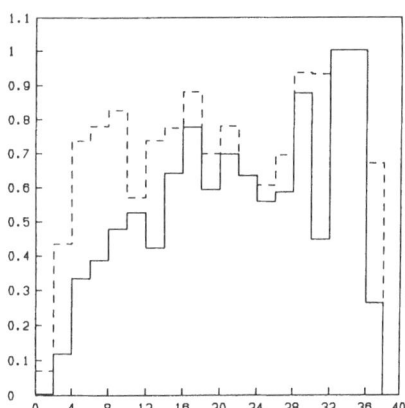

Fig. 6. The $p_{t\mu}$ relative to the hadronic axis distributions for muons from B (solid), C (dashed) and π/K (dash-dot) decays.

Fig. 7. The fraction of $B \to \mu$ events compared to all QCD sources (no π/K decays) after the trigger for all (solid) and with b-cuts (dashed) versus $p_{t\mu}$.

31

Table 2. Muon Rate Summary

| | | σ (μb) | | Fraction |
	π/K decay	$C \to \mu$	$B \to \mu$	$B \to \mu$
$p_{t\mu} > 0$				
Raw	1000			
Trigger	34	2.0	.84	.02
Kink cut	6.6	2.0	.84	.09
B-cut	1.4	.40	.56	.24
Kink + B-cut	0.8	.40	.56	.32
$p_{t\mu} > 2$				
Raw	5.1			
Trigger	2.0	1.1	.70	.18
Kink cut	0.7	1.1	.70	.28
B-cut	.34	.21	.50	.48
Kink + B-cut	.16	.21	.50	.58
$p_{t\mu} > 4$				
Raw	.50			
Trigger	.4	.31	.40	.36
Kink cut	.14	.31	.40	.48
B-cut	.08	.06	.29	.67
Kink + B-cut	.04	.06	.29	.74

III. $\psi \to \mu\mu$ Study

The $\psi \to \mu\mu$ channel has already been seen at UA1.[5] With their mass resolution of 150 MeV, a nice peak with 247 signal events over a background of 26 was found. They estimated that about 80% of these ψ's came from χ decay and the rest from B-decay. Attempts to verify this using the data (related hadronic energy) proved inconclusive. Accepting this, their ratio of $(B \to \psi \to \mu\mu)/(QCD \to \mu\mu)$ was about 2/1.

D0 is different than UA1 in two ways. The first is that we will cover to a much lower muon angle. Figure 8 gives the angular distribution of muons (for events where both muons are outside the toroid and one passes the muon trigger) and Figure 9 gives the acceptance as a function of both the minimum angle and the maximum angle covered. The second is that the muon momentum resolution and thus the dimuon mass resolution will be about twice that in the UA1 ψ sample. Simply scaling from UA1 gives a background level from QCD (BB, CC) of about 1/1 relative to B-decays. As we will see below, an ISAJET calculation confirms this. We can attempt to improve this by:

A. Adding air-gap magnets outside the iron in the forward region

B. Using hadronic energy deposits near the ψ

C. Using vertex information

There seems to be 1 meter of so of "available" space in the forward regions. As can be seen in Figure 9b, about half the accepted ψ events will have both muons in the region from 5-20 degrees (only one of the muons needs to be in this region to improve the mass resolution of $B \to \psi K_S$ events). If we assume an air-gap magnet with a 100 MeV/c field and the current set of drift chambers then the muon momentum resolution would be about 0.003p. This would improve the dimuon mass resolution at the ψ mass by two.

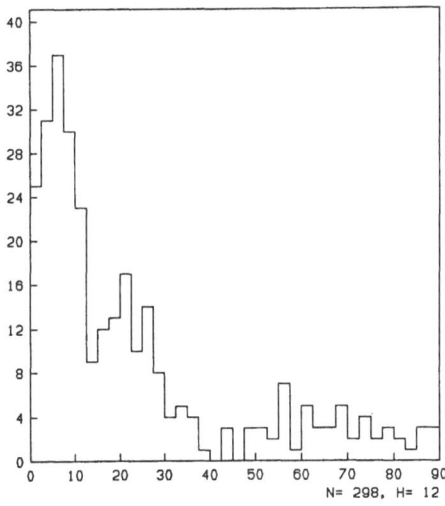

Fig. 8. The θ distribution for muons from $\psi \to \mu\mu$ decays.

The hadronic energy near the muons was studied using a parameter S such that
$$ S = (E_{T1}(\Delta R < .7)/p_{t\mu 1})^2 + (E_{T2}(\Delta R < .7)/p_{t\mu 2})^2. $$
As seen in Figure 10, the signal to background ratio can be improved by about two but with some signal loss.

Vertex detection capability could also improve background rejection. The current D0 microvertex detector only has 2D $(r - \phi)$ readout and is of limited use in rejecting QCD dimuon events (though it can measure the proper time). With a 3D detector, one can measure the distance of closest approach (DCA) of the two tracks. For QCD dimuon events near the ψ mass, the average DCA is 5 mm (figure 11). Thus a vertex detector located in the forward region should have rejections of 5 or more with good signal efficiency. The expection cross sections and backgrounds for ψ's from B-decay are summarized in Table 3 for both single $\psi \to \mu\mu$ events and for those where the other B decays to a muon.

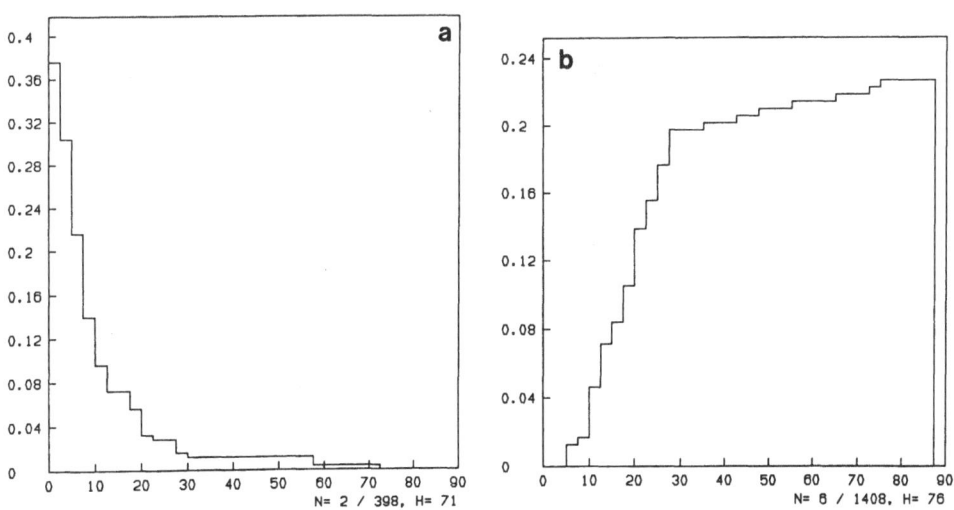

Fig. 9. The geometric acceptance versus a) minimun angle and b) maximum angle (with a 5 degree minimum angle) for $\psi \to \mu\mu$ events from B-decays.

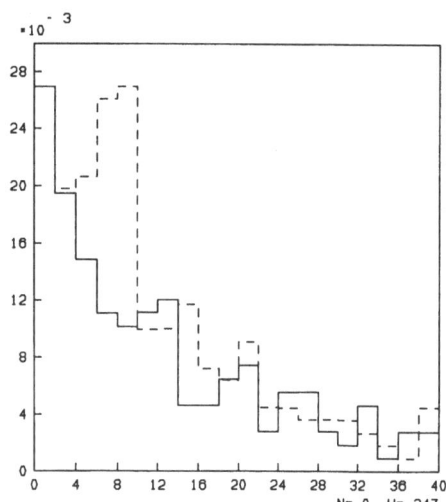

$\cdot 10^{-3}$

N= 0, H= 247

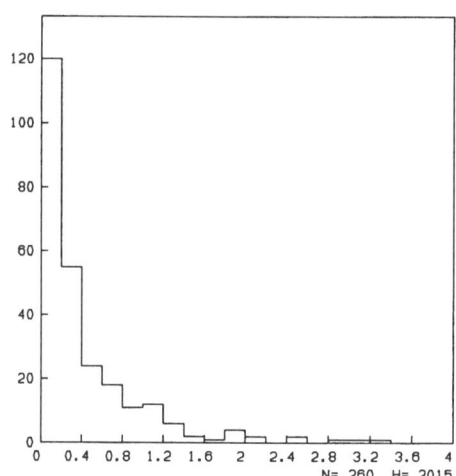

N= 260, H= 2015

Fig. 10. The $S = (E_{T1}/p_{t\mu1})^2 + E_{T2}/p_{t\mu2}^2$ distributions for $\psi \to \mu\mu$ events (solid) and $QCD \to \mu\mu$ events (dashed).

Fig. 11. The distance of closest approach (in cm) of the two muons from $\psi \to \mu\mu$ events (solid) and from $QCD \to \mu\,\mu$ events (dashed).

Table 3a. $B \to \psi \to \mu\mu$ Analysis Chain

| | $B \to \psi \to \mu\mu$ | | $QCD \to \mu\mu$ | Signal/Noise |
	ϵ	σ (nb)	σ (nb) $2 < M_{\mu\mu} < 4$	± 300 MeV
Trigger	.14	5.9	33	.6
$p_{t\mu\mu} > 2.5$	.13	5.5	22	.8
$S < 4$	.09	3.9	12	1.1
DCA < 1 mm	.12	5.5	8	2.4

Table 3b. $BB \to \psi \to \mu\mu\mu$ Analysis Chain

| | $B \to \psi \to \mu\mu$ | | $QCD \to \mu\mu$ | Signal/Noise |
	ϵ	$\sigma(nb)$	$\sigma(nb)$ $2 < M_{\mu\mu} < 4$	± 300 MeV
Trigger	.10	.24	.81	1
$p_{t\mu\mu} > 2.5$	.09	.22	.52	1.4
$S < 4$	.05	.13	.2	2.2
DCA < 1 mm	.09	.22	.2	3.2

$S = (E_{T1}(\Delta R < .7)/p_{t\mu1})^2 + (E_{T2}(\Delta R < .7)/p_{t\mu2})^2$.
DCA = distance of closest approach.
The last two cuts were made independently.

Table 4. $B \to \psi + K_S$ Decay Chain

	rate in μb
$B \to \mu$ trigger rate	0.70
Single muon efficiency (.5)	0.35
Fraction B_d (.4)	0.14
BR $B \to \psi + K_S$ (.0005)	7×10^{-5}
BR $\psi \to ll$ (.14)	1×10^{-5}
BR $K_S \to \pi^+ \pi^-$ (.7)	7×10^{-6}
Acceptance$\times$efficiency (.14)	1×10^{-6}

IV. $B \to \psi + K_S$ Study

As stated above, it appears as if D0 will be able to trigger on $B \to \mu$ events and then used these events to search for some rare B decay modes. Including acceptances and efficiencies (Table 4), the $BB \to (\mu X) + (\psi K_S)$ rate will be about a picobarn. The assumed overall reconstruction efficiency of 10% may be optimistic.

Most of the events come from very low-p_t events. The average muon (electron) and pion energies are 8 and 5 GeV with the particles populating the entire detector. The K_S's average decay path is 33 cm so a large fraction are decaying in the μVTX detector and almost all have decayed before the CDC and FDC. The geometric acceptance for D0 seem to be reasonable (see Figure 12) with a 50% acceptance for $B \to (ee) + (\pi\pi)$ and a minimum angle of 10°. The central muons will range out so that the $B \to (\mu\mu) + (\pi\pi)$ acceptance is 20% for a 5° minimum angle. The events themself are clean. Based upon an eyescan of a small number of events, about 20% will be both accepted and clean with the remaining 30% of the accepted events either being busy or having 'small' 2-pion opening angles. We have begun attempting to reconstruct these events and the initial results look promising.

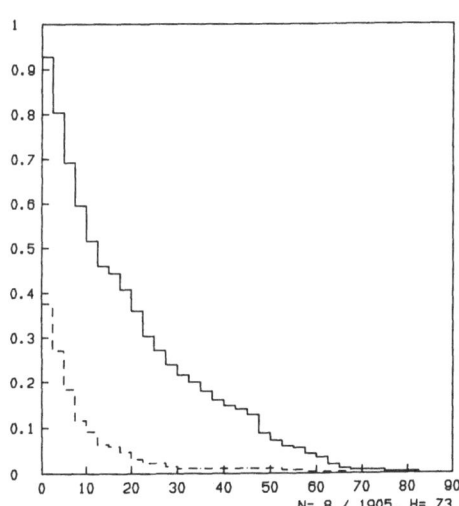

Fig. 12. The geometric acceptance versus minimun angle for four body $B \to ee + \pi\pi$ (solid) and $B \to \mu\mu + \pi\pi$ (dashed) for $B \to \psi + K_S$ events.

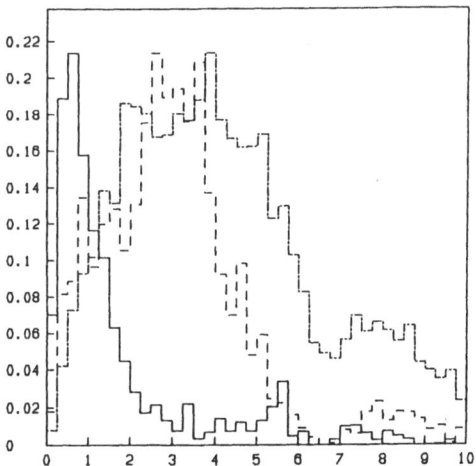

Fig. 13. The $\Delta R_{\mu K}$ for the muons and kaons in $B \to \mu X$ events with the kaon coming from the $B \to \mu$ (solid), the 'other' B (dashed) and a K from the underlying event (dot-dashed).

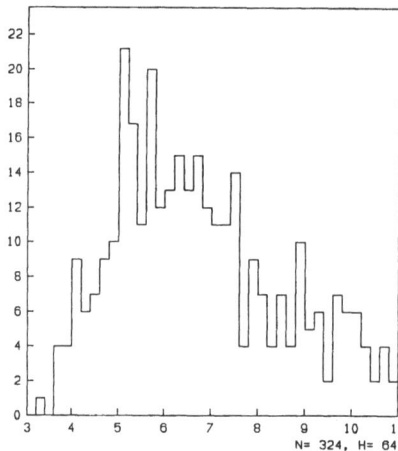

Fig. 14. The $\mu\mu\pi\pi$ mass in GeV for $B \rightarrow \psi + K_S$ events (solid) and the ψ plus a random K from the event (dashed). A 20 pb^{-1} integrated luminosity is assumed.

Most of the leptons and pions are reasonably isolated in the calorimeter. Measuring the pion decay angles and their vertex allows the K_S energy to be determined to $\approx 10\%$. The mass of the ll-pair can be constrained to the ψ mass giving a (ψK_S) mass resolution of 50 MeV RMS ($\psi \rightarrow ee$), 100 MeV RMS ($\psi \rightarrow \mu\mu$), and 30 MeV RMS ($\psi \rightarrow \mu\mu$, one μ in air dipole with .003p resolution).

Only a very preliminary look at backgrounds has been done. The background to ψ decay was discussed above. If we assume that misidentifications in pattern recognition is small, then the dominant K_S background is from other K_S's in the event. For $B \rightarrow \mu + X$ events, ISAJET predicts .3 K_S/event from the $B \rightarrow \mu$, .4 K_S/event from the 'other' B and 3.5 K_S/event from the underlying event. Those from the underlying event will have less p_{tK}: a cut at 1.0 GeV/c eliminates 86% of them but only 53% from B-decay and 20% from $B \rightarrow \psi K_S$. One can also look for spatial correlations. Figure 13 gives the $\Delta R_{\mu K}$ distribution. This clearly helps in separating the source of the K_S. Figure 14 shows the ψK_S mass using the background K_S's in the $B \rightarrow \psi K_S$ events with a p_t cut of 1 GeV/c for an assumed integrated luminosity of 20 pb^{-1}. As is, the signal to background is about 3/1. Hopefully this is improved as backgrounds are eliminated using cuts like $\Delta R_{\psi K}$ (which is equivalent to a cut on the angle in the center of mass).

I would like to thank my collaborators on D0 for their comments on this study. This work was supported by the Department of Energy and the Northern Illinois University Graduate School.

REFERENCES

1. D. Hedin, T. Kramer, K. Roberts, B-Physics at D0, Proceedings of the Workshop on High Sensitivity Beauty physics at Fermilab, November, 1987, p. 265. A. Kernan (p. 43), D. Smith (p. 321) and C. Rahal-Callot, T.Trippe and D. Zieminska (p. 327) submitted related reports to this workshop. Also, D. Hedin, D0 Note 666 (1988).
2. C. Albajar et al., Phys. Lett. 186B, 237 (1987).
3. D0 Design Report, Fermilab (1984).
4. I.I. Bigi and A.I. Sanda, Nuc. Phys. B281, 41 (1987).
5. C. Albajar et al., Phys. Lett. 200B, 380 (1988).

4

B MESON LIFETIME MEASUREMENT

M. Piccolo

Laboratori Nazionali di Frascati dell' I.N.F.N.
Frascati, Italy

1. Introduction

The lifetime of hadrons containing b-quark has been the subject of extensive experimental work and theoretical speculation; its importance is due to implications on some of the fundamental parameters of the Standard Model, such as the top quark mass and the mixing angles.

Since the pioneer measurements of the MAC and MARK II collaborations at PEP in 1983[1][2] the progress has been impressive; but many issues still remain open and await further study.

In this paper we will discuss the field's present status; we will start with an overview of the theoretical motivations for this measurement in the Standard Model framework; we will then review the experimental techniques used, emphasizing the most recent measurements. The following section will be devoted to the results comparison and to the systematic errors discussion; we will conclude with some remarks on the further developments foreseen in the near future.

2. The theoretical implications

In the framework of the Standard Model, the b quark plays a very important role: the study of the B mesons properties can enable us to obtain the values for three out of the four free parameters of the Cabibbo-Kobayashi-Maskawa[3] matrix, which alone determines the structure of the weak charged currents. Two of the three mixing angles are infact linked to the decay properties of the b hadron; the phase which appears in the imaginary part of the matrix elements, and is thought to be responsible for the CP violation, is presently constrained only by K mesons data, but is at least just as sensitive to B physics phenomena.

A brief reminder : the weak charged current quark transitions

$$q_i \longrightarrow q_j \ W$$

proceed with a width proportional to $|V_{ij}|^2$, where V_{ij} is an element of the CKM matrix:

$$V = \begin{pmatrix} V_{ud} & V_{us} & V_{ub} \\ V_{cd} & V_{cs} & V_{cb} \\ V_{td} & V_{ts} & V_{tb} \end{pmatrix}$$

Several parametrizations have been suggested for this matrix, a particularly convenient one by Maiani[4]:

$$\begin{pmatrix} cos\beta cos\theta & cos\beta sin\theta & sin\beta \\ -sin\gamma cos\theta sin\beta e^{i\delta} - sin\theta cos\gamma & cos\gamma cos\theta - sin\gamma sin\beta sin\theta e^{i\delta} & sin\gamma cos\beta e^{i\delta} \\ -sin\beta cos\gamma cos\theta + sin\gamma sin\theta e^{-i\delta} & -cos\gamma sin\beta sin\theta - sin\gamma cos\theta e^{-i\delta} & cos\gamma cos\beta \end{pmatrix}$$

where, in first approximation, θ is the well known Cabibbo angle, γ and β are the equivalent *mixing* angles between the second and the third, the first and the third generations, respectively.

The b quark lifetime, ignoring phase space factors and in the crude approximation where the B semielectronic branching ratio is 1/9, is given by:

$$\frac{1}{\tau_b} = 9(\frac{m_b}{m_\mu})^5 \ \frac{1}{\tau_\mu} \ (|V_{cb}|^2 + |V_{ub}|^2) =$$

$$= 9 \ \frac{G_F^2 m_b^5}{192\pi^3} \ (|V_{cb}|^2 + |V_{ub}|^2) = 9 \times 10^{14} \ (\frac{m_b}{4.9 GeV})^5 \ (|V_{cb}|^2 + |V_{ub}|^2) sec^{-1}$$

and is a measurement of $|V_{bc}|$, since $|V_{bu}|$ is much smaller, as known from the fact that the B mesons decay almost entirely into charmed particles.

In the *naive* spectator model, the B hadron lifetime is the same as the free b quark's, since the light quark doesn't play a significant role in the decay mechanism: all B particles are therefore expected to have the same lifetime. This is of course an oversimplified picture and even if the *spectator* approximation is a valid one, the fact that we must deal with particles, not free quarks, both in the initial and in the final state, must be considered.

Due to the complications deriving from the hadronization, the study of the B-mesons lifetime, which is a measure of the global strength of the decay, is an important probe not only for the weak charged currents, but also for QCD effects. As we consider the various phenomena which may contribute to the real particles total widths, the picture gets increasingly complicated. A partial list of effects to be considered, includes:

a) Mass effects[5]: mostly of a purely kinematics nature, but still important, especially for the B decays into heavy lepton τ and charmed particles.

b) Radiative QCD corrections[6].

c) Short distance[7][8] gluon emission, which leaves the semileptonic widths unchanged, while enhancing the hadronic channels.

More effects could lead to deviations from the *naive* picture: additional non spectator diagrams may contribute to the B_d 's and B_u 's width ; interference effects, when the final state can be reached through different paths, could reduce the partial width into a given channel. These and other contributions could in principle be disentangled, provided that separate branching ratios and lifetimes for the various B mesons species can be measured.

Overall, the spectator model should be a better approximation for the B system than for the D, since the b quark is much heavier than the charm, and the lifetime measurement can be quite reliably translated into determinations of the quantity $|V_{bc}|^2 + |V_{bu}|^2$. The branching ratios measurements would constrain the $|V_{bu}|/|V_{bc}|$ ratio; this is crucial, since the CP violation phenomena can only be explained as a by-product of the quarks mixing in the charged current interaction, if $|V_{bu}|$ is not zero, but we don't yet have any hard evidence to sustain this.

3. The b-lifetime measurement in e^+e^-

All of the experimental data now available on the B lifetime come from e^+e^- annihilation experiments, except one beautiful event with two B decay vertices observed by WA75 at the CERN SPS[9]. The situation is likely to change in the next few years as more fixed target experiments are about to start taking data, with sophisticated solid state vertex detectors.

The b flavored particles *average* lifetime has been measured by several PEP and PETRA experiments ; the first non-zero results were announced in 1983 by MAC and MARK II at PEP and came as a big surprise, since a much shorter value was expected, on the basis that the mixing angle γ was roughly the same as the Cabibbo angle. If in the expression for the b lifetime we enter the value $|V_{bc}| \approx sin\gamma \approx sin\theta_c = 0.22$, we get a value for τ_b of the order of $10^{-14} sec$.

MAC and MARK II quoted values over 1 *psec* for τ_b , thus fixing $sin\gamma$ at a much lower value than $sin\theta_c$, and establishing that the third generation *mixing* with the second is not as strong as the second to the first. This explains why the Cabibbo theory worked so well in describing the weak interactions between the first and second generation particles, and a third generation was not required to explain the data available at the time of the Υ 's discovery. For this reason, the extension to a 3 x 3 matrix introduced by Kobayashi and Maskawa in 1973 went largely ignored until the discovery of the Υ particles.

Even if the B lifetime is *long* for a weak decay, it's still short enough to make the experimental determination difficult for present e^+e^- experiments. Measurements have been performed at PEP and PETRA where B particles travel an average length of 700 to 1000 μm before decaying; no exclusive decay mode has been reconstructed yet at these energies, because of the high background level (b production accounts for only 1/11 of hadronic events), the high decay multiplicity and the low detector acceptance.

B mesons have been completely *reconstructed* up to now only at threshold, at the $\Upsilon(4S)$ mass, where the production mechanism is two-body and the signal to background ratio is much more favorable ; on the other hand, these events are not useful for lifetime measurements since the B's are produced almost at rest.

In order to produce B mesons with a decay path long enough to be measured, the experiment must be performed at a higher energy, in the so called *continuum* region; since B 's cannot be reconstructed, their momentum and velocity can't be determined; this has forced the development of new, unconventional, experimental techniques for measuring the lifetime.

The methods used are based on the following approach: first a sample of *b enriched* events is selected, with a background contamination determined by

Montecarlo simulations; then an estimator of the decay length is found, to a good extent independent from the details of B production for the selected events. The lifetime extracted in this way , is clearly an average quantity which refers to the various species of b particles present in the sample, each one entering in the average with a weight proportional to the production cross-section.

The most widely used technique to obtain a *b enriched* sample consists in selecting multihadronic events containing a lepton with high transverse momentum respect to the thrust axis. About 25 % of the B's decay semileptonically producing leptons with a harder transverse momentum distribution than any competing process; by requiring the presence of a lepton with $p_{\perp} > 1$ GeV/c the b-events percentage in the hadronic sample increases from 9 % to as much as 60 - 70 % , while $\approx 10\%$ of the signal is kept.

The $b\bar{b}$ events produced at PEP/PETRA energies , i.e. at ≈ 30 GeV $c.m.$ total energy , have a rather complex topology: the primary quark-antiquark pair produced in the e^+e^- annihilation, will hadronize as a B mesons pair, plus a few other light mesons, pulling more quarks from the *sea*: following this rather poorly understood process, the energy available is redistributed to the final state particles. The B mesons (b-baryons will be produced in only a few cases, therefore for simplicity we will refer to b-particles as B mesons) will typically carry $\approx 80\%$ of the beam energy, while the remainder is divided into an average of five charged and three neutral light hadrons.

The B mesons travel an average of 0.7 mm before decaying; $\approx 25\%$ of the time a lepton will be produced, together with a charmed particle, with a typical multiplicity of 4 charged prongs; the charmed particle decay will result into a separate vertex and possibly an additional lepton. The B momentum can't be measured, because of the neutrinos which escape detections and the secondary and tertiary vertices cannot be resolved with the available experimental resolution.

The primary vertex is not known a priori: the interaction can occur anywhere within the region of beam crossing; in e^+e^- storage rings the beams are gaussian shaped with r.m.s. of ≈ 500 μm in the horizontal direction and ≈ 50 μm vertically. The beam envelope centroid can be determined with good precision, 10 - 20 μm , by measuring the closest approach distance between Bhabha scattering tracks; This can be monitored on a run by run basis :it is usually stable, and just drifts slowly with time.

Measuring the b-lifetime in e^+e^- annihilation is quite a challenge for the experimentalist : as pointed out before, the interaction point is not known, as well as the B direction and momentum; all these quantities can be determined statistically, not on an event-by-event basis. The available information include the position of the beam spot center, which can substitute the primary vertex, the thrust direction ,*i.e.* the direction which maximizes the projected momenta of the particles in the event; the lepton track which is likely to come directly from the B decay. The method used for extracting the lifetime uses the fact that the B tracks will have an impact parameter distribution not centered at zero respect to the beam centroid, and the shift's value is related to the B lifetime.

4. The detectors

The b-particles lifetime has been studied by several e^+e^- experiments: MAC, MARK II, DELCO and HRS at PEP, TASSO and JADE at PETRA ; most of these collaborations have performed more measurements, based on different data samples and/or different analysis; Following the earlier b-lifetime measurements, several of these detectors have been upgraded with the addition of more precise tracking devices near the interaction region, in order to improve their vertex finding capability.

I will now quickly review the main features of each apparatus, emphasizing the characteristics which are more relevant in the b-lifetime measurement.

MAC is a compact calorimetric apparatus which is designed with emphasis on lepton identification, especially muons, whose tracking and momentum measurement is done twice independently in the central and in the outer regions. The central drift chamber is relatively small, covering the radial region from 10 to 100 cm , and is immersed in a solenoidal magnetic field of $\approx .7\ T$. In 1984 the vacuum pipe radius was reduced to 3.5 cm and a sophisticated smaller chamber was added, to improve the track extrapolation to the origin. The complete calorimetric coverage, unique to this detector, is important for the precise determination of the thrust axis, through the energy flow measurement.

MARK II is a general purpose apparatus with good electron identification capabilities and reasonably good momentum resolution. The tracking system includes a large drift chamber and a smaller high precision vertex detector, which allows a precise measurement of the distance of minimum approach.The thrust axis direction is obtained from the measured momenta of the charged particles only. Electrons are measured in a large acceptance Liquid Argon Calorimeter, while muons are identified over $\approx 55\%$ of the solid angle by chambers preceded by 1 m thick iron absorber.

DELCO is a detector specialized in electron identification; this is achieved by using Cerenkov counters instead of calorimetric techniques. This implies the possibility of identifying electrons over a much wider momentum range, in particular below 1 GeV/c ; since lower momentum tracks have on the average a larger impact parameter, Delco gains noticeably by being able to lower the cut on the electron momentum, and this compensates for the moderate momentum and vertex resolution of the central tracking.

HRS is by far the best detector for what momentum resolution is concerned, thanks to the very large volume dedicated to tracking and to the high field (1.6 T) provided by a superconducting magnet. Part of the data was collected after installing a dedicated vertex detector which further improved the track extrapolation to the origin. The electron identification is obtained with *standard* sampling shower counters.

TASSO is an other general purpose apparatus, with characteristics and performances similar to MARK II ; the electromagnetic calorimeter also uses the liquid argon technique, but its solid angle coverage is not complete since in 2 octants it has been replaced by a hadron identification system. The more recent data has been collected by using a dedicated vertex detector.

JADE is also a general purpose detector; its main feature is a sophisticated jet chamber tracking system, which also provides particle identification using combined dE/dx and momentum measurements. The chamber is inserted in a conventional solenoidal magnet and is followed by a lead glass e.m. calorimeter. This detector provides very good lepton identification over a large solid angle and good momentum resolution for charged tracks.

5. The measurement techniques

I will now discuss the main features of the experimental methods used to extract the b-quark lifetime from the PEP or PETRA data. The *standard* method for measuring a lifetime consists in identifying the decay products, measure their total momentum and the distance of their *vertex* from the interaction point. For the b-particles in PEP/PETRA hadronic events it's not possible to identify all the decay products, so a different approach is needed to convert the measured distances to the *proper* decay time.

The production of a $b\bar{b}$ pair from the e^+e^- annihilation results in hadronic events with peculiar characteristics, which can be exploited to obtain a sample of events *enriched* in B mesons; such a statistical identification can be a substitute for a direct positive one, provided that the background contamination in the sample is known. The B direction can be substituted by the thrust axis, using the fact that the fragmentation is hard and the b-particle follows the b-quark direction quite closely , which in turn is approximated by the thrust axis.

Such a method is clearly somehow dependent on the details of the experiment, on the Montecarlo simulation used to reproduce the data and on the knowledge of the various components present in the selected event sample. It has, however, one clear advantage over the traditional technique used in bubble chamber or emulsion experiments, where in order to sift the interesting events from a very large sample, a cut is made on the distance between the secondary and the primary vertex, thus biasing the lifetime measurement.

B-*enrichment* is obtained with either one of two basic methods that I would classify as **semileptonic** tag and **event shape** tag , respectively; each experiment tailors one or both of these techniques to the characteristics of its own apparatus, so the analysis may differ; I will now briefly describe the general features which apply to all measurements.

The semileptonic tag is widely used to select heavy mesons : the presence of a lepton in a multiprong event is itself a good indicator for the weak decay of a heavy quark object being produced. Other lepton sources in e^+e^- hadronic events include J/ψ decays, Dalitz pairs, τ decays , pions and kaons decays in flight. Since the b-particles are much heavier than any other lepton source, the leptons produced in B semileptonic decays have, on the average, an higher transverse momentum, with respect to the original parent direction. A lower cut on the lepton transverse momentum with respect to the thrust direction is therefore a possible *b-enrichment* tool.

To quantitatively determine such an enrichment, a Montecarlo simulation, reproducing both the characteristics of the collected events and the detector details, is needed. As an example I will describe how this method is implemented in the MAC experiment: both muons and electrons are selected among all tracks having at least a 2 GeV momentum. The transverse momentum distribution of muons, measured respect to the thrust axis, is shown in figure 1 with the breakdown in the different sources: for higher $p_\perp$ values, most of the muons come from direct b decay.

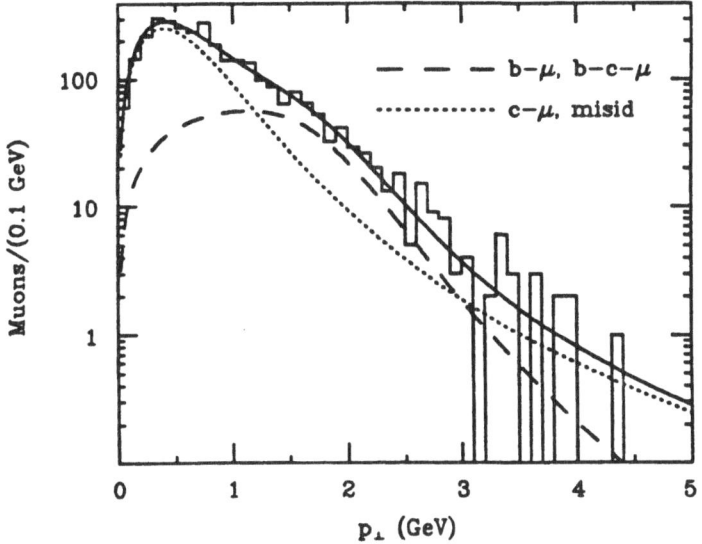

fig 1. MAC $p_\perp$ distribution of muons.

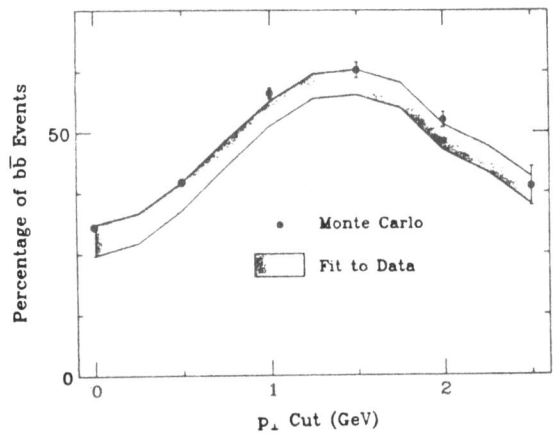

fig. 2 . MAC Percentage of bottom events as a function of
of the $p_\perp$ cut. The band shows the $\pm 1\,\sigma$ of the fit to the data .

The b event fraction in the final event sample, where the b-lifetime is measured, depends on the cut on $p_\perp$ which is used, as shown in figure 2. MAC chooses a cut of 1.5 GeV/c in $p_\perp$, obtaining a purity of (60.3 ± 2.6) % for the muon sample, (59 ± 5) % for the electrons.[10]

Once the b-enriched sample has been selected , the next step is to find an estimator representative of the b quark lifetime. One is given simply by the average value of the leptonic tracks impact parameter; this is basically the technique used for the earlier measurements. Each experiment tried to improve the accuracy by considering the error in the impact parameter measurement for each event,and by using a more accurate statistical procedure than the straight average, to evaluate the distribution center value.

In measuring the impact parameter, it's obviously important to define the zero of the scale and a convention for the sign. The origin is usually taken as the centroid of the beam spot , a precisely known quantity, stable over a time span which exceeds the run duration. The b-lifetime measurement requires a large sample of events, collected over periods of at least several months, in some cases even a few years. The position of the beam centroid can slowly shift with time, but it's continuously monitored, on a run by run basis, using Bhabha scattering events. Shown in figure 3 is the jitter of the interaction point position over a period of about nine months, as determined by the MAC experiment at PEP. While the size of the beam profile is quite large, $\sigma_x \approx 400\mu$, $\sigma_y \approx 50\mu$, its center can be determined with high precision using collinear pairs, the r.m.s. being only 10μ. A good control of the systematic effects is clearly of fundamental importance in order to measure lifetimes in the picosecond range, which result in an average lepton impact parameter of $\approx 100\mu$.

A *signed* impact parameter must be defined, so that a zero lifetime would correspond to a zero mean value for the distribution: the sign is defined positive if the lepton track appears to be emitted forward by a decaying particle travelling along the thrust axis, as shown in figure 4. A small fraction of B decays (less than 10%) will produce a *backward* lepton and contribute to the negative side of the impact parameter distribution; this effect can be taken into account when translating the impact parameter measurement into the lifetime determination.

The average impact parameter is a good lifetime estimator[11]: it can be shown that for massless daughter particles and complete phase space integration, the

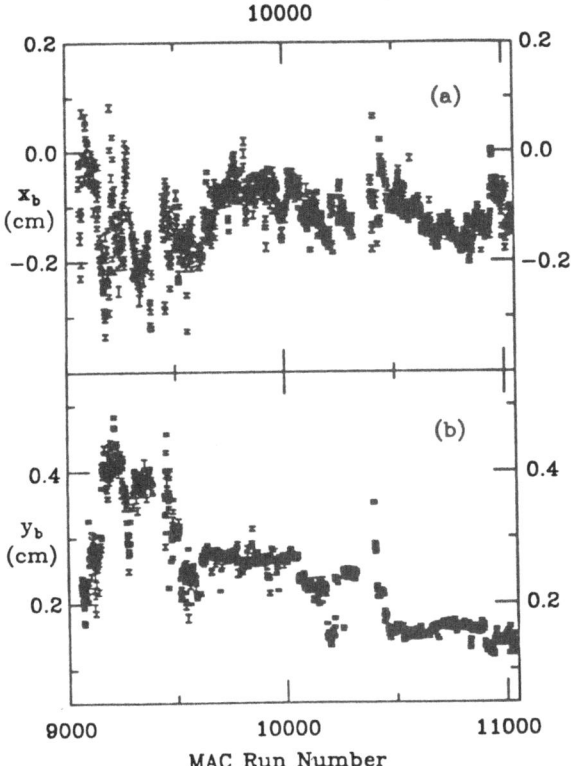

fig. 3 . *MAC jitter of the beam spot centroid over a period of nine months. the x_b coordinate is the radial one , y_b the vertical.*

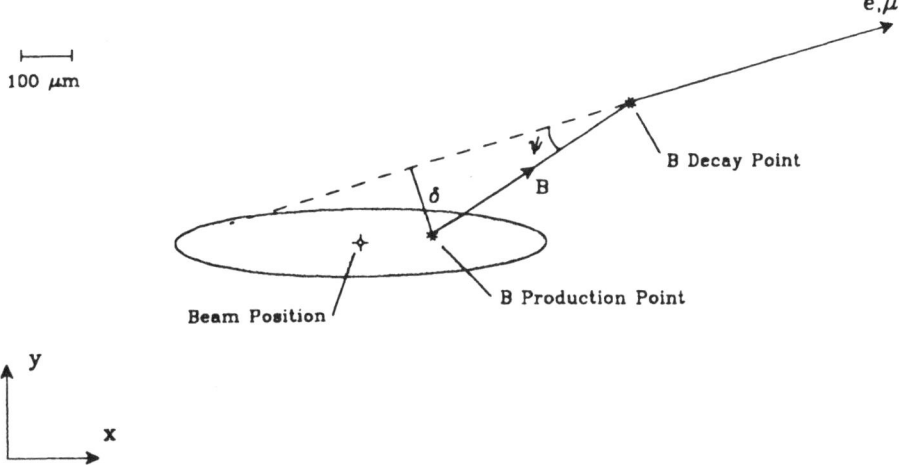

fig. 4 . *Definition of the impact parameter and related quantities .*

47

average value of the tracks impact parameter is independent from the momentum spectrum of the parent particle and is proportional to the lifetime. This is not exactly the case for the typical measurement of b (or c) semileptonic decays, but it's still true in first approximation; using the notation in fig. 4 , we have:

$$< \delta > = < l \sin \psi > = < \beta \gamma \sin \psi > \ \tau$$

the product $\beta \gamma \sin \psi$ is almost a relativistic invariant and can be treated as a constant. This approximation will, however, result into a systematic error, since the constant's value is evaluated by MonteCarlo and is slightly dependent on the momentum distribution of the b particles produced , which in turn is determined by the fragmentation mechanism of the primary quarks.

Another source of systematic error is the less than perfect knowledge of the sample's composition, since the impact parameter's measured average value depends on the sample purity:

$$< \delta_{meas} > = f_b < \delta_b > + f_c < \delta_c > + f_{bkg} < \delta_{bkg} >$$

where f_b , f_c and f_{bkg} indicate the fractions of leptons from b decays and c decays, and background, respectively. These fractions, and the average impact parameters for both charm decays and background , are all evaluated using a MonteCarlo simulation, and are cause of systematic errors when δ_b is derived from δ_{meas} ; these contributions to the systematic uncertainty should be carefully evaluated.

The impact parameter distribution is the convolution of a gaussian whose σ is equal to the experimental resolution, with an exponential whose slope is determined by the lifetime. Several different approaches have been used to disentangle these contributions: maximum likelihood , distribution mean , trimmed mean and median.[12] As an example, I will show the performance of a very simple, yet robust, estimator: the mean value. This is infact an effective quantity to use when determining the slope of an exponential distribution convoluted with a gaussian. The two extreme cases in which the distribution mean retains the complete information coming from the maximum likelihood analysis are:

$$\sigma >> \lambda \ , \ \ \sigma << \lambda$$

where λ and σ are the slope of the exponential distribution and the width of the convoluting gaussian respectively.

In these two limits, the lower bound on the statistical error is not appreciably different if a maximum likelihood fit is performed or the standard deviation of the mean is calculated . When $\lambda \approx \sigma$, the lower bound on the maximum likelihood fit gains in significance over the error on the mean; the maximum effect is 11 % for $\lambda \approx \frac{\sigma}{2}$. On the other hand, in the region where $\lambda \sim \sigma$ the correlation beetween λ and σ causes a non negligible systematic error in the fit, due to the less than perfect knowledge of the experimental resolution. The mean value instead is not affected by this uncertainty, since the error on this quantity can be evaluated in first approximation, as the ratio of the width of the distribution to the square root of the number of entries.

The experimental resolutions however, are not truly gaussian functions, since they all have long non-gaussian tails. This is the main reason why the ability of the simple mean value to reproduce λ is somehow impaired. Other statistical tools, like the median or the trimmed mean, are less sensitive to the distribution's tails and can be used as robust estimators of the lifetime, having a reduced sensitivity to the systematic effects.

The first determinations of the b-flavored particles lifetimes were performed using the technique described above and yielded results around 1-2 psec. (1983-1985) The second generation experiments were carried out during the next couple of years and included improvements both in hardware and software, as more precise tracking chambers and new more sofisticated analysis techniques were employed .

After the first MAC and MARK II results in 1983 proved that the b-lifetime was much longer than anticipated and its measurement was possible in e^+e^- interaction at c.m. energy around 30 GeV, most of the apparata running at PEP and PETRA were equipped with a *vertex detector* and new ideas in data handling started to emerge. The most important improvement on the older analysis was the possibility of attributing a *primary vertex* to each event in the selected sample.

A more precise estimate than the beam centroid as the interaction point was the most important factor in achieving improved lifetime measurements by most experiments. Although the identification of the 4 secondary vertices typically present in a $b\bar{b}$ event was still out of the detectors reach, a significant improvement was obtained by using most of the tracks in each event to determine an *average* vertex; this was then defined as the interaction point in the impact parameter measurement. How this new technique allows a better resolution in the MAC data is shown in figure 5. On the negative side, this method might introduce biases that must be carefully studied in order to understand the systematic errors.

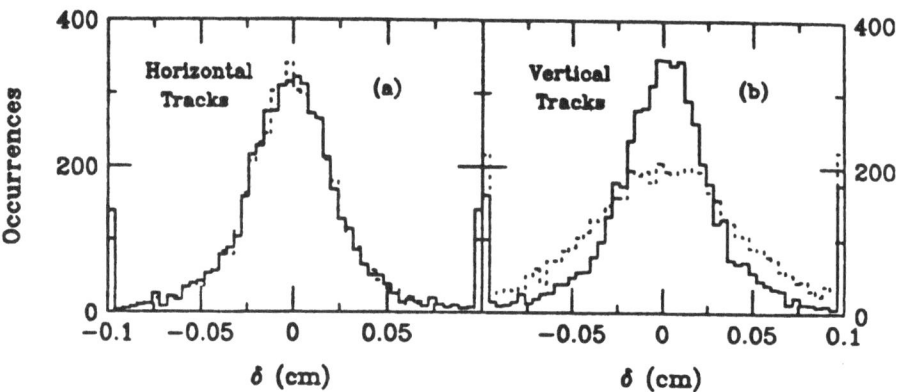

fig. 5 . MAC Impact parameter distribution using the beam centroid (dotted) or the event by event vertex estimate (full).

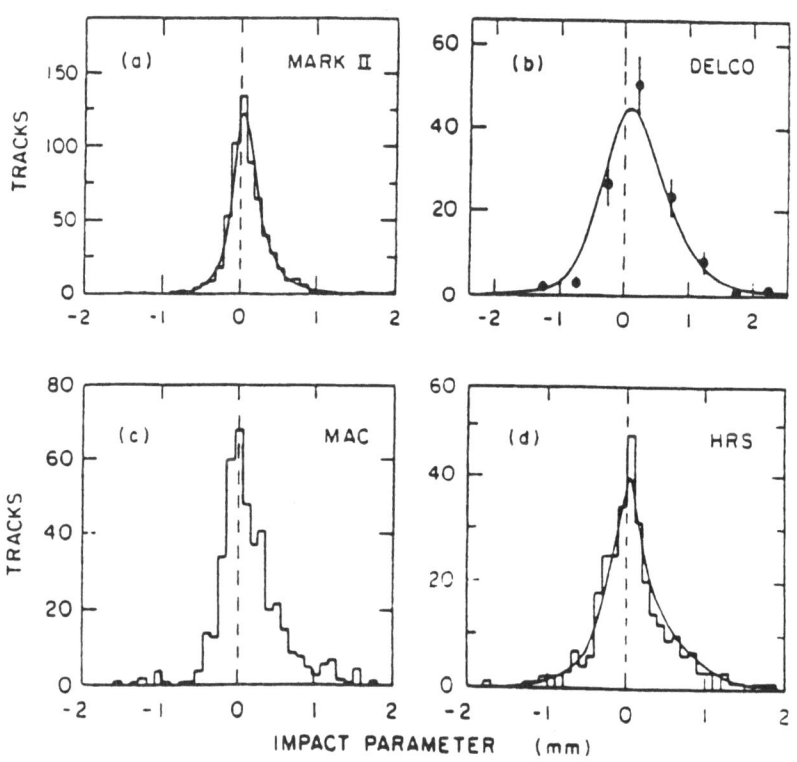

fig. 6 . Impact parameter distributions for the PEP experiments.

The different groups use several variations of the basic analysis technique; in each case a detailed work is necessary , using both real and simulated data, to assess the reliability of the method, its biases and the sources of the systematic uncertainties. MAC, for instance, defines a primary vertex as the one obtained using all tracks in each event , provided that they are well measured, are away from the thrust axis by more than 12 degrees, and the momentum is larger than 0.5 GeV/c.

The average charged multiplicity is larger for B hadronic decays than for the semileptonic modes; a possible concern is that when all tracks are used to determine an *average vertex*, it would not be in the middle of the two B vertices (average interaction point) but shifted toward the side where the B decays hadronically. A detailed Montecarlo study has however proved that this is not an important effect since the bias due to this particular definition of the primary vertex does not exceed 5 microns. Similar conclusions have been reached by other groups when studying their own vertex definition.

The final results of the four experiments running at PEP[13] [14] [15] [16] are shown in figure 6; all the distributions look clearly asymmetric and the b-lifetime values obtained from the analysis of these data are in good agreement .

TASSO[17] and JADE[18] the two PETRA groups which have published results on b-lifetime have each performed several measurements , some using the usual techniques, some with a completely different method, for both the $b\bar{b}$ event enrichment and the lifetime determination. TASSO starts from all its multihadronic events, and selects the b-enriched sample on the basis of a new variable called "boosted sphericity" . The procedure is the following: each multihadronic event is divided into 2 jets by a plane perpendicular to the usual sphericity axis: each jet is then *boosted* to its own center of mass. The β for this transformation is calculated assuming that the jet is from a b quark and that its momentum is the average value obtained from the b fragmentation function. The sphericity for each jet is then computed again in the new reference frame.

The distribution of the product $s_1 \times s_2$ of the two sphericity values is shown in figure 7; a quite clean separation is observed between the lighter quarks and the bottom quark. The major drawback for this technique is that the evaluation of the sample composition, in terms of quark content, is heavily dependent on the Montecarlo assumptions.

The average B lifetime is then determined from the impact parameter distri-

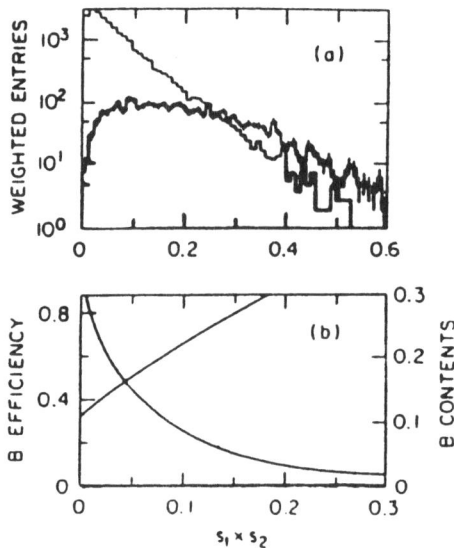

fig. 7 Tasso . Distribution of the product of the
boosted sphericity for each jet : a) weighted entries
for hadronic events: light quarks and b quarks.
b) b-events fraction and b-efficiency as a function of $s_1 \times s_2$.

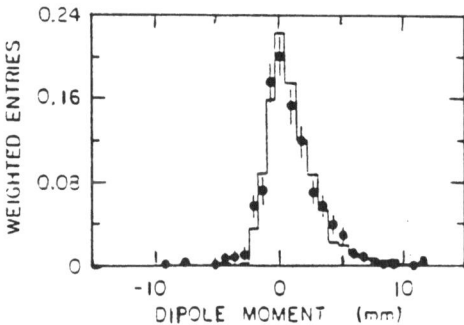

fig 8. JADE : Boosted sphericity weighted distribution
of the dipole moments of hadronic events.

bution, where entries are from all the well measured tracks, after unfolding the background contribution.

The Tasso collaboration also follows a completely different approach: no attempt is made in enriching the sample of b events; all 2-jets events are considered and the decay length of the best 3-prongs vertex in each hemispheres is measured. The b-lifetime is then extracted from this distribution, using the Montecarlo results for lighter quarks. Of course the Montecarlo assumptions play an important role in this method also and this makes it this algorithm prone to systematic uncertainties.

Another TASSO analysis uses a combination of these two methods: the events are first selected on the basis of the boosted sphericity and then the decay length of the best 3 prongs vertex in each jet is measured. One more method that TASSO uses is the so called *dipole method*, based on the measurement of the average vertex separation of the two jets in multihadronic events. The results obtained by each of these analysis, on the same data sample but with rather different biases and other systematic effects, agree completely.

The JADE collaboration has measured the b-particles lifetimes using similar techniques; for example in one analysis the dipole moment is measured for a sample of 30000 multihadronic events : the boosted sphericity is then calculated to assign to each event a probability of being a true $b\bar{b}$, $c\bar{c}$ or light quark event. Their result is shown in figure 8.

6. Summary of results and discussion of the systematic errors

All but one result on the b-lifetime come from e^+e^- experiments; only one fixed target experiment has two B decays, although they are very well reconstructed. In table 6.1 the results from the best measurements performed by each e^+e^- collaboration are summarized, together with information on the analysis technique used and the experimental resolution.

The weighted average of the results gives :

$$\tau_B = (1.18 \pm 0.14) \times 10^{-12} sec$$

Results can be more easily compared using the plot in figure 9.

All available data point to a lifetime between 1 and 1.5 psec , consistent both with each other and with the pioneer MAC and MARK II results; in comparing

Table 1

	MAC	MARK II	DELCO	HRS	JADE	TASSO
L (pb^{-1})	220	200	214	200	70	75
R_{min} Vtx Chamber	4.5	10	12	9	10	8
b$\bar{\text{b}}$ Events (%)	60	65	79	53		
Point res.(μm)	< 50	90	150	100	160	120
Measured λ (μm)	114 ± 13	129 ± 14	249 ± 49	80 ± 27	1080 ± 80	84 ± 5
Lifetime (psec)	1.24 $\pm.2 \pm .17$	0.98 $\pm 12 \pm 13$	1.17 $^{+27}_{-22} \pm .17$	1.02 $^{+0.42}_{-0.42}$	1.46 $\pm .19 \pm .30$	1.39 $\pm .10 \pm .25$

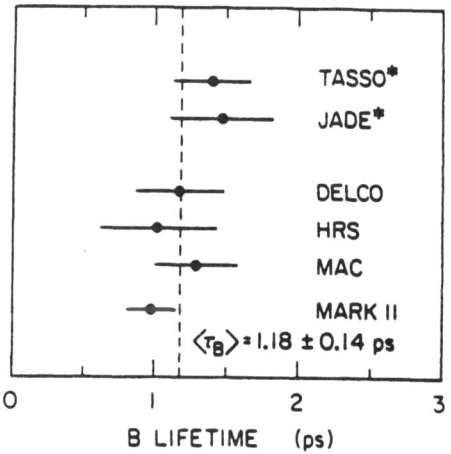

fig. 9 . Summary of the b-lifetime measurements :
for each collaboration only the most accurate result is plotted.
The errors are statistical only.

the data, or when trying to combine the results for quoting a *comprehensive* result, great care should be devoted to the treatment of the systematic errors.

I have already mentioned the fact that systematic errors arise from both the imperfect knowledge of the detector performances (e.g. tails of the spatial resolution function) and the uncertainties in the Physics parameters used in the analysis as a substitute of unknown quantities. Some of these effects are the same in all experiments, some are peculiar to the combination detector - analysis technique.

In a lifetime measurement the systematic errors are the limiting factor, in achieving a good precision rather than statistics. This applies in particular to the experiments we have described, where the impact parameters or the decay distances to be measured are often smaller, or at best of the same order, as the experimental resolution. The presence of unforeseen or not well under control systematic effects could substantially change the measured quantities or even wipe out the effect of a non zero b-lifetime.

Given the large number of detectors involved and the variety of the analysis techniques used, the discussion of the systematic effects should be done separately for each experiment. Even for uncertainties shared by everybody, the resulting systematic effect must be carefully evaluated in each case; consider as an example the b fragmentation function. This influences the lifetime value in more than one respect: the more primary b quark energy goes to the B meson, the fewer tracks from the primary vertex, which all have zero impact parameter. The quantitative dependence of the final result from the uncertainty in the Peterson variable (world data[19] permit $\epsilon_b = 0.012^{+0.019}_{-0.009}$ which correspond to a beam energy fraction of 0.78 ± 0.05 for the B) depends however on the analysis technique (the tracks whose impact parameter has been measured, for example) and on the statistical procedure used in extracting the lifetime from the measured distribution.

As a typical example, I will report here the breakdown of the MARK II measurement's systematic error, which is evaluated to be 13 % overall.[20]

The main instrumental effects are the uncertainties in the impact parameter resolution and in the thrust axis direction, whose contributions are 6 % and 3 % respectively. The resolution function is empirically determined, and affects the lifetime value as the parameters are changed from the minimum value of the likelihood function. The error in the determination of the direction of the parent *b* quark , approximated with the thrust axis, also results in a systematic uncertainty in the lifetime.

Many parameters used in the analysis are unknown and have to be obtained from the Montecarlo simulation: a detailed work is then necessary to understand how much the results vary when the Montecarlo inputs are changed. In the MARK II analysis the uncertainty on the background subtraction contributes a 3 % error to the lifetime measurement, 5 % is due to the b and c fragmentation functions while the biggest effect is due to the uncertainty in the lepton fraction, 9 %.

Effect	$\Delta\tau/\tau$
Lepton fraction	9 %
b and c Fragmentation	5 %
Background (i.e. 2 γ 's)	3 %
Resolution function	6 %
Thrust vs B direction	3 %
Total	$\approx 13\%$

Figure 10 shows how much the b-lifetime determination is affected by systematic uncertainties as the sample purity and the b-quark fragmentation function, which determines the average B meson momentum.

7. Determination of $|V_{cb}|$

We have already pointed out how important the b-quark lifetime is , in the the Standard Model framework, since it's strictly related to the mixing angle between the second and the third quark families, one of the free parameters in the theory to be determined experimentally.

Several assumptions, however, are involved in extracting $|V_{bc}|$ from measured value of B mesons average lifetime; in order to quote a *honest* result from the

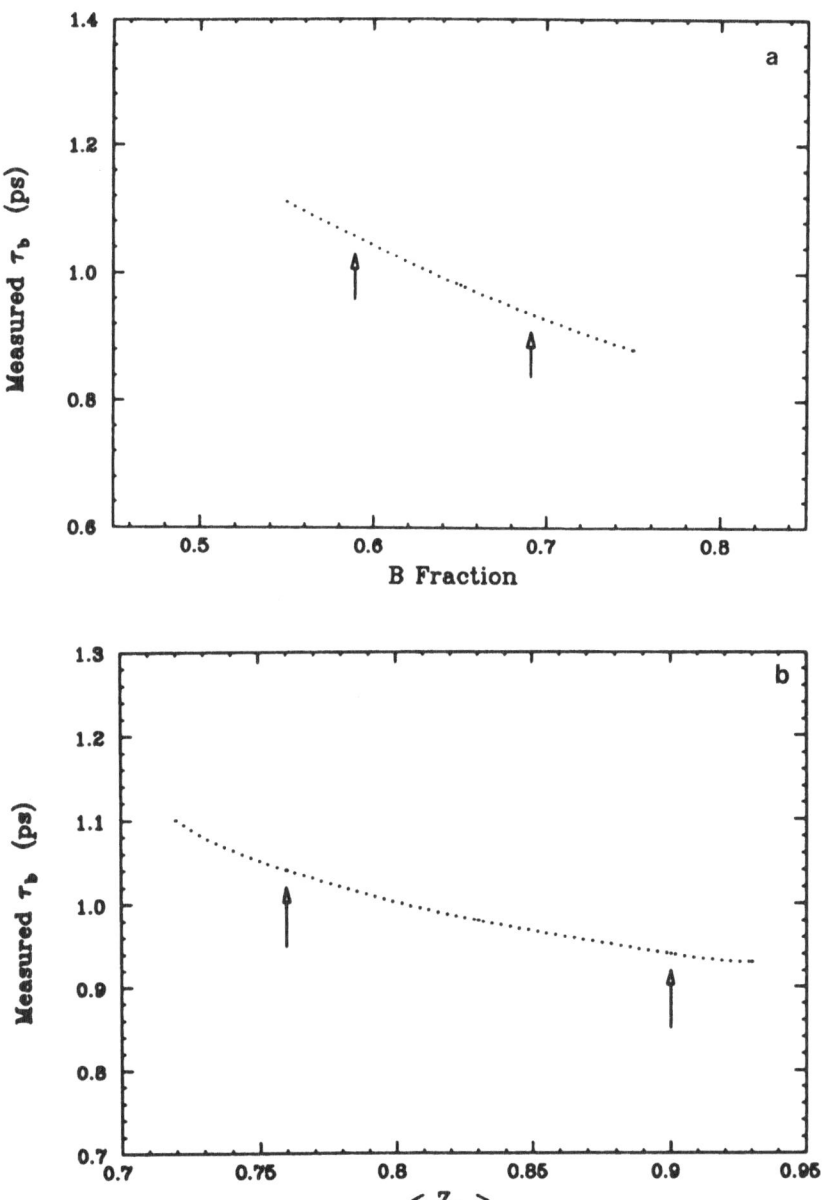

fig. 10 . Dependence of the b-lifetime measurement from :

a) the fraction of b decays in the final sample

b) the $< z_b >$ value of the b quark fragmentation

(Mark II analysis of systematic effects).

available data , theoretical uncertainties have to be taken into account as well as experimental ones.

The coefficients of the C-K-M matrix elements in the lifetime formula are somehow dependent on the model used in the calculation : different kind of approaches give quite consistent results for the $|V_{bc}|$ coefficient but they differ up to 50% for the $|V_{bu}|$.

Using the limits available on the ratio $\frac{|V_{ub}|}{|V_{cb}|}$ [21] [22] , and assuming the *free quark* model to evaluate the functional relation between the measured lifetime and the mixing matrix elements :

$$\tau_b \approx \frac{2.35 \times 10^{-14} \; Br(b \to l\bar{\nu}X)}{2.04|V_{ub}|^2 + |V_{cb}|^2} \; (sec)$$

the value obtained for $|V_{cb}|$ using $Br_{b \to l\bar{\nu}X} = 12.1\%$ is :

$$|V_{cb}| = 0.047 \pm 0.003 \pm 0.005$$

where the first error refers essentially to experimental uncertainties and the second to the theoretical modeling of the $\tau_b \; V_{cb}$ relation.

The b lifetime measurement ,together with the $\frac{|V_{ub}|}{|V_{bc}|}$ ratio and unitarity determine or severely constrain all the absolute values of the C-K-M matrix elements in the hypothesys of three generations : as a matter of fact the matrix becomes almost diagonal and the mixing between the second and third generation of quarks is extremely small . This in turn bears consequences on the mass of the top quark, on the ratio $\frac{\epsilon'}{\epsilon}$ and on the size of CP violation effects expected in beauty Physics.

8. Conclusions

Five years have gone by since the first rough measurements which proved the b-quark lifetime to be much longer than anticipated . A lot of effort by six large collaborations using powerful apparata at PEP and Petra have gone into improving the original measurements . The overall picture which is emerging now shows a high degree of consistency between the various measurements , but the techniques used up to now seem to be limited by systematic effects. Hadro and/or photoproduction experiments are likely to give their contribution in the near future so that a completely new data base of experimental results will be added to the existing ones.

New machines and new apparata will come into operation in the near future, Lep and in particular SLC with its extremely small luminous spot will help understanding many unsolved problems regarding the decay of beauty particles : eventual differences in lifetime between charged and neutral B's , lifetime of the B_s , of the B-baryons and so on. On a longer time scale the *long* b lifetime will be used as a tool to identify and flag b-flavoured particles in order to perform more sophisticated measurements and solve one of the most fascinating problems of our universe : the CP violation phenomemena.

Acknowledgements: It is a pleasure to thank my hosts Dr. Luisa Cifarelli and Dr. Ahmed Ali for a very enjoiable stay in Erice and a superbly organized Workshop.

REFERENCES

1. E. Fernandez *et al.* Phys. Rev. Lett. **51** 1022, (1983)

2. N.S.Lockyer *et al.* Phys. Rev. Let. **51**, 1316, (1983)

3. Kobayashi-Maskawa , Prog. Theor. **652** , 48, 1973

4. L. Maiani ,Journal de Physique , Colloque C3 , 631 , tome 63 (1982)

5. J. L. Cortes, X. Y. Pham ,A. Tounsi Phys. Rev. **D15** 188,(1982)

6. N. Cabibbo , L. Maiani Phys. Lett. **70B** ,109, (1974)

7. G. Altarelli, L. Maiani Phys. Lett. **99B**, 141 (1981)

8. M. K. Gaillard , B. W. Lee Phys. Rev. Lett. **33** ,108 ,(1974)

9. J.P. Albanese *et al.* Phys. Lett. **158B**,186 (1985)

10. H.N.Nelson Ph. D. Thesis SLAC-Rep 322 (1987)

11. S. Petrera and G. Romano Nucl. Instr. and Meth. **174**,61 (1980)

12. A comprehensive discussion on the subject can be found in **H.N.Nelson Ph. D. Thesis** SLAC Report 322 (1987)

13. W.W. Ash *et al.* Phys. Rev. Lett. **58**,640 , (1986)

14. R. Ong *et al.* SLAC-Pub.-4559 (1988)

15. D.E. Klem *et al.* SLAC-Pub.-4025 (1986)

16. J. M. Brom *et al.* Phys. Lett. **195B**,301 (1987)

17. W. Braunschweig *et al.* ,Contribution to the *Int. Symp. on Lepton and Photon interaction at high energy* , Hamburg (1987)

18. D. Muller, Contribution to the *Int. Symp. on the production and decay of heavy flavours* , Stanford , 1987.

19. W. Bartel et al, Z. Physik C, Particles and Fields **33** ,339 , (1987)

20. R. Ong Ph. D. Thesis SLAC Report 320 (1987)

21. Berends *et al.*Phys. Rev. Let. **59** 407 , (1987)

22. W. Schmidt-Parzefall *Proceedings of the Int. Conf. on Lepton and Photon Interaction at High Energy* Hamburg 1987

5

EXPERIMENTAL STATUS OF CP VIOLATION IN $K^0 \to \pi\pi$

A. Nappi

University of Pisa and
INFN, Sezione di Pisa
Pisa, Italy

INTRODUCTION

To the level of precision available in 1985, all observed effects of CP violation (existence of $K_0^L \to \pi\pi$ and the charge asymmetry in K_0^L semileptonic decays) could be explained by the non-zero relative phase of the off-diagonal elements of the hermitian and anti-hermitian part of the $K^0 \bar{K}^0$ mass matrix.

In perturbation theory, one can have first order contribution to the off-diagonal terms of the mass matrix only through an interaction allowing $\Delta S=2$ at tree level, whereas the ordinary weak interaction enters only at second order. Theories trying to explain CP violation are thus naturally divided in two classes ("superweak" and "milliweak") according to whether CP violation is ascribed to the first order or to the second order term. The size of the observed effects is such that in superweak theories[1] no detectable CP violation is expected in the K^0 decay amplitudes, whereas in milliweak theories such effects ("direct" CP violation) are expected to occur, unless suppressed by dynamical mechanisms.

The search for effects of direct CP violation has attracted renewed interest since it was realized that, with three quark generations, the standard electro-weak model can be made into a "milliweak" CP violation theory, by allowing a non-vanishing phase in the quark mixing matrix. This was first pointed out by Kobayashi and Maskawa[2], and explicit calculations[3] showed that, even though no evidence of direct CP violation had yet been seen, this was not sufficient to exclude the Kobayashi-Maskawa model, since the expected effects are small.

Among the various measurements that could be sensitive to effects of direct CP violation the one that has attracted particular experimental interest is the measurement of the ratio of the CP violating rates for $K_0^L \to \pi\pi$ in the two charge states, normalized to the corresponding ratio of the CP allowed rates for $K_S^0 \to \pi\pi$. The formalism for the description of CP violation in $K^0 \to \pi\pi$ will be introduced in the first section, where early measurements will also be recalled. The rest of the paper will describe the three experiments that are presently active in this area, with some emphasis on the description of the experimental techniques. These are sufficiently diversified to make the sources of systematic errors decoupled to a large extent.

Table 1. Earlier measurements of ε'/ε

Year	Authors	$\mid \eta_{00} / \eta_{+-} \mid$
1972	Holder et al.	1.00 ± 0.06
1972	Banner et al.	1.06 ± 0.07
1979	Christenson et al.	1.00 ± 0.09
	Average	1.01 ± 0.04
	$R = \mid \eta_{00} / \eta_{+-} \mid^2 = 1.02 \pm 0.08$	

Year	Authors	ε'/ε
1985	Black et al.	$0.002 \pm 0.007 \pm 0.004$
1985	Bernstein et al.	$-0.005 \pm 0.005 \pm 0.002$
	Average	-0.003 ± 0.005
	$R = \mid \eta_{00} / \eta_{+-} \mid^2 = 1.015 \pm 0.03$	

Two new results have been published recently, by the Fermilab E731 and by the CERN NA31 experiments. A large part of this paper will concentrate on the results of the latter experiment, which shows the first experimental evidence of an effect of direct CP violation .

DEFINITIONS AND EARLY MEASUREMENTS

The ratios of the amplitudes for K^0_L and K^0_S decays to two pions in the two charge states

$$\eta_{+-} = \frac{<\pi^+ \pi^- \mid T \mid K^0_L>}{<\pi^+ \pi^- \mid T \mid K^0_S>} \qquad \eta_{00} = \frac{<\pi^0 \pi^0 \mid T \mid K^0_L>}{<\pi^0 \pi^0 \mid T \mid K^0_S>} \qquad (1)$$

expected to be equal in the superweak hypothesis, become different if the weak decay amplitudes to pure I-spin states are not relatively real. In the Wu-Yang phase convention[4] this is characterized by the parameter

$$\varepsilon' = \frac{1}{\sqrt{2}} \frac{\text{Im } a_2}{a_0} e^{i(\frac{\pi}{2} + \delta_2 - \delta_0)} \qquad (2)$$

where a_2 (a_0) is the weak transition amplitude for K^0 going to the isospin 2 (0) $\pi\pi$ state and δ_2 (δ_0) the corresponding strong $\pi\pi$ phase shift at $\sqrt{s}=m_K$. In the approximation Re $a_2 \ll a_0$, ε' is connected to η_{+-} , η_{00} by

$$\eta_{+-} = \varepsilon + \varepsilon' \qquad ; \qquad \eta_{00} = \varepsilon - 2\varepsilon' \qquad . \qquad (3)$$

Experimental information on ε' is provided by the measurement of the double ratio of decay rates:

$$\frac{\mid \eta_{00} \mid^2}{\mid \eta_{+-} \mid^2} = \frac{\Gamma(K^0_L \to \pi^0 \pi^0) \, \Gamma(K^0_S \to \pi^+ \pi^-)}{\Gamma(K^0_L \to \pi^+ \pi^-) \, \Gamma(K^0_S \to \pi^0 \pi^0)} \approx 1 - 6\,\text{Re}\left(\frac{\varepsilon'}{\varepsilon}\right) \qquad (4)$$

Earlier measurements of ε'/ε[5-9] are summarized in table 1 . The improved precision of the last generation of measurements arises not only from the increased statistics, but also from the use of techniques exploiting cancellation of the flux normalization and (to different extents) of the detection efficiencies. This is also characteristic of the currently active experiments, all aiming at measurements of ε'/ε with errors of the order of 1×10^{-3}.

FNAL E731

This experiment[10] is performed by a Chicago, Elmhurst, Fermilab, Princeton, Saclay collaboration. The elevation view of the apparatus, shown in figure 1, illustrates the double beam technique used. A regenerator is inserted alternately on the path of either of two twin K^0_L beams, thus providing the possibility to collect K^0_L and K^0_S decays at the same time. The rate of events from the "vacuum" beam (V) is proportional to the modulus square of the CP violation parameter η defined in (1)

$$V \propto \mid \eta \mid^2 \tag{5}$$

whereas for the corresponding rate in the "regenerated" beam (R) one can write

$$R \propto \mid \eta + \rho\, e^{(i\Delta m - \frac{\Gamma_S}{2})t} \mid^2 \tag{6}$$

(where ρ is the regeneration amplitude, Δm the $K^0_L - K^0_S$ mass difference, Γ_S the K^0_S decay width). Thus the measurement of R/V, which is performed separately for the charged and neutral decay channels, allows a determination of the ratios $\mid \rho \mid / \mid \eta_{+-} \mid$ and $\mid \rho \mid / \mid \eta_{00} \mid$ of the regeneration amplitude to the CP violation parameters (1). Variation of the phases of ρ and η and of Δm, within the errors with which they are known, do not contribute any significant uncertainty to this determination. ε'/ε is determined from the combination

$$\mid \eta_{00} \mid / \mid \eta_{+-} \mid = \frac{\mid \rho \mid / \mid \eta_{+-} \mid}{\mid \rho \mid / \mid \eta_{00} \mid} \tag{7}$$

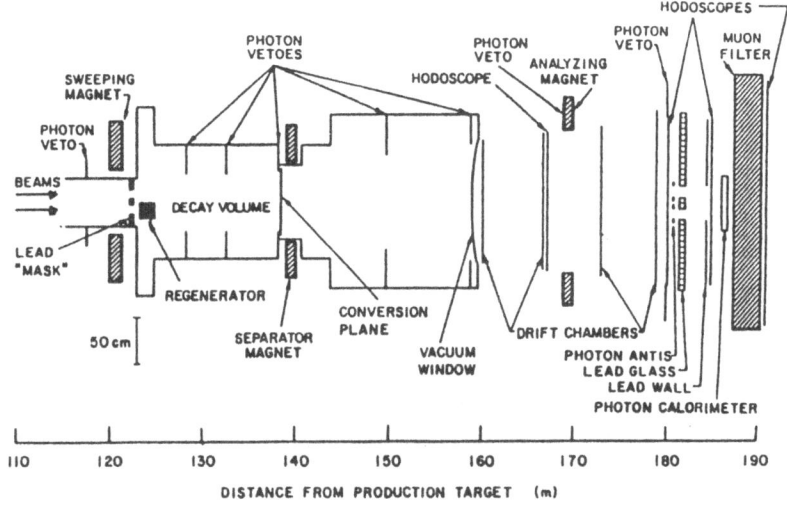

Fig. 1 Elevation view of the detector for FNAL E731

Table 2. Statistics and background subtractions
for E731 test run result

Mode	Events	Background (%)	Systematic error(%)
$K_L \to 2\pi^0$	6747	1.56	0.30
$K_S \to 2\pi^0$	21788	2.90	0.20
$K_L \to \pi^+\pi^-$	35838	1.23	0.18
$K_S \to \pi^+\pi^-$	130025	0.30	0.03

which is insensitive to the regeneration amplitude, since the same regenerator is used for the charged and the neutral run.

The main virtue of this experiment is the fact that ε'/ε is determined from a ratio of rates for K^0_S and K^0_L events which are collected at the same time. Thus the measurement is insensitive to time instabilities or rate dependence of the detector response. Furthermore the alternation of the regenerator between the two beams makes the measured ratio of rates insensitive to differences in flux and acceptance between the two beams as well. Conversely the acceptances for K^0_S and K^0_L events do not cancel in the measured ratio and their estimate gives an important contribution to the systematic error. In fact the data are analyzed in momentum bins but they are integrated over the longitudinal coordinate of the decay vertex which has a different distribution in the "vacuum" and "regenerated" beam.

The experiment has been collecting the bulk of its data in 1987. However a reduced sample collected during a test run in 1985 has been analyzed and the results have been published recently[10]. The statistics available for this result, the background subtractions and the associated systematic uncertainties are summarized in table 2. The background, mainly non $\pi\pi$ decays of K_L and, in K_S, incoherent regeneration, is substantially smaller than in the data of the Chicago-Saclay experiment[9], of which E731 is an outgrowth. This is obtained through many improvements to the apparatus, the most notable being: smaller neutron content in the beam, better calibration and resolution in the lead glass detector, improved photon veto counters, and the use of an active regenerator.

The result for ε'/ε is

$$\varepsilon'/\varepsilon = 0.0032 \pm 0.0028 \text{(statistical)} \pm 0.0012 \text{(systematic)} .$$

The systematic error on the double ratio of rates is dominated by the uncertainty on the acceptance corrections ($\pm 0.5\%$ for the neutral and $\pm 0.25\%$ for the charged channel) and on the background subtraction, as shown in table 2. Other contributions are due to uncertainties in the energy scale ($\pm 0.21\%$) and to possible effects due to accidental events in the detector ($\pm 0.2\%$).

The data taking capability of the experiment in the last run has been substantially improved. Besides the advantage connected with the lower neutron flux in the beam, the experiment also has improved acceptance, trigger and data acquisition. In the 1985 run the main limitation to the statistics that could be achieved was given, in the neutral channel, by the requirement to convert one of the γ's in a 0.1 radiation length converter. This was necessary in order to determine, for each event, whether it originated from the "vacuum" or the "regenerated" beam. In order to overcome this limitation, for the 1987 data, the group will experiment a different technique, based on events collected with an "open" trigger, where the conversion of one of the γ 's is not required. In the analysis, the beam assignment for these events will be made solely on the basis of the "center of gravity" of the energy detected in the lead glass detector. A correction will then have to be applied to the data, to account for cross-overs from the regenerated to the vacuum beam, caused by scattering in the regenerator. The group is confident that this correction can be constrained well, using the charged events, where the decay vertex position is measured and the cross-over problem does not exist.

The data collected last year will yield ~ 280,000 $K_L \rightarrow 2 \pi^0$ events from the "open" trigger. However ~ 30,000 "conversion" triggers have also been collected to allow a comparison of the traditional analysis with the new one. The expectation of the group is a result where both the statistical and systematic error on ε'/ε will be smaller than 1×10^{-3} .

CERN NA31

This experiment[11] is performed by a collaboration[*] of CERN - Dortmund - Edinburgh - Mainz - Orsay - Pisa - Siegen groups. The data described here have been collected in 1986. Only the technique and qualitatively important details of the analysis will be discussed. A more thorough discussion of the experiment and the analysis can be found in references 11-13.

Technique and apparatus

An overall view of the apparatus is shown in figure 2. The technique is complementary to the one of E731, in that $\pi^+\pi^-$ and $\pi^0\pi^0$ decays are collected at the same time, whereas the beam is switched between K^0_L and K^0_S with a periodicity of about one day. Only the charged to neutral ratio $\Gamma(K^0_{L(S)} \rightarrow \pi^+\pi^-)/ \Gamma(K^0_{L(S)} \rightarrow \pi^0\pi^0)$, that does not depend on the beam flux, is measured, for each beam. Approximate cancellation of the acceptances in the double ratio (4) is obtained by a design of the K^0_S beam that makes it as similar as possible to the K^0_L beam and by measuring the double ratio (4) in bins of energy and longitudinal position of the decay vertex.

The beam system is a unique feature of the experiment. The same beam line can either transport K^0_L produced in a target located 120m upstream of the decay region, or bring an attenuated proton beam to the K^0_S target, that can be moved across the decay region in steps of 1.2 m, in order to simulate the flat distribution of the decay vertex position typical of the K^0_L beam. To achieve this, the last elements of the beam (steering magnet, target, sweeping magnet, beam dump and K^0_S collimator) are mounted on a moveable train. The collimation system produces similar beam sizes for K^0_S and K^0_L at detector level. This is a further measure to improve acceptance cancellation, which works to the extent that there are no limiting apertures before the detector. Small deviations from the ideal behaviour are introduced by

---

* The members of the NA31 collaboration are: *CERN:* H.Burkhardt, P.Clarke, D.Coward, D.Cundy, N.Doble, L.Gatignon,V.Gibson, R.Hagelberg, G.Kesseler, J. van der Lans, T.Miczaika, H.G.Sander, A.C.Schaffer, P.Steffen, J.Steinberger, H.Taureg, H.Wahl, C.Youngman; *Dortmund:* G.Dietrich, F.Eisele, W.Heinen;*Edinburgh:* R.Black, D.J.Candlin, J.Muir, K.J.Peach, B.Pijlgroms, I. Shipsey, W.Stephenson; *Mainz:* H.Blümer, M.Kasemann, K.Kleinknecht, B.Panzer, B.Renk, S.Röhn; *Orsay:* E.Auge, R.L.Chase, M.Corti, L.Fayard, D.Fournier, P.Heusse, A.M.Lutz; *Pisa:* L.Bertanza, A.Bigi, M.Calvetti, R.Carosi, R.Casali, C.Cerri, R.Fantechi, S.Galeotti, G.Gargani, I.Mannelli, E.Massa, A.Nappi, D.Passuello, G.Pierazzini; *Siegen:* C.Becker, D.Heyland, M.Holder, G.Quast, M.Rost, W.Weihs, G.Zech.

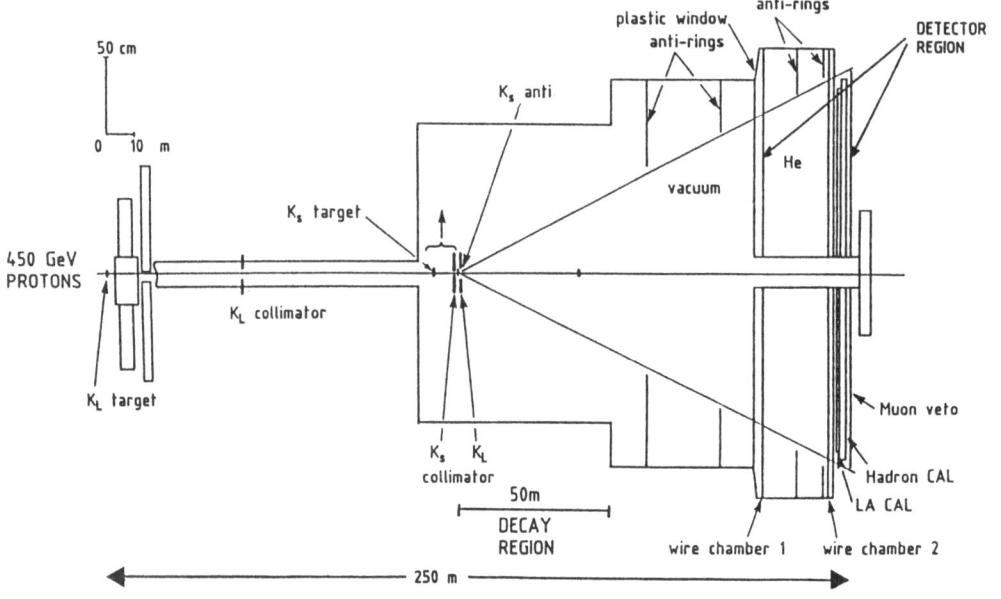

Fig. 2 Overall schematic view of the NA31 experimental set-up

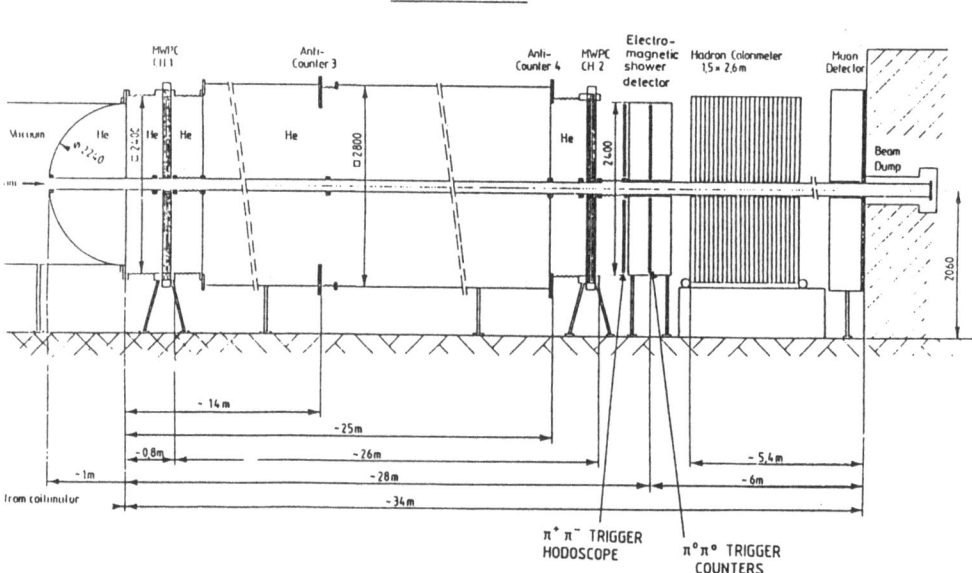

K^0- DETECTOR PART

Fig. 3 The detector region for the NA31 experimental set-up

energy cuts, dependence of the K^0_S beam size on the target position, scattering on material in (or close to) the beam, but the resulting corrections are small.

The detector, shown in figure 3, includes two planes of mini-drift chambers (each providing 4 views), a lead-liquid argon electromagnetic calorimeter with 1.25 cm horizontal and vertical strips, an iron-scintillator hadron calorimeter followed by iron absorbers and μ veto scintillators. Iron-scintillator sandwiches placed at large angles reduce the background from events with additional photons outside the active detector area. When the beam is in K^0_S mode, a veto scintillator, preceded by 7 mm of lead, marks the beginning of the decay region. This is the only material seen by the beam, which is carried in a vacuum pipe through the detector.

The absence of a magnetic field is to be stressed. The energy of the decaying K in the charged channel is determined from the opening angle θ of the two pions, and from the ratio R of their energies measured in the calorimeters

$$E = \frac{1+R}{R\theta} \sqrt{R m_K^2 - (1+R)^2 m_\pi^2} \qquad (8)$$

assuming the kinematics of the ππ decay. Due to the favorable decay kinematics, this determination has a far better resolution than could be obtained by the hadron calorimeter alone. A resolution $\sigma(E)/E \approx 1\%$ is obtained, as long as very asymmetric decays are rejected, by requiring 0.4<R<2.5, a condition which is necessary anyway to reject Λ decays. Moreover the absolute energy scale does not depend on the absolute scale of the hadron calorimeter response. The ππ invariant mass depends more directly on the calorimetric energy measurement and has a typical resolution $\sigma(M)/M \approx 5\%$. The resulting mass spectra in the K^0_S and K^0_L beams, after the analysis cuts, are shown in figure 4. The K^0_S decays have no background, whereas in K^0_L the separation

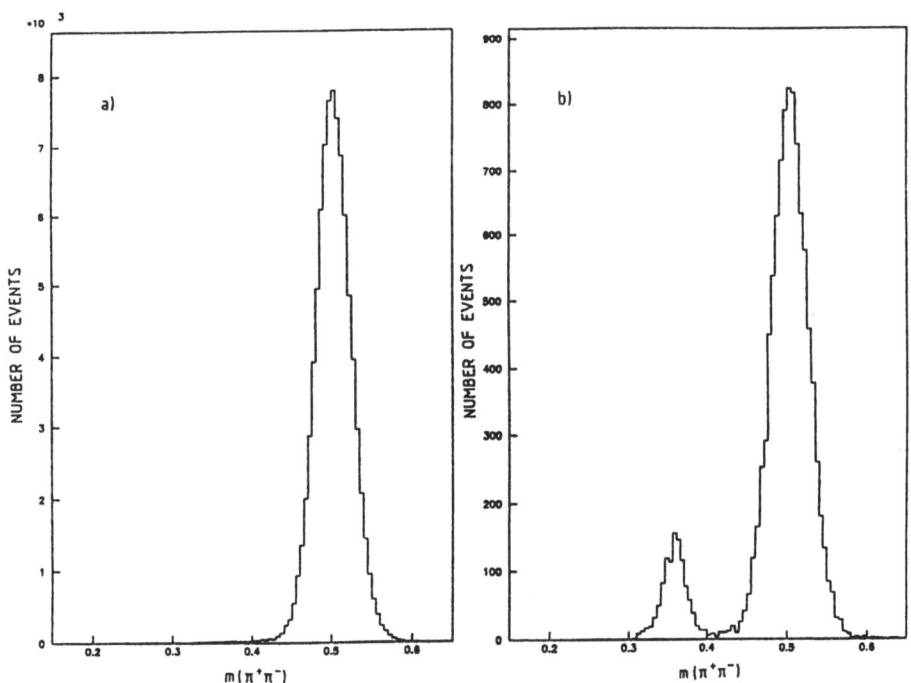

Fig. 4   Invariant mass distribution for $\pi^+\pi^-$ pairs:
a) in KS decays; b) in KL decays

between the signal and the background region is not complete and a subtraction is necessary.

In the neutral channel the transverse coordinates (x_i , y_i) and energies (E_i) of the four γ 's are measured in the electromagnetic calorimeter with resolutions of about 1 mm for the position and $7.5\%/\sqrt{E}$ for the energies. The K^0 mass constraint is used to determine the distance of the decay vertex from the detector

$$z_D - z_V = \frac{1}{m_K} \sqrt{\sum_{i>j} E_i E_j \left[(x_i - x_j)^2 + (y_i - y_j)^2 \right]} \qquad (9)$$

Using this position, the π^0 masses are formed by choosing the two pairs of γ's that give the best values. Resolutions of 2 MeV on the sum and 3.5 MeV on the difference of the two masses are obtained. A Lego plot of this mass distribution for K^0_S data is shown in figure 5 and indicates that also the neutral channel in K^0_S is free from background. The background in K^0_L will be discussed later.

Formula (9) connects the energy scale to the z scale, which is fixed by the anticoincidence at the beginning of the decay region. A fit to the sharp edge in the distribution of the longitudinal vertex position (fig. 6) assures the agreement of the energy scale for charged and neutral events to a precision of 1×10^{-3} . The agreement of the transverse position scale, a necessary requirement for this procedure, is checked to better than 2×10^{-4} using K_{e3} events.

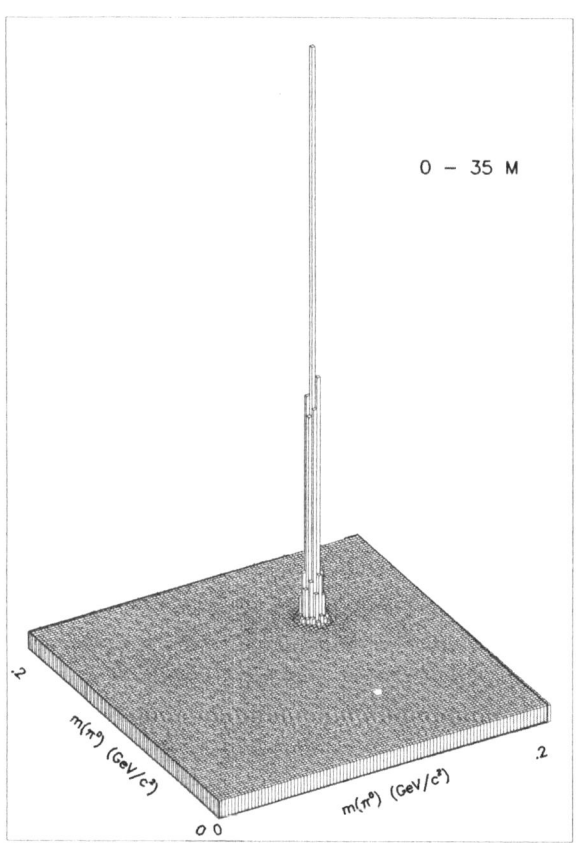

Fig. 5. Lego plot for the invariant mass of the two γ pairs for $K_S \rightarrow \pi^0 \pi^0$

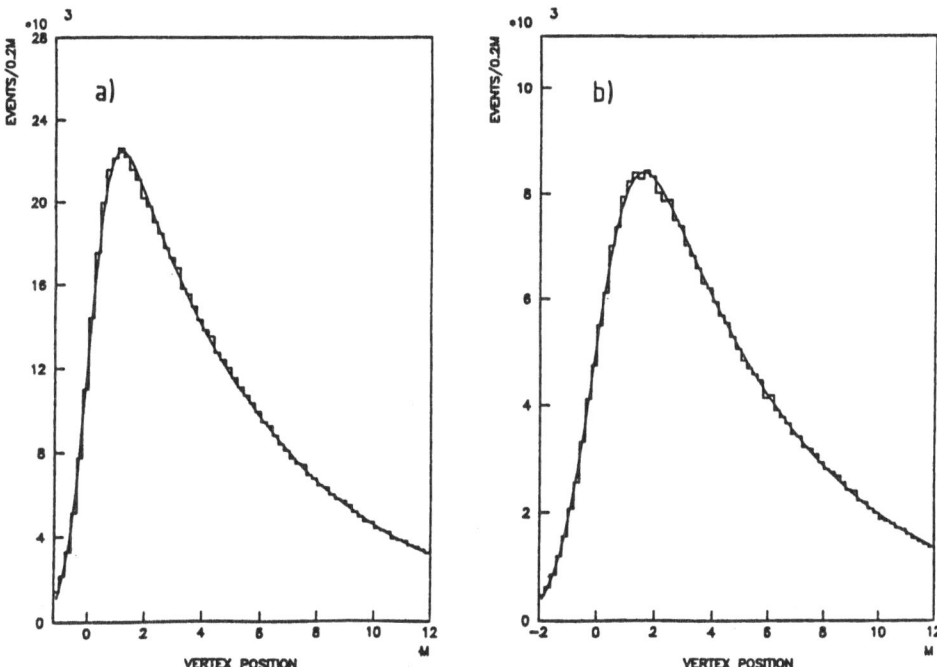

Fig. 6   Distribution of the distance of the reconstructed decay
         vertex from the position of the counter defining the
         beginning of the accepted decay region:
         a) for $K_S \rightarrow \pi^+\pi^-$ and b) for $K_S \rightarrow \pi^0\pi^0$

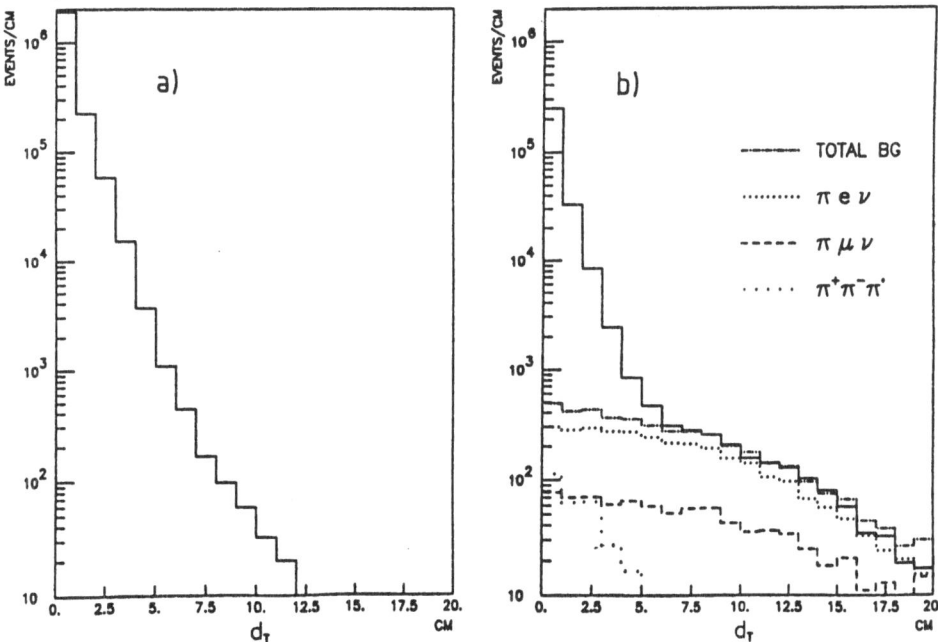

Fig. 7   Distribution of the distance of the K-production target
         from the reconstructed decay plane (d_T) :
         a) for $\pi^+\pi^-$ K_S decays;
         b) for $\pi^+\pi^-$ K_L decays and for various background components

Background subtraction

In the charged mode the background is due to the dominating 3 body K^0_L decays. $K^0_L \rightarrow \pi^+\pi^-\pi^0$, with both photons from the π^0 escaping the detector, normally give an invariant mass well separated from the 2 body peak. However the mass can be pushed to higher values if one of the photons from the π^0 overlaps a charged track. This is a small contribution and it is estimated by extrapolating from events with an additional, well separated, photon. In order to evaluate the contribution from other K^0_L three body decays, the distribution of the distance between the K^0_L target and the decay plane (d_T) is used. This purely geometrical variable, unaffected by the energy measurement, should vanish for coplanar events, except for resolution effects. The presence of a high d_T tail in K^0_L is apparent from figures 7a and 7b, where the distribution of this variable for K^0_L and K^0_S is shown. Suitable cuts on d_T define a "signal" ($d_T < 5$) and "background control" ($7 < d_T < 12$ cm) region.

The percentage of events in the background region rises for small distances of the decay vertex from the K^0_L cleaning collimator, defining the beginning of the decay region. This rise is attributed to decays of K^0_S produced on the lips of the final collimator by neutrons or K^0_L originating from the defining collimator. In order to reduce the uncertainty connected with the subtraction of this background, the accepted range of decay vertex positions is restricted to the region $z > 10.5$ m , where z is the longitudinal vertex position measured from the final K_L collimator. After this cut, the background control region contains 0.36 % of the K^0_L events.

In order to perform the subtraction, one has to know the percentage of events from each background source contributing to the background control region, as well as the shape of the d_T distribution for each source. Since the cuts that are used to suppress the background are quite simple and do not use all the information available on the shower development, the study of the distributions of variables not used in normal cuts allows a determination of the composition of the background sample.[13] $K^0_L \rightarrow \pi\, e\, \nu$ is found to be the dominating source, but small percentages come from $K^0_L \rightarrow \pi\, \mu\, \nu$ (14%), residual K^0_S from the collimator (5%), $K^0_L \rightarrow \pi^+\pi^-\pi^0$ with the photons from the π^0 outside the detector (3%).

The estimated shapes of the d_T distribution for the different sources of background are shown in figure 7b. For the dominating source ($K^0_L \rightarrow \pi\, e\, \nu$) this shape is determined from the data, using a sample of events with reversed electron rejection cuts. For $K^0_L \rightarrow \pi\, \mu\, \nu$ a Monte-Carlo simulation is used, but it is constrained by data collected with a special trigger where the μ veto is released. For the K^0_S produced on the collimator, again a Monte Carlo simulation is used, this time with the experimental constraint provided by the events rejected by the z cut mentioned above . Finally for $K^0_L \rightarrow \pi^+\pi^-\pi^0$, events with a seen additional photon are used, after removing the photon and scaling the energies of the charged tracks to bring the effective mass to the K^0 mass.

The final estimates of the composition of the background control sample and the subtractions applied are shown in table 3, together with the errors associated. The total error includes the systematic error due to uncertainties in the composition and the extrapolation factors.

For the neutral channel the background source is $K^0_L \rightarrow 3\pi^0$, where 2 γ's miss the detector. Monte Carlo studies indicate that in this case the masses of the two γ pairs have a uniform distribution in the vicinity of the signal region. This is confirmed by the data, that are shown in the scatter plot of figure 8. Good events are defined as the ones in the central ellipse shown in the figure and the residual background is estimated by extrapolating

Table 3 . Background composition for $K_L \to \pi^+ \pi^-$ decays
in NA31 data

Background	Control region	Signal region
$K^0 \to \pi^+ \pi^- \pi^0$	$(0.1 \pm 0.1) \times 10^{-3}$	$(1.0 \pm 1.0) \times 10^{-3}$
$K^0 \to \pi e \nu$	$(2.8 \pm 0.2) \times 10^{-3}$	$(4.4 \pm 0.3) \times 10^{-3}$
$K^0 \to \pi \mu \nu$	$(0.5 \pm 0.2) \times 10^{-3}$	$(0.7 \pm 0.3) \times 10^{-3}$
regenerated K_S	$(0.2 \pm 0.1) \times 10^{-3}$	$(0.4 \pm 0.2) \times 10^{-3}$
total	$(3.6 \pm 0.1) \times 10^{-3}$	$(6.5 \pm 2.0) \times 10^{-3}$

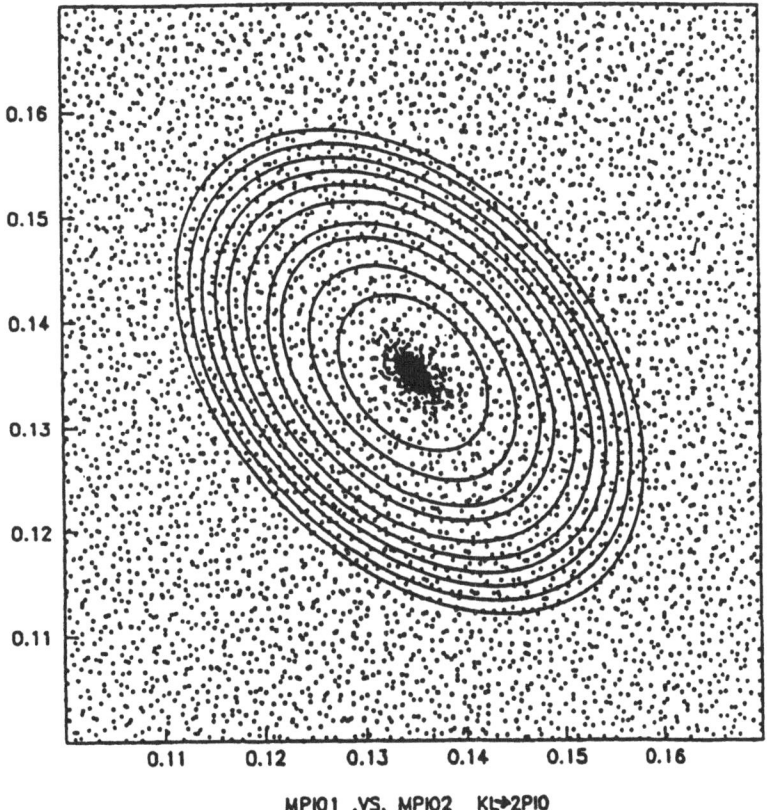

MPIO1 .VS. MPIO2   KL→2PIO

Fig. 8   Scatter plot of the masses of the two best γ pairs
for K_L 4 γ decays

the number of events in elliptical rings having the same area as the central ellipse. This is done by defining a χ^2 variable, whose distribution is shown in figures 9a and 9b for K^0_S and K^0_L respectively. Comparison of the two distributions shows that the distribution is flat outside the signal region, as expected. The total subtraction (actually performed in E,z bins) amounts to 4%. The systematic error is assumed to be 5% of the subtraction, translating into a systematic error on ε'/ε of 0.3×10^{-3}. It should be noticed (figure 10) that, as a result of the K^0 mass constraint (9), this background is strongly z dependent.

Data sample, corrections and systematic errors

The sample of events after all cuts[12,13] contains 109,000 $K^0_L \to \pi^0\pi^0$, 295,000 $K^0_L \to \pi^+\pi^-$, 932,000 $K^0_S \to \pi^0\pi^0$, 2,300,000 $K^0_S \to \pi^+\pi^-$ events. For the ε'/ε analysis, the data are not only binned in 320 rectangular bins of momentum and longitudinal decay vertex position (10 steps in momentum between 70 and 170 GeV/c and 32 steps in z between 10.5m and 48.9m), but also grouped in 16 homogeneous data taking "mini-periods", each including K^0_S runs sandwiched between K^0_L runs, during which the running conditions were kept as stable as possible.

Monte Carlo corrections are applied, in each bin, to account for acceptance effects due to the different beam divergence in K_S and K_L (0.7%) and to the scattering of the K_S beam in the anti-counter and in the collimator (0.3%), as well as for the effects of the finite bin size and of the energy and position resolution. The overall correction to the measured double ratio is 0.3% with an estimated systematic error of 0.1% from the knowledge of the beam divergence and 0.1% from the simulation of the K_S scattering.

Trigger efficiencies throughout the data taking were measured using events that were collected, concurrently with normal triggers, under very

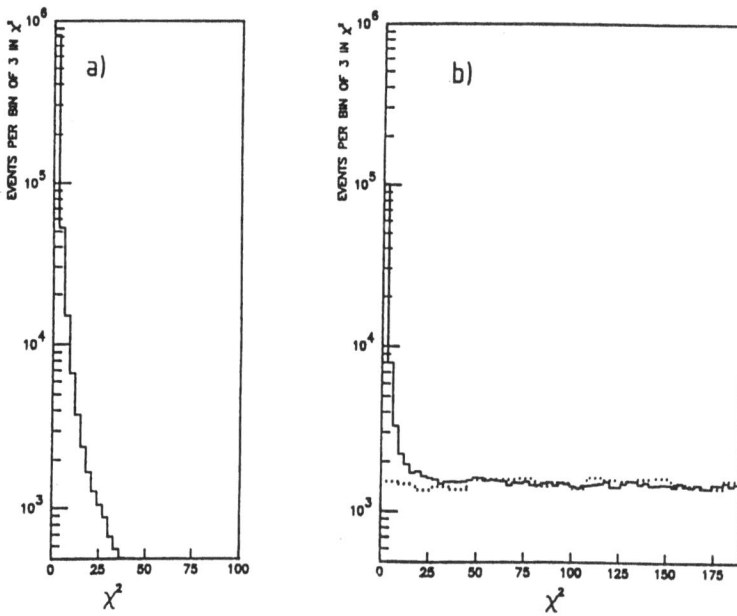

Fig. 9   Number of accepted 4γ events as a function of χ^2
a) for $K_S \to 2\pi^0$;
b) for $K_L \to 2\pi^0$, and a Monte Carlo calculation (dotted line) for background originating from $K_L \to 3\pi^0$ decays

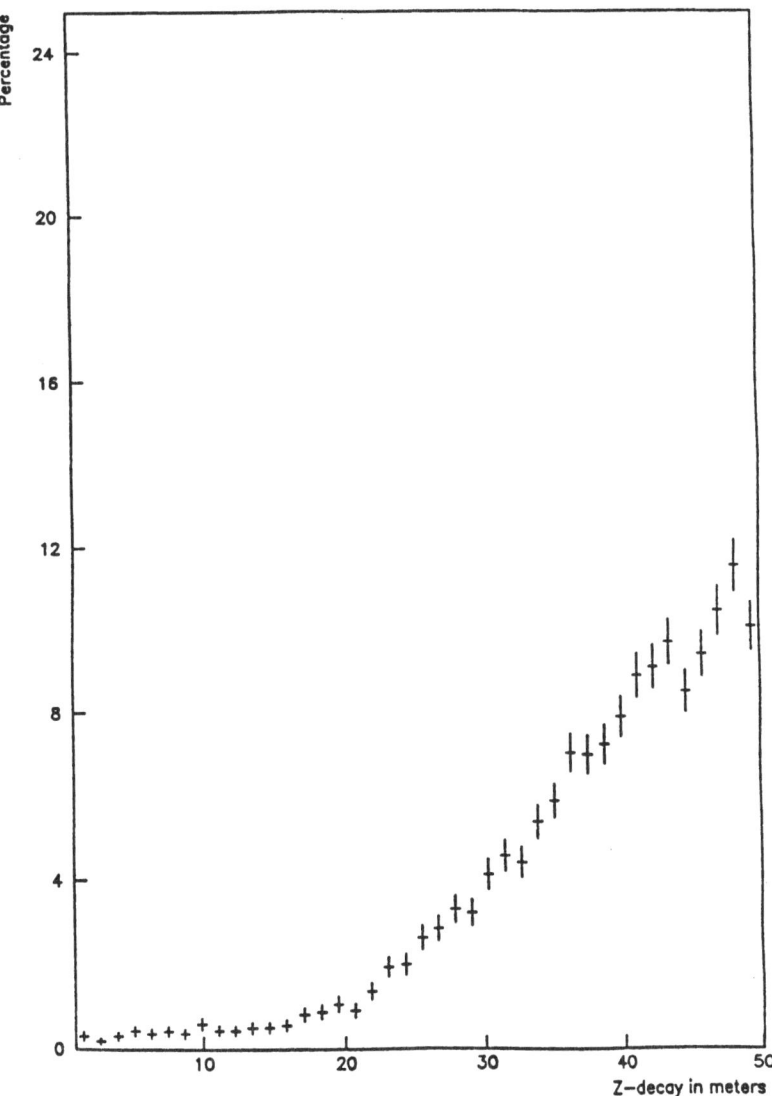

Fig. 10   Dependence of the background from $K_L \to 3\pi^0$ decays on the
distance of the decay vertex from the end of the K_L final
collimator

Table 4. NA31: measurements of trigger inefficiencies and accidental losses

| Channel | Inefficiencies (%) | | Accidental losses (%) |
	Pretrigger	Trigger	
$K_L \to 2\pi^0$	0.06±0.06	0.20±0.10	2.6 ±0.07
$K_S \to 2\pi^0$	0.04±0.02	0.12±0.03	2.5 ±0.05
$K_L \to \pi^+\pi^-$	0.37±0.07	0.05±0.06	2.6 ±0.05
$K_S \to \pi^+\pi^-$	0.48±0.03	0.01±0.01	2.8 ±0.05
Effect on R	-0.12±0.10	-0.03±0.12	-0.34±0.10

Table 5 .   NA31: systematic uncertainties
on the double ratio R (in %)

---

background subtraction for $K_L \to 2\pi^0$	0.2
background subtraction for $K_L \to \pi^+\pi^-$	0.2
$2\pi^0$ /$\pi^+\pi^-$ difference in energy scale	0.3
regeneration in the K_L beam	< 0.1
scattering in the K_S beam	0.1
K_S anticounter inefficiency	< 0.1
difference in K_S/K_L beam divergence	0.1
calorimeter instability	< 0.1
Monte Carlo acceptance	0.1
gains and losses by accidentals	0.2
pretrigger and trigger inefficiency	0.1

---

total systematic uncertainty	± 0.5 %

loose trigger conditions. Table 4 shows the measurements of the inefficiencies separately for the fast pretrigger and for the higher level triggers. As expected, no significant asymmetry between K_S and K_L is found and therefore no correction is applied, but a systematic error of 0.1% from this source is estimated.

Accidental events occurring within the time resolution of the detector could generate asymmetries in the efficiencies for the four decay channels studied. In order to minimize this effect, the beam intensities in K_S and K_L were chosen to give similar single rates in the scintillator hodoscopes. However asymmetries are still possible due to the different nature of the accidental events in K_S and K_L. To estimate this, events collected with a random trigger during normal data taking, have been superimposed to events collected with the normal trigger and passed through the analysis to evaluate the differential inefficiency for the four decay channels. It was found, as shown in more detail in table 4, that accidental losses decrease the double ratio by (0.34±0.1)%. From a study of the stability of this correction versus analysis cuts, a systematic error of 0.2% was estimated from this source.

The uncertainty in the neutral versus charged momentum scale translates into a systematic error on the double ratio, due to the different momentum spectrum of the K_S and K_L beams. By varying the neutral energy scale within the estimated uncertainty of 1×10^{-3} this error is found to be 0.3%.

The time stability of the hadron calorimeter response can have an influence through the offline cuts on the longitudinal shower development (used to reject K_{e3} events) and on the K^0 mass. Possible drifts of the scale between K_S and K_L running are monitored by the peak of the measured K^0 mass. The response of the calorimeter is evaluated to be constant within ±0.5% , leading to an uncertainty on the ratio of K_S to K_L rates smaller than 0.1%.

A summary of the systematic errors discussed is presented in table 5 . Since the errors are mostly uncorrelated, they are summed in quadrature to get a total systematic uncertainty of 0.5%.

Results

The result for the double ratio R defined in (4), after all corrections, is

R = 0.980 ± 0.004(stat.) ± 0.005(syst.)

74

which is significantly different from one. A consistency check of the analysis is obtained by studying the dependence of the double ratio on various parameters. As an example, figure 11 shows that there is no systematic dependence of the double ratio on the data taking miniperiod, on the longitudinal position of the decay vertex, or on the momentum of the decaying K .

If systematic and statistical errors are combined in quadrature, the measured double ratio corresponds to

$$Re (\varepsilon'/\varepsilon) = 0.0033 \pm 0.0011$$

a value that does not agree with the superweak model and is within the range of predictions of the KM model[3].

Future plans

The systematic contribution dominates the error on the result. For this reason, the collaboration is collecting more data in 1988, in order to pin down some of the uncertainties. The improvements to the apparatus include:
a) three stage collimation of the K_L beam to suppress the yield of K_S from the final collimator; this will extend the usable z range to a region which is background free for $K_L \rightarrow 2\pi^0$;
b) improved absorbing power of the K_S beam collimator to suppress the K_S scattering correction;
c) a better coverage for the photon veto's (expecially at small angle) to improve the rejection of 3π events;

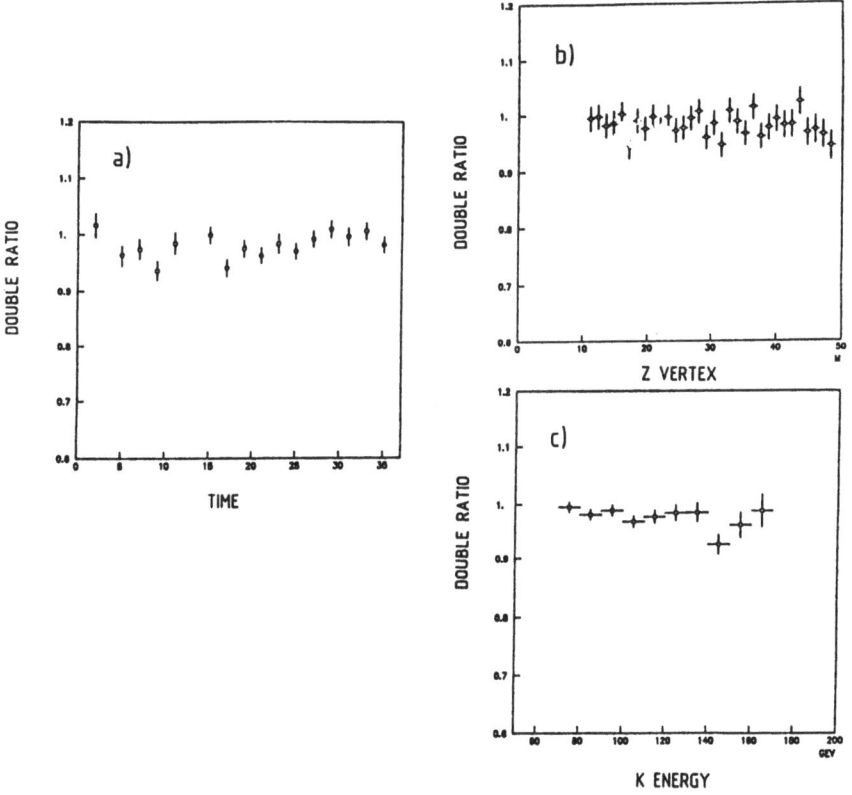

Fig. 11   Double ratio $[N(K_L \rightarrow 2\pi^0)/N(K_L \rightarrow \pi^+\pi^-)]/ [N(K_S \rightarrow 2\pi^0)/N(K_S \rightarrow \pi^+\pi^-)]$
as a function of:
a) time; b) decay vertex position; c) momentum

d) a transition radiation detector (TRD) providing a rejection of K_{e3} events better than a factor 10 for an inefficiency for $\pi\pi$ events smaller than 1%; the additional information, independent from the shower detector information, provided by the TRD, will give the possibility to suppress the K_{e3} background further, in order to verify the present understanding of the charged background and also will allow to relax some of the calorimeter cuts, in order to check the sensitivity to the calorimeter stability;
e) two more planes of iron absorbers + muon counters are added at the end of the apparatus to give a better understanding of the energy loss by muons in the hadron calorimeter and thus verify the present conclusions about the $K_{\mu3}$ background;
f) the primary proton momentum for K_S running will be 360 GeV, instead of 450. This choice will produce a momentum spectrum of the K_S beam more similar to the K_L and thus reduce the sensitivity to the uncertainty on the energy scale by about a factor 2 .

CERN PS195

This experiment[14-15], that will be installed at the CERN LEAR facility by the end of 1988, will provide information on ε'/ε by a novel technique. It will study decay rate asymmetries of tagged $K^0 \bar{K}^0$ produced in $\bar{p}$ interactions at rest by the reactions

$$\bar{p}\, p \rightarrow K^+ \pi^- \bar{K}^0 \qquad ; \qquad \bar{p}\, p \rightarrow K^- \pi^+ K^0 \qquad (10)$$

each with a branching ratio of the order of 2×10^{-3}. All final state particles from reactions (10) and from the subsequent K^0 decay will be fully reconstructed in the detector. The tag of the strangeness of the decaying K^0 relies on the identification of charged particles, which is provided by Cerenkov counters and time of flight measurement.

The principle of the experiment is based on the fact that the decay time distribution of a K^0 produced in a strangeness eigenstate is given by

$$\text{Rate} \left[\begin{array}{c} K^0 \\ \bar{K}^0 \end{array} \longrightarrow f \right] \propto \left(\frac{1}{2} \mp \text{Re}\,\varepsilon \right) \times \qquad (11)$$

$$\times \left[|a_S|^2 e^{-\Gamma_S t} + |a_L|^2 e^{-\Gamma_L t} \pm 2|a_S||a_L| e^{-\frac{\Gamma_S+\Gamma_L}{2} t} \cos(\Delta m\, t + \Phi_S - \Phi_L) \right]$$

(where a_S and a_L are the transition amplitudes for K^0_S and K^0_L going to the final state f, Φ_S and Φ_L their respective phases, Γ_S and Γ_L the K^0_S and K^0_L decay widths). The rate asymmetry arises from two contributions, one connected with $\text{Re}\,\varepsilon$, independent of the decay amplitude, that dominates for $t \rightarrow \infty$, the other which involves decay amplitudes. Information on direct CP violation is contained in this interference term, whose relative contribution is maximum in the region of ≈ 12 K^0_S lifetimes, when f is a CP even state, or at short times, when f is a CP odd eigenstate.

In the measurement of rate asymmetries, the detection efficiencies for K^0 and $\bar{K}^0$ cancel to a very good approximation. However the statistical power is reduced.

The experiment has a very rich program, including detection of CP violation in 3 body K^0 decays. For the 2π channels, information on ε' will be obtained by the measurement of the $K^0/\bar{K}^0$ asymmetry of the decay rates integrated for decay times between 0 and 20 K^0_S lifetimes:

$$I_{+-} = \frac{\text{Rate}\,(\,K^0 \to \pi^+\pi^-\,) - \text{Rate}\,(\,\overline{K}^0 \to \pi^+\pi^-\,)}{\text{Rate}\,(\,K^0 \to \pi^+\pi^-\,) + \text{Rate}\,(\,\overline{K}^0 \to \pi^+\pi^-\,)} = 2\,\text{Re}\,\varepsilon + 4\,\text{Re}\,\varepsilon'$$

$$\tag{12}$$

$$I_{00} = \frac{\text{Rate}\,(\,K^0 \to \pi^0\pi^0\,) - \text{Rate}\,(\,\overline{K}^0 \to \pi^0\pi^0\,)}{\text{Rate}\,(\,K^0 \to \pi^0\pi^0\,) + \text{Rate}\,(\,\overline{K}^0 \to \pi^0\pi^0\,)} = 2\,\text{Re}\,\varepsilon - 8\,\text{Re}\,\varepsilon'$$

giving

$$\frac{\text{Re}\,\varepsilon'}{\text{Re}\,\varepsilon} = \frac{1}{6}\left(1 - \frac{I_{00}}{I_{+-}}\right) \tag{13}$$

An error of 1.5×10^{-3} on this quantity is aimed for. Since both asymmetries are of the order of 3×10^{-3} and the statistical error is dominated by the error on I_{00}, a precision of $\approx 3 \times 10^{-5}$ on this measurement, corresponding to $\approx 10^9 \; \pi^0\pi^0$ decays, is required. In order to collect the statistics needed, the experiment will require $10^{13} \; \overline{p}$, at a rate of 2×10^6 annihilations per second.

At the level of precision required, experimental asymmetries will have to be controlled carefully. In particular, possible asymmetries generated by the different behaviour of the tagging particles, have been considered by the group in the experimental proposal. Their claim is that such asymmetries will only influence the normalization and therefore will not be important for the measurement of ε'/ε.

CONCLUSIONS

The present generation of experiments on ε'/ε have reached the sensitivity of 10^{-3} which is necessary to match the predictions of the Kobayashi - Maskawa model. The first experiment to reach this sensitivity, CERN NA31, has reported a statistically significant evidence of direct CP violation. More measurements with similar sensitivities and different systematics will be available in the next few years to confirm this result.

While the detection of direct CP violation in $K^0 \to \pi\pi$ excludes the super-weak class of models, it is not sufficient to distinguish among different milli-weak models, in view of the uncertainties in the theoretical predictions. Measurements in different channels, either in K^0,[16] or in beauty particle decays, would provide additional constraints to the picture. Undoubtedly, the recent results provide an encouragement to face the experimental challenge that this line of research poses .

REFERENCES

1. L.Wolfenstein, Phys.Rev.Lett. 13:562(1964)
2. M.Kobayashi, T.Maskawa, Prog.Theor.Phys. 49:652(1973)
3. See the reviews:
    L.Wolfenstein, Ann.Rev.Nucl.Part.Sci. 36:137(1986) ;
    J.F.Donoghue, B.R.Holstein & G.Valencia, Int.J.Mod.Phys. A 2:319(1987)
4. T.T.Wu & C.N.Yang, Phys.Rev.Lett.13:380(1964)
5. M.Holder et al., Phys.Lett. 40B:141(1972)
6. M.Banner et al., Phys.Rev.Lett. 28:1597(1972)
7. J.A.Christenson et al., Phys.Rev.Lett. 43:1209(1979)

8. J.K.Black et al., Phys.Rev.Lett 54:1628(1985)
9. R.H.Bernstein et al., Phys.Rev. Lett. 54:1631(1985)
10. M.Woods et al., Phys.Rev.Lett. 60:1695(1988)
11. H.Burkhardt et al., Nucl.Instr.Meth.Phys.Res. A268:116(1988)
12. H.Burkhardt et al., Phys. Lett. 206B:169(1988)
13. M.Kasemann, University of Mainz thesis (1987) , unpublished
    G.Quast, University of Siegen thesis (1988) , unpublished
14. L.Adiels et al.,CERN Experimental proposal PSCC/85-6 (1985)
15. C.Santoni , to appear in the Proceedings of the Stanford symposium on
    Heavy Flavors, Stanford, September 1987
16. For a review see: A Nappi , University of Pisa preprint INFN/PI/AE 87/4,
    to appear in the Proceedings of the Stanford symposium on Heavy Flavors,
    Stanford, September 1987

6

HEAVY FLAVOURS IN

PHOTOPRODUCTION EXPERIMENTS

Bianca Monteleoni

Istituto Nazionale di Fisica Nucleare
Largo E.Fermi 2
50125 Firenze

ABSTRACT

In this review I shall try to cover new experimental results on heavy quark photoproduction. The data have been obtained at CERN, FNAL, and SLAC. The topics covered are: (i) Total cross section measurements, (ii) X_F and P_T distributions, (iii) heavy quark hadronization mechanisms.

1. INTRODUCTION

The first heavy quarks photoproduction experiments have been carried out at FNAL [1], SLAC [2] and CORNELL [3] immediately after the discovery of the J/ψ and ψ' and have helped in establishing the nature of those states as vector mesons associated with a new species of heavy quarks. Those early experiments were generally analyzed in terms of the Vector Meson Dominance Model.

The discovery of the Y and the development of the Photon Gluon Fusion Model in the framework of perturbative QCD have maintained a very high level of experimental interest in heavy flavour production. Large quark masses imply a small α_s and therefore short distance QCD calculations. The physics output of the experiments will consist in the determination of the heavy quarks masses, and the measurements of the gluon structure function and heavy quarks fragmentation functions.

In the past years a large number of experiments has been carried out on real and virtual photoproduction of closed and open charm. The measurements are extremely difficult due to small cross sections, short lifetimes and high multiplicities; only very recently second or third generation experiments are producing the large statistics necessary for detailed studies of production and hadronization mechanisms.

In this review I shall cover recent results on production and hadronization obtained by real photoproduction experiments of open charm. Earlier results have been thoroughly reviewed by T. Nash [5] and by S.Holmes, W.Lee and J. E. Wiss [6]. Results on lifetimes and decay branching ratios are covered by D. Hitlin's [7] report to this workshop.

Section 2 contains a brief introduction to the experiments that have published their final results or are in an advanced analysis stage having completed the data taking. In Section 3 results are presented and discussed in the framework of the Photon-Gluon Fusion Model. Section 4 draws some conclusions and an outlook for the future.

2. THE EXPERIMENTS

Experience has shown that in order to carry out open heavy quark production experiments essential requirements are: i) a high quality vertex detector that allows clean separation between the production and decay vertices, thus reducing the combinatorial background under the signal and ii) a very good downstream spectrometer with particle identification over a wide range of momenta both for charged and neutral particles and excellent energy determination to provide good resolution in the mass distributions.

The four experiments that I am going to describe all fulfill the above requirements, using different techniques. They have also in common a very good determination of the incoming photon energy necessary to measure the energy dependence of the photoproduction cross section; the experiments cover with some overlap the incoming photon energy region from 20 to 230 GeV. In table I are summarized the main features of the experiments.

TABLE I

Details of the Experiments

Experiments	S H F P C [8]	W A 5 8 [9]	N A 1 4' [10]	E 6 9 1 [11]
Laboratory	SLAC	CERN	CERN	FNAL
$< E\gamma >$ (GeV)	20	20 - 70	50 - 150	80 - 230
Target	H_2	Emulsion	Si	Be
Vertex Detector	Bubble Chamber	Emulsion	S M D	S M D
Tracking	9 planes M W P C	30 planes M W P C 8 planes D C	73 planes M W P C	35 planes D C
Number of Triggers	$7 \cdot 10^5$	$6 \cdot 10^5$	$1.7 \cdot 10^7$	$1 \cdot 10^8$

2.1 SHFPC

The SLAC Hybrid Facility Photon Collaboration used as a vertex detector the SLAC rapid cycling 1m. hydrogen Bubble Chamber equipped with a high resolution camera able to resolve 30-50 μm bubbles. The Bubble Chamber operated in a 2.6 tesla magnetic field and was followed by three MultiWire Proportional Chambers, each containing three planes of wires, also operating in the magnet. The momentum resolution obtained for tracks measured both in the B C and in the MWPC was

$$\frac{\sigma_p}{p} = [(0.008)^2 + (0.00085p)^2]^{\frac{1}{2}} \, (p \, in \, GeV/c)$$

For particle identification two atmospheric pressure segmented Cerenkov counters were installed behind the chambers. The first counter was filled with Freon and had a pion threshold around 3 GeV/c and the second was filled with Nitrogen with a pion threshold of 6 GeV/c. An electromagnetic calorimeter built of 204 lead-glass blocks had an energy resolution

$$\frac{\sigma_E}{E} = \left(\frac{0.84 + 4.8}{\sqrt{E(GeV)}}\right)\%$$

The backscattered laser beam spectrum had a central value of 19.6 GeV and a FWHM of 2 GeV. The B C cameras were triggered by requiring either an energy deposition of more than 2 GeV in the electomagnetic calorimeter or a set of hits in the downstream MWPC compatible with a track originating within the fiducial volume of the B C; $3.6 \cdot 10^6$ pictures were taken containing $7 \cdot 10^5$ hadronic events.

2.2 WA58

The CERN WA58 Experiment by the Photon Emulsion Collaboration used emulsion as a vertex detector. The emulsion was placed in the 1.8 tesla Ω magnet and was followed immediately by a high resolution MWPC with 0.5 mm spacing and then by 29 MWPC planes, with 2 mm spacing, divided in 12 chambers. The MWPC operated in the magnetic field; 8 planes of Drift Chambers followed the magnet. A multicell Cerenkov counter filled with Carbon Dioxide at atmospheric pressure with pion threshold around 6 GeV/c and Kaon and proton thresholds at 20 and 38 GeV/c provided particles identification. An electromagnetic calorimeter consisting of 343 lead-glass blocks, 20 radiation lenghts thick, allowed π^0 reconstruction with a standard deviation of 12 MeV/c^2.

The bremsstrahlung beam had a tagged photon spectrum ranging from 20 to 70 GeV. The trigger required four charged particles of which at least one outside the horizontal plane; $6 \cdot 10^5$ triggers were recorded.

2.3 NA14'

The CERN NA14' used as a vertex detector a silicon active target consisting of 34 planes, in 500 μm steps, followed by a telescope of 10 microstrip planes placed in a 1.2 tesla magnet. The downstream spectrometer consisted of 73 MWPC planes. Particle identification was provided by a set of two Cerenkov counters, one filled with Freon with a pion threshold at 3.5 GeV/c, placed in the Goliath Magnet and the other filled with air. Three electromagnetic calorimeters, two of lead-glass and one of lead/scintillator were used to reconstruct η and π^0.

The bremsstrahlung photon beam had a tagged photon spectrum ranging from 50 to 150 GeV. The trigger required at least one charged particle above and one below the horizontal plane and a deposited charge in the active target of at least 2.5 times a minimum ionizing particle; $1.7 \cdot 10^7$ triggers were recorded. The trigger efficiency was 30 % for hadronic events and 60 % for charm events.

2.4 E691

The FNAL E691 Experiment by the tagged Photon Spectrometer Collaboration used a 5 cm. Berillium target followed by a 9 planes silicon microstrip telescope as a vertex detector. The downstream spectrometer consisted of two magnets and 35 Drift Chamber planes, divided in four stations. The momentum resolution for tracks traversing both magnets was

$$\frac{\sigma_p}{p} = 0.5 + 0.05p \, (p \, in \, GeV/c)$$

Particle identification was given by two multicell threshold Cerenkov counters, one filled with Nitrogen with a pion threshold at 5.6 GeV/c and one filled with a Nitrogen-Helium mixture with a pion threshold at 10.4 GeV/c. A segmented lead/liquid scintillator electromagnetic calorimeter had an energy resolution

$$\frac{\sigma_E}{E} = \frac{21\%}{\sqrt{E(GeV)}}$$

An iron/acrylic scintillator calorimeter was used for hadronic energy measurements and had a resolution

$$\frac{\sigma_E}{E} = \frac{75\%}{\sqrt{E(GeV)}}$$

Behind the hadron calorimeter an iron wall was followed by a scintillator wall for muon detector.

The bremsstrahlung beam had a tagged photon spectrum ranging from 80 to 230 GeV. The trigger required a transverse energy deposition in the calorimeters greater than 2.2 GeV together with a total energy deposition greater than 40 GeV and at least one charged particle originating from the target; 10^8 triggers have been recorded. The trigger efficiency was 30 % for hadronic events and 80 % for charm events.

3. RESULTS

3.1 Cross Sections

The results of the earlier charm photoproduction experiments [5,6] have been in qualitative agreement with the predictions of the Photon-Gluon Fusion Model [4]. The model attributes heavy quark pair production to the fusion of the photon with the gluon from one target nucleon.

$$\gamma + g \rightarrow Q + \bar{Q}$$

The cross section for this pointlike process is given by [6]

$$\hat{\sigma}_{\gamma g \rightarrow Q\bar{Q}} = \frac{2\pi\alpha e_Q^2 \alpha_s}{\hat{s}^3} \left\{ \left[\hat{s}^2 + 4m_Q^2 \left(\hat{s} - 2m_Q^2 \right) \right] ln\frac{1+\beta}{1-\beta} - \left[\hat{s}^2 + 4\hat{s}m_Q^2 \right] \beta \right\}$$

where α is the electromagnetic compling constant, α_s is the strong compling constant, $\hat{s}$ is the energy squared in the photon gluon center of mass, m_Q is the heavy quark mass, e_Q is the heavy quark charge and β is $\sqrt{1 - \frac{4m_Q^2}{\hat{s}}}$.

The assumption that the cross section is dominated by the lowest order perturbation theory contribution is justified by the fact that $\alpha_s(4m_Q^2)$ is small.

Recently a full calculation for the first QCD radiative corrections to order $\alpha_s^2\alpha$ has appeared [15].

The total cross section for the inclusive production of a heavy quark pair is obtained by convoluting the elementary cross section with the gluon distribution function G(x).

$$\sigma(s) = \int \hat{\sigma}_{\gamma g}(xs) G(x) dx$$

where s is the square of the photon hadron center of mass energy and $x = \hat{s}/s$.

From the energy dependence of the cross section it is possible to determine $G(x)$. Alternatively assuming a knowledge of the Gluon distribution [16] it is possible to determine the heavy quark mass. The strong dependence of the cross section on the quark mass is due to the very steep decrease of $G(x)$ as a function of x. To a lower value of the quark mass corresponds a lower value for the minimum x allowed by kinematics and therefore a higher value of $G(x)$ and of the cross section. The cross section measurements from the experiments covered in this review are reported in table II.

TABLE II

Cross Sections

Experiments	S H F P C [12]	W A 5 8 [13]	N A 1 4' [10]	E 6 9 1 [14]
$< E\gamma >$ (GeV)	20	45	100	145
Number of Charm decays	100	72	$1.5 \cdot 10^3$	$1 \cdot 10^4$
σ Nucleon (nb)	$62 \pm 8 \, {}^{+15}_{-10}$	230 ± 57	$450 \pm 50 \, {}^{+100}_{-120}$	$580 \pm 10 \pm 60$

Only the SHFPC experiment measures directly the charm production on a photon. The heavy target experiments determine the cross section by relating the number of charm events to the number of hadronic events and then they extrapolate to the per nucleon cross section using slightly different hypotheses on the A dependence of both the charm and hadronic cross sections. WA58 assumes a linear A dependence for charm and $A^{0.9}$ for the hadronic cross section. NA14' assumes the same A dependence for charm and hadronic cross sections, E691 assumes $A^{0.92}$ for hadronic and $A^{0.93}$ for charm; assuming a linear A dependence would lower the result by 14 %. The experiments also find a rising dependence of the cross section from E_γ. The combined results [10,15] from the present and previous experiments favour a charm mass greater than 1.5 GeV/c^2.

3.2 X_F and P_T distributions

The PGFM predictions concerning the X_F and P_T distribution are that the X_F distribution shape should be rather insensitive to the heavy quark masses, but somewhat dependent from the Gluon distribution function, while the P_T distribution should fall rapidly to zero as P_T becomes larger than the heavy quark mass. The average transverse momentum of the heavy quark should be of the order of its mass. Both distributions are little affected by higher order corrections [15]. In table III are reported the values of the average P_T^2 and of the exponent n in the expression

$$\frac{d\sigma}{dx_F} = A(1 + \alpha_F x_F)(1 - x_F)^n$$

for different charm particles obtained by E691.

TABLE III

E691 results

Particles	$< P_T^2 >$ $(GeV/c)^2$	n	$\dfrac{\sigma(200\,GeV)}{\sigma(100\,GeV)}$
D^0, D^+	1.16 ± 0.04	2.95 ± 0.22	1.96 ± 0.24
$D_s{}^+$	1.30 ± 0.26	3.8 ± 1.2	1.3 ± 0.9
$\Lambda_c{}^+$	0.86 ± 0.21	4.1 ± 0.5	1.3 ± 0.9

3.3 Hadronization

The study of heavy quark photoproduction events provides information on the hadronization mechanism. The four experiments described have produced results on the relative abundance of different charm particles. WA58 finds good agreement with the predictions of the Lund Montecarlo with string fragmentation.

The amount of associate production $\bar{D}\Lambda_c^+$ decreases very rapidly with increasing photon energy.

The fraction of associate production over all charm states is $(71 \pm 11 \pm 6)\%$ at 20 GeV, $(28 \pm 13)\%$ at 45 GeV and $(7.5 \pm 2.1)\%$ at 145 GeV.

The results are in good agreement [17] with a hadronization scheme based on the dual parton model in which hadrons are generated along two strings: one baryonic string stretched between the charm quark and the target diquark and one mesonic string stretched between the anticharm quark and the remaining target quark.

4. CONCLUSIONS AND OUTLOOK

The situation for charm photoproduction appears to be quite good. The two high statistics, high energy experiments will soon produce their final results. Together with the QCD calculation by R. K. Ellis and P. Nason they should provide a good determination of the charm quark mass and of the Gluon distribution in the nucleon for x above ~ 0.02. Beauty photoproduction experiments are very difficult. The lower charge and higher mass of the b quark make its detection in photoproduction much more difficult than for charm.

Recently a very high energy experiment, E687, has started taking data at the Tevatron; its contribution will certainly be very important in the charm sector and it will provide first results on beauty photoproduction. For the more distant future the PGFM predicts [18] for Hera $\sim 10^8$ charm and $\sim 10^6$ beauty per year.

A very long time of very hard work still awaits the experimenters interested in heavy quark photoproduction.

ACKNOWLEDGEMENTS

It is a pleasure to thank Luisa Cifarelli and Ahmed Alí for providing the stimulating atmosphere of the 5th Workshop of the INFN ELOISATRON PROJECT on "Heavy Flavours: Status and Perspectives".

I would also like to thank the Staff of the Ettore Majorana Center for Scientific Culture for the superb organization and warm hospitality.

REFERENCES

1. B. Knapp et al., Phys. Rev. Lett. $\underline{34}$ (1975) 1040

2. U. Camerini et al., Phys. Rev. Lett. $\underline{35}$ (1975) 483

3. B. Gittelman et al., Phys. Rev. Lett. $\underline{35}$ (1975) 1616

4. V.A. Novikov et al., Phys. Rep. $\underline{41}$ (1978) 2
   H. Fritzsch and K. H. Streng, Phys. Lett. $\underline{72B}$ (1978) 385
   L. M. Jones and H. W. Wyld, Phys. Rev. $\underline{D17}$ (1978) 759
   J. Babcock, D. Sivers and S. Wolfram, Phys. Rev. $\underline{D18}$ (1978) 62
   M. Glück and E. Reya, Phys. Lett. $\underline{79B}$ (1978) 453

5. T. Nash in Proceedings Int. Symp. on Lepton and Photon Interactions, D. G. Cassel and D. L. Kreinic ed. Cornell (1983) 329

6. S. D. Holmes, W. Lee and J. E. Wiss, Ann. Rev. Nucl. Part. Sci. (1985) 397

7. D. Hitlin this workshop

8. K. Abe et al., Phys. Rev. $\underline{D30}$ (1984) 1

9. A. Forino et al., Nuovo Cimento $\underline{85A}$ (1985) 241

10. M. P. Alvarez et al., Paper N. 739B presented to the XXIV International High Energy Physics Conference Munich 4-10 August 1988

11. J. R. Raab et al., Phys. Rev. $\underline{D37}$ (1988) 2391

12. K. Abe et al., Phys. Rev. $\underline{D33}$ (1986) 1

13. M.l. Adamovich et al., Phys. Lett. $\underline{B187}$ (1987) 437

14. J. C. Anjos et al., Paper N. 465 presented to the XXIV International High Energy Physics Conference Munich 4-10 August 1988

15. R. K. Ellis and P. Nason FERMILAB-PUB-88/54-T, June 1988

16. M. Diemoz et al. Z. Phys. C $\underline{39}$ (1988) 21

17. P. Roudeau Nuclear Physics (Proc. Suppl.) $\underline{1B}$ (1988) 33

18. G. Ingelman and G.A. Schuler DESY 88-020 (1988)

7

HADROPRODUCTION OF TOP FOR $m_t \sim m_W$

D. Denegri

CERN, Geneva, Switzerland
and
CEN, Saclay, France

ABSTRACT

We discuss features of top production and decay at CERN and Fermilab $p\bar{p}$ collider energies for top masses $m_t \sim m_W$. In this mass range the top decays into $t \to W + b$ (physical W) and the $p\bar{p} \to t\bar{t} \to W^+W^-b\bar{b}$ final-state kinematics is rather specific. It is characterized by large transverse momenta $p_T^t \sim p_t^W \sim m_t/2$, highly unbalanced transverse momenta of the two top decay products $p_T^W \gg p_T^b$, making the decay b-jets rather difficult to identify, and with very dispersed and unbalanced W decay products (lepton-neutrino or $q\bar{q}$). In this context, we briefly discuss the few peculiar events observed by UA1 and UA2, as these might well be 'early manifestations' of the top in this mass range. We also discuss the competing single-W + 2-jet background interpretations of these events. Finally, $t\bar{t}$ production up to Large Hadron Collider energies is discussed: cross-sections, kinematical distributions and their mass and energy dependence, and the apparent production of WW final states.

1. INTRODUCTION

We have made a detailed Monte Carlo investigation of $t\bar{t}$ production characteristics for top quark masses $m_t \sim m_W$. In this mass range the top decays into a physical W and a b-quark: $t \to W + b$. In $p\bar{p}$ collisions, $t\bar{t}$ production leads then to $W^+W^-b\bar{b}$ final states as sketched in fig. 1. We give cross-sections and a number of kinematical distributions characteristic of this final state for collision energies from the CERN Collider up to the Large Hadron Collider (LHC) range. If m_t does not appreciably exceed m_W, owing to the large mass difference between the W and b, the decay kinematics is rather peculiar, making the b-jets rather difficult to observe. In the case that $m_t \sim m_W$ the apparent WW final states completely overwhelm electroweak WW production and represent a very severe background to the possible Higgs $\to$ WW signal.

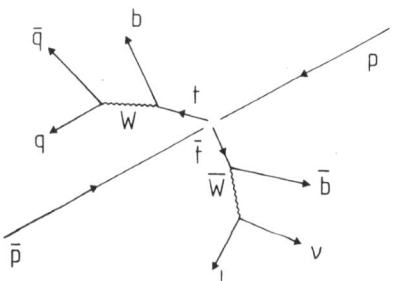

Fig. 1 Sketch of a $p\bar{p} \to t\bar{t} \to W^+W^-b\bar{b}$ final state in the $p\bar{p}$ center of mass.

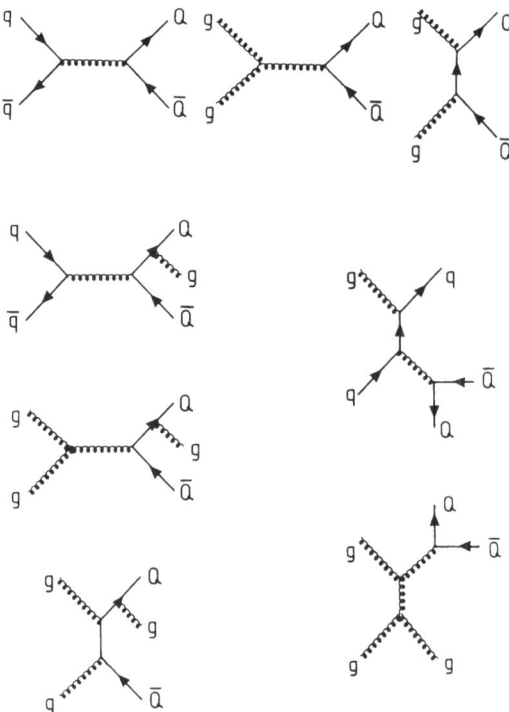

Fig. 2 Some of the Feynman diagrams for heavy flavour production in the EUROJET Monte Carlo [1].

This study of tt̄ production has been done using the EUROJET Monte Carlo [1] appropriately adapted to the régime $m_t \sim m_W$. For the production mechanism this Monte Carlo includes, besides the lowest order α_s^2 (2 → 2) tt̄ production diagrams, also a number of α_s^3 (2 → 3) diagrams (fig. 2) approximating the higher-order QCD contributions. For cross-section estimates as a function of m_t and $\sqrt{s}$, we use either only the 2 → 2 amplitudes, or the sum of 2 → 2 and 2 → 3 contributions with a reasonable K-factor. The K-factor is adjusted through a cut-off on the transverse energy of the jet radiated in the 2 → 3 diagrams. We also study the stability of our results as a function of the Q^2 scale and of the various sets of structure functions available (GHR; DO1,2; EHLQ1,2; DFLM). This procedure is clearly only an approximation, and the real justification and limitations for the K-factors employed can only be found in the recent complete α_s^3 QCD calculations of heavy flavour production by Nason et al. [2], with the accompanying discussion of theoretical uncertainties in Altarelli et al. [3]. Thus the emphasis here is rather on the experimental features of tt̄ production in this top-quark mass range: observables, possible experimental signatures, kinematical distributions and the mass and energy dependence of these, and possible backgrounds. The energy range investigated is from the CERN Collider up to LHC energies.

This study is an extension of an earlier work [4, 5] stimulated by the observation of some peculiar events in UA1/2 [6] which are suggestive of a possible production of a top quark in the ~ 75 to ~ 95 GeV mass range [4]. More recently, this top decay scenario has been studied by a number of authors, with particular emphasis on top production at $\sqrt{s} \sim 2$ TeV and the possible backgrounds [7, 8]. Of course, it may well be that the top mass is significantly below m_W, as suggested by the study of the ratio of partial production cross-sections $\sigma(W \to \ell\nu)/\sigma(Z \to \ell\ell)$, but this is by no means sure at present [9]. Most of our results are also valid for a fourth-generation fermion decaying into a physical (or nearly physical) W and a light quark [4, 5, 8].

2. t̄t PRODUCTION AND DECAY KINEMATICS AS $m_t \rightarrow m_W$

We discuss characteristic kinematical distributions in t̄t production first and cross-sections afterwards. We consider the semileptonic (e or μ) decay of the top, since it is the decay channel which may best allow the observation of top in p̄p collisions [10]. To be specific, we will use the $t \rightarrow be\nu$ mode, since in practice the energy resolution is better for electrons than for muons.

2.1 The lepton-neutrino effective mass in $t \rightarrow b\ell\nu$ decays

The variable that shows the most spectacular variation as m_t approaches m_W is the electron-neutrino effective mass in the $t \rightarrow be\nu$ decay. Figure 3 shows the normalized $e\nu$ effective mass distribution for various top masses. For $m_t = 50$ GeV, the $e\nu$ mass spectrum follows essentially a $t \rightarrow be\nu$ 3-body decay régime. With increasing top mass, however, the $e\nu$ mass becomes more and more skewed owing to the effect of the nearby W pole. The decay kinematics gradually switches to a 2-body $t \rightarrow bW$ decay régime, until for $m_t > m_W + m_b$ the W enters the physical region and there is no further significant evolution of the $e\nu$ mass shape. What is measured, in practice, in the semileptonic decay of the top is not the $e\nu$ effective mass, but rather the transverse mass [11], which shows a less dramatic variation. The effective mass itself can, however, be measured on the purely hadronic (jet) decay of the recoiling top.

2.2 Top-quark transverse momentum in t̄t production

Figure 4 shows the normalized top-quark transverse momentum distribution $(1/\sigma)d\sigma/dp_T^t$ for a range of top masses and at $\sqrt{s} = 0.63$ and 2 TeV. These distributions correspond to the lowest order $(2 \rightarrow 2)$ diagrams only. The higher order $(2 \rightarrow 3)$ diagrams (not shown) give a similar behaviour, with longer tails and a larger average value $\langle p_T^t \rangle$ by 10 to 20%, depending on the low E_T^{jet} cut-off. The main result is that the average top transverse momentum is of the order of m_t, with a long tail extending to

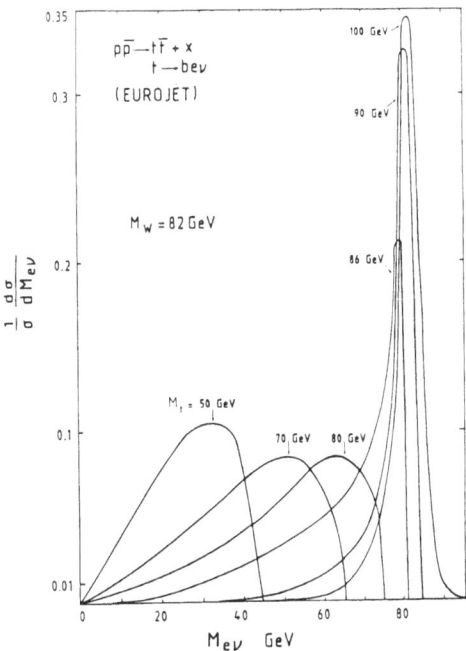

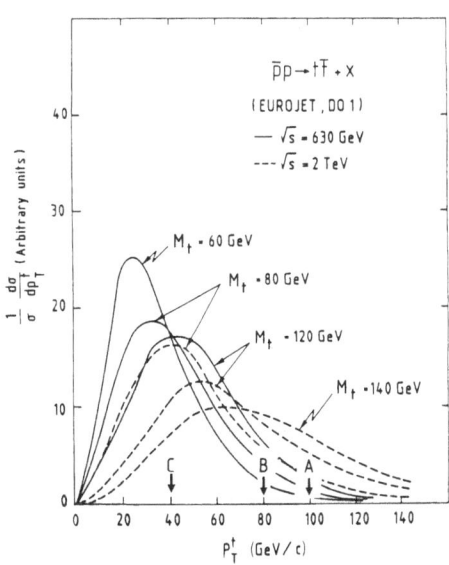

Fig. 3 Effective mass of the lepton-neutrino from the $t \rightarrow be\nu$ decay as m_t approaches m_W [4].

Fig. 4 Normalized top transverse momentum distribution at $\sqrt{s} = 0.63$ and 2 TeV [4].

large values of p_T^t. More precisely, at $\sqrt{s} = 0.63$ TeV we have $\langle p_T^t \rangle \approx m_t/2$, and owing to the limited phase space available the high-p_T^t tail is significantly suppressed in comparison to $\sqrt{s} = 2$ TeV, and the more so the larger is m_t. This result, $\langle p_T^t \rangle \sim m_t$, is expected since both the t-channel exchange gluon–gluon fusion diagram and the (dominant) s-channel $q\bar{q}$ annihilation and gluon–gluon diagrams (fig. 2) have approximately a $d\sigma/dp_T^2 \propto 1/(m_t^2 + p_T^2) \to$ constant behaviour when $p_T^t \to 0$. In fig. 4 are also indicated the p_T^t values for the few peculiar events of UA1/2, if these were interpreted as $t\bar{t}$ production.

2.3 t → Wb decay and the b-quark transverse momentum

The normalized transverse momentum distribution $(1/\sigma)d\sigma/dp_T^b$ for the decay b-quark is shown in fig. 5. It is in this variable that the crossing of the physical W decay threshold in the t → Wb decay generates the most peculiar behaviour. Below $m_t \approx m_W$ the p_T^b spectrum has an extended shape with a long tail (fig. 5). As $m_t \to m_W$ the distribution first shrinks to reach a minimum $\langle p_T^b \rangle \sim 10$ GeV at threshold $m_t = m_W + m_b$. As m_t increases further, the p_T^b distribution starts to expand again with $\langle p_T^b \rangle$ increasing monotonically with m_t. This is to be expected, since from a well-known kinematics (Lorentz boost) effect at large p_T^t and for a small decay Q-value in t → Wb, the momentum is shared between the decay products in the ratio $p_T^b/p_T^W \to m_b/m_W$ as $Q \to 0$. The net result is that most of the large top transverse momentum p_T^t at production is then taken over by the decay W. The shapes of the spectra in fig. 5 show very little dependence on the collision energy $\sqrt{s}$, since they are essentially determined by the top mass dependent decay kinematics. The shapes of p_T^b spectra are also not significantly modified, if the 2 → 3 diagrams are considered. In fig. 6 we show the scatter plot of p_T^b as a function of p_T^W for $m_t = 90$ GeV at $\sqrt{s} = 0.63$ TeV. The unequal sharing of transverse momenta between the decay W and b for $m_t \sim m_W$ is particularly striking in this plot (see also section 6, fig. 21). The effect is even more pronounced at higher energy since $\langle p_T^t \rangle$ is larger. It is however localized to $m_t \sim m_W$, since for $m_t = 150$ GeV (fig. 21b) there is almost equipartition of transverse momenta.

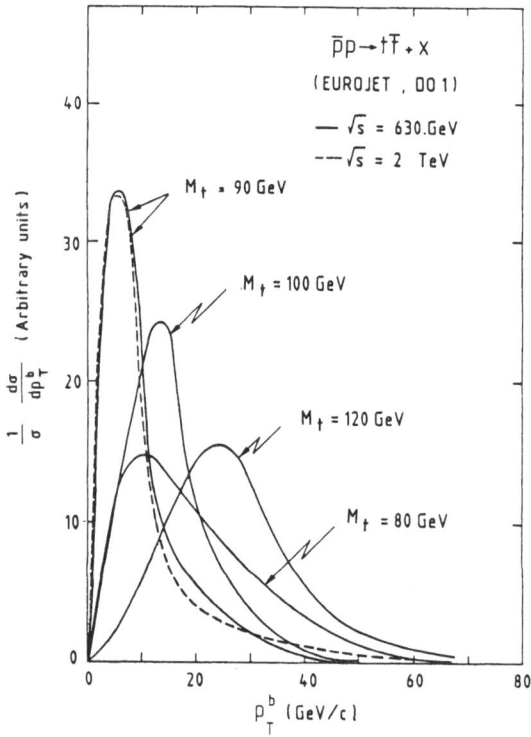

Fig. 5 Normalized b-quark transverse momentum distribution at $\sqrt{s} = 0.63$ and 2 TeV [4].

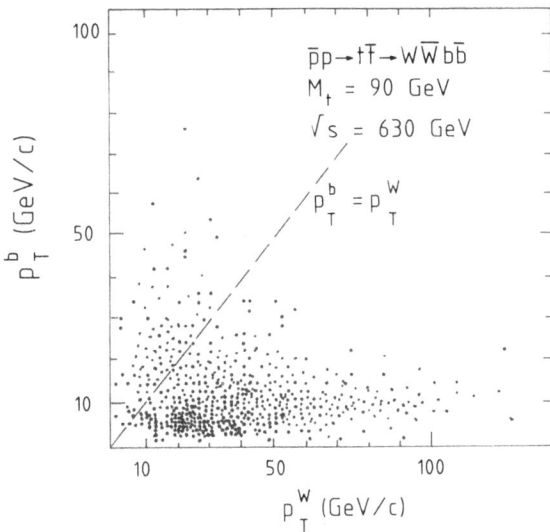

Fig. 6 Correlation between the top decay W and b-quark transverse momenta for $m_t = 90$ GeV at $\sqrt{s} = 0.63$ TeV.

Present experience shows that jet recognition and detection is problematic for E_T^{jet} below ~ 10 GeV. Detectability of jets in this energy range depends on the detector granularity and its energy response and resolution to hadrons in the GeV range. A further difficulty with low-E_T jets is the confusion with abundant gluon bremsstrahlung jets in this E_T range [11], and with accidental jet-like fluctuations of spectator partons in the underlying event. A practical consequence of this unbalance in momentum sharing between W and b means that, for $t\bar{t}$ production with $m_t \sim m_W$, the b-decay jets will be largely lost or confused with bremsstrahlung jets, thus giving rise to essentially WW final states with $\langle p_T^W \rangle \approx \langle p_T^b \rangle \approx m_t/2 \approx 50$ GeV. Since this is a strong interaction production mechanism, at higher energies it overwhelms electroweak WW production by approximately two orders of magnitude [4] (see also fig. 17). This WW final-state configuration, with suppressed and difficult-to-observe b-jets, also means that the main competing physics background in a top search is coming from the higher-order QCD production of W + $\geqslant 2$ jet events [6a, 7]. This QCD background is also significant for $m_t < m_W$ (since only the $\ell\nu$ transverse mass is measurable in the W $\rightarrow \ell\nu$ decay), but the lighter is the top the more important is the $b\bar{b}$–gluon background, which is, in fact, the dominant physics background for $m_t \lesssim 50$ GeV [10].

2.4 W $\rightarrow \ell\nu$ (or W $\rightarrow q\bar{q}$) decay configurations at large p_T^W

Another interesting experimental feature of large transverse momentum W production concerns the W decay products $\ell\nu$ or $q\bar{q}$. In usual (low p_T) single-W production, the W $\rightarrow \ell\nu$ decay products exhibit a strong correlation $E_T^\ell \approx E_T^\nu$ between their transverse energies (or momenta), with a pronounced Jacobian peak in both of these spectra (refs. [11] for example). For W's produced with large p_T, as in $t\bar{t} \rightarrow$ WW$b\bar{b}$, this correlation disappears entirely in the high-p_T^W limit. As visible from fig. 7, for large p_T^W it is replaced by an opposite correlation, where large E_T of the first decay product is correlated with a small E_T of the second one, $\ell\nu$ or $q\bar{q}$. This is a kinematical consequence of a large-mass and large-p_T object decaying into two approximately massless particles, with their transverse momenta being measured with respect to the $p\bar{p}$ beam line, and not the parent W

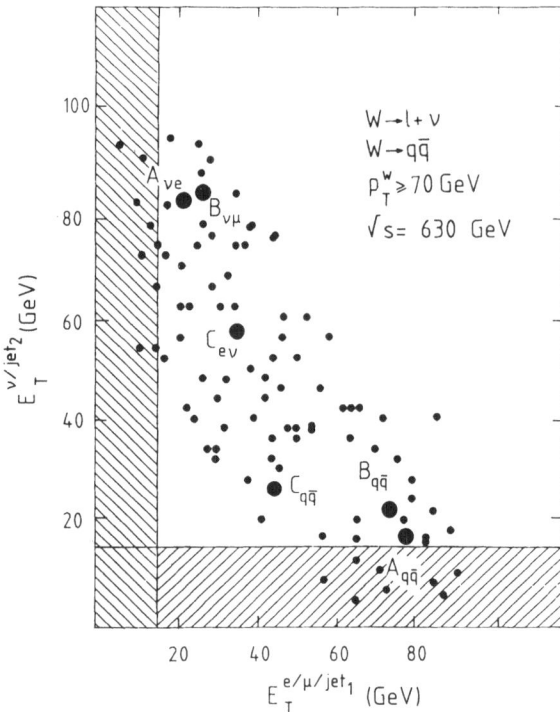

Fig. 7 Correlation between the transverse energies of the two leptons (or the two quarks) from a W decay, for W's produced at large p_T.

line-of-flight. This gives at high p_T^W, as a preferred kinematical configuration, highly unbalanced W decay products, with typically $E_T^{e/jet_1} \sim 70$ GeV and $E_T^{\nu/jet_2} \sim 25$ GeV and vice versa. This kinematical effect is responsible, in particular, for the decreasing W (and thus $t\bar{t}$) detection efficiency with increasing p_T^W (see, for example, refs. [11]). It is due to the increasing fraction of soft W decay products (e, μ, or ν) either below a fixed $E_T^{\ell,\nu}$ detection threshold (the case of UA1 is indicated by the hatched area in fig. 7) or failing the lepton isolation criterion relative to the underlying event [11b]. The quark and lepton kinematical configurations of the apparent WW events of UA1/2 are also indicated in fig. 7.

3.   EXPECTED $t\bar{t}$ CROSS-SECTIONS AND EVENT RATES IN UA1/2 CONDITIONS

3.1 Cross-sections

Figure 8 shows the total $t\bar{t}$ production cross-section as a function of m_t at $\sqrt{s} = 0.63$ TeV for a number of input parton distribution functions (GHR, DO1,2, EHLQ1). The Q^2 scale used is $Q^2 = m_t^2 + p_T^2$. The cross-section shown is the sum of the lowest order and the approximate next order contribution as given by EUROJET, for a relatively natural cut-off at $E_T^{jet} > 5$ GeV. This corresponds at $\sqrt{s} = 630$ GeV to a K-factor of ~ 1.35, slowly decreasing with increasing m_t. This value, in connection with the choice of Q^2, is in a remarkably good agreement with the QCD motivated K-factors of refs. [2, 3]. The various structure functions are used to illustrate the level of uncertainty due to partonic distributions. The full discussion of theoretical uncertainties on $\sigma_{t\bar{t}}$, which are altogether of the order of 50% either way, can be found in ref. [3]. Figure 8 also shows the expected cross-sections at $\sqrt{s} = 2$ TeV in the same conditions (same E_T^{jet} cut-off). The EUROJET K-factor now amounts to ~ 1.9. In view of the results of ref. [2], a value K ~ 1.7 would probably be

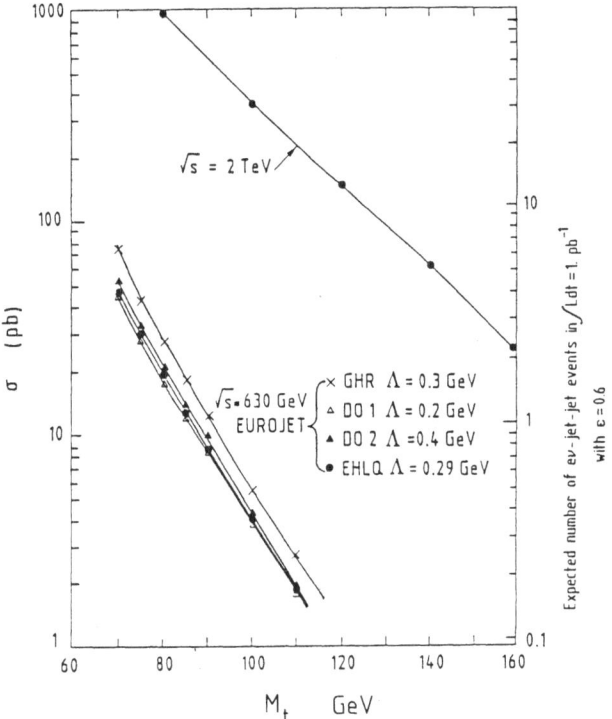

Fig. 8 The $p\bar{p} \rightarrow t\bar{t}$ production cross-section as a function of m_t at $\sqrt{s} = 0.63$ TeV and 2 TeV [4].

better justified and could easily be obtained by adjusting the low E_T^{jet} cut-off. Nonetheless, the main message is clear: for $m_t > 80$ GeV the Fermilab collider has a substantial advantage in $t\bar{t}$ cross-sections (a factor ~ 30) over the CERN Collider. For smaller top masses, $50 \lesssim m_t \lesssim 75$ GeV, top production is in fact dominated by $W \rightarrow t\bar{b}$ production (in particular at $\sqrt{s} = 0.63$ TeV) and the relative advantage of Fermilab is less pronounced, a factor of 4 to 10 depending on m_t.

3.2 Event rates

The right-hand scale of fig. 8 also gives the expected number of $t\bar{t} \rightarrow W\overline{W}b\bar{b} \rightarrow e-\nu +$ jet–jet events (including W branching ratios) for an experimental sensitivity of $\int L \, dt = 1$ event per picobarn, assuming a $W(\rightarrow e\nu)$ detection efficiency of $\epsilon_W = 0.6$. The b-jets are ignored. At $\sqrt{s} = 0.63$ TeV, for 75 GeV $< m_t < 95$ GeV expected event numbers would decrease from ≈ 2 to ≈ 0.4 events with increasing m_t. In the realistic conditions of UA1/2, where $\int L \, dt \approx 0.7$ event per picobarn (data from 1983/85) and $\epsilon_W \approx 0.3$ at large p_T^W [11], from ≈ 0.7 to ≈ 0.2 events per W leptonic mode can be expected in this $m_t \approx 75$ to ≈ 90 GeV range, without any 'stretching' of QCD expectations. Including all theoretical uncertainties on $\sigma_{t\bar{t}}$, this number could vary by a factor of ~ 1.5 [3]. The main message here is that with experimental sensitivities of ~ 1 event per picobarn, under favourable background conditions and thanks to the highly selective experimental signature provided by two on-shell W decays (e/μ-$\nu +$ jet–jet), a top in the $m_t \sim m_W$ range is at the very limit of observability in the UA1/2 data available at present. The forthcoming (1988/89) UA1/2 data-taking periods, where experimental sensitivities of ~ 10 events per picobarn can be expected, could clarify the situation. In view of fig. 8 *it is clear*, however, that the Fermilab collider has a substantial advantage in terms of production cross-sections. The $t\bar{t} \rightarrow WW$ signal to the $W + 2$ jet background ratio may, however, not be as favourable at 2 TeV for $m_t \sim m_W$ (section 5). On the other hand, for $m_t > 120$ GeV, which is beyond the CERN Collider reach, but is within the reach of the Fermilab collider for a sensitivity of more than a few events per picobarn (fig. 8), the observation of the accompanying b-jets might bring a substantial reduction of the QCD background, of the order of α_s per detected jet, thus providing a window of observability for the top at the Fermilab Collider.

4. THE SPECIAL EVENTS OF UA1/2

In this context it is interesting to recall that UA1 has observed two [6a] and UA2 one $W \to \ell\nu$ event [6b] accompanied by two hard jets with $M_{\text{jet-jet}} \sim m_W$. These events are suggestive of $t\bar{t}$ production in the $m_t \sim m_W$ mass range [4].

4.1 Events and the $t\bar{t}$ interpretation

The event of UA2, labelled C, is the only W + 2 jet event within the acceptance of UA2 ($|\eta_{\text{jet}}| < 0.85$ and $p_T^{\text{jet}} > 10$ GeV) [6b]. In this event $p_T^W \approx 40$ GeV/c and with $p_T^W \sim p_T^t$ in the $t\bar{t}$ interpretation, it falls at about the most probable p_T^t according to fig. 4. The jet-jet effective mass is ~ 60 GeV, which is somewhat low in comparison to m_W, but cannot be excluded on a particular event. It is interesting that this event also has a third jet (fig. 9) of $E_T^{\text{jet}} \sim 7$ GeV, which, according to fig. 5, is highly probable for the decay b-jet. According to UA2, however, the production rate and the accompanying 2-(hard)jet kinematical configuration of this event is also entirely consistent with second-order QCD corrections to single-W production [6b].

The UA1 experiment, with a larger jet acceptance ($|\eta_{\text{jet}}| < 3.0$ and $E_T^{\text{jet}} > 7$ GeV), has 10 W + 2 jet events, for a total $W \to e\nu$ sample of ~ 300 events. The only two WW consistent UA1 events (labelled A and B), shown schematically in fig. 9, have been widely discussed [6a]. They are the two largest p_T^W events in the UA1 sample, with $p_T^W \gtrsim 65$ GeV. No additional jets are visible in events A and B, but this is not surprising as discussed in section 2.3. The internal lepton–lepton and jet–jet decay configuration, with its characteristic E_T unbalance, is entirely consistent with expectations for large-p_T^W production (fig. 7). The problem for the $t\bar{t}$ interpretation of these events is the unusually high value of the observed p_T^W. These two events represent a rather unlikely sampling of the expected

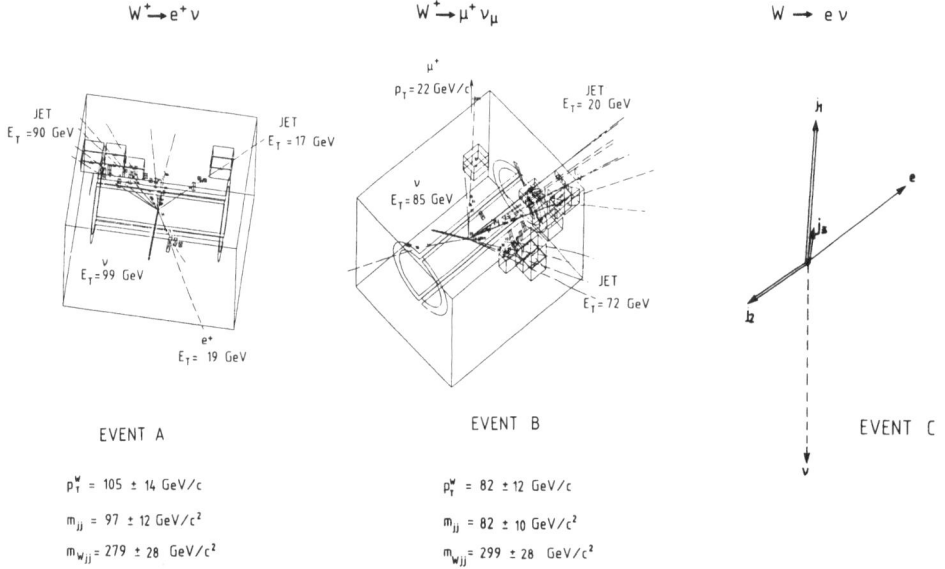

Fig. 9 Sketches of events A and B of UA1 [6a], and C of UA2 [6b].

p_T^W (or p_T^t) distributions for $t\bar{t}$ production (fig. 4). At the observed p_T^W values the expected $t\bar{t}$ production rate is $\leqslant 0.07$ events. However, the resolution on p_T^W is $\sim 15\%$ for large p_T^W, the systematic uncertainty on the absolute energy scale is $\sim 9\%$, and the event probability is a very sensitive function of p_T^W. Displacing p_T^W (measured) $\rightarrow [p_T^W$ (measured) $- 1\sigma]$ increments the $t\bar{t}$ event probability by a factor of ≈ 3.3 for a displacement by 1σ (resolution) and, independently, by ≈ 2.1 for 1σ (absolute energy scale).

There is, in fact, a hint of a possible systematic overestimate of jet energies for $E_T^{jet} \geqslant 35$ GeV. The study of hadronic τ-jets from W $\rightarrow \tau\nu$ decays gives for the W mass $m_W = 89 \pm 3$ (stat.) ± 6 (syst.) GeV [12], as compared to $m_W = 80.2$ GeV expected from the Standard Model for $\sin^2 \theta_W = 0.232 \pm 0.005$. The W mass depends mostly on the hardest τ-jets at $E_T^{\tau\text{-jet}} \geqslant 35$ GeV. An independent fit to m_W, from a somewhat different (partially overlapping) sample of W $\rightarrow \tau\nu$ decays [13], gives also $m_W = 96 \pm 6$ (stat.) ± 6 (syst.) GeV.

4.2 The QCD background interpretation

The single-W + 2-jet background interpretation of these events has been extensively studied by UA1 [6a]. The problem is again the too large values of p_T^W. The situation is best summarized by fig. 10a which shows the scatter plot of the effective mass M(W–jet–jet) as a function of p_T^W for all UA1 W + 2 jet events [6a]. The event population is compared with the expected QCD W + 2 jet production. It is properly clustered in the expected most probable region of the plot, with the two events A and B rather separated from this population. They may be a tail sampling from this QCD population, but the expected number of events in this kinematical configuration is ~ 0.05 [6a]. As already stated, the comparable number of events for top production according to our analysis is ~ 0.07 to 0.02 (a function of m_t), with a factor of ~ 1.5 QCD uncertainty. The contours in fig. 10a correspond to UA1 jet acceptances, and for QCD bremsstrahlung jets the acceptances of UA1 and UA2 differ significantly, so event C of UA2 cannot be meaningfully plotted in fig. 10a. As already said, however, the analysis of UA2 finds event C entirely consistent with a QCD background [6b]. The corresponding plot of M(W–jet–jet) as a function of p_T^W for events A, B, and C for the $t\bar{t}$ hypothesis is shown in fig. 10b. The contours correspond again to UA1 acceptances and detector simulation [14]. Event C can, however, now be meaningfully incorporated in such a comparison, as the effect of the difference between UA1 and UA2 acceptances must now be small. Massive top $t\bar{t}$ production is very central at $\sqrt{s} = 0.63$ TeV, and the large top (and W) mass guarantees a large decay

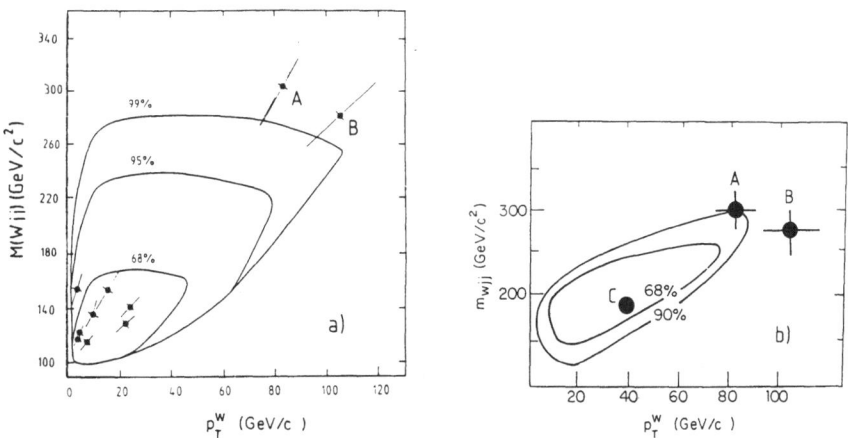

Fig. 10 a) Scatter plot of the W–jet–jet effective mass as a function of p_T^W for the UA1 W + 2 jet events [6a] and the expected QCD distribution;

b) Scatter plot of the W–W effective mass as a function of p_T^W for $t\bar{t}$ production [14].

jet of $\langle E_T \rangle \approx 35$ GeV, well above the experimental jet-detection thresholds. In fig. 10b event C is optimally located for the $t\bar{t}$ interpretation, as already mentioned, while events A and B are marginal. The sampling of the expected $t\bar{t}$ population provided by the three events together is not so bad, however. A 10 to 20% modification of the jet-energy scale for events A and B would of course be equally helpful for both the top interpretation (figs. 4 and 10b) and the W + 2 jet interpretation (fig. 10a).

4.3 A possible way of discriminating between signal and background

The issue is then whether there is at a more detailed level a variable which might help in deciding, at least statistically, between the two competing interpretations. For example, how likely is it to find the second (softer) jet in the observed E_T range? In W + 2 jet production, the observed jets should be part of a monotonically falling bremsstrahlung spectrum $\propto 1/E_T^2$, while in $t\bar{t} \rightarrow WW \rightarrow e\nu$–jet–jet, they are part of a W-decay Jacobian peak smeared by large-p_T^W production. These two jet production spectra should have rather distinct shapes, which might resolve the ambiguity.

Figure 11a shows the jet E_T spectrum from $t\bar{t} \rightarrow WW \rightarrow e\nu$–jet–jet with the expected Jacobian peak, gradually degrading with increasing p_T^W, as already discussed in section 2.4 (fig. 7). Figure 11b shows the jet E_T spectrum of the first (harder) jet and fig. 11c that of the second (softer) jet for W + 2 jet events with $M_{jet\text{-}jet} > 60$ GeV [15], as predicted by the EKS Monte Carlo [16]. The expected shape of the harder and softer jet E_T distributions from $t\bar{t} \rightarrow WW \rightarrow e$-$\nu$ jet–jet is also shown (dashed line). These distributions are for the partons; they do not include apparatus resolution effects. For W + 2 jet events, the dangerous and misleading bump in the $E_T^{jet\,1}$ distribution (fig. 11b) faking a Jacobian peak is the result of the jet-jet mass cut and the 'harder jet' selection bias. In fig. 11c, however, the expected distinctive behaviour for the softer jet is well borne out [15].

The three events consistent with WW production clearly represent a good sampling of the expected $t\bar{t}$ distribution in fig. 11c, but are not grossly inconsistent with a QCD bremsstrahlung

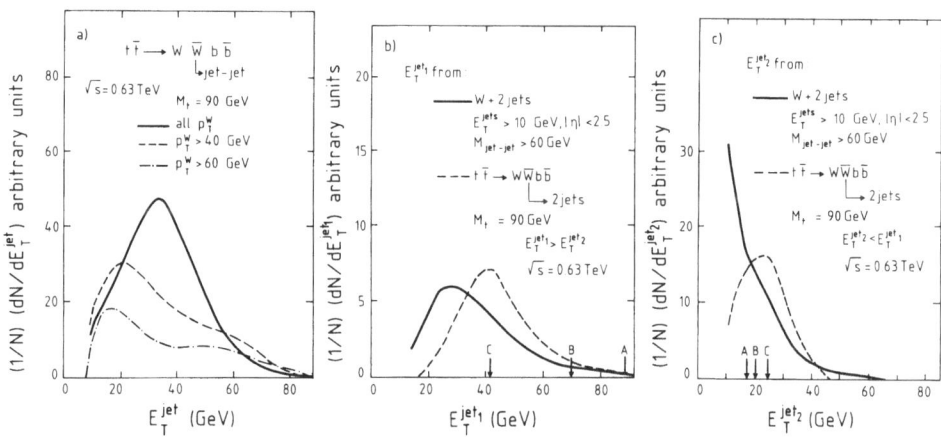

Fig. 11 a) Jet E_T distribution for $t\bar{t} \rightarrow WWb\bar{b} \rightarrow e\nu$ + jet–jet + $b\bar{b}$ for various p_T^W cuts;

b) Jet E_T distribution for the harder jet from W + 2 jets (solid line) and $t\bar{t} \rightarrow WWb\bar{b} \rightarrow e\nu$ + jet–jet + $b\bar{b}$ (dashed line);

c) Jet E_T distribution for the softer jet from W + 2 jets (solid line) and $t\bar{t} \rightarrow WWb\bar{b} \rightarrow e\nu$ + jet–jet + $b\bar{b}$ (dashed line);

W + 2 jet events are obtained with EKS Monte Carlo [16].

spectrum either (jet energy resolution would somewhat smear the bremsstrahlung spectrum). It is clear, however, that it is the bump around $E_T^{jet} \sim m_W/4$ in this distribution of $E_T^{jet 2}$ that could, in the future, with larger statistics, unambiguously reveal a possible top contribution mixed with a QCD background W + 2 jet population. The bump is not a $m_W/2$ owing to the bias of choosing the softer of the W decay jets. The size of the anomaly is of course determined by the relative $t\bar{t}$ and W + 2 jet cross-section ratio, a function mainly of m_t (fig. 8), and to a smaller extent of p_T^W and $\sqrt{s}$ [7, 15]. At $\sqrt{s} = 0.63$ TeV, for example, the W + 2 jet QCD background rate, with $E_T^{jet} > 20$ GeV and $M_{jet\text{-}jet} > 60$ GeV, is comparable with the $t\bar{t}$ rate for $m_t \approx 95$ GeV [15].

The comparison of figs. 11b and c also shows that the E_T unbalance between the two jets is even more pronounced for the QCD background than for the $t\bar{t}$ signal. It is also clear from fig. 11c that jet recognition is needed at low E_T values ($\sim$ 10 GeV) if we wish to distinguish between the two mechanisms. If we want, as a further means of suppressing the QCD background, also to detect the soft b-jets accompanying $t\bar{t}$ production, low-E_T jet recognition is clearly essential.

In conclusion, the probability of the $t\bar{t}$ interpretation of these few events is as good as any other found until now. The sampling of the expected $t\bar{t}$ population provided by the three WW compatible UA1/2 events (fig. 10b) is acceptable, in particular if the p_T^W (and m_{WW}) values of UA1 events are diminished by $\sim$ 15% as suggested by the above discussion. However, the W + 2 jet population sampling in fig. 10a is of comparable probability. The only argument favouring the $t\bar{t}$ interpretation over W + 2 jets is the second (softer) jet E_T spectrum in fig. 11c, where we expect a quasi-Jacobian peak as compared to a steeply falling bremsstrahlung spectrum. Only higher statistics with a good understanding of rates and event configurations of the W + $\geqslant$ 2 jet background could definitely confirm or reject the $t\bar{t}$ production hypothesis. Clearly, a detailed experimental and theoretical understanding of W production at large p_T, with associated jet multiplicities and event configurations, will be needed before top production in this mass range could be ascertained. The forthcoming CERN and Fermilab collider periods, where up to $\sim$ 10 events per picobarn can be expected by the end of 1989, may clarify the situation.

5. $t\bar{t}$ CROSS-SECTIONS FROM CERN TO FERMILAB COLLIDER ENERGIES

Figure 12 shows the energy dependence of the $t\bar{t}$ production cross-section for 40 < m_t < 120 GeV. The sets of structure functions used are EHLQ1 [17] and DFLM [18]. The latter set has a softer gluon distribution than most other parametrizations. As the gluon–gluon fusion diagrams are becoming most important in this m_t range at Fermilab energies, this set probably provides a safe lower limit. The scale is $Q^2 = m_t^2 + p_T^2$. The cross-sections shown are for lowest order (α_s^2) only. With this choice of Q^2, these cross-sections should be rescaled by a K-factor increasing from K $\sim$ 1.3 at 0.6 TeV to $\sim$ 1.5 at $\sqrt{s}$ = 2 TeV for m_t = 100 GeV [2]. The main result of fig. 12 is that for $m_t \approx m_W$, $\sigma_{t\bar{t}}$(Fermilab)/$\sigma_{t\bar{t}}$(CERN) $\sim$ 30. Notice, however, the very steep rise of the $t\bar{t}$ production cross-section in the CERN collider range. A modest increment of $\sim$ 20% (to $\sqrt{s} \approx 800$ GeV) of the CERN collider energy, which is technically feasible but expensive, would increase the top production rate by a factor $\approx$ 3 for $m_t \sim m_W$.

Figure 13 shows the m_t dependence of $\sigma_{t\bar{t}}$ at $\sqrt{s}$ = 0.63 TeV and 1.8 TeV. This is again the lowest order contribution only; the scale is now $Q^2 = m_t^2$ so that the necessary K-factor (of the order of $\sim$ 1.5) can be read off from ref. [2]. At CERN collider energies, for 40 GeV $\leqslant m_t \leqslant m_W$, top production is in fact dominated not by $t\bar{t}$ but rather by W $\rightarrow$ t$\bar{b}$, and $\sigma_{t\bar{b}}$(Fermilab)/$\sigma_{t\bar{b}}$(CERN) $\sim$ 4 [19, 20]. Including both the $t\bar{t}$ and t$\bar{b}$ contributions, for 50 GeV $\leqslant m_t \leqslant m_W$, the advantage that Fermilab has in top production cross-section varies from $\sim$ 4 to $\sim$ 10, and for $m_t > m_W$ it is > 25 when compared to $\sqrt{s}$ = 0.63 TeV, increasing with m_t (see also refs. [7, 8]).

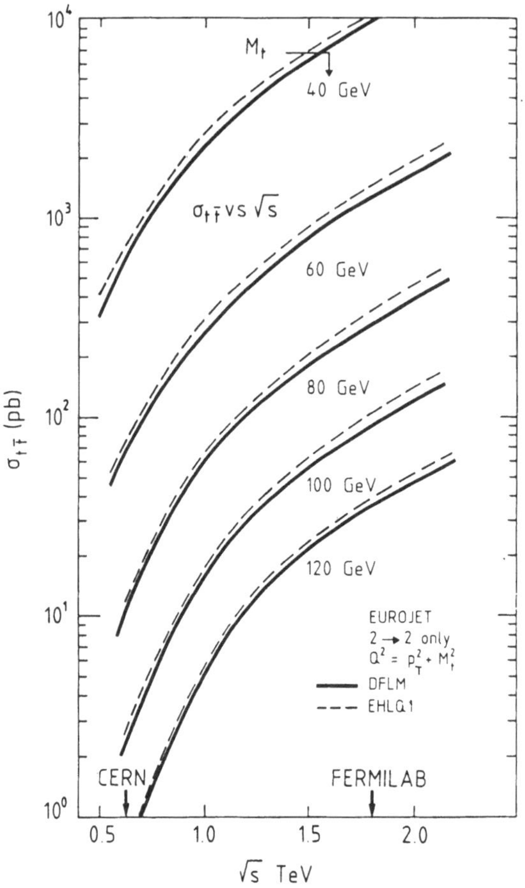

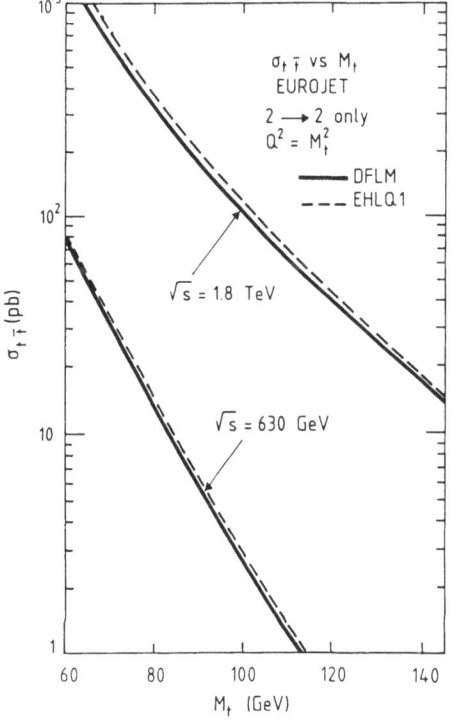

Fig. 12 Variation of the $t\bar{t}$ production cross-section as a function of $\sqrt{s}$ in the CERN–Fermilab $p\bar{p}$ collider range, for $40 < m_t < 120\,\text{GeV}$.

Fig. 13 The $t\bar{t}$ production cross-section as a function of m_t at $\sqrt{s} = 0.63$ and 1.8 TeV.

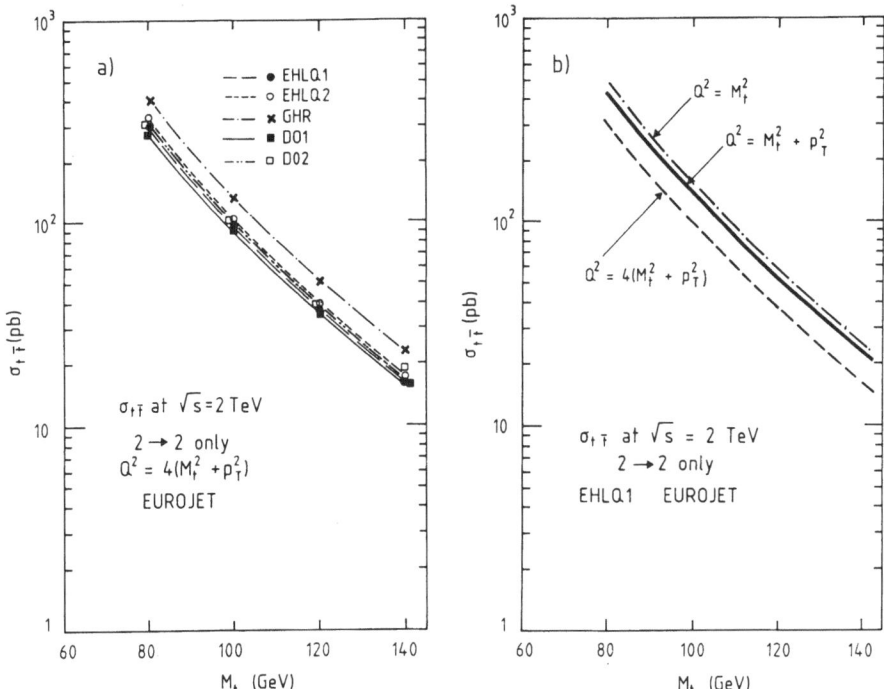

Fig. 14 The $t\bar{t}$ cross-section (at lowest order) as a function of m_t at $\sqrt{s} = 2\,\mathrm{TeV}$,
a) for various choices of structure functions, and b) for various choices of Q^2 scales.

Figures 14a and b illustrate the stability of the $t\bar{t}$ cross-section at $\sqrt{s} = 2\,\mathrm{TeV}$, as a function of various possible sets of structure functions (fig. 14a) and Q^2 scale choices (fig. 14b). As discussed in refs. [2,3], the variation of $\sigma_{t\bar{t}}$ by a factor of ~ 2 at lowest order for various reasonable choices of Q^2 scales is partially compensated by appropriate variations of K-factors when higher order corrections are included.

Figures 15a and b compare the $d\sigma/dp_T^W$ spectra at $\sqrt{s} = 0.63$ and $1.6\,\mathrm{TeV}$ for $t\bar{t} \rightarrow W\bar{W}b\bar{b}$ and the background of single-W production at large p_T^W [20]. At first, it seems that not only the absolute $\sigma_{t\bar{t}}$ rate is more favourable at Fermilab, but also apparently the $t\bar{t}$ signal to single-W background ratio improves when compared to UA1/2 régimes, as the total W production increases by a factor of ~ 4, while the $t\bar{t}$ signal has increased by a factor of ~ 30. What really matters, however, is not the overall W rate, but rather the rate for $W + \geqslant 2$ jets in kinematical configurations similar to those from $t\bar{t}$ production, i.e. with $p_T^W \sim 40\,\mathrm{GeV/c}$, $E_T^{jet} > 10$ to $20\,\mathrm{GeV}$, and $M_{jet\text{-}jet} \sim m_W$. The apparent improvement in the signal-to-background ratio at Fermilab energies suggested by figs. 15a and b may be offset by two effects.

Firstly, the average W transverse momentum $\langle p_T^W \rangle \approx 8\,\mathrm{GeV/c}$ at $\sqrt{s} = 0.63\,\mathrm{TeV}$ [11] is expected to increase to $\sim 14\,\mathrm{GeV/c}$ at $\sqrt{s} = 1.8\,\mathrm{TeV}$ [20]. Secondly, fig. 16a shows the average number of jets (within $|\eta| < 3.0$) as a function of p_T^W for the UA1 sample of W's, and fig. 16b shows the evolution of jet multiplicities as a function of p_T^W [11b]. The average number of jets increases monotonically with p_T^W, at least for $p_T^W \leqslant 40\,\mathrm{GeV}$. If the regularities in figs. 16 extend to higher p_T^W, and in particular if they are also valid as a function of $\langle p_T^W \rangle$, the fraction of background-generating $W + \geqslant 2$ jet events at large p_T^W, say $p_T^W > 40\,\mathrm{GeV}$, should be much larger at Fermilab than at CERN collider energies. A preliminary investigation with the EKS Monte Carlo of $W + 2$ jet production with $E_T^{jet} > 20\,\mathrm{GeV}$

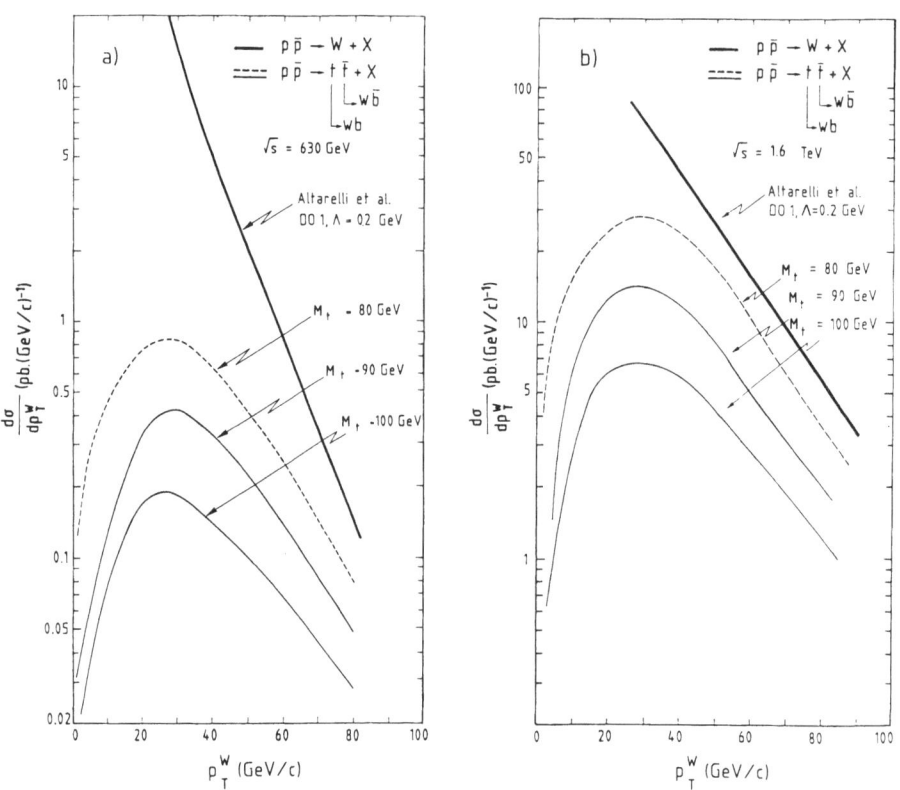

Fig. 15 The inclusive single-W cross-section $d\sigma/dp_T^W$ [20], compared with $d\sigma/dp_T^W$ for $t\bar{t}$ production [4]: a) at $\sqrt{s} = 0.63$ TeV, and b) at $\sqrt{s} = 1.6$ TeV.

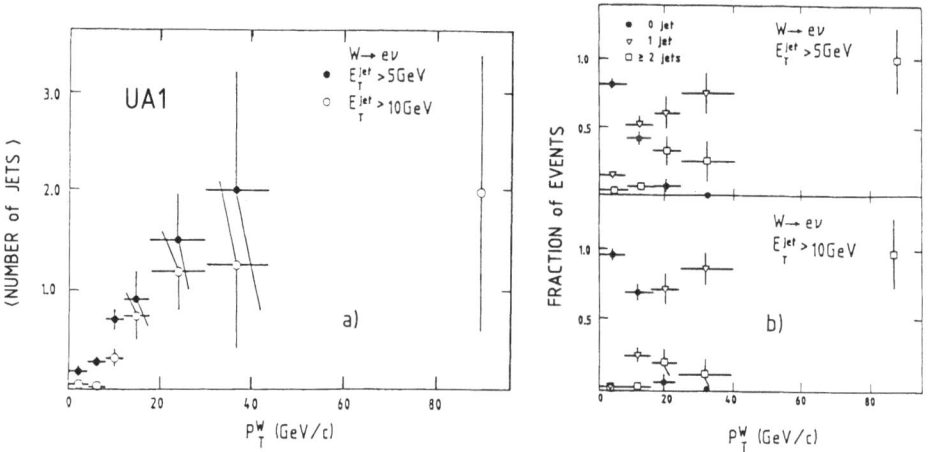

Fig. 16 a) The average number of jets as a function of p_T^W, and

b) jet multiplicity as a function of p_T^W in W production at $\sqrt{s} = 0.63$ TeV [11b]. Jet acceptance is $|\eta| < 3.0$, the distributions are not corrected for acceptance losses.

shows an increase in rate by a factor of $\sim$ 40 between $\sqrt{s}$ = 0.63 and 2 TeV [15]. Particularly disturbing is fig. 16b, if the W + 3 (and 4) jet multiplicity cross-sections become significant in the $p_T^W \sim$ 40 GeV range, where also $d\sigma/dp_T^W$ from $t\bar{t} \rightarrow WWb\bar{b}$ is maximal (few 3- or 4-jet W events have been observed in the UA1 data [11]). These additional soft bremsstrahlung jets might be mistaken or confused with top-decay b-jets.

For m_t > 120 GeV, which is beyond the reach of the CERN Collider, but can be probed at Fermilab with an experimental sensitivity of > 5 events per picobarn, the associated b-jets become easier to detect with increasing m_t. A possible way of distinguishing them from bremsstrahlung jets may be by requiring a b $\rightarrow \mu$ semileptonic signature, but this requires detection of low-p_T muons, or possibly a microvertex detector to detect the displaced b-decay vertices. Another possibility could be to use an isolated dilepton plus missing energy signature, taking advantage of leptonic modes of both W's to suppress the W + $\geqslant$2 jet background. In view of the large E_T unbalance of the W decay products (section 2), the two neutrinos will usually not balance each other, thus giving a significant net missing E_T and characteristic non-back-to-back configurations of the ee, $\mu\mu$, or eμ pairs. However, the high price of requiring two leptonic decays probably limits this option to the region $m_t \lesssim$ 100 GeV. In conclusion, for a quantitative investigation of what is the real top mass reach of Fermilab, a QCD computation of W + $\geqslant$2 jet background production is necessary. Despite the big advantage in top production rate at $\sqrt{s} \sim$ 2 TeV, it seems plausible that again a detailed understanding of W production at large p_T will be needed before top production could be ascertained.

It may also be worth mentioning that for $m_t \sim m_W$, top production kinematics is much less central at $\sqrt{s} \sim$ 2 TeV than at 0.63 TeV, and the larger longitudinal boosts (at fixed m_t) give a significantly larger rapidity spread to the decay quarks and leptons. It is more difficult to unravel W decay products with increasing $|\eta|$, as the isolated lepton signature is increasingly perturbed by spectator spillover and bremsstrahlung jets. In UA1, only the central part $|\eta_e| \leqslant$ 1.5 could be efficiently used for lepton spectrometry in W, Z, and top candidate searches [10, 11, 12, 13]. Thus, the more favourable production kinematics and probably better background conditions at the CERN Collider compensate in part for the unfavourable production rate in the $m_t \sim m_W$ region, but for $m_t \gtrsim$ 100 GeV the rate becomes too low for experimental sensitivities that may reasonably be expected.

6. $t\bar{t} \rightarrow W^+W^-b\bar{b}$ PRODUCTION UP TO LHC ENERGIES

We now extrapolate our study to higher energies. We are aware of the increasing QCD uncertainties affecting this extrapolation [2, 3, 5, 17]. In this energy range top is mainly produced by gluon–gluon fusion and the cross-sections are uncertain by a factor of $\sim$ 2. The model is, however, reliable enough to study basic event features, experimental signatures, and the m_t and $\sqrt{s}$ dependence of averages of kinematical distributions. The reliability of the results can in part be tested by studying the stability of the results for the 2 $\rightarrow$ 2 and 2 $\rightarrow$ 3 amplitudes separately.

6.1 Cross-sections

Figure 17 shows the energy dependence of the $t\bar{t}$ production cross-section over the Fermilab–UNK–LHC energy range, for 90 < m_t < 200 GeV. As the K-factor predicted by EUROJET turned out to be in very good agreement with the QCD expectations of Nason et al. [2], fig. 17 shows the total $t\bar{t}$ cross-section predicted by EUROJET. For the jet E_T cut-off in 2 $\rightarrow$ 3 amplitudes we have taken 10 GeV for $\sqrt{s} \geqslant$ 2 TeV and 15 GeV for $\sqrt{s} \geqslant$ 5 TeV; we used the DO1 structure functions and the scale is $Q^2 = m_t^2 + p_T^2$. The K-factor increases with increasing $\sqrt{s}$ and decreasing m_t. For comparison, fig. 17 also shows the lowest order top production cross-section for

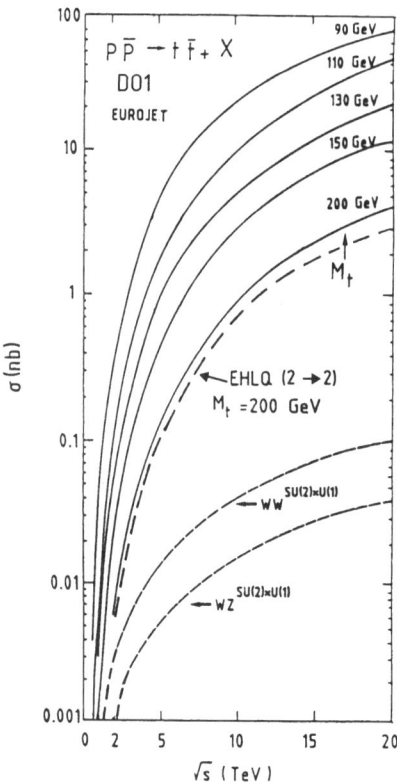

Fig. 17 Variation of the $t\bar{t}$ production cross-section as a function of $\sqrt{s}$ up to the LHC energy range, for $80 < m_t < 200$ GeV [4]. The lowest order calculation for $m_t = 200$ GeV and the electroweak WW cross-section from reference [17] is also shown.

$m_t = 200$ GeV from ref. [17]. There is no significant difference between $p\bar{p}$ and pp cross-sections above $\sqrt{s} \sim 10$ TeV for the top mass range considered. Figure 17 shows that for $m_t \sim m_W$ there is still a very rapid increase in the top production rate in going from Fermilab collider to UNK energies, while there is apparently near saturation of the cross-section in the LHC energy range. However, for large top masses of ~ 200 GeV, and for even higher mass possible 4th-generation quarks, it is very advantageous to go to higher energies still (LHC, SSC, etc.) [5, 8, 17].

As previously mentioned, fig. 17 also shows that if $m_t \sim m_W$, the strong interaction $t\bar{t} \rightarrow W^+W^-b\bar{b}$ apparent WW production is orders of magnitude larger than the electroweak WW production cross-section, also shown in fig. 17 [4, 5, 17]. It is thus likely that for $m_W \lesssim m_t \lesssim 150$ GeV the electroweak WW cross-section is unobservable, being entirely overwhelmed by $t\bar{t}$ production. The electroweak WZ and ZZ final states would, however, not be hampered by this type of background.

6.2 Kinematical distributions at $\sqrt{s} = 10$ TeV

In fig. 18 we show the normalized $(1/N)(dN/dp_T^W)$ W transverse momentum distribution for $m_t = 100$ and 150 GeV at $\sqrt{s} = 10$ TeV. This is the shape from the lowest order diagrams only. It is largely determined by the top mass; the more massive the top the higher is p_T^W. The higher order diagrams give a similar shape with a larger high-p_T^W tail. This is more quantitatively shown in fig. 19,

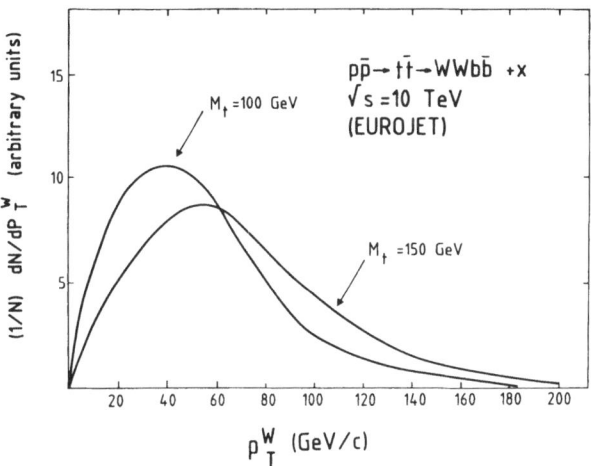

Fig. 18 Shape of the W transverse momentum distribution for m_t = 100 and 150 GeV at $\sqrt{s}$ = 10 TeV (lowest order diagrams only).

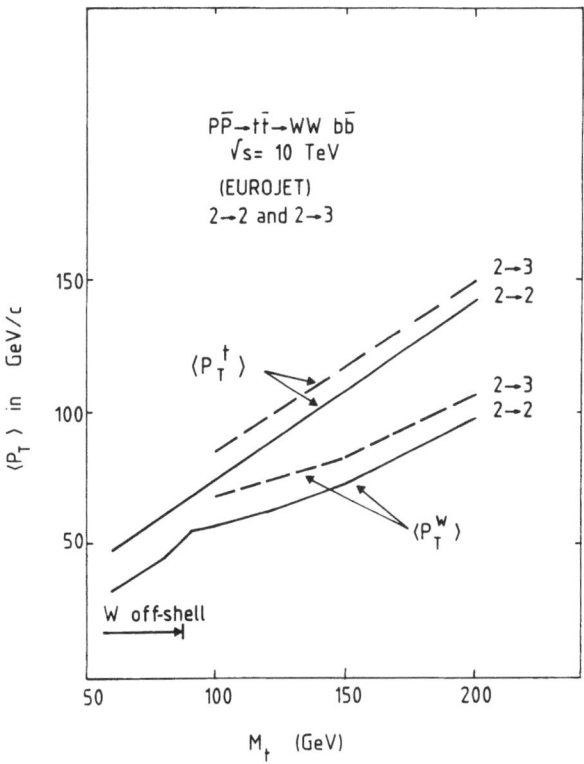

Fig. 19 Variation of the top and W average transverse momenta as a function of the top mass at $\sqrt{s}$ = 10 TeV. The results of 2 → 2 and 2 → 3 diagrams (with E_T cut-off at ⩾ 15 GeV) are shown separately.

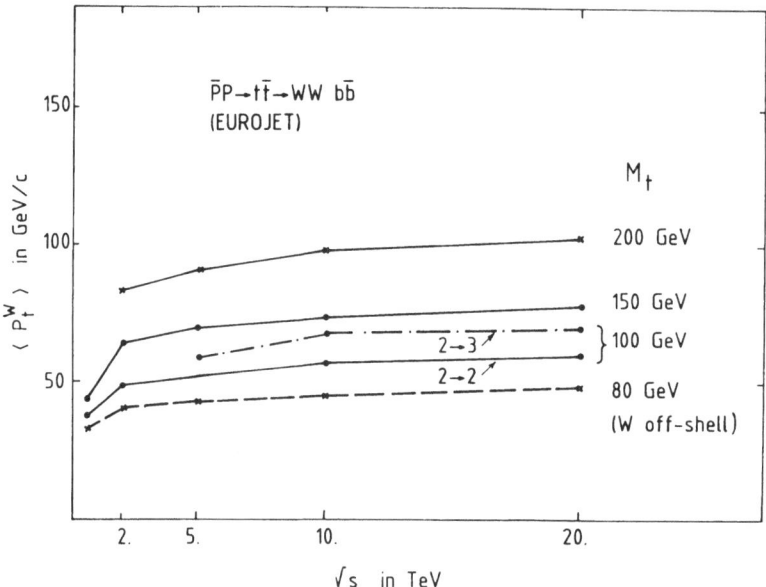

Fig. 20 Energy dependence of the average W transverse momentum for various top masses. For $m_t = 100$ GeV results from the α_s^2 and α_s^3 diagrams are shown.

where the average transverse momenta $\langle p_T^t \rangle$ and $\langle p_T^W \rangle$ are shown as a function of m_t, separately for the lowest order and higher order matrix elements (the E_T^{jet} cut-off is set at $\geqslant 15$ GeV); $\langle p_T^{t,W} \rangle$ increases monotonically with m_t, the kink in $\langle p_T^W \rangle$ at ~ 90 GeV is due to crossing the $m_t = m_W + m_b$ threshold as discussed in section 2. The energy dependence of $\langle p_T^W \rangle$ from SPS to LHC energies is shown in fig. 20. This is $\langle p_T^W \rangle$ from the α_s^2 terms only; for $m_t = 100$ GeV we also show separately the α_s^3 values, which are $\sim 20\%$ larger.

The peculiar kinematics for $m_t \sim m_W$ is well illustrated by the contrast between the two scatter plots in figs. 21a and b showing p_T^b as a function of p_T^W at $\sqrt{s} = 10$ TeV for $m_t = 100$ GeV and 150 GeV respectively. For the low Q-value decay $t \to Wb$ at $m_t = 100$ GeV, p_T^W is essentially always larger than p_T^b, while already at $m_t = 150$ GeV (fig. 21b) there is roughly equipartition of momentum between the two top decay products. Figure 22 shows the decay b-quark transverse momentum spectrum $(1/N)(dN/dp_T^b)$ for several m_t values. The shape is essentially determined by m_t, and for $m_t \leqslant 120$ GeV this p_T^b spectrum is soft in comparison with the gluon bremsstrahlung spectra generated by the $2 \to 3$ diagrams contained in EUROJET. Thus these b-jets will indeed most likely be unrecognizable and $t\bar{t} \to WWb\bar{b}$ would swamp electroweak WW production. The dependence of $\langle p_T^b \rangle$ on m_t over the UNK and LHC energy range is shown in fig. 23. The discontinuity due to crossing the open $t \to Wb$ threshold is evident. The α_s^3 terms do not substantially change this behaviour, $\langle p_T^b \rangle$ is larger by $< 20\%$.

At the UNK and LHC energies it will probably be necessary to have substantially higher jet (and lepton) selection cuts. For jets the reason may be the inflation of the underlying event, the presence of minijets (real nascent jets or accidental fluctuation clusters) which are already present in minimum bias data at a $\sim 10\%$ level at $\sqrt{s} = 900$ GeV, of multiparton interactions producing soft jets, and/or of multiple interactions per beam crossing. Thus, what may be more relevant for the experimental attempts to disentangle electroweak WW production from $t\bar{t} \to WWb\bar{b}$ apparent WW production, is

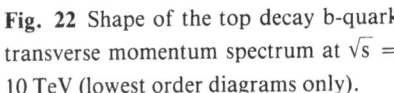
$p\bar{p} \to t\bar{t} \to W\bar{W}b\bar{b}$

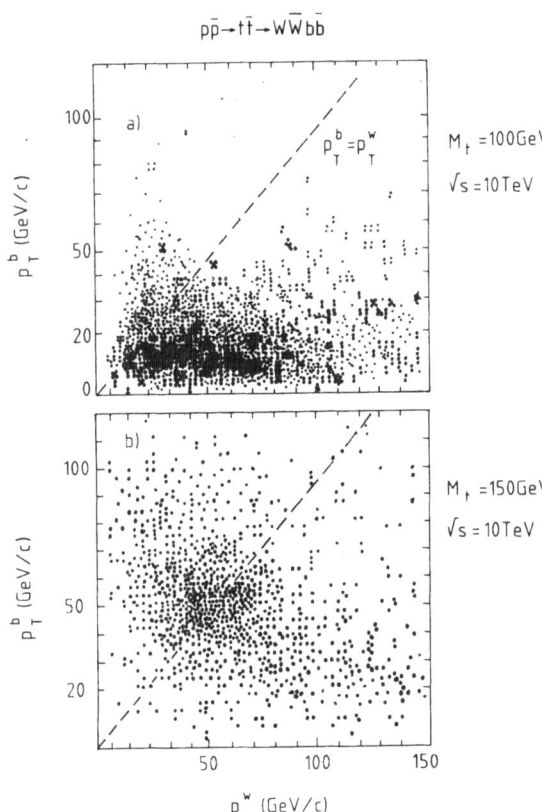

$M_t = 100\,\mathrm{GeV}$

$\sqrt{s} = 10\,\mathrm{TeV}$

$M_t = 150\,\mathrm{GeV}$

$\sqrt{s} = 10\,\mathrm{TeV}$

Fig. 21 Scatter plot of p_T^b as a function of p_T^W in $t \to Wb$ decays at $\sqrt{s} = 10$ TeV, a) for $m_t = 100$ GeV, and b) for $m_t = 150$ GeV.

Fig. 22 Shape of the top decay b-quark transverse momentum spectrum at $\sqrt{s} = 10$ TeV (lowest order diagrams only).

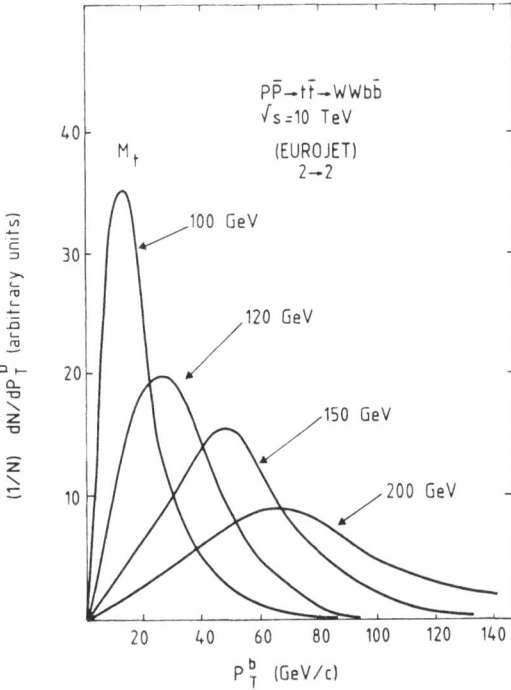

105

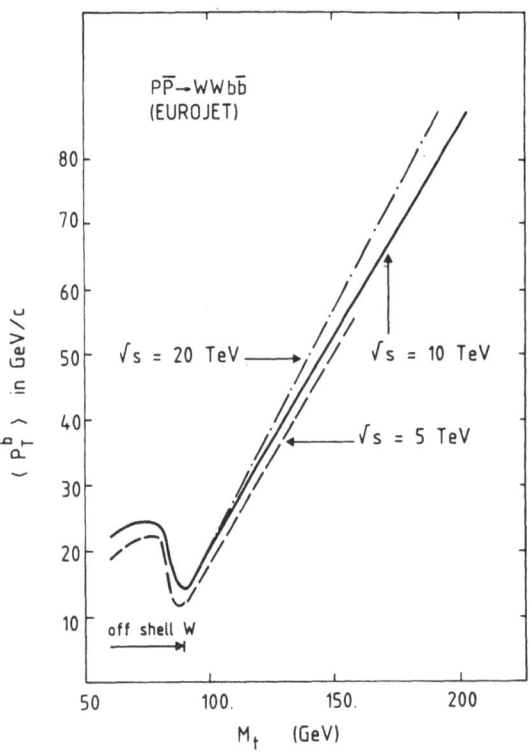

Fig. 23 Variation of the average b-quark transverse momentum with m_t.

Fig. 24 Fraction of b-jets from $t\bar{t} \rightarrow WWb\bar{b}$ below indicated thresholds, as a function of top mass (lowest order diagrams only).

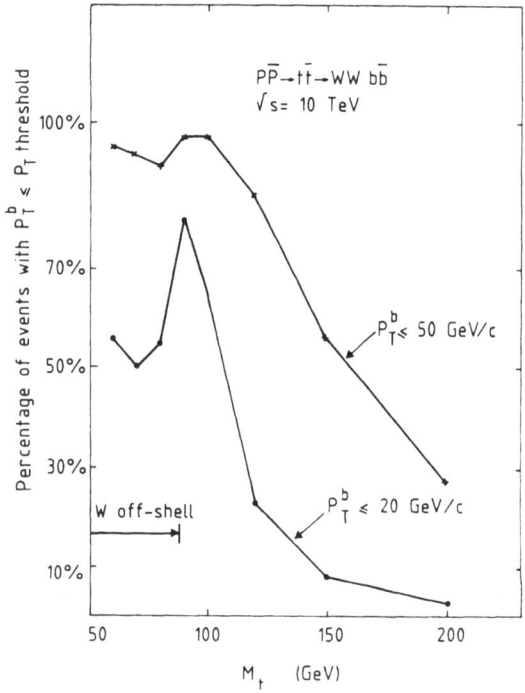

the fraction of WWb$\bar{\text{b}}$ events with a b-jet of p_T^b below a given threshold value. This is shown in fig. 24 for thresholds $p_T^b < 20$ and < 50 GeV/c. The kinematical peculiarity at $m_t \approx m_W$ is evident again, and is localized to $m_t \lesssim 120$ GeV. The threshold above which b-jet recognition may be possible depends on features of the apparatus, the rate of occurrence of hard bremsstrahlung, of multiparton interactions, and of multiple interactions per beam crossing.

In fig. 25 we show the WW effective mass shape, normalized to the appropriate production cross-section at $\sqrt{s} = 10$ TeV for electroweak WW production and for WW from $t\bar{t} \rightarrow$ WWb$\bar{\text{b}}$, at $m_t = 100$ and 150 GeV. In this top mass range there is unfortunately no substantial difference in the WW mass shape either, rendering the task of separating the $SU_2 \times U_1$ component very hard [5].

For completeness, we show in fig. 26 the results from ref. [21] on the electroweak WW cross-section, compared with the expected Higgs $\rightarrow$ WW signal for $m_H = 200, 400, 600$ GeV (shaded histograms), and the other large background source due to QCD W($\rightarrow \ell\nu$) + 2 jet production. There is little hope of easily extracting a Higgs signal in the WW($\rightarrow \ell$-ν-jet-jet) channel, in particular if top production is also present with $m_t \sim m_W$ [4,5]. A possible way out may be by double-tagging the forward-going quark jets (from WW or ZZ fusion into Higgs) accompanying Higgs $\rightarrow$ WW production [5,22]. Probably the best hope of observing the Higgs in hadroproduction at LHC and SSC energies is still through the QCD background-free modes Higgs $\rightarrow$ ZZ $\rightarrow$ leptons, as discussed in refs. [5,23].

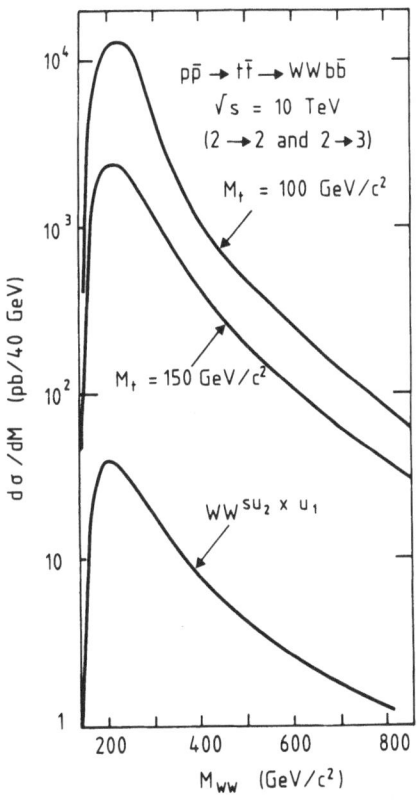

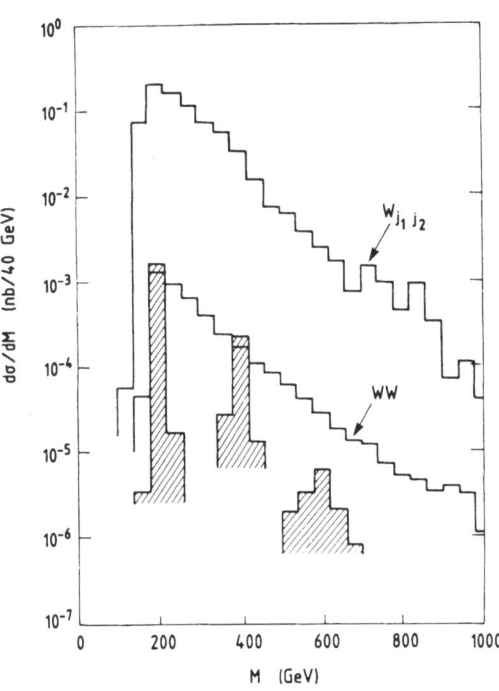

Fig. 25 WW effective mass from $t\bar{t} \rightarrow$ WWb$\bar{\text{b}}$ production at $\sqrt{s} = 10$ TeV compared with electroweak WW production [5].

Fig. 26 Comparison of electroweak WW production, the expected Higgs $\rightarrow$ WW signal (hatched histograms) for $m_H = 200, 400$, and 600 GeV, and the W + 2 jet background from ref. [21].

REFERENCES

[1]   A. Ali et al., Nucl. Phys. **B292** (1987);
      B. Van Eijk, 'EUROJET: a QCD based Monte Carlo program including perturbatively calculated higher order processes in proton-antiproton interactions', Proc. New Particles '85 Conf., Madison, Wisconsin (1985);
      B. Van Eijk, Ph.D. Thesis, University of Amsterdam (1987).

[2]   P. Nason, S. Dawson and K. Ellis, 'The total cross-section for the production of heavy quarks in hadronic collisions', Fermilab–Pub–87/222–T (1987).

[3]   G. Altarelli, M. Diemoz, G. Martinelli and P. Nason, 'Total cross-section for heavy flavour production in hadronic collisions and QCD', preprint CERN–TH 4978/88 (1988).

[4]   P. Colas and D. Denegri, Phys. Lett. **B195** (1987) 295; UA1 TN 86–89, September 1986, unpublished;
      P. Colas, 'Etude de la production d'électrons accompagnés de jets dans l'expérience UA1; recherche du quark top', Thèse de Doctorat d'Etat, Université Paris VI (1987), note CEA–N–2571.

[5]   G. Altarelli et al., Proc. Workshop on Physics at Future Accelerators, La Thuile and Geneva, 1987, CERN 87–07 (1987);
      D. Froideveaux et al., ibid.

[6]   a) C. Albajar et al., UA1 Collaboration, Phys. Lett. **B193** (1987) 389;
      b) P. Bagnaia et al., UA2 Collaboration, Phys. Lett. **B139** (1984) 105;
          R. Ansari et al., UA2 Collaboration, Phys. Lett. **B194** (1987) 158.

[7]   S. Gupta and D.P. Roy, Z. Phys. **C39** (1988) 417;
      R. Kleiss et al., Z. Phys. **C39** (1988) 393;
      S. Geer et al., Phys. Lett. **B192** (1987) 223.

[8]   H. Baer, V. Barger, H. Goldberg and R.J.N. Phillips, 'Top quark signatures at the Tevatron collider', MAD/PH/367, and references therein;
      F. Halzen et al., 'Top quark signatures at the Tevatron collider', MAD/PH/436;
      K.J. Foley et al., 'Bottom and top physics', SLAC–PUB–4426; see also:
      I. Bigi et al., 'Production and decay properties of ultra-heavy quarks, SLAC–PUB–4021;
      A. Martin, Nucl. Phys. B (Proc. Suppl.) **1B** (1988) 133.

[9]   P. Colas, D. Denegri and C. Stubenrauch, Z. Phys. **C40** (1988) 527, and references therein.

[10] C. Albajar et al., UA1 Collaboration, Z. Phys. **C37** (1988) 505.

[11] a) C. Albajar et al., UA1 Collaboration, 'Studies of intermediate vector boson production and decay in UA1 at the CERN Proton-Antiproton Collider', CERN EP/88–168, November 1988; also:
          G. Arnison et al., Lett. Nuovo Cimento **44** (1985) 1;
      b) C. Stubenrauch, 'Etude de la production de bosons W et Z dans l'expérience UA1', Thèse de Doctorat d'Etat, Université de Paris-Sud (1987), note CEA–N–2532.

[12] C. Albajar et al., UA1 Collaboration, Phys. Lett. **B198** (1987) 271.

[13] Y. Giraud-Héraud, 'Mise en évidence de la désintégration du W en $\tau\nu$', Thèse de Doctorat d'Etat, Université d'Orsay, 1988, Collège de France preprint 88–01 (1988).

[14] G. Bauer et al., UA1 TN 87–09 (1987).

[15] B. Andrieu and D. Denegri, UA1 internal report: 'Meeting of the Working Group on UA1 at High Luminosity', 1988;
      B. Andrieu, Ph.D. Thesis, Collège de France, in preparation.

[16] S.D. Ellis et al., Phys. Lett. **B154** (1985) 389.

[17] E. Eichten, I. Hinchliffe, K. Lane and C. Quigg, Rev. Mod. Phys. **56** (1984) 579.

[18] M. Diemoz, F. Ferroni, E. Longo and G. Martinelli, Z. Phys. **C39** (1988) 31.

[19] J. Proudfoot, CDF Collaboration, 'W boson production in $p\bar{p}$ collisions at $\sqrt{s} = 1.8$ TeV', ANL–HEP–CP–88–61 (1988).

[20] G. Altarelli et al., Z. Phys. **C27** (1985) 617.

[21] W.J. Stirling, R. Kleiss and S.D. Ellis, Phys. Lett. **B163** (1985) 261; see also: J.F. Gunion, Z. Kunszt and M. Soldate, Phys. Lett. **B163** (1985) 389.

[22] R.N. Cahn et al., Phys. Rev. **D35** (1987) 1626; R. Kleiss and W.J. Stirling, Phys. Lett. **B200** (1987) 193, and references therein.

[23] G. Altarelli, 'An update of the Higgs search at the LHC', CERN TH-5017/88 (1988). R.N. Cahn, 'Two gauge boson physics at future colliders', LBL–25078 (1988), and references therein.

8

HEAVY FLAVOUR PRODUCTION RESULTS FROM UA1
(p$\bar{\text{p}}$ at $\sqrt{s}$ = 630 GeV)

I. ten Have

NIKHEF-H, Amsterdam

Abstract

The UA1 lepton samples are studied in terms of QCD heavy flavour production. Using the full $O(\alpha_s^3)$ QCD calculations the bottom cross-section is derived: $\sigma(b\bar{b}) = 10.2 \pm 3.3$ μb. Considering the large errors this is in good agreement with the theoretical value: $\sigma(b\bar{b})= 12$ $^{+7}_{-4}$ μb.

The isolated muon and isolated electron samples are used for the top quark search. The two samples together yield a top mass limit: $m_t > 44$ GeV/c^2 at 95 % C.L.. Also a limit is derived on the mass of the b', a fourth generation down-like quark: $m_{b'} > 32$ GeV/c^2 at 95 % C.L..

The possibilities of top searches at the Sp$\bar{\text{p}}$S collider with ACOL and at the Tevatron collider are also discussed.

1 Introduction

Understanding the QCD production of heavy flavour quarks is of both experimental and theoretical interest. Reliable predictions for the production of known heavy flavours, like bottom and charm, are very important. Once the production of known heavy flavours is well-understood, it is possible to search for new heavy objects, such as a new heavy quark.

On the theoretical side important progress has been made recently. P. Nason, S. Dawson and K. Ellis [1] have calculated the cross-sections for QCD heavy flavour production processes up to $O(\alpha_s^3)$. These calculations include the radiative corrections to the $O(\alpha_s^2)$ diagrams and are, therefore, free of divergencies. The uncertainties on the full $O(\alpha_s^3)$ calculation were studied by Altarelli, Diemoz, Martinelli and Nason[2]. The four main sources of uncertainty are the choice of:

- the scale parameter Λ

- the parton structure functions

- the mass of the heavy quarks (4.5 GeV/c² < m_b < 5 GeV/c² and 1.2 GeV/c² < m_c < 1.5 GeV/c²)

- the renormalization scale

Based on data from deep inelastic scattering and measurements of $R_{e^+e^-}$ Altarelli et al. determined Λ to be: $\Lambda = 170 \pm 80$ MeV. They use the DFLM structure functions[3]. This set of structure functions incorporates two important features. Firstly the gluon content falls with $1/(x)^{1+\delta}$ instead of $1/x$. This is in accordance with recent measurements showing a softer gluon content than was suggested by earlier measurements. Secondly the structure functions include a correlation between the gluon density and Λ.

Figure 1 taken from the paper by Nason et al. shows the ratio of the full $O(\alpha_s^3)$ and the $O(\alpha_s^2)$ cross-section as a function of the quark mass for $\sqrt{s} = 0.63$ TeV. For the calculation the Eichten et al. structure functions (set I) has been used. The scale μ has been fixed to $\mu = m$ for the $O(\alpha_s^2)$ calculation. For the higher order cross-section four different μ-scales have been used.

The calculations for the complete $O(\alpha_s^3)$ processes break down when $m/\sqrt{s} \ll 1$ (see figure 1). At the Sp$\bar{p}$S collider, $\sqrt{s} = 0.63$ TeV, the bottom quark lies near the limit beyond which the predictive power of the calculations becomes doubtful.

Altarelli et al. show the cross-section at the Sp$\bar{p}$S collider ($\sqrt{s}$= 630 GeV) for QCD heavy flavour production as a function of the heavy quark mass (see figure 2). The figure also includes predictions for the heavy quark production cross-section at the Tevatron, both for $\sqrt{s} = 1.8$ TeV and for $\sqrt{s} = 2.0$ TeV. The bands shown are due to the uncertainties mentioned above (with exception of the quark mass).

The UA1 experiment studies heavy flavour production at the CERN Sp$\bar{p}$S-collider, which can be considered a heavy flavour factory. For an integrated luminosity of $\int Ldt = 1$ pb^{-1}, 10^8 c$\bar{c}$ and 10^7 b$\bar{b}$ events are produced. The cross-section for open b$\bar{b}$ production is approximately 10^4 nb. For production of the b$\bar{b}$ bound state at the Υ(1s) resonance the cross-section is 1 nb. This can be compared to e.g. the production of b$\bar{b}$ at LEP100 ($\sqrt{s} = 90$ GeV) through the Z°: $\sigma(Z^\circ \to b\bar{b}) \cong 4$ nb.

The first part of this paper describes the study of charm and bottom production in the UA1 single muon and dimuon data (section 3). Information of the available lepton samples is combined to measure the inclusive bottom cross-section (section 4). The result is derived using a recent full $O(\alpha_s^3)$ calculation presented by Nason [4]. Searches for the production of new heavy flavour quarks is presented in the third part of this paper. Mass limits are derived for both the top quark, and for a new fourth generation down-like (b') quark (section 5). Finally in the last part of this paper an outlook for the future is presented (section 6).

Heavy flavour quarks are copious sources of high p_T leptons. The leptons provide a clear signature for a heavy flavour production process. The heavy flavour studies presented in this paper were all carried out using the semi-leptonic decay channels of the heavy quarks. Extensive use has been made of the muon channel as muons can be detected down to low p_T^μ. Moreover the muon isolation is well measurable as it is based on information from both the tracking chamber and the calorimetry.

At the time the heavy flavour studies in UA1 started, the complete $O(\alpha_s^3)$ calculation mentioned above was not available yet. The UA1 studies are based on the then available approximations. The Monte Carlo simulations used will be discussed in the next section.

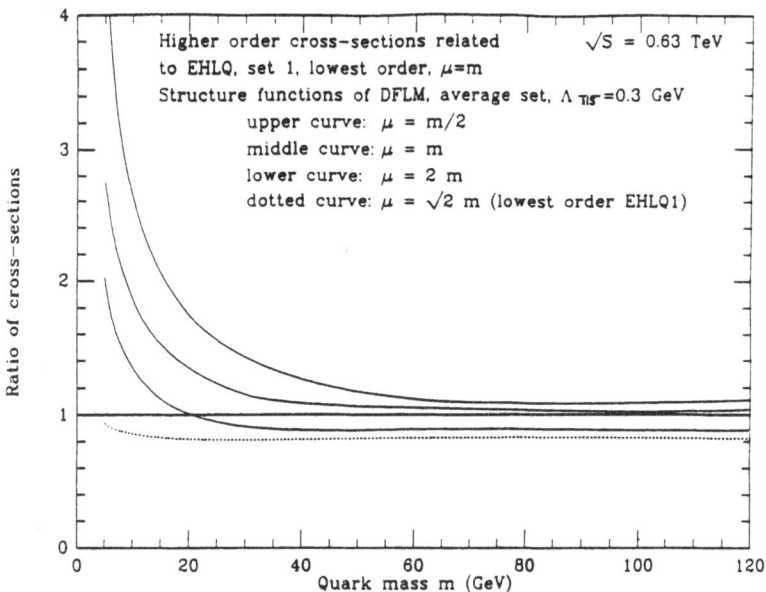

Figure 1. The cross-section ratio of the full $O(\alpha_s^3)$ and the $O(\alpha_s^2)$ calculation, $\mu=m$ at $\sqrt{s} = 0.63$ TeV[1].

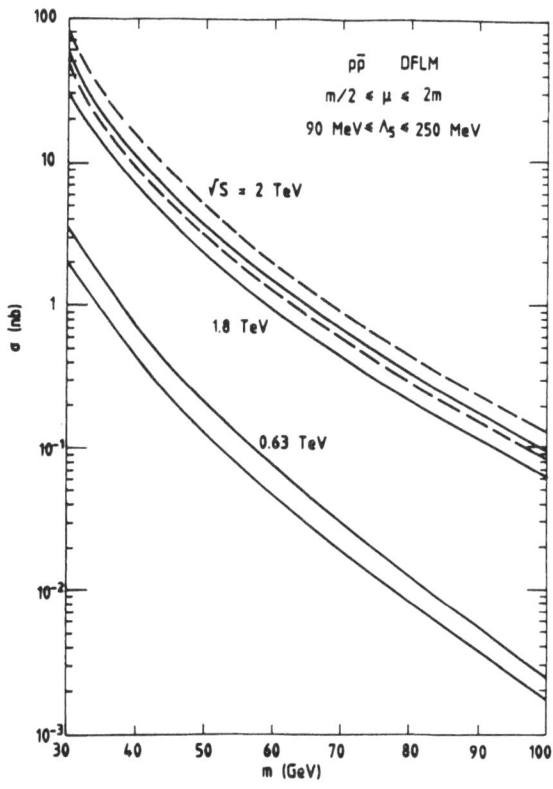

Figure 2. Full $O(\alpha_s^3)$ cross-section as a function of the quark mass for heavy quark production at $\sqrt{s} = 0.63$, 1.8 and 2.0 TeV. The uncertainties on the calculation are shown as bands[2].

2 Monte Carlo Studies

For the heavy flavour studies in UA1 two Monte Carlo packages have been used: ISAJET [5] and EUROJET [6]. The ISAJET Monte Carlo includes the basic processes: gluon fusion, gluon splitting and flavour excitation. Higher orders are incorporated using the parton evolution model [7]. The EUROJET Monte Carlo has a strict separation between $O(\alpha_s^2)$ and $O(\alpha_s^3)$ production processes. The calculations are based on the $O(\alpha_s^2)$ and $O(\alpha_s^3)$ matrix elements. For the $O(\alpha_s^3)$ processes only gluon splitting and gluon bremsstrahlung are included.

ISAJET and EUROJET use different Q^2-scales. ISAJET has a Q^2 definition based on the Mandelstam parameters of the partonic $2 \to 2$ subprocess:

$$Q^2 = \frac{2stu}{s^2+t^2+u^2},$$

while EUROJET uses the Q^2-scale favoured by the UA1 jet data[8]:

$$Q^2 = p_T^2 + m_Q^2$$

Both Monte Carlo's use the Eichten et al. structure functions, set I ($\Lambda=0.2$)[9] as a standard set. The fragmentation is described by the Peterson fragmentation function [10]. The input parameters to this fragmentation function, ϵ_q for charm and bottom were determined by tuning $< z_q >$ to the e^+e^- data. Where z_q is defined as:

$$z_q = \frac{(E + p_{//})^{\text{hadron}}}{(E + p_{//})^{\text{quark}}} \tag{1}$$

The epsilon-parameter for the top quark, ϵ_t has been calculated from ϵ_b using the mass ratio:

$$\epsilon_t = \left(\frac{m_b}{m_t}\right)^2 \epsilon_b$$

3 Heavy Flavour Studies in the Muon Samples

For the study of heavy flavour (b, c) production and the measurement of the bottom cross-section four lepton samples[1] have been used:

- high mass dimuons [11] (also used to study B^o-$\bar{B}^o$ mixing [12].)
  $\int Ldt = 692 \text{ nb}^{-1}$
  cuts: $p_T^{\mu_i} > 3$ GeV/c (i=1, 2), $M_{\mu\mu} > 6$ GeV/c^2
  total number of events: 512

- low mass dimuons [13]
  $\int Ldt = 556 \text{ nb}^{-1}$
  cuts: $p_T^{\mu_i} > 3$ GeV/c (i=1, 2), $2m_\mu < M_{\mu\mu} < 6$ GeV/c^2
  total number of events: 304

---

[1]In total UA1 has five lepton samples. The fifth sample, the single isolated electron[16] sample will be used in the section 5 for the top search.

- J/Ψ sample [14]
  $\int L\,dt = 556$ nb^{-1}
  cuts: $p_T^{\mu_1} > 3$ GeV/c, $p_T^{\mu_2} > 0.75$ GeV/c, $p_T^{\mu\mu} > 4$ GeV/c
  total number of events: 293

- single muon sample [15]
  $\int L\,dt = 556$ nb^{-1}
  cuts: $p_T^{\mu} > 6$ GeV/c
  total number of events: 20,000

QCD heavy flavour production, $p\bar{p} \to Q\bar{Q}$ (Q = c, b), yields a different event topology in these four samples. Generally one expects that the high mass dimuon events have two jets, more or less back-to-back, each jet containing a muon. Low mass dimuons also contain two jets roughly back-to-back. One of the jets contains the $\mu^+\mu^-$ pair. The J/Ψ sample is interesting because of b-production through $p\bar{p} \to b + X \to J/\Psi + X'$. This yields only one jet and on the opposite side a $\mu^+\mu^-$ pair surrounded by some hadronic activity. In the single muon sample the event consists of two jets, only one of which contains a muon. The other quark has decayed hadronically.

3.1 Dimuon Isolation

Besides the QCD process $p\bar{p} \to Q\bar{Q}$ (Q=c, b) X $\to \mu^+ \mu^- $ X', there are other sources of dimuons, like Drell-Yan, J/Ψ or Υ production. These processes in general yield more isolated muons.

To enhance the contribution of QCD heavy flavour production in the dimuon samples, non-isolated events were selected using the dimuon isolation parameter S:

$$S = \left(\sum_{\Delta R=0.7} E_T(\mu_1)\right)^2 + \left(\sum_{\Delta R=0.7} E_T(\mu_2)\right)^2,$$

where $\sum E_T(\mu_i)$ is the sum of the transverse energy in the calorimeter around the muon μ_i. The transverse energy is summed in a cone $\Delta R = 0.7$ in pseudorapidity-azimuthal angle space R:

$$\Delta R = \sqrt{(\eta^{\mu_i} - \eta)^2 + (\phi^{\mu_i} - \phi)^2}.$$

The isolation distribution for the like sign and the unlike sign dimuons from the high mass dimuon sample is shown in figure 3.a and 3.b respectively.

Below $S < 9$ GeV^2 the spectrum for unlike sign dimuons clearly is more peaked than that for the like sign sample. Drell-Yan and Υ production only yield unlike sign dimuons, and these are expected to be isolated. The enhancement of events in the unlike sign dimuons at $S < 9$ GeV^2 is, therefore, interpreted as due to Drell-Yan and Υ production. In the following dimuon events with $S < 9$ GeV^2 are considered isolated and events with $S > 9$ GeV^2 non-isolated.

3.2 High Mass Dimuons

The high mass ($M_{\mu\mu} > 6$ GeV/c^2) dimuon sample has been split into isolated and non-isolated events. QCD heavy quark production is expected to mainly contribute to the non-isolated dimuons. The $c\bar{c}$ and $b\bar{b}$ production will be studied in this sample. The isolated sample is used as a control sample to check that the background to QCD heavy flavour production is well understood.

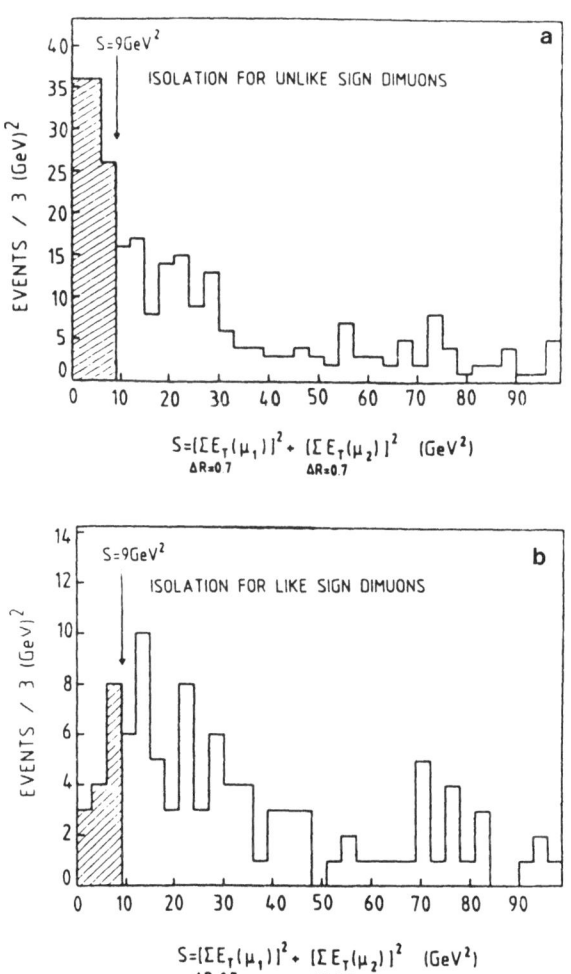

Figure 3. Dimuon isolation for high mass, unlike sign dimuons (3a) and for high mass, like sign dimuons (3b)

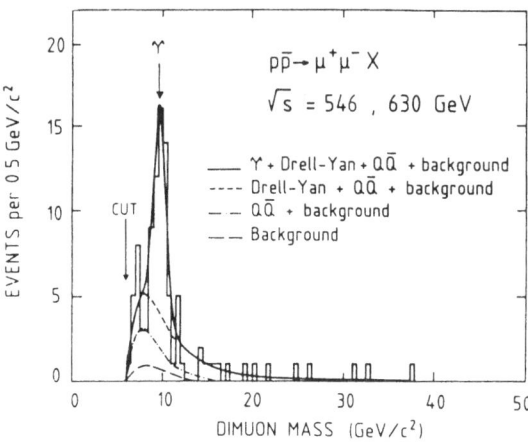

Figure 4. Dimuon mass spectrum for high mass, unlike sign, isolated dimuons. Contributions from decay background, Drell-Yan, Υ and heavy flavour processes are shown.

Figure 4 shows the dimuon mass spectrum for high mass, unlike sign, isolated dimuons. The contribution from π/K decay background (background to prompt muons) is calculated and kept fixed. Based on the different shapes of the dimuon mass distributions the contributions from QCD heavy flavour production ($c\bar{c}$, $b\bar{b}$), Drell-Yan and Υ production are fitted to the data together with the fixed background fraction. In figure 4 the various contributions are shown cumulatively. The total Monte Carlo spectrum is shown as a full line, the data are represented by the histogram. The contributions from these four sources describe the data well.

For the Υ production the assumption was made that the decay of $\Upsilon : \Upsilon' : \Upsilon'' \to \mu^+\mu-$ occurs in the ratio $1 : 0.3 : 0.15$. For the cross-section ISAJET gives:

$$\sigma(\, p\bar{p} \to \Upsilon, \Upsilon', \Upsilon'' \to \mu^+\mu^-) = 0.98 \pm 0.21 \pm 0.19 \text{ nb}$$

This result can be compared to low energy data and to the prediction of Barger and Martin [17](see figure 5). The agreement with theory is found to be very good.

The high mass, non-isolated dimuon events will now be used to study QCD heavy flavour production. A separation between charm and bottom production has been performed. This statistical separation is based on the so-called p_T relative, defined as the transverse momentum of the muon with respect to the axis of the accompanying jet. The mass of the bottom quark is higher than the mass of the charm quark. As a result, a muon originating from a bottom quark will have a higher p_T relative than a muon from a charm quark. The charm/bottom separation has been carried out in both the unlike sign events (figure 6a) and in the like sign events (figure 6b). The curves are the result of a fit of the $c\bar{c}$ and $b\bar{b}$ components. The contribution from π/K-decay is kept fixed. This fit yields that the unlike sign dimuons consist of 66 ± 10 % $b\bar{b}$, 12 ± 9 % $c\bar{c}$ and 22 ± 5 % originating from other, background sources . The like sign dimuons are found to contain 61 ± 12 % $b\bar{b}$ and as expected no contribution from charm. Combining the two results yields:

$$\frac{b\bar{b}}{b\bar{b}+c\bar{c}} \cong 90\%$$

The high bottom content in this dimuon sample is due to the harder bottom fragmentation. Also the bottom quark has a larger mass than a charm quark. Therefore,

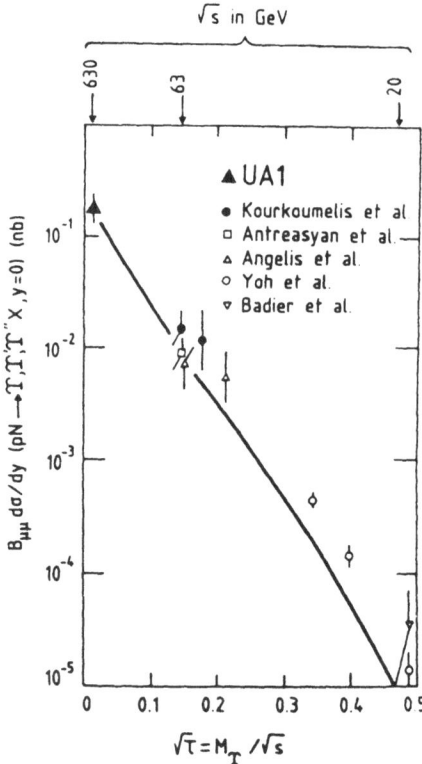

Figure 5. Comparison of the measured cross-section $\sigma(p\bar{p} \rightarrow \Upsilon + X)$ to low energy data and the gluon fusion model of Barger and Martin

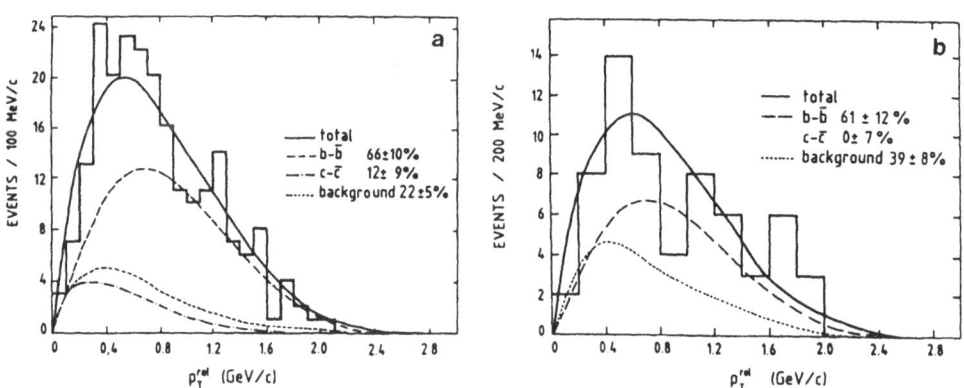

Figure 6. The p_T relative distributions of the high mass, non-isolated, dimuon sample, for unlike sign events (6a) and for like sign events (6b). The curves are the result of a fit of the $c\bar{c}$ and $b\bar{b}$ components with a fixed background fraction

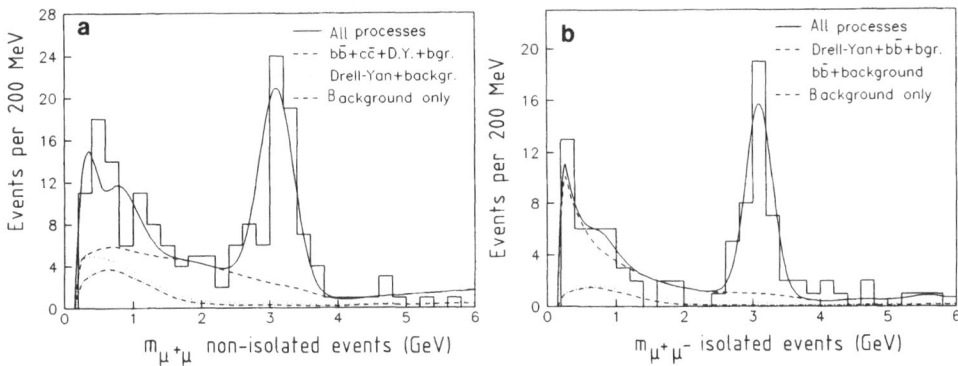

Figure 7. Dimuon mass distribution for the low mass, non-isolated events (7a) and the low mass, isolated events (7b)

a bottom quark can produce, more easily than a charm quark, a high p_T lepton. Both muons are required to have a $p_T^\mu > 3$ GeV/c. Moreover muon pairs, where both muons stem from the same decay branch are suppressed by requiring: $M_{\mu\mu} > 6$ GeV/c^2.

3.3   Low Mass Dimuons

The low mass dimuons ($M_{\mu\mu} < 6$ GeV/c^2) have also been divided into isolated and non-isolated events. The non-isolated dimuons are used to study QCD heavy flavour production. The isolated dimuons form a control sample in which the background to the heavy flavour production is studied. Both samples are restricted to the unlike sign dimuons.

In figure 7 the dimuon mass spectra for the low mass, unlike sign dimuons are shown both for the non-isolated events (7a) and for the isolated events (7b). The contribution from π/K decay background (fixed) together with the fitted contributions from QCD heavy flavour production and Drell-Yan are shown cumulatively in the different curves. The full line that represents all possible processes also includes contributions from light mesons, like $\rho, \eta,\ \omega$ and ϕ ($2m_\mu \leq M_{\mu^+\mu^-} \leq 2$ GeV/c^2) and from J/Ψ (2 GeV/c$^2 \leq M_{\mu^+\mu^-} \leq 4$ GeV/c^2). Both the isolated and the non-isolated events show a clear J/Ψ peak. The J/Ψ and its production mechanisms will be studied in the next section. The contribution from $b\bar{b}$ to the non-isolated sample will be used later on to determine the total $b\bar{b}$ cross-section.

3.4   The J/Ψ Sample

The dimuon mass spectrum for the J/Ψ sample is shown in figure 8. Fitting a gaussian through this distribution yields a mass for the J/Ψ: $M_{J/\Psi} = 3.110 \pm 0.011$ GeV/c^2. This should be compared to the world average: $M_{J/\Psi} = 3.097$ GeV/c^2. The width

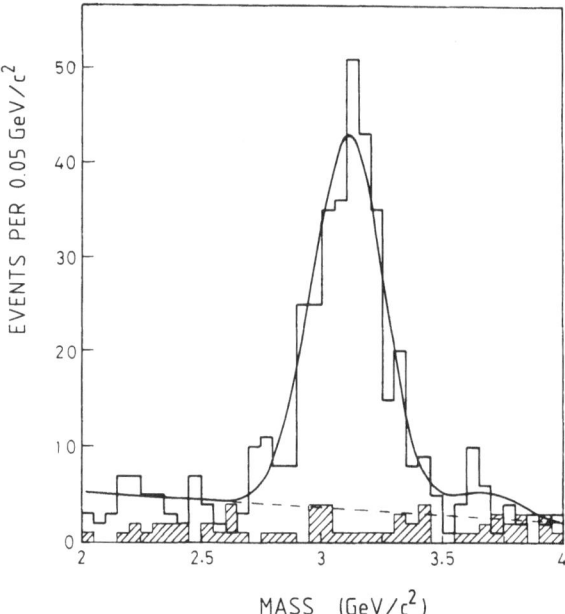

Figure 8. Dimuon mass distribution of the special J/Ψ selection

of the J/Ψ peak is mainly determined by the resolution of the UA1 central tracking chamber.

Two production mechanisms for the J/Ψ production have been studied. The first is the direct production through gluon fusion of a $c\bar{c}$ bound state, χ, which decays into a J/Ψ:

$$p\bar{p} \rightarrow \chi + X$$
$$\quad\quad\hookrightarrow J/\Psi + \gamma$$

The second mechanism is production of a J/Ψ via a B-hadron:

$$p\bar{p} \rightarrow B_{hadron} + X$$
$$\quad\quad\quad\hookrightarrow J/\Psi + X'$$

The branching ratio $B(B_{hadron} \rightarrow J/\Psi)$ is taken to be 1.1 %.

The J/Ψ's produced through these two mechanism exhibit a different behaviour in transverse momentum of the J/Ψ (see figure 9). A curve fit of the two distributions to the data yields:

$$\sigma \,.\, B(p\bar{p} \rightarrow \chi + X \rightarrow J/\Psi + X') = 5.7 \pm 0.8 \pm 1.3 \text{ nb}$$
$$\sigma.\, B(p\bar{p} \rightarrow B_{hadron} + X \rightarrow J/\Psi + X') = 1.8 \pm 0.6 \pm 0.9 \text{ nb}$$
$$(p_T^{J/\Psi} > 5 \text{ GeV/c}, |y| < 2)$$

3.5   The Single Muon Sample

The inclusive muon p_T distribution for the single muon data is shown in figure 10. The data are presented after subtraction of the background from pion and kaon decays.

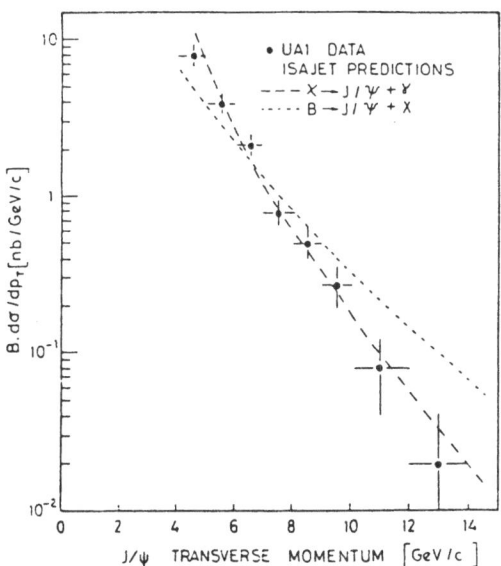

Figure 9. J/Ψ transverse momentum distribution for the two production mechanisms compared to the UA1 data spectrum

At low p_T this decay background is quite large. About 60 % or more of the muons at $p_T^\mu \cong 6$ GeV/c stem from pion or kaon decay. The decay background decreases to approximately 20 % at $p_T^\mu \cong 20$ GeV/c. The p_T^μ band selected to study heavy flavour production should not extend to low p_T^μ. The solid line in figure 10 shows the Monte Carlo prediction for muons from all possible sources: QCD heavy flavour production, Drell-Yan, Υ, J/Ψ, $W^\pm$, Z°. The Monte Carlo predictions have been normalized to the data, where available. The contribution of muons from W-decay is shown separately (dash-dot-dot curve). At high p_T muons from W-decay become dominant. The band chosen to study QCD heavy flavour production should, therefore, also avoid the high p_T region. The band is fixed to $10 < p_T^\mu < 15$ GeV/c.

Again the p_T relative method is used to evaluate the $c\bar{c}$ and $b\bar{b}$ contributions. In the chosen p_T range the measured fraction of $b\bar{b}$ is:

$$\frac{b\bar{b}}{c\bar{c}+b\bar{b}} = 74 \ \%$$

4 Determination of the Inclusive Bottom Cross-section

In this paragraph the $p\bar{p} \to b\bar{b}$ production cross-section will be evaluated. In the single muon sample bottom production is studied in three different p_T^μ bands: $10 < p_T^\mu < 15$, $15 < p_T^\mu < 20$ and $20 < p_T^\mu < 25$. The p_T distribution of the muon's parent bottom quark obtained from the ISAJET Monte Carlo is different for each of the available muon selections (see figure 11a-d). A p_T cut is made on the bottom quark so that $p_T^b > p_T^{min}$ includes 90 % of the muon events in the selection. In the figures this cut-off, p_T^{min}, is indicated by an arrow. The cross-sections $\sigma(p\bar{p} \to b$ or $\bar{b}, p_T^b > p_T^{min}, |y| < 1.5)$ as given by ISAJET are indicated in table 1.

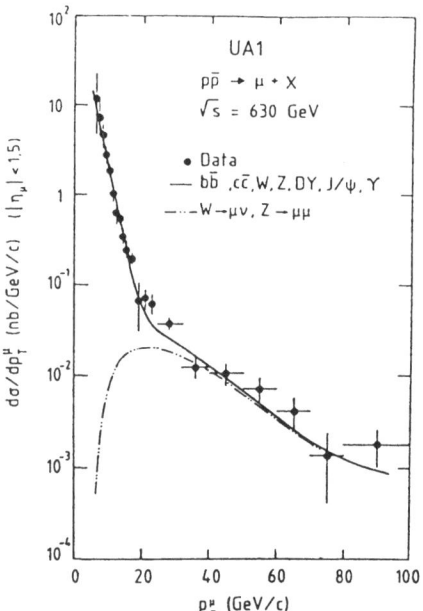

Figure 10. The inclusive muon transverse momentum spectrum, $d\sigma/dp_T^\mu$, versus p_T^μ. The data after subtraction of the decay background are compared to ISAJET Monte Carlo predictions, which include: $b\bar{b}$, $c\bar{c}$, $W^\pm$, Z^o, Drell-Yan, J/Ψ and Υ

Table 1. Summary of cross-sections

Sample	p_T^{min} (GeV/c)	$\sigma(p\bar{p} \to b) + \sigma(p\bar{p} \to \bar{b})$ $p_T > p_T^{min}, \mid y \mid < 1.5$ (μb)
$p\bar{p} \to b \to J/\Psi \to \mu^+ \mu^-$	6	4.5 ($\pm$ 72 %)
$p\bar{p} \to b\bar{b} \to \mu\,\mu$ [high mass]	6.5	2.4 ($\pm$ 55 %)
$p\bar{p} \to b\bar{b} \to \mu^+ \mu^-$ [low mass]	10	0.83 ($\pm$ 48 %)
$p\bar{p} \to b \to \mu$ [10 GeV/c $< p_T^\mu <$ 15 GeV/c]	15	0.42 ($\pm$ 45 %)
$p\bar{p} \to b \to \mu$ [15 GeV/c $< p_T^\mu <$ 20 GeV/c]	23	0.076 ($\pm$ 46 %)
$p\bar{p} \to b \to \mu$ [20 GeV/c $< p_T^\mu <$ 25 GeV/c]	32	0.023 ($\pm$ 48 %)

The measured cross-sections can now be compared to theory. Figure 12 shows the cross-sections as a function of p_T^{min} for both the data and for the $O(\alpha_s^3)$ QCD calculation by Nason[4] [2]. Predictions using the ISAJET all orders calculation and the EUROJET $O(\alpha_s^2) + O(\alpha_S^3)$ calculation are also shown.

---

[2]When the calculations of Nason are concerned $O(\alpha_s^3)$ means $O(\alpha_s^2) + O(\alpha_s^3)$

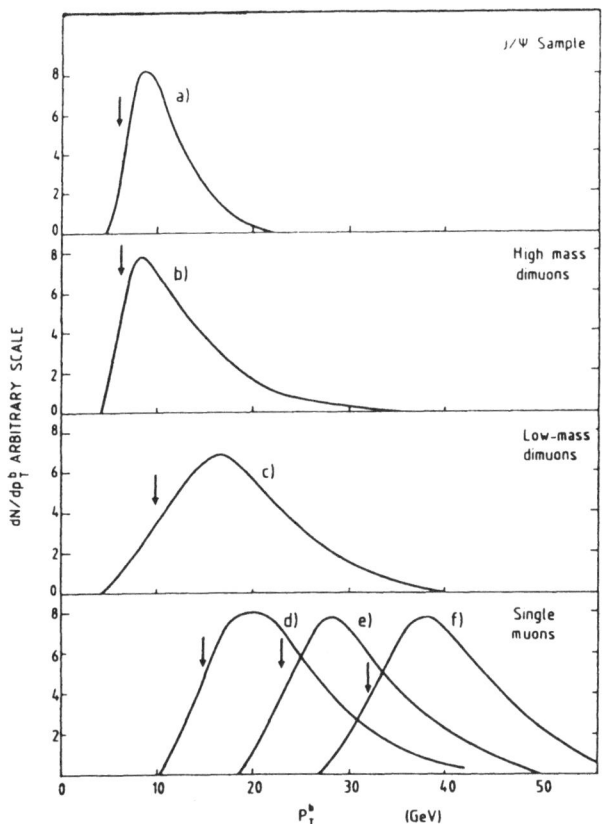

Figure 11. The p$_T$ distribution of the parent bottom quark in the four muon samples

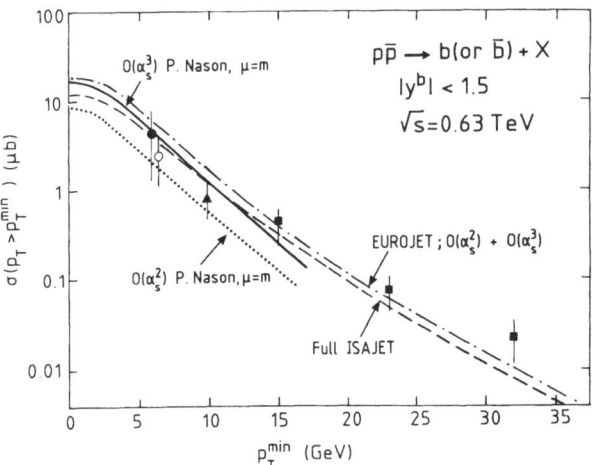

Figure 12. The inclusive bottom cross-section in p$\bar{\text{p}}$ collisions at $\sqrt{s} = 630$ GeV for $p_T^b > p_T^{min}$ and $\mid y^b \mid < 1.5$ as a function of p_T^{min}. The six data points derived from the four muon samples are shown. In the figure also the theoretical predictions of Nason up to $O(\alpha_s^2)$ and $O(\alpha_s^3)$ are plotted. The ISAJET and EUROJET predictions are included too.

The calculation of Nason only gives a reliable prediction of the cross-section up to $p_T^{min} \cong 16$ GeV/c. At high p_T values the bottom quark starts to behave like a light quark and starts to radiate gluons. This multi-gluon radiation is not included in the model used by Nason.

Figure 12 shows that the Nason $O(\alpha_s^3)$ prediction on one hand and the ISAJET and EUROJET predictions on the other hand yield different shapes. ISAJET and EUROJET show a very similar behaviour, that mainly differs in absolute normalization.

The data point measured in the high mass dimuon sample is excluded from this study fit as it does not represent a truly inclusive measurement. One muon comes from the bottom quark the other from the antibottom quark. As none of the theoretical models includes quark-quark correlations, the high mass point cannot be taken into account. The two points above $p_T^{min} = 15$ GeV/c are not used in the curve fit because here the $O(\alpha_s^3)$ QCD calculation is unreliable.

Preserving the shape of the $O(\alpha_s^3)$ calculation by Nason the bottom cross-section $\sigma(p\bar{p} \to b$ or $\bar{b})$ is determined by a fit to the measured partial cross-sections. Integrating down to $p_T^{min} = 0$ yields an inclusive bottom cross-section:

$$\sigma(p\bar{p} \to b \text{ or } \bar{b} + X, \mid y^b \mid < 1.5) = 14.7 \pm 4.7 \ \mu b$$

Going to the $b\bar{b}$ pair production cross-section and extrapolating over all rapidity values yields:

$$\sigma(b\bar{b}) = 10.2 \pm 3.3 \ \mu b$$

This should be compared to the value derived by Altarelli et al. [2]:

$$\sigma(b\bar{b}) = 12 \ _{-4}^{+7} \ \mu b \ (m_b = 5 \text{ GeV/c}^2)$$

Considering the large errors the agreement with theory is very good.

Figure 13 shows the differential bottom cross-section on a double ln-scale. Note that along the x-axis $\ln((p_T^{min})^2 + m_b^2)$ is plotted, so that $p_T^{min} = 0$ lies at 3.22 as indicated in the figure. Besides the prediction of Nason also a simple parametrization, $d\sigma/dp_T = A \cdot (p_T^2 + m_b^2)^{-n}$ is fitted to the data points shown. The best fit yields: A $= 1.1 \times 10^4 \ \mu b/GeV^2$ and n $= 2.79 \pm 0.64$. The mass of the b-quark is fixed to 5 GeV/c^2. Extrapolation of the fitted curve integrated down to $p_T^{min} = 0$ gives:

$$\sigma(p\bar{p} \to b \text{ or } \bar{b} + X, \mid y^b \mid < 1.5) = 19.5 \ _{-9.3}^{+17.5} \ \mu b$$

This simple parametrization gives a remarkably good fit to the data.

5 The search for a New Heavy Flavour Quark

We have studied the production of known heavy quarks (charm and bottom) using our muon data. Our level of understanding of the sources of muon-jet events, which include heavy flavour decays, $W^{\pm}$, Z^o, Drell-Yan, J/Ψ and Υ production allows us to search for other heavy quarks.

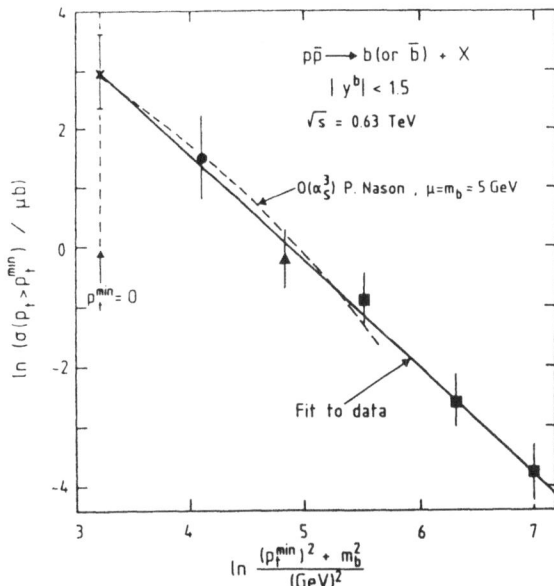

Figure 13. Fit to the measured bottom cross-sections. Besides a fit made using the QCD $O(\alpha_s^3)$ prediction by Nason(dashed line) also a simple parametrization (full line) is fitted to the data points shown: $d\sigma/dp_T = A\,(p_T^2 + m_b^2)^{-n}$, where $A = 1.1 \times 10^4$ $\mu b/GeV^2$ and $n = 2.79 \pm 0.64$. The bottom mass is fixed to 5 GeV/c^2.

5.1 The Top Search

In the top search two production mechanism are considered:

$$p\bar{p} \to t\,\bar{t}$$
$$\quad \rightarrow l^- \,\bar{\nu}\,\bar{b}$$
$$\quad \rightarrow \text{one or more jets}$$

$$p\bar{p} \to W^+ + X$$
$$\quad \rightarrow t + \bar{b}$$
$$\quad \rightarrow \text{jet}$$
$$\quad \rightarrow l^+\,\nu\,b$$

According to existing limits from PETRA and TRISTAN, the top quark is heavy. Therefore, the decay lepton will have a high p_T relative. In other words, the lepton will be well separated from the accompanying jet and will be isolated. Due to the extremely hard top fragmentation and the high top mass the lepton will be produced at high p_T. A top decay will in general also result in a measurable transverse energy (a high p_T neutrino) and one or more jets.

The following processes can fake the top signature:

$$p\bar{p} \to b\,\bar{b}\,g$$
$$\quad \rightarrow \text{jet}$$
$$\quad \rightarrow \text{jet}$$
$$\quad \rightarrow l^- \,\bar{\nu}\,c$$

125

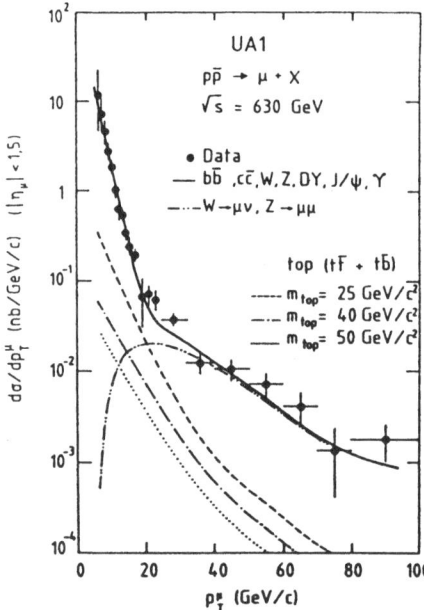

Figure 14. The inclusive muon transverse momentum spectrum, $d\sigma/dp_T^{\mu}$, as a function of p_T^{μ}. The data after subtraction of the decay background are compared to Monte Carlo predictions, which include: $b\bar{b}$, $c\bar{c}$, $W^{\pm}$, Z°, Drell-Yan, J/Ψ and Υ. The predicted contributions from top are shown for three top masses, $m_t = 25$, 40 and 50 GeV/c^2

$$p\bar{p} \rightarrow W^+ + \text{jets}$$
$$\phantom{p\bar{p} \rightarrow W^+ } \hookrightarrow l^+ \nu$$

$$p\bar{p} \rightarrow l^+ l^- + \text{jets}$$
$$(\text{Drell-Yan}, \Upsilon, J/\Psi)$$

The lepton pair produced in Drell-Yan, Υ or J/Ψ is, of course, only background to the top signature if one of the leptons remains undetected.

5.2 Top Search in the Muon Channel

The inclusive muon p_T spectrum for the single muon data after background subtraction is shown again in figure 14 together with the Monte Carlo predictions from all known sources. The expected contribution from top is plotted for three top masses, $m_t = 25$, 40 and 50 GeV/c^2. As can be seen there is little space for a top contribution in this data sample.

However, the inclusive muon p_T spectrum provides only a crude way to search for a top signal. More precise methods have been developed using the expected event topology for top.

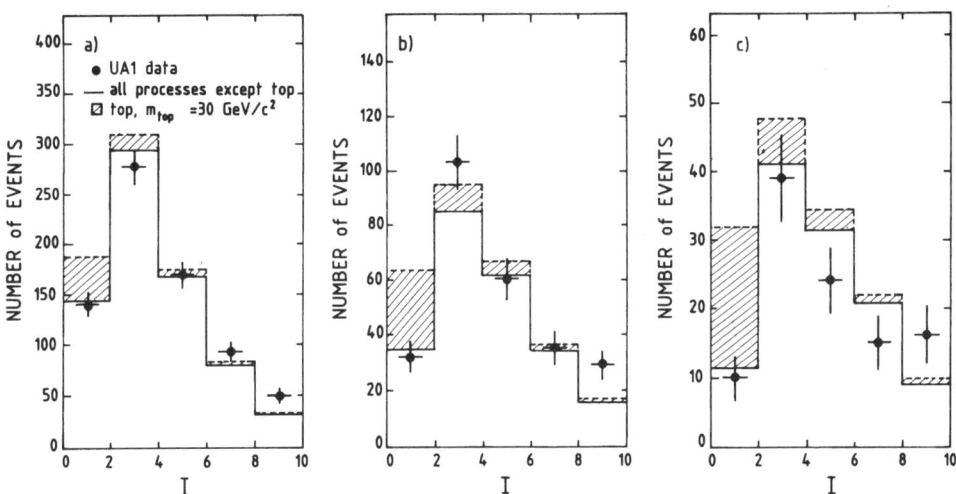

Figure 15. The muon isolation parameter I for three sets of cuts:
a) $p_T^\mu > 10$ GeV/c, $E_T^{jet1} > 12$ GeV, $M_T(\mu\nu) < 40$ GeV/c^2 and no jet 2 requirement.
b) $p_T^\mu > 12$ GeV/c, $E_T^{jet1} > 15$ GeV, $M_T(\mu\nu) < 40$ GeV/c^2 and no jet 2 requirement.
c) $p_T^\mu > 12$ GeV/c, $E_T^{jet1} > 15$ GeV, $M_T(\mu\nu) < 40$ GeV/c^2 and $E_T^{jet2} > 7$ GeV.
The black points with error bars are the UA1 data. The expected contributions from
non-top processes are shown by the solid histogram. The hatched area represents the
predicted contribution from a 30 GeV/c^2 top.

5.3 Muon Isolation

As pointed out before, a top decay is expected to yield events containing an isolated
muon. This is a crucial feature of the event topology. For the top search the muon
isolation is measured using the parameter I:

$$I = \left(\left(\tfrac{\Sigma E_T}{3}\right)^2 + \left(\tfrac{\Sigma p_T}{2}\right)^2\right)^{1/2}$$

where ΣE_T is the sum of the transverse energy in the calorimeters. Σp_T is the sum of
the transverse momenta measured in the UA1 tracking chamber. Both ΣE_T and Σp_T
are summed in a cone of $\Delta R = 0.7$ around the muon. The I-parameter weights the ΣE_T
and the Σp_T to correct for the fact that the tracking chamber detects only charged
particles, whereas the calorimeters measure both charged and neutral particles.

The isolation parameter has been studied under different sets of cuts (see figure
15a, b,c). Figure 15a shows the I distribution when requiring a muon with $p_T^\mu > 10$
GeV/c and at least one jet with $E_T^{jet1} > 12$ GeV. Most of the $W \to \mu\,\nu$ events are
removed from the sample by requiring the transverse mass of the muon-neutrino pair,
$M_T(\mu\nu) < 40$ GeV/c^2. Note that no requirement is made for jet 2.

In figure 15b the I distribution is shown for slightly harder cuts. The thresholds
for the muon and jet 1 are raised to: $p_T^\mu > 12$ GeV/c and $E_T^{jet1} > 15$ GeV. The $M_T(\mu\nu)$
cut remains at the same threshold. In the bin $0 \le I \le 2$ a clear improvement of the
signal-to-background ratio for a top signal is visible.

The signal-to-background ratio can be improved further by requiring in addition a second jet: $E_T^{jet2} > 7$ GeV. The I distribution for this set of cuts is shown in figure 15c.

Requiring the muon isolation to be I < 2 in the third set of cuts, the best signal-to-background ratio has been obtained. Therefore these cuts will be used for the top search.

5.4 Top Search in the Muon + Two Jets Sample

The muon + two jet sample contains 10 events with an isolated muon. First a study has been made of all the known non-top sources contributing to the selected data sample. The main background contribution stems from $p\bar{p} \rightarrow b\bar{b}g$. QCD $c\bar{c}$ and $b\bar{b}$ production yields 6.6 events. Adding the other background sources, π/K decays, $W^\pm$, Z°, Drell-Yan, J/Ψ and Υ production gives a total expected number of events of 11.4 (see table 2a). The data leave very little space for a top contribution. Using Poisson statistics the data are found to be compatible with at most 7 top events at 95 % C.L.. Table 2b shows that e.g. a top of 40 GeV/c^2 would already yield 9.6 events. The isolated muon + one jet selection (second set of selection cuts) is used as a control sample. Also this data sample can only accommodate for a small number of top events.

Table 2a.   Sources of isolated muon + jet events,
with a muon $p_T^\mu > 12$ GeV/c

Sample	Monte Carlo					Data
	K/π Decays	W / Z	D.Y. J/Ψ Υ	b$\bar{b}$ c$\bar{c}$	Total	
μ + 1 jet	7.2 $\pm$ 1.7 $\pm$ 2.2	2.5 $\pm$ 0.5	7.3 $\pm$ 0.7 $\pm$ 3.6	6.3 $\pm$ 0.6	23.3 $\pm$ 2.0 $\pm$ 4.2	22
μ + $\geq$ 2 jets	2.3 $\pm$ 0.4 $\pm$ 0.7	0.6 $\pm$ 0.2	2.0 $\pm$ 0.4 $\pm$ 1.0	6.6 $\pm$ 0.7	11.4 $\pm$ 0.9 $\pm$ 1.2	10

(The first error is statistical, the second one systematic.)

From the event rates alone one can put a limit on the top mass. However, using the shapes of the detailed kinematic properties of the events an even more stringent limit can be derived. Figure 16a-d show the distributions used in the muon channel. A combined likelihood fit of the various distributions is performed. A mass limit will be presented after combining the results from the muon and the electron channel.

Table 2b. Expected top event rates for isolated muon +
≥ 2 jets events with a muon $p_T^\mu > 12$ GeV/c

Sample	Top mass (GeV/c²)			
	25	30	40	50
$t\bar{b}$	2.5 ± 0.3	3.9 ± 0.6	3.6 ± 0.5	3.1 ± 0.5
$t\bar{t}$	21.8 ± 1.1	16.6 ± 0.9	6.0 ± 0.5	2.2 ± 0.3
Total	24.3 ± 1.1	20.5 ± 1.1	9.6 ± 0.7	5.3 ± 0.6

(The errors quoted are statistical only.)

5.5   Top Search in the Electron Channel

The top search in the electron channel is carried out in the sample of isolated electrons with $E_T^{el} > 15$ GeV. The isolation requirement for electrons restricts the hadronic activity in a cone $\Delta R=0.7$ around the electron to $\leq 1/10$ of the electron's p_T (tracking chamber) and to $\leq 1/10$ of the E_T (calorimeters). The p_T and E_T in a cone $\Delta R=0.4$ around the electron are limited to a maximum of 1 GeV. The W $\to$ e ν events are removed by the requirement $M_T(e\nu) < 45$ GeV/c².

The electron has two background sources:

- overlapping $\pi^\pm$ and π^o faking an electron

- photon conversion into an electron-positron pair

The top search in the electron channel concentrates on the isolated electron + one or more jet sample. The background from overlaps and conversions has been calculated. Adding the expected number of events with this topology from all non-top contributions (W$^\pm$, Z^o, Drell-Yan, J/Ψ, Υ, b$\bar{\text{b}}$ and c$\bar{\text{c}}$) yields a total of 23.4 events (see table 3a). The data contain 26 events. At most 12.7 top events (95 % C.L.) could be accommodated by the data. The comparison between Monte Carlo predictions and the data is also made for several other samples (see table 3a). Again there is little room for top. Table 3b shows that e.g. a top of 40 GeV/c² would contribute 13.4 events to the electron, $\geq$ one jet sample.

Also for the electron sample a mass limit can be based on event rates alone, but a more stringent limit can be derived using the shapes of the kinematic distributions. In the electron channel the shapes of the missing transverse energy spectrum and the distribution showing the number of jets in the event are used to make a combined likelihood fit. These distributions are shown in figure 17a, b.

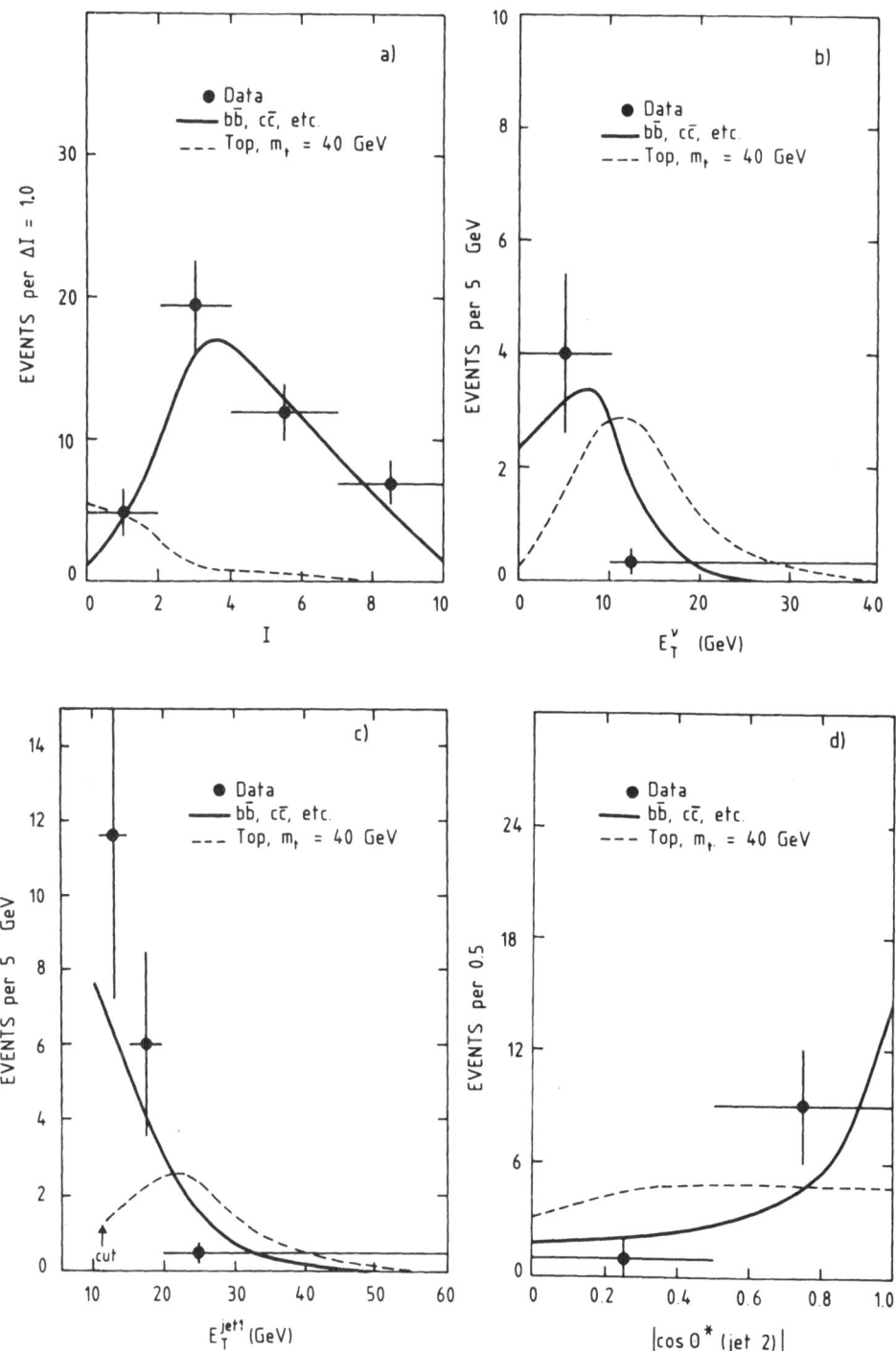

Figure 16. Distributions used in the likelihood fit in the muon channel: a) muon isolation I, b) missing transverse energy, c) E_T of the highest E_T jet and d) angular distribution of the second highest E_T jet

Table 3a. Classification of electron + jet events,
$E_T^e > 15$ GeV, $m_T(e,\nu) < 45$ GeV/c^2

Sample	Monte Carlo					Data
	Overlap Conversions	W / Z	D.Y. J/Ψ Υ	$b\bar{b}$ $c\bar{c}$	Total	
e + 1 jet	5.5 ± 0.3 ± 1.0	5.3 ± 0.3 ± 0.34	4.3 ± 0.5 ± 1.5	1.6 ± 0.35 ± 0.4	16.7 ± 0.7 ± 1.9	19
e + ≥ 2 jets	2.6 ± 0.3 ± 0.5	0.8 ± 0.2 ± 0.06	1.1 ± 0.3 ± 0.38	2.2 ± 0.45 ± 0.55	6.7 ± 0.6 ± 0.7	7
e + ≥ 1 jets	8.1 ± 0.5 ± 1.5	6.1 ± 0.4 ± 0.4	5.4 ± 0.6 ± 1.9	3.8 ± 0.6 ± 0.95	23.4 ± 0.9 ± 2.7	26

(The first error is statistical, the second one systematic.)

Table 3b. Expected top event rates for isolated electron
+ ≥ 1 jet events, with a muon $E_T^e > 15$ GeV

Sample	Top mass (GeV/c^2)					
	25	30	40	45	50	55
$t\bar{b}$	3.3 ± 0.3	5.2 ± 0.5	6.2 ± 0.6	6.3 ± 0.6	6.0 ± 0.6	6.4 ± 0.6
$t\bar{t}$	20.7 ± 0.9	14.1 ± 0.7	7.2 ± 0.5	3.2 ± 0.3	3.3 ± 0.3	2.0 ± 0.2
Total	24.0 ± 0.9	19.4 ± 0.9	13.4 ± 0.8	9.5 ± 0.7	9.3 ± 0.7	8.4 ± 0.6

(All the errors quoted are statistical only.)

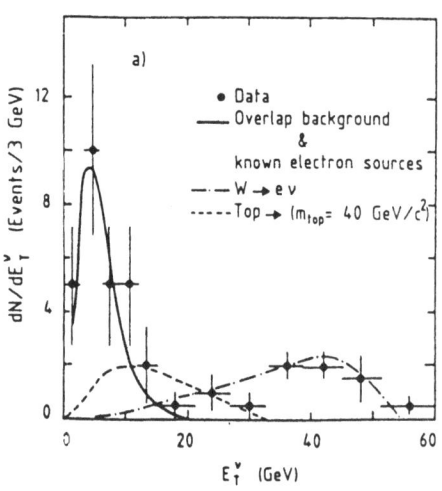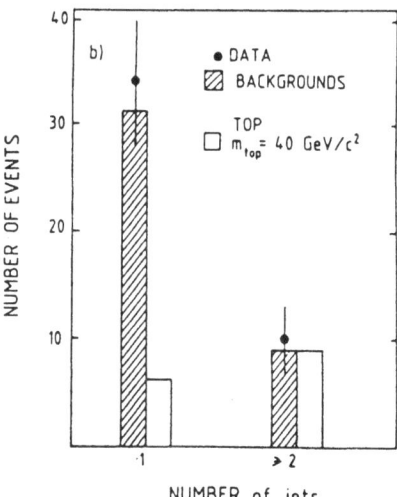

Figure 17. Distributions used in the likelihood fit in the electron channel: a) missing transverse energy, b) number of jets in the event

5.6 Mass Limit from the Combined Electron and Muon Channel

Using the kinematic distributions both from the muon and from the electron channels a limit on the allowed cross-section $\sigma(t\bar{t}+t\bar{b})$ is derived. Figure 18 shows the upper limit on $\sigma(t\bar{t}+t\bar{b})$ as a function of the top mass. Both the 90 % C.L. and 95 % C.L. contours are given. The EUROJET Monte Carlo has been used to predict the $t\bar{t}$ cross-section up to $O(\alpha_s^3)$.

The intersection of the 95 % C.L. contour with the theoretical prediction $\sigma(t\bar{t}+t\bar{b})$ yields a limit on the top mass, $m_T > 56$ GeV/c^2. Note that at 95 % C.L. no limit can be given based on W → $t\bar{b}$ alone.

5.7 Uncertainties on the Top Mass Limit

The uncertainties on the EUROJET cross-section are illustrated by figure 19. The 95 % C.L. contour is now expressed in terms of K:

$$K = \frac{\sigma}{\sigma_o \text{(lowest order)}}$$

where σ is the actual cross-section. The lowest order cross-section, σ_o, is the EUROJET $O(\alpha_S^2)$ calculation using Eichten et al. structure functions, set I and the Q^2 definition: $Q^2 = p_T^2 + m_Q^2$.

The current top mass limit is derived by the EUROJET calculation up to $O(\alpha_s^3)$ shown by the curve marked EUROJET. The other curves shown are EUROJET $O(\alpha_s^2)$ predictions. First the set of structure functions used is varied. Using the Duke and Owens [18] set I structure functions the weakest top mass limit is derived. Changing in addition the Q^2-scale to $Q^2 = \hat{s}$ gives an even more conservative limit. The Duke and Owens structure function together with $Q^2 = \hat{s}$ yields as the most conservative mass limit, $m_t > 44$ GeV/c^2.

132

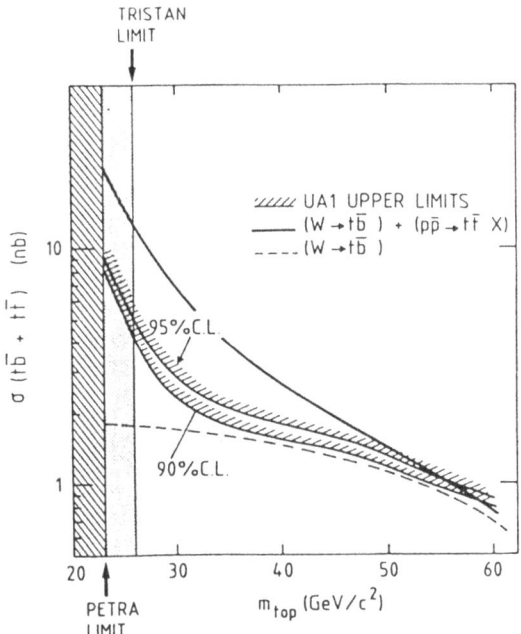

Figure 18. Combined cross-section limits for the top quark production from muon and electron channel

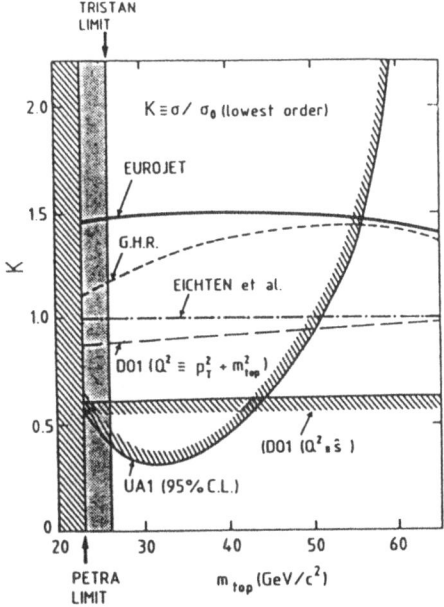

Figure 19. Cross-section normalized to the $O(\alpha_s^2)$ cross-section as given by EUROJET using Eichten et al. structure functions, set I and $Q^2 = p_T^2 + m_Q^2$. The 95 % C.L. contour from the data is shown. The curve labelled EUROJET represents the $O(\alpha_s^2)$ + $O(\alpha_s^3)$ cross-section. The other curves are based on $O(\alpha_s^2)$ calculations. The Q^2 scale is defined as $Q^2 = p_T^2 + m_Q^2$ except in the lower Duke and Owens curve, where $Q^2 = \hat{s}$

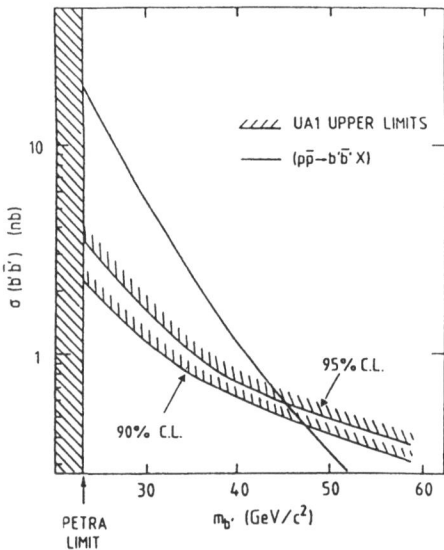

Figure 20. Cross-section limits from for the b' quark from combined muon and electron channel information

5.8   Search for a b' Quark

The b' quark is a fourth generation, down-like quark ($Q_{b'} = -1/3$). The partner of the b' is assumed to be very heavy. As a consequence the $p\bar{p} \rightarrow W^+ \rightarrow t'\,\bar{b}'$ channel is not open. The b' is solely produced in the QCD process $p\bar{p} \rightarrow b'\,\bar{b}'$.

The life time of the b' is assumed to be short, $\tau = O(10^{-13}$ s). Thus the decay muons from a b' are prompt.

Using the same analysis method as has been used for the top quark also a limit on the mass of the b' quark can be derived. Figure 20 shows the 90 % and 95 % C.L. upper limits together with the predicted cross-section $\sigma(p\bar{p} \rightarrow b'\,\bar{b}'$ X) (EUROJET). The limit on the b' mass derived from figure 20 is:

$$m_{b'} > 44 \text{ GeV}/c^2.$$

Error analysis (see section 5.7) yields as the most conservative limit:

$$m_{b'} > 32 \text{ GeV}/c^2.$$

5.9   Mass Limits based on the Full $O(\alpha_s^3)$ calculation

Altarelli et al. [2] have used the complete $O(\alpha_S^3)$ calculation by Nason et al. [1] to derive a limit for both the top and the b' mass from the UA1 measurements (see figure 21). They base their limit solely on QCD processes. Figure 21 shows the central curve of Altarelli et al. together with the band of uncertainties discussed before. Using the 95 % C.L. contour from the combined muon and electron channel information for $t\bar{t}$ and $b'\bar{b}'$, they derive as the most conservative limits:

$$m_t > 41 \text{ GeV}/c^2$$
$$m_{b'} > 34 \text{ GeV}/c^2$$

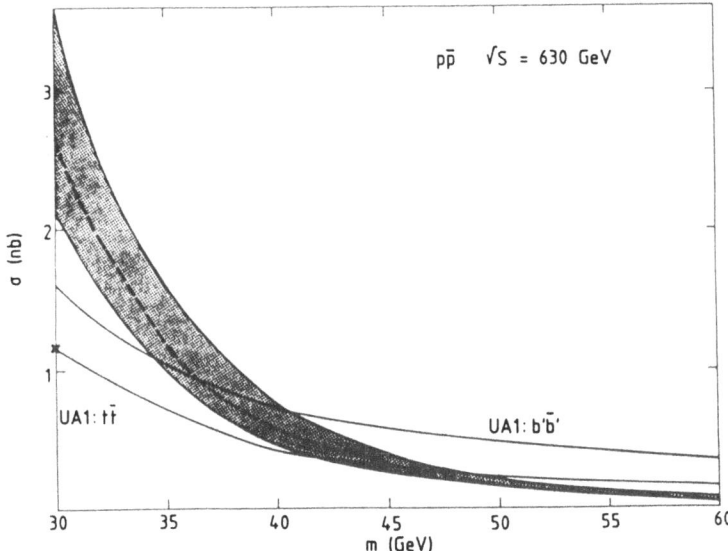

Figure 21. Comparison of the predicted cross-sections for heavy flavour production to the UA1 $t\bar{t}$ and $b'\bar{b}'$ 95 % upper bounds. The uncertainties on the theoretical prediction are shown as a band[2]

Thus Altarelli et al. confirm the UA1 limits.

Despite the fact, that they base their limits on the higher cross-section of the full $O(\alpha_s^3)$ calculation, the limit on the top mass Altarelli et al. derive is slightly weaker than the most conservative limits presented by UA1. The reason for this is, that Altarelli et al. have used the DFLM structure functions, which represent a softer gluon content in the (anti-)proton. As the main $t\bar{t}$ production mechanism is gg $\rightarrow$ $t\bar{t}$, finally the limit derived is weaker.

6 Outlook for the Future

In the future UA1 will continue the top searches using the higher antiproton yield of ACOL. A study of the expected number of top events as a function of the top mass is shown in figure 22. The number of events is given both for the present statistics, 700 nb^{-1} and for the expected integrated luminosity at ACOL, 10 pb^{-1}. The numbers given do not include any selection cuts.

The distribution is split into two distinctly different regions, region I: $m_t < m_W$ and region II: $m_t > m_W$.

The region $m_t < m_W$ has been extensively studied using the present statistics. Top will be produced through p$\bar{\text{p}}$ $\rightarrow$ $t\bar{t}$ and p$\bar{\text{p}}$ $\rightarrow$ $W^\pm$ $\rightarrow$ tb.[3] As shown in section three of this paper the background to the top signature in region I is well understood. To avoid the uncertainties on the $t\bar{t}$ QCD calculation, one would like to study top production in p$\bar{\text{p}}$ $\rightarrow$ $W^\pm$ $\rightarrow$ tb alone. A factor four in statistics is needed to do this. Mark that at the Sp$\bar{\text{p}}$S collider for $m_t \geq 40$ GeV/c^2 the largest contribution comes from p$\bar{\text{p}}$ $\rightarrow$ $W^\pm$ $\rightarrow$ tb.

---

[3]tb = t$\bar{\text{b}}$ or $\bar{\text{t}}$b

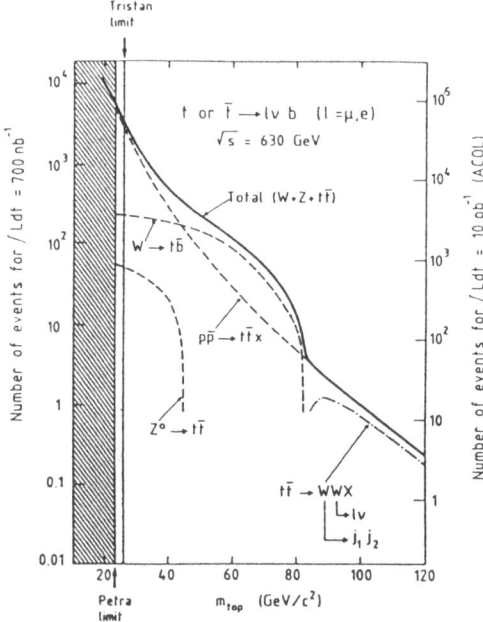

Figure 22. The number of predicted events from top production processes as a function of the top mass for both 700 nb^{-1} (present statistics) and for 10 pb^{-1} (expected statistics at ACOL). The figure can be split into two mass regions: $m_t < m_W$ and $m_t > m_W$. The important top production processes and their backgrounds in the two regions are discussed in the text.

In region II, $m_t > m_W$ top is only produced in QCD $t\bar{t}$ production. As the top mass is higher than the W mass, the top can decay into a real W and a b quark. In this case top production can yield for example the following signature:

$$p\bar{p} \to t\bar{t} \to W^- \; W^+ \; X$$
$$\qquad\qquad\qquad \vert \rightarrow l^+ \; \nu$$
$$\qquad\qquad \vert \rightarrow \text{jet 1, jet 2}$$

The main background to this process comes from $p\bar{p} \to W^\pm + 2$ jets.

The expected number of $t\bar{t}$ events that follow the above decay is small, even for an integrated luminosity of 10 pb^{-1}. Interesting for the the top search in region II are the two high p_T W events [19] seen by UA1. These events (figure 23 shows one example) display the correct event topology for heavy top decay via a real W, as described above. For $m_t \cong m_W$ we expect approximately 0.7 such events per semi-leptonic decay mode, integrated over all expected p_T^W. The observed events, however, seem to have $p_T^W \geq 60$ GeV/c. At these large p_T values we expect between 0.05-0.1 events. In the data two candidates for this heavy top decay have been found.

A comparison of the possibilities to discover top at the Sp$\bar{p}$S and the Tevatron collider is made in figure 24. The upper part of the picture shows the expected cross-section for top production processes. The prediction for $p\bar{p} \to t\bar{t}$ is taken from Altarelli et al. [2]. The band shown is their band of theoretical uncertainties. The measured cross-sections $\sigma(W^\pm \to e^\pm \; \nu)$ at the two colliders are used to predict the

136

$$W^+ \rightarrow \mu^+ \nu_\mu$$

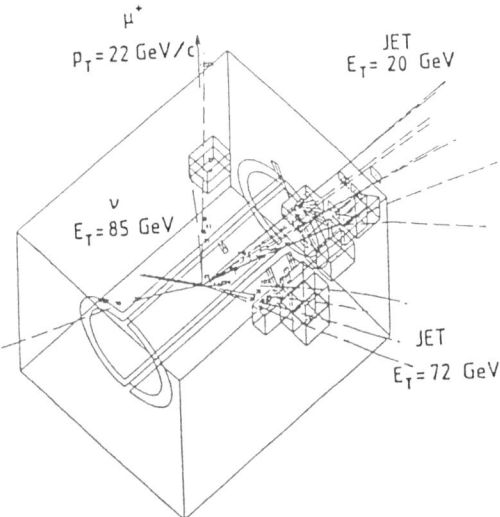

Figure 23. Event A

contribution from $W^\pm \rightarrow tb$. The $W^\pm \rightarrow tb$ contribution at the Tevatron is less pronounced than at the Sp$\bar{p}$S. The cross-section for QCD $t\bar{t}$ production grows faster with energy than the cross-section for $W^\pm \rightarrow tb$ production.

The lower part of the picture shows the ratio of the cross-section $\sigma(p\bar{p} \rightarrow t + X)$ at the Tevatron and at the Sp$\bar{p}$S collider as a function of the top mass:

$$R = \frac{\sigma(p\bar{p} \rightarrow t + X, \sqrt{s} = 1.8 TeV)}{\sigma(p\bar{p} \rightarrow t + X, \sqrt{s} = 0.63 TeV)}$$

In the region $m_t < m_W$ the cross-section at the Tevatron is roughly a factor ten higher than the cross-section at the Sp$\bar{p}$S.

If the top mass lies above the W mass the ratio R becomes much larger than ten and increases with increasing top mass. In this range of top masses the advantage of the Tevatron becomes too large to be compensated by higher luminosity or with a better detector acceptance. The conclusion from this study is, that if $m_t < m_W$ the Sp$\bar{p}$S is competitive with the Tevatron, but for $m_t > m_W$ the Tevatron has a clear advantage over the Sp$\bar{p}$S.

7 Conclusions

The cross-section for heavy flavour bound states as measured in the dimuon data are in good agreement with theoretical predictions. Adding the contributions of J/Ψ, Drell-Yan, Υ and muon background from π/K decays to the predictions for $c\bar{c}$ and $b\bar{b}$ QCD production processes, a good description of both the single muon and the dimuon data is obtained.

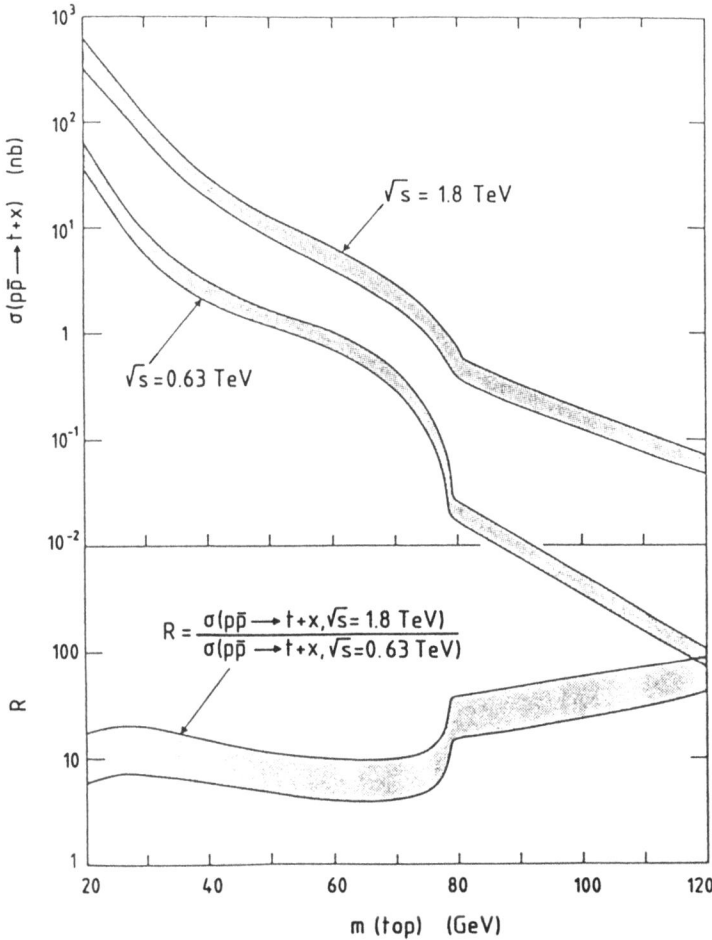

Figure 24. Top production cross-section $\sigma(p\bar{p} \to t + x)$ as a function of the top quark mass is shown in the upper part of the figure. The prediction for the QCD production is taken from Altarelli. The contribution from $W^{\pm}$ is derived from the measured $W^{\pm} \to e^{\pm} \nu$ cross-sections at the two colliders. The ratio of the expected $\sigma(p\bar{p} \to t + x)$ at the Tevatron and at the $Sp\bar{p}S$ collider is shown in the lower part of the figure.

The muon samples (with exception of the high mass dimuons) have been used to derive a bottom cross-section. The cross-section calculation is based on the $O(\alpha_s^3)$ calculation of $\sigma(p_T^b > p_T^{min})$ by Nason [4]. UA1 obtains the value $\sigma(b\bar{b}) = 10.2 \pm 3.3$ μb. A QCD prediction by Altarelli et al. [2] yields $\sigma(b\bar{b}) = 12 ^{+7}_{-4}$ μb. Considering the large errors the agreement between the UA1 measurement and theory is good.

From the study of the single muon and dimuon data it is clear that they are well described by the sum of the non-top processes. Special isolated lepton (muon, electron) selections were used for the top search. The top mass limit based on combined information from the muon and the electron channel is $m_t > 44$ GeV/c^2 at 95 % C.L.. Comparing the cross-sections given by the full $O(\alpha_s^3)$ calculation to the UA1 data, Altarelli et al. derive: $m_t > 41$ GeV/c^2 at 95 % C.L..

The UA1 data have also been used to put a limit on the b'-mass (the b' is a fourth generation, down-like quark). The b'-mass limit is found to be: $m_{b'} > 32$ GeV/c^2 at 95 % C.L.. Altarelli et al. give a limit $m_{b'} > 34$ GeV/c^2 at 95 % C.L..

For $m_t < m_W$ the cross-section for top production processes is approximately a factor ten higher at the Tevatron than at the Sp$\bar{p}$S. In the region $m_t > m_W$ this factor increases to forty or more. If $m_t > m_W$ the advantage of the Tevatron will be difficult to compensate by e.g luminosity or detector acceptance.

Acknowledgements

I should like to thank the organizers for inviting me to this interesting workshop. Also I thank all my colleagues in UA1 who made these results possible. In particular I should like to thank Felicitas Pauss, Michel Della Negra and Daniel Denegri for the interesting discussions.

References

[1] P. Nason, S. Dawson and K. Ellis, ' The Total Cross-section for the Production of Heavy Quarks in Hadronic Collisions', Fermilab-Pub-87/222-T, (1987)

[2] G. Altarelli, M. Diemoz, G. Martinelli and P. Nason,'Total Cross-section for Heavy Flavour Production in Hadronic collisions and QCD', CERN Preprint CERN-TH.4978/88 (1988)

[3] M. Diemoz, F. Ferroni, E. Longo and G. Martinelli,'Parton Densities from Deep Inelastic Scattering to Hadronic Processes at Super Collider Energies', CERN Preprint CERN-TH.4751/87 (1987)

[4] P. Nason, Talk presented at Les Rencontres de Physique de la Vallee d'Aoste, La Thuile, Italy (1988)

[5] F. Paige, S.D. Protopopescu: ISAJET Monte Carlo, BNL 38034 (1986)

[6] A. Ali et al., Nucl. Phys. B292 (1987) 1
B. van Eijk, PhD. Thesis, University of Amsterdam (1987)

[7] G. C. Fox, S. Wolfram, Nucl. Phys. B168 (1980) 285

[8] G. Arnison et al. (UA1 collaboration), Phys. Lett. 172B (1986) 461

[9] E. Eichten et al., Rev. Mod. Phys. 56 (1984) 579
Erratum: Rev. Mod. Phys. 58 (1986) 1065

[10] C. Peterson et al., Phys. Rev. D27 (1983) 105

[11] C. Albajar et al. (UA1 collaboration), Phys. Lett. 186B (1987) 237

[12] C. Albajar et al. (UA1 collaboration), Phys. Lett. 186B (1987) 247

[13] C. Albajar et al. (UA1 collaboration), Preprint CERN-EP/88-46

[14] C. Albajar et al. (UA1 collaboration), Phys. Lett. 200B (1988) 380

[15] C. Albajar et al. (UA1 collaboration), Z. Phys. C37 (1988) 489

[16] C. Albajar et al. (UA1 collaboration), Z. Phys. C37 (1988) 505

[17] V. Barger, A.D. Martin, Phys. Rev. D31 (1985) 1051

[18] D. W. Duke and J. F. Owens, Phys. Rev. D30 (1984) 49

[19] C. Albajar et al. (UA1 collaboration), Phys. Lett. B 193 (1987) 389

9

MASSES OF HEAVY FLAVOURED HADRONS

André Martin

Theory Division, CERN
1211 Geneva 23
Switzerland

I shall not come back on the predictions on the masses of beautiful baryons on which I have lectures many years ago here in Erice[1] and more recently in San Miniato[2].

First I would like to speak of something relevant to the B-$\bar{\text{B}}$ factories which is the masses of the B_s and B_s^*. A very successful fit of $s\bar{s}$, $c\bar{s}$, $c\bar{c}$ and $b\bar{b}$ bound states is given by the non-relativistic effective quark antiquark central potential[3]

$$V = 8.064 + 6.870 \ r^{0.1} \tag{1}$$

when units are powers of GeV, with effective quark masses

$$m_s = 0.518 \text{ GeV}, \ m_c = 1.8 \text{ GeV}, \ m_b = 5.174 \text{ GeV}. \tag{2}$$

One also needs a phenomenological hyperfine splitting

$$C \frac{\vec{\sigma}_1 \vec{\sigma}_2}{m_1 m_2} \ \delta^3(\vec{r}_1 - \vec{r}_2) \tag{3}$$

where C is fitted to the $J/\psi - \eta_c$ mass difference.

With this potential one finds

$$m_{D_s} = 1.99, \ m_{D_s}^* = 2.11 \tag{4}$$

to be compared to the experimental values[4]

$$m_{D_s} = 1.97$$

and

$$m_{D_s}^* - m_{D_s} = 0.140 \text{ GeV}.$$

Therefore we expect deviations of at most 20 MeV. However, the D_s^*-D_s separation is wrong by 15%. Our model predicts

$$M(b\bar{s})_{0^-} = 5354 \text{ MeV}$$

$$M(b\bar{s})_{1^-} = 5408 \text{ MeV} \tag{5}$$

and also

$$M(b\bar{c})_{0^-} = 6250 \text{ MeV}$$

$$M(b\bar{c})_{1^-} = 6318 \text{ MeV} \tag{6}$$

Concerning the B_s's our predictions mean that with a mass 10868 MeV for the T_{ss} state (according to Dan Coffman, this conference), the decays

$$T_{ss} \to B_s \bar{B}_s$$

$$B_s^* \bar{B}_s + B_s \bar{B}_s^*$$

$$B_s^* \bar{B}_s^*$$

are allowed (the latter with a very small phase space).

One might wish to correct the model to account for the experimental hyperfine splitting D_s^*-D_s. If one does so one finds slightly higher masses

$$M_{B_s} = 5374 \text{ MeV}$$

$$M_{B_s}^* = 5410 \text{ MeV.} \tag{7}$$

This, however, does not alter the previous conclusions.

It is amusing to notice that the B_s-B_u mass difference is therefore about 100 MeV, much less than what is believed to be the m_s-m_u mass difference between constituent quarks, about 200 MeV, so that flavour SU(3) symmetry is restored in these heavy mesons.

I would like to remind $B\bar{B}$ factory builders not to forget the B_c mesons. They are only 900 MeV higher and would be very interesting to see. Their production rate could be small.

Now I turn to a different and very interesting animal, the ccc baryon, the analogue of the Ω^-, with charmed quarks replacing strange quarks. This baryon has already been seriously studied by Bjorken[5]. My remarks will be somewhat complementary to those of B.J. In a way if, as some people say, the $c\bar{c}$ system is the hydrogen atom of particle physics, the ccc baryon, with spin 3/2 naturally, is the Helium or Lithium atom. It is a stable particle with respect to strong interactions with a lifetime of 1 to 3×10^{-13} seconds and it produces a doubly ionizing track. Bjorken has tried to estimate its production rate and found that it is not unfeasible.

Now, what is the expected mass of the ground state? First of all, there is the rule of Nussinov[6]

$$M_{QQQ} > \frac{3}{2} M_{Q\bar{Q}} \tag{8}$$

which gives

$$M_{ccc} > \frac{3}{2} M_{J/\psi} = 4.65 \text{ GeV} \tag{9}$$

Looking at

$$M_{\Omega^-} = 1.665 \text{ GeV compared to } 3/2 \ M_\phi = 1530$$
$$M_\Delta = 1232 \text{ GeV compared to } 3/2 \ M_\rho = 1155$$

I would expect something around 4.8 GeV. One could make a calculation of the mass using the potential (1) and the rule

$$V_{QQ} = \frac{1}{2} V_{Q\bar{Q}} \tag{10}$$

which is very reasonable, produces automatically the Nussinov inequality (8), and is, as shown by J.-M. Richard, very successful in the case of the Ω^-[7].

It is important to know the precise mass of the ccc, because for instance, if M_{ccc} is less than 4.773 GeV the ($\overline{ccc}$) antiparticle <u>cannot</u> annihilate in matter. Indeed, then

$$M_{ccc} + M_P < 2M_{D^+} + M_{D^0} .$$

If, on the other hand,

$$4.773 < M_{ccc} < 4.903$$

the only annihilation mode is $2D^+ + D^0$ and this is a complicated re-arrangement collisions.

The spectrum of the ccc system is also very interesting. If we believe that potential (1) is a good potential both for sss and ccc, then the spectrum is the same for both systems with a rescaling of the level spacing:

$$\frac{\Delta E_{ccc}}{\Delta E_{sss}} = \left(\frac{m_s}{m_c}\right)^{1/21} = 0.942$$

For a first guess we can use the Ω^- spectrum calculated in the Chao, Isgur, Karl model[8] and rescale it. We find that the spacing between the first $3/2^+$ excited state and the $3/2^+$ ground state is 360 MeV so that there is a very little phase space for a decay $(ccc)^* \to ccc+2\pi$, less that for the transition $\psi' \to J/\psi+\pi\pi$. We expect therefore that $(ccc)^*_{3/2^+}$ will be narrow (less than 50 KeV). The same applies to the lowest $1/2^-$ and $3/2^-$ states. The other possible decays, when allowed by quantum numbers, are $(ccc)^* \to ccc +\gamma$ and $(ccc)^* + \pi_0$ which is isospin violating. The strong decay $(ccc)^* \to (c\bar{d}) + (ccd)$ occurs only for excitations above $2M_D - J/\psi = 640$ MeV, according to J.-M. Richard.

143

In conclusion the ccc spectrum is very interesting. Decay widths are very small. The observation of the levels would constitute an excellent test of non-relativistic calculations and of the rule (10).

I would like to remind to those who think that it is unrealistic to believe that the production and study of ccc states are possible that, in the particle physics of our times, an experiment impossible at time T is possible at time T+10 (in years!) Who would have thought twenty years ago that antiprotons could be produced and accumulated so abundantly?

REFERENCES

1. A. Martin, in "The Search for Charm Beauty and Truth at High Energy", G. Bellini and S.C.C. Ting editors, Plenum Press, New York (1984), p. 501.

2. A. Martin, Nucl. Phys. B (Proc. Suppl.) 1B (1988) 133, North Holland, Amsterdam.

3. A. Martin, Phys. Lett. 100B (1981) 511.

4. Review of Particle Properties, Particle Data Group, Phys. Lett. 170B (1988) 1.

5. J.D. Bjorken, Hadron Spectroscopy 1985, S. Oneda editor, American Institute of Physics, New York (1985), p. 390.

6. S. Nussinov, Phys. Rev. Lett. 5 (1983) 2081.

7. J.-M. Richard, Phys. Lett. 100B (1981) 515;
   J.-M. Richard, Phys. Lett. 139B (1981) 408.

8. K.-T. Chao, N. Isgur and G. Karl, Phys. Rev. D23 (1981) 155.

NOTE

For the content of the paper on "Order and Spacing of Energy Levels" by André Martin, the reader is referred to the review by A. Martin in Recent Developments in Mathematical Physics, Springer Verlag, Berlin-Heidelberg-New York (1987), p. 53, and to A. Martin, J.-M. Richard and P. Taxil, CERN preprint TH.5273/89, submitted to Nuclear Physics B.

10

MEASURING THE CHIRALITIES OF WEAK b → c(u) TRANSITIONS IN EXCLUSIVE

B MESON DECAYS

Jürgen G. Körner
Institut für Physik
Johannes Gutenberg-Universität
Staudinger Weg 7, Postfach 3980
D-6500 Mainz, West Germany

ABSTRACT

I discuss possibilities to measure the chirality of the weak b → c(u) transitions in exclusive bottom meson decays in (i) the exclusive semileptonic decay process B → D* + l + ν_l and (ii) in the exclusive nonleptonic decays of bottom mesons to baryon-antibaryon pairs.

1. INTRODUCTION

In the standard model of electroweak interactions the weak b → c and b → u transitions are left-chiral, i.e. the c(u)-quark leaves the weak interaction vertex with predominant negative helicity. It is then an interesting and important question to ask whether and how this helicity information is transmitted to the final hadron or hadrons into which the c(u)-quark hadronizes. It is clear that the exclusive one-hadron decays are best suited for such an analysis since the helicity information is least degraded in the exclusive one-hadron hadronization.

In this talk I discuss some aspects of this interesting issue in the context of two different varieties of exclusive b → c(u) bottom meson decays. In Sec.2 I study the exclusive semileptonic (s.l.) bottom meson decays B → D* + l + ν_l where information on the helicity of the c-quark leaving the weak interaction vertex is handed on to the vector meson D*. This then leads to characteristic polar and azimuthal lepton-hadron correllations in the s.l. cascade decay B → D* (→Dπ) + l + ν_l which I will discuss. In Sec.3 I turn to the exclusive nonleptonic (n.l.) bottom meson decays into baryon-antibaryon pairs where the left-chirality of the weak b → c(u) transition leads to a number of interesting sum and selection rules.

2. EXCLUSIVE SEMI-LEPTONIC B → D* + 1 + ν_l DECAY

I begin by discussing the experimental situation. Approximately 2/3 of the total s.l. decay rate of bottom mesons is into the exclusive D* mode. The s.l. branching ratios are $\simeq$ 10% for the e and μ modes and considerably reduced for the τ-mode [1]. With a production rate of 10^6 - 10^8 bottom mesons per year expected at present and future colliders this leaves a healthy sample of s.l. B → D* + 1 + ν_l decays which would allow one to thoroughly analyze the type of lepton-hadron angular correlations that I will be discussing in the following.

Let me start the discussion by defining invariant form factors for the s.l. B → D* transitions. One has

$$\langle D^*(p_2)|A_\mu + V_\mu|B(p_1)\rangle = \varepsilon_2^{*\alpha} T_{\mu\alpha} \tag{1}$$

where

$$T_{\mu\alpha} = F_1^A g_{\mu\alpha} + F_2^A p_{1\mu} p_{1\alpha} + i F^V \varepsilon_{\mu\alpha\rho\sigma} p_1^\rho p_2^\alpha . \tag{2}$$

In the following I shall always work in the zero-lepton-mass limit. Thus I have dropped invariants multiplying $q_\mu = (p_1 - p_2)_\mu$ in (2). The general non-zero lepton mass case is discussed in [1].

In order to fix the $q^2 = 0$ values of the form factors I match the spin properties of the B → D* transitions to the free quark decay b → c transitions. The assumption is that the spectator quark is spin-inert. It neither affects the spin properties of the active quarks in the current-induced heavy quark to light quark transitions nor is its own spin flipped. To this end I first calculate the free quark decay (FQD) helicity amplitudes h^{FQD} in the quark (m_1) rest system with the light quark (m_2) moving along z. One finds (helicity label is that of the current)

$$h_0^{FQD} = \langle c\downarrow|J_0|b\downarrow\rangle = (\sqrt{Q_+} + \sqrt{Q_-})\frac{p+q_0}{\sqrt{q^2}}$$

$$h'_0^{FQD} = \langle c\uparrow|J_0|b\uparrow\rangle = - (\sqrt{Q_+} - \sqrt{Q_-})\frac{\sqrt{q^2}}{p+q_0} \tag{3}$$

$$h_{-(+)}^{FQD} = \langle c\downarrow(\uparrow)|J_{-(+)}|b\uparrow(\downarrow)\rangle = +(-)\sqrt{2}(\sqrt{Q_+} +(-)\sqrt{Q_-})$$

where $Q_\pm = (m_1 \pm m_2)^2 - q^2$, p is the c.m. momentum $2m_1 p = \sqrt{Q_+ Q_-}$ and q_0 is the energy of the virtual W in the c.m. system $q_0 = (m_1^2 - m_2^2 + q^2)/2m_1$.

For the matching procedure I also need the helicity form factors of the mesonic transitions. They are (helicity label is that of the current)

$$H_0^{D*} = \frac{1}{2M_2\sqrt{q^2}} \left((M_1^2 - M_2^2 - q^2) \, F_1^A + 2M_1^2 p^2 F_2^A \right) \tag{4}$$

$$H_\pm^{D*} = - F_1^A \mp M_1 p F^V \quad .$$

I use the minimum momentum transfer point $q^2 = 0$ to do the matching. For $q^2 = 0$ the FQD amplitude $h_0'^{FQD}$ vanishes. Thus one has

$$H_0 \quad = \langle D^{*+}(\text{long.} | J_0 | \overline{B^0} \rangle \simeq - \frac{1}{2} I \langle c\!\downarrow | J_0 | b\!\downarrow \rangle,$$

$$\tag{5}$$

$$H_{-(+)} = \langle D^{*+}\!\downarrow(\uparrow) | J_{-(+)} | \overline{B^0} \rangle \simeq +(-) \frac{1}{\sqrt{2}} I \langle c\!\downarrow(\uparrow) | J_{-(+)} | b\!\uparrow(\downarrow) \rangle ,$$

where the factors $\frac{1}{2}$ $(\pm \frac{1}{\sqrt{2}})$ are spin projection factors for the triplet and singlet spin wave functions of the $(b\bar{q}_{spectator})$ and $(c\bar{q}_{spectator})$ meson bound states. I is to be interpreted as the $B \to D^*$ wave function overlap.

Then by identifying $m_b = M_B$ $(= M_1)$ and $m_c = M_{D^*}$ $(= M_2)$ one obtains

$$F_1^A{}_{(0)} = (M_1 + M_2) \cdot I,$$

$$\tag{6}$$

$$- F_2^A{}_{(0)} = F^V{}_{(0)} = 2/(M_1 + M_2) \cdot I$$

from comparing (4) and (5) at $q^2 = 0$. Equivalently, in terms of the helicity amplitudes one has

$$H_0(0) = \frac{1}{\sqrt{q^2}} (M_1^2 - M_2^2) \, I$$

$$H_+(0) = 2M_2 \, I \tag{7}$$

$$H_-(0) = 2M_1 \, I \quad .$$

In order to be definite I take $I = 0.7$ for the wave function overlap mismatch factor I as e.g. estimated in [3] for the $b \to c$ transitions.

The matching solution Eq.(7) clearly shows the dominance of the negative helicity configuration, i.e. $H_-(0) > H_+(0)$. In fact one can show that $H_-(q^2) \geq H_+(q^2)$ for all values of q^2 by using the q^2-continuation procedure as described in the following. The helicity information of the free quark decay $(|h_-^{FQD}| \geq |h_+^{FQD}|)$ has thus been passed on to the particle decay.

The q^2-dependence of the form factors is fixed by nearest meson-dominance in the appropriate current channel with monopole behaviour (q^{-2})

for F_1^A and dipole behaviour (q^{-4}) for F_2^A and F^V according to the power-counting rules of QCD [5]. For the sake of simplicity I work only with one effective meson ($b\bar{c}$) current mass, for which I take B_c^* (6.34 GeV). The spacing among the various ($b\bar{c}$) bound state levels is presumably so small that one effective mass value is sufficient to set the scale of the q^2-dependence in the range $0 \leq q^2 \leq (M_1 - M_2)^2$.

The q^2-dependence of our form factors is thus given by

$$F(q^2) = F(0) \left(\frac{m_{FF}^2}{m_{FF}^2 - q^2} \right)^n \tag{8}$$

where $n = 1$ for F_1^A, $n = 2$ for F_2^A and F^V and $m_{FF} = 6.34$ GeV.

The matching solutions (6) and the power-behaved form factors (8) completely specify our model of $B \to D^*$ semileptonic decays.* This model was developed in [2] and [4]. For brevity's sake I shall refer to this model as the KS (Körner-Schuler) model in the following.

Let me now turn to the s.l. decay distribution. I shall consider the full $(1 \to 4)$-body s.l. cascade decay distribution $B \to D^* (\to D\pi) + l^- + \bar{\nu}_l$. The $D^* \to D\pi$ decay plane establishes an azimuthal reference plane which is quite useful since it increases the analyzing power of the s.l. decay process. The four-fold decay distribution reads

$$\frac{d\Gamma(B \to D^*(\to D\pi) + l^- + \bar{\nu}_l)}{dq^2 \, d\cos\Theta d\chi \, d\cos\Theta^*} = \frac{G^2}{(2\pi)^4} |V_{bc}|^2 \frac{p}{8M_1^2} L_{\mu\nu} H^{\mu\nu}$$
$$\cdot B(D^* \to D\pi) \tag{9}$$

where Θ is the polar angle of the lepton measured w.r.t. the D^*-direction in the $(l^- - \bar{\nu}_l)$ CM system (see Fig.1). χ is the azimuthal angle between the two decay planes spanned by $(D\pi)$ and $(l^- - \bar{\nu}_l)$ (see Fig.1), and Θ^* is the polar angle of the D relative to the D^* in the D^* rest frame. $B(D^* \to D\pi)$ is the branching ratio $\Gamma_{D^* \to D\pi}/\Gamma_{D^* \to all}$. G is the Fermi coupling constant $G \approx 1.02 \, m_p^{-2} \times 10^{-5}$, and V_{bc} is the $b \to c$ KM matrix element.

---

* Similar $q^2 = 0$ form factor values have been derived in [3] from an infinite momentum frame analysis. However, the authors of Ref.[3] use monopole type form factors also for the higher momentum form factors F_2^A and F^V in disagreement with the power counting rules.

148

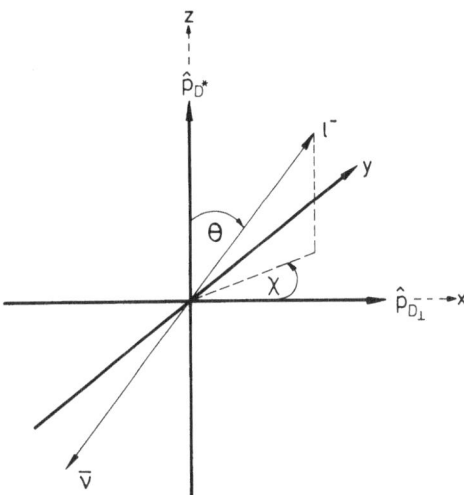

Fig. 1. Definition of the polar and azimuthal angles Θ and χ of the lepton l^- in the $(l^- - \bar{\nu}_1)$ CM frame. z-axis is along $\vec{P}_{D*}$ and x-axis in the $(\vec{p}_{D*}, \vec{p}_D)$ plane with $p_{Dx} \geq 0$.

In our normalization the lepton tensor $L_{\mu\nu}$ is given by

$$
L_{\mu\nu} = \frac{1}{8} \, \mathrm{Tr} \, \rlap{\,/}{l} \, \gamma_\mu (1 - \gamma_5) \, \rlap{\,/}{l}' \, \gamma_\nu (1 - \gamma_5)
$$
$$
= (l_\mu l_{\nu'} + l_\nu l'_\mu - \frac{q^2}{2} g_{\mu\nu} + i \, \epsilon_{\mu\nu\alpha\beta} \, l^\alpha l'^\beta) \tag{10}
$$

The hadron tensor $H_{\mu\nu}$ is related to the current-induced $B \to D^*$ hadronic matrix elements (2) via

$$
H_{\mu\nu} = Z^{\alpha'\beta'} T_{\mu\alpha'} T^*_{\nu\beta'} \tag{11}
$$

where $Z_{\alpha\beta}$ is the decay tensor describing the decay $D^* \to D\pi$ and is given by

$$
Z_{\alpha\beta} = \frac{3}{2} \, \frac{M_2^2}{(p_2 p_D)^2 - M_2^2 M_D^2} \, (-p_{D\alpha} + \frac{p_2 p_D}{M_2^2} p_{2\alpha})(-p_{D\beta} + \frac{p_2 p_D}{M_2^2} p_{2\beta}) \tag{12}
$$

The angular dependence of $L_{\mu\nu} H^{\mu\nu}$ can be made manifest by writing $L_{\mu\nu} H^{\mu\nu}$ in the helicity basis. One has [1]

$$
\begin{aligned}
L_{\mu\nu} H^{\mu\nu} = \frac{2}{3} q^2 \{ & \frac{3}{8} (1 + \cos^2\Theta) \, \frac{3}{4} \sin^2\Theta^* \, \hat{H}_U \\
& + \frac{3}{4} \sin^2\Theta \, \frac{3}{2} \cos^2\Theta^* \, \hat{H}_L \\
& - \frac{3}{4} \sin^2\Theta \, \cos 2\chi \, \frac{3}{4} \sin^2\Theta^* \, \hat{H}_T \\
& - \frac{9}{16} \sin 2\Theta \, \cos\chi \, \sin 2\Theta^* \, \hat{H}_I \\
& + \frac{3}{4} \cos\Theta \, \frac{3}{4} \sin^2\Theta^* \, \hat{H}_P \\
& - \frac{9}{8} \sin\Theta \, \cos\chi \, \sin 2\Theta^* \, \hat{H}_A \} ,
\end{aligned} \tag{13}
$$

where the reduced hadron tensor components $\hat{H}_i$ (i = U,L,T,I,P,A) are bilinear expressions of the three helicity amplitudes H_+, H_- and H_0 of Eq.(4).
They are given by

$$\hat{H}_U = |H_+|^2 + |H_-|^2 \qquad\qquad \text{unpolarized transverse}$$

$$\hat{H}_L = |H_0|^2 \qquad\qquad \text{longitudinal}$$

$$\hat{H}_T = \text{Re}\,(H_+ H_-^*) \qquad\qquad \text{transverse interference}$$

$$\hat{H}_I = \tfrac{1}{2}\,\text{Re}\,(H_+ H_0^* + H_- H_0^*) \qquad \text{transverse longitudinal interference}$$

$$\hat{H}_P = |H_+|^2 - |H_-|^2 \qquad\qquad \text{parity-odd}$$

$$\hat{H}_A = \tfrac{1}{2}\,\text{Re}\,(H_+ H_0^* - H_- H_0^*) \qquad \text{parity asymmetric}$$

(14)

The labelling of the reduced hadron tensor components $\hat{H}_i$ in terms of the
polarization components of the gauge boson $W^-_{\text{off-shell}}$ follows the conventional notation of one-gauge-boson exchange physics. Note that I have dropped angular terms in Eq.(13) that are multiplied by $\text{Im}\,(H_i H_j^*)$, $i \neq j$, assuming that the three helicity amplitudes are relatively real.

It is convenient to define partial helicity rates $d\Gamma_i/dq^2$ in terms
of the reduced hadron tensor components $\hat{H}_i$ via

$$\frac{d\Gamma_i}{dq^2} = \frac{G^2}{(2\pi)^4}\,|V_{bc}|^2\,\frac{pq^2}{12M_1^2}\,\hat{H}_i \qquad\qquad (15)$$

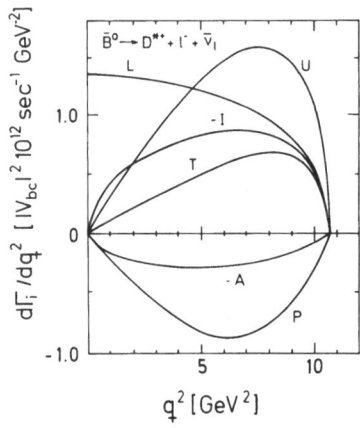

Fig. 2. Helicity rates $d\Gamma_i/dq^2$ for $W^-_{\text{off-shell}}$ polarization components
i = U,L,T,I,P,A in the KS model [2].

Fig.2 shows the q^2-dependence of the helicity rate functions as calculated in the KS model [2]. The unpolarized transverse rate function
$d\Gamma_U/dq^2$ dominates for $q^2 > 4.5$ GeV^2 but is suppressed towards smaller q^2-
values due to the spin-kinematical q^2-factor (see Eq.(4)) where the longi-

tudinal contribution dominates. The interference rates $d\Gamma_i/dq^2$ (i=T,I,P,A) are smaller than $d\Gamma_U/dq^2$ and $d\Gamma_L/dq^2$ over most of the available q^2-range but are nevertheless non-negligible. For the integrated total helicity rates Γ_i one obtains [1]

$$(\Gamma_U, \Gamma_L, \Gamma_T, \Gamma_I, \Gamma_P, \Gamma_A) =$$
$$(12.7, 13.1, 5.3, 8.2, -6.9, -2.6) \cdot |V_{bc}|^2 \; 10^{12} \; \text{sec}^{-1} . \tag{16}$$

In this talk I will mainly concentrate on the two p.v. rate functions $d\Gamma_P/dq^2$ and $d\Gamma_A/dq^2$ which are antisymmetric under the exchange $H_+ \leftrightarrow H_-$ and are thus sensitive to the chirality of the underlying weak $b \to c$ transition. An inspection of the angular decay distribution (13) shows that their contributions can be projected out by defining the following asymmetry ratios [1]:

$$P : \quad A_{FB} = \frac{d\Gamma(\Theta) - d\Gamma(\pi-\Theta)}{d\Gamma(\Theta) + d\Gamma(\pi-\Theta)} \tag{17a}$$
$$\pi/2 \le \Theta \le \pi$$

$$A : \quad A_A = \frac{d\Gamma(\Theta^*,\chi)-d\Gamma(\Theta^*,\pi+\chi)-d\Gamma(\pi-\Theta^*,\chi)+d\Gamma(\pi-\Theta^*,\pi+\chi)}{d\Gamma(\Theta^*,\chi)+d\Gamma(\Theta^*,\pi+\chi)+d\Gamma(\pi-\Theta^*,\chi)+d\Gamma(\pi-\Theta^*,\pi+\chi)}$$

$$0 \le \Theta^* \le \pi/2 \tag{17b}$$
$$-\pi/2 \le \chi \le \pi/2 .$$

I have used a notation in Eq.(17) such that the angles that do not appear in the arguments of the differential rate $d\Gamma$ in (17) have been integrated out over their physical ranges ($0 \le \Theta, \Theta^* \le \pi, 0 \le \chi \le 2\pi$). Integrating over the remaining variables (numerator and denominator separately!) one finally obtains the following values for the asymmetry ratios in the KS model [1]

$$A_{FB} = -\frac{3}{4} \Gamma_P/\Gamma_{U+L} = 0.20 \tag{18}$$
$$A_A = -3 \Gamma_A/\Gamma_{U+L} = 0.15 \tag{19}$$

Present e^+e^--experiments running on the $\Upsilon(4S)$ produce bottom mesons which are practically at rest. Since leptons can only be detected and measured above a certain threshold energy E_1^{cut} in these experiments (which is typically 0.5 GeV for electrons and 1.0 GeV for muons) this excludes the extreme backward region $\cos\Theta \to 1$. The experimentally accessible angular range is then

$$-1 \le \cos\Theta \le \text{Min} (\cos\Theta (q^2, E_1^{cut}); 1) \tag{20}$$

where

$$\cos\Theta (q^2, E_1^{cut}) = -\frac{4M_1 E_1^{cut}-M_1^2-q^2+M_2^2}{2M_1 p} \tag{21}$$

151

It is interesting to consider asymmetries subject to the experimental constraint (20). One needs to, however, symmetrize the angular range (20) for the asymmetry definitions, Eq.(17). I shall accordingly define the symmetries A_{FB} and A_A also in the symmetrized restricted angular range

$$- \text{Min} (\cos\Theta \ (q^2, E_1^{cut});1) \leq \cos\Theta \leq \text{Min} (\cos\Theta \ (q^2, E_1^{cut});1) \quad (22)$$

This is illustrated in Fig.3 where I show a plot of the double decay distribution $d\Gamma_p/dq^2 \ d\cos\Theta$ as a function of q^2 and $\cos\Theta$. The forward-backward asymmetry is clearly evident in Fig.3a. Fig.3c and Fig.3d show the same rate distribution but within the restricted symmetric Dalitz plot domain according to Eq.(22). The asymmetric domain Eq.(20) is also indicated in the plots by blacking out the relevant backward phase space region.

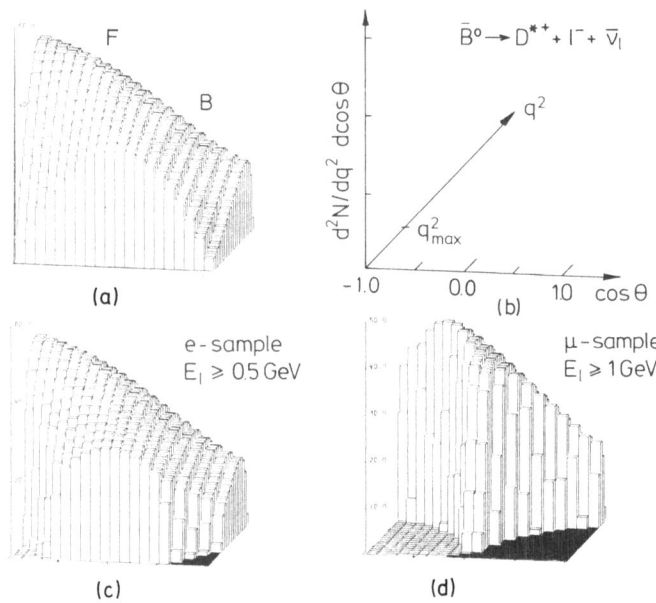

Fig. 3. Unnormalized double differential rate $d\Gamma/dq^2 \cos\Theta$ for $\overline{B}^0 \to D^{*+} +$
    $1^- + \overline{\nu}_1$ in the KS model [2]. (a) complete phase space, (b) coordinate scales, (c) and (d) symmetrized restricted phase space domains for (c) e-sample E_1^{cut} = 0.5 GeV, (d) μ-sample E_1^{cut} = 1.0 GeV.
    F: forward hemisphere $-1 \leq \cos\Theta \leq 0$, B: backward hemisphere
    $0 \leq \cos\Theta \leq 1$.

In Fig.4 we show a plot of the two asymmetry ratios A_{FB} and A_A as a function of E_1^{cut}. E_1^{cut} defines the symmetrized restricted phase domain (22) through Eq.(21). The asymmetries rise from zero at $E_1^{cut}=(M_1-M_2)/2$ = 1.63 GeV (zero phase space) to their largest values A_{FB} = 0.20 and A_A = 0.15 (see Eqs.(18,19)) at E_1^{cut} = 0 when the whole phase space domain is accessible for the asymmetry measurement.

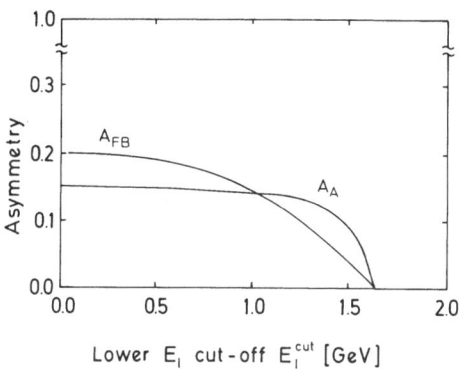

Fig. 4. Forward-backward asymmetry A_{FB} and asymmetry A_A as a function of
lower energy cut E_1^{cut} in KS model [2].

The forward-backward asymmetry A_{FB} drops to 0.19 and 0.15 from its
maximal value 0.20 at the two cut values E_1^{cut} = 0.5 GeV and E_1^{cut} =
1.0 GeV for the e- and μ-samples, respectively. The asymmetry A_A remains
practically flat at its maximal value of - 0.15 up to cut values of E_1^{cut} =
1.1 GeV which includes the two above E_1^{cut} values relevant for the e- and
μ-samples.

Taking the KS model as a measure, only = 5% and = 40%, respectively,
of the total s.l. rate $B(b) \to D^*(c) + 1^- + \bar{\nu}_1$ have been forfeited at these
two cut values. The two suggested asymmetry measurements with their expected asymmetry values in excess of 0.15 for realistic experimental cuts should
be experimentally feasible in the near future.

A measurement of the sign of the asymmetries alone would be a significant experimental achievement as it would allow one to check on the chirality of the underlying b → c transition irrespective of the details of
the underlying quark model that is used to describe the exclusive hadronization phase. Measuring in addition the magnitude of the asymmetries
would provide a probe of the details of the bound state quark dynamics involved in the description of the current-induced B → D* transition.

Naturally one could also attempt to measure the signs of the p.v. rate
functions $d\Gamma_P/dq^2$ and $d\Gamma_A/dq^2$ in the rest frame of the decaying bottom meson. It is well-known that e.g. the energy spectrum of the lepton 1^- is
shifted towards the forward direction for a left-chiral b → c transition
as compared to a right-chiral b → c transition in this frame. This is quite
evident in Fig.5 where I have plotted the KS model predictions for the lepton's energy spectrum for left-chiral (V-A) and right-chiral (V+A) b → c
transitions in the s.l. B(b) → D*(c) decay. However, the shape of the lepton spectrum depends on the details of the hadron dynamics that is being

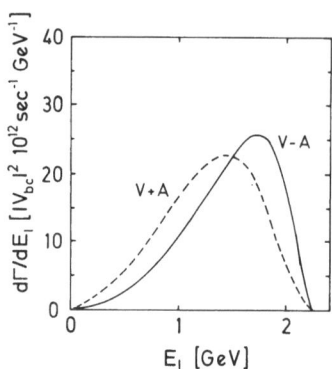

Fig. 5. Lepton energy spectrum in the s.l. decay $\overline{B^0} \to D^{*+} + 1^- + \overline{\nu}_1$ in KS
model [2] for left-chiral (V-A; full line) and right-chiral (V+A;
dashed line) b → c transition, in $\overline{B^0}$ rest frame.

used to describe the current-induced hadronic B(b) → D*(c) transition. This
will complicate the experimental analysis of the sign of $d\Gamma_P/dq^2$ in such
a measurement. In this sense the $(1^- - \overline{\nu}_1)$ CM frame and the asymmetry observables A_{FB} and A_A defined in this frame provide an optimal framework to
analyze the sign of the p.v. contributions $d\Gamma_P/dq^2$ and $d\Gamma_A/dq^2$.

I would like to conclude this section by remarking that a similar analysis of the exclusive s.l. decays of bottom mesons into the light vector
mesons ω and ρ, and the charmed mesons D into the mesons K* (ρ,ω) would
allow one to similarly conclude for the chirality of the fundamental b → u
and c → s (c → d) current-induced quark transitions [1].

3. EXCLUSIVE NON-LEPTONIC DECAYS OF BOTTOM MESONS INTO BARYON-ANTIBARYON
PAIRS

Recently, the ARGUS collaboration has presented evidence for the ex-
clusive b → u modes $B^+ \to p\overline{p}\pi^+$ and $\overline{B^0} \to p\overline{p}\pi^+\pi^-$ [6]. They are quoted to occur with branching ratios of $(5.2 \pm 1.4 \pm 1.9) \times 10^{-4}$ and $(6.0 \pm 2.0 \pm 2.2) \times$
10^{-4}, respectively [6].* The (pπ) invariant masses in the latter decays
lie in the Δ(1236) mass region indicating that the b → u decay of bottom
mesons into ground state baryon-antibaryons could occur with a branching
ratio of $O(10^{-4})$.

A rough estimate of the baryon matrix element involved in the above
b → u *baryonic decays points* to a ratio of $|V_{bu}/V_{bc}| \simeq 0.3$ [8]. This would

---

* The ARGUS rates have been contested by recent findings of the CLEO col-
   laboration who quote $B_{B^- \to p\overline{p}\pi^-} < 1.4 \cdot 10^{-4}$ and $B_{\overline{B^0} \to p\overline{p}\pi^+\pi^-} < 2.9 \cdot 10^{-4}$,
   [7].

imply that the b → c decays of bottom mesons into ground state baryon-anti-baryon pairs could occur with a branching ratio of $O(10^{-3})$.*

Given that exclusive baryon decays occur at rates of $O(10^{-4})$ and $O(10^{-3})$ for b → u and b → c transitions, respectively, this opens up the exciting prospect to experimentally investigate the details of the weak structure exhibited in these nonleptonic decay processes, at present and next gene-ration bottom meson facilities.

In this talk I will concentrate on **relative** baryon-antibaryon rates. In particular, I will show how the left-chirality of the weak b → c and b → u transitions lead to a number of sum and selection rules among the various baryonic B meson decay modes [10].

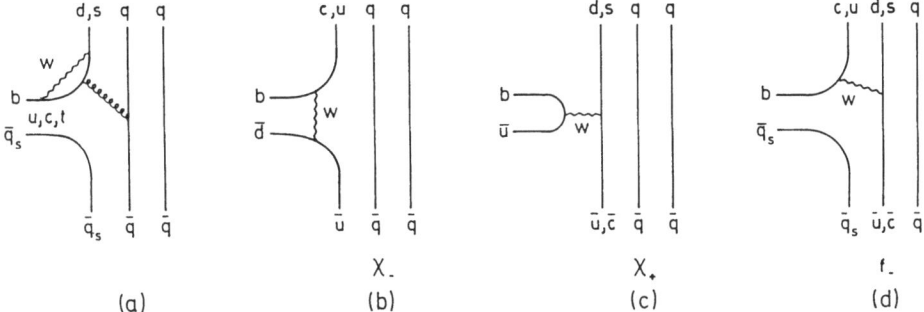

Fig. 6. Charged current contributions to baryonic bottom meson decays,
(a) penguin contributions, (b) and (c) W-exchange contribution,
(d) W-decay contribution. Spectator quark is denoted by q_s.

In Fig.6 we have drawn the various flavour diagrams that contribute to the baryon decays of bottom mesons. These involve the penguin contri-butions (Fig.6a), W-exchange contributions (Figs.6b and 6c) and the W-de-cay contributions (Fig.6d). Cabibbo suppressed transitions involving u → s and c → d transitions will be disregarded. Note that the penguin contri-butions (Fig.6a) have been calculated to be strongly suppressed [9].

---

* A more refined theoretical calculation now predicts the baryon-antiba-ryon rates to occur with branching ratios of $O((0.3 - 3) \cdot 10^{-3})$ and $O((0.5 - 2) \cdot 10^{-4} \ (V_{bu}/V_{bc})^2)$ for b → c and b → u transitions, respecti-vely [9]. Rough estimates in [10] point to similarly small branching ra-tios. We shall return to this point at the end of this section.

Let us begin by writing down the standard model effective weak Hamiltonian that induces the basic weak nonleptonic (n.l.) quark decay processes $b \rightarrow c + d + \bar{u}$, $b \rightarrow c + s + \bar{c}$, $b \rightarrow u + d + \bar{u}$ and $b \rightarrow u + s + \bar{c}$. One has (see e.g. [4])

$$H_{n.l.}^{eff} = \frac{G}{\sqrt{2}} V_{bc} \frac{1}{2} \{ \frac{1}{2} (f_+ + f_-)(V_{du}(\bar{c}b)_L (\bar{d}u)_L + V_{sc}(\bar{c}b)_L (\bar{s}c)_L)$$
$$+ \frac{1}{2} (f_+ - f_-)(V_{du}(\bar{c}u)_L (\bar{d}b)_L + V_{sc}(\bar{c}c)_L (\bar{c}b)_L) \} \qquad (23)$$
$$+ \frac{G}{\sqrt{2}} V_{bu} \{ \bar{c} \rightarrow \bar{u} \} .$$

The bracket notation $(\bar{q}_a q_b)_L$ is the usual short hand notation for the left-chiral colour singlet combination of (mass eigenstate) quark fields with flavours a and b and V_{ab} are the corresponding Kobayashi-Maskawa matrix elements. In particular, note that the weak $b \rightarrow c$ and $b \rightarrow u$ currents entering the Hamiltonian Eq.(23) are in their left-chiral form as predicted by the standard model. The $f_\pm$ are the renormalized current-current coefficients. For the b-sector one estimates [4,11].

$$f_+ = 0.85$$
$$f_- = 1.4 \qquad (24)$$

For the W-exchange contributions Fig.6b and Fig.6c the relevant colour-flavour factors resulting from (23) are (see e.g. [4])

$$\chi_\pm = \frac{1}{2} (f_+ \pm f_-) + \frac{1}{N_c} \frac{1}{2} (f_- \mp f_-) \qquad (25)$$

where N_c is the number of colours ($N_c = 3$ in QCD) and χ_+ and χ_- are associated with the "colour enhanced" B^- and the "colour suppressed" $\bar{B}^0$ W-exchange decays as indicated in Fig.6. However, by using dipole type form factors for the baryons one can show that the W-exchange contributions are negligibly small [10]. The W-exchange contributions will be neglected in the following. We shall assume that the n.l. baryon-antibaryon decays of bottom mesons are dominated by the W-decay contributions Fig.6d.

Concerning the W-decay contributions in Fig.6d one finds that only the antisymmetric flavour combination in (23) (proportional to f_-) contributes to these according to the Körner-Pati-Woo (KPW) theorem [12,13].

The KPW theorem is a simple consequence of the fact that a local product of left-chiral currents with (V-A)(V-A) structure is invariant under Fierz transformations.* Since the quarks in baryons are colour antisymmetrized one then remains only with the flavour-antisymmetric piece proportional to f_- in the Hamiltonian (23).

---

* In the following we shall always assume that the Cabibbo currents $(\bar{d}u)$ and $(\bar{s}c)$ are left-chiral.

Another consequence of the KPW theorem is that the (ud) diquark in the W-decay $b \to$ (ud) $+ \bar{u}$ is flavour and spin antisymmetric, i.e. the (ud) has isospin and spin 0. This then immediately leads to the $\Delta I = 1/2$ sum rules among the amplitudes [10]

$$A_{\overline{B^0} \to p\bar{p}} = A_{\overline{B^0} \to n\bar{n}} - A_{B^- \to n\bar{p}} \tag{26a}$$

$$\sqrt{2}\, A_{\overline{B^0} \to \Lambda \overline{\Sigma^0}} = A_{B^- \to \Lambda \overline{\Sigma^+}} \tag{26b}$$

Further one has the selection rules [10]

$$\overline{B^0},\ B^- \not\to \Sigma^0 + X \tag{27a}$$

$$\overline{B^0},\ B^- \not\to 3/2^+ + X \tag{27b}$$

where the $3/2^+$ is a decuplet ground state baryon.

If the $b \to u$ transitions were right-chiral, one then would have a $(V+A)(V-A)$ current x current structure. The (ud) diquark would then be in an isospin and spin 1 state. In such a case the decays (27) would be allowed, but the decays (26b) would be forbidden. In the case that the sum and selection rules (26,27) are experimentally verified, one could then directly draw a conclusion on the left-chiral nature of the underlying $b \to u$ transitions.

The situation for the decays $b \to$ (us) $+ \bar{c}$, $b \to$ (cd)$+\bar{u}$ and $b \to$(cs)$+\bar{c}$ is more involved since one also has to take flavour symmetry breaking effects into account. At the level of the basic decay process $b \to (q_1 q_2)+ \bar{q}$ one notes that the matrix elements $b \to (q_1\uparrow q_2\downarrow) + \bar{q}$ and $b \to (q_1\downarrow q_2\uparrow)+ \bar{q}$ acquire different weights if $m_{q_1} \neq m_{q_2}$ due to chirality suppression effects [10]. However, this flavour symmetry breaking effect is mostly compensated for by the explicit flavour symmetry breaking effects in the baryon wave function where the ratio of longitudinal momentum fractions of the quarks q_1 and q_2 is given by $= m_{q_1}/m_{q_2}$ [10]. Thus the KPW theorem also applies in this more general flavour breaking situation, i.e. the light-heavy diquark $(q_1 q_2)$ in the n.l. decay process $b \to (q_1 q_2) + \bar{q}$ is again approximately flavour and spin asymmetric.

Again, this leads to a number of sum rules which are too numerous to be recorded here [10]. Also one has the selection rule

$$\overline{B^0},\ B^- \not\to 3/2^+(c) + X\ , \tag{28}$$

similar to the selection rule, Eq.(27b) where the $3/2^+$ is now a ground state spin $3/2^+$ charmed baryon. Again, if the selection rule (28) would be experimentally verified, one could then conclude for the left-chiral nature of the $b \to c$ transitions.

Before concluding this section we endeavour to attempt a very rough order of magnitude estimate of the absolute scale of the exclusive baryonic W-decay contributions. Let us remark that in a perturbative context one has the power law behaviour

$$B_{W\text{-decay}} \sim F_B^2 \, m_B^{-10} \qquad\qquad (29)$$

where F_B is a dimensional $[m^1]$ constant that characterizes the meson wave function. For a heavy-light quark meson as the B-meson the scaling law $F_B \sim m_B^{-1/2}$ was proposed in [14]. This would then imply an overall scaling law of $B_{W\text{-decay}} \sim m_B^{-11}$. Two inverse powers of m_B result from a helicity suppression factor. The source of this suppression is the helicity suppressed quark in the helicity zero diquark system. Compared to the scaling law $B_{W\text{-exchange}} \sim f_B^2 \, m_B^{-12}$ for the W-exchange contributions [10] one gains two powers of m_B from the fact that the perturbative gluon propagator that attaches to the light quark in the bottom meson provides only for a m_B^{-1} suppression [9].

From the experience one has gained from the analysis of the exclusive mesonic B-meson decays one knows that the W-exchange contributions are quite suppressed relative to the W-decay contributions [4]. It appears that the time-like off-shell propagators appearing in the W-decay contributions do not suppress exclusive decays as compared to the time-like propagators in the W-exchange contributions. This can be explicitly followed through in the factorization approach of Ref.[4]. The suppression of W-exchange is essentially given by the penalty of having to create a (time-like) $q\bar{q}$-pair. For the exclusive mesonic $b \to c$ and $b \to u$ decays the suppression is determined by power behaved time-like form factors which give rate suppression factors

$$F_W\text{-exchange} \approx \left(1 - \left(\frac{m_B}{2\text{GeV}}\right)^2\right)^{-2} \quad \text{and} \quad \approx \left(1 - \left(\frac{m_B}{1\text{GeV}}\right)^2\right)^{-2},$$

respectively, [4]. The W-decay contributions to the mesonic B-meson decays, on the other hand, are enhanced due to the squared time-like form factor factors

$$F_W\text{-decay} \approx \left(1 - \left(\frac{2\text{GeV}}{m_B}\right)^2\right)^{-2} \quad \text{and} \quad \approx \left(1 - \left(\frac{1\text{GeV}}{m_B}\right)^2\right)^{-2},$$

respectively. Thus, the W-decay rate contributions are enhanced relative to the W-exchange rate contributions by the factors $\approx (m_B/2\text{GeV})^4$ and $\approx (m_B/1\text{GeV})^4$, respectively.

For the baryonic B-meson decays W-decay again involves one ($q\bar{q}$-pair) creation less than W-exchange. Let us therefore assume that the ratio of W-decay and W-exchange contributions are the same for the baryonic and mesonic decays although there is no direct form factor interpretation in the baryonic case. In addition one has to consider the relative colour-flavour factors $(f_-/\chi_+)^2$ in this ratio. As mentioned above, one can attempt to estimate the W-exchange contributions to the baryon-antibaryon decays by using a dipole type form factor behaviour for the baryon current matrix elements. Using the W-exchange estimates of [10] one then estimates that the $b \to c$ and $b \to u$ W-decay branching ratios should be of the order $O(5 \cdot 10^{-4})$ and $O(10^{-4} (V_{bu}/V_{bc})^2)$, respectively. It is clear that the latter rate estimates are smaller than the more optimistic rate projections given at the beginning of this section. It could very well be that the above order of magnitude estimate is too simplistic. Only future experiments will be able to tell.

ACKNOWLEDGEMENTS

Part of this work was done while I was a visitor at the DESY theory group. I would like to thank R. Peccei for his hospitality and the DESY directorate for supporting my visit. The results on s.l. B-decays presented in Sec.2 have been obtained together with G.A. Schuler. I would like to acknowledge informative discussions with A. Ali and V.L. Chernyak.

REFERENCES

1. J.G. Körner and G.A. Schuler, Exclusive Semileptonic Bottom Meson Decays Including Lepton Mass Effects, Mainz preprint MZ-TH/88-14 (1988)
2. J.G. Körner and G.A. Schuler, Z.Phys. C38(1988)511
3. M. Wirbel, B. Stech and M. Bauer, Z.Phys. C29(1985)637
4. A. Ali, J.G. Körner, G. Kramer and J. Willrodt, Z.Phys. C1(1979)269
5. S.J. Brodsky and G.P. Lepage, Phys.Rev. D22(1980)2157
6. H. Albrecht et al. (ARGUS Collab.), Phys.Lett. B209(1988)119
7. C. Bebek et al. (CLEO Collab.), Phys.Rev.Lett. 62(1989)8
8. M. Shifman, Invited Talk given at "1987 International Symposium on Lepton and Photon Interactions at High Energies",
    Nucl.Phys. B (Proc.Suppl.) 3(1988)289
9. V.L. Chernyak and I.R. Zhitnitsky, Novosibirsk preprint 1988
10. J.G. Körner: Spin, Helicity and Flavour Patterns in Exclusive Decays of Bottom Mesons into Baryon-Antibaryon Pairs, preprint MZ-TH/88-01, Mainz (1988), to be published in Z.Physik C
11. J. Ellis, M.K. Gaillard, D.V. Nanopoulos and S. Rudaz,
    Nucl.Phys. B131(1977)285
12. J.G. Körner, Nucl.Phys. 25B(1970)282

13. J.C. Pati and C.H. Woo, Phys.Rev. D3(1971)2920

14. E. Shuryak, Nucl.Phys. B198(1982);
    M.A. Shifman and M.B. Voloshin, Sov.Nucl.Phys. 45(2)(1987)292;
    H.D. Politzer and M.B. Wise, Phys.Lett. 208B(1988)504

RARE B DECAYS: A WINDOW ON THE STANDARD MODEL AND BEYOND IT FOR THE 90's

A. Masiero

Istituto Nazionale di Fisica Nucleare - Sezione di Padova
Via Marzolo, 8 - 35131 Padova, Italy

Flavour changing neutral currents (FCNC) have played a major role in leading to the present formulation of the standard model (SM) of the electroweak interactions. In particular the Glashow-Iliopoulos-Maiani (GIM)[1] mechanism and the prediction of the charm quark find their origin in the requirement of suppressing FCNC effects in the kaon system. The FCNC test (in particular, in kaon physics) have helped SM also "indirectly" by severely constraining (if not ruling out at all) extensions of it.

How about the FCNC constraints which are imposed on B physics? We all know how important the $\Delta B = 2$ oscillations of the $B_d - B_{\bar{d}}$ system have been in this last year. Here I want to focus on other FCNC tests of B physics: the rare B decays where the $\Delta B = 1$ $b \to s$ transitions take place. I shall show that some of these processes are already on the verge of yielding constraints on SM and extensions of it of a relevance comparable to those which are provided by FCNC in kaon physics. For other rare charmless B decays with strange particles in the final state the comparison between theoretical expectations and experimental possibilities is not so rosy: in some cases, the only wayout seems to be the construction of dedicated machines (B factories).

The plan of my talk is the following. In the first part I'll examine how SM stands in front of the challenge of the FCNC tests in rare $b \to s+$ charmless states decays. In the second part I'll consider some extensions of SM. Much of the time will be devoted to spontaneously broken $N = 1$ supergravity theories whose low energy limits yields SM, but I shall not "forget" also three other classes, left-right symmetric, two Higgs doublets and fourth generation models on which new light can be shed by the rare B decays that I discuss here.

In SM the transition $b \to s+$ charmless particles can occur at the tree level (Fig. 1) or at the one-loop level (Fig. 2). In the example of fig. 1 the amplitude for $b \to u\bar{u}s$ is strongly Cabibbo suppressed ($\sim (sin\theta_c)^4$), so that it may be more convenient to pay the price of going to one-loop (Fig.2) but with a less severe Cabibbo suppression ($\sim (sin\theta_c)^2$). The diagrams of the kind depicted in fig. 2 are called <u>penguin diagrams</u> and they constitute the major source of the $\underline{\Delta B = 1}$ <u>b decay</u> that we consider here.

It is crucial to distinguish between <u>two "species" of penguins</u>: a) those which are "complete" with two legs (like that in Fig. 2) and b) those that are "truncated", namely the radiative decays where a real photon or a gluon with $q^2 = 0$ are emitted. These two species show two completely different GIM attitudes. This is best seen by looking at the general structure of the electromagnetic current matrix element.

$$< q_2(p - q)|j^\lambda (=)q_1(p) \equiv \bar{u}_2(p - q)T^\lambda u_1(p), \tag{1a}$$

where

$$T^\lambda = F_1(q^2)i\sigma^{\lambda\mu} q_\mu + F_2(q^2)(q^\lambda q^\mu - q^2 g^{\lambda\mu})\gamma_\mu \tag{1b}$$

For the decay in Fig. 2 the relevant contribution arises from the monopole from factor $F_2(q^2)$ in (1b); indeed, for this term, the GIM cancellation manifests itself with a typical $\ell n(m_q^2/m^2)$ <u>enhancing factor</u>, where q is the heaviest quark circulating in the loop (the top in our case) and m denotes the typical hadronic mass scale in the process under consideration (m_B in the case depicted in Fig. 2).

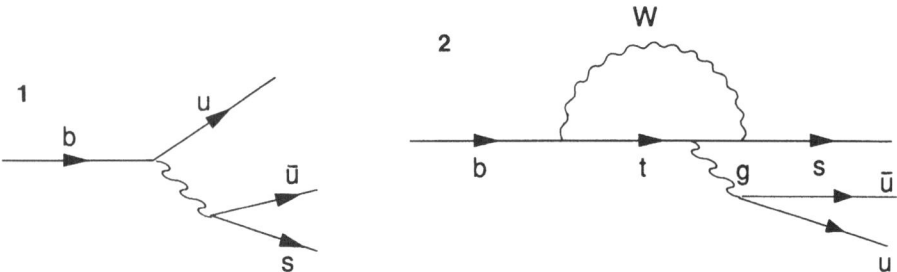

Figures 1 and 2. Tree- and one-loop level diagrams giving rise to b→s u ū, respectively.

On the contrary, if we cut the legs of the penguin in Fig. 2, i.e. we discuss the $b \rightarrow s + \gamma$ decay, <u>only</u> the magnetic transition form factor $F_1(q^2)$ in (1b) gives a nonvanishing contribution, since in this case a helicity flip in the $b \rightarrow s$ transition must be present. Due to the different "infrared" behaviour of $F_1(q^2)$ and $F_2(q^2)$, in this case no large $\ell n(m_q^2/m^2)$ factors appear in the GIM cancellation, but rather we are in the presence of the more classical GIM <u>"power" suppression</u>, $(m_q^2 - m_{q'}^2)/M_w^2$, where q and q' are two quarks running in the loop.

Summarizing, the penguin contributions to a $b \rightarrow s$ transition present <u>two quantitatively very different GIM factors</u>:

i) <u>"mild" GIM</u>, with enhancing factors $0(\ell n(m_q^2/m^2))$, for the decays $b \rightarrow sq\bar{q}$ or $b \rightarrow s\ell\bar{\ell}$, where the γ or the g are virtual (Fig. 2);

2) <u>"hard" GIM</u>, with suppressing factors $0(m_q^2/M_w^2)$, for the radiative decays $b \rightarrow s\gamma$ or $b \rightarrow sg$ with real γ or g at $q^2 = 0$.

The above remark on the two different kinds of GIM manifestation is crucial for the analysis that I present in my talk both in the SM context or in extensions of it, in particular in the SUSY case. Armed with the above general notions on FCNC in b physics we can now proceed to a more detailed study of the above two classes i) and ii) of $b \to s$ transitions in SM and SUSY.

RARE B DECAYS IN SM

I think that in SM the most interesting charmless b decay with strange particles in the final products is by far the radiative decay $b \to s + \gamma$. Here we witness a fascinating process of "metamorphosis". As I said in the previous section, "truncated" penguin diagrams, like the one responsible for $b \to s+\gamma$, belong to the "hard" GIM species. This is readily seen by computing the amplitude $b \to s + \gamma$ from the diagrams in Fig. 3:

$$A(b \to s + \gamma)_{SM} = [\epsilon_\mu i \bar{s} \sigma^{\mu\nu} q_\nu (m_b P_R + m_s P_L) b] \cdot$$
$$\cdot \frac{G_F}{2\sqrt{2}} \frac{e}{2\pi^2} \{ V_{ts}^* \, V_{tb} [F_2(x_t) - F_2(x_u)] + V_{cs}^* \, V_{cb} [F_2(x_c) - F_2(x_c) - F_2(x_u)] \} \quad (2)$$

where $x_j = m_j^2/m_w^2$, $j = u, c, t$, $P_{R,L} = (1 \pm \gamma_s)/2$ and the V_s' being elements of the Kobayashi-Maskawa (KM) matrix. The function $F_2(x)$ can be found in Inami and Lim [2], eq. (B3). The essential feature of (2) is the presence of the GIM mechanism [1] which leads to the cancellation of contributions of order $G_F e/\pi^2$, resulting in an extra power suppression of the type $(m_i^2 - m_u^2)/m_w^2$ from $F_2(x_i) - F_2(x_u) (i = c, t)$.

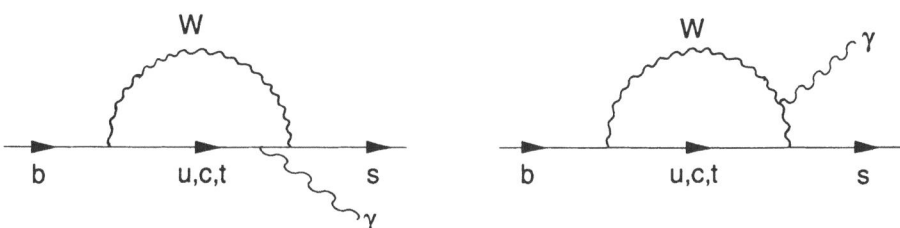

Figure 3. One-loop diagram for b→s+γ in SM.

Thus, as expected, the one-loop contribution to $b \to s + \gamma$ shows a honest to goodness "hard" GIM suppression. The surprise comes when we compute the <u>QCD corrections</u> to the operator $\bar{s} i \sigma^{\mu\nu} q_\nu (m_b P_R + m_s P_L) b$. Their expression in the leading-logarithmic approximation in first order in α_s can be immediately derived from the results of Kogan and Shifman [3]. In our paper of ref. 4] we explain why we are confident that taking the leading-logarithmic approximation at the first order in α_s, at least for $m_t < m_w$, can be a reliable procedure [5]. Neglecting the small charm contribution, the inclusion of these QCD corrections modifies (2) as follows [4]:

$$A(b \to s + \gamma)_{QCD\ corr} = \{\epsilon_\mu \bar{s} i \sigma^{\mu\nu} q_\nu (m_b P_R + m_s P_L) b\} \frac{G_F}{2\sqrt{2}} \frac{e}{2\pi^2} \cdot$$

$$\cdot V_{ts}^* V_{tb} [F_2(x_t) + \frac{4}{3} \frac{\alpha_s}{\pi} \ell n(m_t^2/m^2)], \tag{3}$$

where $m \sim 5\ GeV$ and we take $\alpha_s = 0.15$. Eq. (3) shows that the two-loop level entails a drastic change in the GIM cancellation. Instead of $F_2(x_t)$ leading to $0(m_t^2/m_w^2)$ we have now a "mild" GIM with the typical large log factor $\ell n(m_t^2/m^2)$. For any value of $m_t (m_t < m_w)$, the two-loop factor $\frac{4}{3} \frac{\alpha_s}{\pi} \ell n(m_t^2/m^2)$ dominates over the one-loop factor $F_2(x_t)$. In fig. 4 we show the resulting $BR(b \to s + \gamma)$ as a function of m_t. The solid line is computed using the QCD corrected value of the $A(b \to s + \gamma)$ as given by eq. (3), while the dashed line shows the result for $BR(b \to s + \gamma)$ when the QCD uncorrected amplitude of eq. (2) is used.

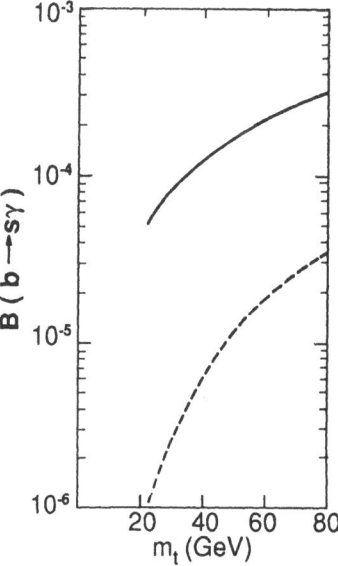

Figure 4. BR(b→s+γ) as a function m_t in SM. The solid (dashed) line denotes the QCD corrected (uncorrected) values.

The QCD corrections enhance $BR(b \to s\gamma)$ by almost one order of magnitude with respect to the one-loop contribution even for $m_t = 80 GeV$ (the enhancement encreases for decreasing values of m_t, as it is obvious comparing eq. (12) and (3)).

This effect of impressively large QCD corrections in the $b \to s + \gamma$ decay was first pointed out, independently, in the papers of ref. 4] and 6]. Both these works are based on the computations of QCD corrections to $s \to d + \gamma$ by the russian authors in ref. 3] and 5]. Recently, an entirely new computation based on a somewhat different approach has been performed by the authors of ref. 7]. Quantitatively their results essentially confirm what is presented here in Fig. 4.

Thus, taking into account QCD corrections, SM predicts $BR(b \to s + \gamma) \sim 0(10^{-4})$ (for instance, for $m_t = 80 GeV$ we obtain $BR(b \to s + \gamma) \simeq 3 \cdot 10^{-4}$). How does this compare with experimental situation ? The best piece of experimental information on these decays comes from Argus. They have recently improved the upper bound of $BR(B \to K^* + \gamma)$:

$$BR(B \to K^* + \gamma) < 2.4 \cdot 10^{-4} \ (Argus)^{8]} \qquad (5)$$

Unfortunately, the relation between the exclusive $B \to K^* + \gamma$ and the inclusive $b \to s + \gamma$ modes is plagued by relevant theoretical uncertainties. Taking the two most recent (and maybe most reliable) estimates [6,9]:

$$\frac{\Gamma(B \to K^* + \gamma)}{\Gamma(b \to s + \gamma)} = (4.5 \div 7)\% \qquad (6)$$

the Argus limit (5) would then imply:

$$BR(b \to s + \gamma) < (3.4 \div 5.3) \cdot 10^{-3} \qquad (7)$$

Hence we need an improvement of roughly one order of magnitude on $BR(B \to K^* + \gamma)$ to start testing the presence of the expected large QCD corrections in SM. For the experimentalists who think that I am asking too much when I demand a bound on $BR(B \to K^* + \gamma)$ at the level of 10^{-5} I can propose an alternative avenue which, obviously, has also its own problems, though. Since $B \to K^* + \gamma$ is only few percent of the whole $b \to s + \gamma$, one can consider higher K resonances $(K^{**}(1400), K^{***}(1800), ...)$ with a mass less than m_D^*, in order to impose a cut on the energy of the photon to eliminate the background represented by $B \to D^* + \gamma$. In ref. 9] it was estimated that

$$\Gamma(B \to X_s(m_{X_s} < m_{D^*}) + \gamma)/\Gamma(b \to s + \gamma) \simeq 37\%, \qquad (8)$$

where X_s denote these higher K resonances, and, so, summing over all of them (or at least, the first ones, like K^{**} and K^{***}) might represent an interesting alternative to circumvent the smallness of $BR(B \to K^*\gamma)$.

The situation is much less exciting for the other radiative decay, $b \to s + g$. The dominant decay mode is not that with a "real" gluon carrying $q^2 = 0$, but rather the one in which the gluon momentum satisfies $q^2 > 0$. The most important channels are $b \to sgg$ and $b \to sq\bar{q}^{10]}$. In any case the BR's cannot be more than few times 10^{-3}. The QCD radiative corrections for the case of "real" gluon emission have not been computed yet. However, even for values as large as those found in the $b \to s + \gamma$ case, the $0(10^{-3})$ for $BR(b \to sg)$ cannot be significantly exceeded. Needless to say, what makes me rather pessimistic is the formidable challenge that one has to meet to detect such a decay at values of the $BR's$ of $0(10^{-3})$. At the moment both the Argus and Cleo collaborations agree that we cannot even rule out the existence of charmless b decays at the level of 20% and so it is clear that there is quite a long way to go to reach a threshold sensitive to the SM expectation for $b \to s + g$. We shall see that $b \to s + g$ may be remarkably enhanced in SUSY models; there, indeed, even $BR(b \to s + g)$ of 10% or more are attainable.

Having discussed the "truncated penguins" of the radiative $b \to s + \gamma, g$ decays, we turn now our attention to the "complete" penguins where the "mild" GIM is operating. The most interesting example is provided by the semileptonic decays [2,11] $b \to s + \ell^+ + \ell^-$ (a representative diagram is obtained from that in Fig. 2 with the virtual photon converting into $\ell^+ \ell^-$ instead of $q\bar{q}$). Obsviously, being the large logs of the "mild" GIM present already at the one loop level, there is no chance for the QCD corrections to be large. For the full computation of the process also box diagrams with up quarks, neutrinos and two W exchanges have to be included. The final result yields a $BR(b \to s\ell^+\ell^-)$ in the range $10^{-6} \div 10^{-5}$. For instance, taking ℓ to be the electron, one obtains $BR = 2.4 \cdot 10^{-6}$ for $m_t = 45~GeV$ and $BR = 10^{-5}$ for $m_t = 165~GeV$.

Experimentally the best piece of information on these processes comes from CLEO [12], their inclusive bound on FCNC $BR(B \to \ell^+\ell^- + X)$ being $1.2 \cdot 10^{-3}$. I think that probing SM in the interesting $b \to s\ell^+\ell^-$ decays might represent one of the issues for B factories. That we have to wait for these machines is even clearer if one wants to consider exclusive channels like $B \to K + \ell^+\ell^-$. It has been estimated [13] that

$$\Gamma(B \to K\ell^+\ell^-)/\Gamma(b \to s\ell^+\ell) = (2.3 \text{ to } 9.3)\% \qquad (9)$$

Obviously even more problematic is the detection of b semileptonic decays with a "two-generation gap", namely $b \to d + \ell^+ + \ell^-$. Indeed [13]:

$$\Gamma(b \to d\ell^+\ell^-)/(b \to s\ell^+\ell^-) \simeq 0.005 \text{ to } 0.1 \qquad (10)$$

for m_t from 200 GeV to 40 GeV, respectively, and

$$\Gamma(B \to \pi\ell^+\ell^-)/\Gamma(b \to d\ell^+\ell^-) = (1.7 \text{ to } 6.7)\% \qquad (11)$$

Finally, the other FCNC b semileptonic decay, $b \to s + \nu + \bar{\nu}$, has a BR which is very slightly larger than $BR(b \to s + \ell^+ + \ell^-)$. [2,11,13]

I want to conclude this review of rare b decays which involve the $b \to s$ transition by mentioning that this class of processes may help us in establishing a lower bound on the mass of the physical neutral Higgs scalar h^o of SM[14]. The process one considers is $b \to s + h^o$. Its amplitude is proportional to m_t^2: one power comes from the Yukawa coupling of h^o to the quark t, whilst the other one originates from the helicity flip which is realized on the internal t line. Due to this fact the $BR(b \to s + h^o)$ is quite large. For instance for $m_t = 50~GeV$ one obtains $BR(b \to s + h^o) \sim 4\%$ (for $m_{h^o} < 1~GeV$) and remember that $BR(b \to s + h^o)$ grows with the fourth power of m_t!

PROBING SUSY IN RARE B DECAYS

It is by now well known that the major interest in discussing FCNC in the SUSY context lies not so much in the mere supersymmetrization of the FCNC SM contribution with W and up-quarks replaced by their s-partners, the $\tilde{W}$ and up-squarks, but rather is the presence of an entirely new source of FCNC in the strong interaction sector of SUSY theories [15]. This point is obviously crucial for my present discussion and so I shall try to briefly recall here the major ingredients of this new FCNC source.

We know that in SM the vertices $gq\bar{q}$ do not induce any flavour change. Their SUSY version is $\tilde{g} - q - \tilde{q}^+$, where $\tilde{g}$ and $\tilde{q}$ denote the gluino and squark respectively. I want to show that these vertices can yield a flavour change, for instance the $\tilde{g}$ line can enter a vertex where d_L and $\tilde{s}_L$ (the scalar partner of the left-handed quark s) are present. To prove this one must show that it is not possible to diagonalize the q and $\tilde{q}$ mass matrices simultaneously. This is the analogue of the lack of simultaneous diagonalization of the up- and down- quark mass matrices which is ultimately responsible for the flavour changes in the $\bar{u}\,Wd$ vertices.

Let me first focus on the left-handed sector of the theory. After the spontaneous breaking of local SUSY, the $N = 1$ supergravity lagrangian that I am considering here consists of two parts [16] :a global N=1 supersymmetrization of SM and a set of terms which break softly this residual $N = 1$ global SUSY. The 3×3 mass matrix $m^2_{\tilde{d}_L \tilde{d}^*_L}$ receives two contributions, one from each of these two parts:

$$m^2_{\tilde{d}_L \tilde{d}^*_L} = m_d m_d^+ + \mathbf{1} m^2, \tag{12}$$

where m_d denotes the down quark mass matrix and m is the universal mass contribution to all the scalars of the theory from the SUSY softly broken sector of the theory. Obviously, at this stage, there is no flavour change in the $\tilde{g} - d_L - \tilde{d}_L^+$ vertices since $m^2_{\tilde{d}_L \tilde{d}^*_L}$ is diagonalized by the same unitary rotation which diagonalizes $m_d m_d^+$.

However the mass relation (12) needs to be renormalized from the superlarge scale of supergravity breaking to the M_W scale if we want to use it in confronting our experimental data. In this process of renormalization $m^2_{\tilde{d}_L \tilde{d}^*_L}$ acquires terms which are proportional to $m_d m_d^+$ and (what is relevant for us!) to $m_u m_u^+$. These latter contributions should not surprise. Indeed, the supersymmetrization of the usual Yukawa coupling $h_u Q H u^c$ which gives rise to the up-quark masses yields a term in the superpotential where the d_L superfield interact with the superfields H^+ and u^c through the Yukawa coupling h_u. From this term it is relatively easy to construct contributions to $m_{\tilde{d}_L \tilde{d}_L^+}$ which are proportional to $m^2 h_u h_u^+$ and, hence, to $m_u m_u^+$. Thus, taking into account these renormalization effects, eq. (12) modifies into [15]:

$$m_{\tilde{d}_L \tilde{d}^*_L} = m_d m_d^+ + \mathbf{1} m^2 + c\,m_u m_u^+, \tag{13}$$

where the parameter c is computable in terms of the inputs of the theory by solving their set of renormalization group equations.

Eq. (13) contains two precious pieces of information [15]:

a) due to the presence of the $cm_u m_u^+$ term, the $g - d_L - \tilde{d}_L^+$ vertices can give rise to flavour changes;

b) the angles which are present in these flavour changes are the same angles which appear in the lack of simultaneous diagonalization of the $m_d m_d^+$ and

$m_u m_u^+$ matrices in the $W - u - d$ vertices, i.e. they are the familiar Cabibbo-Kobayashi- Maskawa angles.

Although the above discussion contains the essential points which intervene in any gluino mediated FCNC process, the actual computation of rare b decays requires a much more detailed knowledge of the low energy limit of spontaneously broken $N = 1$ supergravity theories. In particular, in the above I focused only on $m^2_{\tilde{d}_L \tilde{d}_L^*}$, but this is just a submatrix of the general 6×6 down squark squared mass matrix which includes also the $m^2_{\tilde{d}^\circ \tilde{d}^{\circ*}}$ and $m^2_{\tilde{d}\tilde{d}^\circ}$ submatrices. Fortunately the error one commits by neglecting these two other submatrices is not so large: $m^2_{\tilde{d}^\circ \tilde{d}^{\circ*}}$ keeps an expression like in eq. (12) essentially unaltered by the renormalization procedure and the terms $m^2_{\tilde{d}\tilde{d}^\circ}$, being proportional to m_d are rather small. Another delicate point concerns the actual values of the parameter c in eq. (13) (remember that the rates of the b decays are proportional to the square of c and so a precise determination of this parameter is not only a matter of academical interest!). When I speak of "a" value for the parameter c one should recall that c varies if one changes the inputs of the theory, so, in particular, the value of the squark and gluino masses which enter the computation of the rare b decays. For each point of the squark-gluino mass plane there is a different value of c and its knowledge requires the solution of the complicate set of renormalization group equations of the theory.

Having pointed out some more sophisticated ingredients which are needed for a complete computation, I urge the reader not to feel overwhelmed by the above "technical problems". As it often happens in physics, also in this case a reasonable choice of approximations (restrictions to the $m^2_{\tilde{d}_L \tilde{d}_L^*}$ submatrix, use of the current mass eigenstates with mass insertions to account for flavour changes, "average" value of c,...) leads to a result which differs only unessentially from what one obtains in the full rigorous elaboration, as we shall see when considering each rare b decay process in detail. Before doing that, we still need a general discussion concerning the fate of the GIM mechanisms in the SUSY context [17].

In the SUSY loop we sum over the exchange of the down squarks of the three generations. At the vertices we have still the elements of the KM matrix and so now we are in the presence of a superGIM mechanism which assures the vanishing of the b decay amplitudes for degenerate down squark masses.

We have still to distinguish between hard and mild GIM cancellations, with factors $0(\tilde{m}_q^2/m^2)$ and $0(ln(\tilde{m}_q^2/\tilde{m}_{q'}^2))$, respectively. Here $\tilde{m}_q$ denotes the mass of the heaviest squark circulating in the loop, $\tilde{q}'$ is another $Q = -1/3$ squark mass eigenstate and m sets the scale for the global SUSY breaking. The point is that the universal mass m (eq.12) represents the dominant contribution to all the squark mass and so $ln(\tilde{m}_q^2/\tilde{m}_{q'}^2) = ln(1 + \epsilon) \sim \epsilon$ where ϵ is $0(\tilde{m}_q^2/m^2)$. Hence, the distinction between mild and hard superGIM is completely immaterial from a quantitative point of view. The large logs of the mild GIM situation in SM are now replaced by logs which, quantitatively, behave like the usual power suppression.

The above comparison GIM-superGIM is quite instructive: if we consider a process with a mild GIM suppression there is no chance for any significant SUSY enhancement. The SUSY advantage due to the presence of the strong couplings in the $\tilde{g}$ exchange in place of the usual weak couplings of the W exchange is swamped by the different GIM factors.

To reassure the readers who may find the above arguments somewhat hand-waving I am ready to provide a quantitative result. Take the typical "mild" GIM decay $b \rightarrow s + \ell^+\ell^-$. By comparing the SUSY with the SM amplitude we obtain [17]:

$$\frac{A_{SUSY}(b \rightarrow s\ell^+\ell^-)}{A_{SM}(b \rightarrow s\ell^+\ell^-)} = 0.25 \div 0.40 \qquad (14)$$

for $m_t \simeq 80 GeV$ and $\tilde{q}$ and $\tilde{g}$ in the 70-80 GeV range.

The b radiative decays $b \rightarrow s + \gamma$ and $b \rightarrow s + g$ have all the right credentials to be the places for conspicuous SUSY enhancemens: strong instead of weak couplings and no particular reason for a penalizing GIM since the GIM suppression is "hard" also in the SM case. This line of thought prompted S. Bertolini, F. Borzumati and myself to compute $b \rightarrow s + \gamma$[17] and $b \rightarrow s + g$[18] in the SUSY context. We used some simplifying assumptions that I previously mentioned and we found that, indeed, a relevant enhancement was present. Given the importance of the results and the rapid experimental improvement in the bound for $BR(B \rightarrow K^* + \gamma)$, G. Ridolfi and I [19] took the pain of repeating the computation without any unnecessary assumptions. We worked in the squark mass eigenstate basis (i.e. we diagonalized the full 6 × 6 down squark mass matrix) and we made use of the complete solution of the renormalization group equations. As I said, we could essentially confirm the previously results of Bertolini et al. In any case, I report here the results of Ridolfi and myself, being the computation more accurate.

The $N = 1$ supergravity lagrangian after SUSY spontaneous breakdown depends on five SUSY parameters [16]:

a) m that I have already introduce (see eq. 12);

b) M, the common gaugino mass term at the grand unification scale;

c) μ, the dimensional coupling of the superfield bilinear term in the superpo-tential which mixes the two Higgs doublets;

d,e) the dimensionless parameters A and B which enter the trilinear and bilinear term, respectively, of the SUSY soft breaking sector. The whole set of renormalization group equations whose solution provides the low energy SUSY version of SM depends on the above set of five SUSY parameters plus m_t. The request of a flat Kähler metric further reduces the number of independent SUSY parameters, since $B = A - 1$. We impose now a stringent (and theoretically most appealing) constraint, namely the radiative breaking of $SU(2) \times U(1)$. In this way a combination of SUSY parameters is fixed in such a way as to reproduce the correct value of M_z. We are left with three independent SUSY parameters. We choose them to be $m_{\tilde{g}}$, the mass of the lightest $Q = -1/3$ squark (that I shall denote by $m_{\tilde{b}}$, although it does not coincide with the partner of the b quark which, in general, is not even a mass eigenstate) and, finally, the ratio v_2/v_1 of the two Higgs vacuum expectation values. The figures below present our results in the $m_{\tilde{b}} - m_{\tilde{g}}$ plane for fixed values of v_2/v_1 and m_t.

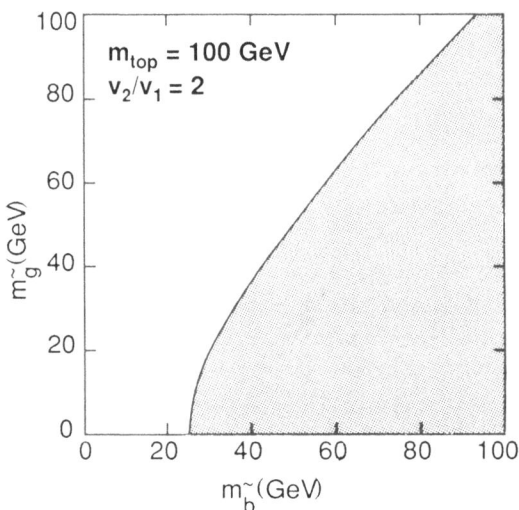

Figure 5. In the dotted region spontaneous breaking of the SU(2)xU(1) symmetry takes place at the correct scale.

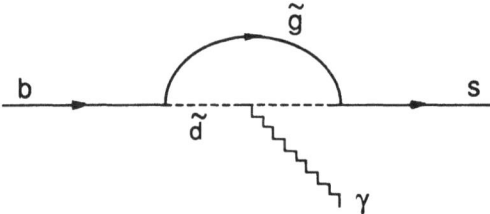

Figure 6. Feynman diagram for the super-symmetric contribution to b→s+γ.

First, in Fig. 5 we show for which values of $m_{\tilde{l}}$ and $m_{\tilde{g}}$ the spontaneous breaking of the $SU(2) \times U(1)$ symmetry takes place at the correct scale (dotted area). I examine now the further constraints on the dotted region which can be derived from $b \to s + \gamma$ and $b \to s + g$.

The dominant contribution in SUSY to $b \to s + \gamma$ comes from the one-loop diagrams where gluinos and down squarks are exchanged (Fig. 6).

The decay amplitude is

$$A(b \to s + \gamma) = \frac{e\alpha_s}{9\pi} \frac{m_b^2}{m_{\tilde{g}}^2} \bar{s} \frac{\sigma_{\mu\nu} q^\nu \epsilon^\mu}{m_b + m_s} (F_2^V + F_2^A \gamma_5)b \tag{15a}$$

where

$$F_2^{V,A} = (\Gamma_{L2i}^\dagger \Gamma_{Li3} \pm \Gamma_{R2i}^\dagger \Gamma_{Ri3})P(x_i) + \frac{m_{\tilde{g}}}{m_b}(\Gamma_{R2i}^\dagger \Gamma_{Li3} \pm \Gamma_{L2i}^\dagger \Gamma_{Ri3})Q(x_i) \tag{15b}$$

$$P(x) = \frac{1}{12(x-1)^4}(x^3 - 6x^2 + 3x + 2 + 6x \ log \ x), \tag{15c}$$

$$Q(x) = \frac{1}{2(x-1)^3}(x^2 - 1 - 2x \ log \ x) \tag{15d}$$

and $x_i = (\tilde{m}_i/m_{\tilde{g}})^2$, where $\tilde{m}_i^2$ is the i-th eigenvalue of the full 6×6 down squark squared mass matrix. Finally the mixing 6×3 matrices $\Gamma_{L,R}$ in (15b) are defined as in ref.20.

The amplitude (15) yields the decay rate:

$$\Gamma(b \to s + \gamma) = \frac{\alpha_s^2 \alpha}{81\pi^2} \frac{m_b^5}{2m_{\tilde{g}}^4}(|F_2^v|^2 + |{}_2^A|^2). \tag{16}$$

Imposing the very conservative experimental bound $BR(b \to s + \gamma) < 10^{-2}$ (it would correspond to take $\Gamma(B \to K^* + \gamma)/\Gamma(b \to s + \gamma) = 2 \cdot 10^{-2}$) we further limit the allowed region in the $m_{\tilde{l}} - m_{\tilde{g}}$ plane, as shown in Fig. 7a. Looking at the bound in eq. 7 it might be more interesting to impose the condition $BR(b \to s + \gamma) < 10^{-3}$ (a bound which should be possible to reach in the near future). Then, the disallowed region enlarges by the amount shown in Fig. 7b.

Repeating the same kind of calculation for the case b→s+g and imposing the (conservative?) bound BR(b→s+g)<20% leads to the disallowed region shown in Fig. 8.

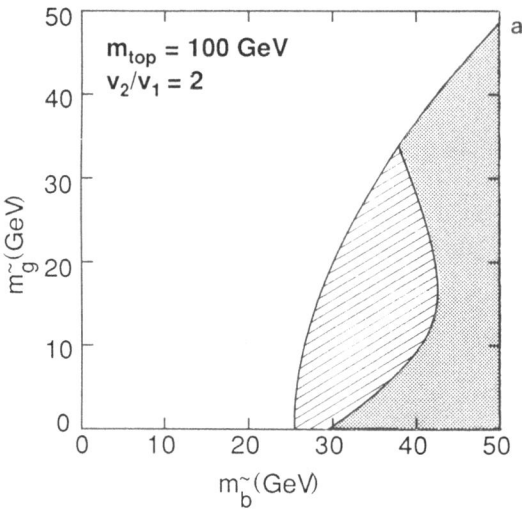

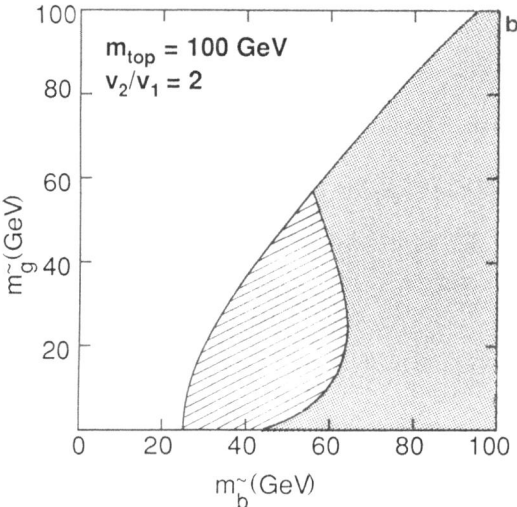

Figure 7. Region of the $m_{\tilde{b}}$-$m_{\tilde{g}}$ plane where a) BR(b→s+γ)>10^{-2} and b) BR (b→s+γ)>10^{-3}, for m_t=100 GeV and v_2/v_1=2 (shaded area). The region allowed by SU(2)xU(1) breaking is also shown (dotted area).

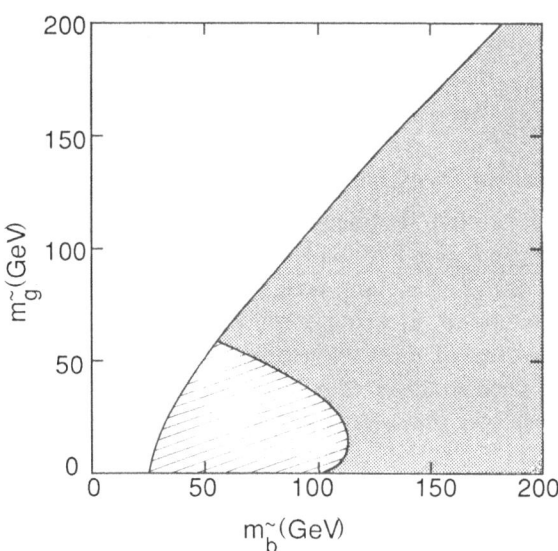

Figure 8. Region of the $m_{\tilde{b}}$-$m_{\tilde{g}}$ plane where BR(b→s+γ)>20%, for m_t=100 GeV and v_2/v_1=2 (shaded area). The region allowed by SU(2) xU(1) breaking is also shown (dotted area).

I believe that the bounds given here on $m_{\tilde{g}}$ and $m_{\tilde{b}}$ are so far the most stringent ones to be derived from virtual effects of SUSY particles in rare low energy processes. It is true that the total area forbidden by UA1 [21] (but allowed by the $SU(2) \times U(1)$ radiative breaking) remains larger than that disallowed by the rare radiative b decays that I consider here. However, if we take $BR(b \to s + \gamma) < 10^{-3}$ then there is already some (small) part of the $m_{\tilde{b}} - m_{\tilde{g}}$ plane which is ruled out by $b \to s + \gamma$, but is otherwise allowed by the UA1 limits. In any case I believe that what is important is the complementarity of the direct and indirect searches for SUSY particles. The bounds that $b \to s + \gamma$ (and, hopefully, $b \to s + g$) may provide reach an interval of values of squark and gluino masses comparable to what UA1 has explored and, even more important, the two procedures to derive these bounds are based on methods and assumptions which are quite different. In the case of UA1 there might be some problem in dealing with the complicated "jettology" and some assumptions are made concerning the lightest supersymmetric particle and the degeneracy of squark masses. In our case we have two free parameters, v_2/v_1, and m_t, that we have to fix in order to give bounds on $m_{\tilde{b}}$ and $m_{\tilde{g}}$.

Finally, QCD corrections are expected to be small in the SUSY case, since all the particles in the loop are heavy and so no large log can appear. However, no estimate has been produced so far.

I conclude this section by briefly reporting the results of the analysis of the SUSY contributions to $b \to s + h^o$ that Bertolini, Borzumati and I have recently performed [22]. In SUSY there are three neutral massive Higgs fields: a pseudoscalar, h_3^o and two scalars, h_1^o and h_2^o. Quite severe conditions must be respected if one wants at least one of these three bosons to remain light, i.e. with a mass $< 4\ GeV$. Barring accidental cancellations, the only way to have $m_{h_3^o} < 4\ GeV$ is that μ (i.e. the parameter which determines the mixing of the two Higgs doublets) be quite small, $\mu \sim 0(10^{-1} GeV)$. Although it is technically possible to obtain such a small μ, one has to worry about its phenomenological and cosmological implications. The radiative breaking of $SU(2) \times U(1)$ requires m_t as large as $\sim 100\ GeV$, one chargino is certainly lighter than m_w and to ensure that it is heavier than 23 GeV one has to impose restrictions on the parameter space, and, finally it might be difficult to respect the lower bound on the mass of stable neutralinos which is based on cosmological considerations. The leading contribution to $b \to s + h_3^o$ comes from the exchange of gluino and squarks. Analogously to what we found for the previous radiative decays, also in this case a large enhancement is present. Even for m_t as low as $45 GeV$, if the squarks and gluino masses are below 100 GeV one can infer a lower bound of $\sim 4\ GeV$ on $m_{h_3^o}$. If $m_t \geq m_w$, then $m_{h_3^o}$ must saturate the phase space available ($m_{h_3^o} \simeq 4.7\ GeV$), unless large SUSY masses and/or contrively chosen soft breaking parameters provide the needed suppression.

What surprised us (and also the reader at this point, I am sure) was the study of the decays $b \to s + h_2^o, h_1^o$. There are two ways to obtain a light scalar: h_3^o is also very light and/or $v_1 \simeq v_2$ (v_2 and v_1 must differ by no more than 7%). The former option has its own potential problems, as we have seen and so I'll focus here on the latter one. In ref.22 we have shown that for $v_1 = v_2$ the sum of the gluino mediated one-loop diagrams which contribute to $b \to s + h_{1,2}^o$ has an extra suppression factor $m_s^2/m_b^2 \sim 10^{-3}$ with respect to the single terms. The leading contribution is due to one-loop diagrams where charginos and up-squarks are exchanged. Still interesting bounds can be derived, although more parameters appear and so the conclusions are not as neat as in the gluino mediated $b \rightarrow s + h_3^o$ decay. The interested reeder is invited to consult ref. 22] for a sufficiently thorough discussion of this more complicate situation.

A final remark is in order. When discussing the SUSY contributions to rare B decays I have systematically considered <u>only</u> the exchange of s-particles, namely gluinos, squarks or charginos, which are partners of ordinary particles. However, I mentioned that the implementation of SUSY dictates a major change in the <u>Higgs sector</u>: instead of the usual Higgs doublet of SM we must now introduce two doublets. Consequently one charged Higgs scalar is physical and it contributes to all the rare B decays that I considered. Obviously these contributions are significant only if it is not too heavy. The SUSY context is a particular case of the so-called two Higgs doublet models. In particular in SUSY the important requirement that one Higgs doublet is responsible for the up quark masses, whilst the shown quarks get a mass from the other doublet is respected and so no tree level FCNC transitions can occur. A series of papers by W.-S. Hou and collaborators deals with the charged Higgs contributions to rare B decays [23]. We refer the reader to them and I simply mention that in some cases, like for instance $b \rightarrow s + \gamma$, the charged Higgs exchange can produce enhancements which are of the same order of those produce by s-particle exchange. For instance taking $(v_2/v_1)^2 = 10^{-2}, m_t = 150 \, GeV$ and $m_H^{\pm} = 155 \, GeV$ one obtains $BR(b \rightarrow s + \gamma)_{2H} \simeq 5 BR(b \rightarrow s + \gamma)_{SM}$. Thus, the prospects for enhancements in the SUSY version of SM may be even richer than those that I presented here.

RARE B DECAYS IN THREE EXTENSIONS OF SM

In this Section I would like to present a few remarks concerning the implications of the study of charmless b decays with strange particles in the final state in the following three extensions of SM: models with more than one Higgs doublet, models with a fourth fermionic generation and left-right symmetric theories.

At the end of the previous Section I have already mentioned that SUSY models need the presence of at least two Higgs doublets in order to provide a mass for all the fermions of the theory. This is just a particular case of models where two Higgs doublets are present. An important distinction must be made. If one wants to enforce the property of absence of tree level FCNC one doublet must give mass to the up quarks while the other one is responsible for down quark masses.* Alternatively, one can give up this requirement and then deal with the tree-level FCNC problem in a different way, for instance with a large mass for some physical neutral scalar. In this latter situation, large enhancements in the rare B decays that I consider here are possible, but I think that these models with tree level FCNC must be rather contrived in any case and so I don't think

---

*A second possibility to avoid tree level FCNC would be to allow only one Higgs doublet to couple to both types of quarks. This option is less severely constrained and, consequently, larger enhancements for rare b decays are possible. I don't find it particularly attractive, however, and so I invite the interested reader to see further details on this option in the paper of ref.10

that they deserve my further attention. Thus I focus on the former class which contains the SUSY case, in particular. If we denote by v_1 and v_2 the vacuum expectation values which provide the up- and down- quark masses, respectively, one would immediately identify the hierarchy $v_1 > v_2$ as the "natural" one given that $w_t >> m_b$ and $m_c > m_s$. In this "natural" case it is still possible to have some rare b decay enhanced, but certainly not by tremendously high factors [10]. $BR(b \rightarrow s\ell^+\ell^-)$ remains $< 10^{-5}$, but $b \rightarrow s + \gamma$ can reach a BR between 10^{-4} and 10^{-3}, thus reaching almost the SUSY prediction. Nothing prevents, however, a more "perverse" choice of the Yukawa couplings so that $v_1 < v_2$. In this "unnatural" case large enhancements are indeed possible. The decay rate of $b \rightarrow s + \gamma$ can become comparable to the one we found in the SUSY case. The $BR(b \rightarrow s + \gamma)$ can be even larger than in SUSY and, moreover, now also the semileptonic decay $b \rightarrow s\ell^+\ell^-$ can present a BR which is one order of magnitude larger than in SM.

Coming to the left-right models, there are two parameters which determine their potentiality in enhancing rare b decays: the W_R mass and the $W_L - W_R$ mixing that I'll denote by ξ. If one takes $M_{W_R} > 1.6 TeV$ and $\xi \leq 4 \cdot 10^{-3}$, then the contributions to rare b decay which arise when exchanging a W_R boson are always smaller than the SM ones [24]. However those who still "love" the LR models can object that the above bounds on M_{W_R} and ξ rely on some theoretical assumptions and are not derived from direct experimental evidence. As for M_{W_R}, the corresponding lower bound of 1.6 TeV was derived by imposing that the $W_L - W_R$ contribution to the box diagram which provides $\Delta m_{K - \bar{K}}$ does not exceed the experimental result. However, crucial is the assumption in the evaluation of this box diagram that the mixing angles in the right-handed sector are the same as those that appear in the left-handed sector. This situation actually occurs only in a special class of LR models known as models with "manifest" left-right symmetry. Also the upper hand on ξ, $\xi < 4 \cdot 10^{-3}$ is plagued by some theoretical uncertainty. Indeed, it was obtained by analyzing non-leptonic K decays by making use of current algebra, PCAC and bag model assumptions. The direct bound on ξ from β-decay experiments is more than one order of magnitude larger, $\xi < 0.06$. In a recent analysis [24], Cocolicchio, Costa, Fogli and I showed that if ξ is allowed to be between $4 \cdot 10^{-3}$ and $6 \cdot 10^{-2}$ considerably large enhancements of all the rare b decays occur. We refer the interested reader to our analysis for a more quantitative discussion.

Finally, a lot of attention has been given in the literature to the possibility that a fourth generation may lead to conspicuous enhancements of some rare b decays [25]. For this to occur we need a rather heavy fourth up-quark, t', say $m_{t'} > 150 GeV$ and/or large mixing angles between t' and b, s say $|V_{t'b}V_{t's}| > 0.2 \div 0.3$. Of particular interest is the situation concerning $b \rightarrow s\ell^+\ell^-$. As I said no SUSY enhancement can occur, while on the contrary the fourth generation models can lead to a $BR \sim 10^{-4}$ if the above requirements are met. The enhancement for $b \rightarrow s + \gamma$ and $b \rightarrow s + g$ are comparable to those that I presented in the SUSY case.

A more detailed analysis of these three classes of extensions of SM is beyond the scope of my talk [26]. However, I hope that the few remarks that I reported here can convince the reader that the study of rare b physics presents an extreme relevance in establishing bounds on the couplings and masses of the new particles that all these extensions predict.

CONCLUSIONS

There is an important message that I hope my talk has made clear enough. When we speak of using rare b decays in probing SM and possible new physics most people think that this concerns only B factories and it is not that relevant in a very near future. This is only partially correct. It is true that for some processes, for instance the semileptonic b decays $b \to s\ell^+\ell^-$ and $b \to s\nu\bar{\nu}$, it is likely that we have indeed to wait for B factories to have some hope of observing them. However, this is not the case for other rare b decays, in particular for my favourite radiative $b \to s+\gamma$ decay. Here the possibility of testing SM (and in particular the presence of large QCD corrections) and new physics (notably SUSY models) lays around the corner. I urge a great attention to this process not only is an obvious context like CLEO II, but also in accelerator physics. There exist some studies of feasibility at LEP where a bound of 10^{-4} on $BR(B \to K^* + \gamma)$ seem achievable. At the same time, radiative decays provide an interesting challenge also from the theoretical side. The problems of QCD radiative corrections, computation of exclusive channels and interpretation of the experimental data related to these processes are far from being solved. I think that also some more work is needed in trying to be more specific on the constraints that rare b decays can impose on extensions of SM. In particular, these decays represents an ideal place for probing SUSY models in the low energy precision tests. We are planning a more detailed investigation of these implications where all the SUSY contributions (including the exchange of the physical charged Higgs scalar) are considered in the context of spontaneously broken N=1 supergravity theories with the radiative breaking of SU(2)xU(1).

"Se sono rose fioriranno": I think that rare B decays have a reasonable chance of blossoming, in the context of the B physics of the 90's.

NOTE ADDED

After accomplishing this written version of my talk, I received two papers which are relevant for the above analysis. The paper of R. Grigjanis et al. 27] presents a new computation of the QCD corrections to $b \to s + \gamma$ and $b \to s + g$. The authors claim that the previous calculations in refs. 4-7] used a regularization scheme which does not take into account the peculiarities of loop integrals containing γ_5 terms. Using a different method of handling γ_5 terms-dimensional reduction - they conclude that the $BR(b \to s + \gamma)$ is roughly one-half of the value reported in the solid line of fig. 4. Moreover, they find that the processes $b \to s + g$, with the gluon on-shell is strongly suppressed by the QCD corrections.

C.A. Dominguez et al.28] report a new evaluation of $\Gamma(B \to K^* + \gamma)/\Gamma(b \to s+\gamma)$ in the framework of QCD sum rules combined with Vector Meson Dominance. They obtain $\Gamma(B \to K^* + \gamma)/\Gamma(b \to s + \gamma) = 0.28 \pm 0.11$, thus a value much larger than that I reported in eq. 6 which was obtained in the framework of the Constituent Quark Model (CQM).

Acknowledgements

It is with great pleasure that I thank my collaborators S. Bertolini, F. Borzumati, D. Cocolicchio, G. Costa, G. Fogli, Jick Kim and G. Ridolfi with whom I shared the excitement (and, sometime, the disappointment) of the study of rare b decays in SM and beyond it.

I benefited a lot from stimulating discussions on different topics of this talk with A. Ali, G. Altarelli, R. Barbieri, J. Ellis, K. Enqvist, Paula Franzini, G. Giudice, L. Hall, W.-S. Hou, W. Marciano, A. Sanda, A. Soni and J. Trampetić.

REFERENCES

1] S. Glashow, J. Iliopoulos and L. Maiani, Phys. Rev. D2, 1285 (1970).

2] T. Inami and C.S. Lim, Progr. Theor. Phys. 65, 297 (1981); E65 (1981) 1772.

3] Ya. I. Kogan and M.A. Shifman, Yad. Fiz. 38, 1045 (1983) [Sov. J. Nucl. Phys. 38, 628 (1983)].

4] S. Bertolini, F. Borzumati and A. Masiero, Phys. Rev. Lett. 59, 180 (1987).

5] For an evaluation of QCD corrections which are obtained by summing up all the leading-logarithmic contributions in a four- quark model see M.A. Shifman, A.I. Vainshtein and V.I. Zakharov, Phys. Rev. D18, 2583 (1978).

6] N.G. Deshpande, P. Lo, J. Trampeti G. Eilam and P. Singer, Phys. Rev. Lett. 59, 183 (1987).

7] B. Grinstein, R. Springer and M.B. Wise, Phys. Lett. 202B, 138 (1988).

8] H. Albrecht et al., ARGUS Collab., preprint DESY 88-062 (1988).

9] T. Altomari, Phys. Rev. D37, 677 (1988).

10] W.-S. Hou, Univ. of Pittsburgh report PITT-87-03 (1987).

11] W.-S. Hou, R.S. Willey and A. Soni, Phys. Rev. Lett. 58 (1987) 1608.

12] CLEO Collaboration, A. Bean et al., Phys. Rev. 35, 3533 (1987).

13] This result reported by J. Ellis and P.J. Franzini, Proceedings of the Workshop on Heavy Quark Factory and Nuclear Physics Facility with Superconducting Linacs, E. De Sanctis, M. Greco, M. Piccolo and S. Tazzari eds. (Courmayeur, 1987) p. 341 was obtained using three different models for hadronic bound states of quarks: M. Wirbel, B. Stech and M. Bauer, Z. Phys. C29, 637 (1985); S. Nussinov and W. Wetzel, Phys. Rev. D36 130; B. Grinstein, M.B. Wise and N. Isgur, Phys. Rev. Lett. 56, 298 (1986.

14] R.S. Willey and H.L. Yu, Phys. Rev. D26, 3086 (1982); B. Grzadkowski and P. Krawczyk, Zeit. für Phys. C18, 43 (1983); T.N. Pham and D.G. Sutherland, Phys. Lett. 151B, 444 (1985); R. Ruskov, Phys. Lett. 187B, 165 (1987), B. Grinstein, L. Hall and L. Randall, preprint LBL-25095, UCB-PTH-88/6 (1988).

15] M.J. Duncan, Nucl. Phys.B221, 285 (1983); J.F. Donoghue, H.P. Nilles and D. Wyler, Phys. Lett. 128B, 55 (1983); A. Bouquet, J. Kaplan and C. Savoy, Phys. Lett. 148B, 69 (1984).

16] E. Cremmer, B. Julia, J. Scherk, P. van Nieuwenhuizen, S. Ferrara and L. Girardello, Phys. Lett. 79B (1978) 23; Nucl. Phys. B147 (1979) 105. E. Cremmer, S. Ferrara, L. Girardello and A. van Proeyen, Phys. Lett. 116B,

231 (1982); Nucl. Phys. B212, 413 (1983). For a review, see, for instance, H.P. Nilles, Phys. Rep. 110, 1 (1984).

17] S. Bertolini, F. Borzumati and A. Masiero, Phys. Lett. 192B, 437 (1987).

18] S. Bertolini, F. Borzumati and A. Masiero, Nucl. Phys. B294, 321 (1987).

19] A. Masiero and G. Ridolfi, preprint CERN-TH 5081/88 (1988), to appear in Phys. Lett. B.

20] J.M. Gerard, W. Grimus, A. Raychauduri and G. Zoupanos, Phys. Lett. 140B (1984) 349.

21] C. Albajar et al., UA1 Collaboration, Phys. Lett. 198B, 261 (1987).

22] S. Bertolini, F. Borzumati and A. Masiero, preprint CMU-HEP 88-07 (1988).

23] W.-S. Hou and R.S. Willey, preprint MPI-PAE/PTh 81/87 (1987). R.K. Ellis, G.C. Joshi and M. Matsuda, Phys. Lett. 179B (1986) 119; D. Cocolicchio, G. Costa, G.L. Fogli and A. Masiero, Proceedings of the Workshop on Heavy Quark Factory and Nuclear Physics Facility with Superconducting Linacs, E. De Sanctis, M. Greco, M. Piccolo and S. Tazzari eds. (Courmayeur, 1987), p.355.

24] D. Cocolicchio, G. Costa, G. Fogli, J.H. Kim and A. Masiero, preprint CERN-TH 5110/88, BARI-TH/88-30, DFPD 88/TH/6 (1988).

25] W.-S. Hou, A. Soni and H. Steger, Phys. Lett. 192B (1987) 441; J.L. Hewett, Phys. Lett. 193B (1987) 327.

26] For a detailed discussion see D. Cocolicchio et al. in ref. 24.

27] R. Grigjanis, P.J. O'Donnell, M. Sutherland and H. Navelet, preprint UTPT-88-11 (1988).

28] C.A. Dominguez, N. Paver and Riazuddin, preprint DESY 88-110 (1988).

12

CP VIOLATION IN THE *B* SYSTEM

David London

Deutsches Elektronen Synchrotron - DESY
Hamburg, Fed. Rep. Germany

1. INTRODUCTION

For more than 20 years, the origin of CP violation has been one of the fundamental questions in particle physics. To date, CP violating phenomena have only been seen in the kaon system. First of all, the observation of the decay $K_L \rightarrow \pi\pi$ [1] is evidence that ϵ, the CP-violating mixing parameter, is non-zero. More recently, the measurement of ϵ'/ϵ by the NA31 experiment at CERN [2] indicates that CP is also violated in kaon decays. This latter result, if confirmed, already rules out some models for CP violation, the superweak model for example. It is likely that the B system will provide us with more clues, and perhaps even tell us whether or not the standard model explanation for CP violation, the Cabibbo-Kobayashi-Maskawa (CKM) matrix, is correct. The results of the ARGUS [3] and CLEO [4] collaborations have indicated that $B_d\overline{B_d}$ mixing is significantly larger than expected. This has led to an enormous amount of interest in the implications for the CKM matrix and, as we shall see, has suggested that CP violating asymmetries in the B system could be quite large. In this paper, I will discuss the prospects for CP violation in the B system.

I will start off with a review of CP violation in the kaon system (Sec. 2). In Sec. 3, I discuss mixing in the B system, contrasting the effects with those found in the K system. CP violating phenomena in the B system are examined in Sec. 4. After briefly discussing the possibilities for seeing a CP asymmetry in the semi-leptonic decay mode, and for seeing CP violation via final state interactions, I then turn to a more promising possibility, that of CP violating asymmetries in hadronic decay modes. The size of these asymmetries depends crucially on the CKM matrix elements, and I review the limits which result from both the K system and from the ARGUS/CLEO result. Experimental problems in seeing such asymmetries are discussed. I then examine the possibility for observing CP violation via time-dependent measurements. Sec. 5 contains a summary and conclusions.

2. THE K SYSTEM - MIXING AND CP VIOLATION

There are two ways in which CP is violated in the kaon system. The first is through CP violation in the state, characterized by the mixing parameter ϵ. This enters when the weak states K_S, K_L are expressed as linear combinations of the strong (electromagnetic) states:

$$|K_S^0\rangle = \frac{1}{\sqrt{2(1+|\epsilon|^2)}}\left[(1+\epsilon)|K^0\rangle - (1-\epsilon)|\overline{K^0}\rangle\right]$$
$$|K_L^0\rangle = \frac{1}{\sqrt{2(1+|\epsilon|^2)}}\left[(1+\epsilon)|K^0\rangle + (1-\epsilon)|\overline{K^0}\rangle\right] , \tag{1}$$

where I have used the convention $CP|K^0\rangle = -|\overline{K^0}\rangle$; $CP|\overline{K^0}\rangle = -|K^0\rangle$. If ϵ were zero, the weak states would be CP eigenstates: K_S would have CP $+$; K_L would have CP $-$. Therefore a non-zero ϵ is evidence for $\Delta S = 2$ CP violation. In the standard model, CP violation is explained by an imaginary phase in the CKM mixing matrix, which enters into the vertices in the box diagram for mixing in the kaon system (Fig. 1). ϵ is (in principle) calculable from this diagram. However, as we shall see, there are large theoretical uncertainties. Experimentally, it has the value [5]

$$\epsilon = (2.275 \pm 0.021) \times 10^{-3} . \tag{2}$$

It is also possible to have CP violation in the decays of kaons, parametrized by the $\Delta S = 1$ CP violating parameter ϵ', which arises from different isospin phases in the amplitudes for the decays $K \to 2\pi$:

$$A_0 = \langle \pi\pi, I = 0|H_W|K^0\rangle$$
$$A_2 = \langle \pi\pi, I = 2|H_W|K^0\rangle , \tag{3}$$

and

$$\epsilon' \propto \mathrm{Im}\left(\frac{A_2}{A_0}\right) . \tag{4}$$

In the standard model, it is expected that $\epsilon' \ll \epsilon$, and the NA31 group at CERN [2] recently measured ϵ'/ϵ to be

$$\left(\frac{\epsilon'}{\epsilon}\right) = (3.3 \pm 1.1) \times 10^{-3} . \tag{5}$$

Therefore, in the kaon system, CP violation with $\Delta S = 2$ is much larger than that with $\Delta S = 1$.

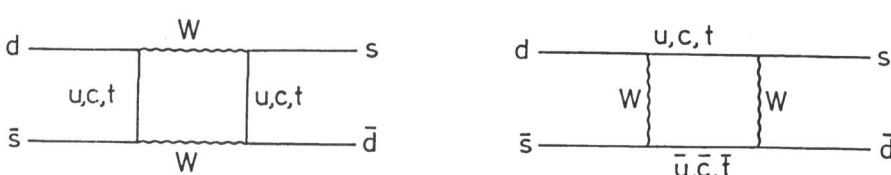

Fig. 1. Diagrams contributing to ϵ in the CKM model.

3. THE B SYSTEM - MIXING

In the B system, there are also two possibilities for CP violation. First of all, as in the kaon system, CP can be violated in the mixing between B and $\overline{B}$ mesons (Fig. 2). However, the B system differs from the kaon system in one crucial respect. Since B-mesons are so heavy, the phase space for their decays is quite large. Therefore both B and $\overline{B}$ have essentially the same lifetime, i.e. $\Delta \tau_B \ll \tau_B$. The calculation which yields the lifetime difference comes from the box diagram. and is similar to that which gives ϵ_B, the $\Delta B = 2$ CP-violating parameter in the B system [6]. Therefore ϵ_B is expected to be quite small in the standard model. This calculation has been done [7], and yields

$$\epsilon_B = \begin{cases} O(10^{-4}), & B_d, \\ O(10^{-5}), & B_s. \end{cases} \tag{6}$$

It therefore seems that the prospects for observation of $\Delta B = 2$ CP-violating phenomena are essentially hopeless. However, in the B system. the situation is reversed with respect to the kaon system, namely, CP violation in B decays ($\Delta B = 1$) can be large. For such phenomena, the relevant parameter is x, the ratio of the energy of the oscillation (i.e. the mass difference) and the total width for the B mesons:

$$x = \frac{\Delta M}{\Gamma} \quad \frac{\text{(transition energy)}}{\text{(mean total width)}} . \tag{7}$$

After all, mixing hardly matters if the particle decays before it has a chance to oscillate into its antiparticle. This can be seen explicitly - because of B-$\overline{B}$ mixing, a state which starts out as a pure B^0 or $\overline{B^0}$ will evolve in time to a mixture of B^0 and $\overline{B^0}$:

$$\begin{aligned} |B^0(t)\rangle &= f_+(t)|B^0\rangle + \frac{1 - \epsilon_B}{1 + \epsilon_B} \, f_-(t)|\overline{B^0}\rangle \\ |\overline{B^0}(t)\rangle &= \frac{1 + \epsilon_B}{1 - \epsilon_B} \, f_-(t)|B^0\rangle + f_+(t)|\overline{B^0}\rangle . \end{aligned} \tag{8}$$

Here, $|B^0\rangle$ represents a pure B^0 state at $t = 0$, $|\overline{B^0}\rangle$ represents a pure $\overline{B^0}$ state at $t = 0$, and

$$\begin{aligned} f_+(t) &= e^{-imt} e^{-\Gamma t/2} \cos(\Delta mt/2) \\ f_-(t) &= e^{-imt} \epsilon^{-\Gamma t/2} i \sin(\Delta mt/2) . \end{aligned} \tag{9}$$

From Eqn (9), it is clearly seen that the competition between Δm and Γ is the important consideration for seeing CP violation in B decays. For B_d-$\overline{B}_d$ mixing, the combined ARGUS and CLEO measurements give

$$x_d = 0.70 \pm 0.13 . \tag{10}$$

As we shall see, this is a large number, and leads to the possibility of substantial CP violating asymmetries in B decays. to which I now turn.

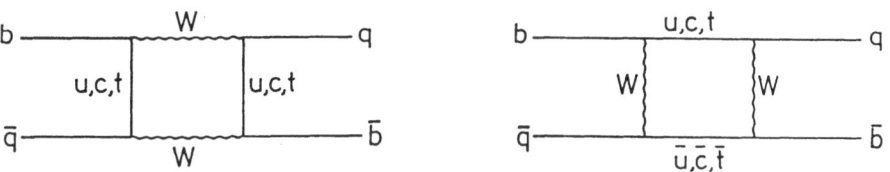

Fig. 2. Diagrams contributing to B^0-$\overline{B^0}$ mixing.

4. THE B SYSTEM - CP VIOLATION

In the B system, CP violation is indicated by a difference in the rates for $B \to f$ and $\overline{B} \to \overline{f}$. In this section, I examine four possible scenarios for such a CP violating asymmetry.

4.1 Lepton Asymmetries

The first possibility is via a lepton asymmetry [8]. B-$\overline{B}$ mixing is measured by looking for same sign dileptons; CP violation is indicated by a difference in the cross sections for producing positively and negatively charged dileptons:

$$A_l = \frac{N(l^+ l^+) - N(l^- l^-)}{N(l^+ l^+) + N(l^- l^-)} \ . \tag{11}$$

However. this is precisely the CP violation in the mixing matrix referred to earlier, i.e.,

$$A_l \simeq -4 \operatorname{Re} \epsilon_B \ . \tag{12}$$

As was pointed out earlier, ϵ_B is expected to be very small in the B system, which leads to very small predicted asymmetries [9]:

$$A_l \leq \begin{cases} O(10^{-3}), & B_d, \\ O(10^{-4}), & B_s. \end{cases} \tag{13}$$

To see such a small asymmetry requires 10^9-10^{10} B's. However, it is clearly still worth looking for, since the observation of a larger CP violating asymmetry would be clear evidence of physics beyond the standard model.

4.2 $B^\pm$ - Final State Interactions

Another possibility is to look for CP violation in the decays of charged B's [10]. CP can be violated if two different amplitudes contribute to the decay of a B^- (B^+) into a final state f ($\overline{f}$).

$$
\begin{aligned}
A(B^- \to f) &= |A_1| e^{i\delta_1} e^{i\phi_1} + |A_2| e^{i\delta_2} e^{i\phi_2} \\
A(B^+ \to \overline{f}) &= |A_1| e^{i\delta_1} e^{-i\phi_1} + |A_2| e^{i\delta_2} e^{-i\phi_2} \ .
\end{aligned}
\tag{14}
$$

Here, δ_i are the strong phase shifts (for example. isospin phases), and ϕ_i are the weak phases. The CP asymmetry is then

$$A_f^{+-} \propto \sin(\phi_1 - \phi_2) \sin(\delta_1 - \delta_2) \ . \tag{15}$$

As can be seen, a non-zero asymmetry requires (i) that there be a difference in the weak CP phases, and (ii) that the strong phases of the two amplitudes be different. Such an asymmetry is relatively easy to see experimentally. However, the calculations are unreliable, so that the theoretical interpretation of a positive signal would be difficult. Nevertheless, these CP asymmetries should still be searched for, although the implications for the CKM matrix would require more theoretical analysis.

I now turn to the most likely prospect for CP violation in the B system - that of hadronic decay asymmetries. If we consider a non-leptonic final state f such that both B^0 and $\overline{B^0}$ can decay both to it and its CP conjugate state $\overline{f}$, then CP violation is manifested in a non-zero value of [11]

$$A_f = \frac{\Gamma(B^0(t) \to f) - \Gamma(\overline{B^0}(t) \to \overline{f})}{\Gamma(B^0(t) \to f) + \Gamma(\overline{B^0}(t) \to \overline{f})} \ . \tag{16}$$

Using Eqn (8), and integrating over time, we obtain

$$A_f = -\frac{2x \, \mathrm{Im}\lambda_f}{2 + x^2 + x^2 |\rho_f|^2} \ , \tag{17}$$

where

$$\rho_f = \frac{A(\overline{B^0} \to f)}{A(B^0 \to f)} \ , \tag{18}$$

and

$$\lambda_f = \frac{1 - \epsilon_B}{1 + \epsilon_B} \, \rho_f \ . \tag{19}$$

Therefore a nonzero value of the imaginary part of λ_f would lead to a CP asymmetry. Note that λ_f, which is a product of $\Delta B = 2$ and $\Delta B = 1$ pieces, can be nonzero even if $\epsilon_B = 0$. Now, if f is not a CP eigenstate, then A_f depends on hadron dynamics in the ρ_f term, which leads to some theoretical difficulties in calculating the asymmetry. This is avoided by taking f to be a CP eigenstate. Furthermore, when only one combination of CKM matrix elements contributes to $B^0 \to f$, and another to $\overline{B^0} \to f$, then $|\rho_f| = 1$, i.e. ρ_f is a pure phase [12]. An example of this is shown in Fig. 3, where I have given the diagrams for B^0 and $\overline{B^0}$ decaying to ΨK_S. There,

$$A(B^0 \to f) \sim V_{bc}^* V_{cs} \ . $$
$$A(\overline{B^0} \to f) \sim V_{bc} V_{cs}^* \ . \tag{20}$$

In general, $|\rho_f|$ will be equal to 1 whenever the (quark level) decays $b \to u\bar{u}d$, $b \to u\bar{u}s$, $b \to c\bar{c}d$, or $b \to c\bar{c}s$ occur. For these cases, A_f takes the familiar form

$$A_f = -\frac{x}{1 + x^2} \, \mathrm{Im}\lambda_f \ . \tag{21}$$

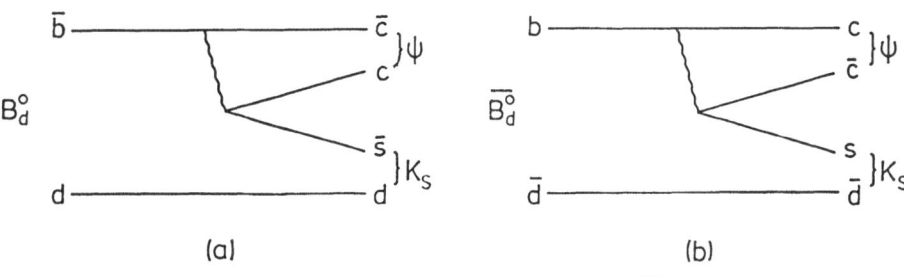

Fig. 3. Diagrams for (a) $B_d^0 \to \Psi K_S$ and (b) $\overline{B_d^0} \to \Psi K_S$.

It can also be shown that $\frac{1-\epsilon_B}{1+\epsilon_B}$ is pure phase, assuming that ϵ_B is small, and that the box diagram is dominated by the t-quark [13]:

$$\frac{1-\epsilon_B}{1+\epsilon_B} = \frac{\xi_t}{\xi_t^*} \,, \qquad \xi_t = V_{tb}V_{t\alpha}^* \,, \alpha = d, s \,. \tag{22}$$

As mentioned earlier, the standard model explanation for CP violation is that it is due to a non-zero phase in the CKM matrix. How are the phases ρ_f and $\frac{1-\epsilon_B}{1+\epsilon_B}$ related to the CKM phase? A parametrization of the CKM matrix which is convenient when discussing B physics is the following [14]:

$$V = \begin{pmatrix} 1 - \frac{1}{2}\lambda^2 & \lambda & A\rho\lambda^3 e^{i\delta} \\ -\lambda & 1 - \frac{1}{2}\lambda^2 & A\lambda^2 \\ A\lambda^3(1 - \rho e^{-i\delta}) & -A\lambda^2 & 1 \end{pmatrix} . \tag{23}$$

This form of the CKM matrix is an approximation, accurate to $O(\lambda^3)$, where λ is the Cabibbo angle, $\lambda \simeq 0.22$. With this parametrization, the only two elements which have "large" $(O(\lambda^3))$ phases are V_{ub} and V_{td}. It is these phases which will give rise to substantial CP violating asymmetries in B decays. (There are other elements with smaller phases which are relevant to the kaon system, but I shall ignore them here.) CP violating asymmetries measure one or the other of these phases, or both. Using this parametrization, $\frac{1-\epsilon_B}{1+\epsilon_B}$ takes on two values:

$$\frac{1-\epsilon_B}{1+\epsilon_B} = \begin{cases} \frac{V_{tb}^* V_{td}}{V_{tb} V_{td}^*} = \frac{V_{td}}{V_{td}^*} \equiv e^{2i\phi} & (B_d) \\ \frac{V_{tb}^* V_{ts}}{V_{tb} V_{ts}^*} = 1 & (B_s). \end{cases} \tag{24}$$

ρ_f also takes on two values, depending on how the b-quark decays:

$$\rho_f = \begin{cases} \frac{V_{ub}}{V_{ub}^*} \equiv e^{2i\delta} & \text{(Cabibbo suppressed)} \\ \frac{V_{cb}}{V_{cb}^*} = 1 & \text{(Cabibbo allowed)}. \end{cases} \tag{25}$$

Since λ_f is the product of ρ_f and $\frac{1-\epsilon_B}{1+\epsilon_B}$, there are three classes of model independent asymmetries which can be sizeable [15]. Each measures different combinations of KM matrix elements, all related to δ:

1) Cabibbo allowed B_d decays (e.g. $B_d \to \Psi K_S$)

$$\mathrm{Im}\lambda_1 \simeq \mathrm{Im}\frac{V_{td}}{V_{td}^*} = \sin 2\varphi = \frac{2\rho\sin\delta(1-\rho\cos\delta)}{1+\rho^2-2\rho\cos\delta} \tag{26}$$

2) Cabibbo suppressed B_d decays (e.g. $B_d \to \pi^+\pi^-$)

$$\mathrm{Im}\lambda_2 \simeq \mathrm{Im}\frac{V_{td}}{V_{td}^*}\frac{V_{ub}}{V_{ub}^*} = \sin 2(\phi + \delta) = \frac{2\sin\delta(\cos\delta - \rho)}{1+\rho^2-2\rho\cos\delta} \tag{27}$$

3) Cabibbo suppressed B_s decays (e.g. $B_s \to \rho^0 K_S$)

$$\mathrm{Im}\lambda_3 \simeq \mathrm{Im}\frac{V_{ub}}{V_{ub}^*} = \sin 2\delta = 2\sin\delta\cos\delta \tag{28}$$

These are the only CP asymmetries in B decay which are completely calculable from the KM matrix, without additional (unreliable) hadronic information. I will now discuss the ranges of these asymmetries which are allowed by current data.

In addition to the phase, δ, and the Cabibbo angle, λ, there are two other parameters in the CKM matrix in Eqn (23), A and ρ. These are, in principle, obtainable from B decay. The B lifetime fixes A (through V_{bc}) to be $A = 1.05 \pm 0.17$ [16]; B semileptonic decays give (conservatively) $\rho < 0.9$ [17], and the preliminary ARGUS result for V_{ub} [18] yields $\rho > 0.3$. Note that CLEO has not confirmed this ARGUS result [4], so the lower bound is not firm. In any case, we do not take any fixed value of ρ in our analysis, but allow it to vary in the range $0 \leq \rho \leq 0.9$.

Now, δ is constrained by the measurements of ϵ and x_d (the current measurement of ϵ' does not yield constraints better than those of ϵ and x_d [19]). The theoretical expression for ϵ is given by [20]

$$
|\epsilon| = \frac{G_F^2 f_K^2 M_K M_W^2}{6\sqrt{2}\pi^2 \Delta M_K} B_K \left(A^2 \rho \lambda^6 \sin \delta \right) \left(y_c \left\{ \eta_3 f_3(y_c, y_t) - \eta_1 \right\} \right.
$$
$$
\left. + \eta_2 y_t f_2(y_t) A^2 \lambda^2 (1 - \rho \cos \delta) \right) , \tag{29}
$$

where $y_i = m_i^2 / M_W^2$, f_2 and f_3 are weakly dependent functions of the top and charm masses, and the η_i are QCD corrections. The two important unknowns in Eqn (29) are the top quark mass and the bag parameter, B_K. The mass of the top quark is constrained by both experimental and theoretical considerations. Direct searches [21] put a lower limit $m_t > 41$ GeV, while the upper bound $m_t \leq 180$ GeV results from the study of radiative corrections within the standard model [22]. B_K encapsulates our present ignorance of the matrix element of $\left(\bar{d}\gamma^\mu(1 - \gamma_5)s \right)^2$ between K^0 and $\overline{K^0}$. A reasonable range is $1/3 \leq B_K \leq 1$, with $B_K = 1$ corresponding to the vacuum insertion approximation.

Theoretically, x_d receives its dominant contribution from the presence of top quarks in the box diagram and one finds [23]

$$
x_d = \tau_B \frac{G_F^2}{6\pi^2} M_B M_W^2 \left(f_{B_d}^2 B_{B_d} \right) \eta_B y_t f_2(y_t) \left\{ A^2 \lambda^6 \left(1 + \rho^2 - 2\rho \cos \delta \right) \right\} , \tag{30}
$$

where η_B is a QCD correction factor. Here the hadronic uncertainty is hidden in the factor $f_{B_d}^2 B_{B_d}$, whose meaning is analogous to that of the corresponding quantities in the kaon system, except that here also f_B is not measured. There are a large number of estimates for this quantity, most of which differ from one another, but $(100 \text{ MeV})^2 \leq f_{B_d}^2 B_{B_d} \leq (200 \text{ MeV})^2$ includes most of them. We now fit the theoretical expressions to the experimental numbers, at 90% c.l. Fig. 4 shows the allowed region in the ρ-δ plane for the values $B_K = 1/3, 2/3$ and 1, but allowing m_t and $f_{B_d}^2 B_{B_d}$ to vary over their entire ranges. As is clear from the figures, the allowed area for the CKM matrix parameters is quite sensitive to the value of B_K taken.

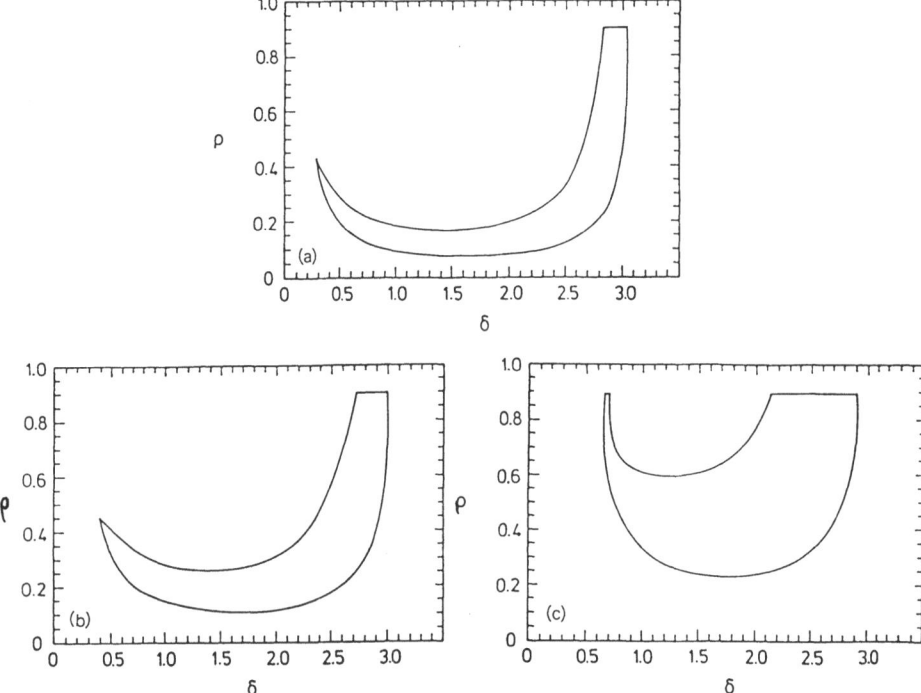

Fig. 4. The domain in ρ-δ space (δ in radians), within which the standard model is compatible with the measurements of ϵ and x_d (90% c.l.). We allow m_t to vary between 41 GeV and 180 GeV, and $(f_{B_d}^2 B_{B_d})^{1/2}$ between 100 MeV and 200 MeV: (a) $B_K = 1$, (b) $B_K = 2/3$, (c) $B_K = 1/3$.

For the CP violating asymmetries (Eqs (26)-(28)), we will take B_K to be 2/3 for the purpose of illustration. In this case the allowed regions are shown in Fig. 5. The region for $\mathrm{Im}(\lambda_1)$ is the smallest, with a maximum value of about 0.4 allowed. On the other hand, $\mathrm{Im}(\lambda_2)$ and $\mathrm{Im}(\lambda_3)$ are allowed quite sizable values. However, recall that the asymmetries are reduced by a factor $x/(1 + x^2)$ (Eqn (21)). For B_d mesons, this is a factor of about 0.5. But for B_s mesons, this factor can be quite a bit smaller. Within the standard model, one expects

$$\frac{x_s}{x_d} \simeq \left| \frac{V_{ts}}{V_{td}} \right|^2 , \tag{31}$$

which is $O(\lambda^{-2})$, leading to a conservative lower limit of $x_s > 3$ [15]. This gives a reduction factor of 0.3, but it is likely to be considerably smaller. Therefore, unless x_s is smaller than expected in the standard model, B_d decays, both Cabibbo allowed and Cabibbo suppressed, appear to be the best prospects for seeing CP violation in the B system via time-integrated methods.

There are more complications when considering the experimental side of things. Let us first consider threshold e^+e^- machines, such as DORIS, CESR, or a possible B-factory. The hadronic CP asymmetries were calculated assuming that one knew

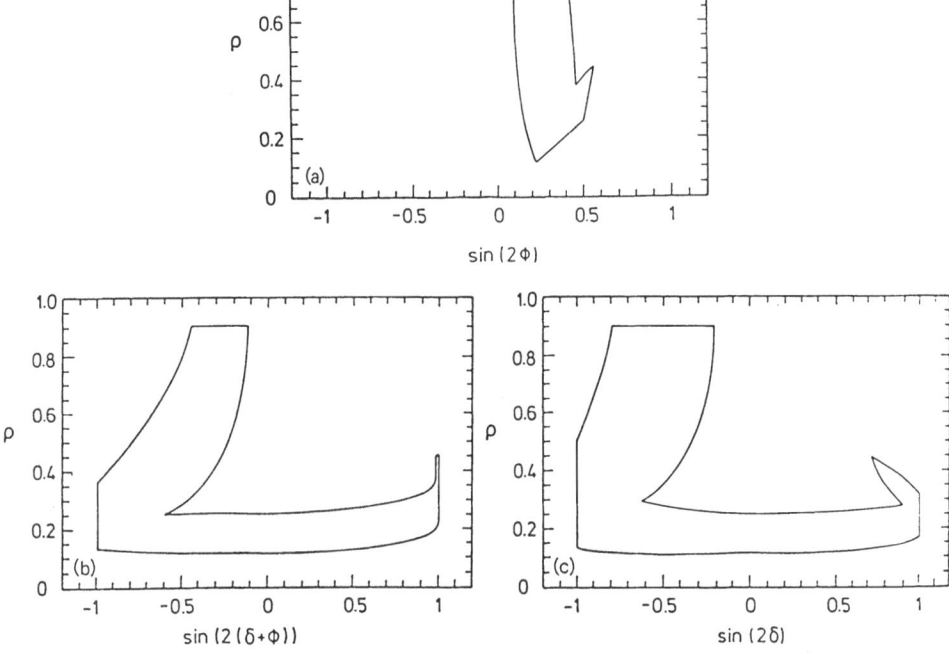

Fig. 5. Varying $(f_{B_d}^2 B_{B_d})^{1/2}$ and m_t, 100 MeV $\leq (f_{B_d}^2 B_{B_d})^{1/2} \leq$ 200 MeV and 41 GeV $\leq m_t \leq$ 180 GeV, and fixing $B_K = 2/3$, the areas within which the standard model is compatible with the measurements of ϵ and x_d (90% c.l.) are shown for the following parameter spaces: (a) $(\sin 2\varphi, \rho)$, (b) $(\sin 2(\delta + \phi), \rho)$, (c) $(\sin 2\delta, \rho)$.

whether it was a B or $\overline{B}$ which decayed. However, since pure B or $\overline{B}$ beams do not exist, one has to "tag" on the other B in order to know whether it was B or $\overline{B}$ which decayed. This is done at such machines by looking for its semileptonic decay. Using the semileptonic tag, the (pre-time-integrated) branching ratios for a $B\overline{B}$ pair to decay into a final state f and a leptonic tag are [24]

$$B.R.(B(t)\overline{B}(\bar{t})|_{C=\mp 1} \rightarrow f + (Dl\bar{\nu}X)_{\text{tag}})$$
$$\propto e^{-\Gamma(t+\bar{t})}\{1 - \sin(\Delta m(t \mp \bar{t}))\, \text{Im}\lambda_f\} \ . \tag{32}$$

$$B.R.(\overline{B}(t)B(\bar{t})|_{C=\mp 1} \rightarrow f + (\overline{D}\,l\nu\overline{X})_{\text{tag}})$$
$$\propto e^{-\Gamma(t+\bar{t})}\{1 + \sin(\Delta m(t \mp \bar{t}))\, \text{Im}\lambda_f\} \ . \tag{33}$$

The important point is that for $C = -1$, i.e. for the $B\overline{B}$ pair in an odd relative angular momentum state, the asymmetry vanishes when the times t and $\bar{t}$ are treated symmetrically. This is precisely the case at the $\Upsilon(4s)$, so that the $\Upsilon(4s)$ *cannot be used to see time-integrated CP asymmetries*. Above the $\Upsilon(4s)$, an L-even state can be produced from the decay of a $B\overline{B}^* + c.c.$ state. Unfortunately, the cross section is quite a bit smaller at this energy. There has also been a suggestion to use asymmetric e^+e^- beams at the $\Upsilon(4s)$. This would have the effect of separating the decay vertices, so that it might be possible to avoid integrating over the whole time distribution, and thereby partially evade the suppression of CP asymmetries from the L-odd state [25]. Although it is not clear whether or not this can be done, we will see that an asymmetric $\Upsilon(4s)$ machine could be quite useful for the measurement of time-dependent CP violating effects.

At hadron machines, there is an additional possibility for tagging. If the B^0 or $\overline{B^0}$ is produced along with a charged B, then the charge of the $B^\pm$ tags the neutral B. However, this requires full reconstruction of the charged B, which is quite difficult. For seeing asymmetries, fixed target machines, such as TEV II or UNK, may have some advantages over hadron colliders. Both types of machines produce enormous numbers of B's, and both have quite large backgrounds. However, in fixed target experiments, the produced B's are quite boosted, so that they travel a considerable distance before decaying. This will help in reducing some of the background. For example, experiment E771 at Fermilab has suggested looking for $B_d \to \Psi X$ [26]. One background problem comes from Ψ's produced at the vertex. However, since the B's are boosted, it is possible to look for $\mu^+\mu^-$ pairs coming from a secondary vertex, which is a clear signal of a Ψ coming from a B decay. Nevertheless, it is not yet certain whether the background can be reduced sufficiently to see clear evidence for CP violation.

Regardless of which machine is used, the number of B's required to see CP violation is quite large. The effective branching ratios (which include efficiencies for detecting secondary decays) are all of order 10^{-5}. To see CP violation at a 3σ level requires at least 100 B's. And the tagging will cost at least a factor of 10. Therefore the observation of CP violation will require at least 10^8 $B\overline{B}$ pairs. Although this number is quite large, it is quite typical of all of this type of time-integrated CP violating asymmetries. The one consolation is that the situation would be worse if the asymmetries were smaller than 10%. Fortunately, as Fig. 5 shows, the standard model appears to favour large asymmetries.

4.4 Time-dependent CP Asymmetries

Finally, there is the possibility of time-dependent CP asymmetries [27]. The probability of obtaining a final CP eigenstate f at time t for a beam which at $t = 0$ was pure B^0 is

$$N_f(t) = N_f(0)e^{-\Gamma t}\left[1 - \mathrm{Im}\lambda_f \sin \Delta m t\right] ; \qquad (34)$$

For $\overline{B^0}(t)$, it is

$$N_f(t) = N_f(0)e^{-\Gamma t}\left[1 + \mathrm{Im}\lambda_f \sin \Delta m t\right] . \qquad (35)$$

First of all, if the time development of such a decay, for either case, were measured, a simple deviation from an exponential would signal CP violation. More importantly, however, there can be some quite spectacular effects. In Fig. 6a are shown the two curves for $B_d(\overline{B_d}) \to f$ ($f = \Psi K_S$, for instance), for the ARGUS/CLEO result $x_d = 0.70$, and $\mathrm{Im}(\lambda_f)=0.3$. Depending on the detector, it might be possible to separate the two curves, although this could be difficult. However, for the B_s system, the effects can be extremely large. Fig. 6b contains the graphs for $B_s(\overline{B_s}) \to f$ (e.g. $f = \rho K_S$) for $x_s = 15$ and for

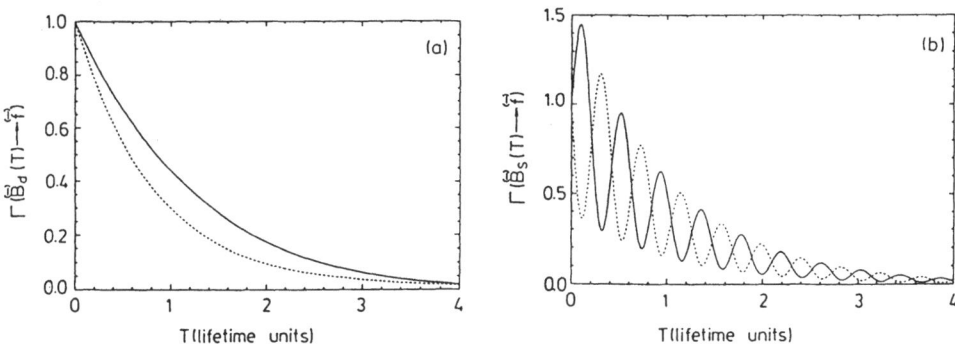

Fig. 6. The time dependence for $B \to f$ (dotted line) and $\overline{B} \to f$ (solid line) are shown: (a) $B_d(\overline{B_d}) \to f_d$, $x_d = 0.70$, $\text{Im}\lambda_f = 0.3$, (b) $B_s(\overline{B_s}) \to f_s$, $x_s = 15.0$, $\text{Im}\lambda_f = -0.6$.

$\text{Im}(\lambda) = -0.6$. Here the differences between the two curves are enormous, and because of the large value of x_s, there are many oscillations within a few lifetimes.

Experimentally, there are some quite difficult problems, of course. To get the time dependence of the B's, extremely good vertex resolution is required. Furthermore, as written, these decays occur in the rest frame of the decaying B. The transformation to this frame requires complete reconstruction of the B, and with very good energy resolution. Finally, these decays also need tagging, and the rates are, as ever, quite small. Therefore, for the same reasons as in the time-integrated case, the number of $B\overline{B}$ pairs needed to see CP violation via time-dependent measurements is at least 10^8.

There are several possibilities for such experiments. The most important requirement is simply that the B travel a significant average distance before decaying, in order to be able to obtain reasonable time resolution. One possibility is to look at $B\overline{B}$ pairs produced at a Z factory [28]. Here the average decay length is a few millimeters. In addition, with polarization, the forward-backward asymmetry could be used to separate B and $\overline{B}$. Secondly, fixed target machines could also be used for such time-dependent measurements. For example, at TEV II, a B travels an average of 7000 microns before decaying, while at UNK, it could travel up to several centimeters. Finally, there is the possibility of using an asymmetric e^+e^- collider at the $\Upsilon(4s)$ [29]. Although the $B_d\overline{B_d}$ pair is produced essentially at rest in the centre-of-mass frame, the asymmetric energies give the B's a boost in the lab frame. Therefore, time-dependent measurements may be possible.

5. SUMMARY AND CONCLUSIONS

In conclusion, there are a number of possibilities for the observation of CP violating phenomena in the B system. Lepton asymmetries are predicted to be quite small in the standard model but should still be searched for as evidence of new physics. CP violation due to final state interactions is rather easy to see experimentally but, because of hadronic uncertainties, would be difficult to interpret in terms of implications for the CKM matrix. Within the standard model, the most promising possibility is looking for CP asymmetries in nonleptonic decays. There are two ways to search for these, using time-integrated and time-dependent means. In the time-integrated case, there are three classes of such asymmetries – B_d decays, both Cabibbo allowed and Cabibbo suppressed, and Cabibbo suppressed B_s decays. The CP asymmetries for B_d decays are predicted to be rather large. However, due to the (expected) large value of x_s, the asymmetries for the B_s decays should be quite small. Therefore, in the standard model, decays of B_d's are the best place to search for such CP asymmetries. Time-dependent methods look for differences in the time distributions of B and $\overline{B}$ mesons decaying to a final state f. Depending on the parameters of the CKM matrix, it may or may not be possible to see such differences in the decays of B_d mesons. However, decays of B_s mesons should yield spectacular effects. Regardless of whether time-integrated or time-dependent methods are used, the number of $B\overline{B}$ pairs required to see a CP asymmetry at the 3σ level is at least 10^8. Therefore the observation of CP violation in the B system will take a lot of work. But the reward, namely the possibility of determining whether or not the CKM matrix explains CP violation, is most certainly worth it.

REFERENCES

[1] J. H. Christenson, J. W. Cronin, V. L. Fitch, and R. Turlay, Phys. Rev. Lett. **13** (1964) 138.

[2] NA31 Collaboration: H. Burkhardt et al, Phys. Lett. **206B** (1988) 169. See also A. Nappi, these proceedings.

[3] ARGUS Collaboration: H. Albrecht et al, Phys. Lett. **192B** (1987) 245.

[4] CLEO Collaboration: A. Jawahery, *Proceedings of the XXIV International Conference on High Energy Physics*, Munich, W. Germany (1988).

[5] M. Aguilar-Benitez et al, (Particle Data Group), Phys. Lett. **170B** (1986) 1.

[6] For more details, see R. Peccei, *Proceedings of the Workshop on the Experimental Program at UNK*, Protvino, 1987, and references therein.

[7] See, for example, A. Buras, H. Steger, and W. Slominski, Nucl. Phys. **B238** (1984) 529.

[8] L. Okun, V. Zakharov, and B. Pontecorvo, Lett. Nuovo. Cim **13**, (1975) 218; A. Pais and S. B. Treiman, Phys. Rev. **D12**, (1975) 2744.

[9] J. Ellis et al, CERN preprint CERN-TH.4816/87 (1987); P. J. Franzini, CERN preprint CERN-TH.4846/87 (1987).

[10] M. Bander, D. Silverman, and A. Soni, Phys. Rev. Lett. **43** (1974) 242; J. Bernabeu and C. Jarlskog, Z. Phys. **C8** (1981) 233; L. L. Chau and H. Y. Cheng, Phys. Rev. Lett. **53** (1984) 1037; **59** (1987) 958; I. I. Bigi and A. I. Sanda, Nucl. Phys. **B281** (1987) 41.

[11] A. Carter and A. I. Sanda, Phys. Rev. Lett. **45** (1980) 952; Phys. Rev. **D23** (1981) 1567; I. I. Bigi and A. I. Sanda, Nucl. Phys. **B193** (1981) 85; **B281** (1987) 41; Y. Azimov, V. Khoze, and M. Uraltsev, Yad. Fiz. **45** (1987) 1412; D. Du, I. Dunietz, and D. Wu, Phys. Rev. **D34** (1986) 3414.

[12] I. Dunietz and J. Rosner, Phys. Rev. **D34** (1986) 1404; See also D. Du. I. Dunietz, and D. Wu, Phys. Rev. **D34** (1986) 3414 and I. I. Bigi and A. I. Sanda, Nucl. Phys. **B281** (1987) 41.

[13] See for example, I. Dunietz and J. Rosner. Phys. Rev. **D34** (1986) 1404.

[14] L. Maiani, Phys. Lett. **62B** (1976) 183: L. Wolfenstein, Phys. Rev. Lett. **51** (1983) 1945.

[15] P. Krawczyk, H. Steger, D. London, and R. Peccei. Nucl. Phys. **B307** (1988) 19.

[16] G. Altarelli and P. Franzini, Z. Phys. **C37** (1988) 271.

[17] M. Gilchriese, *Proceedings of the XXIII International Conference on High Energy Physics*. Berkeley. USA (1986).

[18] ARGUS Collaboration: W. Schmidt-Parzefall. *1987 International Symposium on Lepton and Photon Interactions at High Energies*, ed. W. Bartel and R. Rückl (North-Holland. Hamburg, 1987) p. 257.

[19] G. Altarelli and P. J. Franzini, CERN preprint CERN-TH.4914/87 (1987).

[20] A. J. Buras, W. Slominski and H. Steger, Nucl. Phys. **B238** (1984) 529; **B245** (1984) 369.

[21] UA1 Collaboration: C. Albajar et al, Z. Phys **C37** (1988) 505; G. Altarelli, M. Diemoz, G. Martinelli, and P. Nason. CERN preprint CERN-TH-4978/88 (1988).

[22] U. Amaldi et al, Phys. Rev. **D36** (1987) 1385.

[23] A. J. Buras, W. Slominski and H. Steger, Nucl. Phys. **B238** (1984) 529; **B245** (1984) 369; J. Hagelin, Nucl. Phys. **B193** (1981) 123.

[24] I. I. Bigi and A. I. Sanda. Nucl. Phys. **B281** (1987) 41.

[25] J. L. Rosner. A. I. Sanda, and M. P. Schmidt, *Proceedings of the Workshop on High Sensitivity Beauty Physics at Fermilab*, ed. A. J. Slaughter, N. Lockyer, and M. P. Schmidt (Fermi National Accelerator Laboratory, Batavia, 1987) p. 165.

[26] B. Cox et al, "A Proposal to Study Beauty Production and Other Heavy Quark Physics Associated With Dimuon Production in 800-925 GeV/c PP Interactions", Fermilab Proposal 771 (April 1986). See also B. Cox, these proceedings.

[27] I. Dunietz and J. Rosner, Phys. Rev. **D34** (1986) 1404; Y. Azimov, V. Khoze, and M. Uraltsev, Yad. Fiz. **45** (1987) 1412;

[28] W. B. Atwood, I. Dunietz, and P. Grosse-Wiesmann, SLAC preprint SLAC-PUB-4544 (1988).

[29] R. Aleksan, J. E. Bartelt, P. Burchat. and A. Seiden, SLAC preprint SLAC-PUB-4673 (1988).

13

A THEORETICAL DETERMINATION OF THE PC VIOLATING PHASE ϕ_{00}

Michel Gourdin

Université Pierre et Marie Curie
Paris - France

I. INTRODUCTION

The problem of the violation of PC in the $K^{\circ}\bar{K}^{\circ}$ system is now 24 years old and many efforts have been made, from both the experimental and the theoretical sides, in order to know the magnitude and to understand the origin of the violation of PC in Particle Physics.

A reasonable approach of the problem will proceed, in my opinion, into three steps. The first one is to construct a framework for the analysis of the experimental data. This framework is based on quantum mechanics using, in particular, the superposition principle and the conservation of probabilities and on the isotopic spin symmetry of strong interactions broken by weak interactions responsible of the decay of neutral K mesons.

The second step is a phenomenological analysis of the experimental data taking into account the two following facts occuring simultaneously in the decay : the violation of PC and the violation of the empirical non leptonic $|\Delta I| = 1/2$ rule in the $K \to 2\pi$ decay.

The last step is a comparison between experiment and theory which essentially means for the present time the one provided by the standard model in its simplest version with three generations of quarks and leptons.

After a brief review of the necessary basic definitions and relations of parameters and a quotation of the most relevant experimental results we restrict ourselves, in this lecture, to the step two of the discussion.

II THEORETICAL FRAMEWORK

1°) We first write the eigenstates of the mass matrix having well defined lifetimes. Assuming TCP invariance we introduce only one complex parameter ε

$$|K_S> = \frac{1}{[2(1+|\varepsilon|^2)]^{1/2}} \, [\, (1+\varepsilon) \, |K^0> + (1-\varepsilon)| \, \bar{K}^{\circ}> \,] \tag{1}$$

$$|K_L> = \frac{1}{[2(1+|\varepsilon|^2)]^{1/2}} \, [\, (1-\varepsilon) \, |K^0> + (1+\varepsilon)| \, \bar{K}^{\circ}> \,] \tag{2}$$

The hermitian product of these two eigenstates is non zero and observable. With TCP invariance it is real

$$\langle K_S | K_L \rangle = \frac{2 \operatorname{Re} \varepsilon}{1 + |\varepsilon|^2} \tag{3}$$

2°) PC violation in the 2π decay modes.
It is usual to introduce two complex parameters as follows

$$\eta_{+-} = \frac{\langle \pi^+\pi^- | T | K_L \rangle}{\langle \pi^+\pi^- | T | K_S \rangle}$$

$$\tag{4}$$

$$\eta_{00} = \frac{\langle \pi^0\pi^0 | T | K_L \rangle}{\langle \pi^0\pi^0 | T | K_S \rangle}$$

We also experimentally know the partial decay widths $\Gamma(K_S \to \pi^0\pi^0)$, $\Gamma(K_S \to \pi^+\pi^-)$ with their ratio R defined by

$$R = \frac{\Gamma(K_S \to \pi^0\pi^0)}{\Gamma(K_S \to \pi^+\pi^-)} \tag{5}$$

and the 2π decay mode for the charged K meson $\Gamma(K^\pm \to \pi^\pm\pi^0)$

3°) Because of the Bose Einstein statistics the 2π final states have only possible total isotopic spins I=0 or I=2. We then define

$$\varepsilon_0 = \frac{\langle 0 | T | K_L \rangle}{\langle 0 | T | K_S \rangle} \tag{6}$$

$$\varepsilon_2/\omega = \frac{\langle 2 | T | K_L \rangle}{\langle 2 | T | K_S \rangle} \tag{7}$$

$$\omega = \frac{1}{\sqrt{2}} \frac{\langle 2 | T | K_S \rangle}{\langle 0 | T | K_S \rangle} \tag{8}$$

and the related quantity

$$\varepsilon' = \varepsilon_2 - \omega \, \varepsilon_0 \tag{9}$$

A trivial algebra leads to the relations

$$\eta_{+-} = \frac{\varepsilon_0 + \varepsilon_2}{1 + \omega} \tag{10}$$

$$\eta_{00} = \frac{\varepsilon_0 - 2 \varepsilon_2}{1 - 2 \omega} \tag{11}$$

$$R = \frac{1}{2} \left| \frac{1-2\omega}{1+\omega} \right|^2 \text{(PS)}$$

where (PS) is a phase space correction factor due to the $\pi^\pm$-π^0 mass difference

$$\text{(PS)} = 1,0148$$

In what follows instead of R we shall use the parameter ρ^2 defined by

$$\rho^2 = \frac{2R}{(PS)} = \left| \frac{1-2\omega}{1+\omega} \right|^2 \qquad (12)$$

4°) The analysis of the experimental date is conveniently made after introducing the reduced amplitudes A_I free of the strong interaction phases δ_I for the 2π system of isotopic spin I taken at the K meson mass

$$< I \mid T \mid K^\circ > = A_I\, e^{i\delta_I} \qquad\qquad I=0,2$$

Using TCP invariance

$$< I \mid T \mid \overline{K^\circ} > = A_I^*\, e^{i\delta_I} \qquad\qquad I=0,2$$

The complex nature of the reduced amplitudes A_I is of course due to the violation of PC in the phase convention for $|K^\circ>$ and $|\overline{K^\circ}>$ we have chosen.

We now obtain the expression of ε_0, ε_2/ω and ω

$$\varepsilon_0 = \frac{\varepsilon\, \mathrm{Re}\, A_0 + i\, \mathrm{Im}\, A_0}{\mathrm{Re}\, A_0 + i\, \varepsilon\, \mathrm{Im}\, A_0}$$

$$\varepsilon_2/\omega = \frac{\varepsilon\, \mathrm{Re}\, A_2 + i\, \mathrm{Im}\, A_2}{\mathrm{Re}\, A_2 + i\, \varepsilon\, \mathrm{Im}\, A_2} \qquad (13)$$

$$\omega = \frac{1}{\sqrt{2}}\, e^{i(\delta_2-\delta_0)}\, \frac{\mathrm{Re}\, A_2 + i\, \varepsilon\, \mathrm{Im}\, A_2}{\mathrm{Re}\, A_0 + i\, \varepsilon\, \mathrm{Im}\, A_0}$$

From these formulas we immediately get the TCP constraints

$$\frac{\mathrm{Re}\, \varepsilon_0}{1+|\varepsilon_0|^2} = \frac{\mathrm{Re}\, \varepsilon_2/\omega}{1+|\varepsilon_2/\omega|^2} = \frac{\mathrm{Re}\, \varepsilon}{1+|\varepsilon|^2} \qquad (14)$$

Let us notice that the complex parameters ε_0, ε_2 and ω are independent of any phase convention. This is not the case for ε where only the real part has a physical meaning.

5°) Finally anticipating on the smallness of the violation of PC in neutral K meson decay $O(10^{-3})$ a first order calculation seems to be fully justified. The previous expressions take simpler forms

$$\varepsilon_0 \simeq \varepsilon + i\, \frac{\mathrm{Im}\, A_0}{\mathrm{Re}\, A_0} \qquad (15)$$

$$\varepsilon_2/\omega \simeq \varepsilon + i\, \frac{\mathrm{Im}\, A_2}{\mathrm{Re}\, A_2} \qquad (16)$$

$$\omega = \frac{1}{\sqrt{2}}\, e^{i(\delta_2-\delta_0)}\, \frac{\mathrm{Re}\, A_2}{\mathrm{Re}\, A_0} \qquad (17)$$

and the TCP constraints reduce to

$$\mathrm{Re}\, \varepsilon_0 = \mathrm{Re}\, \varepsilon_2/\omega = \mathrm{Re}\, \varepsilon \qquad (18)$$

For the parameter ε' we obtain the expression

197

$$\varepsilon' \simeq \frac{1}{\sqrt{2}} e^{i(\delta_2 - \delta_0)} \frac{\text{Re } A_2}{\text{Re } A_0} \left[\frac{\text{Im } A_2}{\text{Re } A_2} - \frac{\text{Im } A_0}{\text{Re } A_0} \right] \qquad (19)$$

III EXPERIMENTAL RESULTS

1°) Let us first consider the PC violating parameters η_{+-} and η_{00}. From the Particle Data Tables we get

$$|\eta_{+-}| = (2,275 \pm 0,021)10^{-3} \qquad \phi_{+-} = 44°6 \pm 1°2$$
$$|\eta_{00}| = (2,299 \pm 0,036)10^{-3} \qquad \phi_{00} = 54° \pm 5°$$

From the experiment NA 31 at CERN we have

$$\left| \frac{\eta_{00}}{\eta_{+-}} \right|^2 = 0,980 \pm 0,004 \pm 0,003$$

Therefore we have three independent experimental results involving $|\eta_{+-}|$ and $|\eta_{00}|$. We now try to extract from these data the best information concerning the difference $|\eta_{+-}| - |\eta_{00}|$.

Using the Particle Data Tables we have

$$\frac{|\eta_{00}|}{|\eta_{+-}|} = 1,0105 + 0,0251 \, \varepsilon_M \qquad (20)$$

$$|\eta_{+-}| - |\eta_{00}| = (-2,4 - 5,7 \, \varepsilon_M) \, 10^{-5} \qquad (21)$$

where, in the one standard deviation limit for $|\eta_{+-}|$ and $|\eta_{00}|$, the error parameter ε_M is restricted by

$$-1 \leqslant \varepsilon_M \leqslant +1 \qquad (22)$$

On the other hand we have a direct measurement of the ratio $|\eta_{00}| / |\eta_{+-}|$ and adding the errors in quadrature we get

$$\frac{|\eta_{00}|}{|\eta_{+-}|} = 0,9899 \pm 0,0032 \qquad (23)$$

Comparing now the equations (20) and (23) we can solve in ε_M

$$\varepsilon_M = -0,82 \pm 0,13 \qquad (24)$$

which clearly satisfies the bounds (22) showing the compatibility of the two sets of data.
Inserting now this value of ε_M in the equation (21) we obtain a more occurate estimate of the differences of the moduli

$$|\eta_{+-}| - |\eta_{00}| = (2,27 \mp 0,74) \, 10^{-5} \qquad (25)$$

Such a difference is non zero and positive by three standard deviations

2°) We now consider the ratio R. From the particle data tables we have the following branching ratios for $K_S \rightarrow 2\pi$ decays

$$\text{BR } (Ks \rightarrow \pi^+ \pi^-) = 68.61$$
$$(\qquad \pm 0,24 \,) \,\%$$
$$\text{BR } (Ks \rightarrow \pi°\pi°) = 31.39$$

and we deduce

$$R = 0,4575 \pm 0,0051$$

Taking into account the phase space correction factor (PS) we finally get

$$\rho^2 = 0{,}9017 \pm 0{,}0100$$

3°) The information for the phase difference $\delta_2 - \delta_0$ is coming from π-π phase shift analysis and from K_{e_4} decay.

A weighted average, quoted by Devlin and Dickey gives

$$\delta_2 - \delta_0 = -41°4 \pm 8°1$$

The analysis of $K \rightarrow 2\pi$ and $K \rightarrow 3\pi$ data by the same authors produces a similar value

$$\delta_2 - \delta_0 = -45°3 \pm 5°1$$

A more recent analysis of π-π scattering by Biswas et al gives a different value

$$\delta_2 - \delta_0 = -29°2 \pm 3°$$

In what follows we shall use the first quoted weighted average value.

IV THE INCONSISTENCY PROBLEM

1°) Let us solve the equations (10) and (11) in ε_0 and ε_2

$$\varepsilon_0 = \frac{1}{3} (2\eta_{+-} + \eta_{00}) + \frac{2}{3} \omega (\eta_{+-} - \eta_{00}) \qquad (26)$$

$$\varepsilon_2 = \frac{1}{3} \omega (\eta_{+-} + 2\eta_{00}) + \frac{1}{3} (\eta_{+-} - \eta_{00}) \qquad (27)$$

At first order in the violation of PC the constraint (18) due to TCP invariance is equivalently written

$$Re\ (\frac{\varepsilon_2}{\omega} - \varepsilon_0) = 0 \qquad (28)$$

With the equations (26) and (27) we get

$$Re\ [\ \frac{\eta_{+-} - \eta_{00}}{\omega}\ (1+\omega)\ (1-2\omega)\ ] = 0 \qquad (29)$$

Let us consider first the approximation where the violation of the non leptonic $|\Delta I| = \frac{1}{2}$ rule can be neglected in equation (29). From equation (17) the phase of ω is $\phi_\omega = \delta_2 - \delta_0$ and the TCP relation (29) implies

$$Phase\ (\eta_{+-} - \eta_{00}) = \delta_2 - \delta_0 + \frac{1}{2} \pi \equiv \phi \qquad \text{modulo } \pi \qquad (30)$$

Using first the particle table data for η_{+-} and η_{00} we obtain, within one standard deviation limit

$$\frac{Im\ (\eta_{+-} - \eta_{00})}{Re\ (\eta_{+-} - \eta_{00})} = -0{,}98 \qquad (31)$$

On the other hand with the estimate $\delta_2 - \delta_0 = -41°4 \pm 8°1$ we have $\phi = 48°6 \pm 8°1$ and we get

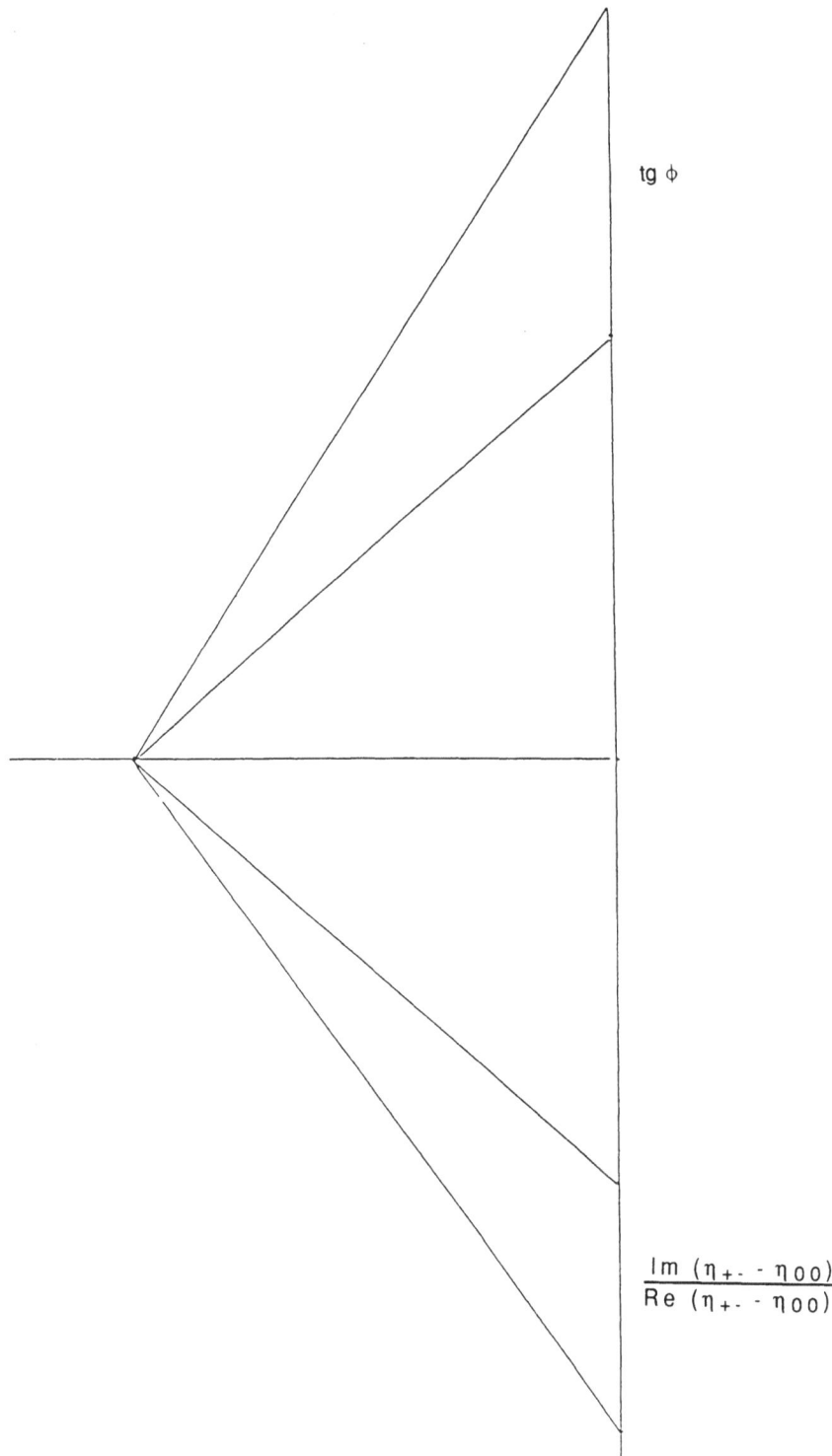

tg ϕ

$$\frac{\text{Im } (\eta_{+-} - \eta_{00})}{\text{Re } (\eta_{+-} - \eta_{00})}$$

<u>Figure 1</u>   the two domains for $\dfrac{\text{Im } (\eta_{+-} - \eta_{00})}{\text{Re } (\eta_{+-} - \eta_{00})}$ and tg ϕ allowed within one standard deviation limits.

$$\text{tg } \phi = 1,13 \begin{array}{c} +0,39 \\ -0,28 \end{array} \tag{32}$$

We clearly observe an inconsistency between the two numbers (31) and (32) which have to be identical because of TCP. It is an inconsistency between the experimental data on the two PC violating parameters η_{+-} and η_{00} on one side and the value of the difference of phases $\delta_2 - \delta_0$ due to strong interactions. Figure 1 illustrates the situation

2°) A first explanation of such an inconsistency is that the constraint (28) due to TCP invariance is not satisfied. Nevertheless before to conclude into a violation of TCP invariance in neutral K meson decay it is probably more reasonable to look more carefully at the data.

The experimental values of η_{+-} and η_{00} are very close to each other and, as a consequence, the difference $\eta_{+-} - \eta_{00}$ is a third complex number whose phase and modulus are not accurately measured.

We shall adopt this point of view in our analysis of the experimental data and in particular we shall look at which value of the phase difference $\phi_{00} - \phi_{+-}$ the previous inconsistency will disappear.

V PARAMETER ω

1°) Before to proceed into our program it is convenient to study first the complex parameter ω which describes the violation of the non-leptonic $|\Delta I| = \frac{1}{2}$ rule in $K_S \rightarrow 2\pi$ decay. Let us remember that, in the approximation where only the first order effects in the violation of PC are retained, the parameter ω is free of such effects.

2°) The basic formula is

$$\rho^2 = |\frac{1-2\omega}{1+\omega}|^2 \tag{33}$$

The quantity $\rho 2$ is deduced from the experimental ratio R for $K_S \rightarrow 2\pi$ decay and, from equation (17) we already know that the phase of ω is simply $\phi_\omega = \delta_2 - \delta_0$.

Equation (33) leads to a second order equation in $|\omega|$. For obvious physical reasons related to the $|\Delta I| = \frac{1}{2}$ rule we retain only the small $|\omega|$ solution which is given by

$$|\omega| = \frac{2+\rho^2}{4-\rho^2} \text{ Cos } \phi_\omega - [\frac{(2+\rho^2)^2}{(4-\rho^2)^2} \text{ Cos}^2 \phi_\omega - \frac{1-\rho^2}{4-\rho^2}]^{\frac{1}{2}} \tag{34}$$

The quantity $|\omega|$ as a function of the angle $\delta_0 - \delta_2$ is represented on Figure 2 where the two curves correspond to one standard deviation for ρ^2. For $\delta_2 - \delta_0 = -41°4 \pm 8°1$ the value of $|\omega|$ is

$$|\omega| = (2,3 \begin{array}{c} +0,7 \\ -0,5 \end{array}) \ 10^{-2} \tag{35}$$

It corresponds to the dashed region of Figure 2 and the error on $|\omega|$ depends strongly on the error on ϕ_ω and weakly on the error on ρ^2.

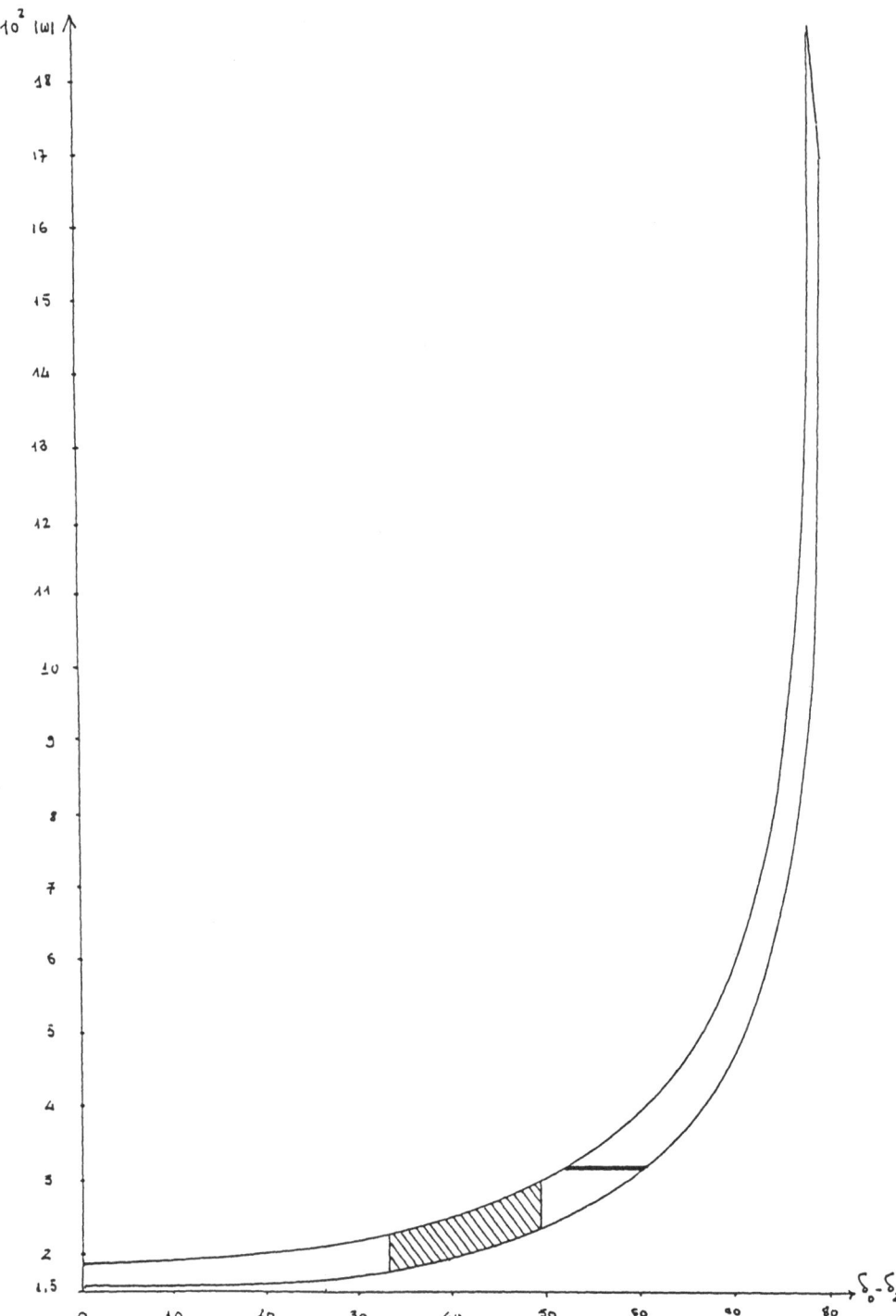

<u>Figure 2</u>   the modulus of ω as a function of the phase difference δ_0 - δ_2 ; the two curves correspond to one standard deviation limit for the ratio R.

202

3°) For values of ϕ_ω in the experimental range an approximate expression of $|\omega|$ is

$$|\omega| \simeq \frac{1}{2} \frac{1 - \rho^2}{2 + \rho^2} \frac{1}{\text{Cos } \phi_\omega} \qquad (36)$$

and the quantity Re ω is essentially independent of ϕ_ω and given by

$$\text{Re } \omega \simeq (1.72 \mp 0.19) \, 10^{-2} \qquad (37)$$

VI THEORETICAL ESTIMATE OF THE PHASE DIFFERENCE $\phi_{+-} - \phi_{00}$

1°) Let us write the difference $\eta_{+-} - \eta_{00}$ as

$$\eta_{+-} - \eta_{00} = \Delta \, e^{i\phi_\Delta} \qquad (38)$$

A first estimate of ϕ_Δ has been given in Section IV, $\phi_\Delta = \phi$, and the difference between these two angles is due to a violation of the non-leptonic $|\Delta I| = \frac{1}{2}$ rule. Therefore it is small and of the order $O(|\omega|)$.

Let us define such a difference as $\delta\phi$

$$\phi_\Delta = \phi - \delta\phi \qquad (39)$$

From equation (29) we have

$$\delta\phi = \text{Phase } (1 - \omega - 2\omega^2) \qquad (40)$$

Using our previous result on $|\omega|$ we obtain a small correction $\delta\phi$

$$0°62 < \delta\phi < 1°42 \qquad (41)$$

which is well inside the error of 8°,1 on ϕ. In what follows we shall use

$$\phi_\Delta = 47°65 \pm 8°1 \qquad (42)$$

2°) We now represent the complex parameters η_{+-} and η_{00} as two vectors of the complex plane. They define a triangle whose third side is obviously $\eta_{+-} - \eta_{00}$. A possible situation with $\phi_{00} < \phi_{+-}$ is represented on Figure 3. The differences between two of the three angles ϕ_Δ, ϕ_{+-} and ϕ_{00} are small and, as a consequence, the triangle is extra flat and we get the first equality

$$\Delta \simeq \frac{[\, |\eta_{+-}| - |\eta_{00}| \,]}{\text{Cos } (\phi_\Delta - \phi_{+-})} \qquad (43)$$

Using now the elementary relations in a triangle between the length of one side and the sine of the opposite angle we obtain the second relation

$$\text{Sin } (\phi_{+-} - \phi_{00}) = \left(\frac{|\eta_{+-}|}{|\eta_{00}|} - 1 \right) \text{Sin } (\phi_\Delta - \phi_{+-}) \qquad (44)$$

Inserting the experimental values of $|\eta_{+-}| / |\eta_{00}|$ and ϕ_{+-} and our estimate (42) of ϕ_Δ we obtain the following bounds for the phase difference $\phi_{+-} - \phi_{00}$

$$-0°08 < \phi_{+-} - \phi_{00} < 0°16 \qquad (45)$$

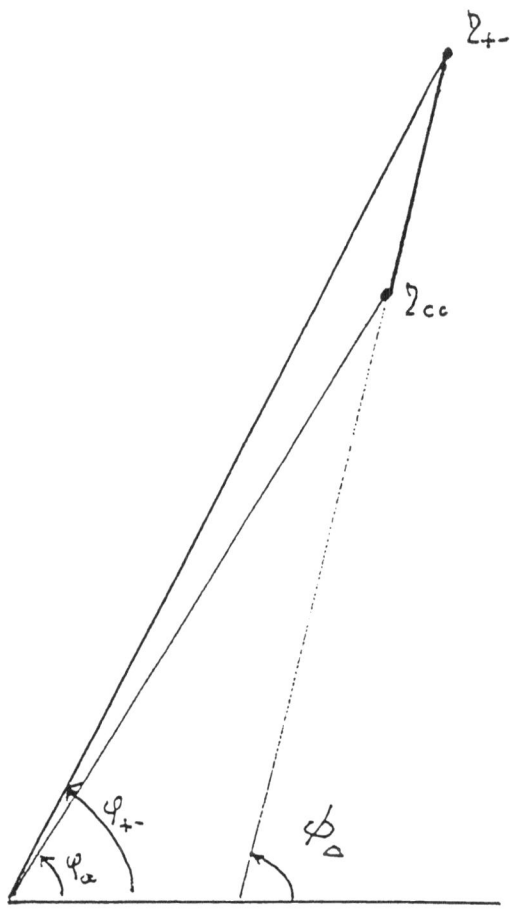

Figure 3   the vectors η_{+-}, η_{00}, and $\eta_{+-} - \eta_{00}$ in their complex plane.

Comparing with the particle data table result

$$\phi_{+-} - \phi_{00} = -9°6 \pm 6°2$$

we see that our estimate (45) differs by less than two standard deviations from the actual experimental value. Therefore an accurate measurement of $\phi_{+-} - \phi_{00}$ is urgently needed and it could solve the inconsistency problem.

VII PARAMETER ε_0

1°) We use the identity

$$\eta_{00} \equiv \eta_{+-} - (\eta_{+-} - \eta_{00}) \tag{46}$$

and we get, from equation (26) the following exact expression of ε_0

$$\varepsilon_0 = \eta_{+-} - \frac{1}{3} (1 - 2\omega) \Delta \; e^{i\phi} \tag{47}$$

The value of ε_0 is strongly dominated by that of the first term η_{+-} and because of the smallness of Δ (equation 25) the second term brings a correction which, in relative value is less than $4 \; 10^{-3}$.

On the other hand noticing that the angles ϕ_{+-} and ϕ have similar values the phase ϕ_{ε_0} of ε_0 is expected to be close of ϕ_{+-}.

2°) Let us point out that a simple exact formula can be derived for Re ε_0 by using, in equation (47), the precise definition of ϕ_{Δ} . The result is

$$\text{Re } \varepsilon_0 = \text{Re } \varepsilon = \text{Re } \eta_{+-} - \frac{1}{3} \rho \Delta \cos \phi \tag{48}$$

3°) The numerical estimates for the parameter ε_0 are the following

a) $$\text{Re } \varepsilon_0 = (1,615 \mp 0,034) \; 10^{-3}$$
$$\text{Im } \varepsilon_0 = (1,592 \pm 0,034) \; 10^{-3} \tag{49}$$

where the errors are dominated by that on the phase $\phi_{+-} = 41°6 \pm 1°4$

b) $$\left| \varepsilon_0 \right| = (2,268 \pm 0,021) \; 10^{-3} \tag{50}$$

here the error is entirely due to that on the modulus $\left| \eta_{+-} \right| = (2,275 \pm 0,021)10^3$

c) The phase difference $\phi_{\varepsilon_0} - \phi_{+-}$ is very small and positive

$$\phi_{\varepsilon_0} - \phi_{+-} \leqslant 0°02 \tag{51}$$

4°) It is interesting to add one piece of information coming from the PC violation asymmetry observed in the semi-leptonic decay of the K_L state. Defining

$$\delta = \frac{\Gamma(K_L \to \pi^- e^+ \nu_e) - \Gamma(K_L \to \pi^+ e^- \bar{\nu}_e)}{\Gamma(K_L \to \pi^- e^+ \nu_e) + \Gamma(K_L \to \pi^+ e^- \bar{\nu}_e)} \tag{52}$$

the experimental result is

$$\delta_{exp} = (3.33 \pm 0,14) \; 10^{-3} \qquad (53)$$

Assuming the validity of the $\Delta S = \Delta Q$ rule and TCP invariances we simply get

$$\delta_{th} = \langle K_S | K_L \rangle \qquad (54)$$

which in the first order of the violation of PC is nothing but

$$\delta_{th} = 2 \; \mathrm{Re} \; \varepsilon \qquad (55)$$

From our previous analysis of the 2π decay mode the corresponding value of the asymmetry $\delta_{2\pi}$ is given by

$$\delta_{2\pi} = (3,23 \mp 0,068) \; 10^{-3} \qquad (56)$$

The two values of $\delta_{2\pi}$ and δ_{exp} are perfectly compatible within errors.

From these results we can obtain an estimation of the violation of the $\Delta S = \Delta Q$ rule in the semileptonic decay of neutral K mesons. By defining

$$x = \frac{A(\Delta S = -\Delta Q)}{A(\Delta S = +\Delta Q)} \qquad (57)$$

we replace equation (55) by

$$\delta_{th} = 2 \; \mathrm{Re} \; \varepsilon \; S(x) \qquad (58)$$

where

$$S(x) = \frac{1 - |x|^2}{|1-x|^2} \qquad (59)$$

From the comparison of δ_{exp} and δ_{th} we obtain

$$S(x) = 1,03 \pm 0,06 \qquad (60)$$

which is essentially equivalent to

$$\mathrm{Re} \; x = 0,015 \pm 0,030 \qquad (61)$$

5°) A third source of information concerning $\mathrm{Re} \; \varepsilon$ is provided by the so-called Bell-Steinberger unitarity relation which, in the framework of TCP invariance can be written as

$$(\gamma_L + \gamma_S) \; \mathrm{Re} \; \varepsilon \; [1 + i \; \mathrm{tg} \; \alpha_N] = \sum_F \langle F |T| K_S \rangle^* \; \langle F |T| K_L \rangle \qquad (62)$$

where the natural angle α_N is given by

$$\alpha_N = \mathrm{Arc} \; \mathrm{tg} \frac{2(m_L - m_S)}{\gamma_L + \gamma_S} = 43°, 62 \pm 0°, 12 \qquad (63)$$

The 2π meson contribution to the unitarity relation (62)

$$\sum_{2\pi} = \sum [\langle \pi^+\pi^- |T| K_S \rangle^* \langle \pi^+\pi^- |T| K_L \rangle + \langle \pi^0\pi^0 |T| K_S \rangle^* \langle \pi^0\pi^0 |T| K_L \rangle] \qquad (64)$$

can be written by using the definitions of η_{+-} and η_{00} as

$$\sum_{2\pi} = \eta_{+-} \; \Gamma\,(K_S \rightarrow \pi^+\pi^-) \; + \; \eta_{00} \; \Gamma\,(K_S \rightarrow \pi^0\pi^0) \tag{65}$$

Assuming that the K_S width is saturated by the 2π meson modes

$$\Gamma\,(K_S \rightarrow \pi^+\pi^-) + \Gamma\,(K_S \rightarrow \pi^0\pi^0) = \gamma_S$$

and introducing the ratio R of the two modes previously defined we get

$$\sum_{2\pi} = \gamma_S \; [\frac{\eta_{+-} + R\eta_{00}}{1+R}]$$

Using again the trick $\eta_{00} \equiv \eta_{+-} - (\eta_{+-} - \eta_{00})$ we obtain

$$\sum_{2\pi} = \gamma_S \; [\; \eta_{+-} \; - \; \frac{R}{1+R} \; (\eta_{+-} - \eta_{00}) \;] \tag{66}$$

Neglecting now the phase space correction factors due to the $\pi^\pm$ - π^o mass difference we also have

$$\frac{R}{1+R} \rightarrow \frac{\rho^2}{1+\rho^2} = \frac{|\,1-2\omega\,|^2}{3\,(1+2\,|\,\omega\,|^2)}$$

and the 2π meson contribution to the unitarity relation takes the form

$$\sum_{2\pi} = \gamma_S \; [\; \eta_{+-} \; - \; \frac{1}{3} \; \frac{|\,1-2\omega\,|^2}{1+2\,|\,\omega\,|^2}(\eta_{+-} - \eta_{00}) \;] \tag{67}$$

From

$$\eta_{+-} - \eta_{00} = \Delta \; e^{i\phi}$$

and the determination of ϕ_Δ as given by the TCP constraint it is straightforward to obtain a simple expression for the real part of $\displaystyle\sum_{2\pi}$

$$Re \sum_{2\pi} = \gamma_S \; [Re \; \eta_{+-} \; - \; \frac{1}{3}\rho \; \Delta \, Cos\, \phi \;]$$

which is nothing but

$$Re \sum_{2\pi} = \gamma_S \; Re \; \varepsilon \tag{68}$$

Comparing now this result with the left hand side of the relation (62) we immediately obtain

$$\gamma_L \; Re \; \varepsilon \; = \; Re \sum_{F \neq 2\pi}$$

We then have the dominance of the 2π contribution in the unitarity relation and as far as the real parts are concerned we get

$$\frac{\mathrm{Re}\displaystyle\sum_{F\neq 2\pi}}{\mathrm{Re}\displaystyle\sum_{2\pi}} = \frac{\gamma_S}{\gamma_L} \simeq 1{,}7 \quad 10\text{-}3 \tag{69}$$

As a last remark the natural angle α_N, due to the mass difference and the width sum, turns out to be compatible within errors with ϕ_{+-}. Such a result also supports strongly the dominance of the unitarity relation by the 2π meson contributions. It can also be, in this respect, considered as a test of TCP invariance which predicts the hermitian product $\langle K_S | K_L \rangle$ to be real.

VIII PARAMETER ε'

1°) In the literature the parameter ε' is defined from ε_0, ε_2 and ε by

$$\varepsilon' = \varepsilon_2 - \omega\,\varepsilon_0 \tag{70}$$

Using the experimental parameters η_{+-} and η_{00} we get

$$\varepsilon' = \frac{1}{3}(1 - \omega - 2\omega^2)(\eta_{+-} - \eta_{00}) \tag{71}$$

Taking into account the TCP constraint (22) we obtain the equivalent expression

$$\varepsilon' = \frac{1}{3}\left| 1 - \omega - 2\omega^2 \right| \Delta\, e^{i\phi} \tag{72}$$

The first factor is close to $\frac{1}{3}$ because of the non leptonic $\left| \Delta I \right| = \frac{1}{2}$ rule and with the determination of ω given in Section V we have

$$\frac{1}{3}\left| 1 - \omega - 2\omega^2 \right| = 0{,}3276 \pm 0{,}00007 \tag{73}$$

Inserting our previous estimate of Δ in equation (25) we obtain

$$\varepsilon' = (0{,}744 \pm 0{,}242)\,10^{-5}\,e^{i\phi} \tag{74}$$

2°) It is convenient to compare ε' and ε_0. With

$$\varepsilon_0 = (2{,}268 \pm 0{,}0021)\,10^{-3}\,e^{i\phi} \tag{75}$$

we first have

$$\left| \frac{\varepsilon'}{\varepsilon_0} \right| = (3{,}28 \pm 1{,}07)\,10^{-3} \tag{76}$$

On the other hand the phases ϕ and $\phi_{\varepsilon 0}$ are very close to each other

$$\phi - \phi_{\varepsilon 0} = 4° \pm 8°2$$

and the ratio $\varepsilon'/\varepsilon_0$ in quasi real, positive with a numerical value given by equation (76).

3°) As a last remark if we neglect the violation of the non leptonic $|\Delta I| = \frac{1}{2}$ rule we can obtain a simple approximate form for the ratio $\varepsilon'/\varepsilon_0$

$$\frac{\varepsilon'}{\varepsilon_0} \simeq \frac{|\eta_{+-}| - |\eta_{00}|}{2|\eta_{+-}| + |\eta_{00}|} \qquad (77)$$

Using the measured value of $|\eta_{+-}| / |\eta_{00}|$ we obtain

$$\varepsilon'/\varepsilon_0 = (3,34 \pm 1,00) \ 10^{-3}$$

which is clearly in agreement with the more accurate estimate (76) because of the large experimental error on the relevant parameter $1- |\eta_{00}| / |\eta_{+-}|$.

IX DETERMINATION OF THE $K \rightarrow 2\pi$ DECAY AMPLITUDES

1°) The analysis of the experimental decay widths for the $K \rightarrow 2\pi$ transitions is made by using the formula

$$\Gamma (K \rightarrow 2\pi) = \frac{K_{CM}}{8\pi} \ |\frac{<2\pi|T|K>}{m_K}|^2 \qquad (78)$$

where K_{CM} is the centre of mass momentum of the 2π meson system.

2°) Let us start with the $K_S \rightarrow 2\pi$ decay amplitudes. The input are the partial decay width for the $\pi^+\pi^-$ and $\pi^0\pi^0$ modes and the phase difference $\delta_2 - \delta_0$ due to final state interaction. The particle data tables give the K_S lifetime

$$\tau_S = 0,8923 \ (23) \ 10^{-10} \ s$$

We remark that the modulus $|A_I|$ of the reduced amplitudes are affected by PC violation only at second order. The angle ϕ_ω being located between $-\pi/2$ and O, the ratio ReA_2/ReA_0 is positive and we simply have

$$\frac{A_2}{A_0} = \frac{ReA_2}{ReA_0} = \sqrt{2} \ |\omega| \qquad (79)$$

The numerical results are the following

$$A_0 = (334,27 \pm 0,56) eV \qquad (80)$$

$$A_2 = (10,86 \ ^{+3,11}_{-2,20}) \ eV \qquad (81)$$

3°) For the charged K meson decay we use the data

$$\tau_K = 1,2371 \ (26) \ 10^{-8} \ s$$

$$BR \ (K^+ \rightarrow \pi^+\pi^0) = (21,17 \pm 0,15) \ \%$$

The reduced amplitude A_C is determined up to its sign to be

$$A_C = \pm (18,34 \pm 0,006) eV \qquad (82)$$

4°) The decay of a K meson of isotopic spin $I = \frac{1}{2}$ into a 2π meson system of total isotopic spin $I=0$ or $I=2$ can proceed through 3 types of transitions, $|\Delta I| = \frac{1}{2}$, $|\Delta I| = \frac{3}{2}$ and $|\Delta I| = \frac{5}{2}$. The corresponding reduced amplitudes $a_{1/2}$, $a_{3/2}$, and $a_{5/2}$ are related to A_0, A_2 and A_C by using Clebsch Gordan coefficients

$$a_{1/2} = \sqrt{2}\ A_0$$

$$a_{3/2} = 2/5\ \sqrt{2}\ A_2 + 2/5\ \sqrt{3}\ A_C$$

$$a_{5/2} = 3/5\ \sqrt{2}\ A_2 - 2/5\ \sqrt{3}\ A_C$$

We compute the numerical value of $a_{1/2}$, $a_{3/2}$, $a_{5/2}$ by choosing A_C and A_0 of the same sign in order to minimize $a_{5/2}$. The result is

$$a_{1/2} = (472.7 \pm 0,7)\ eV$$

$$a_{3/2} = (18,85 \begin{array}{c} +1,76 \\ -1,24 \end{array}) eV$$

$$a_{5/2} = (-3,49 \begin{array}{c} +2,64 \\ -1,85 \end{array}) eV$$

As expected we observe the dominance of the $|\Delta I| = \frac{1}{2}$ transition but our $|\Delta I| = \frac{5}{2}$ amplitude is not zero by 1,5 standard deviation.

5°) Let us compare our analysis with a previous one of Devlin and Dickey. These authors choose $a_{5/2} = 0$. As a consequence they determine A_2 from A_C by using the relation $A_2^2 = 2/3\ A_C^2$. With our numerical value of A_C we get

$$A_2 = (14,97 \pm 0,05) eV \qquad (83)$$

In our analysis A_2 is obtained from $K_S \rightarrow 2\pi$ decay only and the value of $\delta_2 - \delta_0$. If we insist on the value (83) for A_2 this will determine a range of allowed value for $\cos\phi_\omega$ shown by the horizontal bar on Figure 2. The corresponding value of $\delta_2 - \delta_0$ is then

$$\delta_2 - \delta_0 = -56°5 \pm 4° \qquad (84)$$

which is outside the averaged value previously quoted of $-41°4 \pm 8°1$ and definitely inconsistent with the value $-29°2 \pm 3°$ proposed by Biswas and al. Surprisingly the fit of Devlin and Dickey gives a value $-45°3 \pm 5°1$ which is also incompatible with (84) within one standard deviation.

6°) In conclusion the phenomenological analysis of the data is not definitely conclusive concerning the two closely related quantities $\delta_2 - \delta_0$ and A_2/A_0. More experimental information is needed.

1°) The phenomenological knowledge of the parameter ε_0 is very satisfactory and the errors are of the order of 2% for both the real and the imaginary parts.

Concerning the parameter ε' a decisive improvement has recently been made with the obtention of a non-zero value by three standard deviations. Obviously more data are needed in order firstly to confirm that measurement and secondly to reduce both statistical and systematic errors.

For the third complex parameter ω the situation appears to be a little confusing especially for the phase difference $\delta_2 - \delta_0$ coming from strong interactions. Let us just point out that from a phenomenogical point of view the $|\Delta I| = 5/2$ transition reduced amplitude $a_{5/2}$ has no a priori reason to be strictly zero and only experiment will decide what is the the value of $a_{5/2}$. If it turns out to be different from zero the next step will be to understand the origin of such a non vanishing value.

2°) It is clear that TCP invariance is one of the basic ingredient of our analysis and the inconsistency problem described in section IV has to find a solution. It seems that an accurate determination of the phase difference $\phi_{+-} - \phi_{00}$ is the cleanest way to clarify the situation and the range of allowed values given in equation (45) is a very strong constraint.

3°) Assuming now that the two complex PC violating parameters ε_0 and ε' have been experimentally determined the next step will be to compare theory and experiment in order to understand the origin of the violation of PC and its magnitude.

The basic formalism is the complex 2x2 hamiltonian H for the $K^\circ \bar{K}^\circ$ system which, as usual, is separated into it hermitian M and skew- hermitian Γ parts

$$H = M - \frac{1}{2} \ i\Gamma$$

The imaginary parts of the non diagonal elements m_{12} and γ_{12} of M and Γ measure the violation of PC in the mass matrix of the $K^\circ \ \bar{K}^\circ$ system.

Making, for the computation of $Im \gamma_{12}$, the two following assumptions

i - the 2π states dominate the K_S decay and phase space corrections due to the $\pi^\pm - \pi^\circ$ mass difference can be neglected ;

ii - contribution to the order $O \left(|\omega|^2 \right)$ can be neglected

Using formula (15) we obtain the following expression for ε_0

$$\varepsilon_0 = e^{i\alpha} \ Sin \ \alpha_K \ [\frac{Im \ m_{12}}{m_L - m_S} + \frac{Im \ A_0}{Re \ A_0}] \qquad (85)$$

where the phase α_K is given by

$$\alpha_K = Arc \ tg \frac{2 \ (m_L - m_S)}{\gamma_L - \gamma_S} = 43°,72 \pm 0°,12 \qquad (86)$$

The first trivial prediction of the formula (85) is that the phase of ε_0 is simply α_K. Experimentally the value of ϕ_{+-} -close to ϕ_{ε_0} by equation (51) - is consistent with α_K within errors.

The second consequence of equation (85) is a theoretical expression for $Re \ \varepsilon$ which has two independent experimental determinations as discussed in section VII.

$$Re \ \varepsilon = \frac{1}{2} \ Sin \ 2\alpha_K \ [\frac{Im \ m_{12}}{m_L - m_S} + \frac{Im \ A_0}{Re \ A_0}] \qquad (87)$$

Let us point out that the separation of the two contributions in the bracket of equation(85) or (87) depends on the phase convention chosen for the states $|K° >$ and $|\bar{K}° >$. With the usual choice, Im m_{12} is computed from the box diagram and the second term from the penguin diagram in the framework of the standard model with three generations of quarks. Unfortunately the theoretical uncertainties on the parameters used are such that a sensible theoretical determination of Re ε with acceptable errors is not yet possible. The only weak statement we can made is that the experimental value of Re ε is not in contradiction with the large domain allowed by the standard model. An analogous conclusion can be obtained for the second parameter ε'.

BIBLIOGRAPHY FOR EXPERIMENTAL DATA

J.H. Christenson, J.W. Cronin, V.L. Fitch and R. Turlay Phys. Rev. Lett. 13, 138 (1964)
R.H. Bernstein et al. Phys. Rev. Lett. 54, 1631 (1985)
J.K. Black et al. Phys. Rev. Lett. 54, 1628 (1985)
M. Woods et al. Phys. Rev. Lett. 60, 1695 (1988)
H. Burkhart et al. Phys. Lett. B 206, 169 (1988)

Particle Data Tables Phys. Lett. B 170 (1986)

BIBLIOGRAPHY FOR PHENOMENOLOGICAL ANALYSIS

T.J. Devlin and J.O. Dickey Rev. Mod. Phys. 51, 237 (1979)
V.V. Barmin et al. Nucl. Phys. B247, 293 (1984)
N.N. Biswas et al. Phys. Rev. Lett. 47, 1378 (1981)

14

POSSIBLE SEARCHES FOR CP NON-CONSERVATION IN Z BOSON DECAYS *

W. Bernreuther

CERN - Geneva

U. Löw**, J.P. Ma, and O. Nachtmann

Institut f. Theoretische Physik, Universität Heidelberg, FRG

Abstract: We investigate how the decays $Z \to l^+ l^- \gamma (l = e, \mu, \tau)$, $Z \to 2jets + \gamma$, $Z \to 3$ jets can be used to search for possible new sources of CP violation beyond the standard model (SM). CP tests can be made with these decays using CP-odd angular correlations. Within the SM these correlations are unobservably small. We parametrize the CP-violating interactions which may affect these decays by an effective Lagrangian L_{CP} and calculate the CP-odd correlations in terms of the parameters of L_{CP}.

INTRODUCTION

Soon large numbers of Z bosons will be produced at the $e^+ e^-$ colliders SLC and LEP1. In particular, LEP1 is expected to provide several millions of Z bosons per year. The decays of the Z bosons will allow for a detailed study of interactions at a distance scale which is not easily accessible otherwise. It should be natural to look also for CP violation in Z decays. One may use the Z resonance to search for CP–violating effects which are predicted by the 3 generation standard model (SM) where CP violation is induced by the charged weak quark currents[1].The expected CP violation for $B\bar{B}$ mesons (for recent reviews see Refs. 2, 3, 4) produced at the Z peak may offer some hope to find CP-odd effects. For instance, a CP–odd asymmetry which may serve this purpose was proposed recently[5]. Here we are concerned with effects due to possible new sources of CP violation. Several proposals for CP tests in this context

---

* Talk given by W. Bernreuther.

** supported by Bundesministerium f. Forschung u. Technologie, F.R.G.

have already been made[6,7,8]. We have investigated[9] in a systematic way whether and how the main Z decays, i.e. the decays $Z \to l^+ l^-, Z \to l^+ l^- \gamma (l = e, \mu, \tau), Z \to 2 \; jets, Z \to 2 \; jets + \gamma, Z \to 3 \; jets$ can be used for CP tests. The main results will be presented in the following. We always consider the situation where polarization of the final states is not observed. Appropriate observables are CP–odd correlations among the momentum direction vectors of the outgoing states, i.e. CP–odd angular correlations. Within the SM these correlations turn out to be unobservably small. Rather than investigating specific "non–standard" models of CP violation we describe the CP–violating interactions which may affect the above–mentioned decays by a local effective Lagrangian L_{CP}, including operators of mass dimension $d \leq 6$. This allows for a fairly model–independent analysis. The correlations are then calculated by using this Lagrangian; i.e., our results will depend on the parameters of L_{CP}.

It is straightforward to show[9] that tests of CP and/or T invariance cannot be made with the 2–body decay modes $Z \to l^+ l^-, Z \to 2 \; jets$ if no polarizations of the outgoing particles are measured. (Polarization measurements are thought to be feasible for τ leptons.) Therefore, we confine ourselves here to the 3–body decay modes stated above.

THREE BODY DECAYS OF THE Z BOSON

We discuss in some detail how the discrete symmetries CP, T, and CPT can be tested in the decays $Z \to l^+ l^- \gamma$. This analysis also applies to $Z \to 2 \; jets + \gamma$ where the flavour of the jets is tagged. Then we give some results of a similar analysis[9] of $Z \to 2 \; jets + \gamma, Z \to 3 \; jets$ where no flavour identification is made. As stated above we always consider the situation where no polarizations of the outgoing particles are observed. For the CP–odd correlations introduced below it is however crucial that the Z bosons are polarized to a certain degree. This will always be the case for Z bosons produced at SLC and LEP. Let us first consider the spin density matrix of the Z in its rest system. In the Cartesian basis

$$|Z, i > \equiv |Z(\mathbf{p} = 0), \hat{\mathbf{e}}_i > \quad (i = 1, 2, 3) \tag{1}$$

where $\hat{\mathbf{e}}_i$, being the i–th Cartesian unit vector, is the polarization vector of the state (1) the spin density matrix ρ_{ij} can be expanded as follows:

$$\begin{aligned} \rho_{ij} &= \frac{1}{3}\rho_{ij} + \frac{1}{2i}\epsilon_{ijk}s_k - s_{ij}, \\ s_{ij} &= s_{ji}, \quad s_{ii} = 0, \end{aligned} \tag{2}$$

where s_k, s_{ij} describe the vector and the tensor polarization of the Z, respectively. Z bosons produced in $e^+ e^-, p\bar{p}$ or pp collisions are in general polarized to a certain degree. For instance, if the Z's are produced by unpolarized $e^+ e^-$ beams the vector and tensor polarization of the Z in the c.m. system are, respectively:

$$\begin{aligned} \mathbf{s} &= \frac{2g_{Ve}g_{Ae}}{(g_{Ve})^2 + (g_{Ae})^2}\hat{\mathbf{p}}_+, \\ s_{ij} &= \frac{1}{2}(\hat{p}_{+i}\,\hat{p}_{+j} - \frac{1}{3}\delta_{ij}), \end{aligned} \tag{3}$$

where $\hat{\mathbf{p}}_+ = \mathbf{p}_+/|\mathbf{p}_+|$ and $\mathbf{p}_+$ is the e^+ c.m. momentum. The neutral current couplings g_{Ve}, g_{Ae} of the electron are given in the SM by $g_{Ve} = -1/2 + 2\sin^2\theta_w$, $g_{Ae} = -1/2$. Using $\sin^2\theta_w = .23$ and choosing $\hat{\mathbf{p}}_+ = \hat{\mathbf{e}}_3$ then:

$$\mathbf{s} = .16\hat{\mathbf{p}}_+,$$
$$s_{ij} = \mathrm{diag}(-1/6, -1/6, 1/3). \tag{4}$$

Consider now the decays

$$Z \to l^+(k_+) + l^-(k_-) + \gamma(k) \tag{5}$$

where $l = e, \mu, \tau$. We shall always work in the rest system of the Z, i.e., we have $\mathbf{k}_+ + \mathbf{k}_- + \mathbf{k} = 0$. It is very convenient to analyze (5) in terms of its decay matrix. It is defined in the Cartesian polarization basis (1) by

$$R_{ij}(\mathbf{k}_+, \mathbf{k}_-, \mathbf{k}) = \Gamma_{l+l^-\gamma}^{-1}$$
$$\sum_{r,s,\epsilon} \langle l^+(\mathbf{k}_+, r), l^-(\mathbf{k}_-, s), \gamma(\mathbf{k}, \epsilon)| T|Z, i\rangle^* \tag{6}$$
$$\langle l^+(\mathbf{k}_+, r), l^-(\mathbf{k}_-, s)\gamma(\mathbf{k}, \epsilon)|T|Z, j\rangle,$$

where r, s are the spin indices of $l^\pm$ and ϵ is the polarization vector of the photon. The matrix R_{ij} is normalized such that

$$\frac{1}{3} \int d\Gamma \, R_{ii}(\mathbf{k}_+, \mathbf{k}_-, \mathbf{k}) = 1, \tag{7}$$

where $d\Gamma$ is the usual phase space measure. In computing the decay width $\Gamma_{l+l^-\gamma}$ and the integral (7) energy and angular cuts appropriate for a given experiment have to be taken into account. Cuts are also necessary on theoretical grounds in order to avoid the usual infrared and collinear singularities. We will assume such cuts to be $C-$ and $P-$blind; i.e. the cuts should apply equally well to l^+ and l^- and equally well to particles of momenta $\mathbf{k}$ and $-\mathbf{k}$. In this way no $CP-$bias will be introduced by the cuts.

The following decomposition of R_{ij} can be made:

$$R_{ij}(\mathbf{k}_+, \mathbf{k}_-, \mathbf{k}) = \delta_{ij} a(E_+, E_-)$$
$$+ \frac{1}{i}\varepsilon_{ijl} B_l(\mathbf{k}_+, \mathbf{k}_-, \mathbf{k}) \tag{8}$$
$$- C_{ij}(\mathbf{k}_+, \mathbf{k}_-, \mathbf{k}),$$

where $E_\pm = k_\pm^0$ and $C_{ij} = C_{ij}$, $C_{ii} = 0$. Using $\hat{\mathbf{k}}_\pm = \mathbf{k}_\pm/|\mathbf{k}_\pm|$ and the vector $\hat{\mathbf{n}} = (\mathbf{k}_+ \times \mathbf{k}_-)/|\mathbf{k}_+ \times \mathbf{k}_-|$ which is orthogonal to the decay plane as a basis we can decompose $\mathbf{B}$ and the traceless symmetric tensor C_{ij} as follows:

$$\mathbf{B}(\mathbf{k}_+, \mathbf{k}_-, \mathbf{k}) = \hat{\mathbf{k}}_+ b_1(E_+, E_-)$$
$$+ \hat{\mathbf{k}}_- b_2(E_+, E_-) + \hat{\mathbf{n}} b_3(E_+, E_-), \tag{9}$$

$$C_{ij}(\mathbf{k}_+, \mathbf{k}_-, \mathbf{k}) = (\hat{k}_{+i}\hat{k}_{+j} - \frac{1}{3}\delta_{ij})c_1(E_+, E_-)$$
$$+ (\hat{k}_{-i}\hat{k}_{-j} - \frac{1}{3}\delta_{ij})c_2(E_+, E_-)$$
$$+ [\hat{k}_{+i}\hat{k}_{-j} + \hat{k}_{-i}\hat{k}_{+j} - \frac{2}{3}(\hat{\mathbf{k}}_+\hat{\mathbf{k}}_-)\delta_{ij}]c_3(E_+, E_-)$$
$$+ (\hat{k}_{+i}\hat{n}_j + \hat{n}_i\hat{k}_{+j})c_4(E_+, E_-)$$
$$+ (\hat{k}_{-i}\hat{n}_j + \hat{n}_i\hat{k}_{-j})c_5(E_+, E_-). \tag{10}$$

Since R is hermitean $a, b_i, c_j (i = 1, 2, 3, j = 1, \ldots, 5)$ are real functions.

Next we give the constraints on R which arise if CP, T, and CPT invariance holds. The decay mode (5) and the decays (18), (19), (20) considered below allow for CP tests because the Z boson states (1) in the Z rest system are eigenstates of CP — as are the final states in these decays. CP invariance implies

$$R_{ij}(\mathbf{k}_+, \mathbf{k}_-, \mathbf{k}) = R_{ij}(-\mathbf{k}_-, -\mathbf{k}_+, -\mathbf{k}). \tag{11}$$

It is well known that final state interactions can fake T and CPT violations even if the basic interactions involved conserve these symmetries. However, since the SM gauge couplings are small at the scale of the Z mass m_Z and the couplings of possible new CP-violating interactions should also be small for phenomenological reasons, we neglect in the following final state rescattering effects in the decays (5) and (18), (19), (20) below. (See also the comment below.) Then we obtain from CPT invariance

$$R_{ij}(\mathbf{k}_+, \mathbf{k}_-, \mathbf{k}) = R_{ji}(\mathbf{k}_-, \mathbf{k}_+, \mathbf{k}), \tag{12}$$

whereas from T invariance we have

$$R_{ij}(\mathbf{k}_+, \mathbf{k}_-, \mathbf{k}) = R_{ji}(-\mathbf{k}_+, -\mathbf{k}_-, -\mathbf{k}). \tag{13}$$

A more general discussion is given in Refs. 9,10. The constraints on the functions a, b_i, c_j due to (11), (12), and (13) are collected in Table 1. According to the work of Ref. 11, QED final state interactions – respectively, QCD final state interactions in $Z \to q\bar{q}$ gluon – do not generate to one-loop approximation non–zero functions b_3, c_4, c_5.

Suppose now that a CP-violating but CPT-conserving interaction affects the decays (5). We see from Table 1 that in this case the CP-symmetry relations involving $a, b_1, b_2, c_1, c_2, c_3$ can only be violated if final state interactions are present. This implies for instance that in a calculation where both for the CP-conserving and the CP-violating amplitudes only Born diagrams are considered only the CP relations for b_3, c_4, c_5 of Table 1 can be violated. This is exemplified by explicit calculations below.

Integral observables which are sensitive to CP violation in the decays (5) can be easily constructed. Examples are:

i) Asymmetry in the distribution of $\hat{\mathbf{k}}_+ \times \hat{\mathbf{k}}_-$: If rotational invariance holds for the cuts then using (2) and (8) one obtains

$$\langle \hat{\mathbf{k}}_+ \times \hat{\mathbf{k}}_- \rangle = I\mathbf{s} \tag{14}$$

where

$$\begin{aligned}
I &= \frac{1}{3} \int d\Gamma \, |\hat{\mathbf{k}}_+ \times \hat{\mathbf{k}}_-| b_3(E_+, E_-) \\
&= \frac{1}{6} \int d\Gamma \sqrt{1 - (\hat{\mathbf{k}}_+ \hat{\mathbf{k}}_-)^2} \\
&\quad \cdot [b_3(E_+, E_-) + b_3(E_-, E_+)].
\end{aligned} \tag{15}$$

ii) Asymmetry in the distribution of the tensor $\hat{k}_{+i}\hat{n}_j - \hat{k}_{-i}\hat{n}_j + (i \leftrightarrow j)$: Again for cuts respecting rotational invariance we get

$$\langle \hat{k}_{+i}\hat{n}_j - \hat{k}_{-i}\hat{n}_j + (i \leftrightarrow j) \rangle = K \, s_{ij} \tag{16}$$

Table 1   Constraints on the functions a, b_i, c_j of (8)-(10)
from CP invariance and, neglecting absorptive parts,
from CPT and T invariance.

CP	CPT	T
$a(E_+,E_-) = a(E_-,E_+)$	$a(E_+,E_-) = a(E_-,E_+)$	
$b_1(E_+,E_-) = -b_2(E_-,E_+)$	$b_1(E_+,E_-) = -b_2(E_-,E_+)$	
$b_3(E_+,E_-) = -b_3(E_-,E_+)$	$b_3(E_+,E_-) = b_3(E_-,E_+)$	$b_3(E_+,E_-) = 0$
$c_1(E_+,E_-) = c_2(E_-,E_+)$	$c_1(E_+,E_-) = c_2(E_-,E_+)$	
$c_3(E_+,E_-) = c_3(E_-,E_+)$	$c_3(E_+,E_-) = c_3(E_-,E_+)$	$c_4(E_+,E_-) = 0$
$c_4(E_+,E_-) = c_5(E_-,E_+)$	$c_4(E_+,E_-) = -c_5(E_-,E_+)$	$c_5(E_+,E_-) = 0$

where (2) and (8) were used and

$$K = \frac{2}{5} \int d\Gamma(1 - \hat{\mathbf{k}}_+ \cdot \hat{\mathbf{k}}_-)[c_4(E_+, E_-) - c_5(E_-, E_+)]. \tag{17}$$

Experimental observation of $I \neq 0$ and/or $K \neq 0$ would be evidence for CP-violation. Observables which require CP-violating final state interactions for being non-zero can also be constructed[9].

The above analysis and the observables (14), (16) also apply to the decays $Z \rightarrow 2\,jets + \gamma$, where the jets arising from the fragmentation of a quark and an antiquark are tagged.

$$Z \rightarrow \bar{q}(\hat{\mathbf{k}}_+) + q(\hat{\mathbf{k}}_-) + \gamma(\hat{\mathbf{k}}) \rightarrow 2\,jets + \gamma \tag{18}$$

Here $q = u, d, s, c, b$. According to the UA1 lower limit[12] on the mass of the t quark, $m_t > 41\,\mathrm{GeV}$ (see Ref. 13) 3-body decays of the Z involving $\bar{t}t$ are not very likely and we do not consider them in the following.

Finally we briefly discuss the decays $Z \rightarrow 3\,jets, 2\,jets + \gamma$ where no flavour identification of the jets is made. Consider

$$Z \rightarrow jet(\mathbf{k}_1) + jet(\mathbf{k}_2) + jet(\mathbf{k}_3) \tag{19}$$

and

$$Z \rightarrow jet(\mathbf{k}_1) + jet(\mathbf{k}_2) + \gamma(\mathbf{k}_3). \tag{20}$$

In order to use these decays for CP tests the jets have to be ordered according to a CP-blind criterion. Several such criteria can be given[8,9]. In the calculations below we have chosen for definiteness

$$|\mathbf{k}_1| \geq |\mathbf{k}_2| \geq |\mathbf{k}_3| \tag{21}$$

for the decay (19) and

$$|\mathbf{k}_1| \geq |\mathbf{k}_2| \tag{22}$$

for the decay (20). Of course, the remarks made above on the CP-blindness of the cuts apply also here.

A decay matrix formalism analogous to the one outlined above can be set up for (19) and for (20). The essential difference compared to (5) and (18) is that the jets in (19), (20) are to be treated as identical particles in phase space. Details are given in Ref. 9 where also the constraints arising from CP, T, and CPT invariance are derived. It turns out that appropriate observables being sensitive to CP violation in the decays (19) or (20) are the tensors

$$\hat{k}_{ai}\hat{n}_j + \hat{k}_{aj}\hat{n}_i \qquad (a = 1, 2, 3), \tag{23}$$

where $\hat{\mathbf{k}}_a = \mathbf{k}_a/|\mathbf{k}_a|$ and $\hat{\mathbf{n}} = (\mathbf{k}_1 \times \mathbf{k}_2)/|\mathbf{k}_1 \times \mathbf{k}_2|$. For cuts respecting rotational invariance the expectation values of the 3 tensors (23) take the following form:

$$\langle \hat{k}_{ai}\hat{n}_j + (i \leftrightarrow j) \rangle = H_a s_{ij} \qquad (a = 1, 2, 3) \tag{24}$$

where

$$H_a = \frac{2}{5} \int d\Gamma [(\hat{\mathbf{k}}_a \cdot \hat{\mathbf{k}}_1) c_4(E_1, E_2) + (\hat{\mathbf{k}}_a \cdot \hat{\mathbf{k}}_2) c_5(E_1, E_2)] \tag{25}$$

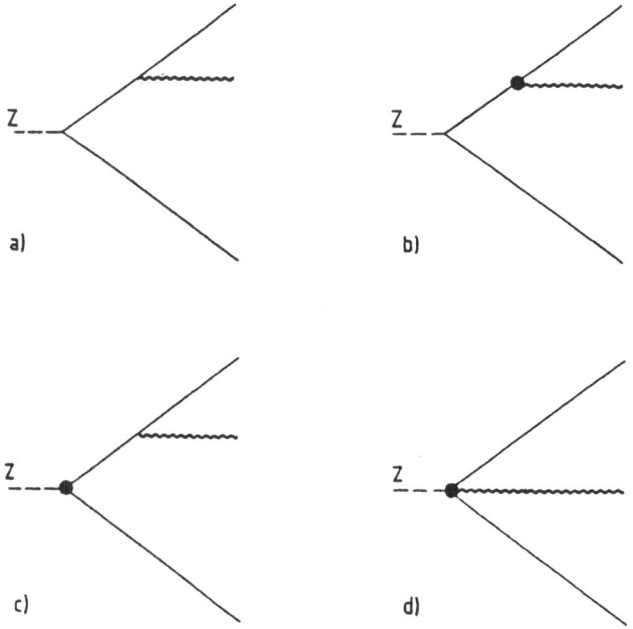

Fig. 1 a) SM Born diagram for $Z \to \bar{f}f\gamma$(f=1,q) and $Z \to \bar{q}qG$. Solid lines represent fermions, the wiggly line represents a photon or a gluon, respectively. b),c),d): Diagrams depicting the CP-violating interactions (full circles) given by (27).

218

and the functions c_4, c_5 in (25) are specific to the respective decay matrices associated with (19) and (20). If CP is a good symmetry then[9] $c_4 = c_5 = 0$. That is, the correlations (24) provide genuine CP tests for the decays (19) and (20) – like the correlations (14) and (16) for the decays (5) and (18). These correlations cannot be faked by final state rescattering arising from CP-conserving interactions. The above CP tests — and others[8] — can be applied also to corresponding e^+e^- reactions away from the Z peak and to $p\bar{p}$ collisions.

CP-VIOLATING EFFECTIVE LAGRANGIAN

Which interactions could generate CP-violating effects in the 3–body decays considered above? Consider the SM Born diagram Fig. 1a which represents these decays at the level of the elementary particles of the SM: $Z \rightarrow \bar{f}f\gamma (f = $ lepton or quark) and $Z \rightarrow \bar{q}qG$ ($G = $ gluon). CP-violating interactions could affect the following 1–particle irreducible vertices contributing to these decays: $ff\gamma$ respectively qqG (Fig. 1b), ffZ (Fig. 1c), $ffZ\gamma$ respectively $qqZG$ (Fig. 1d). (Flavour-off-diagonal vertices are irrelevant on phenomenological grounds.) Within the SM there is no CP violation in the lepton sector if the neutrino masses are zero. As to the quark sector of the SM it is straightforward to show that the charged weak quark currents do not generate any CP-violating contribution to the above vertices to one-loop approximation. Therefore, within the SM, one expects the CP-odd correlations introduced above for the decays (18), (19) and (20) to be at most of order

$$\left(\frac{g_w^2}{4\pi}\right)^2 c_1 c_2 c_3 s_1^2 s_2 s_3 s_\delta \leq 10^{-7}. \tag{26}$$

In (26) g_w is the SU(2) gauge coupling and $c_1 c_2 c_3 s_1^2 s_2 s_3 s_\delta (s_i = \sin \vartheta_i, c_i = \cos \vartheta_i)$ is the combination of mixing angles[1] to which all CP violation in the SM is proportional[14]. That is, the CP-odd correlations discussed above are sensitive to possible new forms of CP violation only. Within the gauge theory context many extensions of the SM – e.g., left–right symmetric models, supersymmetric models, SU(2)×U(1) models with several Higgs doublets – which provide such new sources were discussed in the literature. (For reviews, see Ref. 2,3). Typically these models generate CP-violating contributions to the above vertices already to 1–loop approximation. However, an analysis of these models is in general hampered by the occurrence of many unknown parameters. It seems therefore appropriate for our purposes to use the method of effective Langrangians in order to keep the analysis rather model-independent.

In constructing the relevant CP-violating effective Lagrangian L_{CP} we restrict ourselves to operators of mass dimension $d \leq 6$. This is justified if the mass scale Λ_{CP} characterizing possible new interactions is markedly larger than the Z mass m_Z. Only flavour-diagonal neutral current interactions are considered. The Lagrangian L_{CP} can be represented by various sets of independent operators. The following form[9] is convenient for our purposes:

$$
\begin{aligned}
L_{CP}(x) = \sum_{\psi=q,l} \{ &-\frac{i}{2} d_\psi \bar{\psi}(x) \sigma^{\mu\nu} \gamma_5 \psi(x) F_{\mu\nu}(x) \\
&-\frac{i}{2} \tilde{d}_\psi \bar{\psi}(x) \sigma^{\mu\nu} \gamma_5 \psi(x) [\partial_\mu Z_\nu(x) - \partial_\nu Z_\mu(x)] \\
&+ [f_{V\psi} \bar{\psi}(x) \gamma^\nu \psi(x) + f_{A\psi} \bar{\psi}(x) \gamma^\nu \gamma_5 \psi(x)] Z^\mu(x) F_{\mu\nu}(x) \} \\
+ \sum_q \{ &-\frac{i}{2} d'_q \bar{q}(x) T^a \sigma^{\mu\nu} \gamma_5 q(x) G^a_{\mu\nu}(x) \\
&+ [h_{Vq} \bar{q}(x) T^a \gamma^\nu q(x) + h_{Aq} \bar{q}(x) T^a \gamma^\nu \gamma_5 q(x)] Z^\mu(x) G^a_{\mu\nu}(x) \}
\end{aligned}
\tag{27}
$$

where

$$
F_{\mu\nu} = \partial_\mu A_\nu - \partial_\nu A_{\mu\nu} \qquad G^a_{\mu\nu} = \partial_\mu G^a_\nu - \partial_\nu G^a_\mu - g_s f_{abc} G^b_\mu G^c_\nu.
\tag{28}
$$

In (27), (28) $\psi(x)$ denotes a quark field $q(x)$ (colour indices being suppressed) or a lepton field $l(x)$; $A_\mu(x), Z_\mu(x), G^a_\mu(x)$ is the photon, Z, and gluon field, respectively; g_s is the QCD coupling, and T^a are the generators of $SU_c(3)$ satisfying $[T^a, T^b] = i f_{abc} T^c$. The interaction Hamiltonian corresponding to (27) is odd under CP. The dimension 5 operators with the couplings $d_\psi, \tilde{d}_\psi, d'_q$ correspond to electric dipole, weak dipole, and chromoelectric dipole interactions, respectively. These operators are fermion chirality flipping. The dimension 6 operators in (27) describe fermion-chirality-conserving contact interactions. Anomalous[15] CP-violating $Z\gamma\gamma, ZZ\gamma$, and ZGG interaction vertices could also exist. However, since these vertices must vanish if all gauge bosons are on-shell, such CP-violating interactions are described (up to $d \le 6$) by the $Zff\gamma$ and $ZqqG$ operators of (27). (For comprehensive lists of CP-conserving and violating $d \le 6$ operators investigated in other contexts see Ref. 16.)

The real parameters $d_\psi, \tilde{d}_\psi, d'_q, f_{V\psi}, f_{A\psi}, h_{Vq}$, and h_{Aq} correspond to constant real form factors and vertex functions, respectively; i.e., possible imaginary parts of these form factors and vertex functions are neglected. Higgs models of CP violation[17,18] - for recent investigations see Ref. 19 - are especially interesting for our considerations. Because Higgs couplings are proportional to fermion masses, sizeable CP-violating form factors and 4-particle vertex functions involving heavy flavours - i.e. the τ lepton, the c or b quarks - might be generated if this source of CP violation exists in nature.

CALCULATION OF CP-ODD CORRELATIONS

The respective decay matrices R - i.e. the functions a, b_i, c_j - for the decays (5), (18), (19), and (20) can now be calculated in lowest order perturbation theory using the SM ffZ, $ff\gamma$, qqG couplings and the CP-violating couplings given by (27). The CP-violating terms in the decay matrices arise from the interference between the CP-conserving and CP-violating Born amplitudes depicted in Fig. 1. The functions a, b_i, c_j of the various decay matrices are listed in Ref. 9. These functions allow for a calculation of the above CP-odd correlations which can be searched for in the

Table 2   Decay widths for (5), (18) and coefficients appearing
in the correlation integrals (31), (32). Cuts are
specified in the text.

ψ	decay width (MeV)	$w_{1\psi}$	$w_{2\psi}$	$w_{3\psi}$	x_ψ
e	2.54	.018	.018	7.6×10^{-7}	.068
μ	3.15	.014	.014	1.2×10^{-4}	.056
τ	2.20	.020	.020	2.9×10^{-3}	.076
u	2.00	.029	.029	2.4×10^{-5}	.097
c	1.96	.029	.029	3.1×10^{-3}	.097
d	.64	.023	.023	1.8×10^{-5}	.075
s	.64	.023	.023	3.2×10^{-4}	.075
b	.57	.025	.024	8.1×10^{-3}	.079

3-body Z decays. Instead of using the parameters appearing in (27) it is convenient to introduce the following dimensionless parameters $\hat{d}_\psi$, $\hat{f}_{V\psi}$, $\hat{f}_{A\psi}$, $\hat{h}_{Vq}$, $\hat{h}_{Aq}$ through

$$\tilde{d}_\psi = \frac{e}{\sin\theta_w \, \cos\theta_w \, m_Z} \hat{d}_\psi,$$

$$f_{V\psi} = \frac{-e^2 Q_\psi}{\sin\theta_w \cos\theta_w m_Z^2} \hat{f}_{V\psi}, \qquad f_{A\psi} = \frac{-e^2 Q_\psi}{\sin\theta_w \cos\theta_w m_Z^2} \hat{f}_{A\psi}, \qquad (29)$$

$$h_{Vq} = \frac{e g_s}{\sin\theta_w \cos\theta_w m_Z^2} \hat{h}_{Vq}, \qquad h_{Aq} = \frac{e g_s}{\sin\theta_w \cos\theta_w m_Z^2} \hat{h}_{Aq},$$

where $\psi = l, q$; $e > 0$ is the proton charge, and Q_ψ is the electric charge of the fermion ψ in units of e. The (chromo) electric dipole form factors $d_\psi(d'_q)$ actually drop out in the calculations of the CP-violating functions of the various decay matrices - due to a spin mismatch between the respective CP-conserving and CP-violating amplitudes. Furthermore, in the CP-odd correlations given below the following SM neutral current couplings of a fermion ψ appear:

$$g_{V\psi} = T_{3\psi} - 2Q_\psi \sin^2\theta_w, \quad g_{A\psi} = T_{3\psi}, \qquad (30)$$

where $T_{3\psi} = -1/2 (\psi = e, \mu, \tau, d, s, b)$, $T_{3\psi} = 1/2 (\psi = u, c)$. In calculations we have used $\sin^2\theta_w = .23$ and $m_Z = 92 \ GeV$. For the calculations of the decay widths we took for the electromagnetic and strong couplings at the Z mass: $\alpha = 1/128, \alpha_s = .11$.

The necessary energy and angular cuts which we use (see below) for calculating the CP-odd correlations do not violate rotational invariance. Therefore the formulae (14), (16), and (24) hold. The results of our calculations can then be represented as follows:

i) $Z \to l^+ l^- \gamma$, $Z \to \bar{q} q \gamma \to 2 jets + \gamma$ (tagged jets):

For a given fermion ψ, the integrals I_ψ, K_ψ determining the correlations (14) and (16), respectively, take the form:

$$I_\psi = \hat{f}_{V\psi} g_{V\psi} w_{1\psi} - \hat{f}_A g_{A\psi} w_{2\psi} + \hat{d}_\psi g_{A\psi} w_{3\psi}, \qquad (31)$$

$$K_\psi = (\hat{f}_{A\psi} g_{V\psi} - \hat{f}_{V\psi} g_{A\psi}) x_\psi, \qquad (32)$$

and the numbers $w_{i\psi}$ ($i = 1, 2, 3$), x_ψ, which depend on the cuts, are given in Table 2. In calculating these numbers we used the following cuts: Denoting the minimal angle required between two particles by δ and the minimal energy required for each particle by E, we chose: $\delta_{e^+e^-} = \delta_{e\gamma} = \delta_{\tau^+\tau^-} = \delta_{\tau\gamma} = 7^0$, $\delta_{\mu^+\mu^-} = \delta_{\mu\gamma} = 3^0$, $\delta_{q\bar{q}} = \delta_{q\gamma} = 30^0$, $E_e = 1\ GeV$, $E_\mu = E_\tau = 3\ GeV$, $E_q = 2\ GeV\ (q = u, d, c, s)$, $E_b = 4.3\ GeV$, and $E_\gamma = 10\ GeV$. For the quark masses we used $m_u = m_d = 10\ MeV$, $m_s = 175\ MeV$, $m_c = 1.3\ GeV$, and $m_b = 4.3\ GeV$. If one lowers the minimal photon energy for instance to $1\ GeV$, the numbers for $w_{i\psi}$, x_ψ given in Table 2 decrease by approximately a factor of 3 basically because the respective decay widths increase by about the same factor. (This reflects the infrared singularity in the soft photon limit.) The small numbers for $w_{3\psi}$ reflect the fact that the interference of the helicity conserving amplitude Fig. 1a with the amplitude Fig. 1c involving helicity flips results in a suppression factor m_ψ / m_Z (m_ψ = fermion mass).

ii) $Z \to 3 jets$, $Z \to 2 jets + \gamma$:

It turns out that the tensor correlations (24) for these decays are generated only by the helicity conserving $d = 6$ interactions in (27) — like the tensor correlation (16). Therefore we put the quark masses to zero in the following. For the decay (19) the integrals H_a of the 3 correlations (24) then take the form:

$$H_a^{3jet} = y_a \sum_q \left(\hat{h}_{Aq} g_{Vq} - \hat{h}_{Vq} g_{Aq} \right) \qquad (a = 1, 2, 3). \qquad (33)$$

Calculating the correlations (24) for the decay (20), one gets for the integrals (25):

$$H_a^{2jet} = z_a \sum_q Q_q^2 \left(\hat{f}_{Aq} g_{Vq} - \hat{f}_{Vq} g_{Aq} \right) \qquad (a = 1, 2, 3). \qquad (34)$$

The numbers y_a and z_a in (33) and (34) depend on the cuts and jet-ordering criteria. For the decays (19), (20) we used the ordering criteria (21), (22), respectively. The following cuts were chosen: $E_{jet} = E_\gamma = 5\ GeV$, $\delta_{jet,jet} = \delta_{jet,\gamma} = 30^0$. We then obtain

$$(y_1, y_2, y_3) = (-2.1 \times 10^{-3}, 2.5 \times 10^{-3}, 2.8 \times 10^{-4}) \qquad (35)$$

$$(z_1, z_2, z_3) = (-.026, .028, .011). \qquad (36)$$

The respective decay widths are $\Gamma_{3jet} = 745\ MeV$, $\Gamma_{2jet+\gamma} = 8.9\ MeV$. Increasing the minimal energies and/or minimal angles leads to larger numbers y_a, z_a, but, of course, to smaller decay widths. It remains to be investigated whether other jet-ordering criteria lead, for given cuts, to stronger correlations (24).

222

Which level of sensitivity can be reached for the dimensionless parameters (29) by measuring our CP-odd correlations? We give only crude order-of-magnitude estimates assuming 5×10^6 Z's and assuming the widths of the distributions of the (dimensionless) CP-odd observables to be .1. Measurements of the tensor correlations (16) and (24) should be sensitive to $\hat{f}_{Vl}$, $\hat{f}_{Al}$, $\hat{f}_{Vq}$, $\hat{f}_{Aq}$, $\hat{h}_{Vq}$, $\hat{h}_{Aq} \sim O(10^{-1})$. The existing upper bounds on the electric dipole moments of e, μ, and of the neutron presumably preclude such rather large values for $\hat{f}, \hat{h}$ associated with e, μ, u, d. (A conclusion without caveats is, however, not possible.) In any case measurement of the CP-odd tensor correlations will provide new information about CP-odd vertex functions for heavy flavours. Due to helicity suppression the vector correlations (14) are insensitive to the weak dipole form factors for e, μ, u, d, s. If sizeable weak dipole form factors exist for τ, c, b – i.e., $\hat{d}_\tau, \hat{d}_c, \hat{d}_b \sim O(1)$ – detection may be possible by means of (14). However, such large values would lead to a considerable deviation of the total Z width from its SM value.

SUMMARY

We have shown that the standard 3-body decay modes of the Z boson can be used to search for new sources of CP violation. The CP-odd correlations which we introduced and calculated in a model-independent way, serve this purpose. They should be fairly easily measurable in an experiment. These CP tests can also be applied to the corresponding $e^+ e^-$ reactions away from the Z resonance and to the corresponding $p\bar{p}$ reactions, although our calculations are specific to Z boson decays. Given the expectation that several millions of $Z's$ will be produced at LEP1 measurements of these correlations will provide new information – at least; useful upper bounds – on some CP-violating flavour-diagonal vertex functions and weak dipole form factors for heavy flavours.

REFERENCES

1. M. Kobayashi and T. Maskawa, Progr. Theor. Phys. **49**, 652 (1972).

2. J. F. Donoghue, B. Holstein, and G. Valencia, Int. J. Mod. Phys. **A2**, 319 (1987).

3. W. Grimus, Fortschr. d. Physik **36**, 201 (1988).

4. I. I. Bigi, V. A. Khoze, N. G. Uraltsev, and A. I. Sanda,
   SLAC preprint SLAC-PUB-4476(1987);
   see also the reports given by A. Ali, D. London, B. Cox, and D. Cline at this workshop.

5. W. B. Atwood, I. Duinez, and P. Grosse-Wiesmann,
   SLAC preprint SLAC-PUB-4544 (1988).

6. L. Stodolsky, Phys. Lett. **150B**, 221 (1985).

7. W. S. Hou, N. G. Deshpande, G. Eilam, and A. Soni,
   Phys. Rev. Lett. **57**, 1406 (1986);
   J. Bernabéu, A. Santamaria, and M. B. Gavela, Phys. Rev. Lett. **57**, 1514 (1986).

8. J. F. Donoghue and G. Valencia, Phys. Rev.Lett. **58**, 451 (1987).

9. W. Bernreuther, U. Löw, J. P. Ma, and O. Nachtmann, "CP violation and Z boson decays", Heidelberg preprint HD-THEP–88–26 (1988).

10. W. Bernreuther, U. Löw, J. P. Ma, and O. Nachtmann, "How to test CP, T, and CPT invariance in the three photon decay of polarized $^3S_1$ positronium", HD-THEP-87-25, to be published in Z. Phys. C (1988).

11. J. G. Körner and H. Schuler, Z. Phys. C **26**, 559 (1985).

12. C. Albajar et al., Z. Phys. C **37**, 505 (1988);
    D. Denegri, talk given at this workshop.

13. G. Altarelli, M. Diemoz, G. Martinelli, and P. Nason, CERN preprint CERN-TH. 4978 (1988).

14. C. Jarlskog, Phys. Rev. Lett. **55**, 1039 (1985).

15. F. M. Renard, Nucl. Phys. **B196**, 93 (1982);
    G. Gounaris, R. Kögerler, and D. Schildknecht, Phys. Lett. **137B**, 261 (1984);
    Y. Tomozawa, Phys. Lett. **139B**, 455 (1984);
    V. Barger, H. Baer, and K. Hagiwara, Phys. Rev. **D3**, 1513 (1984);
    A. Barroso et al., Z. Phys. **C28**, 149 (1985); **C33**, 243 (1986);
    K. Hagiwara, R. D. Peccei, D. Zeppenfeld, and K. Hikasa,
    Nucl. Phys. **B282**, 253 (1987).

16. W. Buchmüller and D. Wyler, Nucl. Phys. **B268**, 621 (1985);
    C. N. Leung, S. T. Love, and S. Rao, Z. Phys. **C31**, 433 (1986).

17. T. D. Lee, Phys. Rev. **D8**, 1226 (1973).

18. S. Weinberg, Phys. Rev. Lett. **37**, 657 (1976).

19. G. C. Branco, A. J. Buras, and J. M. Gérard,
    Nucl. Phys. **B259**, 306 (1985);
    H. Y. Cheng, Phys. Rev. **D34**, 1397 (1986);
    J. Liu and L. Wolfenstein, Nucl. Phys. **B289**, 1 (1987);
    G. Ecker, W. Grimus, and H. Neufeld, Phys. Lett. **194B**, 251 (1987).

15

HEAVY FLAVOUR PRODUCTION IN HIGH ENERGY ELECTRON-PROTON COLLISIONS:
Theoretical Issues and Expectations at HERA and Beyond

G. A. Schuler

II. Institut für Theoretische Physik
Universität Hamburg

Abstract

I review the status of the theory of heavy quark production in ep collisions. The various production modes are discussed and cross-section estimates given. Uncertainties in the theoretical cross-section calculation are investigated. The characteristics of heavy quark production in ep collisions are outlined.

1   Introduction

I review the status of the theory of heavy quark production in ep collisions. Heavy flavour physics in ep collisions is first of all important for "practical" reasons like the interpretation of charged lepton signals, search for new heavy quarks, study of B physics or background estimates. For these reasons it is necessary to have a reliable description of heavy flavour events in ep collissions. Boson gluon fusion into a heavy quark-antiquark pair is the dominant production mechanism and many characteristics of heavy flavour events can be understood in this model. Yet there is also the theoretical interest to aim a precise quantitative test of the parton approach and of the QCD predictions in a highly non-trivial dynamical situation. In this context it is important to discuss the various possibilities of heavy quark production in ep collisions as well as to critically examine the related theoretical sources of uncertainty. After having presented total production cross-sections in section 2 I deal with these problems in Section 3. The general characteristics of heavy flavour events in ep collisions are then outlined in Section 4.

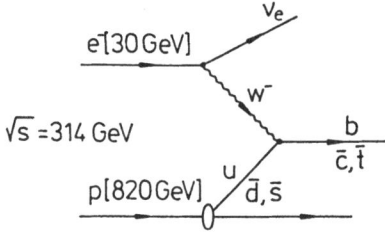

Figure 1. *Quark-Parton Model diagram contributing to the charged current interaction giving a single heavy quark.*

2 Total production cross-sections

In lepton-nucleon collisions, heavy quark production can already occur in the Quark-Parton Model (QPM) via the charged current process

$$e^{\pm}(l_e) + q(p) \rightarrow \nu(l') + Q(p_Q) \tag{1}$$

shown in Figure 1. Here the $W^{\pm}$ emitted from the incoming electron (positron) picks a light quark out of the proton and turns it into a heavy one resulting in a cross-section proportional to the square of the appropriate Cabbibo-Kobayashi-Maskawa (CKM) matrix element. It turns out that the tiny mixing angles between the light quarks and the heavy quarks cannot be compensated by the abundance of the former in the proton. In fact, the QPM contribution to heavy quark production is negligble compared to the other production mechanisms [1] making it extremely difficult to measure the CKM matrix elements of interest V_{cd}, V_{cs}, V_{td}, V_{ts}, V_{tb}, V_{cb}, V_{ub} at HERA through the CC processes (1). For a given cms energy $\sqrt{s}$ the cross-section for the QPM reaction (1) is specified by two kinematical variables, e.g. x and y, related to the boson momentum, $q = l_e - l'$, and the total hadronic energy, W, via the relations

$$Q^2 = -q^2 = xys \quad , \quad W^2 = (P + q)^2 = (1 - x)ys. \tag{2}$$

Furthermore, x_q, the momentum fraction of the proton momentum, P, carried by the incoming quark equals Bjorken x, $x_q^{QPM} = x$.

In QCD the leading order contribution to heavy quark production is due to the $O(\alpha_s)$ boson-gluon fusion (BGF) mechanism shown in Figure 2.

$$V(q) + g(p) \rightarrow Q_f(p_f) + \bar{Q}_{f'}(p_{f'}). \tag{3}$$

In the charged current (CC) case V is the $W^{\pm}$ boson whereas in the neutral current (NC) case it corresponds to γ/Z^0 exchange and the produced quark and antiquark have the same flavour f. The cross-section

$$\sigma(e^{\pm}p \rightarrow Q\bar{Q}'X) = \int dy \int dQ^2 \int dx_g \int dz \int d\Phi \, g(x_g, \mu_F^2) \, h(y, Q^2, x_g, z, \Phi; \mu_R^2) \tag{4}$$

is a convolution of the gluon density $g(x_g, \mu_F^2)$ and a QCD part h for the subprocess. The latter [2] depends on the heavy quark masses, the electroweak charged/neutral

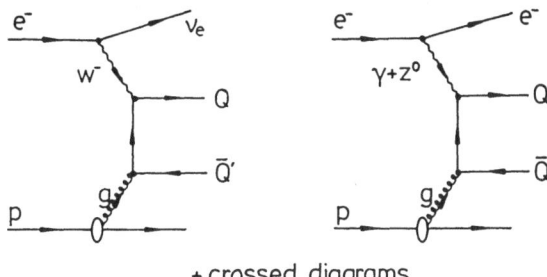

+ crossed diagrams

Figure 2. *Lowest order QCD diagrams for boson-gluon fusion into a heavy quark-antiquark pair.*

current structure including γ/Z^0 interference and polarization of the $e^\pm$ beam. The cross-section also depends on the two mass scales, the factorization mass in the gluon density, μ_F, and the renormalization mass in the strong coupling, μ_R. In addition to the normal deep inelastic scattering (DIS) variables (y, Q^2), three new independent variables enter: (i) the gluon momentum fraction x_g; (ii) the variable $z = P \cdot p_f / P \cdot q$ related to the angle θ between the $Q\bar{Q}$-axis and the boson-gluon axis in this subsystem cms (e.g. in the NC case we have: $2z = 1 - \beta \cos \theta$ where $\beta = \sqrt{1 - 4m^2/\hat{s}}$); and (iii) the azimuthal angle Φ between the lepton and hadron planes. I note that the gluon momentum fraction, x_g, is now related to the usual DIS variable x by

$$x_g = x + \frac{\hat{s}}{ys} \geq \max \left\{ x, \frac{(m_f + m_{f'})^2}{s} \right\} \tag{5}$$

Here $\hat{s} = (p_f + p_{f'})^2$ is the invariant mass square of the $Q\bar{Q}'$ subsystem. Also, as opposed to the QPM subprocess (1), the BGF processes (3) need the gluon density in a proton as input. Thus ep colliders like HERA offer the possibility of extracting the gluon density by measuring specific heavy flavour distributions. To judge such prospects, however, a critical examination of the uncertainties in the cross-section calculation of eq. (4) is required. Also the possibilities of heavy quark production besides the BGF mechanism have to be investigated. I shall return to these problems in the next section after having presented total production rates based on BGF.

The inclusive heavy quark cross-section, $\sigma_Q = 2\sigma(ep \rightarrow eQ\bar{Q}X) + \sum_{Q'} \sigma(ep \rightarrow \nu Q\bar{Q}'X)$, is given in Figure 3 in terms of the ep (or μp) cms energy $\sqrt{s}$. The results are based on the BGF cross-sections evaluated in the leading order, $O(\alpha_s)$, using $\mu_F = \mu_R = \hat{s}$, $\Lambda = 200\,\mathrm{MeV}$ and $n_f = 6$ in the one-loop formula for α_s, and the gluon density parametrization set 1 of [3]. The charm cross-section is also shown using the the softer MRS [4] alternative. The quark masses are taken as $m_c = 1.5\,\mathrm{GeV}$, $m_b = 5\,\mathrm{GeV}$ and the top quark mass varied between $m_t = 50$ and $100\,\mathrm{GeV}$. Concentrating on the HERA configuration of 30 GeV electrons on 820 GeV protons, the charm and bottom cross-sections are comparable to (or larger than) those expected at SLC/LEP. On the contrary, at the HERA energy, $\sqrt{s} = 314\,\mathrm{GeV}$, top quark production is still at threshold, i.e. the top cross-section varies strongly with the top quark mass. E.g. we find $\sigma_t = 4.1, 0.31, 0.05\,\mathrm{pb}$ for $m_t = 40, 60, 80\,\mathrm{GeV}$, respectively. With an integrated luminosity of 200 pb^{-1}, corresponding to ~1 year of running at HERA, one would

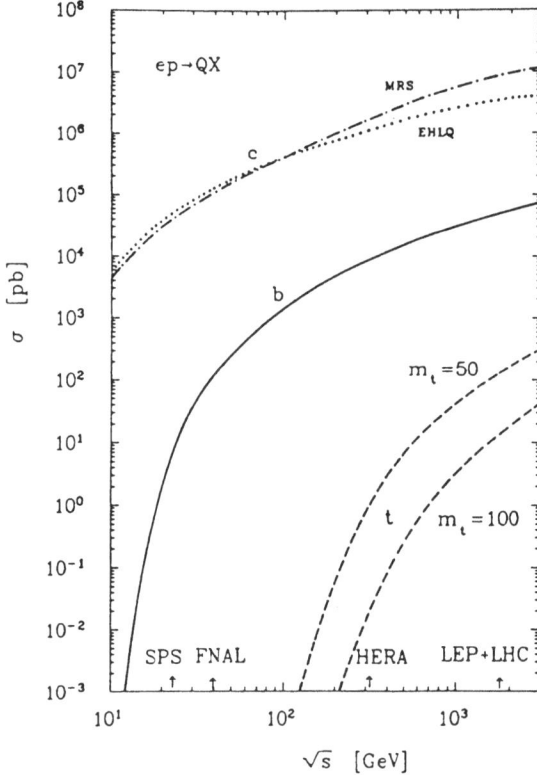

Figure 3. *Inclusive heavy flavour cross-section based on lowest order BGF. (EHLQ [3] gluon density, for charm also the softer MRS [4] alternative.)*

produce about 10^8 charm, a million bottom particles and still about 10 top events with $m_t = 80\,\text{GeV}$ indicating the prospects of B physics at HERA (HERA a "bottom factory") and the limit up to which a top search at HERA seems to be feasible.

To understand the origin of these heavy quark cross-sections one has to distinguish between a light top ($m_t \leq 40\,\text{GeV}$), charm and bottom on the one hand and a heavy top ($m_t > 55\,\text{GeV}$) on the other hand. The large cross-sections in the former case are provided by neutral current processes at very low Q^2, i.e. γ exchange giving essentially real photoproduction. For a heavy top the CC process dominates due to the smaller threshold associated with single top quark production ($\bar{t}b$) in comparison to the pair production ($t\bar{t}$) in NC. Since heavy quarks with masses up to about $40\,\text{GeV}$ are mainly photoproduced the Weizsäcker-Williams approximation (WWA) can be used to simplify their cross-section evaluation. Yet note that, as is the case for all leading approximations, there is some ambiguity on the choice of the upper and lower limits of Q^2. Thus the use of the WWA introduces an additional (unnecessary) uncertainty in the theoretical cross-section calculation.

I close this section by considering the contributions of heavy quark production to the DIS structure functions, $F_i(x, Q^2)$, in the NC and CC sectors,

$$F_i^{NC/CC} = F_i^{u,d,s} + F_i^c + F_i^b + F_i^t \quad . \tag{6}$$

Here u, d, s denotes the contributions from the light (massless) quarks. The structure

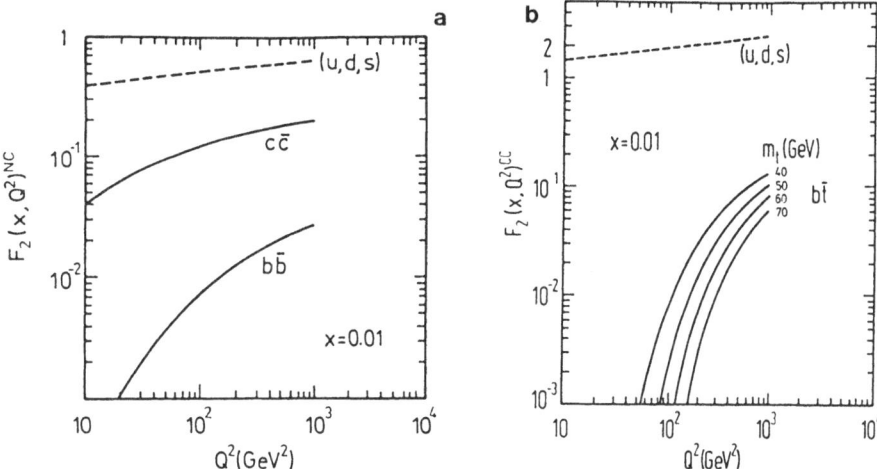

Figure 4. *Predicted Q^2 dependence of $F_2(x,Q^2)$ for NC events (a) and CC events (b) for $x = 0.01$ using $m_c = 1.3\,GeV$ and $m_b = 4.5\,GeV$. The standard leading $\ln Q^2$ scaling violating contributions of the light quarks (u,d,s) to F_2 are shown by the dashed curves. The parton distributions are taken from [7].*

functions, $F_i^Q(x,Q^2)$, are defined in the usual way via [2,5,6]:

$$\frac{d^2\sigma^Q}{dx\,dy} = \frac{4\pi\alpha^2}{Q^2 xy}\left\{g_1\left[xy^2 F_1^Q(x,Q^2) + (1-y)F_2^Q(x,Q^2)\right] + g_2 xy(2-y)F_3^Q(x,Q^2)\right\} \quad (7)$$

where $g_{1,2}$ are electroweak couplings suitable for the NC and CC case, respectively, which can be found e.g. in [2]. In the NC sector, Figure 4a [5], the $t\bar{t}$ contribution in the at HERA observable Q^2 range (given by $y \geq 0.01$) is unobservable small, less than 10^{-4}. The much larger $c\bar{c}$ component reaches its asymptotic leading logarithmic Q^2 dependence already at $Q^2 = 100\,\text{GeV}^2$, from there on the charm component of the proton behaves the same way ($\sim \ln Q^2$) as the light u, d and s quarks. This is clear since for $Q^2 \geq 4m_c^2$ only the logarithmic terms in the cross-section formula survive and we are left with the dominant $\ln Q^2$ terms as for massless quarks. For $b\bar{b}$ production, however, the powerlike-like m_b^2/Q^2 terms dominate over the whole Q^2 range accessible at HERA. But their absolute size is small as compared to the leading $\ln Q^2$ contributions from the light quarks. Thus it will be hard to observe a deviation from the logarithmic scaling violation due to the heavy quarks in the NC sector at HERA.

In the CC sector, Figure 4b [5], the contribution from top production ($\bar{t}b$) is sizeable at high values of Q^2 in the small x region. Since for kinematic reasons we are here close to the $\bar{t}b$ threshold, the top contributions strongly vary with Q^2 which gives rise to power-like deviations from the leading $\ln Q^2$ scaling violating terms due to all lighter quark contributions to F_i. The observation of such deviations from the logarithmic scaling violations could in principle be used as an indication for heavy quark (top) production. Yet, it turns out [5] that a resolution better than 5% would be needed which rules out this possibility. On the other hand, one can conclude that heavy quark production thresholds will not contaminate the leading $\ln Q^2$ scaling violation tests of QCD at HERA using $F_i(x,Q^2)$.

229

3 Uncertainties of the cross-sections

The total error on the heavy quark cross-sections is in part due to our ignorance of the different input quantities and in part to intrinsic theoretical ambiguities. Insufficient known input quantities are the heavy quark masses, the QCD scale Λ (and the number of flavours n_f), and the parton densities both inside the proton and inside the photon. Theoretical uncertainties arise from the choice of the renormalization and factorization mass scales, μ_R and μ_F. The effect of higher order terms should be included as well as the other channels of heavy quark production. Finally there is an ambiguity of attributing certain contributions as part of higher order QCD corrections to BGF or as resolved photon (see below) contributions.

A major source of uncertainty for charm and bottom production is the proper definition of the heavy quark masses (current or constituent) and the exact values to use. A decrease of these masses by 0.3 GeV from the assumed ones, $m_c = 1.5\,\text{GeV}$ and $m_b = 5.0\,\text{GeV}$, leads to an increase of the charm and bottom cross-sections by 68% and 22%, respectively, whereas an increase by 0.3 GeV gives a reduction by 36% and 17%, respectively.

Next there is an ambiguity due to our ignorance of the input structure functions. Yet there is a correlation between the value of Λ and the form of the gluon density in the QCD analysis of scaling violations in deep inelastic scattering. The extraction of the input gluon density and the subsequent QCD evolution have been performed using a definite value of Λ. Thus if one wants to change Λ, the parametrization of the gluon density must be chosen accordingly (or vice versa). I account for this fact by taking that value of Λ (and of n_f) that was used in the respective parametrization and estimate the resulting error by the use of different parametrizations [3,4,7,8]. Both the top and the bottom cross-sections are rather insensitive, they change by less than 25%. On the other hand, charm production is changed by up to a factor of two. The reason is not so much the change of α_s but the fact that charm production probes the gluon density, $g(x_g, \mu_F^2)$, down to $x_g \sim 10^{-4}$ where the gluon density is quite uncertain, whereas the range in x_g does not extend to such small values for top and bottom, see eq. (5).

Another important source of errors is introduced by the scale dependence of the result. In principle, one can distinguish the renormalization scale, μ_R, and the factorization scale, μ_F, c.f. eq. (4). On physical grounds the scales should be of the order of the heavy quark mass. In the following I identify the two scales and estimate the uncertainty in μ by varying it from half the heavy quark mass to twice its value, $m/2 \le \mu \le 2m$. The result is that the top cross-section increases by a factor two when lowering μ from $2m$ to $m/2$, whereas charm and bottom cross-sections are suprisingly stable. Depending on the respective parametrization of the gluon density they increase or decrease slightly by less than 20%. This last fact is due to an accidental cancellation in the lowest order BGF cross-section between the scale dependence of the gluon density and that of $\alpha_s(\mu)$. While the strong coupling constant increases as μ becomes smaller, the gluon density, $g(x_g, \mu)$, decreases if $x_g \le 0.07$ as it is usually the case for charm and bottom production. For top production, on the other hand, $x_g \ge 0.07$, typically and thus the gluon density amplifies the increase of the cross-section through α_s as μ gets smaller.

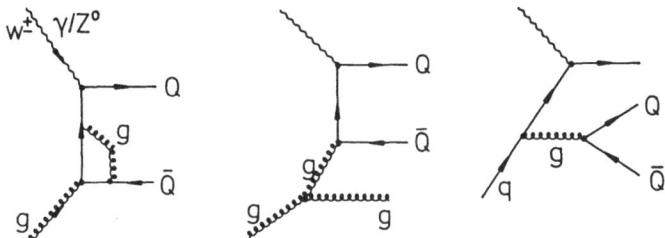

Figure 5: $O(\alpha_s^2)$ diagrams contributing to heavy quark production in ep collisions.

Based on the lowest order, $O(\alpha_s)$, alone it is not possible to rescue from the mass scale dependence, all values for μ are equally valid. A change of μ around m introduces an error of order $\alpha_s^2(m)$. At next-to-leading accuracy, the variation of μ in the leading term is automatically compensated at order α_s^2 by the corresponding change in the correction, so that the resulting error is in this case of order $\alpha_s^3(m)$. Thus, the ambiguity in the result for a given change of μ is normally decreased when going from the leading to the next-to-leading accuracy. This can of course not be expected for charm and bottom production due to the subtle cancellations of the mass scale dependences in these cases in the lowest order. A full $O(\alpha_s^2)$ calculation for ep collisions is still outstanding. It requires the calculations of diagrams shown in Figure 5 for nonzero Q^2. I use the result of the inclusive $O(\alpha_s^2)$ calculation of photoproduction of heavy quarks in [9] to estimate the next-to-leading effects in electroproduction using the WWA. In fact, the results I obtain are similar to those found for heavy quark production in $p\bar{p}$ collisions [10]. According to this approximation, top production increases by roughly a factor of two and becomes less sensitive to changes in μ with a maximum at $\mu \sim m/2$. Using the parton density parametrization set 1 of [3] (the one of [7]) the top cross-section varies between 1.8 pb and 2.6 pb (1.45 pb and 2.3 pb). Thus top production can be quite reliable computed. The maximum error due to changes in the parton densities and the mass scale is about 60%. On the other hand, bottom and charm cross-sections are more uncertain. Here $\sqrt{s}/m_Q$ is already so large that the corrective terms are dominant over the lowest order cross-section at all values of order m of the scale μ. The most important contributions to the cross-section come from the corrections to the BGF mechanism, in particular from the diagrams with a spin-1 gluon in the t-channel which dominate at the large values of $\sqrt{\hat{s}}/m_Q$ available at HERA for charm and bottom production. Since higher order corrections not only change the absolut normalization but also influence the event shape, a realistic event simulation should include these terms at least in an approximate way as it is e.g. done in [11].

Charm and bottom cross-sections are dominated by almost real photon exchange. Yet, at low Q^2 there are important contributions coming from subprocesses involving the photon structure function, see Figure 6. This latter which describes the quark and gluon content of the photon is given by the sum of a VDM component and of an anomalous component [12]:

$$p_{a/\gamma}(x,\mu^2) = p_{a/\gamma}^{VDM}(x,\mu^2) + p_{a/\gamma}^{AN}(x,\mu^2) \quad . \tag{8}$$

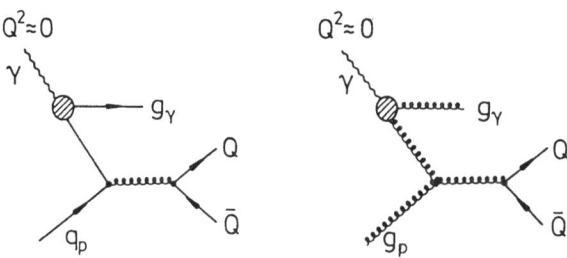

Figure 6. *Resolved photon contributions to heavy quark production in ep collisions.*

The anomalous component, proportional to $\log \mu^2 / \Lambda^2$, is calculable in perturbative QCD. The other one may be estimated using the VDM approach [13]. At higher values of the quark $p_\perp$ it is expected to be negligible. The anomalous component, in the following referred to as "resolved photon" contributions in contrast to the "direct" contribution through BGF, gives rise to gg and $q\bar{q}$ initiated hard scattering subprocesses (and to qg ones in next-to-leading order in α_s). The gg initial state is expected to give the dominant corrections [14]. The authors of [15] estimate the resolved photon corrections to charm and bottom production to about 20%. The results, however, rely strongly on the not yet well known parametrizations of the parton structure functions in the photon. Experimentally a separation of heavy quark production through the direct mechanisms and the resolved photon processes seems to be feasible. Resolved photon contributions drop off rather quickly with increasing $p_\perp$. Also the rapidity distributions are different. Thus cuts in rapidity and $p_\perp$ (or even in Q^2) allow to isolate the resolved photon contributions from the direct BGF contributions to charm and bottom production. I note that there is potentially a problem of double counting. The splitting of the photon into collinear quark pair, Figure 6, is partially included in the contribution of the higher order QCD subprocess $\gamma + q \rightarrow Q + \bar{Q} + q$, see Figure 5. A cut in the $p_\perp$ of the $Q\bar{Q}$ system might allow a separation of the two contributions, but more work is needed for a clean separation.

A fair fraction of charm and bottom events is also expected in high energetic leptoproduction where the proton of the beam is scattered quasi-elastically with only very small energy loss and emerges at very small momentum transfer. Hard QCD scatterings in these diffractive reactions at HERA can be described by the exchange of a pomeron. Assuming that the pomeron can be characterized by a momentum density, $G_{\mathcal{P}}(x)$, describing gluons in the pomeron, the charm and bottom cross-sections at HERA in diffractive production were estimated [16] to $O(150\,\mathrm{nb})$ for $c\bar{c}$ and $O(0.8\,\mathrm{nb})$ for $b\bar{b}$. These cross-sections amount to about 10% of the total charm and bottom cross-sections.

Finally there are uncertainties particularly relevant for charm production. First there are higher twist effects, suppressed by a power of $1\,\mathrm{GeV}/m_Q$, thus non negligible for charm production. A possible higher twist contribution is the concept of intrinsic charm [17] where one imagines that the heavy quark exists as part of the hadron wave function. Another possibility is the effect of coalescence enhancement [18] based on final state interactions of the heavy quark with the proton remnants. Secondly, there is the problem of considering the charm quark as a "parton". Evolved parton distributions containing the charm quark require a proper subtraction in the QCD corrections not to double count the contribution where the exchanged virtual quark is nearly on mass-shell and collinear to the incoming gluon [19].

For top production in ep collisions there exists still another source, namely the production of a $W^\pm$ boson which subsequently decays into $\bar{t}b$ (or $t\bar{b}$). The W production cross-section at HERA is estimated to be [20], $\sigma_W \sim 0.5$–1 pb, resulting in 5–10 top events per year ($\int L dt = 200\,\text{pb}^{-1}$). This corresponds to an increase of about 5–30% of the total top cross-section for $m_t = 50$–70 GeV and is thus a non negligible contribution in that mass range.

4 Characteristics of heavy flavour production ep collisions

Having found that the production of heavy quarks is dominated by BGF I now study their characteristics based on this mechansim. The basic features of heavy quark production can be derived from the lowest order formula. Since the full $O(\alpha_s^2)$ calculation of differential heavy quark distributions is still outstanding I account for the higher order effects on the event shape by the use of a parton cascade simulation algorithm [21] incorporated in the Monte Carlo event generator [11] which was used to obtain the following differential distributions. In this programme also the full hadronic final state is simulated including the target remnants [22] using the Lund string model [23] for the hadronization step.

I first discuss the overall kinematics specified by x, y and $Q^2 = xys$. A light top quark ($m_t \leq 40$ GeV) and also charm and bottom quarks are mainly photoproduced, i.e. we find the typical $1/Q^2$ behaviour with an exceedingly small lower limit in Q^2, $\sim m_e^2(4m^2/s)^2$, which, however, increases with the quark mass. Equivalently, the scattered electron is dominantly at very small angles and thus very hard to measure. Hence the kinematics must be reconstructed from the hadronic system. The Weizsäcker-Williams approximation can be used as a reasonable approximation [2], but let me emphasize that still about 1.7% of charm, 5% of bottom and 23% of top events are deep inelastically produced with $Q^2 \geq 10$ GeV^2. As a reflection of the Q^2 dependence the NC processes are also dominantly at small x. The cross-section is also peaked at small y for charm, but shifts to larger y with increasing mass through the increased threshold, $x_g ys \sim \hat{s} \geq 4m_Q^2$.

A heavy top quark ($m_t \geq 55$ GeV), on the other hand, is dominantly produced by the CC reaction (3). The exchange of the massive W boson leads to a rather uniform distribution in Q^2 and also in x and y. In fact, the Q^2 distribution of $\bar{t}b$ production is quite similar to the one of light quark production with a rather large mean value of Q.

The heavy quark antiquark system is specified by two of the following variables: The invariant mass of the $Q\bar{Q}'$ system, $\hat{s}$, the momentum fraction of the gluon, x_g, the transverse momentum of the heavy quark, $p_\perp$, and its rapidity, η, both measured in the $Q\bar{Q}'$ cms system or in the laboratory (Lab) frame (besides the azimuthal angle Φ, see eq. (4)). The distribution in the invariant mass of the $Q\bar{Q}$ system, $\hat{s}$, shown

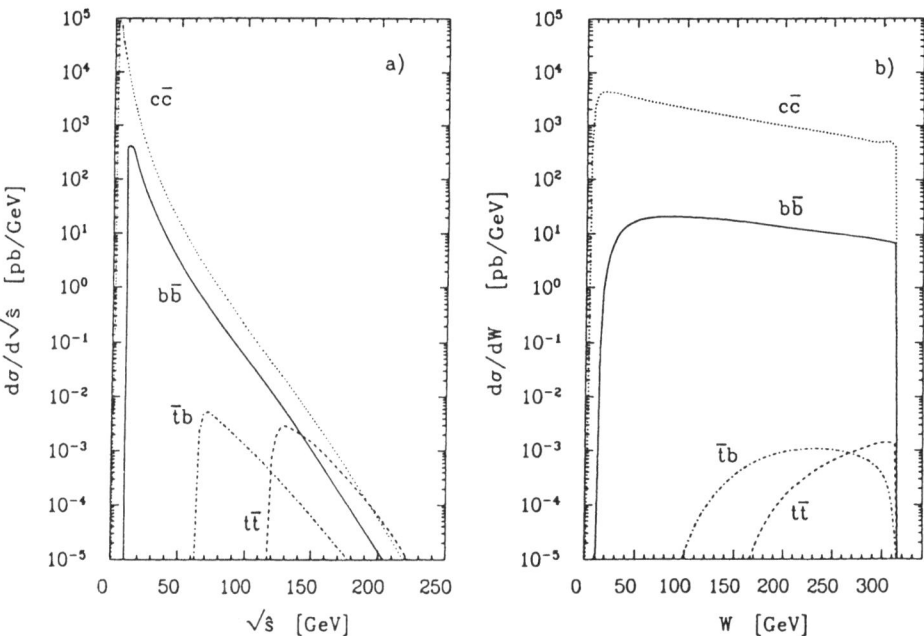

Figure 7. *Dependence of the BGF cross-section on (a) the invariant mass, $\sqrt{\hat{s}}$, of the heavy quark-antiquark system, and (b) the invariant mass, W, of the total hadronic system including the target remnant.*

in Figure 7a, dominates close to threshold and falls steeply above. The reason is the combined effect of the $1/\hat{s}^2$ dependence of the partonic cross-section and the strong fall-off of the gluon density with increasing x_g. The invariant mass W of the complete hadronic system, $W^2 = (1 - x)Q^2/x$, has a rather different distribution, Figure 7b, due to the addition of the target remnant. A cut in W is not effective in separating the heavier flavours bottom and top from the dominant charm production whereas a cut in $\hat{s}$ would be better. On the other hand $\hat{s}$ is more difficult to measure experimentally.

Closely related to $\hat{s}$ is the momentum fraction, x_g, of the gluon, see eq. (5). From eq. (5) it is clear that both the lower limit on x_g and its mean value increase with the heavy quark mass. This is illustrated in Table 1. As was shown in [1] the x_g distributions of the heavy quark cross-sections follow closely the input gluon density which make an extraction of the gluon density at HERA feasible. From the experimental point of view, however, the variable x_g is not directly accessible and its reconstruction through eq. (5) depends on the ability to determine $\hat{s}$ and the possiblity to separate the BGF mechanism where the gluon density enters at the Born level from all other heavy quark production processes.

The $p_\perp$ and rapidity behaviour of the heavy quark cross-sections can be understood most easily in the $Q\bar{Q}$ cms (with the boson gluon axis defining the z-axis). I shall discuss the corresponding distributions in the Lab frame afterwards. Consider the NC lowest order BGF cross-section in the WWA:

Table 1. *Mean and minimum value of the momentum fraction of the gluon, x_g, and average values of the momentum fraction of the boson, y, the longitudinal velocity of the $Q\bar{Q}$ sytem w.r.t. the Lab-system, β_{long}, and the rapidity of the heavy quark in the Lab-sytem, η_{cms}^{Lab}.*

	$c\bar{c}$	$b\bar{b}$	$\bar{t}b$	$t\bar{t}$
$< x_g >$	0.025	0.034	0.19	0.29
$x_{g,\mathrm{min}}$	9×10^{-5}	1×10^{-3}	0.04	0.14
$< y >$	0.18	0.29	0.55	0.77
$< \beta_{\mathrm{long}} >$	-0.58	-0.52	-0.81	-0.82
$< \eta_{cms}^{Lab} >$	-0.67	-0.58	-1.1	-1.2

$$\frac{d^3\sigma}{dy\,dx_y\,dz} = P_\gamma(y)g(x_g,\mu_F^2)\frac{\mathrm{const}}{yx_gs}\left[\frac{z}{1-z} + \frac{1-z}{z} + \frac{4m^2}{z(1-z)\hat{s}} - \left(\frac{2m^2}{z(1-z)\hat{s}}\right)^2\right] \quad (9)$$

Here $P_\gamma(y)$ is the probability of finding a photon in the electron. Let us assume that the parton number densities are constant,

$$yP_\gamma(y)x_gg(x_g,\mu_F^2) \approx \mathrm{const.} \quad (10)$$

Then introducing the transverse mass square of the heavy quark in the $Q\bar{Q}$ cms

$$m_\perp^2 \equiv p_\perp^2 + m^2 = z(1-z)\hat{s} \quad (11)$$

I can rewrite the cross-section in the following way:

$$\sigma \approx \mathrm{const} \ln\left(\frac{s}{4m^2}\right)dz\,\frac{dm_\perp^2}{m_\perp^4}\left[1 - 2z(1-z)\left(1 - \frac{2m^2}{m_\perp^2} + \frac{2m^4}{m_\perp^4}\right)\right] \quad (12)$$

The integration limits are $m^2 \leq m_\perp^2 \leq z(1-z)s$ and $z_- \leq z \leq z_+$ where $z_\pm = (1 \pm \beta_0)/2$, $\beta_0^2 = 1 - 4m^2/s$. From eq. (12) we conclude that the $p_\perp$ beahaviour of the heavy quark in the $Q\bar{Q}$ cms is given by:

$$\frac{d\sigma}{dp_\perp^2} \propto \frac{\sqrt{1 - 4m_\perp^2/s}}{m_\perp^4} \quad (13)$$

Thus, in the $Q\bar{Q}$ cms we expect, $< p_\perp > \approx m$, and a fall-off at large $p_\perp$ as $1/p_\perp^4$. To obtain the $p_\perp$ distribution in the Lab frame we need to know the transformation between the $Q\bar{Q}$ sytem and the Lab frame. The transverse momentum of the scattered lepton has to be balanced by the $Q\bar{Q}$ sytem. Its modulus is given by

$$p_\perp^{DIS} = \sqrt{(1-y)Q^2} \quad (14)$$

Hence for the low Q^2 NC events the $Q\bar{Q}$ system and the Lab frame are only connected via a longitudinal boost along the beam axis. Thus the $p_\perp$ distributions of the heavy quarks in both frames will look quite similar. I show the $p_\perp$ (w.r.t. the beam

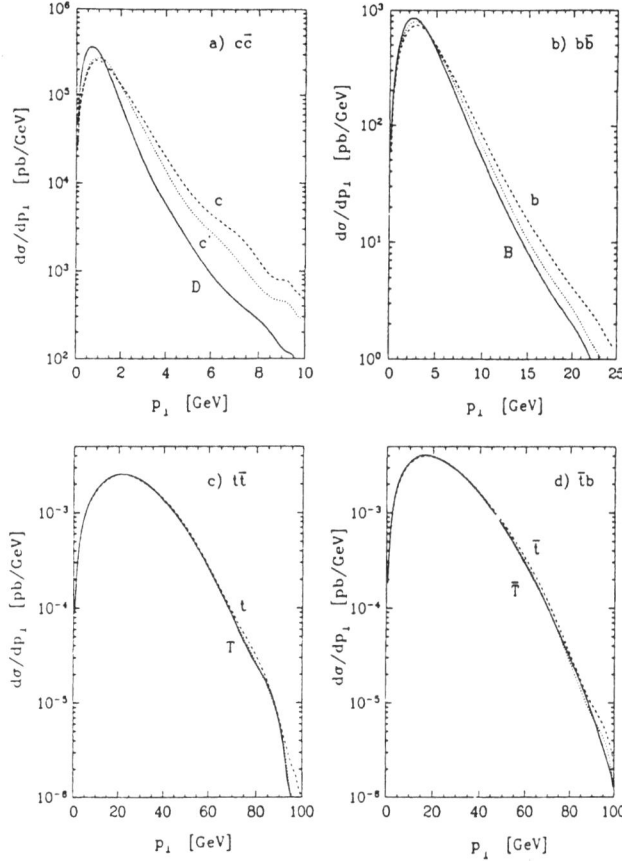

Figure 8. *Transverse momentum in the HERA lab frame for the heavy quark before (dashed) and after parton shower gluon emission (dotted), and for the corresponding hadrons after string fragmentation (full).*

axis) distributions of the heavy quarks in the HERA Lab frame in Figure 8. The distributions for NC heavy quark production indeed have the predicted characteristics with $< p_\perp > \approx m/2$. Only when $p_\perp \gg m_Q$, the curves are a bit steeper than the $1/p_\perp^4$ expected because the photon and gluon number densities $x f(x)$ are not constant, but decrease with increasing x. The effect of multiple gluon emission on the event shape can also be seen in Figure 8 as well as the distributions of the heavy hadrons. Their distributions essentially follow those of the corresponding heavy quarks.

The $p_\perp$ behaviour of the CC $\bar{t}b$ production cannot be explained in the foregoing way. Here Q^2 is typically so large that the $Q\bar{Q}$ cms and the Lab frame are no longer related by just a longitudinal boost. In fact, $< \sqrt{Q^2} >$ is larger than the $p_\perp$ of the hard scattering process, i.e. $< p_\perp^{DIS} > = O(Q)$ is larger than $< p_\perp^{QCD} > = O(m)$. Thus contrary to the NC production of light quarks, the $p_\perp$ of the heavy quarks in CC reactions is essentially given by the DIS p_- eq. (14). This also explains why the $p_\perp$ distribution of $\bar{t}b$ events in Figure 8 is flater than the NC distributions.

I now come to the rapidity distributions of heavy quarks. In the $Q\bar{Q}$ cms the rapidity, η, is related to the variable z via (NC case):

$$\eta = \frac{1}{2} \ln \left(\frac{1-z}{z} \right) \tag{15}$$

236

Integrating eq. (12) over $m_\perp^2$ I find:

$$\frac{d\sigma}{d\eta} \propto \frac{e^{2\eta} + e^{-2\eta}}{\left(e^\eta(1 + e^{2\eta})\right)^4} \tag{16}$$

Thus in the $Q\bar{Q}$ cms we find that the rapidity distribution is central and approximatively constant for small values of η. Furthermore it falls off as the modulus of η becomes large.

The rapidity distribution in the Lab frame is given by:

$$\eta^{Lab} = \eta^{cms} + \eta_{cms}^{Lab} \tag{17}$$

Here η_{cms}^{Lab} is the rapidity of the $Q\bar{Q}$ cms in the Lab frame. Its distributions can be understood once the distributions in x_g, y (and Q^2) are known. x_g, the momentum fraction of the gluon, is defined by the relation, $p = x_g P$, where p is the gluon and P the proton momentum. A similar relation can be found for the variable y. In the Lab frame we have:

$$y = \frac{E^\gamma + p_{long}^\gamma}{E^e + p_{long}^e} \tag{18}$$

Here E^a and p_{long}^a are the energy and the longitudinal (along the beam axis) momentum of particle a in the Lab-system, respectively. Now, in the limit $Q^2 = 0$, eq. (18) simplifies to

$$q = yl_e \quad , \tag{19}$$

i.e. the photon is emmitted collinearly from the electron moving along the beam axis. Thus for photoproduction the heavy quark cms and the Lab frame are related by a longitudinal boost only. The velocity of this boost is given by

$$\beta_{long} = \frac{yE^e - x_g E^p}{yE^e + x_g E^p} \tag{20}$$

The rapidity, η_{cms}^{Lab}, entering eq. (17) can now be written as:

$$\eta_{cms}^{Lab} = \ln\left[\gamma(\beta + 1)\right] = \frac{1}{2}\ln\left(\frac{yE^e}{x_g E^p}\right) \tag{21}$$

For example at HERA, $E^e = 30\,\mathrm{GeV}$ and $E^p = 820\,\mathrm{GeV}$. Since the distributions of the cross-section in x_g and y are quite similar we expect the rapidity eq. (21) to be rather constant. In the limit $Q^2 \ll s, 4m^2 \ll s$, the variables x_g and y are limited symmetrically from below by the heavy quark mass, $yx_g \geq 4m^2/s$, and their distributions are essentially independent of the heavy quark mass. Thus in the NC case the cms velocity and the cms rapidity should approximately be independent of the heavy quark mass. These ideas are confirmed in Table 1 where I give the average values of β_{long} and η_{cms}^{Lab}. Furthermore we observe that β_{long} and η_{cms}^{Lab} are quite small, i.e. the $Q\bar{Q}$ cms is almost at rest in the Lab frame, in mean it is slowly moving in the proton direction.

According to eq. (17) the rapidity of the heavy quark in the Lab frame is the sum of its rapidity in the $Q\bar{Q}$ cms and the rapidity of the latter in the Lab system. From the previous arguments we expect a broad distribution in η^{Lab} which is slightly shifted in proton direction. This behaviour can in fact be seen in Figure 9 where I show

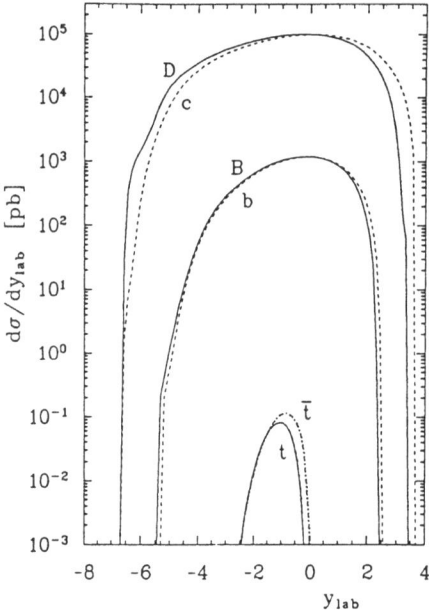

Figure 9. *Rapidity distribution of heavy quarks and hadrons (sum of mesons and baryons of the corresponding flavour) in the laboratory frame of HERA. The separate sets of curves are for $c\bar{c}$, $b\bar{b}$, $t\bar{t}$ and $\bar{t}b$ production, respectively. (Top quark and hadron curves overlap.)*

the rapidity distributions of the heavy quarks and hadrons in the Lab frame. Also shown are the distributions of the heavy hadrons fragmented with the LUND string fragmentation. Due to similar losses in both energy and momentum the rapidity is essentially unchanged by gluon radiation and fragmentation. We find that the asymmetric electron and proton momenta are thus almost completely compensated by a suitable choice of the photon and gluon energy fractions. It is only for top production that the heavy quark cms moves faster and thus the top quark is slightly more forward (in proton direction) produced in average. This can be explained by the fact that for top production such a high partonic energy $\hat{s}$ is needed that even with a maximum energetic photon ($y \rightarrow 1$) an even faster gluon is required.

I close this section by commenting the two modes of heavy quark production in ep collisions. NC heavy quark production in ep collisions is similar to the situation at the $p\bar{p}$ collider in the sense that the initial state of the hard scattering subprocesses consists out of two collinear partons. In ep collisions, the incoming electron with momentum l_e emitts collinearly a photon with momentum $q \approx yl_e$, and similarly the proton with momentum P a gluon with momentum $p = x_g P$, see Figure 10a. The scale μ of the whole process is thus set by the $p_\perp$ of the hard scattering subprocess which is of the order of the heavy quark mass:

$$low \ Q^2 \ events: \quad \mu \sim p_\perp^{QCD} = O(m_Q) \tag{22}$$

Heavy top quark ($m_t \geq 55\,\mathrm{GeV}$) events, on the other hand, are dominantly produced through the CC reaction (3). The Q^2 distribution of $\bar{t}b$ production is essentially

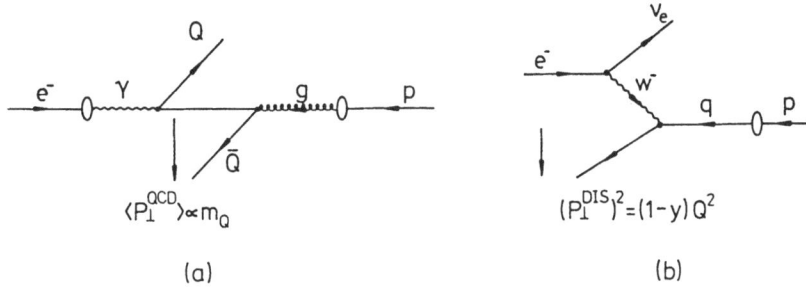

Figure 10. *The two modes of an ep collider: (a) low Q^2 events with a scale $\mu \sim p_\perp^{QCD}$, and (b) truely DIS events with $\mu \sim p_\perp^{DIS}$.*

the same as the one of light quark production with a rather large mean value of Q. Here the situation is complementary to the $p\bar{p}$ collider, the $p_\perp$ of scattered lepton which is balanced by the hadronic side sets the scale of the process rather independently of the quark mass, Figure 10b:

$$\text{high } Q^2 \text{ events:} \qquad \mu \sim p_\perp^{DIS} = O(\sqrt{Q^2}) \tag{23}$$

According to these two heavy quark production modes, charm and bottom signatures go along the lines of similar analyses in hadron collisions exploiting e.g. the $p_\perp$ of the lepton w.r.t. the beam axis [24]). In the contrary, a top search at HERA requires new strategies, in particular when investigating the nonleptonic top decay sample where the DIS CC background from light quarks is crucial [25].

Acknowledgements. I am grateful to G. Ingelman for interesting collaboration on this study. I thank A. Ali for organizing an interesting workshop and a pleasant stay in Erice.

References

[1] G. Ingelman and G.A. Schuler, Z. Phys. C40 (1988) 299.

[2] G.A. Schuler, Nucl. Phys. B299 (1988) 21.

[3] E. Eichten, I. Hinchliffe, K. Lane and C. Quigg, Rev. Mod. Phys. 56 (1984) 579, ibid. 58 (1986) 1047.

[4] A.D. Martin, R.G. Roberts and W.J. Stirling, Phys. Rev. D37 (1988) 1161.

[5] M. Glück, R.M. Godbole and E. Reya, Z. Phys. C38 (1988) 441.

[6] U. Baur and J.J. van der Bij, Nucl. Phys. B304 (1988) 451.

[7] M. Glück, E. Hoffman and E. Reya, Z. Phys. C13 (1982) 119.

[8] D.W. Duke and J.F. Owens, Phys. Rev. D30 (1984) 49.

[9] R.K. Ellis and P.Nason, FERMINLAB-Pub-88/54-T.

[10] G. Altarelli, M. Diemoz, G. Martinelli and P. Nason, Nucl. Phys. B308 (1988) 724.

[11] G. Ingelman and G.A. Schuler, AROMA 1.2 – A Generator of Heavy Flavour Events in *ep* Collisions, DESY preprint in preparation.

[12] P. Aurenche et al., Nucl. Phys. B286 (1987) 553.

[13] R.P. Feynman, Photon-hadron interactions, W.A. Benjamin Inc. 1972.

[14] R.K. Ellis and Z. Kunszt, Nucl. Phys. B303 (1988) 653.

[15] M. Drees and R. M. Godbole, MAD/PH/419 and DO-TH 88/13 preprint 1988.

[16] K. H. Streng, in Proceedings of the HERA Workshop, Hamburg 1987, Editor: R. D. Peccei.

[17] S. J. Brodsky, C. Person and N. Sakai, Phys. Rev. D23 (1981) 2745;
S. J. Brodsky, P. Hoyer, C. Person and N. Sakai, Phys. Lett. 93B (1980) 451.

[18] S. J. Brodsky, J. F. Gunion and D. E. Soper, Phys. Rev. D36 (1987) 2710.

[19] F. I. Olness and Wu-Ki Tung, Nucl. Phys. B308 (1988) 813.

[20] K. J. F. Gaemers et al., in Proceedings of the HERA Workshop, Hamburg 1987, Editor: R. D. Peccei.

[21] T. Sjöstrand and M. Bengtsson, Comput. Phys. Common. 43 (1987) 367.

[22] B. Andersson, G. Gustafson, G. Ingelman and T. Sjöstrand, Z. Phys. C13 (1982) 361.

[23] B. Andersson, G. Gustafson, G. Ingelman and T. Sjöstrand, Phys. Rep. 97 (1983) 31.

[24] A. Ali et al., in Proceedings of the HERA Workshop, Hamburg 1987, Editor: R. D. Peccei.

[25] G. Ingelman, G. A. Schuler and J. F. de Trocóniz, DESY 88-143 (1988).

16

THE LEPTON–QUARK MASS SPECTRUM–
A GUIDE TO THE PHYSICS BEYOND THE STANDARD MODEL[*]

Harald Fritzsch

Sektion Physik der Universität München
and
Max–Planck–Institut für Physik und Astrophysik
– Werner Heisenberg Institut für Physik
München, Germany

1. INTRODUCTION

The standard model of quarks, leptons and their gauge interactions gives a complete description of all phenomena observed in particle, nuclear and atomic physics. Nevertheless it is a model based on about twenty free parameters and can, for this reason, hardly be regarded as the final theory of the basic constituents and forces of nature. A closer inspection of these parameters, however, reveals that the majority of these parameters (13) refer to the lepton–quark mass spectrum and to the associated weak interaction mixing parameters. Thus any further insight into the dynamics which is responsible for the spectrum of leptons and quarks would at the same time be a step beyond the standard model and an important step towards the construction of a theory which would allow theoreticians eventually to calculate all parameters of the standard model in terms of one free mass or scale parameter, e.g. the Planck mass.

In this talk I shall present a few ideas and speculations about the mechanism which might be responsible for the generation of the quark and lepton masses. Let me first present the experimental data of the lepton and quark mass spectrum, the first new and yet unexplained mass spectrum which has entered physics after the discovery of the hadronic mass spectra about 30 years ago. The latter was the starting point of the first speculations about the composite structure of the hadrons and of a path which led eventually to the final theory of the hadrons and their strong interactions, the quark–gluon gauge theory (QCD). Likewise it would not be far–fetched to believe that the lepton–quark mass spectrum could be the starting point of theoretical ideas related to a dynamical substructure of the lepton and quarks, e.g. of composite models[1].

---

[*]Supported in part by DFG–contract Fr. 412/7–2

We shall suppose that there are three electroweak doublets of quarks and leptons:

u [5.1]   c [1350]   t [?]
d [8.9]   s [175]   b [5300]

ν_e[0] ν_μ[0] ν_τ[0]
e [0.5] μ [106] τ [1784]

The numbers in the brackets denote the mass values (in MeV). The neutrino masses are taken to be zero. The quark masses are those normalized at a scale $\mu = 1$ GeV. Only the central values are shown; the quark mass values are subject to large uncertainties (see e.g. ref. (2)).

It is instructive to look at the lepton–quark mass spectrum in two different ways. First we consider the masses on a linear scale (see Fig. (1)). The largest mass values (m_t, m_b, m_τ) are normalized to one. For the t–quark we shall assume a mass value of m_t[1 GeV] = 100 GeV, which is savely within the present experimental bounds (see e.g. ref. (3)).

Despite the uncertainties within the quark mass sector we can make the following comments:

a) The mass spectra of the three flavor channels displayed in Fig. (1) are almost entirely dominated by the mass of the member of the third generation.

b) The relative importance of the second generation decreases as we proceed upwards in the charge ($\mu \rightarrow s \rightarrow c$). In the lepton case the muon contributes about 5.6 % to the sum of the masses ($m_e + m_\mu + m_\tau$), while in the charge –1/3 –channel the s–quark contributes only 3.2 % , and in the charge + 2/3 – channel the c–quark contributes only 1.3 % (for m_t= 100 GeV).

c) The relative importance of the masses of the members of the first generation is essentially zero.

Another instructive way to display the spectrum is to use a logarithmic mass scale and the generation index as the x–coordinate (Fig. (2)). Although it is by no means clear that a mass plot of this type is interesting from a dynamical point of view, it seems that a certain regular pattern persists. If one connects the mass values of each flavor channel, the connecting lines seem to be approximately straight and parallel, apart from the lines which end at the u–quark and the electron (both seem to be too light).

The spectrum exhibits clearly a hierarchical pattern: The masses of a particular generation of leptons or quarks are small compared to the masses of the following generation, if there is any, and large compared to the previous one if there is any.

Furthermore a hierarchical pattern emerges also if we consider the phenomenon of weak interaction mixing. The experimental data combined with the constraints obtained from unitarity give the following form of the weak interaction mixing matrix V[4]:

Fig. 1.   The lepton and quark masses on a linear scale (each flavor separately, the quark masses are defined at $\mu = 1$ GeV). The largest mass eigenvalues (m_t, m_b, m_τ) are normalized to one. For the t–quark a mass value m_t [1 GeV] $= 100$ GeV was assumed.

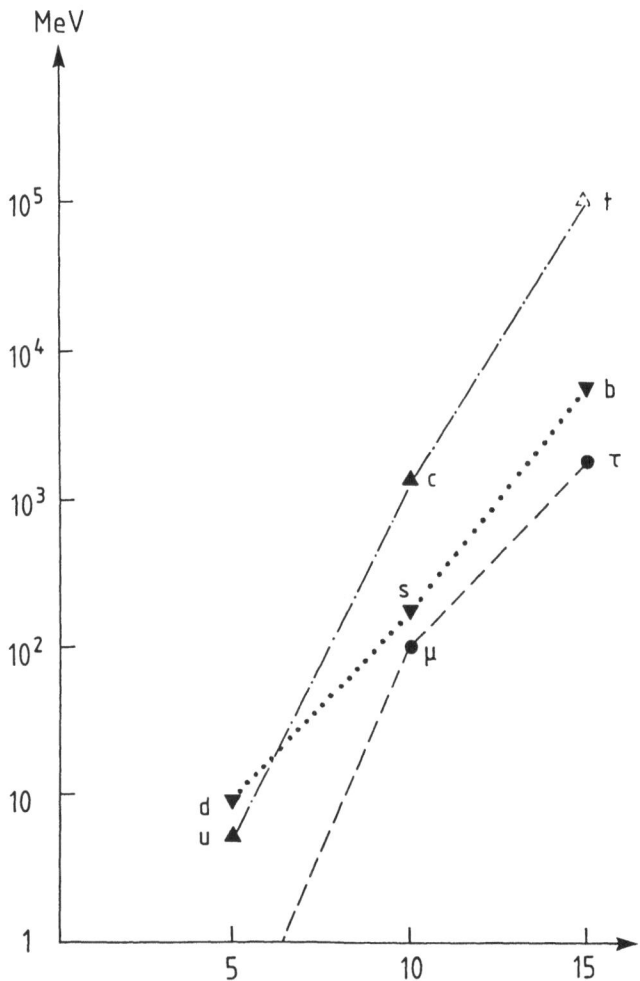

Fig. 2.   The lepton–quark mass spectrum on a logarithmic scale. The generation
index is used as the x–coordinat. The electron mass is not shown.

$$V = \begin{bmatrix} V_{ud} & V_{us} & V_{ub} \\ V_{cd} & V_{cs} & V_{cb} \\ V_{td} & V_{ts} & V_{tb} \end{bmatrix} \qquad (1.1)$$

$$= \begin{bmatrix} 0.9748 \ldots 0.9761 & 0.217 \ldots 0.223 & 0.003 \ldots 0.008 \\ -0.217 \ldots 0.223 & 0.9733 \ldots 0.9754 & 0.028 \ldots 0.08 \\ 0.00 \ldots 0.02 & -0.028 \ldots 0.06 & 0.9980 \ldots 0.9999 \end{bmatrix}$$

(Only the absolute values of the matrix elements are given.)

Clearly this matrix is not far away from a diagonal matrix. Furthermore the matrix elements V_{cb}, V_{ts} are small compared to V_{us}, V_{cd}, and the elements V_{ub}, V_{td} are yet significantly smaller than V_{cb}, V_{ts}. Qualitatively this pattern is displayed in Fig. (3). Subsequently I shall discuss various implications of the observed hierarchical pattern.

2. MASS HIERARCHIES AND CHIRAL SYMMETRIES

In view of the observed hierarchies it seems useful to consider the various stages of mass generation given in the subsequent table, following the arguments given in ref. (5).

Stage	Masses different from zero
I	none.
$\downarrow$(t,b massive)	
II	t,b.
$\downarrow$(c,s. massive)	
III	t,b; c,s.
$\downarrow$(u,d massive)	
IV	t,b; c,s; u,d.

If all six quark masses vanish, the Lagrangian of the standard model is invariant under the chiral symmetry $U(3)_L \times U(3)_R$, where $U(3)$ is the symmetry group connecting the three generations ($U(3)$ stands as usual for the direct product $SU(3) \times U(1)$, and the index L(R) refers to the left handed or righthanded quark fields q_L (q_R).

At the present time it is completely unknown what dynamical mechanism is responsible for generating the masses of the quarks (and leptons). Nevertheless it is a fact that the chiral symmetry $U(3)_L \times U(3)_R$ in the space of the generations is exact

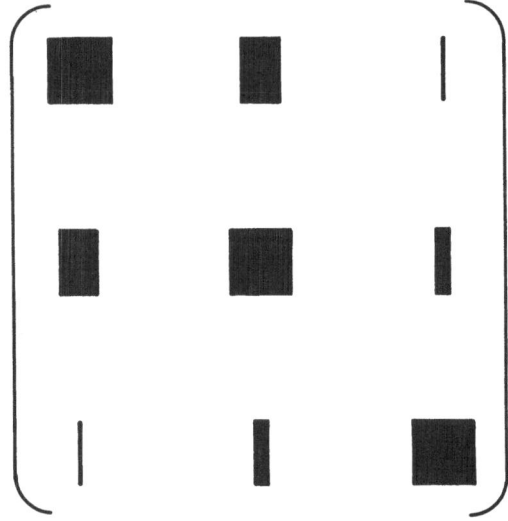

Fig. 3. The qualitative pattern of the weak interaction mixing matrix V.

only if all masses vanish, and vice versa. Furthermore in this limit all weak mixing angles vanish. Of course, in the real world this symmetry is broken. Nevertheless the chiral generation symmetry does play a relevant rôle. It guarantees that the symmetry limit $m_q = 0$ (q= u,d...) is a natural limit, i.e. it can be realized without posing artifical constraints on the underlying yet unknown dynamical framework (whatever this may be).

Stage II: $m_u = m_d = 0$, $m_c = m_s = 0$, m_t, $m_b \neq 0$

The (u,d)–quarks as well as the (c,s)–quarks are massless. The (t,b)–quarks are massive (their masses assume the values realized in nature).

Obviously this stage is not very far from the case given in nature, since the masses of the (u,d; c,s)–quarks are indeed very small compared to the t and b masses and might be regarded as perturbations. In terms of mass eigenstates, the mass matrices for the (2/3) and (−1/3) charged quarks are proportional to the matrix

$$\begin{bmatrix} 0 & 0 & 0 \\ 0 & 0 & 0 \\ 0 & 0 & 1 \end{bmatrix} \tag{2.1}$$

Such a situation can be obtained in a natural way only if it follows from a specific breaking of the original chiral generation symmetry. Since only the t and b quarks are massive, the original chiral symmetry $U(3)_L \times U(3)_R$ is not completely violated, but it is reduced to the subsymmetry $U(2)_L \times U(2)_R$ acting on the (u,d; c,s)–system.

What is the weak interaction mixing in the stage II? In principle the stage II mass pattern allows a non–trivial weak mixing described by just one angle, as described by the following weak doublets:

$$\begin{bmatrix} u & \vdots & c & \vdots & t \\ d & \vdots & s\cos\alpha + b\sin\alpha & \vdots & -s\sin\alpha + b\cos\alpha \end{bmatrix} \tag{2.2}$$

It can easily be seen that any parametrization of the weak interaction mixing at the stage II can be reduced to the one given above. Note that here the (u,d)–pair remains unmixed; this can always be arranged by a suitable unitary transformation among the two massless doublets.

The weak interaction mixing given in eq. (2.2) must be the consequence of a particular structure of the quark mass matrix. The quark mass term describing such a mixing is as follows:

$$(\overline{d}_o, \overline{s}_o, \overline{b}_o)_R \begin{bmatrix} 0 & 0 & 0 \\ 0 & 0 & 0 \\ 0 & B & A \end{bmatrix} \begin{bmatrix} d_o \\ s_o \\ b_o \end{bmatrix}_L + \text{h.c} \tag{2.3}$$

(d_o, s_o, b_o: quark states in which the weak currents are diagonal). The parameters A and B can be arranged to be real and positive.

It is obvious from the matrix given in eq. (2.3) that the mixing parameter does not influence the mass of the s–quark; the determinant of the mass matrix in eq. (2.3) is zero, independent of the magnitude of the mixing. The mass term can be written explicitly as

$$B \cdot \overline{b}_{o_R} s_{o_L} + A \cdot \overline{b}_{o_R} b_{o_L} + \text{h.c.}. \tag{2.4}$$

We should like to interpret the mass pattern at stage II as a result of a specific breaking of the chiral $U(3)_L \times U(3)_R$ symmetry which is broken down to $U(2)_L \times U(2)_R$. This reduction of the symmetry establishes a hierarchical pattern of the fermion masses (hierarchical chiral symmetry breaking).

The masslessness of the quarks of the first two generations in the stage II can only be guaranteed if the subgroup $U(2)_L \times U(2)_R$ of the original group $U(3)_L \times U(3)_R$ is preserved. However the mass term given in eq. (2.4) does not respect this symmetry if the mixing term $B \cdot \overline{b}_{o_R} s_{o_L}$ is present; this term transforms under $U(2)_L \times U(2)_R$ as (1/2, 0). For example, as a result of the chiral transformation $s_{o_L} \rightarrow e^{i\alpha} s_{o_L}$, $s_{o_R} \rightarrow e^{-i\alpha} s_{o_R}$ the first term in eq. (2.4) is multiplied by $e^{i\alpha}$, violating the chiral $U(2)_L \times U(2)_R$ symmetry. The weak interaction mixing and the chiral symmetry cannot both be implemented. Thus there is no weak interaction mixing in the limit of stage II. The weak interaction doublets are simply given by the mass eigenstates:

$$\begin{bmatrix} u & \vdots & c & \vdots & t \\ d & \vdots & s & \vdots & b \end{bmatrix} \tag{2.5}$$

In reality it is observed that the (t,b) system mixes only very slightly with the other quarks, which in our view is related to the fact that the (c,s)–masses are small compared to the (t,b)–masses, and the hierarchical symmetry limit of stage II is not far away.

Stage III: $m_u = m_d = 0$, m_c, m_s; m_t, $m_b \neq 0$.

In order to arrive at stage III, the chiral symmetry is broken down to $U(1)_L \times U(1)_R$. In terms of weak eigenstates the general mass term for the charge $(-1/3)$–quarks respecting this symmetry has the structure

$$(\overline{d}_o, \overline{s}_o, \overline{b}_o)_R \begin{bmatrix} 0 & 0 & 0 \\ 0 & C & B' \\ 0 & B & A \end{bmatrix} \begin{bmatrix} d_o \\ s_o \\ b_o \end{bmatrix}_L \tag{2.6}$$

The parameters A, B, B', C can be made real by suitable phase rotations of the quark fields. The mass matrix is Hermitian $(B = B')$ after a rotation of the right–handed quark fields (s_{o_R}, b_{o_R}).

The mass term (eq. (2.6)) can be diagonalized by a rotation in the (s_o, b_o)–plane. Only one mixing angle α enters, and we arrive at the same weak doublets as given in eq. (2.2). The weak mixing matrix denoted here by $V^{(o)}$ is given by:

$$V^{(0)} = \begin{bmatrix} 1 & 0 & 0 \\ 0 & \cos\alpha & \sin\alpha \\ 0 & -\sin\alpha & \cos\alpha \end{bmatrix} \qquad (2.7)$$

The weak mixing matrix at stage III is orthogonal – no complex phases are present. We conclude that in the limit $m_u = m_d = 0$ there is no CP violation. All CP violating effects must be proportional to positive powers of the u or d quark masses. Furthermore the u and d quarks remain unmixed in the limit $m_u = m_d = 0$. The Cabibbo angle vanishes in this limit, as well as the transitions u $\longleftrightarrow$ b or t $\longleftrightarrow$ d.

This result is, of course, not valid in general, but follows in our special approach from the underlying hierarchical structure.

<u>Stage IV:</u> All quark masses different from zero.

At the last stage IV the (small) masses of the u and d quarks are generated. The mixing matrix V takes its final form. Since the u and d masses are small compared to the other masses, the weak mixing matrix V can be regarded as a perturbation of the matrix $V^{(0)}$ denoted in eq. (2.8). This implies that the elements V_{us}, V_{ub}, V_{cd} and V_{td}, which vanish in the limit $m_u = m_d = 0$, can be expanded in positive powers of the ratios m_d/m_s, m_d/m_b, m_u/m_c and m_u/m_t. Of course, the corresponding expansion coefficients cannot be predicted in the general approach discussed here. This can only be done if one has a specific algebraic structure for the mass matrix. For example, in the ansatz of ref. (6) the chiral symmetry constraints discussed here are fulfilled. One finds:

$$V_{us} \approx -\sqrt{m_d/m_s} + e^{i\alpha}\sqrt{m_u/m_c} \qquad (2.8)$$

$$V_{ub} \approx \sqrt{m_d/m_b}\ m_s/m_b + \sqrt{m_u/m_c}\ (e^{i\alpha}\sqrt{m_s/m_b} - e^{i\beta}\sqrt{m_c/m_t})$$

$$V_{cb} \approx \sqrt{m_u/m_c} + e^{i\alpha}\sqrt{m_d/m_s}$$

$$V_{td} \approx \sqrt{m_u/m_t}\ m_c/m_t + \sqrt{m_d/m_s}\ (e^{i\alpha}\sqrt{m_c/m_t} - e^{i\beta}\sqrt{m_s/m_b})$$

(α, β phase parameters).

All these matrix elements vanish in the limit m_u, $m_d \to 0$, in accordance with the general conclusions drawn above.

As the u and d quarks acquire their masses, the b–s mixing coefficients (V_{cb}, V_{ts}) are essentially not changed (apart from very small corrections due to unitarity).

The considerations made above can easily be extended to include further quark doublets, provided the masses of the new quarks fit into the hierachical mass pattern as observed for the first three doublets. Suppose there exists a fourth doublet denoted by (h,l) (h: "high", l: "low"). If the masses of these quarks are large enough, e.g. $m_l > 80$ GeV, $m_h > 200$ GeV, the set of chiral symmetries discussed above can be extended by another stage at which all masses except the h and l masses are zero. At this stage the (h,l)–doublet remains unmixed. The mixing of this doublet with the (t,b)–doublet would arise once the t and b masses are introduced.

3. THE MASS GAP AND THE RANK ONE MASS MATRIX

The observed hierarchy of the lepton and quark masses suggests that it may be useful to consider a specific limit in which the masses of the first two generations vanish, i.e. the limit $m_\nu = m_e = 0$, $m_\nu = m_\mu = 0$, $m_u = m_d = 0$ and $m_c = m_s = 0$ (the stage II, discussed above). Of course, it depends on yet unknown details of the mass generation mechanism whether such a limit can be achieved in a consistent way. We simply assume that this is the case. Furthermore we shall assume that in this limit all weak mixing angles vanish as a result of a chiral symmetry, acting on the fermions of the first and second generation.

In the limit where the mass matries are proportional to the matrix given in eq. (2.1) there exists a mass gap: The third generation is split from the massless first two generations.

In ref. (7) the author has emphasized that a mass matrix like the one given in eq. (2.1) follows always if the general mass matrix has rank one. Especially it is obtained if one starts from a matrix in which all matrix elements are equal, e.g. the lepton mass matrix (see also refs. (8,9,10):

$$M = \begin{bmatrix} 1 & 1 & 1 \\ 1 & 1 & 1 \\ 1 & 1 & 1 \end{bmatrix} \cdot m_\tau / 3. \tag{3.1}$$

It is easily seen that the diagonalization of this matrix leads to the mass term given in eq. (1). More generally any 3 x 3–matrix M of rank one can be written as

$$M = \text{const. } h_i h_k \tag{3.2}$$

where h_i is a 3–vector. (For simplicity we assume M to be real, since in the absence of the masses of the first two generations the general mass matrix can always be made real and symmetric by suitable unitary transformations of the fermions).

More recently a mass matrix of the type given in eq. (3.2) was considered in ref. (11), where it was used as a starting point to construct the full mass matrices of the quarks, including the weak interaction mixing terms. It was also emphasized that such a matrix plays an important rôle in other fields of physics, where mass gap phenomena are observed:
a) In the BCS theory of superconductivity the energy gap is related to a matrix of the type (3.1) in the Hilbert space of the Cooper pairs (see also ref. (12)).
b) The pairing force in nuclear physics which is introduced in order to explain large mass gaps in nuclear energy levels has the property that the associated Hamiltonian in the space of the M= 0 pairs (M: angular momentum) has equal matrix elements, i.e. a structure of type (3.1).

We ask ourselves the question whether the mass gap in the lepton–quark mass spectrum could have a similar dynamical explanation as in the two phenomena mentioned above. This would mean that a rank one mass matrix like the one given in eq. (3.1) would play a significant rôle, and the question arises whether a dynamical explanation of such a matrix can be found. A specific model of this type has been discussed in 1983[13]. There the mass gap was related to the electromagnetic self energies of composite leptons and quarks. In lowest order of α one obtains a rank one matrix. However one consequence of this model, the prediction $m_t / m_b \approx 4$, is ruled out by the experiments. Nevertheless it will become clear subsequently that some of the features of this model might be correct.

Before we enter the discussion of the lepton–quark situation, we mention another gap phenomenon in physics which in our view may be more closely connected to the lepton–quark mass problem than the BCS phenomenon and the nuclear pairing force: the mass pattern of the pseudoscalar mesons in QCD.

The QCD Mass Gap in the Pseudoscalar Channel

It is well–known that in the absence of the quark mass m_u, m_d and m_s the nonet of pseudoscalar mesons is split in a very characteristic way: there exist eight massless pseudoscalar mesons, while the ninth one, the η'–state, acquires a mass of order 1 GeV. This mass gap was one of the phenomena which led originally to the idea that the forces among quarks are generated by the exchange of color octet gluons[14], since in this case the SU(3) singlet axial current acquires an anomalous divergence.

We are especially concerned with the mass spectrum of the neutral pseudoscalar mesons, i.e. linear superpositions of the quark bilinears $\bar{u}u$, $\bar{d}d$ and $\bar{s}s$. The mass and mixing pattern of the neutral pseudoscalar mesons π^0, η' is well reproduced by the ansatz[15].

$$
M^2_{\bar{q}q} = \begin{bmatrix} M^2_{\bar{u}u} & & \\ & M^2_{\bar{d}d} & \\ & & M^2_{\bar{s}s} \end{bmatrix} + \lambda \begin{bmatrix} 1 & 1 & 1 \\ 1 & 1 & 1 \\ 1 & 1 & 1 \end{bmatrix} \tag{3.3}
$$

$(q = u, d, s)$

where $M^2_{\bar{u}u}$ etc. stands for the masses generated by the breaking of the chiral $SU(3)_L \times SU(3)_R$ symmetry which are linear in m_u etc., and λ stands for the strength of the transition of the $\bar{q}q$ – configuration into another $\bar{q}q$ – configuration. In QCD the matrix describing these transitions is precisely of the type (3.1), provided one uses the basis of the bilinears $\bar{u}u$, $\bar{d}d$ and $\bar{s}s$.

Numerically the strength of the transition term is large[15]: $\lambda \approx 0.5$ GeV^2. Presumably the magnitude of this term is related to the action of instantons, which breaks the associated $U(1)$ – axial symmetry[16]. In the absence of the quark masses the mass of the SU(3) singlet meson, i.e. the coherent state $1/\sqrt{3}\,(\bar{u}u + \bar{d}d + \bar{s}s)$, is massive, while the two other coherent states $1/\sqrt{2}\,(\bar{u}u-\bar{d}d)$ and $1/\sqrt{6}\,(\bar{u}u + \bar{d}d - 2\bar{s}s)$ remain massless.

The effective mass term of eq. (3.3) arises in QCD as a second order effect, involving the mixing of the $\bar{q}q$ states with 0^{-+}–gluonium states. If one were able to turn off this mixing, there would exist nine massless pseudoscalar mesons instead of eight in the chiral limit $m_u = m_d = m_s$. Thus the mixing

$\bar{q}q \leftrightarrow \bar{q}'q'$ proceeds via intermediate glue states as follows: $(\bar{q}q) \leftrightarrow 0^{-+} (\text{glue}) \leftrightarrow (\bar{q}'q')$.

The SU(3) flavor symmetry requires that the resulting mass matrix, the second term in eq. (3.3), has the same entry λ everywhere, i.e. it is of rank 1. Thus the mass gap in the pseudoscalar channel appears. We believe that the way this mass gap is generated, can teach us not only details about the QCD interactions, but may provide us with an interesting analogy to the lepton–quark mass spectrum.

The Lepton–Quark Mass Gap

We believe that the dynamical mechanisms which lead to the mass gaps in the spectrum of the neutral pseudoscalar mesons and in the lepton–quark mass spectrum are somewhat related. In fact, in both cases the chiral limits $m_{\pi^0} = m_\eta = o$ and $m_e = m_\mu = o$ etc. are obtained if a chiral SU(2)–symmetry, acting either on the u– and d–quarks or on the fermions of the first two generations, is imposed. Even after the symmetry breaking the resulting spectra are quite similar:

$$m^2_{\pi^0} << m^2_\eta << m^2_{\eta'}$$

$$m_e << m_\mu << m_\tau , \quad m_u << m_c << m_t , \quad m_d << m_s << m_b.$$

Of course, there exists also a major difference between the two cases: the leptons and quarks are fermions, while the pseudoscalar mesons are bosons. The latter are acting in the chiral limit $m_u = m_d = m_s = 0$ as massless Goldstone bosons while correspondingly the leptons and quarks are massless due to a supposedly exact chiral symmetry. In view of our nearly total ignorance about the physics beyond the standard model we find it interesting to speculate that at least the generation of the η'–mass in QCD and the generation of (m_t, m_b, m_τ) are analogous. At this moment it would be premature to assume a specific underlying theoretical framework, e.g. a composite model, a specific technicolor framework or an extended symmetry scheme, based on a supersymmetric extension of the standard model or on a superstring model – doing so, would be almost certainly wrong.

Nevertheless we feel that such an underlying dynamical theory should be somewhat similar to QCD, which, after all, represents a composite model of the hadrons, Thus we think, a composite model of the leptons and quarks has the best chances to be able to explain the mass pattern of the leptons and quarks and in particular the mechanism to create the mass gap, discussed subsequently.

Let us recall that it is crucial for the mass gap of the pseudoscalar mesons to appear that there exists a continuum of gluonium states above a certain threshold of the order of 1...2 GeV. The lowest of these states may be an isolated resonance, although this is not necessary for the argument – in any case the mixing is caused by a matrix of the type given in eq. (3.3).

Likewise we suppose that the mass gap for the leptons and quarks in the limit is due to the mixing of the fermions with heavy fermion states, which are analogous to the QCD gluonium states. We leave it open at which energies these states appear. If one associates their mass scale with the scale of compositeness of the fermions, they must have a mass of at least a few hundred GeV.

For simplicity we represent the heavy states by isolated heavy fermions H, although a continuum of states would serve the same purpose:

$$H = \begin{bmatrix} N & \vdots & U \\ \cdots & & \cdots \\ E & \vdots & D \end{bmatrix}$$

(N: heavy neutral state, E: heavy charged lepton state of charge –1; U, D: heavy quark states of charge 2/3, –1/3 respectively).

First we discuss the mass generation for the charged leptons. In the absence of mixing all three leptons denoted by l_1, l_2, l_3 (1,2,3: family index) are massless; the symmetry of the lepton mass term is $SU(3)_L$ x $SU(3)_R$. Once the mixing of the massless lepton states with the heavy fermion is introduced, the mass eigenstates are (in lowest order of the small mixing angles α_i and β_i, taken to be real)

$$E'_L = (\alpha_1 l_1,\ \alpha_2 l_2,\ \alpha_3 l_3,\ E_L) \tag{3.4}$$

$$E'_R = (\beta_1 l_1,\ \beta_2 l_2,\ \beta_3 l_3,\ E_R)$$

The resulting mass matrix for the previously massless fermions is of rank one and given by:

$$(l_1,\ l_2,\ l_3)_L \begin{bmatrix} \alpha_1\beta_1 & \alpha_2\beta_1 & \alpha_3\beta_1 \\ \alpha_1\beta_2 & \alpha_2\beta_2 & \alpha_3\beta_2 \\ \alpha_1\beta_3 & \alpha_2\beta_3 & \alpha_3\beta_3 \end{bmatrix} \begin{bmatrix} l_1 \\ l_2 \\ l_3 \end{bmatrix}_R m_E + \text{h.c.} \tag{3.5}$$

Analogously the mass matrices of the quarks are constructed. The associated mixing parameters α_i and β_i are, of course, different in each case. It is assumed that the neutrinos remain massless.

By suitable unitary $SU(3)_L$ x $SU(3)_R$ transformations these matrices M can always be brought to the following specific forms:

a)
$$M = \begin{bmatrix} 0 & 0 & 0 \\ 0 & 0 & 0 \\ 0 & 0 & 1 \end{bmatrix} m \tag{3.6}$$

(m stands generically for m_τ, m_t, m_b).

There is, however, an important constraint for the quarks. If one achieves this form for the quarks of charge 2/3, at the same time the mass matrix for the quarks of charge $(-1/3)$ must have the same form after applying the same unitary transformations in the family space. Only if this is the case, the weak mixing angles vanish, as they should. This is guaranteed, for example, if the two mass matrices are proportional to each other: $M(2/3) = \text{const.} \cdot M(-1/3)$.

b)
$$M = \lambda \begin{bmatrix} 1 & 1 & 1 \\ 1 & 1 & 1 \\ 1 & 1 & 1 \end{bmatrix} \tag{3.7}$$

This basis, which we shall denote as the coherent state basis, is of special interest. Note that the mass term is simply given by $\lambda \sum_{i,j} \bar{f}_i f_j$ (f: lepton, quark). The mass matrix is identical to the induced λ–term in the pseudoscalar meson matrix, given in eq. (3.3). We recall that in the meson case this basis is especially suited: only in this basis the mesons are represented by <u>pure quark states</u>, not by mixtures. Of course, this is of no relevance, if the quarks are massless, but it becomes very important, once the chiral symmetry is broken by small mass terms.

Something similar is expected to happen in the lepton–quark case. Thus the matrix (3.7) might be a good starting point for introducing symmetry breaking effects. An attempt in this direction was made recently[11].

The coherent state basis is one in which the mass eigenstates are represented by superpositions:

$$R = 1/\sqrt{3}\,(l_1 + l_2 + l_3) \quad \mu = 1/\sqrt{6}\,(l_1 + l_2 - 2l_3) \quad e = 1/\sqrt{2}\,(l_1 - l_2) \qquad (3.8)$$

$$t = 1/\sqrt{3}\,(u_1 + u_2 + u_3) \quad c = 1/\sqrt{6}\,(u_1 + u_2 - 2u_3) \quad u = 1/\sqrt{2}\,(u_1 - u_2)$$

$$b = 1/\sqrt{3}\,(d_1 + d_2 + d_3) \quad s = 1/\sqrt{6}\,(d_1 + d_2 + d_2) \qquad d = 1/\sqrt{2}\,(d_1 - d_2).$$

In view of the scarce information we have at present about the internal dynamics of the leptons and quarks we do not know, whether this description of the fermions in terms of coherent states in more than a specific mathematical representation.

Definitely it is so in the meson case: it is the basis in which the SU(3)–invariant nature of the mixing with the gluonia–states is manifest.

We expect something similar to be the case in the lepton–quark case. Thus the fermion states f_1, f_2, f_3 would be those states which are "pure" in a dynamical sense, e.g. they have simple wave functions in a composite model.

We remind the reader that also in the case of superconductivity and of the nuclear pairing force the mass eigenstates are coherent superpositions of "physical" states which are described by simple wave functions (e.g. the Cooper pairs in superconductivity).

For the leptons one deduces the coherent states in terms of mass eigenstates:

$$l_1 = 1/\sqrt{2}\,e + 1/\sqrt{6}\,\mu + 1/\sqrt{3}\,\tau$$

$$l_2 = -1/\sqrt{2}\,e + 1/\sqrt{6}\,\mu + 1/\sqrt{3}\,\tau \qquad (3.9)$$

$$l_3 = -\sqrt{2}/3\,M + 1/\sqrt{3}\,\tau$$

(Analogous relations with the same coefficients are valid for the quarks).

Within our approach we see a solution to a problem, which has plagued many models of the physics beyond the standard model, the problem of the near masslessness of the first and to some extent also of the second generation. In the coherent state basis this is easily understood. For example, the electron state $e = 1/\sqrt{2}\,(l_1 - l_2)$ is nearly massless, since there is a nearly complete cancellation of the l_1– and l_2– mass terms, as a consequence of the rank one structure of the dominant lepton–quark mass term.

Below we discuss two simple and useful applications of the scheme discussed above.

a) The origin of the induced λ–term and the mass of the t–quark

At this stage we are unable to give a detailed dynamical description of the origin of the mixing term. Nevertheless one may speculate about some of its properties. Apart from small corrections, the ratio m_t/m_b is given by the ratios of the λ–terms for the charge 2/3–sector and the charge (−1/3) sector. On the other hand the λ–terms are the products of certain mixing angles, multiplied by m_U and m_D respectively. In many composite models and in models of the technicolor type one expects the threshold mass parameters m_U and m_D to be identical, due to an underlying SU(2) flavor symmetry. Thus the ratio m_t/m_b is directly related to the strength of the mixing interaction.

From the experimental limits on m_t one deduces that m_t/m_b is at least of the order of 10. Since the essential difference between the charge (2/3) and the charge (−1/3) sector is the electric charge, one possible conclusion is that the mixing is caused by the electromagnetic interaction, acting together with nonperturbative effects of the underlying hyper– or technicolor forces, e.g. instantons. In this case an expansion of the λ–terms in the finestructure constant should be possible:

$$\lambda = \text{const. } \alpha Q^2 + \text{const. } \alpha^2 Q^4 + ... \tag{3.10}$$

(Q: 2/3 or (−1/3) respectively).

The first term in this expansion must be absent; otherwise one would have $m_t/m_b = 4$, which is excluded by experiments. However the next term could well be present, in which case it would dominate, and one finds:

$$\frac{m_t}{m_b} = 16. \tag{3.11}$$

Using m_b (1 GeV) $\approx$ 5300 MeV, this relation leads to:

$$m_t(1 \text{ GeV}) \approx 85 \text{ GeV}. \tag{3.12}$$

Rescaling this value due to the QCD effects, one finds for the physical value of m_t, i.e. $m_t(m_t)$:

$$m_t \approx 50...55 \text{ GeV}, \tag{3.13}$$

a value, not yet excluded by the experiments, but not far from the experimental limit.

It is interesting to note that experimental limits on m_t require that the first term in eq. (3.10) must vanish. The same is true in the case of the mixing of the pseudoscalar mesons, where $\lambda = O(\alpha_s^2)$. This is related to the fact that the mixing interaction must involve at least two gluons, which originate from the annihilation of a $\bar{q}q$–pair.

In our case the situation might be analogous in the sense that the λ–term is caused by the electromagnetic annihilation of fermion–antifermion pairs in the wave functions of the leptons and quarks. Of course, such an interpretation would make sense only in composite models. Indeed it has been speculated that the nearly massless states of the leptons and quarks may involve many pairs of the underlying constituents[17].

b) <u>Flavor conservation and nonconservation</u>

In our approach the flavor problem is directly related to the existence of three massless generations of leptons and quarks in the absence of the λ-term. By definition these states can be chosen to be orthogonal to each other. Thus each lepton and quark flavor is separately conserved. Of course, this is not a solution of the flavor problem, but relates it to the spectrum of the leptons and quarks, i.e. a feature of the lepton–quark physics which may soon be clarified.

Once the λ-term is introduced, the third generation acquires a mass. The first two generations remain massless, and we still have separate conservation laws for the first two generations, however no such law for the third generation. Due to the mixing with the heavy states transitions like $E \rightarrow \tau + \gamma$
or $U \rightarrow t + g$ (g: gluon) are allowed.

Although the generation of the masses of the first two generations is not a topic which we have studied here, it is expected that conservation laws for the quark and lepton flavors of the first two generations are broken slightly as soon as their masses are introduced. Thus transitions like $\mu \rightarrow e\gamma$ should accur. Likewise an excitation of the electron to a muon or a γ–lepton in an electromagnetic interaction like $e+p \rightarrow \mu + X$, i.e. the process studied with the future HERA maschine, should be possible. Of course, these flavor violating terms are proportional to powers of small mass ratios like m_e/m_E and m_μ/m_E and therefore very small, but perhaps large enough to be detectable soon.

4. CONCLUSIONS

In this talk I have described a number of ideas which one might consider after looking at the pattern of masses exhibited in the lepton–quark mass spectrum. I have emphasized the rôle of chiral symmetries in the space of the generations of the quarks in providing relations between the various mass eigenvalues and the mixing angles.

An approach to the flavor problem and to the hierarchical mass spectrum of the leptons and quarks, based on the introduction of coherent states, was discussed. It was argued that the mass generation for the third lepton–quark generation is nothing but a gap phenomenon and is rather similar to the mass generation for the pseudoscalar mesons. In an analogy to the latter case the masses of the members of the third generation are due to a mixing with a continuum of heavy states. Thus the third lepton–quark generation is somewhat distinct from the other ones. If this mixing is due to electromagnetic effects, one expects $m_t \approx 50...55$ GeV. The same mechanism which leads to the mass generation causes the appearance of flavor changing effects; only in the absence of the lepton and quark masses the various quark and lepton flavors are exactly conserved.

If our interpretation of the mass gap seen in the lepton–quark spectrum is correct, it would mean that all mass gap phenomena seen in physics – superconductivity, nuclear pairing forces, QCD mass gap, lepton–quark spectrum – are due to an analogous underlying dynamical mechanism. The exploration of further details of this mechanism could lead soon to a deeper understanding of the physics beyond the standard model.

REFERENCES

1. See e.g: Composite W–Bosons and their Dynamics, Proceedings of the Int. School on Subnuclear Physcis, Erice 1984.
    Schrempp, B., 1986, Proc. of the VIth. Int. Conf. on Proton Physics, Aachen, p. 642.

2.   Gasser, J., and Leutwyler,H., 1982, Phys. Rep. 87, 77.
3.   Shochet, M.J., 1988, Proc. of the 24. Int. Conference on High Energy Physics, Munich.
4.   For a recent review see: Kleinknecht, K., Proceedings of the 24. Int. Conference on High Energy Physics, München 1988.
5.   Fritzsch, H.,1987, Phys. Lett. 184 B, 3091.
6.   Fritzsch, H., 1979, Nucl. Phys., B 155, 189.
7.   Fritzsch, H., 1984, Proc. Europhysics Conference on Flavor Mixing in Weak Interactions, Erice 1984. L.L. Chau ed.
8.   Harari, H., Haut, H., and Weyers, J., 1978, Phys. Lett. 78 B, 459.
9.   Koide, Y., 1983, Phys. Rev. D 28, 252.
10. Jarlskog, C., 1986, Proc. Int. Conf. on Production and Decays of Heavy Flavors, Heidelberg, Schubert, K.R., and Waldi, R. edts..
11. Kaus, P., and Meshkov, S., 1988, Caltech preprint, to appear in Europhysics Letters.
     Fritzsch, H., 1988, preprint MPI–PAE/PTh 22/88.
12. Nambu, Y., these proceedings.
13. Baur, U., and Fritzsch, H., 1984, Phys. Lett. 134 B, 105.
14. Fritzsch,H., Gell–Mann, M., and Leutwyler, H., 1973, Phys. Lett. 47 B 3) 365.
15. Fritzsch H., and Minkowski,P., 1975, Nuovo Cimento 30 , 393.
16. t'Hooft, G., 1976, Phys. Rev. D 14, 3432.
17. See e.g.: Harari, H., and Seiberg, N., 1981, Phys. Lett. 98 B, 269.
     Fritzsch, H., and Mandelbaum, G., 1981, Phys. Lett. 102 B, 319.

17

THRESHOLD B̄B FACTORIES

David B. Cline

Physics Department
University of California, Los Angeles
405 Hilgard Avenue
Los Angeles, California 90024-1547

ABSTRACT

We review the requirements of B̄B Factories to facilitate the observation of rare B decays and the study of CP violation. Two advances are required over present colliding beam machines: (1) much higher luminosity, (2) collisions that increase the reconstruction efficiency and facilitate proper time measurements. The latter requirement can likely be met by using asymmetric colliding beams (either circular or linear). The limitations on achievable luminosity in circular machines likely limits the maximum luminosity to 10^{33}cm^{-2}sec^{-1} (SIN Machine). Linear collider B̄B Factories offer the possiblitiy of greater luminosity but are still in the R&D stage. Examples of B̄B Factory designs at UCLA and Frascati are described.

Introduction

There are two promising directions of study in K or B Physics: Rare Decays and CP Violation. A considerable effort is underway at BNL, CERN, and FNAL to study rare K decays and study CP violation in the K_L° system. Kaon factories at LAMPF and Canada are being designed to carry on this work. On the other hand the B system has hardly been studied but may have a greater sensitivity in both the rare decays and CP violation.

About two years ago at the Heidelberg Conference the concept of B̄B Factories was born[1,2,3]. This occured at about the same time as the announcement of substantial B° mixing by the UA1 group[4]. Later the ARGUS group reported large mixing in the B_d° system to the surprise of almost all theorists[5]. During the past two years several B factory design studies have started and one machine has been formally proposed for construction. Two workshops devoted to this idea have been held.

In this report we review the progress to date in these designs and indicate the areas where substantial uncertainty still exists and where R&D efforts are still required.

1. Rare B Decays

The large mass of the b quark and the strong coupling to the massive t quark increase the sensitivity of rare B decays to new physics such as the existence of a 4th family of quarks. There are numerous examples of rare B decay calculations in the literature. One illuminative case is the decays

$$b \rightarrow s + \ell + \bar{\ell} \tag{1}$$
$$b \rightarrow s + \nu + \bar{\nu} \tag{2}$$

These decays can be contrasted with the decays

$$s \rightarrow d + \ell + \bar{\ell} \tag{3}$$
$$s \rightarrow d + \nu + \bar{\nu} \tag{4}$$

which are studied as

$$K^+ \rightarrow \pi^+ + e^+ + e^- \tag{5}$$
$$K^+ \rightarrow \pi^+ + \nu + \bar{\nu} \tag{6}$$

It is well known that the GIM breaking decay branching action for 6 is expected to appear at the level of about 10^{-10}. In contrast reactions (1) and (2) can display GIM breaking decay at the level of $10^{-5} - 10^{-7}$. Thus the B decays much more sensitive to non GIM effects than the corresponding K decays. A list of the current estimate for rare B decays is given in Table 1. [7]

2. CP Violations in B Decay

The coupling of the b quark to off diagonal K-M matrix elements is expected to enhance the level of CP violation in certain processes. (Those that favor the off diagonal elements such as $B^\circ \rightarrow \pi^+\pi^-$.) In addition it appears that B_d° mixing is larger than was originally expected, futher enhancing the chance for observing CP violation. Many theorists now believe that the future study of CP violation will be largely concentrated in the B decay sector.

Rare B decays may exhibit the largest CP violation. Thus high luminosity $\bar{B}B$ production is required. Table 1 indicates some B decays that may exhibit large CP violation and the estimated level. Another way to study CP violation is to observe the proper time behavior of the B_d° or B_s° system. This is described in some detail in Reference 8.

In order to derive an estimate of the number of $\bar{B}B$ pairs required to observe CP violation, we find

Table 1. Estimated Rare B Decay Branching Ratio

MODE	ESTIMATED BRANCHING RATIOS	CP ASYMMETRY
$B \to D^+D^-$	$\sim 10^{-3}$	
$B^\circ \to \Psi K_s$	$(4 \times 10^{-4})^* \to 2 \times 10^{-5}$	~ 0.2
$B \to K\pi$ **	$\sim 10^{-5}$	?
$B \to \pi\pi$	$\sim 10^{-5}$	> 0.05
$B \to \bar{p}p$	$\sim 10^{-5}$	~ 0.3
$B \to K\nu\bar{\nu}$	$\sim 10^{-7} - 10^{-6}$	–
$B \to Ke^+e^-$	$\sim 10^{-7} - 10^{-6}$	–
$B^\circ \to \mu^+\mu^-$	$\sim 10^{-8}$	
$B \to K\mu^\pm e^\pm$	?	

* Detection of $4K_s$ final state in the mode $4 \to \ell\bar{\ell}$, $K_s \to \pi\pi$ introduces an additional reduction in the effective branching ratio

** $B\bar{B}$ tagging may not be necessary – a rate difference between $K^+\pi^-$ and $K^-\pi^+$ could be sufficient to detect CP violation

$$N \simeq \frac{1}{\alpha^2} \cdot \frac{1}{B_r} \cdot \frac{1}{Q} \cdot 10$$

where α is the CP violating asymmetry expected, B_r is the branching ratio of the decay mode and Q is a quality factor that includes the reconstruction efficiency and tagging efficiency for the other B (or $\bar{B}$). For present colliders Q is very small. It might be hoped that Q could reach 10^{-1} using a linear collider with a very small beam pipe and asymmetric energies. For example, with $\alpha = 10^{-1}$, $B_r = 10^{-4}$, $Q \sim 10^{-1}$ we find the number of events needed is 10^8. a more careful calculation has been carried out by Rosner, Sanda and Schmidt and the results are shown in Figure 1.[8]

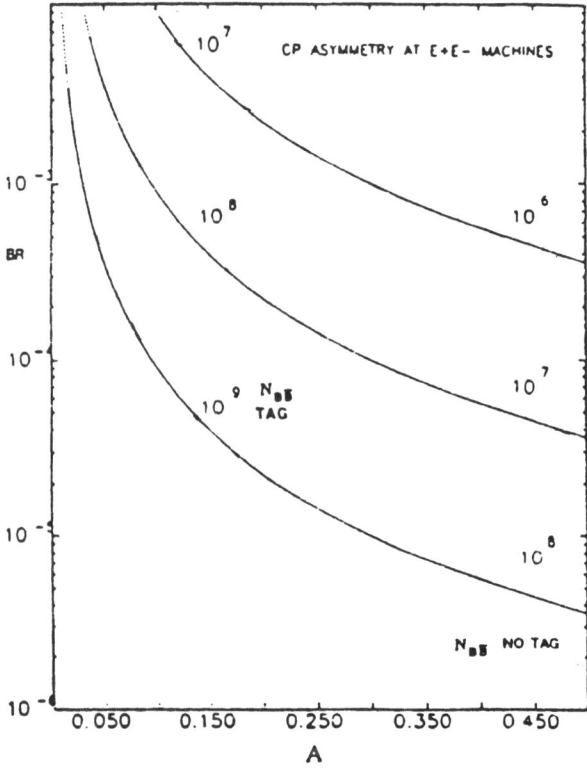

Figure 1. Number of B decays needed to see a CP violating asymmetry A in a process with branching ratio BR. Larger numbers assume a 10% tagging efficiency.

3. Requirements in Luminosity and Reconstruction Efficiency

The currently proposed strategy for observing CP violation in the B sector is to select improbable or rare B decays and to search for a rate asymmetry or a deviation in the proper time distribution from the expected distribution with no CP violation. The former method required identification of the B or $\bar{B}$ in the same event, thus requiring good reconstruction efficiency. The second method requires accurate proper time measurement. Both methods constrain the parameters of the $\bar{B}B$ system. It is now recognized that the best solution to these problems is to operate colliding beam machines in asymmetric energy mode to give the $\bar{B}B$ system a net laboratory momentum. This will increase the γ of each B and thus the path length before decay, enhancing the proper time measurements. Furthermore, the Lorentz boost should increase reconstruction efficiency.

The major advantage of the asymmetric energies is to give a boost to the B and $\bar{\text{B}}$ systems as shown in figure 2.[9] In Figure 3 we show the decay lengths for symmetric colliding beam energies. In the vicinity of the Υ (4S) the decay length is very small. At higher energy the decay length increases, however, unfortunately the production $\bar{\text{B}}\text{B}$ cross section falls rapidly. Figure 4 shows the same decay lengths for the asymmetric case. Note that for the case of 2.5 GeV and 10 GeV the decay length increases to about 350 μm. The conbination of this long decay length and the very small beam pipe for a linear collider could result in a large Q factor.

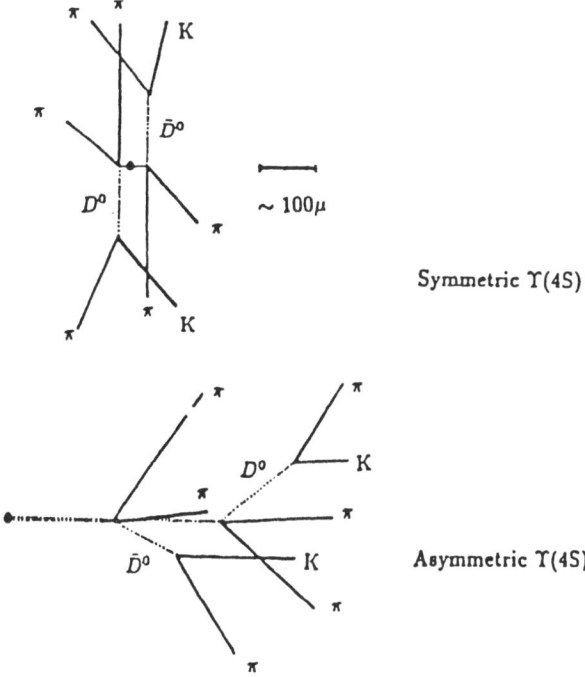

Figure 2. Lorentz Boost of the $\bar{\text{B}}\text{B}$ System for the Symmetric and Antisymmetric $\bar{\text{B}}\text{B}$ Factories

The observation of the rare B decay (see Table 1) can only be carried out at an e^+ e^- machine. Studies are underway to determine if the boosted system offers an advantage.

A promising technique to measure the B_d°, B_s° mixing and to search for CP violation effects is to study time distribution. A recent study of the detector parameters has been carried out by Gratta, Schwarz and Zaccardelli. They define the necessary resolution in decay length measurements [9]

$$\sigma_{\text{mix}} \leq \frac{1}{\text{n}}\frac{2\pi}{X_{\text{s,d}}}[\beta\gamma c\tau]$$

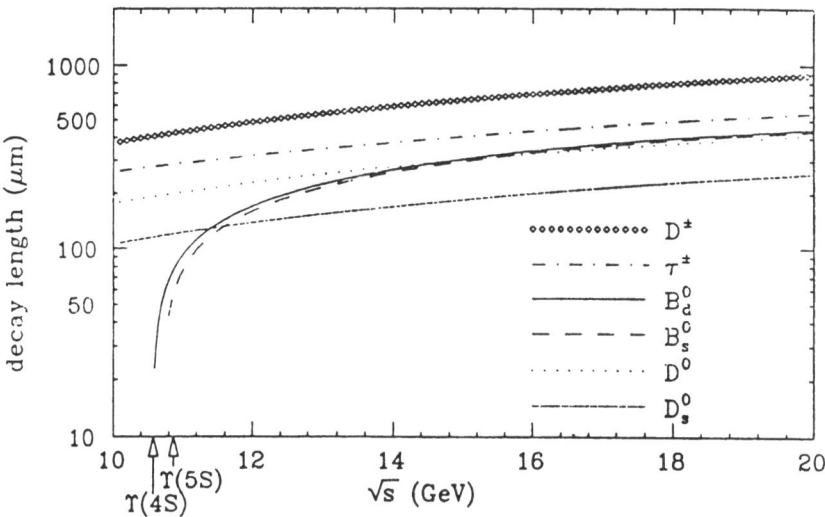

Figure 3. The mean decay length of D mesons, τ leptons and B mesons as a function of $\sqrt{s}$ using the values for their proper lifetimes and masses. The different fragmentation functions for the charm and the bottom quark have been taken into account. Note that for the B_s^0, the values obtained for the mean decay length are very model dependent in the vicinity of the Υ [5S] resonance. (Ref 9)

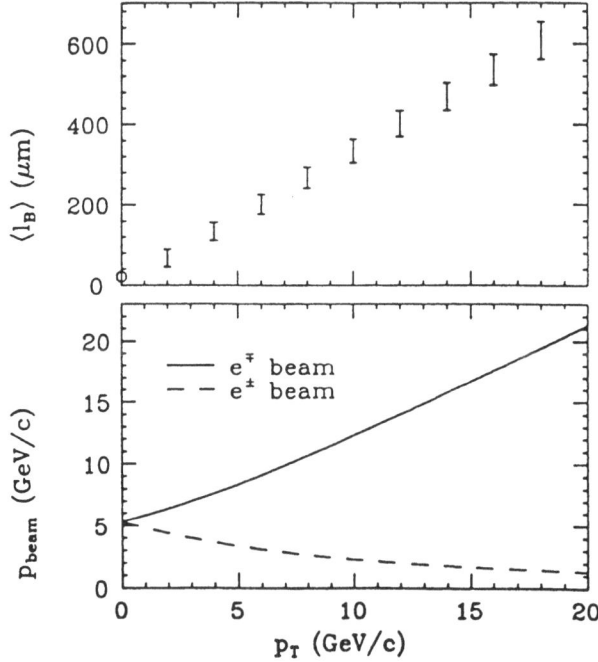

Figure 4. The expected decay length range of B mesons from a moving Υ [4S] as a function of the momentum of the Υ [4S] (p_T). the two boundaries for the decay length range are given by the two extreme cases where the B meson is emitted from the Υ [4S] antiparallel or parallel to the boost direction. Also shown are individual e^+ and e^- beam momenta necessary to create the Υ [4S] resonance in the center of mass. (Ref 9)

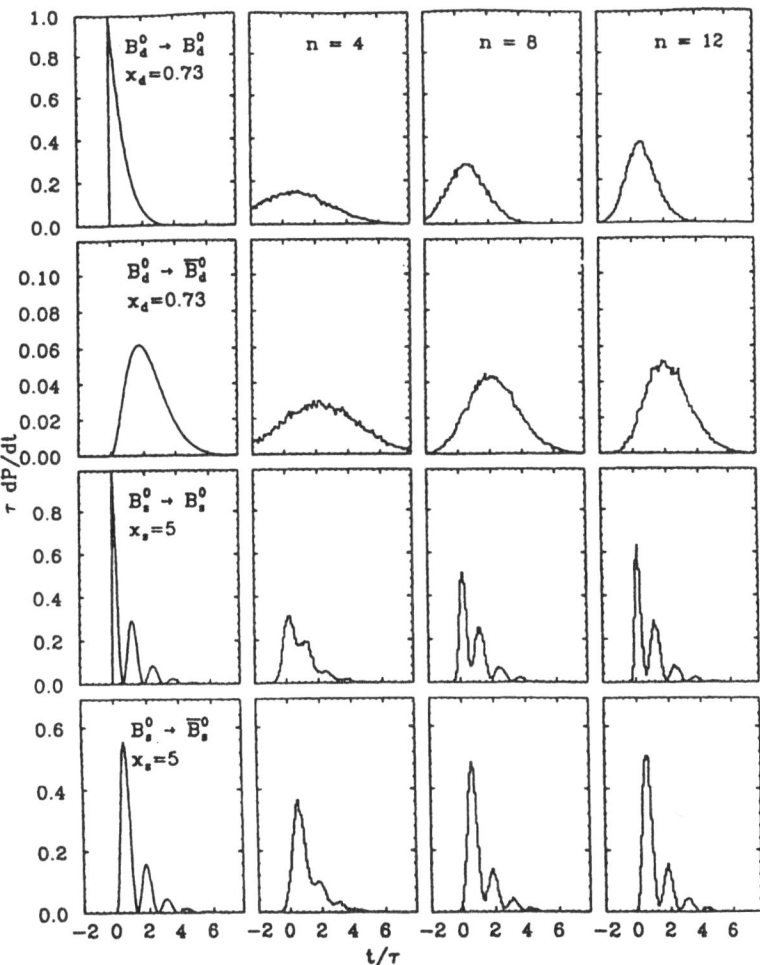

Figure 5. Monte Carlo generated mixing distributions for B_d^o and B_s^o mixing, smeared with thee different values of the assumed decay length resolution σ_{mix} (see text) using n=4, n=8, and n=12 for x_d=0.73 and x_s=scales for the individual distributions. (Ref 9)

where $X_{s,d}$ are the mixing parameters, n is a quality of measurement factor and $\beta\gamma c\tau$ are well known. Figure 4 shows the results of a monte carlo generation of decays. For the boosted case they find

$$\sigma_{\text{mix}} \leq 320\mu\text{m} \text{ for } B_d^\circ - \bar{B}_d^\circ \text{ on the } \Upsilon(4S)$$

and

$$\sigma_{\text{mix}}^{5S} \leq 40\mu\text{m} \text{ for } B_s^\circ - \bar{B}_s^\circ \text{ for the } \Upsilon(5S)$$

These values are obtainable with present micro vertex detectors.

Table 2.   Parameters of the B Meson System and Figure of Merit for Different Types of $\bar{\text{B}}\text{B}$ Sources (From P. Oddone, Ref 9)

	$\Upsilon(4S)$ Symmetric	$\Upsilon(4S)$ Boosted	PEP (25 GeV)		Z°	
	R = 1 cm	R = 1 cm	R = 1 cm	R = 4 cm	R = 1 cm	8 cm
$< P_\beta >$ GeV/c	0.3	5	9	9	20	20
$< \gamma\beta >$	0.08	1	1.7	1.7	20	20
$< \bar{b}b \text{ separation} >$	35	300	1020	1020	4200	4200
Impact parameter relative to beam collision	17	150	280	280	500	500
$< P_\pi >$ GeV/c	0.5	0.7	1.0	1.0	2.0	2.0
δ microns	21	18	15	60	11	11
Figure of merit M	1.6	8.3	18	4.5	45	5.7

4. Limitations on Luminosity in Circular Colliders

We now describe the various approaches that have been considered for threshold $\bar{\text{B}}\text{B}$ factories. A summary of the limitations and possible goals in luminosity for these machines is shown in Figure 6. At present it seems that only a linear collider $\bar{\text{B}}\text{B}$ factory has the potential for leading to the observation of CP violation in the B system. However, as we shall see, this is still subject to a great deal of uncertainty which calls for a strong R&D program in this field.

Circular colliders have many advantages and several disadvantages for the production of large quantities of $\bar{\text{B}}\text{B}$ pairs. The major advantage is the ease of construction of circular machines after two decades of experience with such machines. The disadvantages are (1) relatively large beam pipe at the IP making proper time measurements difficult, (2) limitations on luminosity due to the maximum time shift allowed by the machine, (3) difficulty of operating the machines in an asymmetric energy mode. We note that the CESR machine at Cornell would eventually reach a luminosity of $5 \times 10^{32}\text{cm}^{-2}\text{sec}^{-1}$.[11]

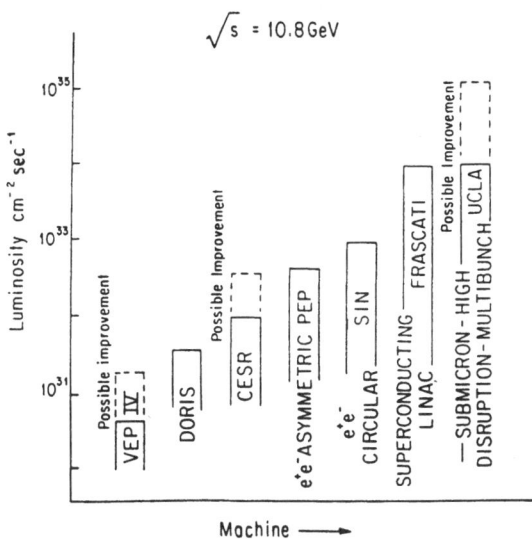

Figure 6. Overview of the Possible Luminosities For Circular and Linear Collider $\bar{B}B$ Factories compared with some existing colliders.

The maximum luninosity that can be obtained in a circular machine is given by[10]

$$\mathcal{L}^{\mathrm{max}} \simeq (\mathrm{nf}) \times \epsilon_{\mathrm{x}} \times \frac{(\Delta\nu)^2}{\beta^*_{\mathrm{y}}}$$

where nf is the number of bunches, σ_{x} is the natural horizontal emittance of the beam, $\Delta\nu$ is the linear tune shift and β^*_{y} is at the IP of the collider. There is strong evidence of a limitation of $\Delta\nu$ in existing machines. This likely limits $\mathcal{L}^{\mathrm{max}}$. $\Delta\nu$ will be partially limited by bunch crossing at regions other than the IP and this is one reason the SIN project will have two rings.[10] It is now generally considered that $\mathcal{L}^{\mathrm{max}}$ can not exceed $\sim 10^{33} \mathrm{cm}^{-2}\mathrm{sec}^{-1}$ due to these limitations.

5. Circular Collider $\bar{B}B$ Factory : SIN Example

A complete design of a circular $\bar{B}B$ factory has been carried out by a SIN – West Germany group.

In order to overcome part of the time shift limitations the SIN machine will operate with two rings.[10] In addition, electrons and positrons will be injected into the collider at the final energy, thus avoiding the necessity of accelerating the electrons and positrons in the storage ring. This should also help raise the luminosity. A schematic of the machine is shown in Figure 7. The current design parameters are given in Table 3. a recent calculation of the luminosity prospects is shown in Figure 8. The initial luminosity of the machine would be $5 \times 10^{32} \mathrm{cm}^{-2}\mathrm{sec}^{-1}$ and this assumes 10 bunches per beam and a tune shift of $\Delta\nu \simeq 0.03$. Using 10 cavities per ring it will be possible to reach 6 GeV in each beam. A later improvement in the Beta function at the IP magnet alignment and assuming $\Delta\nu \sim 0.04$ and using 20 bunches can bring the luminosity to 10^{33} $\mathrm{cm}^{-2}\mathrm{sec}^{-1}$. From the recent SIN study it appears that this will likely be the maximum luminosity possible.[10]

Table 3. Parameters for the SIN $\bar{\text{B}}$B Factory Project

			Standard Optics	Micro beta
Circumference	L [m]	:	648.0	
Number of bends		:	56	
Bending radius	R [m]	:	41.443	
Energy loss/turn	ΔE [MeV]	:	1.389	
Hor. betafunction	β_x^* [m]	:	1.00	1.0
Vert. betafunction	β_z^* [m]	:	0.03	0.015
Tune	Q_x	:	7.779	
	Q_z	:	9.279	
Chromaticity	ξ_x	:	-12.8	-12.9
	ξ_z	:	-19.3	-30.6
Number of sextupole families		:	2	
Compensated chromaticity	ξ_x	:	+1	
	ξ_z	:	+1	
Mom. comp. factor	α	:	$2.5 \cdot 10^{-2}$	
hor. emittance	ϵ_x [m·rad]	:	$5.5 \cdot 10^{-7}$	
min. vert. emittance	ϵ_z [m·rad]	:	$9.9 \cdot 10^{-10}$	
vert. emittance with 3% coupling	ϵ_z [m·rad]	:	$1.7 \cdot 10^{-8}$	
Energy spread	ΔE/E	:	$6.6 \cdot 10^{-4}$	
Damping times	τ_x [msec]	:	16.5	
	τ_z [msec]	:	15.6	
	τ_s [msec]	:	7.6	
RF-frequency	f_{RF} [MHz]	:	500	
# of cavities		:	10	
max. number of bunches		:	20	
current	I [mA]	:	485	
RF-power	P_{RF} [kW]	:	1050	

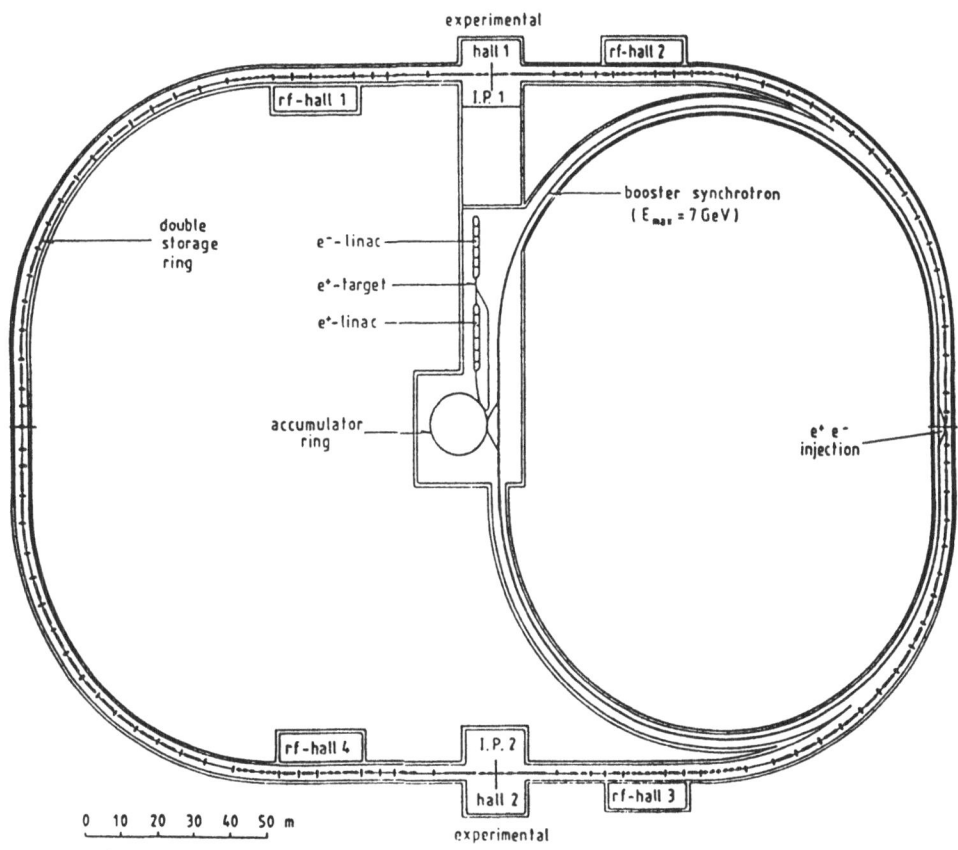

experimental
hall 1

rf-hall 2

rf-hall 1

I.P. 1

booster synchrotron
($E_{max} = 7\,GeV$)

double
storage
ring

e^--linac

e^+-target

e^+-linac

accumulator
ring

$e^+\,e^-$
injection

rf-hall 4

I. P. 2

hall 2

rf-hall 3

experimental

0 10 20 30 40 50 m

Figure 7. View of the SIN Circular Collider $\bar{B}B$ Factory

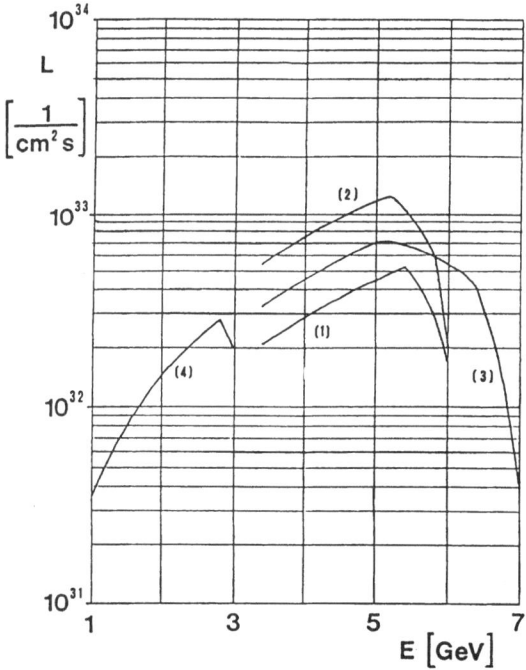

Figure 8. The luminosity of the SIN B meson factory. Curve (1): standard optics, ΔQ=0.03, 10 cavities, 10 bunches, P_{RF}=1050 kW. Curve (2): β_z^*=2cm, ΔQ=0.04, 10 cavities, 10 bunches, P_{RF}=1050 kW. Curve (3): β_z^*=2 cm, ΔQ=0.04, 14 cavities, 6 bunches, P_{RF}=1050 kW. Curve (4): standard opt., ΔQ=0.03, 2 one cell cavities, P_{RF}=129 kW: Above 5.3 GeV the luminosity can be increased through the installation of more RF power or superconductive cavities.

A study of an asymmetric energy two storage ring circular collider has been carried out by a SLAC/LBL group.

Table 4. Parameters for an Asymmetric Collider (from A. Garren)

APIARY 1 Parameters				26 February 1988	
		2 GeV		12 GeV	
Bunch spacing	S_B		25.88		
Number of bunches	B	6.0		85.0	
Particles/bunch	N_B	2.90×10^{11}		1.45×10^{11}	
Current	I	538.0		269.0	mA
Emittances	ϵ_x	0.3		0.01	μm
(design)	ϵ_y	0.03		0.01	μm
3-functions at IP	β_x^*	0.254		0.762	m
	β_o^*	0.0254		0.0762	m
	η^*	0		0	m
Beam-beam tune shifts	$\Delta\nu_x$	0.05		0.05	
	$\Delta\nu_x$	0.05		0.05	
Luminosity	$\mathcal{L}$		5×10^{32}		$cm^{-2}s^{-1}$
Peak β function values	$\hat{\beta}_x$	24.0		360.0	m
	$\hat{\beta}_y$	24.0		388.0	m
	$\hat{\eta}$	2.9		1.9	m
Tunes (unadjusted)	ν_x	4.60		22.95	
	ν_y	4.35		16.06	
Natural emittance	ϵ_{xo}	0.3214		0.301	μm
Circumference	$2\pi R$	155.3		2200.0	μm
Rigidty	B_ρ	6.671		40.03	T-m
Critical photon energy from first dipole	E_c	0.4		15.0	keV
Bend angle of first dipole	θ_1	50.0		8.0	mrad

A preliminary design of this machine has been carried out by A. Garren from LBL.[12] A set of parameters for the 2 GeV or 12 GeV circular collider is given in Table 4. Note the maximum luminosity in this design is $5 \times 10^{32} cm^{-2} sec^{-1}$ and this assumes $\Delta\nu \simeq 0.05$ (a value larger than the SIN design). Bringing the beams into collision is a non trivial matter and a schematic of the Garren solution is shown in Figure 9. [12] One problem with this machine may be the large β that occures in parts of the machine due to the ultra low β^* at the IP. The lattice β function of the small machine is shown in Figure 10. While this machine may not exceed $5 \times 10^{32} cm^{-2} sec^{-1}$ luminosity, it may be valuable in the study of B decays due to the potentially larger reconstruction efficiency.

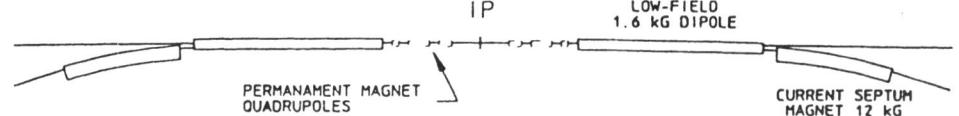

INTERACTION REGION

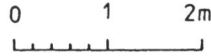

Figure 9. IR for the asymmetric e$^+$ e$^-$ circular collider study (A. Garren)

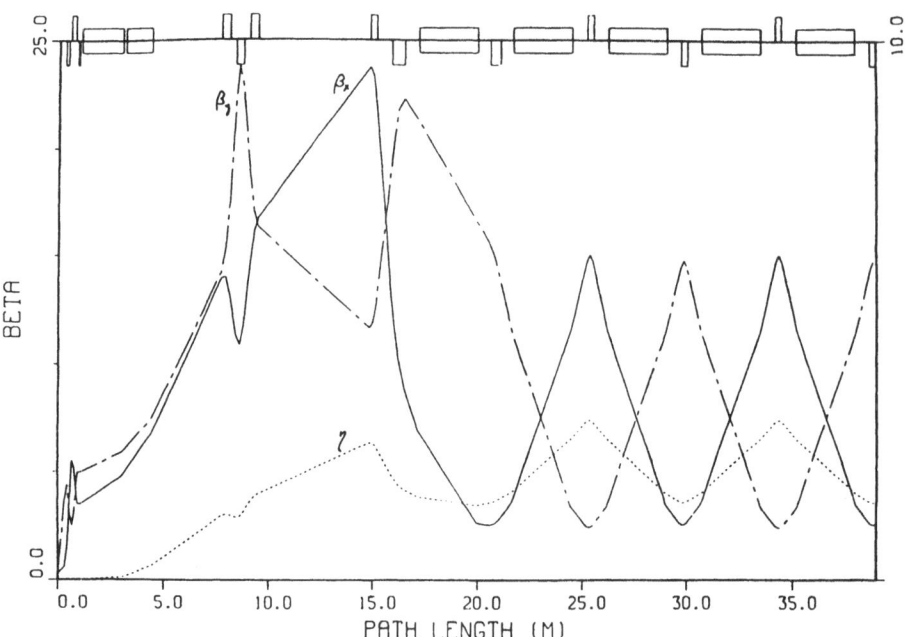

Figure 10. β and η Functions for the small ring in the asymmetric circular collider study (A. Garren)

7. Linear Collider $\bar{B}B$ Factories : R&D Questions

Linear collider $\bar{B}B$ Factories seem to offer the possibility of very high luminosity (perhaps $10^{35}\text{cm}^{-2}\text{sec}^{-1}$ under very favorable circumstances). However, these machines are still in the R&D phase. Many questions need to be answered before we can be certain that high average luminosity is possible.

Two specific machine designs have been carried out

(1) A Superconducting recirculating linac $\bar{B}B$ Factory by a CERN – Frascati group
     2

(2) An Asymmetric submicron – High Gradient $\bar{B}B$ Factory by the UCLA group
     3, 13

These machines emphasize different parameters to achieve high luminosity. In the Frascati design the repetition rate of the machine is assumed to be 12 kilo Hertz putting a strain on the positron source.[2] In the UCLA design the emphasis is put on achieving very small spot size (about 0.1 μm) putting a severe constraint on the electron and positron source emittance and the final focus system.

At the 1987 UCLA $\bar{B}B$ Factroy Workshop, P. Wilson from SLAC worked out a set of parameters for a high luminosity $\bar{B}B$ factory.[14] Table 5 gives a list of the parameters. In this example it would be possible to reach a luminosity of $10^{34}\text{cm}^{-2}\text{sec}^{-1}$. The possibility of reaching this luminosity is perhaps a little greater now due to the recent calculation of Chen and Yokoyo on the disruption luminosity enhancement.

Table 5. Parameters for a 10 GeV B Factory Linear Collider

$$E_o = 10\text{GeV}$$
$$\mathcal{L} = 10^{34}\text{cm}^{-2}\text{s}^{-1}$$
$$L = 100 \text{ m}$$
$$b = 4$$
$$\frac{\sigma_p}{\rho} = 2.0 \times 10^{-3}$$
$$\sigma_z = 0.30 \text{ mm}$$
$$\eta_b = 0.028$$
$$N = 2.2 \times 10^{10}$$
$$\sigma_\perp^* = 0.32 \ \mu\text{m}$$
$$D = 9.0, \text{H}_D = 6$$
$$\Upsilon = 5 \times 10^{-3}$$
$$\delta_{c\ell} = 9 \times 10^{-3}$$
$$f_r = 11.1 \text{ KHz}$$
$$f_b = 44.4 \text{ KHz}$$
$$P_b = 1.6 \text{ MW}$$
$$\epsilon_n = 3.0 \times 10^{-6} \text{ rad}$$
$$\beta^* = 0.7 \text{ mm}$$
$$\langle N_p \rangle = 4.0$$
$$\frac{\sigma_w}{W} = 5 \times 10^{-3}$$

8. The Frascati Recirculating Superconducting B Factory Design

A schematic of the Frascati design is shown in Figure 11. The current design parameters are given in Table 5. Note that the luminosity of $10^{34} \mathrm{cm}^{-2} \mathrm{sec}^{-1}$ can be reached in the "low resolution mode". The machine will use superconducting cavities to accelerate the $e^{\pm}$. It is not clear if this can operate in an asymmetric energy mode. The machine will also serve the nuclear physics groups in Italy. To do this a complicated set of beam gymnastics are to be used, as shown in Figure 12. [15]

The Frascati collider is a very interesting type of $\bar{\mathrm{B}}\mathrm{B}$ factory. The use of Superconducting cavities allows a high repetition rate and this allows for high luminosity. There are several problems with this approach, one of which is the positron source and the target heating.[16]

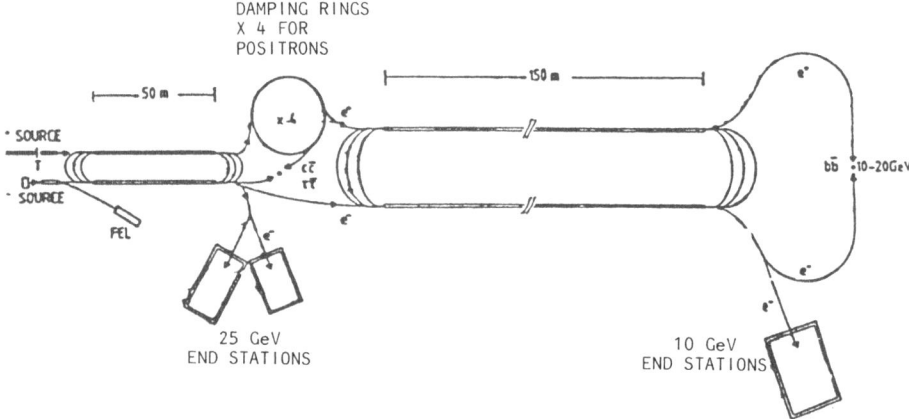

Figure 11. Early design of the Frascati $\bar{\mathrm{B}}\mathrm{B}$ Factory

A strong R&D program has been started at Frascati – CERN to solve many of these problems.

9. The UCLA Asymmetric – Submicron – High Gradient Linear Collider B Factory Design

The early $\bar{\mathrm{B}}\mathrm{B}$ factory design from the UCLA group is shown in Figure 13. a later design differs from this previous design by two changes

(1) The beam energies are asymmetric
(2) The positron problem is alleviated by recirculation of the e^{+} and production using the 10 GeV e^{-} beam and is shown in Figure 14.

The parameters of the symmetric version of the UCLA $\bar{\mathrm{B}}\mathrm{B}$ factory are given in Table 6.[13] Note that the collider assumes very small spot size at the IP. We will now describe several aspects of the machine design.

274

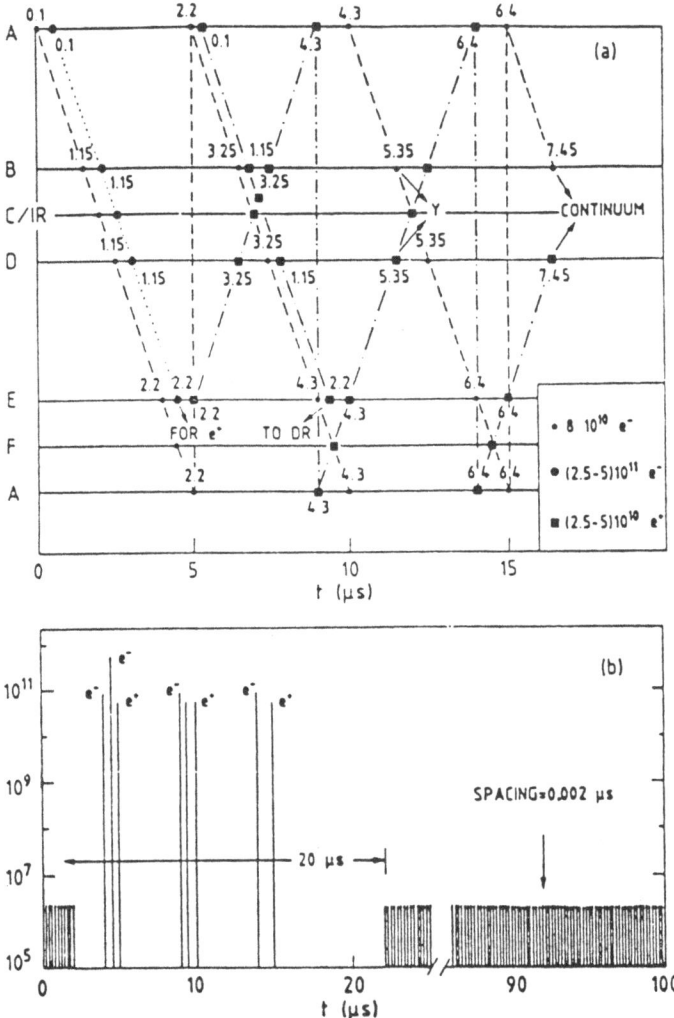

Figure 12. Some of the beam gymnastics for a later design of the Frascati machine.

Table 6. Parameter List for the One-Racetrack ARES Project

Mode:	High-resolution Υ(4S)	Medium-resolution Υ(5S)	Low-resolution 'continuum'
E_o (GeV)	5.29	5.43	7.5
W (GeV)	10.58	10.86	15.
$N^+(10^{10})$	2.5	2.5	5.0
$N^-(10^{10})$	8.0	8.0	8.0
ϵ_n (10^{-6}m)	2.0	2.0	2.0
ϵ_L^+ (10^{-2}m)	4.0	4.0	6.0
ϵ_L^- (10^{-2}m)	1.0	1.0	1.0
σ_z^+ (mm)	3.0	1.5	0.7
σ_z^- (mm)	1.0	0.5	0.5
β^* (mm)	5.0	2.0	2.0
$\sigma_x = \sigma_y$ (μm)	1.0	0.6	0.5
D^+	21	27	25
D^-	22	28	28
H_D a)	8.5	8.5	8.5
f_r (kHz)	10.	10.	10.
ΔW_b (MeV)	4	15	75
ΔW_L (MeV)	9	17	45
P^+ (MW)	0.2	0.2	0.6
P^- (MW)	0.7	0.7	1.
P_T^- (MW)b)	0.9	0.9	1.8
P_{mains} (MW)c)	18	18	20
L (10^{33}cm^{-2}s^{-1})	. 1.4	3.5	10

a) Computed using the expression given in Ref.
b) Power of a 2.2 GeV electron beam needed to produce the positrons.
c) Total power absorbed by the SC linacs.

276

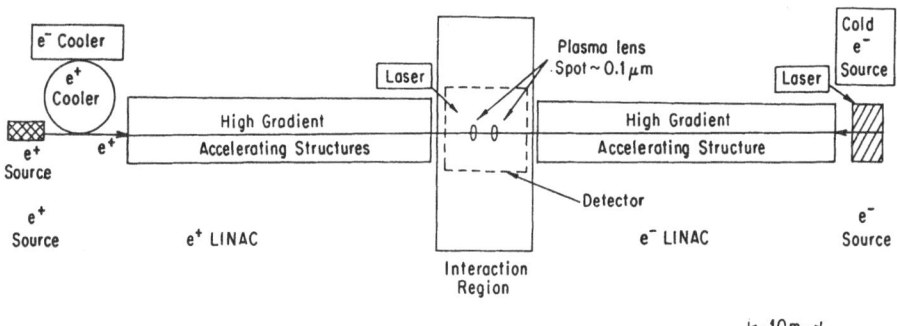

Figure 13. Early schematic design of the UCLA Linear Collider $\bar{B}B$ Factory

$$10 \text{ GeV } e^- \text{ on } (2-4) \text{ GeV } e^+$$

$$E_{cm} = 2\sqrt{E_{e^+}E_{e^-}}$$

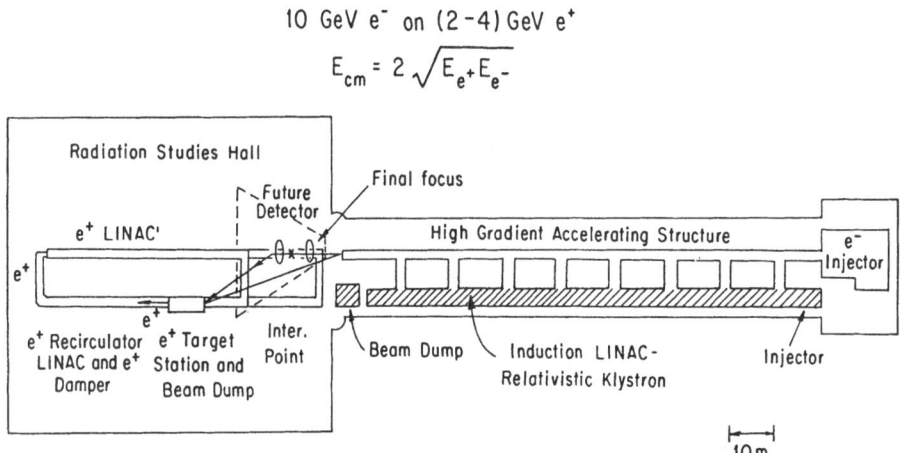

Figure 14. Most recent conceptual design of the UCLA $\bar{B}B$ Factory with Asymmetric Beam and a New Positron Source

There is reason to be optimistic about the high luminosity of a submicron -low energy collider. Recent calculation by Chen and Yokoya show that increasing the beam-beam disruption can produce a large enhancement in the luminosity provided the parameter[17]

$$A = \frac{\sigma_z}{\beta*}$$

is kept small. Results from the calculations are shown in Figure 15a and 15b.[17] Note that $H_D \sim 25$ may be possible a factor of 4 above the earlier estimated limit. In order to keep A small and D large, it will likely require a very short focal length final focus. Such a lens may be provided by a plasma lens as shown in Figure 16.[18] An ANL/SLAC/UCLA group is developing such a lens for the SLC and this could be used for the UCLA $\bar{B}B$ factory.

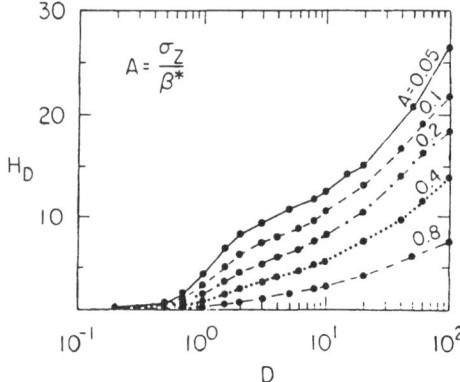

Figure 15a. Luminosity enhancement factor as a function of D, computed with five different values of A. The values are so chosen that they are equally separated on the logrithmic scale.

It is likely that the achievement of utlra small spot sizes will require the shortest possible accelerator to avoid emittance blowup. High gradient Cu acceleration is possible as shown in Figure 17. A possible RF driver that could achieve the required power at 11.4 GHz is the Relativistic Klystron shown in Figure 18.[19] There have been recent breakthroughs in this technique by the SLAC/LLNL/LBL groups where a gradient of 130 MeV/ has recently been obtained at LLNL.[19]

Finally, the solution to the positron problem for the UCLA $\bar{B}B$ Factory may consist of partially reusing the positrons that have been produced by the 10 GeV e^- beam used after the collision.[13, 20] A schematic of this e^+ collider-source is shown in Figure 19. The details of this linac-damper are being worked out[13, 20].

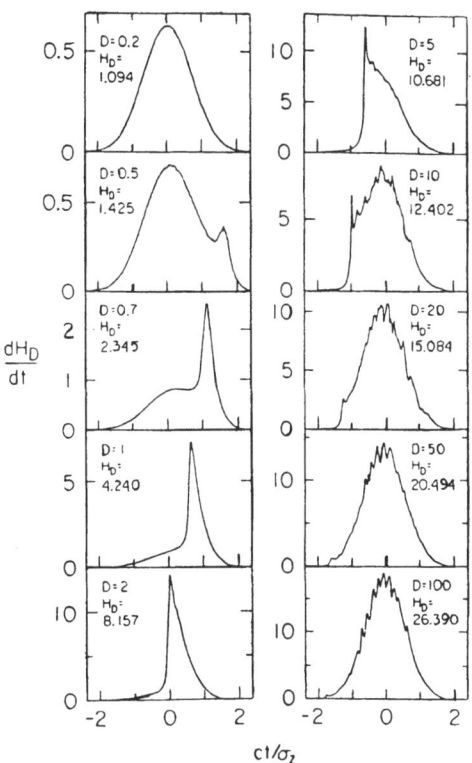

Figure 15b. Computer analysis on the time evolution of the luminosity enhancement factor H_D, at various different values of D. For very small and very large D's, dH_D/dt varies as a Gaussian function (although for large D regime there are small wiggles superimposed), while for medium values of D there is an obvious spike. (Ref 17)

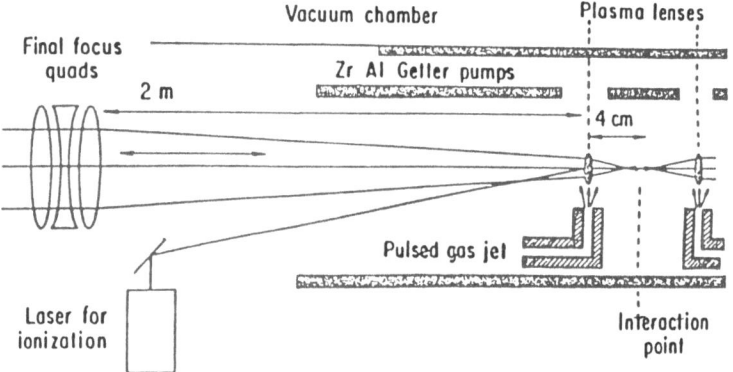

Figure 16. Design of a Plasma Lens by the ANL/SLAC/UCLA group (Ref 18)

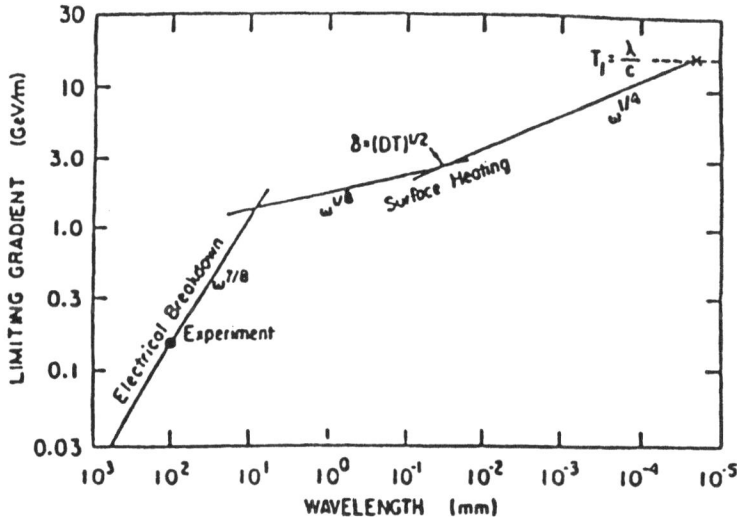

Figure 17. Limiting gradient for a Cu structure as a function of the RF wavelength of the driver source

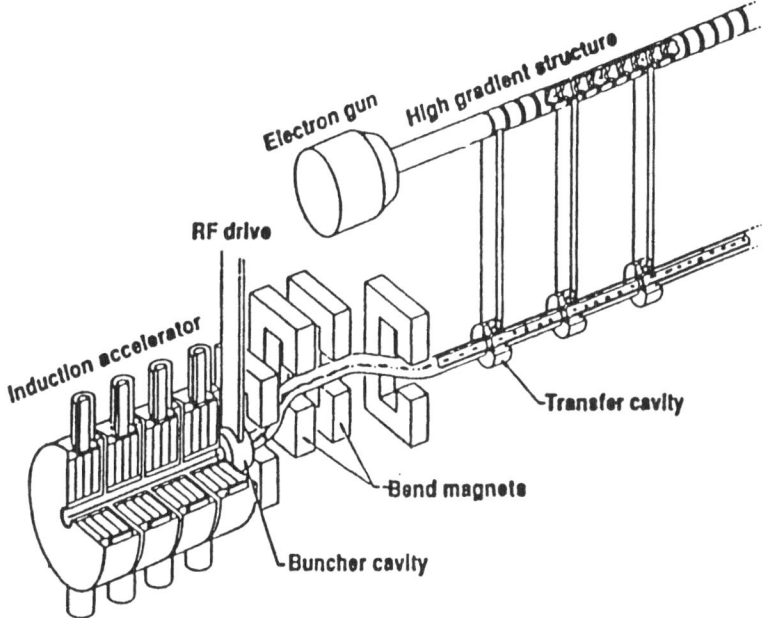

Figure 18. Schematic of the Relativistic Klystron being tested by the SLAC/LLNL /LBL groups

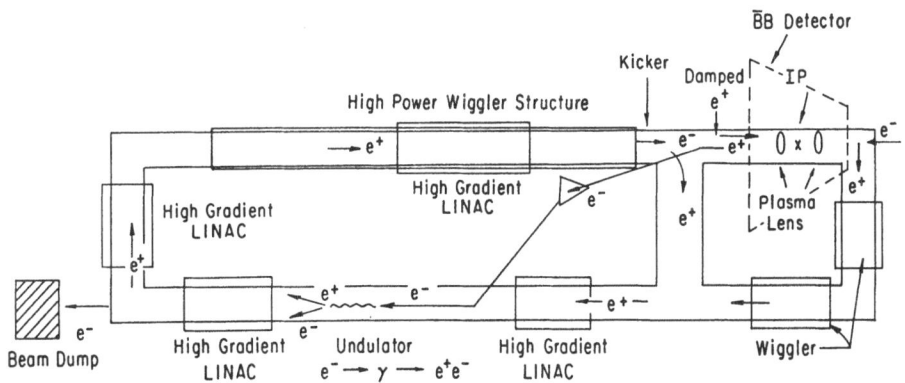

Figure 19. An e^+ Recirculating-Damping Linac for the UCLA $\bar{B}B$ Factory that partially solves the positron source problem.

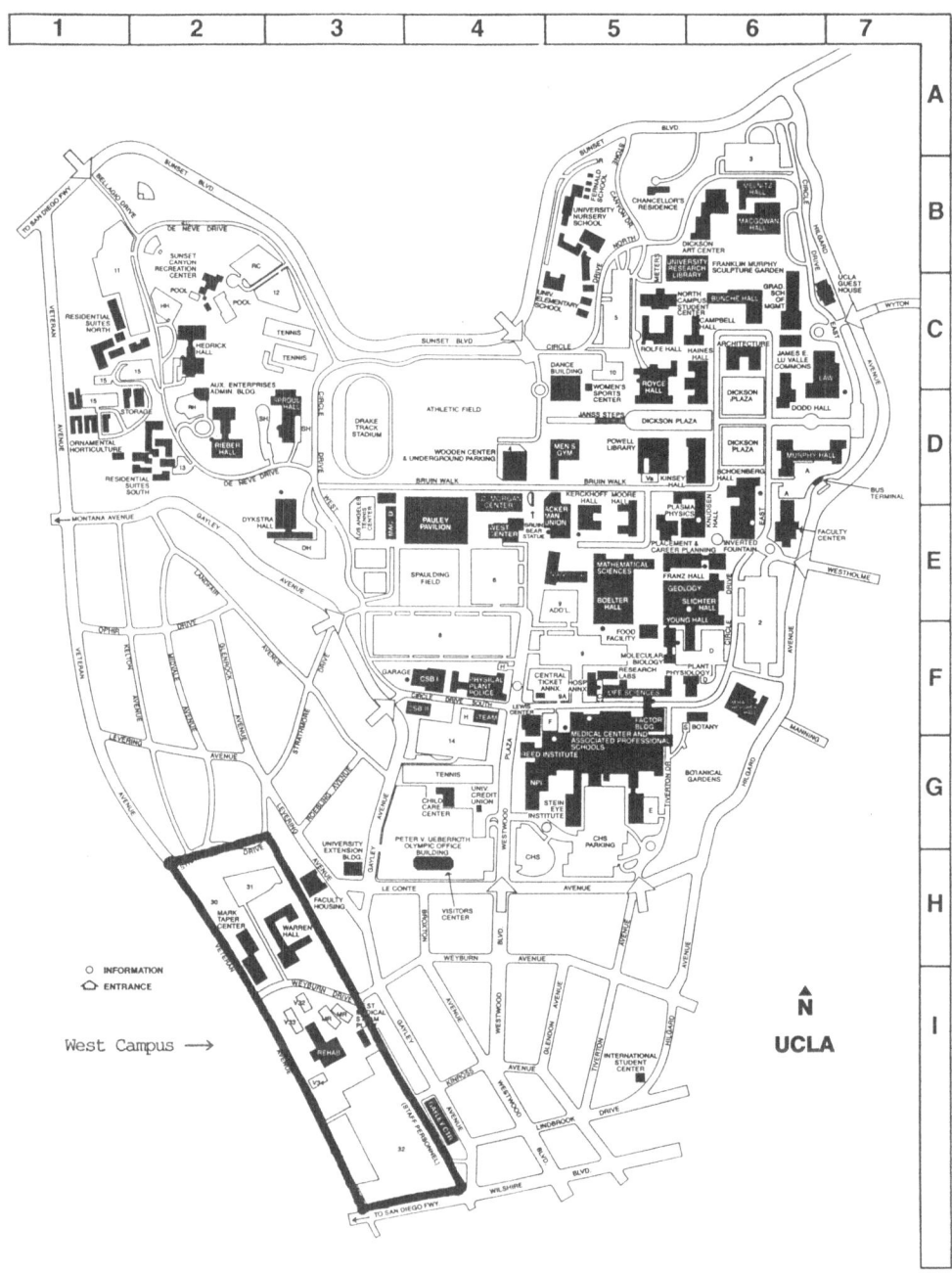

Figure 20. Possible site for the UCLA B̄B Factory in Westwood, Los Angeles

Finally, the UCLA machine is imagined to be constructed on the University campus in Westwood, LA (see Figure 20). This mode of operation is likely appropriate to such an R&D machine.[13]

Table 7. Parameters of a Submicron Spot $\bar{B}B$ Factory Collider

$$E_o = (5 \rightarrow 7)\text{GeV per beam}$$
$$\mathcal{L} = 10^{34}\text{cm}^{-2}\text{sec}^{-1}$$
$$L = 50 \rightarrow 75 \text{ m}$$
$$b = 4$$
$$\frac{\sigma_p}{\rho} = 3 \times 10^{-3}$$
$$\sigma_z = (0.4 \rightarrow 0.02) \text{ mm}$$
$$\eta_b = 0.03$$
$$N = 3 \times 10^{10}$$
$$\sigma_\perp = (0.14)\mu\text{m}$$
$$D = 36, \ H_D = 10$$
$$\Upsilon = 5 \times 10^{-3}$$
$$\delta = 8 \times 10^{-3}$$
$$f_r = 1 \text{ kHz}$$
$$f_b = 4 \text{ kHz}$$
$$P_b = 0.2 \text{ MW}$$
$$\epsilon_n = 3 \times 10^{-6}\text{m} - \text{RAD}$$
$$\beta^* = 0.07 \text{ mm}$$
$$\frac{\sigma_w}{w} = 4 \times 10^{-3}$$

10. Conclusions

It is clear that the future study of rare B decays and CP violation require a very large luminosity $\bar{B}B$ factory. The e^+ e^- production of $\bar{B}B$ near threshold offers the cleanest environment to study B physics. In order to increase the reconstruction efficiency it will be necessary to operate the collider in an asymmetric mode. This is possible for both circular and linear colliders. Very likely the circular collider will have the same luninosity limitations as symmetric colliders.

Linear colliders appear to offer the greatest luminosity. They can be operated in an asymmetric mode as shown by the UCLA design. Furthermore, this is rather natural if the high energy beam is to be used to produce positrons as well.

There are several other important ideas in this field. The Novosibirsk group is designing a $\bar{B}B$ factory that is similar to the SLAC/SLC in that it would use a single linac to accelerate both e^+ and e^- and arcs to bring the particles into collision.[21]

A recent idea to use the Inverse Free Electron Laser as the accelerator has been put forward by C. Pellegrini. The IFEL is shown in Figure 21.[20] The potential advantage of this approach is that the IFEL can help keep the $e^\pm$ emittance small during the acceleration.

These and other ideas are likely to keep this field very lively and the R&D program very interesting.

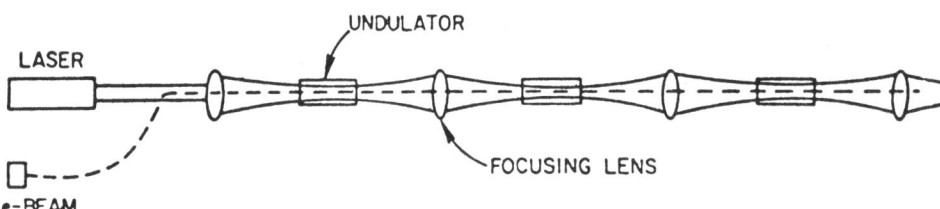

Figure 21. An IFEL Driven for a $\bar{B}B$ Factory (C. Pelligrini)

We can compare the possible rare B decay physics that could result from the realization of a $\mathcal{L} \sim 10^{34} \mathrm{cm}^{-2} \mathrm{sec}^{-1}$ $\bar{B}B$ factory in Italy or at UCLA in Figure 22. The natural comparison is with π and K factories. Note , however, that the theoretical rates for many B decay processes are enhanced by a ratio of

$$(\frac{m_t}{m_c})^2 \sim \mathcal{O}(2000)$$

or more, which partially compensates for the expected ratio of K to B particles that can be produced per year of $\sim \frac{10^{14}}{10^9} \sim 10^5$. Furthermore, the detector and background limits could be different for the rare B and K decays partially overcoming this large rate advantage. Clearly the future of $\bar{B}B$ factories is assured.

Acknowledgments

I wish to thank the following people for discussions on $\bar{B}B$ factories: U. Amaldi, E. Bloom, W. Barletta, G. Coignet, A. Garren, P. Chen, A. Fridman, R. Sheffield, D. Stork, C. Pellegrini, A. Soni, P. Wilson, K. Wille. I wish to thank the Aspen Center for Physics for the hospitality where this report was written.

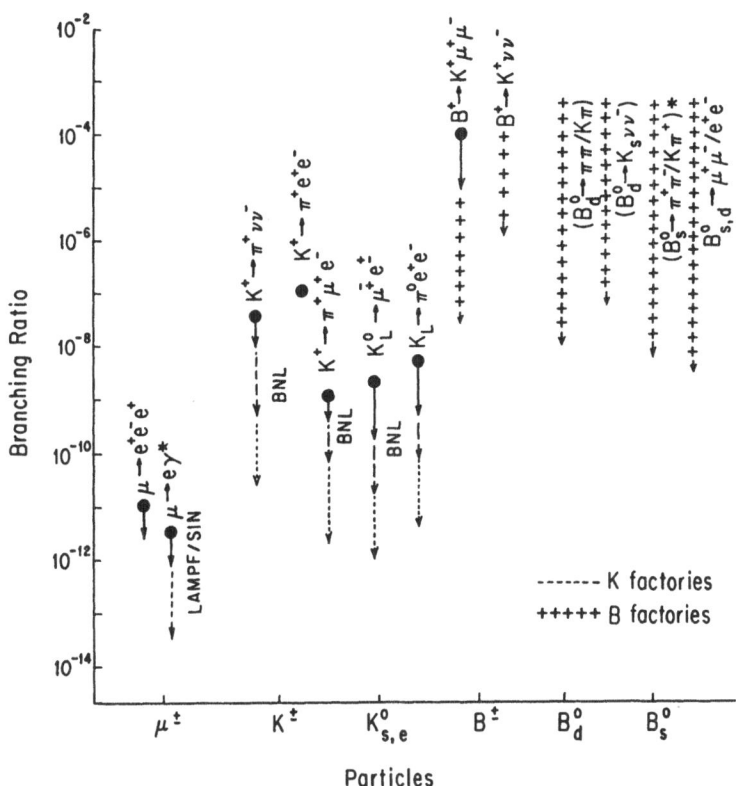

Figure 22. Comparison of the Range of Sensitivity of π, K and $\bar{B}B$ Factories and some of the Interesting Decay Modes.

References

1. K. Wille, proceedings of the 1986 Heidelberg Heavy Flavor Conference. DESY Report (1986)

2. U. Amaldi and G. Coignet, Nucl. Instr Meth. A 260 , 7 (1987)

3. D. Cline, Proceedings of the 1986 Snowmass Workshop and the Madison Workshop on Advanced Accelerator Physics Concepts (1987) and Proceedings of the UCLA Linear Collider $\bar{B}B$ Factory Conceptual Design Workshop, editor D. Stork World Scientific 386 (1987)

4. D. Cline, Invited talk at the 1986 Heidelberg Conference on Heavy Flavors, 1986 and Proceedings of the Workshop on Collider Physics, Madison 1986

5. See for example: the Proceedings of the SLAC Heavy Flavor Workshop, 1987, edited by E. Bloom and A. Fridman, to be published in the New York Academy of Sciences (1988)

6. For a recent summary see the talk of E. Bloom, B Factories, SLAC Pub - 4604 - 1988

7. See for example: J. Ellis and P. Franzini, Rare B Decays within the Standard Model, CERN - TH - 4952/88 and the references therein. See also the papers by A. Soni (UCLA preprints)

8. See for example: J. Rosner, A. Sanda and M. Schmidt, E. Fermi Institute 88-12 Report NO. DOE/ER/40325-30-Task B and the references therein.

9. See for example: G. Gratta, A. Schwarz, and C. Zaccardelli, Vertex Detectors at e^+ e^- B Meson Factories, (UCSC) SCIPP - 88/04 and references therein. – See also the talk of P. Oddone in the UCLA Linear Colloder Workshop Proceedings, D. Stork editor, World Scientific 423, (1988)

10. See K. Wille Feasibility Study of a B Meson Factory – SIN Report PR-88-01 for references to these limits. See also reference 6

11. Private communication, K. Berkelman, Cornell

12. Private communication, A. Garren, LBL

13. D. Cline, An Asymmetric Linear Collider $\bar{B}B$ Factory Using a Recirculating Damping e^+ Linac UCLA Center for Advanced Accelerators 6-1988 and 1-1988 and the report at the SLAC B Meson Factory Workshop, SLAC Report 324 (1988)

14. P. Wilson, report in the Proceedings of the UCLA $\bar{B}B$ Factory Workshop D. Stork Editor, World Scientific 423 (1988)

15. For a recent report on the Frascati design see the talk of U. Amaldi at the SLAC B Meson Factory Workshop, SLAC, edited by A. Fridman, SLAC Report 324 (1988)

16. Private communication, P. Sievers, CERN

17. P. Chen and Yokoya, SLAC Publication 4339 (1987)

18. See for example: the Proceedings of two workshops at SLAC on the Plasma Lens – UCLA Center for Advanced Accelerators Reports Plasma Lens 1 and 2, 1988

19. Private communication from W. Barletta (LLNL) and R. Miller (SLAC)

20. Private communication, C. Pellegrini, BNL

21. Private communiction, V. Siderov, Novosibirsk

18

EXPERIMENTAL ASPECTS OF HEAVY QUARK PHYSICS AT HERA[1]

F. Barreiro[2], **M.A. García, and J. F. de Trocóniz**

Universidad Autónoma de Madrid[3]

Madrid. Spain

1.Introduction

HERA is the $e - p$ collider now under construction at DESY, Hamburg. It will provide collisions between 30 GeV electrons and 820 GeV protons and is expected to become operational by 1990. Two detectors have been approved for data taking at HERA, H1 and ZEUS. A schematic layout of ZEUS is shown if Fig. 1.1. This figure illustrates the angular coverage of the high resolution calorimeter in the (w.r.t. the incoming p direction) forward (FCAL), central (BCAL) and rear (RCAL) directions. It also shows the position of the vertex detector (VXD), the central tracking device (CTD) as well as the forward and rear tracking chambers (FTD and RTD). The forward. barrel and rear calorimeters are planned to be instrumented with two planes of Si pads for improved electron/hadron separation. Muons are detected in the forward direction in a spectrometer using drift and limited streamer tube chambers plus scintillator counters interspersed between the magnetized iron yoke and toroid. The barrel and rear muon detectors are based on limited streamer tube chambers before, in between and behind the backing calorimeter and behind the concrete shield.

With the purpose of sharpening our thinking on the physics potential offered by HERA, DESY devoted the 1987 annual Theory Workshop to the topic "Physics at HERA". Twelve study groups were formed to cover the various aspects under this title. We are going to present in this report some of the most important results obtained in the study group "Heavy Quark Physics at HERA". Particular attention is paid to the problem of reconstructing the original quark direction of flight as well as to the importance of quantitatively assessing the amount of gluon radiation. Procedures for tagging charm. bottom and top hadron events are discussed as well as the implications for experimentation at HERA. A more comprehensive review can be found in [1]. See also the lectures given at this Meeting by A. Ali [2] and G. Schuler [3].

---

[1]Based on lectures given by F.B. at the VI ELOISATRON Workshop on Heavy Flavours, Erice. Italy

[2]Alexander von Humboldt Fellow, now at DESY.

[3]Supported by CICYT. Spain.

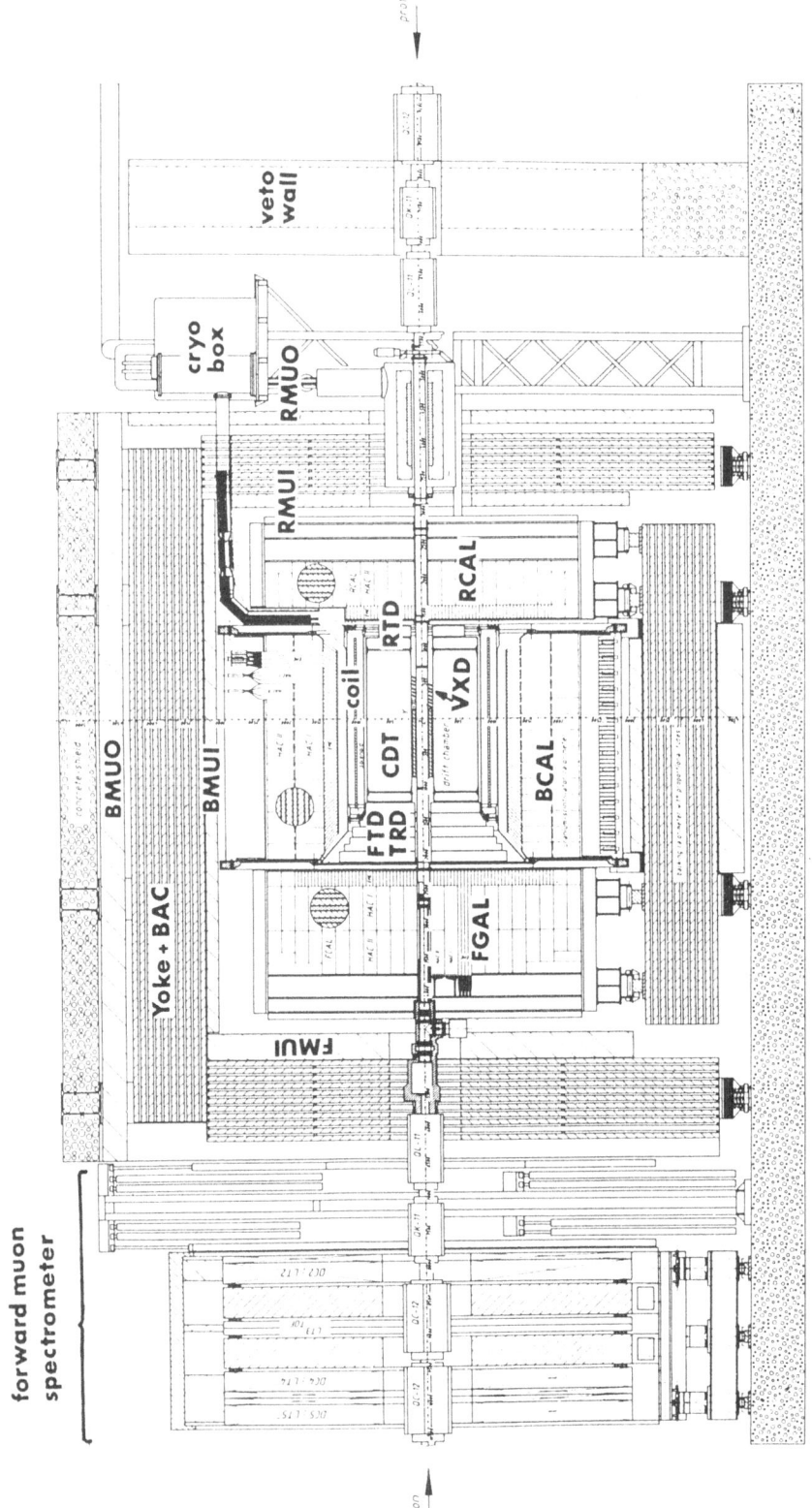

Fig. 1.1. View of the ZEUS detector

288

2.Heavy Quark Production in ep Collisions

In high energy $e - p$ collisions, the main source of heavy flavour production is the boson-gluon fusion (BGF) mechanism. The leading order Feynman diagrams are shown in Fig. 2.1 and correspond to the charged current (CC) and neutral current (NC) processes

$$CC : W^- + g \to Q + \bar{Q}' \qquad (2.1)$$

$$NC : \gamma/Z^0 + g \to Q + \bar{Q} \qquad (2.2)$$

The cross sections for these processes can be calculated convoluting the gluon density in the proton with the QCD parton level cross-sections obtained from the diagrams shown in Fig. 2.1.

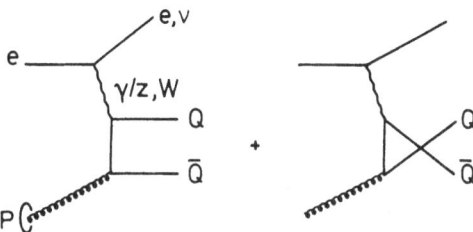

Fig. 2.1.Lowest order QCD diagrams for boson-gluon fusion into a heavy quark-antiquark pair

These calculations to $O(\alpha_S \alpha^2)$ [4-8] are subject to uncertainties due to the choice of

- the gluon density parametrisation

- the scales needed to define the running electromagnetic and strong coupling constants

- the values of the heavy quark masses

We will show results obtained [6] with the parametrisation of Eichten et al. [9]. Uncertainties due to alternative parametrisations of the gluon structure function are very large, $O(100\%)$. for charm production. large. $O(25\%)$. for bottom and small for top production. Uncertainties due to the choice of scales are comparatively smaller, $O(10\%)$, while those coming from the choice of heavy quark masses are also important for charm and bottom cross-sections.

For the values of the parameters in the Standard Model which are quoted in Table 2.1. the cross sections for charm, bottom and top production are given in Tables 2.2. 2.3 and 2.4. respectively. For completeness. the contributions coming from the Quark Parton Model. which are small compared to the ones stemming from the BGF mechanism. are also given.

Table 2.1. Assumed values of the standard model parameters
used in the estimates of cross-sections

Quantities	Values
CKM matrix elements	$V_{ud}^2 = 0.95$ $V_{us}^2 = 0.05$ $V_{ub}^2 = 10^{-4}$ $V_{cd}^2 = 0.05$ $V_{cs}^2 = 0.948$ $V_{cb}^2 = 0.002$ $V_{td}^2 < 4.10^{-4}$ $V_{ts}^2 = 0.002$ $V_{tb}^2 = 0.998$
Quark masses	$m_s = 0.5\ GeV$ $m_c = 1.5\ GeV$ $m_b = 5.0\ GeV$
Weak boson masses and angle	$m_W = m_z cos\theta_W = \frac{38.68 GeV}{sin\theta_W}$ $sin^2\theta_W = 0.226$
Mass scales	$M_s^2 = M_g^2 = \hat{s}$
QCD scale	$\Lambda_{QCD} = 0.2\,\text{GeV}$
No. of flavours in $\alpha_S(Q^2)$	$n_f = 3$

Table 2.2. Charm cross sections at HERA

$\sigma(ep \to cX)$ [pb] at HERA ($m_c = 1.5\,\text{GeV}$)								
	CC			NC $c\bar{c}$				inclusive
e^-p	$\bar{c}d$	$\bar{c}s$	$\bar{c}b$	NC	γ	Z	γ-Z	$c + \bar{c}$
BGF	0.46	8.2	0.012	5.1×10^5	5.1×10^5	0.60	4.7	1.0×10^6
e^-p	$\bar{d} \to \bar{c}$	$\bar{s} \to \bar{c}$						
QPM	0.26	3.3						3.6
e^+p	$d \to c$	$s \to c$						
QPM	0.75	3.3						4.1

Table 2.3. Bottom cross sections at HERA

$\sigma(ep \to bX)$ [pb] at HERA ($m_b = 5\,\mathrm{GeV}$)								
	CC			NC $\quad b\bar{b}$				inclusive
$e^- p$	$\bar{u}b$	$\bar{c}b$	$\bar{t}b$	total	γ	Z	γ-Z	$b+\bar{b}$
BGF	$\leq 0.96 \times 10^{-3}$	0.012	0.13	4.2×10^3	4.2×10^3	0.35	0.59	8.4×10^3
$e^- p$	$u \to b$							
QPM	$\leq 0.56 \times 10^{-2}$							$\leq 0.56 \times 10^{-2}$
$e^+ p$	$\bar{u} \to \bar{b}$							
QPM	$\leq 0.12 \times 10^{-2}$							$\leq 0.12 \times 10^{-2}$

Table 2.4. Top cross sections at HERA for $m_t = 60\,\mathrm{GeV}$

$\sigma(ep \to tX)$ [pb] at HERA ($m_t = 60\,\mathrm{GeV}$)								
	CC			NC $\quad t\bar{t}$				inclusive
$e^- p$	$\bar{t}d$	$\bar{t}s$	$\bar{t}b$	total	γ	Z	γ-Z	$t+\bar{t}$
BGF	$\leq 0.14 \times 10^{-3}$	$\leq 0.62 \times 10^{-3}$	0.13	0.09	0.09	0.4×10^{-3}	0.6×10^{-4}	0.31
$e^- p$	$\bar{d} \to \bar{t}$	$\bar{s} \to \bar{t}$						
QPM	2×10^{-3}	8×10^{-3}						1×10^{-2}
$e^+ p$	$d \to t$	$s \to t$						
QPM	4×10^{-3} QPM	8×10^{-3}						1.2×10^{-2}

The following comments are in order:

- NC cross-sections for charm and bottom quarks dominate over the CC processes.

- Top quark production through CC reactions cannot be neglected, actually this is the dominant contribution for m_t above 55 GeV, see Fig. 2.2.

If one takes into account that at LEP-I/SLC one expects [10]

$$\sigma(e^+e^- \to Z_0 \to cX) = 6.5 \ nb \tag{2.3}$$

$$\sigma(e^+e^- \to Z_0 \to bX) = 9.0 \ nb \tag{2.4}$$

one can conclude that the bottom pair cross-section at HERA is comparable to that at LEP-I/SLC, while the charm cross-section is two orders of magnitude larger at HERA. After five years of running at the expected yearly luminosity of $200pb^{-1}$, we expect to have accumulated $O(10^9)$ charmed hadron events and $O(10^7)$ bottom hadron events.

Summarizing this section, HERA has the potential to be a powerful charm and bottom factory, and will have a chance to see top hadron events if the top mass lies below 80 GeV.

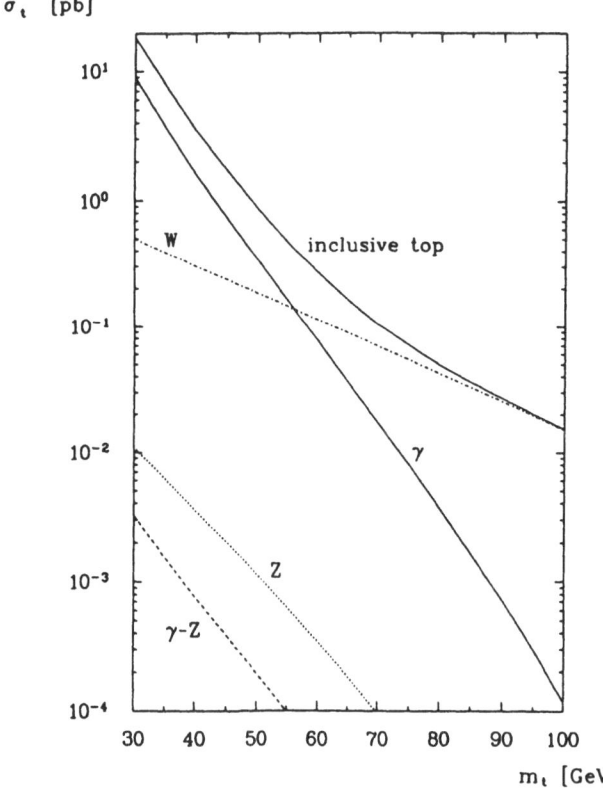

Fig. 2.2. Inclusive cross-section for quark top production via boson gluon fusion as a function of the top quark mass. Curves are for charged current interactions (W) and in the neutral current case separately for pur γ, pure Z and their interference $\gamma - Z$. From [6].

292

3. General characteristics of heavy flavour production at HERA

In the preceeding section we have seen that the cross sections for heavy quark production at HERA are quite large. The next question we have to address is obviously, how large are the detection efficiencies. Let us first discuss the cross-sections for heavy quark production in the x, Q^2 plane, with x and Q^2 defined as usual in deep inelastic scattering. i.e. Q^2 as the squared momentum transfer at the lepton vertex and x as the fractional momentum carried by the gluon in the proton. For the sake of illustration we show in Tables 3.1 and 3.2 the normalized distributions $\frac{1}{\sigma} \frac{d^2\sigma}{dxdQ^2}$ for $t\bar{b}$ and $b\bar{b}$ production respectively, for an assumed value of the top quark mass of 60 GeV.

Notice that NC processes are dominated by very low x, very low Q^2 values. On the other hand. CC processes have a much broader distribution both in x and Q^2. These are consequences of the fact that NC processes are γ exhange dominated, while CC processes are due to the exchange of heavy W bosons. Thus, the experimental signature for BGF events through NC processes. the dominant source of "light" heavy quark production, is simple to remember:

- the scattered electron goes undetected down the beam pipe, as a consequence of the low-Q^2 dominance discussed above;

- the proton fragments carry also very little transverse momentum and essentially go undetected in the direction opposite to the scattered electron;

- a $Q - \bar{Q}$ pair is produced back-to-back in the transverse or $r - \phi$ plane, as illustrated in fig. 3.1.

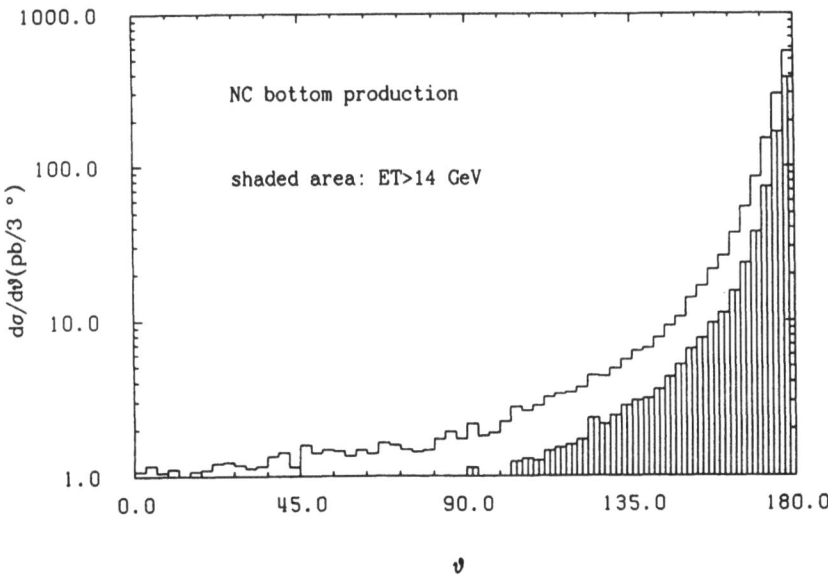

Fig. 3.1. Angular distribution between the b and $\bar{b}$ in the transverse or $r - \phi$ plane in the reaction $ep \to b\bar{b}X$. The shaded area indicates the results when a cut $\sum E_T \geq 20 \, GeV$ is applied.

Table 3.1. Percentages of $t\bar{b}$ pairs from boson gluon fusion relative to total $t\bar{b}$ cross section

	$Q^2 \leq 10$	$10 \leq Q^2 \leq 10^2$	$10^2 \leq Q^2 \leq 10^3$	$10^3 \leq Q^2 \leq 10^4$	$10^4 \leq Q^2 \leq 4.10^4$
$x \leq 10^{-3}$	0.40	1.41	–	–	–
$10^{-3} \leq x \leq 10^{-2}$	0.01	2.59	13.93	–	–
$10^{-2} \leq x \leq 5.10^{-2}$	0.0	0.05	15.58	32.08	–
$5.10^{-2} \leq x \leq 10^{-1}$	0.0	0.0	0.51	21.38	–
$10^{-1} \leq x \leq 2.10^{-1}$	0.0	0.0	0.04	8.30	1.20
$2.10^{-1} \leq x \leq 5.10^{-1}$	0.0	0.0	0.0	0.75	0.99
$5.10^{-1} \leq x \leq 1$	0.0	0.0	0.0	0.0	0.0

Table 3.2. Percentages of $b\bar{b}$ pairs from boson gluon fusion relative to total $b\bar{b}$ cross section

	$Q^2 \leq 1$	$1 \leq Q^2 \leq 10$	$10 \leq Q^2 \leq 10^2$	$10^2 \leq Q^2 \leq 10^3$	$10^3 \leq Q^2 \leq 10^4$
$x \leq 10^{-3}$	81.154	8.978	3.065	–	–
$10^{-3} \leq x \leq 10^{-2}$	0.019	0.999	3.399	1.037	–
$10^{-2} \leq x \leq 5.10^{-2}$	0.0	0.014	0.517	0.638	0.031
$5.10^{-2} \leq x \leq 10^{-1}$	0.0	0.0	0.018	0.086	0.014
$10^{-1} \leq x \leq 2.10^{-1}$	0.0	0.0	0.003	0.023	0.006
$2.10^{-1} \leq x \leq 5.10^{-1}$	0.0	0.0	0.001	0.0	0.001
$5.10^{-1} \leq x \leq 1$	0.0	0.0	0.0	0.0	0.0

Up to now, we have restricted our discussion of heavy quark production at the parton level. It is clear that for a realistic estimate of the efficiency to tag charm, bottom or top hadrons, the hadronization of quarks and gluons as well as the weak decays of heavy quarks have to be taken into account. To incorporate these features we have used the Monte Carlo simulation model AROMA [11], which is based on the following ingredients:

- the complete $O(\alpha^2 \alpha_s)$ matrix elements for the BGF processes discussed earlier,

- hadronization of the quarks and gluons using the LUND string model [12],

- weak decays of the charm, bottom and top quarks according to the standard V-A matrix elements.

Furthermore, AROMA has built in the possibility of gluon emission from the $Q - \bar{Q}'$ system in a parton shower approach[13]. Gluon emission has clearly to be considered in order to realistically estimate charm and bottom backgrounds for top search. However, it is not yet clear that the parton cascade formalism for time-like processes, which is presently built in AROMA, is adequate for space-like processes considered at HERA.

Since we have seen that $Q - \bar{Q}$ pairs are preferentially produced back-to-back in the transverse plane, and since the transverse momentum of the parton produced in the hard scattering subprocess is essentially given by its mass, we find it convenient to look at the distribution of the total transverse energy $\sum E_T$ deposited by charm, bottom and top events in a quasi-ideal calorimeter, in the sense that only a beam pipe cut of 100 $mrad$ has been considered, but no energy resolution smearing. The total transverse energy has the additional advantage that experimentally it is very easy to implement it in the trigger logic for hadronic events. The distributions obtained are shown in Fig. 3.2. Notice that they peak at values which are roughly the sum of the quark masses, and fall off steeply for higher values.

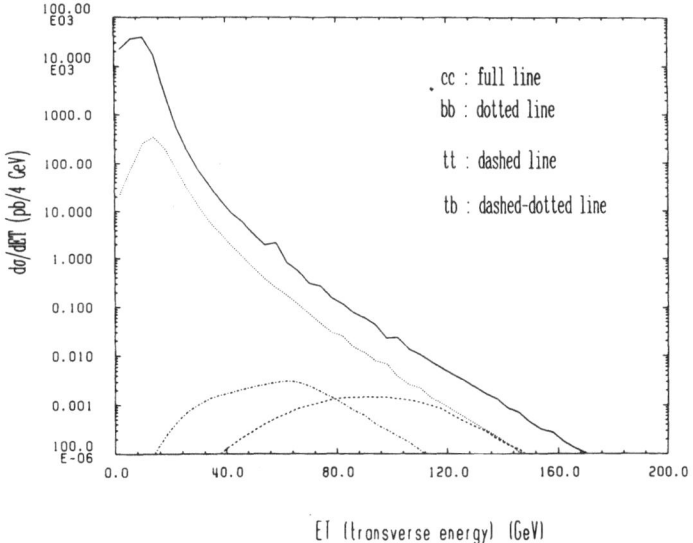

Fig. 3.2. Distribution of the total transverse energy $\sum E_T$ of all stable particles outside a beam pipe cut of 100 $mrad$, with curves corresponding to $t\bar{t}$, $t\bar{b}$ and the various backgrounds

One is therefore tempted to conclude that as far as quark pair production (in the transverse plane) is concerned, HERA is a γ g collider and similar to an e^+e^- storage ring with variable centre of mass energy, the luminosity peaking at centre of masses suitable for production of $Q - \bar{Q}$ pairs at rest.

In order to reconstruct the quark direction of flight, one could therefore apply the jet algorithms with which we are familiar in e^-e^- physics [14]. One such possibility is to divide the final state hadrons into two jets, in such a way as to maximize the sum of the longitudinal momenta along these two directions in space. This is the so called twoplicity method, a suitable generalization of thrust to not exactly back-to-back configurations. That one can reliably reconstruct the quark direction of flight is illustrated in fig. 3.3a, where we plot the angle between the generated quark direction and the reconstructed one, for the case of NC bottom pair production. The dotted (solid) line indicates the possibility that the $Q - \bar{Q}$ system does (not) radiate gluons according to the parton shower discussed in the preceeding section. The average angle between the generated and the reconstructed quark direction turns out to be 27° (resp. 26°).

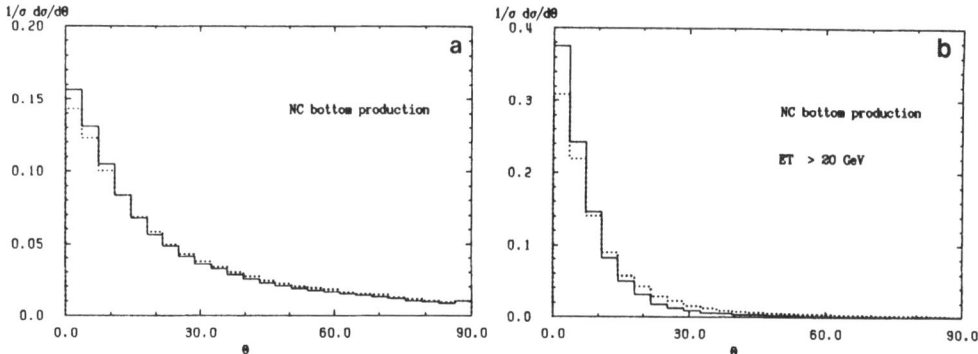

Fig. 3.3. Distribution of the angle between the Monte Carlo generated and reconstructed quark direction of flight, for the process $ep \rightarrow e b\bar{b}X$. The dotted line indicates that gluon radiation has been taken according to a parton shower formalism. The full line stands for no gluon radiation out of the $b\bar{b}$ pair. Two cases have been considered a) no cut on $\sum E_T$ and b) a cut on $\sum E_T \geq 20\,GeV$.

Furthermore one can make use of the standard procedures in e^+e^- jet physics, to select jet-like configurations. It is clear that by increasing the E_T-threshold for triggering, we would select final states with a more pronounced jet structure, and this will ease the problem of reconstructing the original quark direction. As illustrated in fig. 3.3b, the average angle between the generated b direction of flight and the reconstructed axis goes down to 11° (resp. 8°) after imposing a cut on $\sum E_T \geq 20\,GeV$.

We would like to finish this Section, by presenting, see Table 3.3, some average properties characterising the heavy quark induced hadronic final states. Along with the cross sections for the various channels, mean values for $\sum E_T$, charged multiplicity and circularity are given.

Table 3.3 . Characteristics of heavy quark production at HERA

general cut: beam hole angle $\theta = 0.1$ rad				
for DIS events: $Q \geq 2\,\mathrm{GeV}$, $x \geq 0.001$, $W^2 \geq 5\,\mathrm{GeV}^2$				
for light quark events: $p_\perp^q \geq 1\,\mathrm{GeV}$				
process	$\sigma[\mathrm{pb}]$	$\langle \sum E_\perp \rangle [\mathrm{GeV}]$	$\langle C \rangle$	$\langle n_{ch} \rangle$
NC				
$b\bar{b}$	4.2×10^3	14.	0.44	16
$c\bar{c}$	5.1×10^5	7.2	0.51	12
DIS	1.3×10^5	8.6	0.12	5
qG	7.3×10^5	3.4	0.42	5
$q\bar{q}$	7.8×10^5	6.2	0.50	10
CC				
$\bar{c}s$	8.3	25.	0.18	13
light quarks	71.	42.	0.14	15
$\bar{t}b \quad m_t =$				
40 GeV	0.36	48.	0.30	21
60 GeV	0.13	58.	0.28	21
80 GeV	0.05	70.	0.26	22
$t\bar{t} \quad m_t =$				
40 GeV	1.9	66.	0.36	26
60 GeV	0.09	93.	0.37	28
80 GeV	4.3×10^{-3}	117.	0.38	29

4. Comments on top quark searches at HERA

If one tries to further exploit the analogy discussed in the previous section between $Q - \bar{Q}$ production in e^+e^- annihilation and in $e - p$ collisions at HERA, one would think that imposing stringent cuts on $\sum E_T$ would result in a suppression of the charm and bottom background in a top search. As one can see from fig. 3.1 this will not suffice to isolate a clean top sample. However the jet configuration of the surviving charm and bottom pair final states will be very different from those exhibited by top pair events. The difference will be similar to that between quark pair production in e^+e^- annihilation at centre of mass energies near and much above threshold. Thus, one could think that by judiciously choosing cuts on both $\sum E_T$ and a jet measure like sphericity, circularity or twoplicity, one could select samples with enriched top hadron contents. This illustrated in Fig. 4.1 where we show the circularity distribution for $c\bar{c}$, $b\bar{b}$ and $t\bar{t}$ after imposing a cut on $\sum E_T \geq 100\ GeV$ and demanding at least one lepton in the corresponding final states. Two cases have been considered, namely (a) with and (b) without gluon radiation. The former has been taken into account in the parton shower formalism discussed in previous sections [13]. Notice that the signal over background ratio is significantly larger in case (b) as expected. Just as in top searches at e^-e^- machines one would expect a shoulder in the tail of the sphericity distribution. This illustrated in Fig. 4.2 where the mass of the top quark has been assumed to be 60 GeV.

An alternative possibility to search for top hadron events will be to

- divide the event into two jets with the twoplicity method,

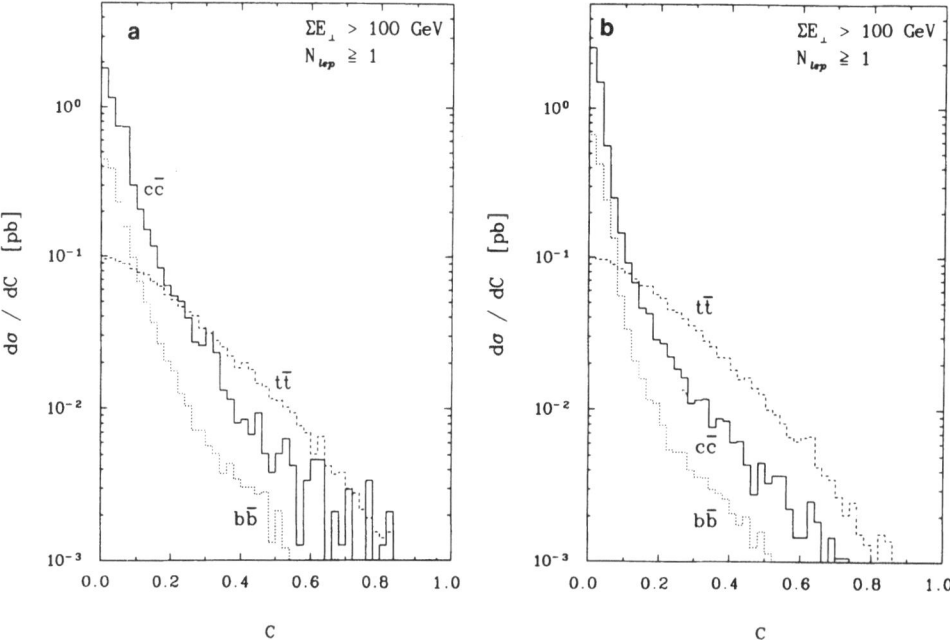

Fig. 4.1. Circularity distribution for $ep \rightarrow eQ\bar{Q}X$, with $Q = c, b, t$, a) gluon radiation in a parton shower is considered and b) no gluon radiation at all.

- plot the light versus the heavy jet mass.

The corresponding distributions for $c\bar{c}$, $b\bar{b}$ and $t\bar{t}$ production with and without parton cascading are shown in Fig. 4.3. The top quark mass has been assumed to be 60 GeV. Again. a clear separation is possible if gluon radiation were overestimated in the parton shower algorithm presently built in AROMA.

We would like to finish this section with two comments:

- A thorough discussion of top quark search strategies both in NC and CC processes can be found in [15]

- It is imperative to include in current Monte Carlo fragmentation models for heavy quark production at HERA a parton shower formalism appropriate for space like processes as well as the exact $O(\alpha_S{}^2\alpha^2)$ corrections in order to assess the importance of gluon radiation in the heavy quark pair final states at HERA.

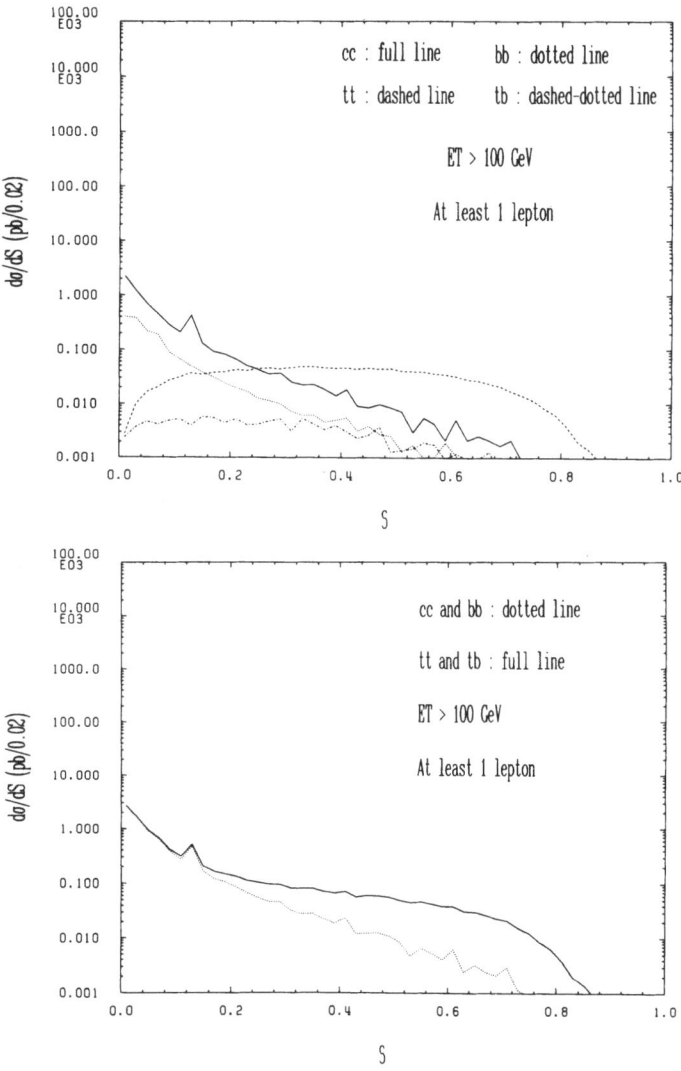

Fig. 4.2. Similar to Fig. 4.1 but for sphericity distributions

299

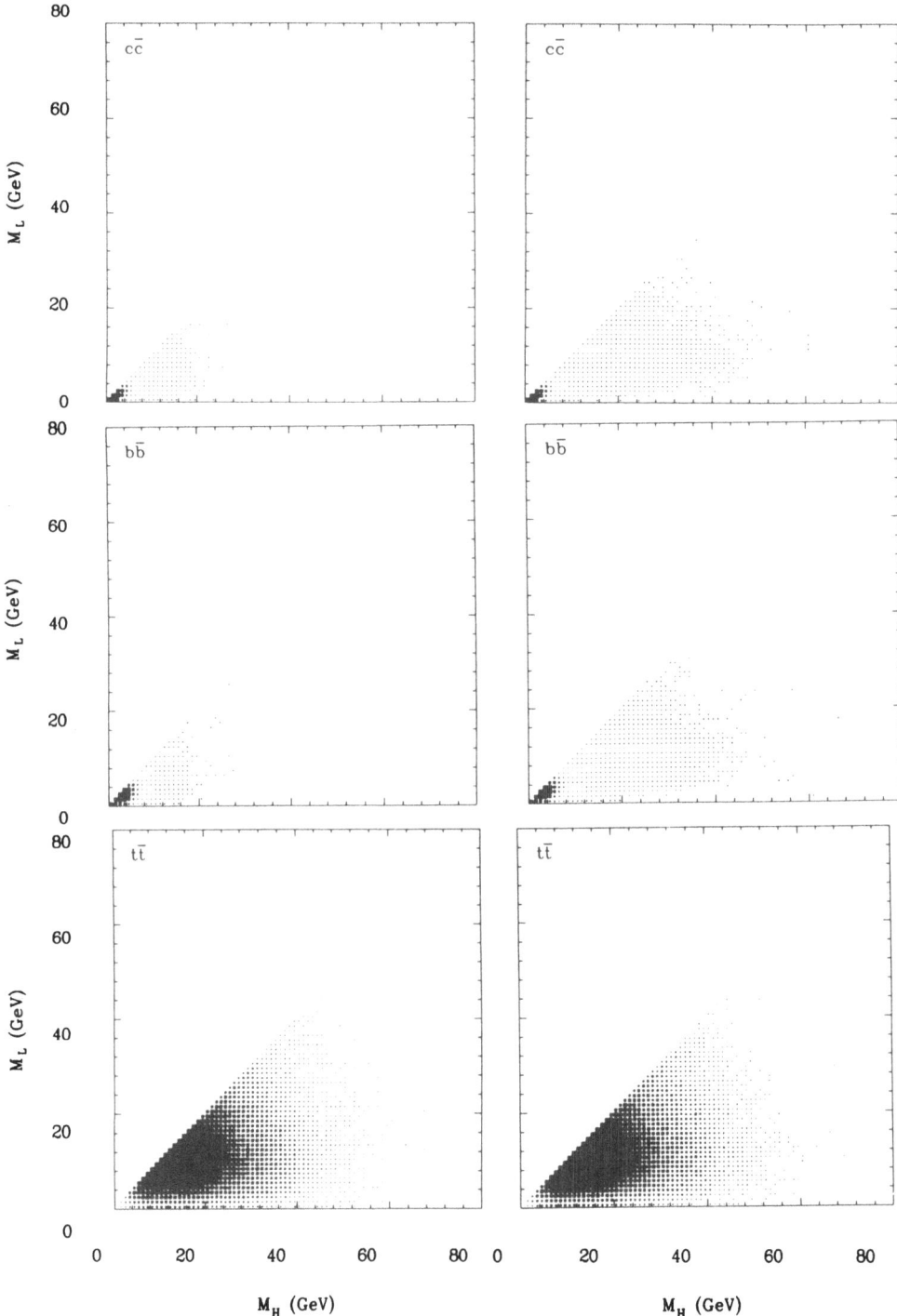

Fig. 4.3. Light vs heavy jet mass in the process $ep \rightarrow Q\bar{Q}X$ with $Q = c, b, t$. Left column: no gluon radiation considered, right column: gluon radiation in a parton cascade considered

5. Charm and Bottom Physics at HERA

We would like to start this section by discussing the energy-angle profiles for charm and bottom jets. As already shown in Sections 2 and 3, charm and bottom production at HERA is dominated by the NC process $\gamma g \to Q\bar{Q}$ with $Q = c, b$ at very low momentum transfers. The scattered electron in $ep \to eQ\bar{Q}X$ is in most of the cases lost in the beam pipe and the Q and $\bar{Q}$ are produced almost back-to-back in the $r - \varphi$ plane. The scatter plots $\frac{d^2\sigma}{dE_Q d\theta_Q}$ (θ_Q measured w.r.t. the incoming electron direction) for the heavy mesons produced in the processes $ep \to ecX \to eDX$ and $ep \to ebX \to eBX$ are shown in Fig. 5.1.

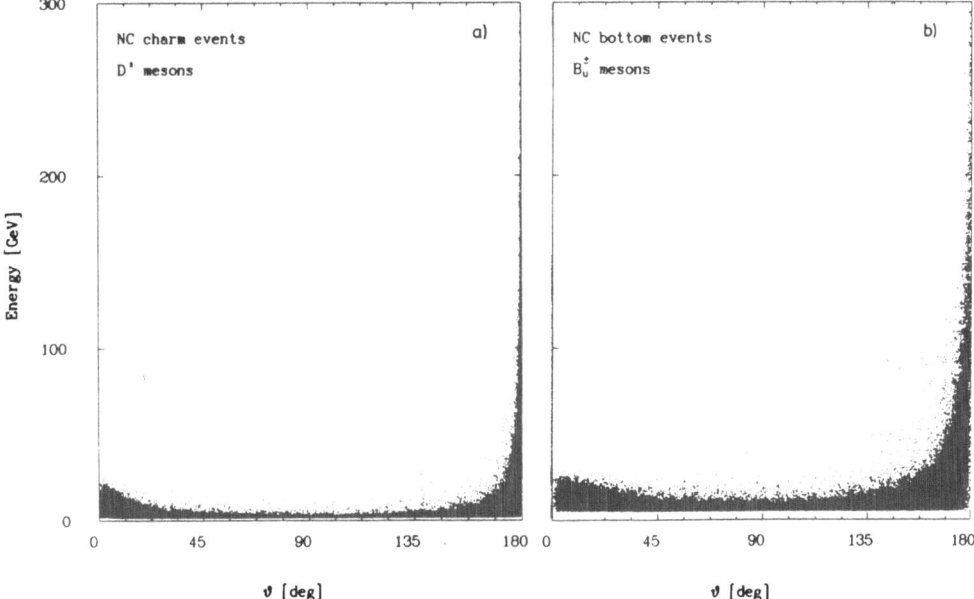

Fig. 5.1. Scatter plot showing the correlation between the energy and polar angle of the heavy mesons produced in the NC process $ep \to eQ\bar{Q}X$, a) $Q = c$, b) $Q = b$. From [16].

This is one of the most interesting results obtained in our study group since it shows that the Lorentz boost is not large enough so as to overcome the effects of the u-channel exchange depicted in Fig. 2.1.

Thus, although the most energetic heavy hadrons fly along the incoming proton beam direction, the fraction emerging along the incident electron is not negligible, see Fig. 5.2. They are even more energetic than those centrally produced and they offer an additional advantage namely their decay products are not contaminated by the proton remnants.

It is interesting to see how a beam pipe cut of approximately 100 *mrad* affects the energy spectrum of charmed and beauty hadrons, as well as their decay leptons and kaons. This is illustrated in Fig. 5.3. The following comments are in order

- heavy hadrons and their decay leptons and kaons have very similar energy distributions

- the effect of the beam pipe is very drastic, despite of which charmed and beauty hadrons with Lorentz factors up to $\gamma_D = 40$, $\gamma_B = 15$, respectively, will be measurable at HERA

- the mean decay lepton and kaon energies are in the 2 GeV region

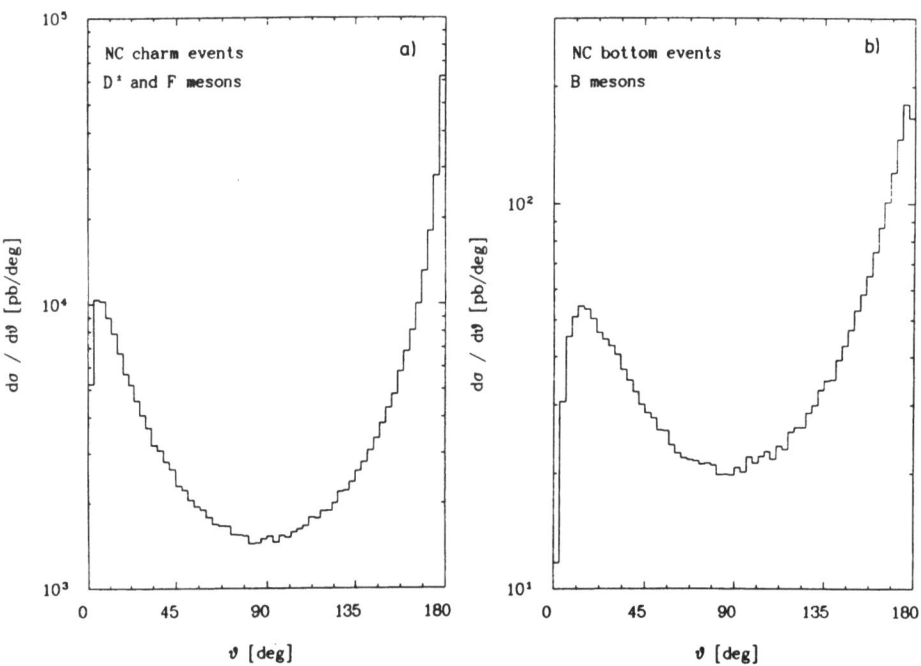

Fig. 5.2. Polar angle distribution of the heavy mesons produced in the NC process $ep \rightarrow eQ\bar{Q}X$ at HERA $\sqrt{s} = 310\ GeV$, a) $Q = c$, b) $Q = b$. From [16]

The separation of charm and bottom events has been a long standing issue in e^+e^- annihilation and $p\bar{p}$ collisions. This separation is neccesary for a number of experimental measurements, like those of

- lifetimes,

- branching ratios.

While the reconstruction of charmed mesons, D^* in particular, is a matter of routine, the efficiency of reconstructing beauty hadrons is of $O(10^{-3})$, mainly due to the large multiplicities involved. Our poor knowledge of exclusive and inclusive B decays makes the separation between charmed and bottom hadrons difficult. Two tags could be used at HERA to select bottom events inclusively:

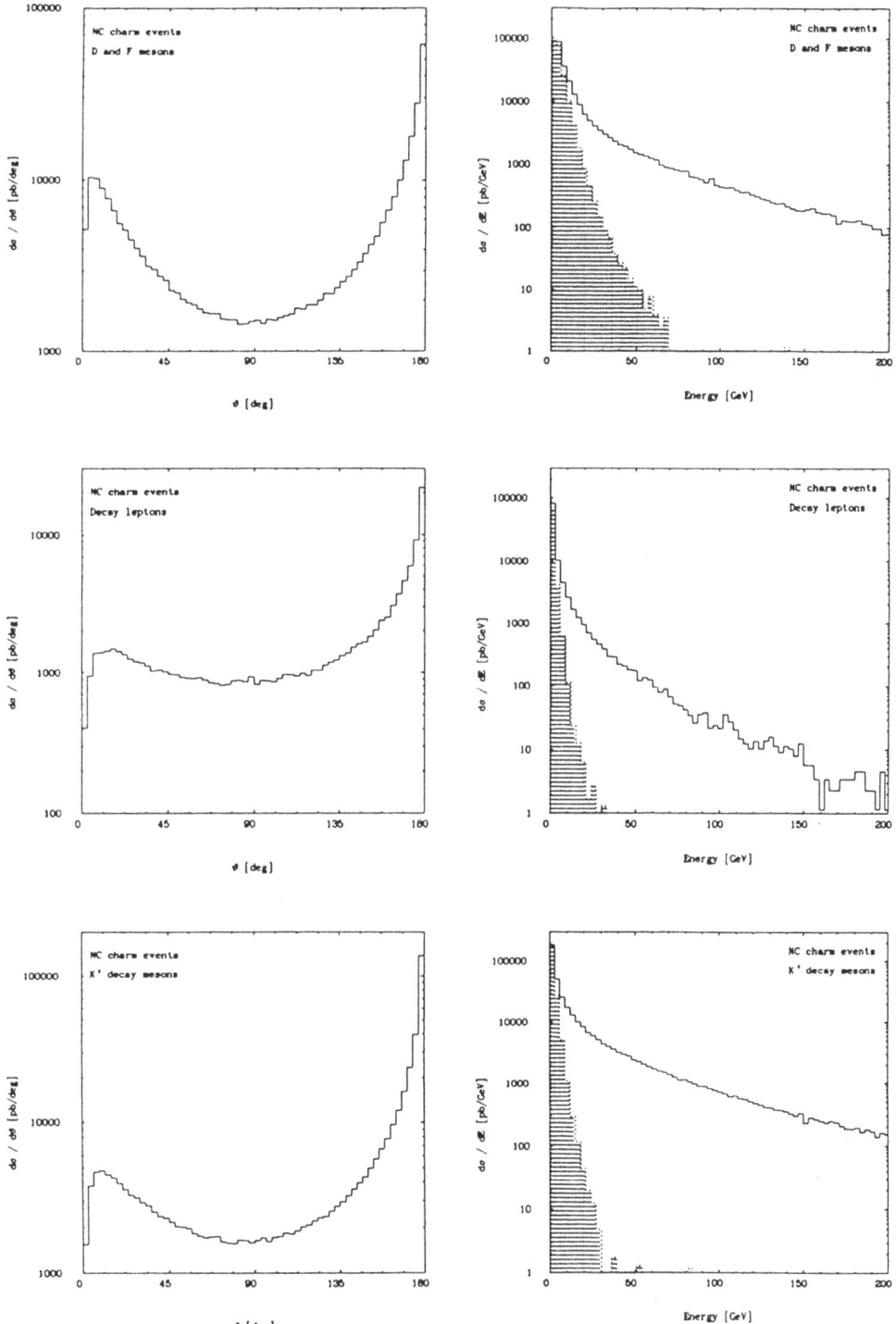

Fig. 5.3. a) Energy and polar angle distribution for heavy mesons as well as their decay leptons and kaons produced in the NC process $\epsilon p \to \epsilon c\bar{c}X$ at HERA, $\sqrt{s} = 310\ GeV$. From [16].

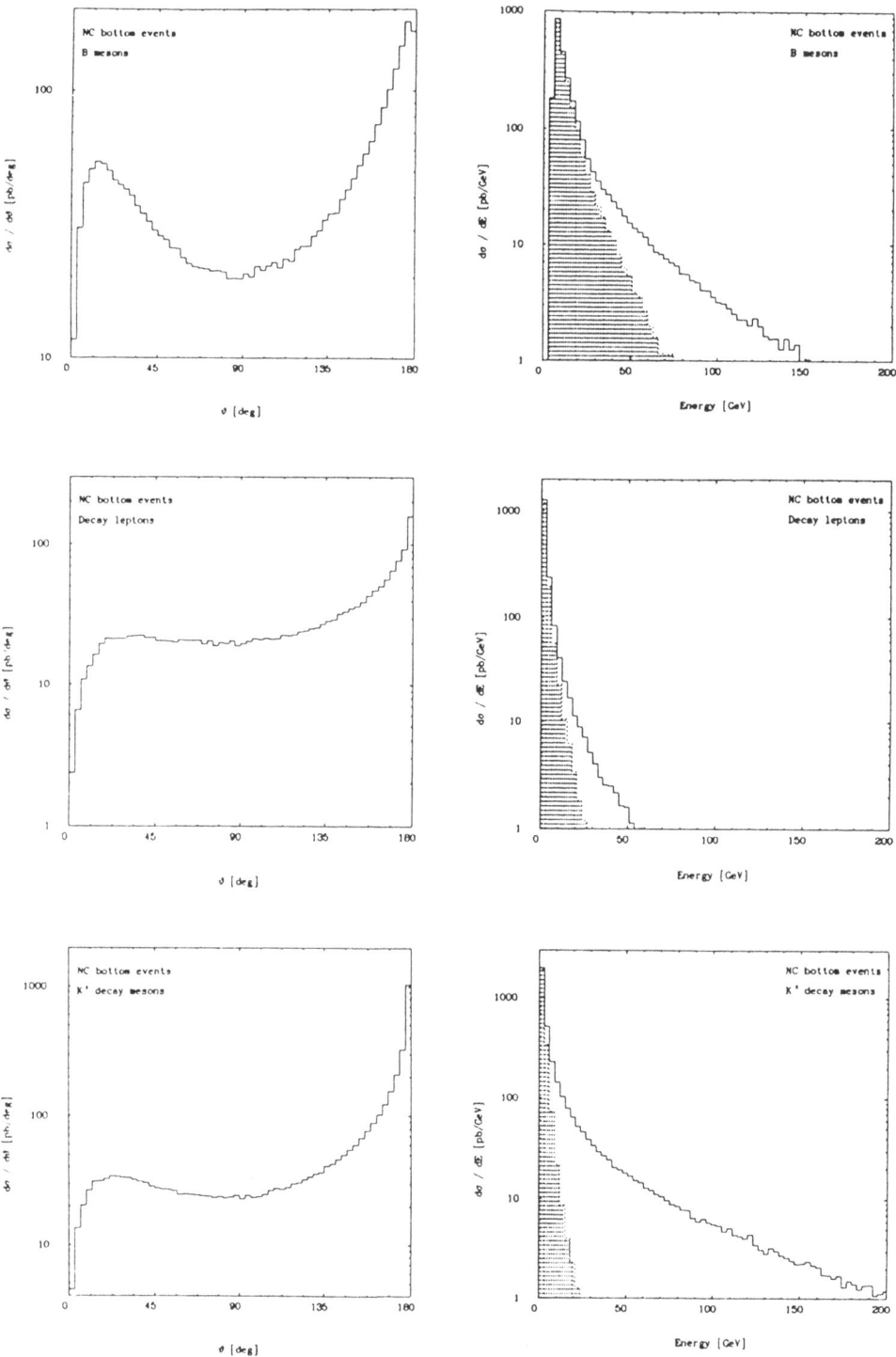

Fig. 5.3. b) Energy and polar angle distribution for heavy mesons as well as their decay leptons and kaons produced in the NC process $ep \to eb\bar{b}X$ at HERA, $\sqrt{s} = 310\ GeV$. From [16]. From [16]

- large p_T leptons from the semileptonic decay $b \to cl\nu_l$;

- J/Ψ from the decay $B \to J/\Psi X$ which experimentally has a branching fraction of 1.5%.

The main background source for large p_T leptons comes from the much larger charm cross-section and the subsequent semileptonic decay $c \to sl\nu_l$. If gluon radiation were absent in the $Q\bar{Q}$ system, a well suited method to reconstruct the quark direction of flight, like twoplicity discussed in Section 3, will permit us to properly measure the p_T of the lepton w.r.t the jet axis and exploit the very different end points in the spectra for both decays. This is illustrated in Fig. 5.4. Here again we have considered two cases with and without gluon radiation. If the $Q\bar{Q}$ is subject to a parton showering, the separation is rendered more difficult, see Fig. 5.4b.

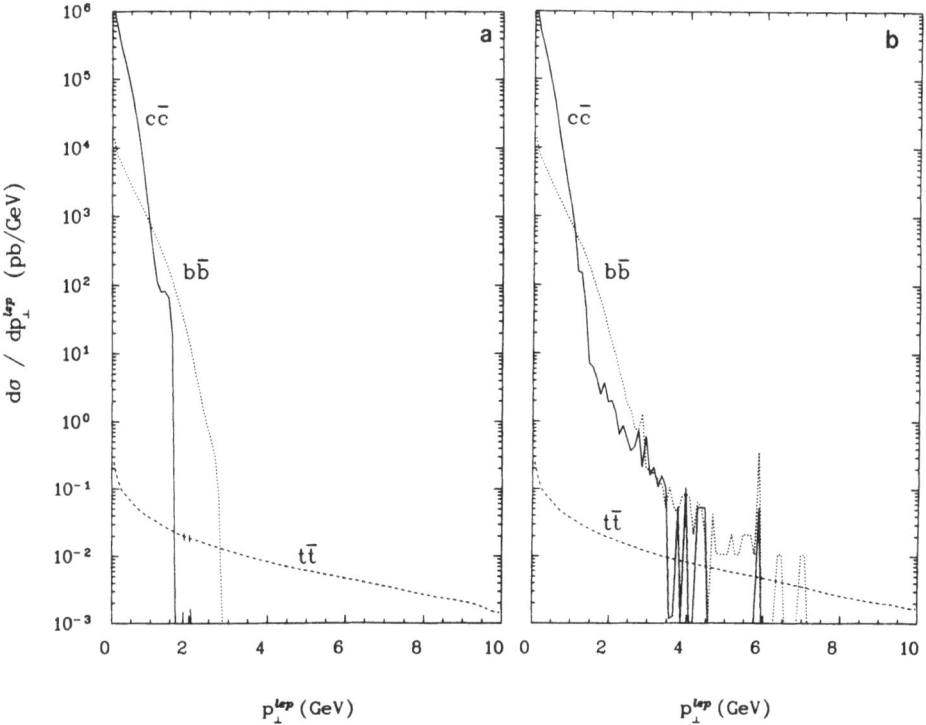

Fig. 5.4.Lepton p_T distribution in the process $ep \to lX$ from semileptonic decays of heavy quarks produced in $ep \to Q\bar{Q}X$. The p_T is measured w.r.t. the quark direction as reconstructed using the twoplicity method. a) no gluon radiation, b) right gluon

One has then to think of more sophisticated methods like the one discussed in [1] based on imposing a cut on the hadronic energy accompanying the trigger lepton. We estimate $O(10^4)$ bottom events which could be tagged this way per $100pb^{-1}$ integrated luminosity, see Fig. 5.5. This will yield a sizeable sample of bottom events which could be used to undertake more detailed studies like $B - \bar{B}$ oscillations.

6. Summary and Conclusions

We have discussed some aspects of heavy quark pair production at HERA. The cross-sections and relevant distributions were all based on perturbative QCD, including the BGF mechanism. Two extreme cases were considered i.e. with and without gluon radiation in the quark pair final state, the gluon radiation according to a parton shower formalism . Hadronisation and weak decays were taken into account in the LUND string picture. Beam pipe acceptance cuts were imposed but no other detector effects were taken into account.

After five years of running we expect $O(10^9)$, $O(10^7)$ and $O(10^2)$ charm, bottom and top quark jets (for $m_T = 60 GeV$) be produced at HERA. A substantial fraction of these events will be measurable but an effort has to be done to

- keep the $\sum E_T$ trigger threshold to a minimum,

- make VXD detectors as long as possible so that they also cover a large fraction of the backward hemisphere.

- keep calorimeters as hermetic as possible and instrumented in 4π with powerful electron-hadron separators.

ZEUS has taken the neccesary steps to implement the points discussed above. Fig. 1.1 shows how the VXD has been enlarged, see dashed area, in the backward direction as a consequence of the results presented so far. This study also prompted the group to equip the rear calorimeter RCAL with silicon diodes for electron identification.

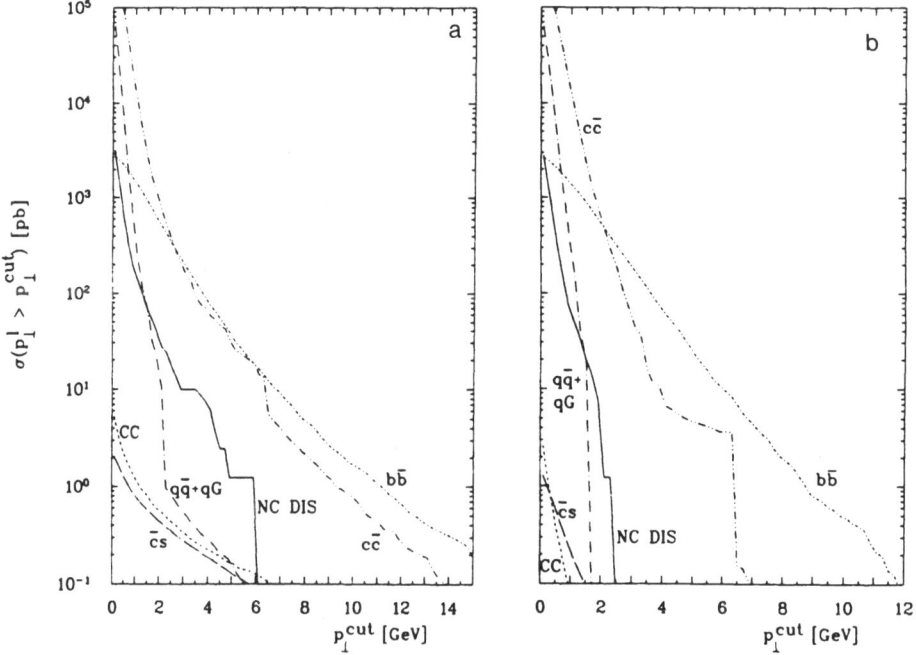

Fig. 5.5.a) integrated cross section for the process $\epsilon p \rightarrow (\epsilon, \nu_e) l X$ as a function of a cut on the p_T of the hardest lepton measured w.r.t. beam axis. Beam pipe cut is 100 $mrad$. b) same but for an isolated lepton defined as $E_{acc} \leq 2\ GeV$.

Acknowledgements

We expresss our thanks to our colleagues in the working group on Heavy Quark Physics at HERA. One of us (FB) gratefully acknowledges L. Cifarelli and A. Ali for their kind invitation to the VI-th INFN ELOISATRON Project Workshop on Heavy Flavours. and the Alexander von Humboldt Stiftung for supporting his stay at DESY during summer of '89. when this report was written.

Two of us (MAG and JFT) acknowledge the binational exchange programm CICYT-Kfz Karlsruhe. for supporting their visists to DESY.

Last but not least, we would like to thank G. Wolf for his interest in this study and for a critical reading of the manuscript.

References

[1] Heavy Quark Physics at HERA

A. Ali. G. Ingelman, G.A. Schuler, F. Barreiro, M.A. Garcia, J.F de Troconiz. R.A. Eichler and Z. Kunszt, *DESY 88-119 and FTUAM-EP 88-05, published in the Proceedings of the DESY Workshop on HERA Physics, R. Peccei editor*

[2] A. Ali, see contribution to these Proceedings

[3] G.A. Schuler. see contribution to these Proceedings

[4] U. Baur, J.J. van der Bij. CERN-TH.4875/87 (1987) preprint

[5] M. Glück, R.M. Godbole. E. Reya. Dortmund Univ. preprint, DO-TH-87/17 (1987)

[6] G.A. Schuler. *Nucl. Phys. B299 (1988) 21*

[7] R.A. Eichler. Z. Kunszt. ETHZ-IMP P/88-1 (1988) preprint

[8] R.K. Ellis and Z. Kunszt. *Nucl. Phys. B303 (1988) 633*

[9] E. Eichten. I. Hinchliffe. K. Lane. C. Quigg, Rev. Mod. Phys. 56 (1984) 579, ibid. 58 (1986) 1047

[10] A. Ali, proc. workshop on Physics at LEP, Eds. J. Ellis, R. Peccei, CERN 86-02, Vol. 2, p. 220

[11] G. Ingelman, G.A. Schuler. AROMA 1.2 – A Generator of Heavy Flavour Events in *ep* Collisions, DESY preprint in preparation

[12] B. Andersson, G. Gustafson, G. Ingelman. T. Sjöstrand, Phys. Rep. 97 (1983) 31

[13] M. Bengtsson. T. Sjöstrand, Comput. Phys. Common. 43 (1987) 367

[14] F. Barreiro. S. Brandt and M. Gebser, *Nucl. Inst. and Meth. A240 (1985) 237*

[15] G. Ingelman. G. Schuler and J.F. de Troconiz. *DESY 88-143 and FTUAM-EP 88-07. accepted in Nucl. Phys. B*

[16] A. Ali et al.. *to be published*

19

PROSPECTS IN TOP PHYSICS

Peter Igo-Kemenes

Heidelberg University
Heidelberg, Fed. Rep. Germany

INTRODUCTION

During the last two decades our understanding of elementary particles grew at a remarkable pace. This period has seen the development of the standard model and its spectacular success. The most important benchmarks were the discovery of weak neutral currents and, more recently, the confirmation of the existence of the intermediate vector bosons $W^{\pm}$ and Z^0 at the predicted masses. Today the standard model describes with remarkable precision most physical phenomena down to distances of the order of 10^{-16} cm and, conversely, no phenomenon has yet been found which would cast serious doubt on the validity of the standard model [1].

The standard building blocks of the universe are elementary fermions, quarks and leptons, which belong to three distinct families. All members are known today with the exception of the top quark, the 'up' member of the third family or the isopartner of the bottom quark. Forces between the elementary constituents are transmitted by the exchange of intermediate vector bosons: gluons for the strong interaction, $W^{\pm}$, Z^0, and γ for the electroweak force. Particle masses are generated by the Higgs mechanism which predicts the existence of one neutral scalar particle, the Higgs boson—yet to be discovered. All these 'standard particles' are known today with the exception of the top quark and the Higgs boson, and we are witnessing a world-wide effort to discover them.

The interest of these discoveries goes far beyond the limits of the standard model. It is more and more clear that, despite all its successes, the standard model cannot be regarded as the ultimate truth; its main weakness is the inability to explain its own complex phenomenology. The spectrum of fermion masses, the deeper meaning of flavours and of the family structure, and other fundamental features have to be taken today as 'facts of life' instead of having a simple and coherent explanation! Today's interest is therefore oriented beyond the standard model, towards a more global theory which would account for the standard model phenomenology in a natural way [2]. Speculations go in many directions and both theorists and experimentalists seek for hints as to which direction to follow.

New insight is expected, of course, from the coming generation of accelerators which will, within the next decades, extend the accessible energy range into the multi-TeV region. Supersymmetric particles, leptoquarks, or an extended Higgs sector would be spectacular manifestations of physics beyond the standard model, but these very high energy machines still seem remote. A less spectacular, but more easily accessible way towards new physics is to detect small but systematic deviations from standard model predictions, where the discovery of the top quark and the Higgs boson would be of prime importance. Today the masses of these particles are two almost free parameters which, by their uncertainty, constitute a severe limitation to the predictive power of the standard model. In this context, higher-order rare phenomena, for example rare kaon [3, 4] or B-meson decays [5, 6], are of fundamental interest. Within the standard model these processes are described by virtual heavy quark loops, including the top quark. If the top-quark mass were known, the standard model would provide a firm prediction for their rate, and any deviation from the prediction could be interpreted as a sign for new physics. Thus, beyond the challenge of discovering another new particle, the deeper motivation of the world-wide searches for the top quark is to improve the predictive power of the standard model and, by the same token, to increase the sensitivity of the scrutiny for new physics.

Today the hope for a rapid discovery of the top quark is well founded. It is worth mentioning that there cannot be any doubt concerning the existence of the top quark, the isodoublet nature of the bottom quark [7, 8] being experimentally proven. The top-quark mass is confined by experiment, both from below and from above, and current accelerators are in the process of scanning more and more extended parts of the allowed mass range. The lower limit on the top mass is provided by detailed searches carried out at e^+e^- colliders: the PETRA limit of 23 GeV has recently been raised at TRISTAN to 27 GeV, where the search will continue up to 33 GeV. At the $S\bar{p}pS$ hadron collider, the early UA1 'signal' has been successively revised and finally transformed into a solid lower limit of 41 GeV [9, 10]. An indirect lower bound has also been derived from the recent observation of $B^0\bar{B}^0$ mixing by the ARGUS and CLEO collaborations; the combined result reads [11–13] $m_t > 45$ GeV. It is important to note that the limits from hadron colliders and from $B^0\bar{B}^0$ mixing only hold in the context of the standard model, where it is assumed that the top quark has standard couplings and decays via $t \rightarrow bW^*$. Many non-standard decays can be imagined (e.g. $t^+ \rightarrow bH^+$ in supersymmetric models with an extended Higgs sector), towards which the experimental searches would have been ineffective. The upper limit of $m_t < 180$ GeV is obtained [14], also within the context of the standard model, from the size of the electroweak radiative correction to $\sin^2 \theta_w$. By combining the experimental information on $B^0\bar{B}^0$ mixing, the B-meson lifetime, and the recent observation of 'direct' CP violation in the K-meson system [15], the author of ref. [16] claims a preferred range for the top-quark mass from 70 to 90 GeV.

Today both the $S\bar{p}pS$ and the Tevatron are producing data and hope to push the top search in 1989 up to 70 GeV. In the same year both LEP and the SLC will come into operation and will continue the search in e^+e^- collisions. In its first phase, LEP will be able to extend the search to 55 GeV and later, after the energy upgrade, to 100 GeV. If the top mass is larger than 100 GeV, only hadron colliders will have the opportunity to discover the top quark in the foreseeable future, until a new e^+e^- collider, like the one discussed in ref. [17], becomes available.

At the time of writing there is no way to tell whether the top mass is below or above 100 GeV. If it is above, the discovery and the whole realm of top physics will be a matter of the far future; we therefore concentrate here on the other alternative, more interesting at present, and assume that the top quark is lighter than 100 GeV. In this case the LEP collider will contribute in an outstanding way to the study of the top-quark properties through the annihilation reaction $e^+e^- \rightarrow t\bar{t}$. Today it is even possible that the top quark is pair-produced on the Z^0 resonance, in which case top physics at LEP will start as soon as 1989! Assuming for the first year of operation a luminosity of 1 pb^{-1}, one may count on 30,000 Z^0 decays, of which 400 ($m_t = 44$ GeV) to 800 ($m_t = 40$ GeV) could be $Z^0 \rightarrow t\bar{t}$ events. The distinctive topological properties of $t\bar{t}$ events will permit a clean separation from the background of light quark pairs. In the following years the LEP experiments will register 10^7 Z^0 decays, and the high-statistics study of the top-quark properties will include the measurement of the exact mass, the electroweak couplings, and possibly the search for rare decays. If the top mass is higher than half the Z^0 mass, it has to be pair-produced in the e^+e^- continuum, where the $t\bar{t}$ cross-section is much lower, typically 1 to 10 pb. The yearly rate will still be of 200 to 2000 $t\bar{t}$ events (assuming a luminosity of 200 pb^{-1} per experimental year), but for most physics results the precision will in this case be limited by statistics.

A highly important aspect of top physics is the search for toponium, the $t\bar{t}$ bound state. The discovery of the ground state will be followed by a search for the lowest radial excitations. The cross-section and the level spacing will give information lacking on the $q\bar{q}$ potential at very small distances (below 0.1 fm), in the perturbative region. Toponium production and decay provide supplementary information on the top-quark properties and finally, under favourable circumstances, toponium decay may be the place to look for the standard Higgs particle and for less standard particles such as charged Higgs bosons or supersymmetric particles. The various studies have already proven [18–20] that the LEP machine, because of its high energy resolution, is in a unique position to carry out the toponium research programme.

The next chapter will be devoted to 'open top' physics at LEP. The discussion will include aspects of the top search, the measurement of the mass, the determination of the weak couplings, and some aspects of the search for the Cabibbo-forbidden t → s decay. The following chapter will treat the broad subject of toponium physics, initially in the standard model scenario, followed by some simple extensions, such as supersymmetry or an extended Higgs sector. The most trivial extension beyond the standard model is to admit a fourth family of fermions. The corresponding 'down' quark, b', may well be within reach at LEP. As the search for b' and for the corresponding 'bottomonium' states is an important part of the LEP programme, the discussion will be extended to b' whenever appropriate.

ASPECTS OF OPEN TOP PHYSICS

Essentially all earlier studies agree that in e^+e^- annihilation the threshold for top production could be observed without difficulty, either below or above the Z^0 resonance. The events have a particular, spherical shape when the c.m.s. energy is near the production threshold, owing to the high transverse momentum of the top fragments with respect to the original top-quark direction [$p_T(max)/p = m_t/E_{cms}$]. Any topological quantity such as the event sphericity, thrust, aplanarity (fig. 1), transverse mass, etc., will help to distinguish between the 'fat' top events and the 'narrow' light quark events. However, if the top mass is higher than the W mass, the main background after the topological selection will come from W^+W^- pairs, and a more sophisticated analysis is needed to establish the signal. Typical cross-sections, for the signal and for the competing processes, and some 'case studies', both for t and for b', are described in Chapter 3 of ref. [19]. It appears that, even in the least favourable case of b' with mass above the W mass, a signal can be established from a modest data sample of 50 pb^{-1} luminosity. Other properties that will corroborate the t (b') signal are the increased numbers of jets (typically six), of charged and neutral particles and of prompt leptons, due to the longer cascade of top events: t → b → c → u. Also, high-p_T leptons from the semileptonic decay t → b$\ell\nu$ (ℓ = e, μ, τ) will play an important role. In fact, semileptonic decays may prove to be the best 'discovery channel': the probability for at least one of the two quarks to decay into a muon or electron plus jets is 35%, only slightly below the probability for purely hadronic decays [21]; on the other hand, this channel simultaneously incorporates various distinctive features, namely the high sphericity and an energetic, isolated lepton [21, 22]. (For an efficient use of the lepton isolation to discriminate top events against the background, see fig. 2, taken from ref. [22].)

A new and complementary method to tag top quarks takes advantage of the refined vertex-detection capabilities of the LEP experiments which will measure lifetimes as short as 5×10^{-14} s, and will recognize 'long-lived' b quarks from the t (b') cascade. This method is of particular interest in the difficult case when one is running at a c.m.s. energy higher than the W^+W^- threshold. If $m_t < m_W - m_b$ holds, a sizeable fraction of the W bosons decay into t$\bar{b}$, with the t decaying into another b quark. Another case where vertex detection could be of interest is for $m_{b'} < m_t$. The b' then decays to a c quark, a jump of two families, with a correspondingly long lifetime of about 10^{-12} s. The vertex-detection technique together with all the methods discussed above, will contribute to establish the new threshold due to t or b'.

The precise *determination of the top mass* (a precision of about 1 GeV is the prerequisite for an efficient search for toponium) is a more elaborate problem. Of course, all the topological distributions mentioned earlier 'carry' the top-mass information; however, the precise event shape depends on the yet unknown fragmentation properties of the top quark. Various other ways to measure the top mass have been proposed, and we discuss them briefly here, mainly to draw attention to the difficulties that will be encountered when applying them.

The t$\bar{t}$ cross-section near the production threshold is a sensitive function of the top mass [23, 24]; however, *the event selection through topological cuts was found to considerably attenuate this sensitivity*; it also introduces systematic uncertainties through the unknown

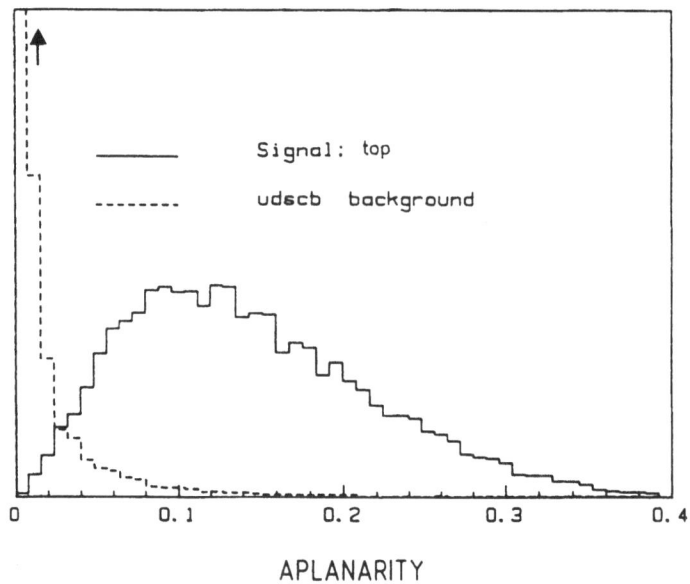

APLANARITY

Fig. 1 Comparison of the aplanarity distribution for $e^+e^- \to t\bar{t}$ ($m_t = 65$ GeV) and for the background from light quarks. The c.m.s. energy is 160 GeV. Taken from ref. [19].

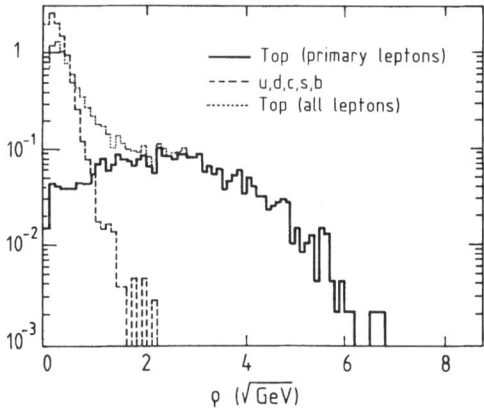

Fig. 2 Comparison of the distribution in ϱ, a separation variable based on the isolation of the prompt lepton in semileptonic decays, for top with mass of 50 GeV and for light quarks; the c.m.s. energy is 110 GeV; ϱ is defined [22] as $\varrho = \min[E_\ell(1 - \cos\theta_{j,\ell})]^{1/2}$, where E_ℓ is the lepton energy and $\theta_{j,\ell}$ the angle between the lepton and jet j. (The index j is running over all identified jets.)

top-quark fragmentation properties. In addition, QCD corrections to the cross-section are substantial [24] and essentially unknown to higher orders.

The top mass can be inferred from the shape of the lepton energy spectrum in semileptonic top decays. In practice, the method is useful only if the $t\bar{t}$ pairs are produced at the Z^0 resonance; otherwise the method lacks statistical accuracy. The various systematic uncertainties, from bremsstrahlung, QCD effects, top fragmentation and top-spin-lepton-spin correlation have been estimated [24]; they add up to only 0.5 GeV for $t\bar{t}$ produced at the Z^0 resonance.

One can attempt, of course, to reconstruct the top mass from its decay products. If, however, both the t and the $\bar{t}$ decay into hadrons, the association of the many jets (six or more) to one or the other of the top quarks is by no means trivial, and the many possibilities introduce a high combinatorial background. It seems therefore more promising to select those events where one of the top quarks decays via the semileptonic channel; the number of possible associations of the jets then reduces to 4. Various algorithms have been tried out to select the correct association [21, 22], and the method seems to work, at least for $m_t < m_W$. They require, however, high standards in terms of pattern recognition, association of tracks to jets, and a full geometrical coverage to reconstruct the missing-momentum vector associated to the escaping neutrino.

From the above discussion it should appear that the determination of the top mass with the desired accuracy of 1 GeV is by no means trivial, especially if the mass is high. With time, however, the various systematic uncertainties will be understood, and the many redundant methods will contribute to pin down the top mass with the desired precision. Let me mention in this context that the most precise determination of the top mass should come from the detection of the toponium ground state. The theoretical uncertainty in relating the toponium mass to the top mass is estimated [20] to 300 MeV only!

Once the top quark (or the b') is discovered and its mass is measured, the next task would be to determine the *weak vector and axial-vector couplings* (v_t, a_t) and to verify the standard model prediction of 'universality'. This is usually done by combining the total cross-section, proportional to $v_t^2 + a_t^2$, with the forward–backward charge asymmetry, proportional to $v_t \cdot a_t$. For the latter, semileptonic decays have to be selected ($t^{\pm} \rightarrow b\ell^{\pm}\nu$; BR = 10% per lepton flavour), in which case the charge of the top quark is given by the charge of the daughter lepton. If the top quark is too heavy to be produced on the Z^0 resonance, the yield of semileptonic $t\bar{t}$ events (20 to 160 events per year) will be too low for a statistically significant measurement of the charge asymmetry; the total yield of $t\bar{t}$ events (200 to 1600 events per year) will, however, be sufficient to measure the cross-section and to distinguish between the 'up' or 'down' nature of the discovered quark. (The standard model cross-section for 'up' is roughly two times higher than for 'down'.) If, on the other hand, the top quark is abundantly produced at the Z^0 resonance, the forward–backward charge asymmetry of the daughter lepton can be measured with a high statistical accuracy. The difficult task will then be to relate the measured lepton asymmetry to the weak couplings. The relation is difficult to establish for various reasons, the most important being [25] the low velocity of the top quarks (which makes it possible for the lepton to show up in the 'wrong' hemisphere), and the partial depolarization of the top quark during the fragmentation process. One cannot tell *a priori* if the daughter lepton arises from the decay of T^* vector-mesons, or from pseudoscalar T-mesons, or from both in given proportions. This may influence significantly the top-quark daughter-lepton angular correlation and, under unfavourable circumstances, may even reverse the sign of the lepton asymmetry with respect to the original top asymmetry (fig. 3). Methods to deal with this problem are proposed by the authors of ref. [25]; it is clear, however, that the interpretation of the lepton asymmetry in terms of the weak couplings of the top quark will require a broader experimental approach and a good theoretical understanding.

The *decay properties* of the top quark, related to the Cabibbo-Kobayashi-Maskawa mixing matrix are, of course, of basic interest. The matrix elements which are relevant to top decay, V_{tb}, V_{ts}, and V_{td}, are currently obtained [13] from the experimental knowledge of all other elements, using the standard model assumption of 'unitarity'. The direct measurement of the ratio $|V_{ts}/V_{tb}|^2$ via the inclusive branching ratio $\Gamma(t \rightarrow s)/\Gamma(t \rightarrow b)$ is a difficult task since its value is expected to be low, of about 0.003. Observation of the b $\rightarrow$ u transition, the

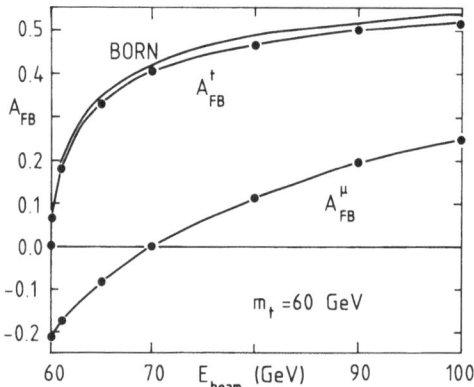

Fig. 3 The forward–backward charge asymmetry of the top quark with 60 GeV mass and of daughter muon (t → bμν), as a function of the beam energy; taken from ref. [25].

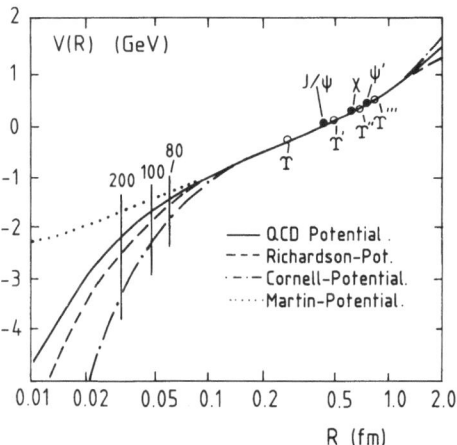

Fig. 4 Quarkonium potentials 'adjusted' to fit J/ψ and Υ data; taken from ref. [20]. For illustration, we have indicated on the figure the radial size (R) of the various J/ψ and Υ states, as well as the size of the toponium ground state, for mass equal to 80, 100, and 200 GeV.

analogue of the t → s decay for the b system, has recently been claimed by ARGUS [26], but contradicted by CLEO [27]. Both experimental groups search for low-multiplicity exclusive B-meson decays without charm, such as $B^+ \rightarrow p\bar{p}\pi^+$ and $B^0 \rightarrow p\bar{p}\pi^+\pi^-$, to infer from there the inclusive b → u rate. This method, already difficult in the case of the b system, will be even more so for top-quark decays, due to the higher track and jet multiplicities in the final state. An alternative method [24] takes advantage of the slight difference in Q-value of the two decays t → s$\ell\nu$ and t → b$\ell\nu$, which causes a slight difference in the end-point energy of the two inclusive lepton spectra. The size of the effect, of the order of 1%, is comparable to the energy resolution of the LEP detectors. Because of the very low branching ratio of t → s$\ell\nu$ and the low population of the lepton spectra in the end-point region, the method can only be attempted if the top is copiously produced at the Z^0 resonance. Even there it seems marginal; nevertheless, the special topology of an isolated, high-p_T lepton, plus a leading K-meson in the same hemisphere, might allow the selection of a small sample of t → s transitions.

Let us conclude the discussion of 'open top' aspects by simply stating that the top lifetime, which decreases like the fifth power of the top mass, is too short (10^{-18} to 10^{-20} s) to be directly measured; it can, however, be inferred from the decay properties of the toponium ground state [20]. Also, essentially because of the short lifetime of T-mesons, there is no hope of observing $T^0\bar{T}^0$ oscillations.

TOPONIUM PHYSICS

The search for toponium and the study of the toponium spectrum is the logical continuation of charmonium (J/ψ) and bottomonium (T) spectroscopy, which have contributed in the past to a very large extent to our knowledge of the q$\bar{q}$ potential and of perturbative QCD.

According to present understanding, the *q$\bar{q}$ potential* is flavour-independent, and consists essentially of a long-distance 'confining' part, proportional to the distance of separation (r), and a short-distance Coulomb-like potential (1/r). The singularity of the latter, at r = 0, is logarithmically attenuated; this is expressed by the Richardson potential model [28]: $V(r) \rightarrow [r \log(a/r)]^{-1}$ for r → 0, a potential of order α_s, which already incorporates the QCD feature of 'asymptotic freedom'. Many experimental data suggest that terms of order α_s^2 have an important contribution and have to be included; in this case the q$\bar{q}$ potential can be expressed in terms of the scale parameter of the strong interaction, $\Lambda_{\overline{MS}}$. A comprehensive discussion of the precise form of V(r) and the various constraints arising from J/ψ and T spectroscopy, is presented in ref. [29]. The reader can also find there the exact expression for the particular QCD potential (order α_s^2, variant 'J', $\Lambda_{\overline{MS}}$ = 200 MeV), which was used in the LEP 200 Workshop studies [19], from which most of the following discussion is drawn. An excellent review of the whole 'toponium scenario' can be found in ref. [20].

It is a fact that the measured properties of the J/ψ and T spectra are adequately described by a great variety of potential models (fig. 4), ranging from the simplest 'power law' (Martin) and the singular Coulomb potential (Cornell) to the most sophisticated QCD potentials of first (Richardson) and second ('QCD') order. This becomes clear from the figure, where the J/ψ and T states are indicated at the appropriate radial distances. These states, by their size, probe the q$\bar{q}$ potential only at intermediate distances (r > 0.1 fm), where all models can be 'adjusted' by a judicious choice of the long- and short-distance contributions. At distances below 0.1 fm, however, the various potential models have diverging predictions, and a measurement of the properties of the toponium spectrum would be able to discriminate between them. Toponium states, by their smaller radial dimensions, would probe the q$\bar{q}$ potential in its deep regions, where the contribution from the long-distance part is really negligible, and where the QCD prediction of asymptotic freedom is the dominant feature.

Both the production cross-section and the level spacing are sensitive to the exact shape of the q$\bar{q}$ potential. In e^+e^- annihilation the toponium cross-section depends crucially on the electroweak couplings, especially close to the Z^0 resonance. The 'effective' cross-section (which is shown in fig. 5 for the 1S and 2S states and compared with the relevant competing processes) is further influenced by initial-state radiation and by the energy spread of the colliding beams; therefore the cross-section is less appropriate to fix the parameters of the q$\bar{q}$

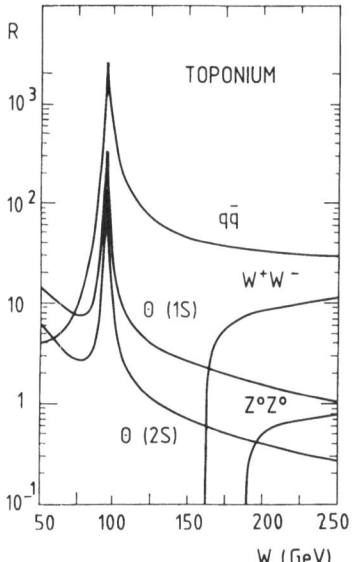
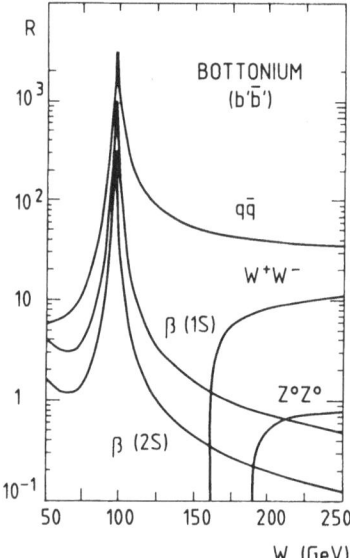

Fig. 5 The 'effective' cross-section of toponium and heavy bottomonium states, as a function of the mass (which equals the c.m.s. energy W), compared with the relevant sources of background; $R = \sigma/\sigma_0$, where σ_0 is the 'point' cross-section $4\pi\alpha^2/3W^2$. Taken from ref. [19].

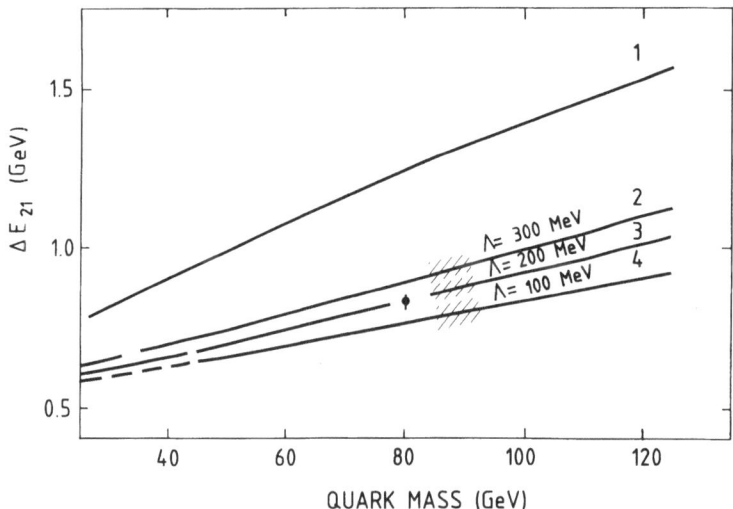

Fig. 6 The energy separation ΔE_{21} for toponium, as a function of the top mass, for the 'Richardson' potential (curve 1) and for the second-order QCD potential (curves 2 to 4) with various values of $\Lambda_{\overline{MS}}$. Taken from ref. [19]. The experimental point at m_t = 80 GeV indicates the precision one can aim to achieve at LEP.

potential. Experimentally, the quantity to measure is the energy separation $\Delta E_{21} = E(2S) - E(1S)$ between the 1S ground state and the first radial excitation (2S); its sensitivity to various potential models, as a function of the top mass, can be read off fig. 6, taken from ref. [19]. The figure sets the scale for the precision which is required in order to be sensitive to second-order QCD effects (compare 'Richardson' and 'QCD'), and to the precise value of $\Lambda_{\overline{MS}}$. The 'experimental' point is the result of a Monte Carlo case study and will be discussed later.

Before turning to the purely experimental aspects of detecting toponium states and measuring ΔE_{21}, we first review some of the basic properties of the *toponium spectrum* and of *toponium decays*. With the particular choice of the potential mentioned above, there are 10 to 18 radial excitations (S-states) and a similar number of orbital or P-states below the open-top production-threshold (see fig. 7). The separation ΔE_{21} is of the order of 0.6 to 1 GeV, which is large compared with the design energy spread of LEP. The latter varies between 45 MeV near the Z^0 resonance and 250 MeV at the maximal LEP energy of 200 GeV; the lowest S-states can thus, in principle, be resolved. (P-states are only weakly produced in e^+e^- annihilation, via the axial current, where the cross-section, proportional to v^2/c^2 of the top quark, is very low.) The most relevant toponium decay channels are the annihilation into fermion–antifermion pairs via the charged and the neutral electroweak current, and the 'single-quark decay', where one of the quarks within the toponium undergoes a charged-current β-decay. While near to the Z^0 resonance the annihilation channel is the most important one, at higher masses the single-quark decay soon becomes the dominant mode. Strong, gluonic decays ($\rightarrow$ ggg, $\rightarrow$ ggγ), and the Wilczek process $\rightarrow$ $H^0\gamma$, have a low branching ratio, owing to the predominance of single-quark decays. Partial and total decay widths are shown in fig. 8, and the corresponding mathematical expressions can be found in Appendix C of ref. [19].

At this point it is worth comparing (fig. 8) toponium decays with those of *heavy bottomonium* (b' of a fourth generation), for which the decay pattern is very different. Under the assumption of $m_{b'} < m_{t'}$, the single-quark decay of bottomonium (b' $\rightarrow$ t or b' $\rightarrow$ c) implies a 'jump' of at least one family and is therefore heavily suppressed. The decay into fermion–antifermion pairs will thus dominate over the whole LEP range, and the suppression of the gluonic and Higgs channels will be less severe. The two figures also show that the bottomonium states remain narrow, whereas the width of toponium states increases rapidly with mass. Indeed, around 250 GeV, the toponium decay width becomes comparable to the level spacing ($\Delta E_{21} \approx 1$ GeV), which means that the toponium resonances merge into the open top threshold, and talking of isolated toponium states becomes meaningless. (Physically, the lifetime of the top quark becomes short compared to the time it takes to form a $t\bar{t}$ bound state.) This merging of toponium states occurs for $m_t > 125$ GeV, which is beyond the range accessible at LEP.

The *experimental aspects* of the search for toponium states for m_t below and above 55 GeV have been worked out in refs. [18] and [19], respectively. The prerequisite for an efficient, high-resolution energy scan is that the open-top threshold has already been localized, and the top mass fixed with an accuracy of about ± 1 GeV. The methods to achieve this goal have been discussed in the previous section.

If toponium happens to fall into the region of the Z^0 resonance, its dominant decay mode, annihilation into fermion–antifermion pairs, interferes with the same final state from Z^0 decay. A peculiar interference pattern arises [30, 20], with a close succession of maxima and minima, both in the toponium cross-section and in the forward–backward charge asymmetry as a function of the c.m.s. energy (fig. 9). The precise measurement of this pattern, a sensitive test of the interplay between the electroweak couplings and QCD, would put very stringent requirements on beam stability and on the spread in c.m.s. energy.

Single-quark decays offer themselves naturally for the detection of toponium states, especially for masses above the Z^0 resonance, where they soon become the dominant decay channel. Although the relevant backgrounds are 20 to 50 times higher than the signal, they can be reduced by simple cuts [19]. The largest background arises from $e^+e^- \rightarrow Z^0\gamma$, where the high-energy gamma escapes along the beam pipe; it communicates, however, a longitudinal boost to the event, which can be exploited for background rejection. The most distinctive feature of single-quark decays is the very high event sphericity, as the top quarks decay almost at rest. For toponium masses below the W^+W^- threshold, a selection based on

317

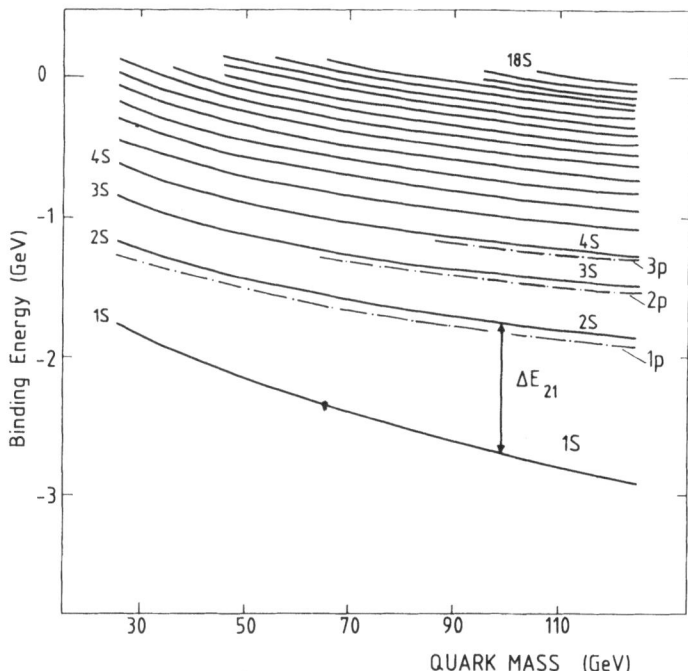

Fig. 7 Quarkonium energy level spectrum as a function of the quark mass, for a typical QCD potential (ref. [29], variant 'J', $\Lambda_{\overline{MS}}$ = 200 GeV). The separation ΔE_{21} = E(2S) − E(1S) is indicated.

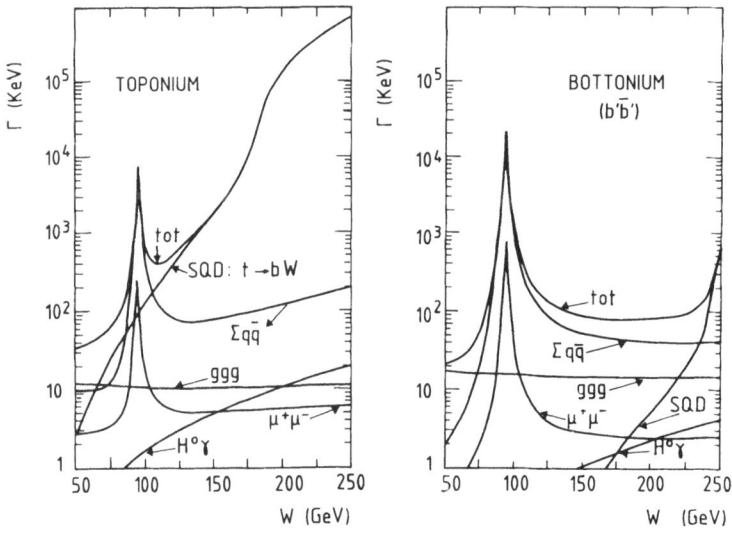

Fig. 8 Partial and total decay widths for the toponium and heavy bottomonium ground states, as a function of the mass. Taken from ref. [19].

topological criteria and on longitudinal momentum balance is sufficient to reduce the backgrounds from $e^+e^- \to q\bar{q}$ and $Z^0\gamma$. For higher masses, a more sophisticated analysis is required to eliminate the W^+W^- pairs; for example, one may attempt [19] to reconstruct the events under the W^+W^- hypothesis, and require the two W's not to be back-to-back.

The luminosity needed to detect toponium in a high-resolution energy scan for single-quark decays has been estimated [18, 19], assuming a scanning region of 2 GeV and a step size of two times the c.m.s. energy spread. It appears that a significant toponium signal (3 standard deviations above background) only requires the modest integrated luminosity of 25 to 50 pb^{-1}, over the whole mass range from 110 to 200 GeV. For the first radial excitation (2S), the lower production cross-section makes its detection more difficult, but still perfectly possible, especially once the 1S state has been localized. Recently it has been shown that a scan for semileptonic single-quark decays is even more economic [21].

After the 1S and 2S states have been detected, the energy difference ΔE_{21} can be measured with high precision. The absolute c.m.s. energy of LEP is accurate to 50 MeV only, but the difference of two nearby energy settings is known with better precision, estimated to be 15 MeV. A matching statistical accuracy can be obtained at the price of about 300 pb^{-1}, a good year of running. (The estimate was done [19] for a toponium mass of 160 GeV.) The combined accuracy (22 MeV) is very respectable: with such a precision, the measurement would of course be sensitive to QCD terms of order α_s^2 but, beyond that, would also significantly contribute to the precise knowledge of $\Lambda_{\overline{MS}}$, as indicated by the 'experimental' point in fig. 6.

Heavy bottomonium ($b'\bar{b}'$) is much more difficult to detect, as its single-quark decay has no significant rate (fig. 8). The dominant decay mode, annihilation into fermion–antifermion pairs, has a contribution 30 to 50 times higher from the e^+e^- continuum, and there is no way of reducing this background. A scan (in the total cross-section) is only worth while if the 'effective' signal is enhanced with respect to the background, by decreasing significantly (at least by a factor of 3) the c.m.s. energy spread, without prejudice in terms of luminosity.

The search for the *Higgs particle* via toponium decay into $H^0\gamma$ has been discussed in great detail [18] for m_t below 120 GeV. For higher toponium masses this channel is severely suppressed by the dominant single-quark decays, and one arrives at the conclusion that toponium decays do not add significantly to the Higgs search via the process $e^+e^- \to Z^0H^0$ (Z^0 real or virtual).

Let us now turn to *non-standard decays* of toponium (heavy bottomonium). Superstring phenomenology predicts [31, 32] several neutral and charged Higgs bosons on the electroweak energy scale. Since the Higgs coupling to fermions increases with the fermion mass, the single-quark decay with Higgs in the final state (e.g. toponium $\to H^+H^-b\bar{b}$) would play a dominant role, provided it is kinematically allowed (fig. 10). If supersymmetric particles are relevant to the LEP energy range, quarkonium may again deviate in a spectacular way from standard expectations. The detailed pattern (see e.g. the relevant part of ref. [18]) depends upon the relative masses of all particles involved (t, b', $\tilde{t}$, $\tilde{b}'$, $\tilde{\gamma}$, $\tilde{g}$, ...), and only a few qualitative guidelines can be discussed here; they hold equally for toponium and heavy bottomonium.

The single-quark decay $t \to \tilde{t}\tilde{g}$ would increase the toponium decay width above 1 GeV, and would wipe out the resonance structure. The electromagnetic analogue, $t \to \tilde{t}\tilde{\gamma}$, has less dramatic consequences (see fig. 10 taken from [19]), and can readily be observed if $m_{\tilde{\gamma}} < m_{\tilde{g}}$ and if the toponium mass is between $m_{\tilde{t}} + m_{\tilde{\gamma}}$ and $m_{\tilde{t}} + m_{\tilde{g}}$. If the single-quark decays with supersymmetry, $t \to \tilde{t}\tilde{g}$ and $\tilde{t}\tilde{\gamma}$, are kinematically excluded, t-channel exchange with $\tilde{t}\tilde{t}$ and $\tilde{g}\tilde{g}$ final states may still appear with a weak but detectable rate. In any case, toponium (heavy bottomonium) decays may turn out to be a clean source of supersymmetric particles, which can be detected via the usual signatures of high missing energy. By measuring on- and off-resonance, most backgrounds are easily subtracted.

A FINAL COMMENT

Rumours are often generated by very weird processes and should not be taken to have *serious* scientific value; nonetheless, the scientific world listens to them with relish and even nourishes them. While writing this text, in January 1989, the author was exposed to many

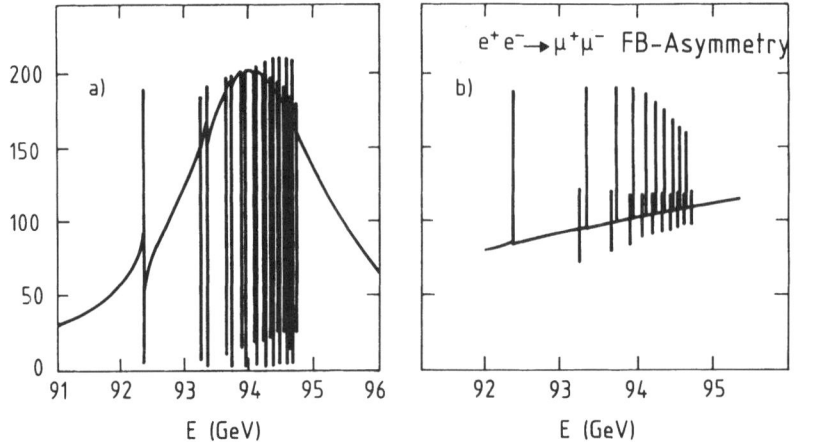

Fig. 9 The interference in the $\mu^+\mu^-$ final state, from toponium and Z^0 decays. a) The cross-section; b) the forward–backward charge asymmetry. The top mass is 1/2 m_{Z^0}. The pattern is strongly reduced and almost invisible if it is folded with the c.m.s. energy resolution of the LEP machine. Taken from ref. [20].

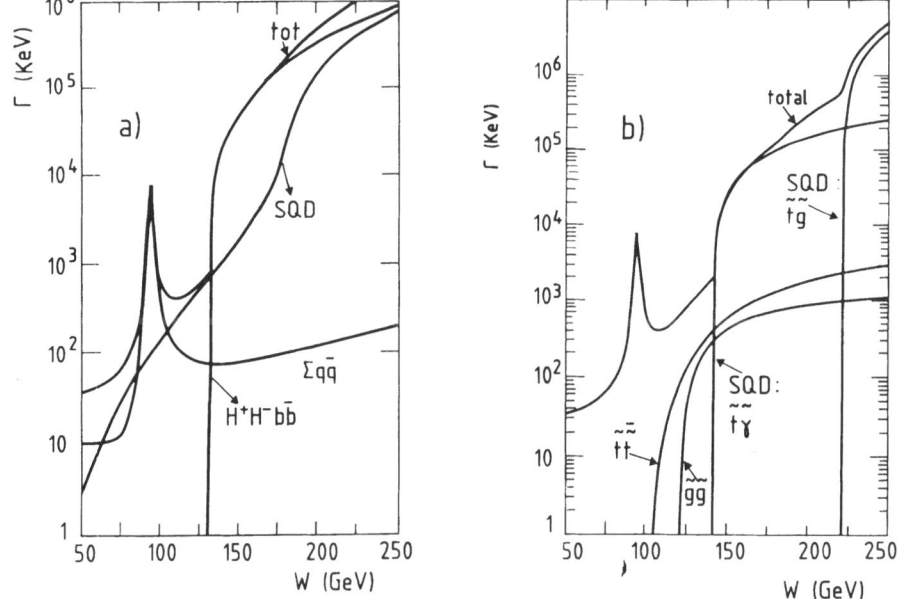

Fig. 10 Partial and total decay widths of the toponium ground state, for simple extensions of the standard model. In part a) the decay $t \rightarrow H^+b$, in part b) various decays into supersymmetric particles are allowed. Taken from ref. [19].

haunting whispers about the imminent discovery of the top at hadron colliders. Today 'rumour has it' that the top quark might have been spotted at 65 GeV. Great!! This would mean a beautiful future for LEP: the rich field of Z^0 and W^+W^- physics would be further enriched by the no less exciting realm of precision measurements of 'open top' and of toponium! Of course, the good news may vanish as easily as it came, in which case one will have to continue living without the top quark for a while. The author can only hope that the present rumour soon gets a solid material support, and wishes the LEP machine and the associated experiments a long and happy existence!

References

1. P. Darriulat, Proc. EPS Conf. on High Energy Physics, Uppsala, Sweden, 1987 (Univ. Uppsala, 1987), Vol. 2, p. 738.

2. G.L. Kane, ibid., p. 761.

3. J. Ellis and J.S. Hagelin, Nucl. Phys. **B217** (1983) 189.

4. W.J. Marciano, Proc. Int. Conf. on High Energy Physics, Berkeley, 1986 (World Scientific Publ., Singapore, 1987), Vol. 2, p. 815.

5. P.J. O'Donnell, Phys. Lett. **B175** (1986) 369.

6. N.G. Desphande et al., Phys. Rev. Lett. **59** (1987) 183.

7. A. Bean et al., Phys. Rev. **D35** (1987) 3533.

8. B. Naroska, Phys. Rep. **148** (1987) 67.

9. S. Geer, same Proc. as Ref. [1], Vol. 1, p. 219.

10. G. Altarelli et al., Nucl. Phys. **B308** (1988) 724.

11. H. Albrecht et al., Phys. Lett. **192B** (1987) 245.

12. J. Spengler, Nucl. Phys. B (Proc. Suppl.) **1B** (1988) 145.

13. K. Kleinknecht, Proc. 24th Int. Conf. on High Energy Physics, Munich, 1988 (Springer Verlag, Berlin, 1989), p. 98.

14. U. Amaldi et al., Phys. Rev. **D36** (1987) 1385.

15. H. Burkhardt et al., Phys. Lett. **206B** (1988) 169.

16. A.J. Buras, Munich preprint MPI-PAE/PTh 70/88 (1988), invited talk given at the 19th European Symposium on Proton–Antiproton Interactions and Fundamental Symmetries, Mainz, 1988.

17. Proc. Workshop on Physics at Future Accelerators, La Thuile and Geneva, 1987 (CERN 87-07, Geneva, 1987). Therein: P. Igo-Kemenes, Vol. 2, p. 69.

18. Physics at LEP (CERN 86-02, Geneva, 1986). Therein: W. Buchmüller et al., Vol. 1, p. 203.

19. Proc. ECFA Workshop on LEP 200, Aachen, 1987 (ECFA 87-108, Geneva, 1987). Therein: P. Igo-Kemenes et al., Vol. 2, p. 251.

20. J.H. Kühn and P.M. Zerwas, Phys. Rep. **167** (1988) 321.

21. B. Gobbo, Ph. D. Thesis, Univ. Trieste (1987).
    B. Gobbo and L. Rolandi, Trieste preprint INFN–AE 88-11 (1988).

22. A. Simon, Diploma work, Univ. Bonn (1988).

23. S. Gusken et al., Phys. Lett. **155B** (1985) 185.

24. A. Ali, Proc. Ref. [18], Vol. 2, p. 220.

25. S. Jadach and J.H. Kühn, Phys. Lett. **191B** (1987) 313.

26. H. Albrecht et al. (ARGUS Collab.), same Proc. as Ref. [13], p. 505.

27. C. Bebek et al. (CLEO Collab.), same Proc. as Ref. [13], p. 545.

28. J.L. Richardson, Phys. Lett. **82B** (1979) 272.

29. K. Igi and S. Ono, Phys. Rev. **D33** (1986) 3349.

30. J.H. Kühn and P.M. Zerwas, Phys. Lett. **154B** (1985) 446.

31. J. Ellis and S. Rudaz, Phys. Lett. **128B** (1983) 248.

32. J.H. Kühn, Acta Physica Polonica **B16** (1985) 969.

20

PHOTONS, ELECTRONS AND MUONS IN PAST, PRESENT AND FUTURE

Ulrich Becker

European Organization for Nuclear Research (CERN)
Geneva, Switzerland and MIT, Cambridge, U.S.A.

The "art of detection" and the strategy of physics will be the focus of this presentation. Firstly, why γ, e, and μ ? Because they are fundamental; directly detectable in unique ways, hence at high energies easy to distinguish from hadrons. The physical background is small and the production rate rare. If we consider :

$$\begin{bmatrix} \text{Good physics} \\ \text{Choice} \\ \text{Strategy} \end{bmatrix} * \begin{bmatrix} \text{Good Detection} \\ \text{type} \\ \text{Resolution} \end{bmatrix} = \begin{bmatrix} \text{Potential for} \\ \\ \text{New Physics} \end{bmatrix}$$

γ, e and μ play a central role both in physics strategy and possible detector resolution. The later point being the key to "seeing new things" just as Galileo's telescope or Leeuwenhoek's microscope.

PAST

The potential of pp $\rightarrow$ e^+e^- and $\mu^+\mu^-$ was clearly recognized by Conversi, Massam, Müller and Zichichi [1]. Figure 1 shows their apparatus, a Lorentz boosted collider detector as we would say now. The recognition of electrons involved the new 1965 technology of shower spark chambers, clearly displaying the narrow electron showers with many hits opposed to pion interactions. A rejection of $5 * 10^{-4}$ was achieved, enabling the measurement of the timelike form factor of the proton at relative low q^2. We imagine with amazement the course physics would have taken if this experiment could have been carried out with much higher energies.

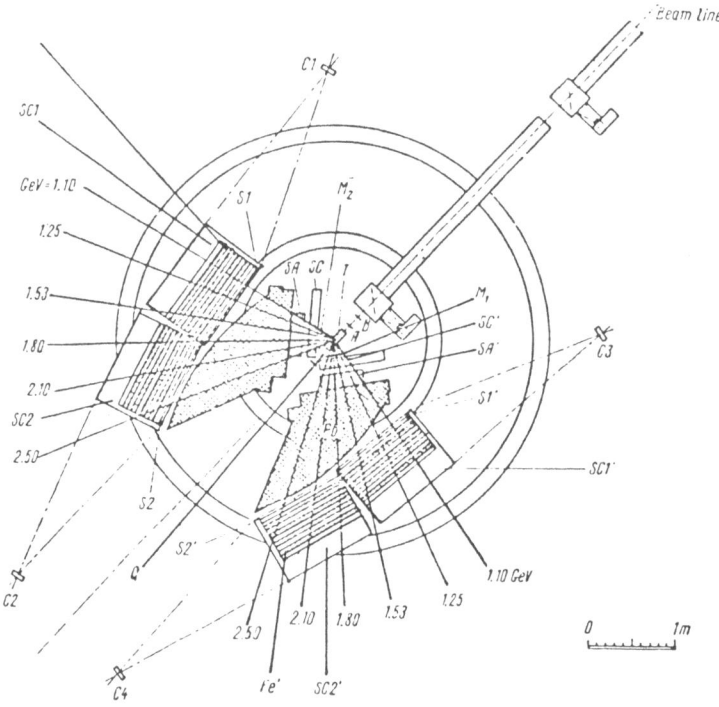

Fig. 1   One of the first detecors for systematic lepton pair studies from Ref. 1.

Starting 1965, S.C.C. Ting et al.[2] developed double arm spectrometers, to detect :

$$\gamma A \rightarrow e^+e^-A$$

and confirmed Q.E.D. However, in order to detect electron pair masses of $\approx$ 750 MeV from QED and $\rho \rightarrow e^+e^-$, the high rejection of 10^{-8} by four Cerenkov counters was needed to suppress a 10^{+5} $\pi^+\pi^-$ background. To

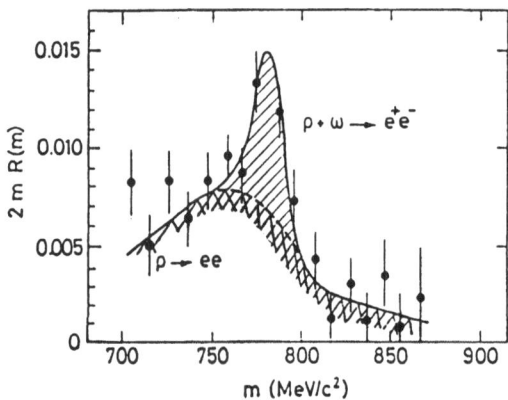

Fig. 2   The observation of interference in ρ and ω decaying into the same final state required 5 MeV mass resolution.

324

further see $\omega \to e^+e^-$ in interference with ρ required a mass resolution of 5 MeV (Figure 2).

By this decay mode and the demonstration of the coherent diffraction production, ρ, ω, and φ were shown to behave just like heavy photons. But why three only?

Higher energies (30GeV) were available at Brookhaven for the study of

$$p + Be \to e^+e^- + X(Be)$$

at the price of outnumbering the e^+e^- by 10^6 genuine hadron pairs and many accidentals. The spectrometer [3] in Figure 3 demonstrates the "art of detection" which in a different form, but with similar principles may be necessary for future experiments at hadron colliders.

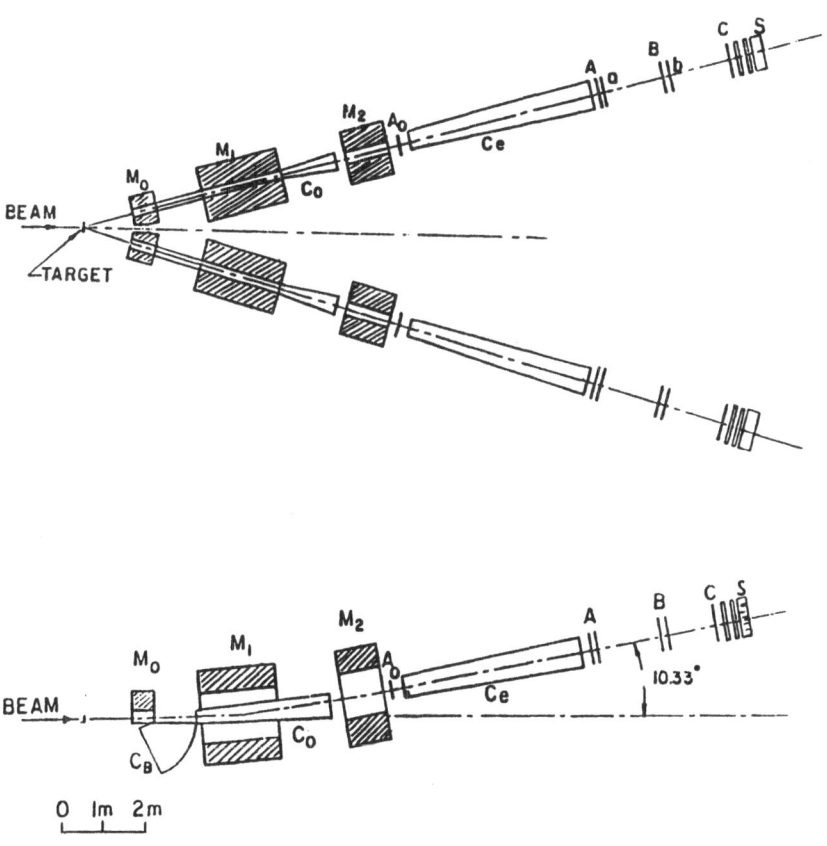

Fig. 3   Double arm spectrometer of S.C.C. Ting et al., at BNL. M=Dipole magnets, C=Cerenkov counters, A_o, A, B, C, D=Proportional chambers, and S=shower counter hodoscopes.

Four Cerenkov counters identify e^+e^- with full efficiency and 10^{-10} rejection. Hydrogen as radiator produces the least knock on electrons (10^{-3}) as false signals from passing pions. Locating the counters inside the momentum analyzing dipole magnets prevents knock on electrons to enter the next counter. However, the radiated signal is only 40 photons per good electron track. High quality spherical and parabolic mirrors transferred those with little loss on the cathode of the new RCA 31000 quantacron producing eight photoelectrons. Due to the high amplification of the first GaAs dynode these could be resolved individually (Figure 4a). As expected, the counter proved to be 99,9% efficient.

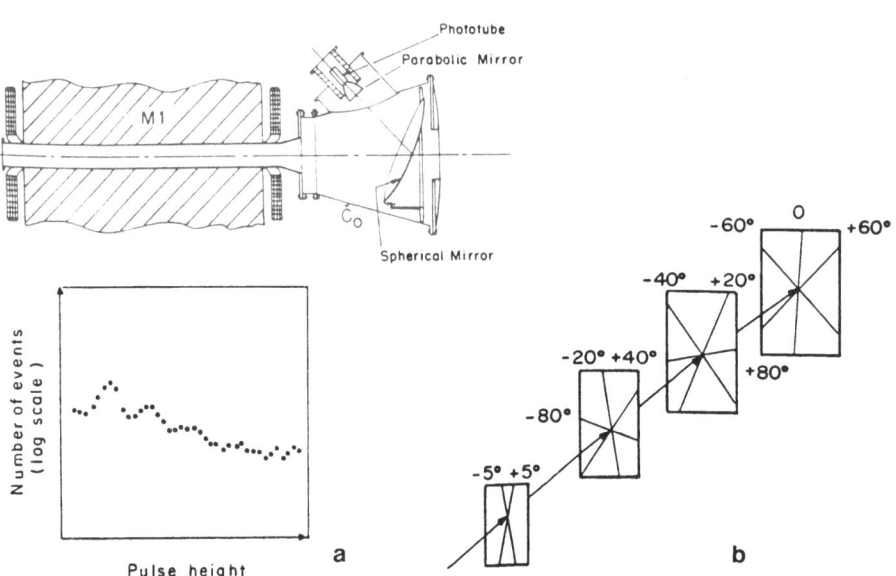

Fig. 4   a) Pulse-height spectrum from the phototube of the C_o Cerenkov counter with He as radiator. Clearly visible are three photoelectron peaks.
          b) Relative orientation of the planes of wires in the proportional chambers.

Four proportional chambers in each arm with high rate gas of Ar: Methylal developed a superb mass resolution of

$$\frac{\Delta m_{ee}}{m_{ee}} = 0.3\% \ ,$$

much to the criticism of the experiment approval committee not seeing the need from the ρ, ω, φ widths known in 1972. Having three 120°rotated wire planes, the copious neutrons could be rejected on the spot by requiring the wires to add up to a constant in each chamber (Figure 4b).

Figure 5 gives the mass spectrum of the measured electron pairs [3]. Being a total surprise in 1974 the high redundancy of the experiment (shower counter to overidentify electrons, timing cuts, change of pair geometry by change of magnet current) could prove the discovery of the J particle, which opened a new field of physics (S.C.C.Ting, Nobel Prize 1976).

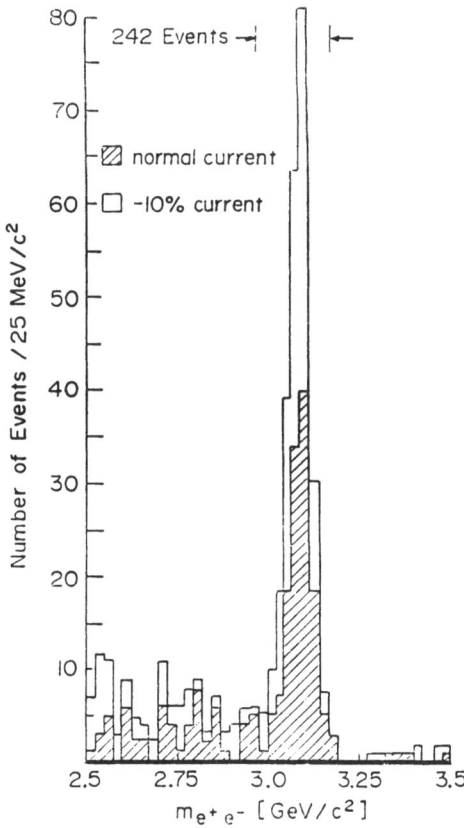

Fig. 5   Discovery of J particle [3]. The unusually good mass resolution led to the clear observation.

To determine the structure of Charmonium a precision tool like the crystal ball with good photon resolution of
$$\Delta E/E = 2.6\%/\sqrt[4]{E}$$
was necessary.

An example for the importance of lepton detection is given by the experiment shown in Figure 6a. The 1967 proposal [4] was set up to look for new sequential leptons by the μ-e signal

- exactly in the way they were found [5] in 1975 (Figure 6b) when higher energies became available. The search for acoplanar muons, μ-e or μ − h combinations at still higher energies at PETRA did not yield evidence [6] for the leptons of a fourth family, see Figure 6c.

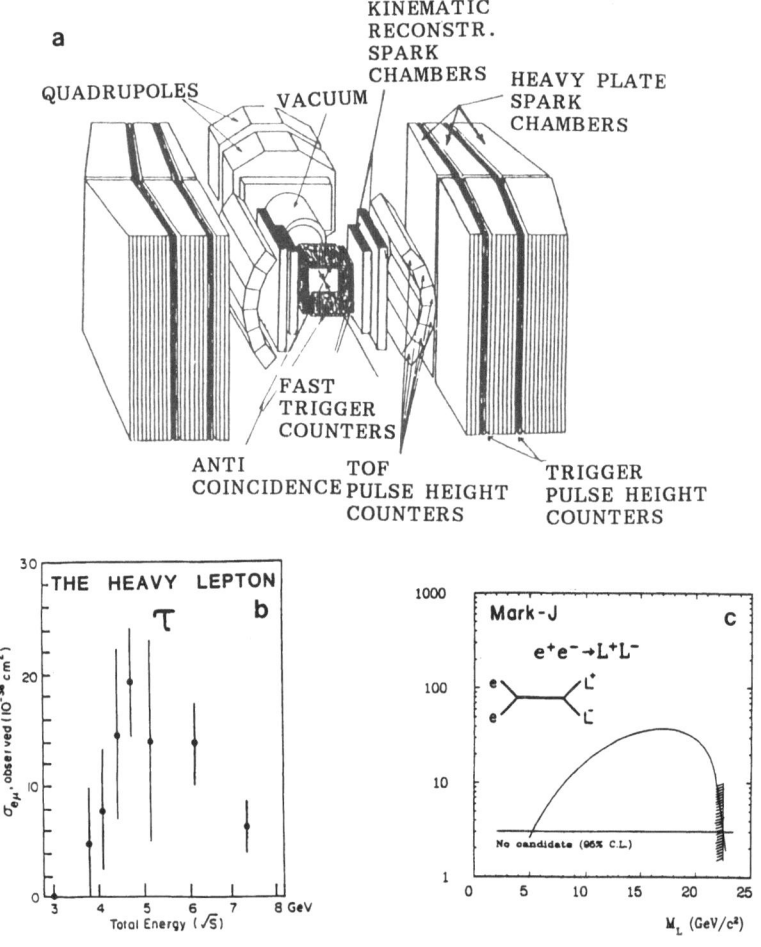

Fig. 6   a) Detector for proposed search of e-μ events by Zichichi [4].
            b) Discovery of τ in e-μ events by Perl [5].
            c) No more heavy leptons with m < 23 GeV were seen by Ref. [6].

Turning back to the role of resolution in the past, Figure 7a shows the discovery of the γ [7] in

$$p\ N \rightarrow \mu^+\mu^-\ X$$

which was enabled by good mass resolution of 1.5%. Figure 7b gives the signal of γ → μ⁺μ⁻ detected with 11% mass resolution by an experiment [8] at the ISR using iron toroids. The message for the future is clear:

Resolution is absolutely necessary.

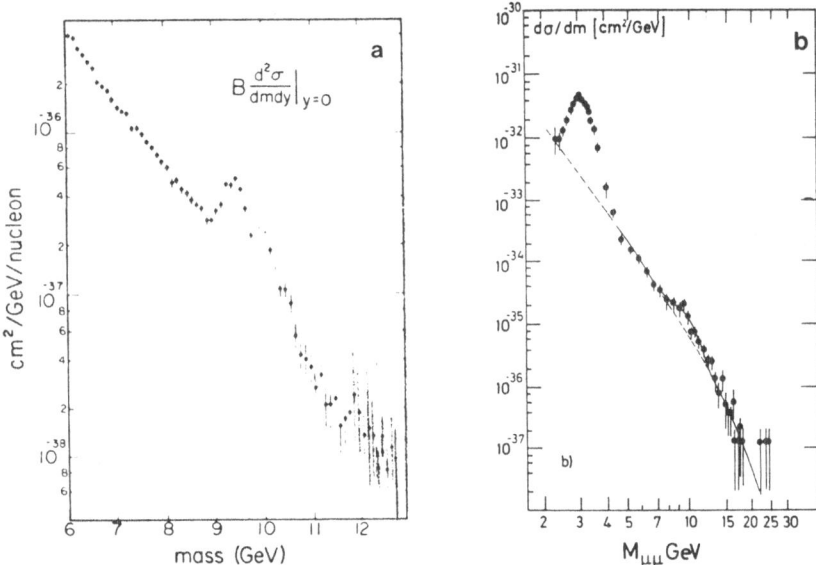

Fig. 7   a) Discovery [7] of $T \rightarrow \mu\mu$ with 1,5% resolution
              b) Iron toroid spectrometer with 11% resolution

PRESENT

Some of the experimental questions along known guidelines presently
are: How many types of leptons exist?

          How large are they?

          Can they be divided in smaller parts?

The same questions are to be repeated for quarks. Even more important, however, is the quest for the totally unknown areas, in particular:

          What is the origin of mass?

          What is the quantization rule generating the masses?

          Is the Higgs mechanism existent or the "ether" of the late 20th
          century?

Attacking this merely from the experimental point of view we note
that the physics of e, μ, γ had been most successful in making discoveries [3-9]:

          1) $pN \rightarrow J \rightarrow ee + X$

          2) $e^+e^- \rightarrow cc \rightarrow \gamma + X$

          3) $pN \rightarrow T \rightarrow \mu\mu + X$

          4) $e^+e^- \rightarrow (\tau \rightarrow \mu\nu\bar{\nu}) + (\tau \rightarrow e\nu\bar{\nu}) \rightarrow e\mu + X$

          5) $pp \rightarrow W \rightarrow \mu + X$

               $\rightarrow Z \rightarrow ee + X$

Although the hadronic branching ratio of all new particles was much larger than the leptonic one, the fact is that all discoveries were made by observing leptonic decays with good resolution/rejection. The detector I wish to describe now emerged from these considerations.

L3 DETECTOR [10]

Meant for LEP, it is a large magnetic hall of a 5 kgauss field with 5.9 m radial and 14 m longitudinal dimensions. Led by Prof. S.C.C. Ting, the detector is under construction by the collaboration of 33 institutes. The design is solely oriented to achieve

1% Resolution

detecting multi-GeV γ, μ, and e.

In addition, hadron jet energies are measured over an almost hermetically closed surface to detect neutrino induced missing p_T for example. Figure 8 shows the L3 detector surrounding the interaction region of LEP providing 80-160 GeV e$^+$e$^-$ annihilations. Emerging charged particles are tracked in the Time Expansion Vertex Chamber (TEC) and neutrals leave energy deposits in the BGO electromagnetic shower counter and the uranium calorimeter while muons penetrate all to reach the three layers of muon chambers.

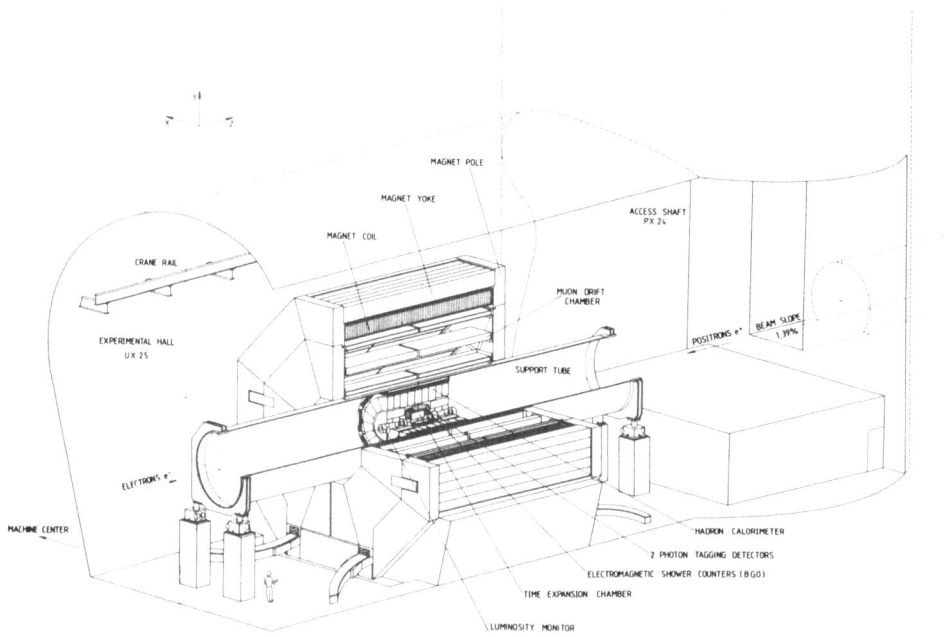

Fig. 8　View of the L3 detector in the I2 region of LEP.

Figure 9 gives a very schematic view of different events. Electrons are seen in the TEC and deposit all energy in the BGO. Muons penetrate all detectors, leaving MIP (minimal ionizing) signals in the e.m. and hadron calorimeter. Photons are absorbed in the BGO having no TEC track. Opposed to these clear signals, hadrons show a complexity of tracks with energy in the hadron calorimeter and BGO.

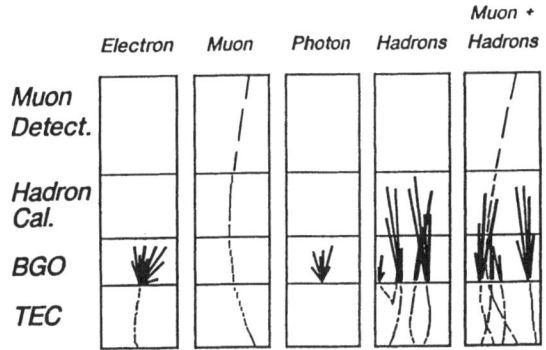

Fig. 9   Appearance of different particles in the L3 detector. Schematic only.

Magnet

With 30kA in 168 turns of 12 m inner diameter the magnet generates 5kgauss over 12 m length. The stored power is 160 MJ, the time constant of change 11 sec. Using 1000 tons of Aluminum conductor the ohmic losses are only 4 MW. Figure 10 shows the coil assembled - dwarfing a person. The impressive engineering, e.g. the 6 turn packages of the coil all got located within 2 mm accuracy, which is to be compared to a 4 mm lift of the center by thermal expansion, when the coil reaches the operating temperature of 55°. The magnetic field is very homogeneous (< 5%) in the TEC region, and varies only 5% over the muon detector region. The field is measured by 960 probes to <18 gauss accuracy and checked by 12 nuclear resonance probes to 1 gauss.

Vertex Chamber

Firstly, electrons are distinguished from photons by leaving a track in the TEC. Also, to measure lifetimes up to 10^{-13} sec a resolution of 30 to 40 μm is needed which is also necessary to distinguish the charge of 50 GeV tracks to 3 σ, enabling measurements of forward-backward asymmetries.

Fig. 10   Assembly of 1000 to coil for the L3 Magnet.
CERN Photo Jan. 88.

The chamber works in the time expansion mode [11]. That means that ionization electrons drift at

$$6 \ \mu m/ns,$$

which is about 1/8 of the normal speed and allows the Flash ADC electronics to record the structure of the incoming ionization. The centroid of this is evaluated later and expected to be accurate to

$$30 \ \mu m/\sqrt{cm}$$

for a "cool" gas like the CO_2:i·Butane = 80:20 used at 2 atm. Figure 11a shows a cosmic shower event in a test chamber. Figure 11b gives the resolution achieved under "battle" conditions with a TEC in the Mark J Detector at PETRA. The L3 chamber depicted in Figure 12 consists of three chambers. The inner surrounds in 12 segments of 8 sense wires of 20 μm each the 1.5 mm thick Beryllium beam pipe. Two of eight wires determine the Z coordinate to 1% by charge division. The outer chamber has 54 TEC wires in 24 sectors, 8 wires of which have charge division. Finally the outmost layer of 24 mm radial thickness constitutes the Z chamber. 1152 wires in 1 atm of $Ar:CO_2= 80:20$ are parallel to the beam pipe in two layers and generate the signals for the cathode pads of -68°, +68°, 0°, 90° orientation with 4.4 mm pitch. They determine the coordinate along the beam, Z, to less than 0.5 mm.

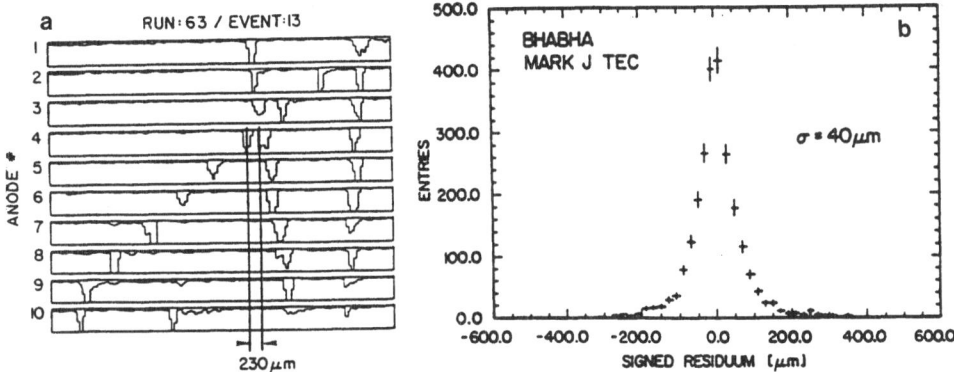

Fig. 11   a) Charge on 8 wires as recorded by 100 MHz Flash ADC's. Three tracks of a shower are visible.
b) Reconstruction of 23 GeV electrons from Bhabha events at PETRA.

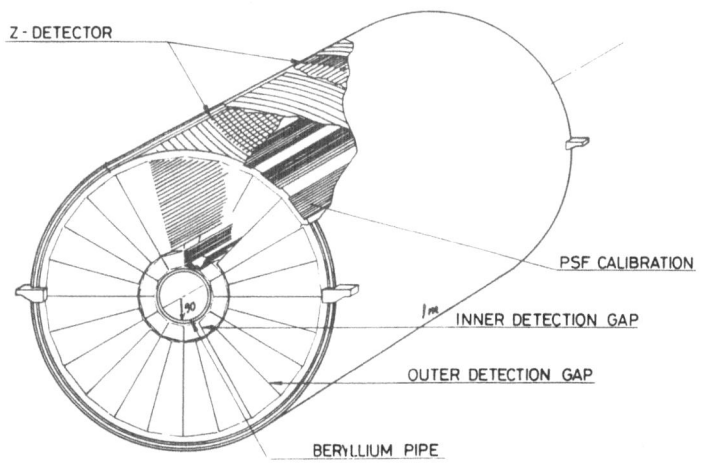

Fig. 12   TEC vertex chamber of L3 having 24(12) segments to measure momenta. Cathode strip chambers measure the coordinate along the beam.

The TEC tracks are recorded by 2000 Flash ADC's with 100 Mc and processed by 1000 microprocessors TMS99/05. Calibration in situ will be provided by 200 rectangular scintillation fibers of 0.5 mm thickness covering each segment.

Figure 13 gives a physics example. Measuring the $\mu^+\mu^-$ asymmetry, one can distinguish a single Z from the case of having two. Both, a precision measurement close to m_z or, better: a measurement at 160 GeV would clearly signal the advent of a second Z assumed with a mass of 214 GeV.

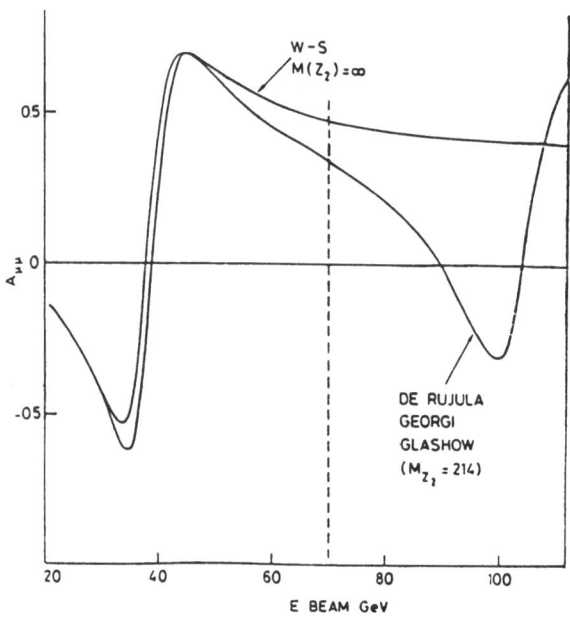

Fig. 13 Predicted forward-backward asymmetry of muon pairs for Z_0 region with and without an additional Z (214 GeV).

Bismuth Germanate (BGO) Shower Counter [12]

Figure 14 shows the most unique component of L3 : 12000 crystals of BGO (11.6 tons) surround the TEC and measure e and γ to

$$\Delta E/E < 1\% \qquad \text{for } E > 2 \text{ GeV}$$

$$\Delta(x \text{ or } y) < 3\text{mm for } E > 1 \text{ GeV}$$

Since the detector operates in a 5k gauss field, the light of the showers is read out by 24000 Si photodiodes of 3.6m² total area. Bismuth Germanate, $Bi_4Ge_3O_{12}$, must be grown in the correct one of many phases with little impurities. The **Shanghai Institute of Ceramics** has achieved to grow 24 cm flawless and therefore radiation hard crystals. Table 1 gives the features.

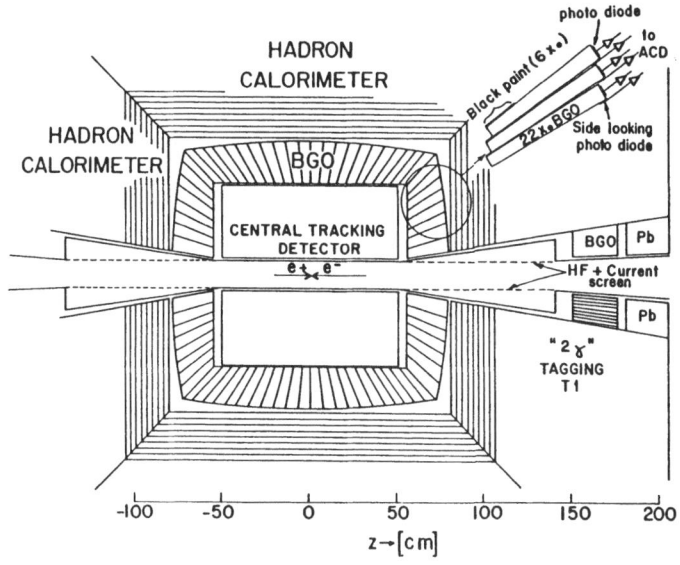

Fig. 14 Electromagnetic shower counter of 12000 BGO crystals to measure
the energy 1.4% /$\sqrt{E}$ and the position of electrons and photons.

Table 1

Properties of BGO

Molecular formula	: $Bi_4Ge_3O_{12}$
Density	: 7.13 g/cm^3
Hardness	: 5 Mho
Melting point	: 1050° C
Non-hydroscopic	
Interaction length	: <u>22 cm</u>
Radiation length	: <u>1.12 cm</u>
Moliere radius	: 2.7 cm
dE/dx (MIP)	: 9.2 MeV/cm
Refractive index at 486 nm	: 2.15
Temperature gradient of light output	: -1.5%/° (at 20° C)
Light output decay constant	: 300 ns

Besides high density of 7.13g/cm³, BGO has a short radiation length
and is not hygroscopic. Compared to NaI these features allow building a
more compact detector with better resolution. For NaI gaps in detection
are caused by containers which are needed to keep water out.

The achievement of the Shanghai Institute of Ceramics to produce
($22X_o$) long crystals is absolutely essential: A 50 GeV shower has only
0.6% rear leakage to be compared with 10% in the Crystal Ball having
$16X_0$.

335

Figure 15a shows a BGO crystal in the thin Carbon fiber container. Reflective paint enhances the light output to 1500 electrons/MeV on the two photodiodes and at the same time increases the homogeneity along the crystal. Two light fibers from different Xenon flash tubes monitor the calibration and cross-check at the same time. A real crystal is shown in Figure 15b with a readout microprocessor. Since the light output diminishes 1.5% for 1°C increase the whole array is immersed in freon liquid of a temperature stabilized system which is monitored by many sensors.

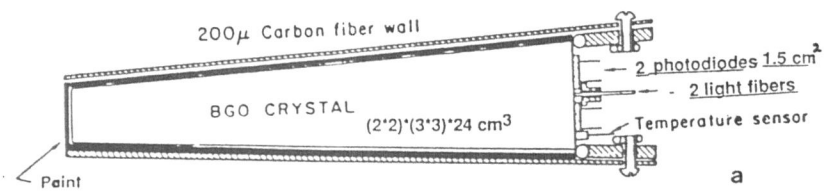

Fig. 15 a) Cross-section of BGO crystal, mounted in the C-fiber structure.
b) Crystal with one diode; at left, micro processor 6305 Hitachi.

The crystals are cut accurately to 200 μm as to enable a tight assembly. Checking the spectral transmission to be above 55, 60, 65% at 400, 480, 630 nm selects good crystals from light output and radiation hardness. The scintillation light from e.m. showers is registered by 2 silicon diodes of 3 cm². The amplified signal is integrated and stored in two channels, < 2 GeV and < 200 GeV. For accurate dynamic range a 20 bit ADC under control of a 6305 microprocessor (one per crystal) digitizes in 300μs and stores up to 40 events. 200 processor units 68010 in a VME based system constitute the energy cluster trigger and enable the read out.

Figure 16 shows the Carbon fiber 'structure partially filled with crystals and preamplifiers. 140 kg of structure carry ~ 10 tons of crystals with deflections of < 1mm! In the back the fully assembled first half barrel is visible.

Fig. 16   Left: finished first half barrel with electronics and temperature stabilization.
Right: second half barrel with 40% of crystals inserted.

Test results are given in Figure 17. A resolution of~5% is achieved at 100 MeV and $\leqslant$ 1% for energies > 2 GeV. These results have been confirmed by measurements with the first half barrel where all crystals were calibrated with electrons of 2, 10 and 50 GeV.

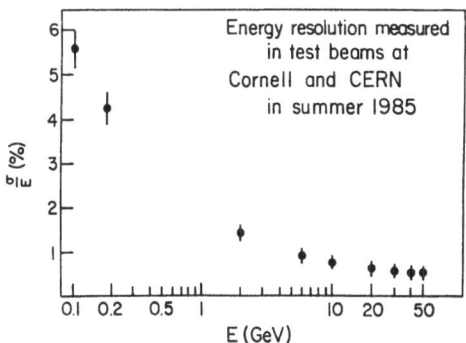

Fig. 17   Measured Energy resolution of BGO crystals with Silicon Diodes.

Opposed to sodium iodide BGO is self curing after radiation damage, i.e. after krad exposures it will restore the light output in a few days. The expected levels for the main barrel are < 1 rad/day and should not cause noticeable effects. Nevertheless, the transmission is continuously checked by two independent Xenon light pulsers which in turn are compared to reference ^{241}Am sources. Cosmic rays leave 80 - 200 MeV (pending on geometry) while traversing the crystals. In two days this provides a 1% calibration considering the underground position at LEP. Using a RFQ (radiofrequency quadrupole) accelerator to bombard an Li target with 1.5 MeV Hydrogen atoms causes a nuclear reaction liberating 17.6 MeV photons which can calibrate the detector in situ. To name a physics application, obviously with such good resolution and calibration it will be a pleasure to register the γγ transitions of toponium!

Hadron Calorimeter [13]

The central part consists of 9 rings with 144 modules and the forward part is made of 12 half rings. Half a million cells of proportional chambers are sandwiched between 7824 Uranium plates of 5 mm (see Figure 18). Finer "electronic towers" which are matched to the BGO crystals are formed by connecting subgroups of tubes within the mechanic modules, thus providing a resolution of

$$\Delta\Theta = \Delta\varphi = 2° \text{ in angle and}$$
$$\Delta E = 0.5\sqrt{E} \text{ in energy.}$$

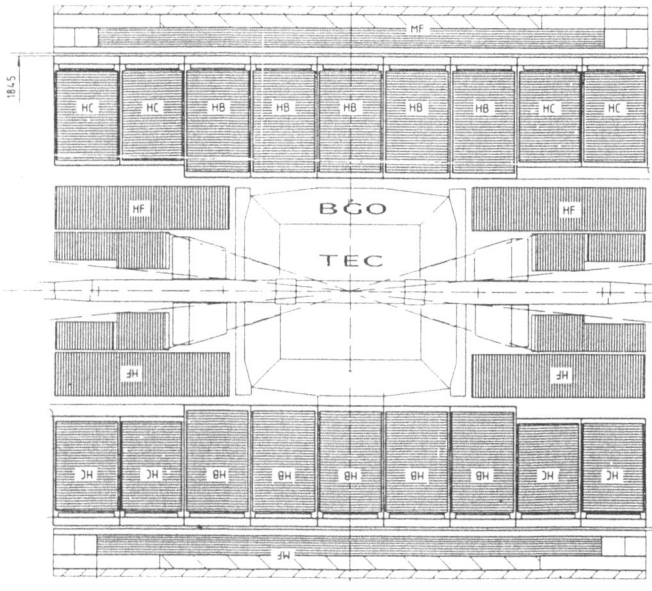

Fig. 18   Side view of the hadron calorimeter.

Figure 19 shows one central barrel module. The penetration thickness is 4λabs, and a muon will loose 1.6 GeV by ionization. The natural radio-activity of Uranium is suppressed to 10% by the 1 mm brass walls of the proportional tubes; ageing posing no problem. The proportional tubes with Argon/CO_2 = 80/20 show no change in amplification after a dose equivalent to several hundred "Uranium" years. In fact, the spectrum of pulse heights from the Uranium is used for intercalibration of all modules and serves as monitor for the performance. Test beam results confirmed the resolution to be (see Figure 20)

$$\Delta E/E \simeq 59\%/\sqrt{E} \text{ for } \pi\text{'s of } 4 - 50\text{GeV}$$
$$\Delta E/E \simeq 33\%/\sqrt{E} \text{ for } e\text{'s of } 2 - 22\text{GeV.}$$

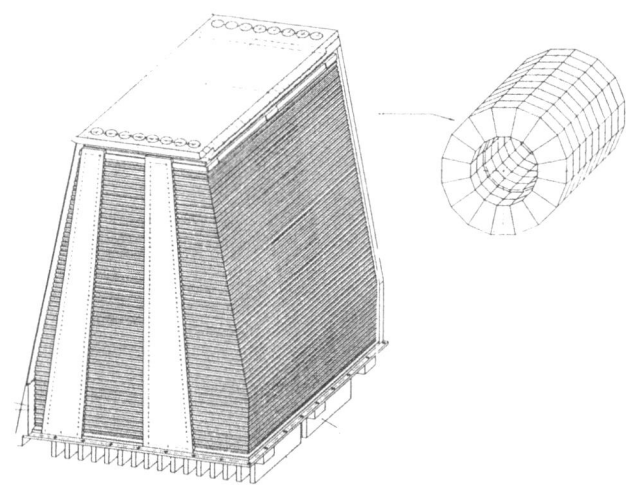

Fig. 19 Hadron calorimeter module. 58 Uranium plates alternate with 30 φ and 30 Z measuring proportional tube layers of 8 mm thickness.

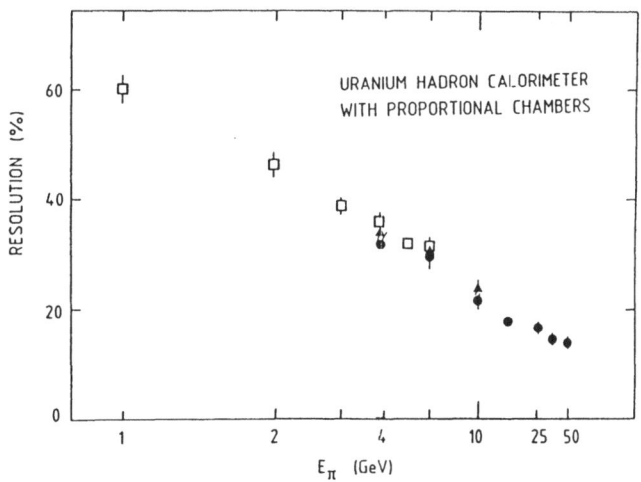

Fig. 20 Resolution for pions.

339

Figure 21 presents the finished barrel calorimeter in the gigantic support tube of the L3 experiment. The spatial resolution for the end-cap calorimeter is about 1,7 cm for the rφ coordinate. As expected, the showers from 10 GeV electrons show a more narrow distribution than the hadronic cascades of 10 GeV pions.

The good spatial energy cluster resolution enables studies like the one shown in Figure 22 for 4 jet reconstruction for the hypothetical reaction :   ee → H⁺ H⁻ → 4 jets,
to find new particles H(20 GeV) decaying into 2 quarks. Of course, an analogous scenario applies for ee → W⁺W⁻.

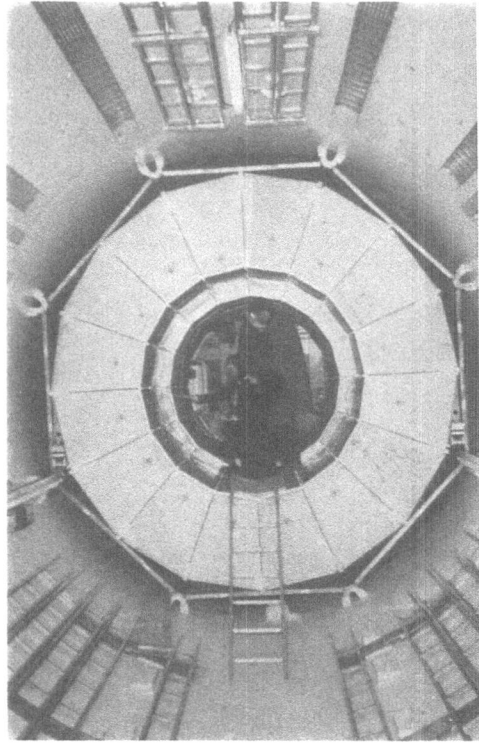

Fig. 21. Hadron Calorimeter in pos-   Fig. 22. Reconstruction of hypothet-
iion in the support tube.   ical particle H with 20 GeV
                                             mass from 4 jet topology.

Precision Muon Detector

Conceptually suited for the much higher Eloisatron or SSC energies this detector has no parallel in present and planned experiments. Using precision drift chambers with minimal multiple scattering, it aims at a dimuon mass resolution of :

$$\frac{\Delta m}{m} = 1.4\% \quad \text{at } m = 100 \text{ GeV}$$

340

Figure 23 gives the schematic end view showing the three chamber layers for muon detection. Although the magnetic analysis power B.l² is very large in the L3 experiment, the sagitta of a 45 GeV muon is only 3.7 mm. Scaling this, we note that practically all muons (> 3GeV) will remain in **one** octant module. To achieve 1.4% dimuon mass resolution, the sagitta must be measured to < 2% by the chambers. Systematic alignment errors should be < 30 μm and the UV laser calibration must be commensurate.

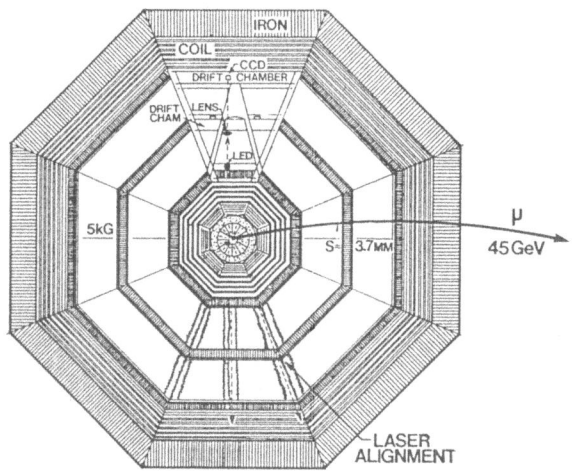

Fig. 23   End view of L3 detector with 45 GeV muon track. Curvature exaggerated.

Figure 24 shows the sixteen precision modules ("octants") which by optomechanical monitors and feedback controlled servo mechanisms are aligned to < 30μm. The detector is huge, with the following overall parameters :

Table 2

Volume	1000 m³
Weight	160 tons
Chambers	80
Area	900 m²
Wires, all	250000
Signals, TDC's	24000
Argon, Ethane 62/38	250 m³
ΔΩ/4π, precision	64%
ΔΩ/4π, all	76%

Figure 25 explains the principle of accuracy. Multi-sampling drift chambers [14] measure the magnetic deflection over 2.9 m in 5 kgauss. Having $N_{1,2,3}$ = 16, 24, 16 wires, the average coordinate is determined to $\varepsilon_i = \sigma /\sqrt{N_i}$, where σ = 200 μm is the single wire accuracy.

The sagitta is then measured to $\Delta S = \sqrt{\varepsilon_1^2/_2 + \varepsilon_2^2} = 50$ μm by the chambers. Adding in quadrature 40 μm for multiple scattering and < 30 μm for alignment we have :

$$\frac{\Delta p}{p} = \frac{\Delta s}{s} = 1.9\% \text{ for a 45 GeV ray.}$$

The dimuon mass resolution is then $\frac{\Delta m}{m} = \frac{1}{\sqrt{2}} \frac{\Delta p}{p} = 1.4\%$.

Scattering in the hadron calorimeter does not change this result for dimuon opening angles $\Theta > 100°$ For $\Theta < 100°$ one uses the angle measured by the vertex chamber.

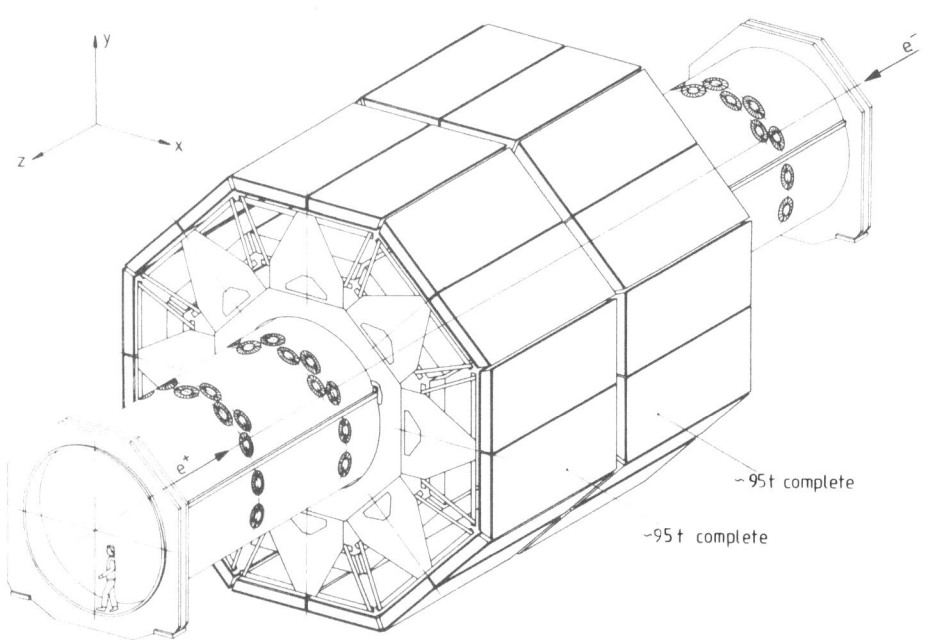

Fig. 24   Sixteen units containing 80 muon detecting drift chambers forming 3 concentric surfaces surrounding the interaction point. Assembled by using rotation on air bearings (front). The 80 to wheels are railed into the magnet.

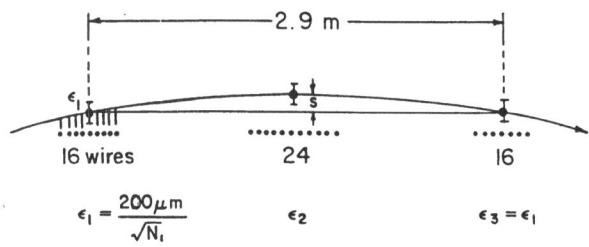

Fig. 25   The necessary accuracy is achieved by averaging 16, 24, 16 bias-free wire measurements.

Figure 26 shows how the three chamber layers are aligned. The light of a diode is projected by a lens onto a quadrant diode receiver. The three elements were adjusted on an optical bench, so that all quadrants receive equal light. Moving the middle chambers by an amount δ shifts the image on the quadrant diode by 2 δ. The imbalance of light on the four cells measures the x and y displacement. This set up will be referred to as a "Straightness monitor". Servo motors put the middle chambers back in a few seconds [15].

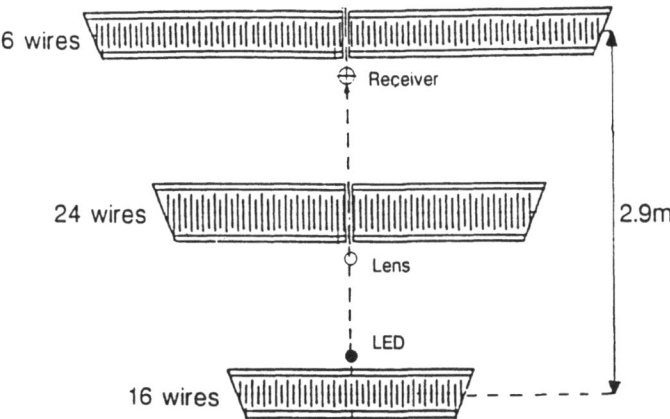

Fig. 26 The 24 wire middle chambers can be moved by a servo motor until LED, lens, and receiver form an ideal line.

The accuracy of wire location within the chamber is achieved by pulling the wires over the corners of optically flat (2 μm) Pyrex glass pieces which are glued on Carbon fiber rods, see Figure 27. This way the 16(24) wires of a cell are planar to 2 μm, and the cells are located to an accuracy of 4.9 μm by the gluing those structures on a precision template. They have almost no thermal expansion and are the origin of locations for computer reconstruction. To reduce the sag of the 5 m long wires from 380 μm to 95 μm, they are supported in the middle. The middle structure is attached to actuators which remotely from the chamber end can be moved as to center the three "straightness" monitors. The residual sag is known to < 3% by the measured tension of the wires and corrected for. Figure 28 shows the monitored deviations of the middle support over a week, which are mostly caused by temperature gradients. They are small and are taken into account.

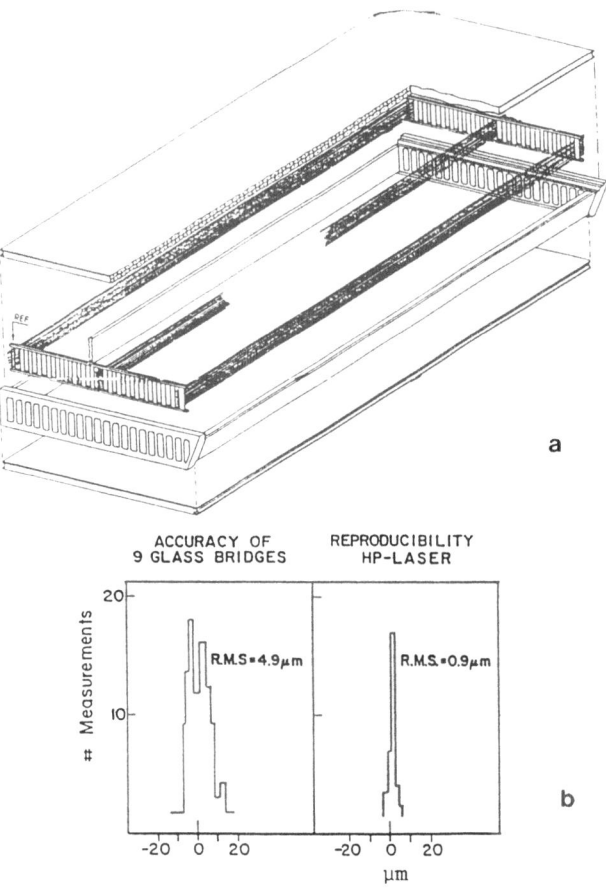

ACCURACY OF
9 GLASS BRIDGES

REPRODUCIBILITY
HP-LASER

R.M.S = 4.9 μm

R.M.S = 0.9 μm

Measurements

Fig. 27 a) Exploded view of a chamber with "Bridge" structures supporting the wires.
b) Measured accuracy of the equidistant spacing of glass in
these bridges.

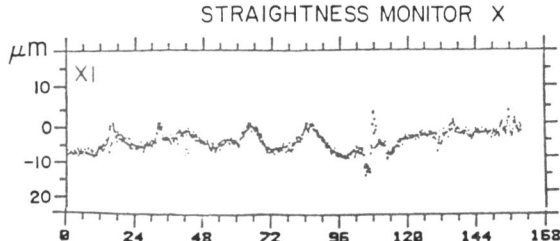

STRAIGHTNESS MONITOR X

Fig. 28 Deviations of the wire supports from absolute straightness
measured by a system similar to the one in Fig 26.

Electronically $\leq$ 30 μm at the 50 mm/μs drift speed mean 0.6 ns. The
TDC electronics was checked and found stable to 0.2 ns. Figure 29 gives a
summary of chamber parameters and operating values. The calculated drift
path lengths are given for one chamber cell. Note that by using only
electrostatically "good" wires the arrival times are equal to 0.1 ns.

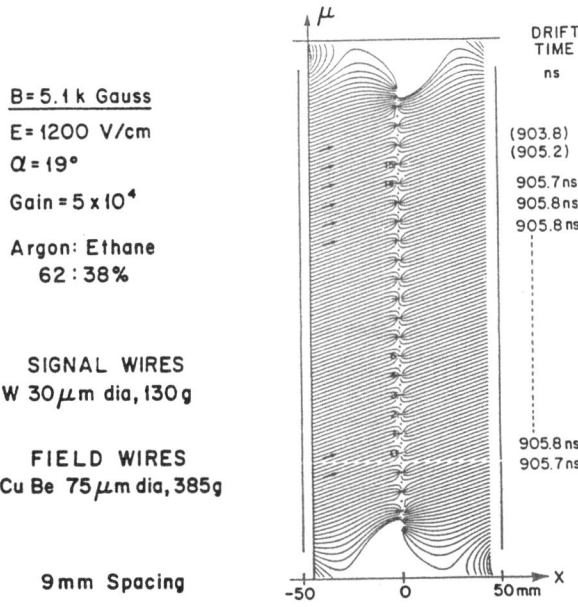

B = 5.1 k Gauss

E = 1200 V/cm

α = 19°

Gain = 5 x 10^4

Argon : Ethane
62 : 38%

SIGNAL WIRES
W 30 μm dia, 130 g

FIELD WIRES
Cu Be 75 μm dia, 385 g

9 mm Spacing

Fig. 29 Chamber operation parameters and
calculated drift times for muon 45 mm
from the wires of one chamber wall.

For optimization of the single wire resolution of the chamber the
signal pulse threshold should be reached after 4 - 11 incoming electrons
from the ionization of the track. Requiring higher numbers means waiting
for electrons coming in on curved electrostatic trajectories with detours
large compared to the ~ 200 μm accuracy required. Figure 30 gives the
residuals for cosmic muon tracks for 20 mV (or nine incoming electrons)
threshold. It was obtained by fitting a line to 14 of the 16 wire measurements in a chamber. The 136 μm accuracy exceeds design value by almost
a factor 2.

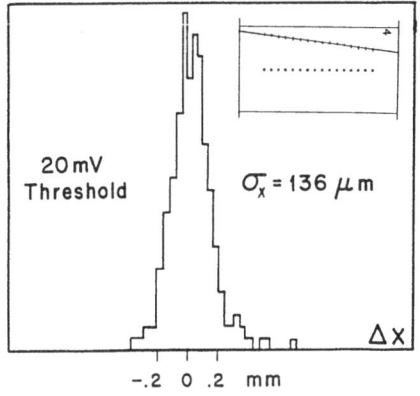

Fig. 30 Distribution of residuals from a
muon track, see insert.

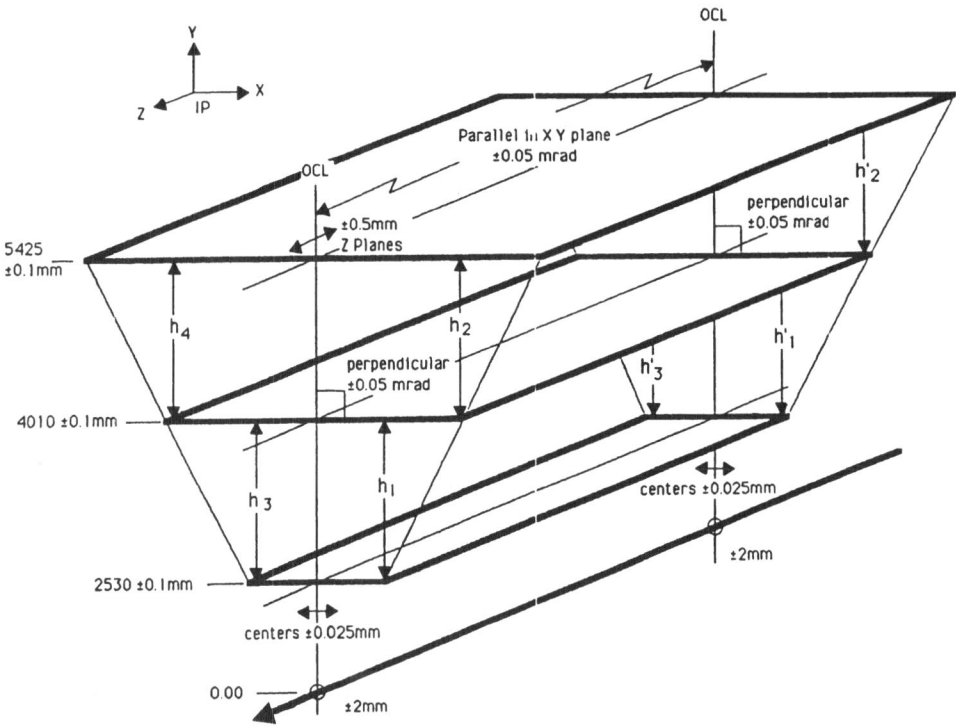

Fig. 31   Alignment accuracies within one module for systematic error
less than 30 μm.

Two outer, two middle and one inner chamber form an octant module
(see Figure 26). The positioning requirements, due to < 30 μm error in
the muon sagitta are given in Figure 31. In particular, we note the
required accuracy of <25 μm in aligning the three chamber layers and
< 50 microradian torque of the module. To minimize the propagation of
errors, the straightness monitor for the three chamber layers refers
directly to the wires by precision pins (see Figure 32a). This is done by
measuring the contact by ohm meters and can be done accurately to a few
microns. The conducting tips of the pins are later disconnected.
Figure 32b shows temperature induced octant deformations over a day,
monitored by the straightness monitor. Planarity is achieved by adjusting
all six reference points to the plane of a laser beacon. The 90° deflec-
ted beam of a laser is whirled around by a rotating mirror to form a
plane. Sensors mounted or the 2 x 3 elements of the straightness monitors
enable us to manually eliminate the torque to < 50 μ rad.

346

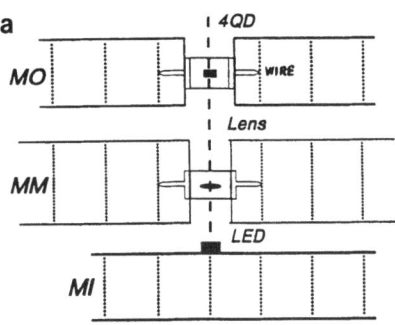

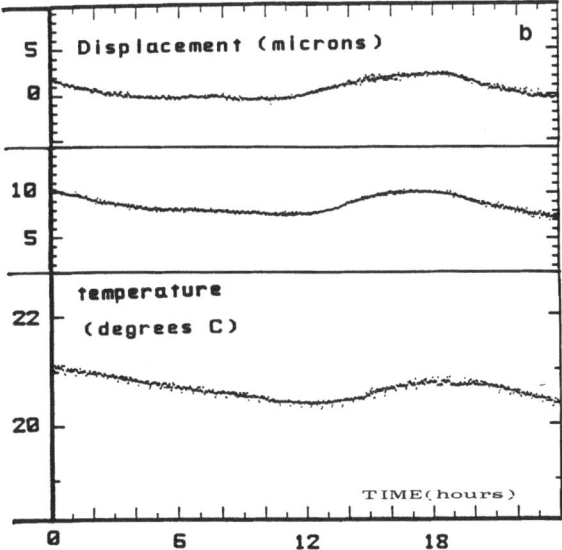

Fig. 32 a) Straightness monitor referencing directly to the wires of the three chamber layers(schematically).
b) Temperature induced deviations registered over one day for both octant ends.

Being optomechanically aligned in this way [16] the modules are checked by 8 UV laser rays traversing the octant, see Figure 33. All measurements should result in "0" sagitta. Figure 34 shows results from many runs of 200 laser shots each. We note that :

a) Octant 3 was aligned to $\approx$ 26 μm within specification;

b) the reproducibility of the laser is sufficient.

Independently we check the adjustment by cosmic rays. Again, having no magnetic field the resulting sagitta should vanish. Figure 35a gives the computer reconstructed track with a magnification of the middle chamber.

347

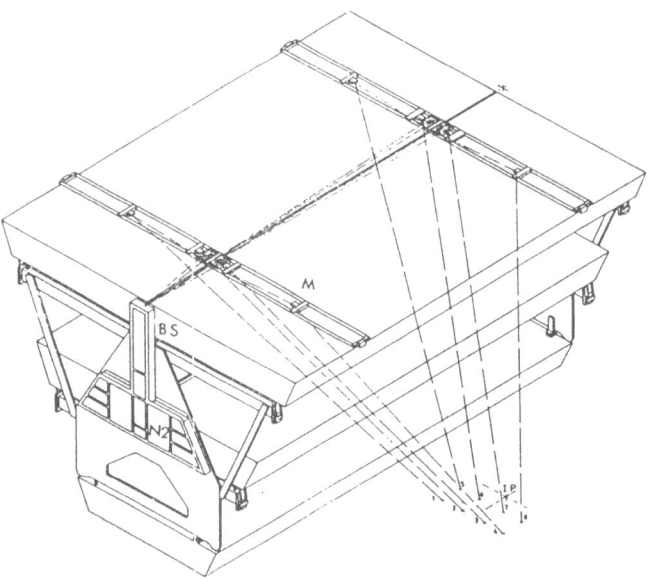

Fig. 33 The beam of UV Laser (N2) under computer control can be steered by (BS) into 8 rays via mirrors (M) toward the interaction point.

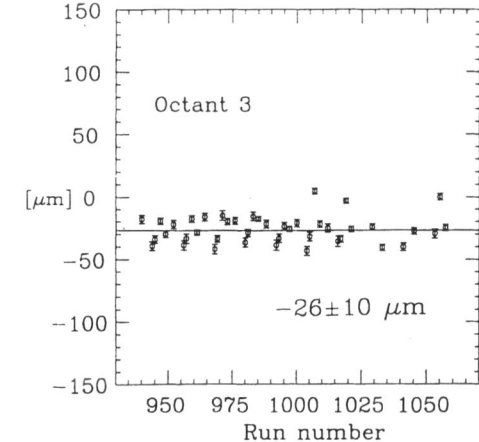

Fig. 34 Sagitta measured with UV laser in 42 different runs.

Admitting all cosmics the residual distribution is wide (1.2 mm) due to low energy muons scattering in the air and honeycomb material of the middle chamber (Figure 35b). The mean value of 7 μm is consistent with the alignment. However, the multisampling L3 chambers allow to eliminate bad multiscattering because each chamber measures the track slope to ~1mr. Eliminating all events scattered more than 2 m r a Gaussian distribution is obtained (Figure 35c) with a centroid of :

$$2\mu m \ \pm \ 400 \ \mu m \ / \ \sqrt{740} \ = \ 2 \pm 15 \ \mu m$$

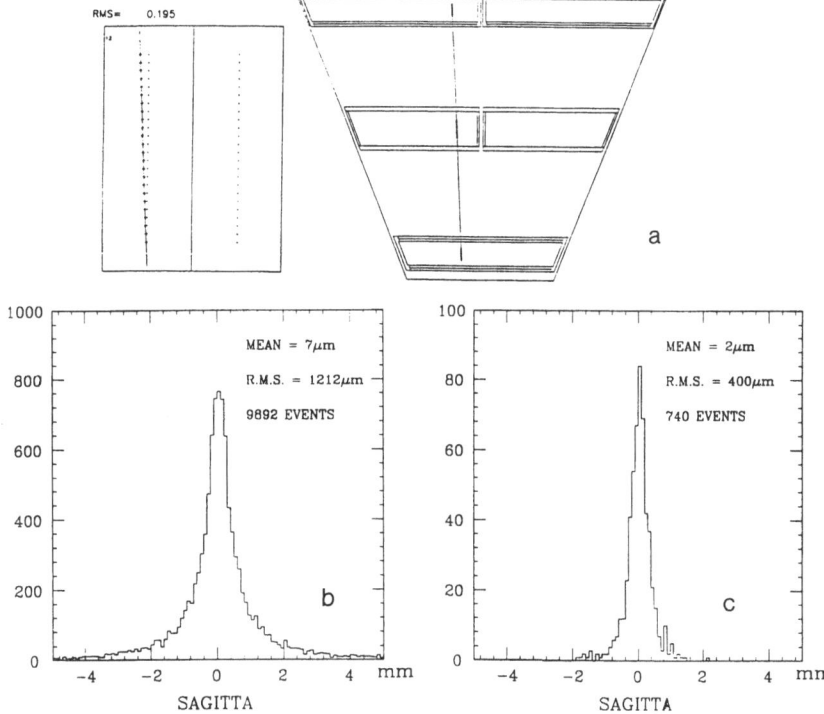

Fig. 35 a) Cosmic ray, computer reconstruction with detail of middle chamber.
b) Distribution of residuals-all cosmics.
c) Events with all slopes in chambers agreeing to the track slope to < 2 mr.

This measurement confirms the optomechanical alignment. Figure 36 depicts the good agreement of the three independent methods for 7 octants measured. We learn that :

<u>alignment < 30 μm is well possible</u>.

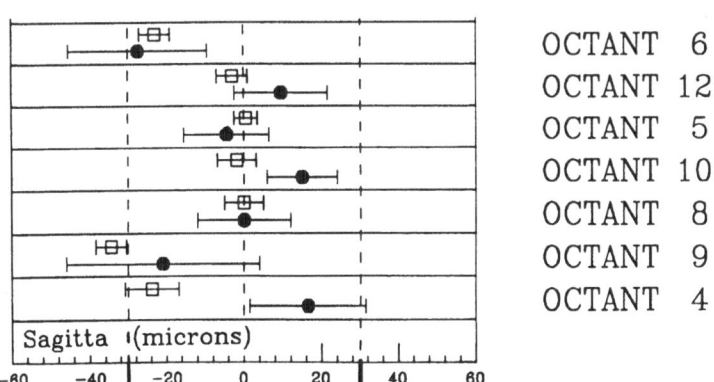

Fig. 36 *Comparison of opto mechanical alignment (___line,0) with Cosmic- (●) and UV laser (□) measurements for seven different octants.*

Figure 37 displays the final check. Each octant is rotated 360° to demonstrate that all deformations are elastic. It is then put to its final orientation in the experiment and adjusted. Installation is foreseen at the start of December 1988. Note that after transport the alignment can be quickly restored by the servo motors.

Fig. 37
Octant rotation.

FUTURE

It will immediately bring the measurement of the Zo peak at SLC and LEP I, but more fundamental would be the discovery of a Higgs particle.

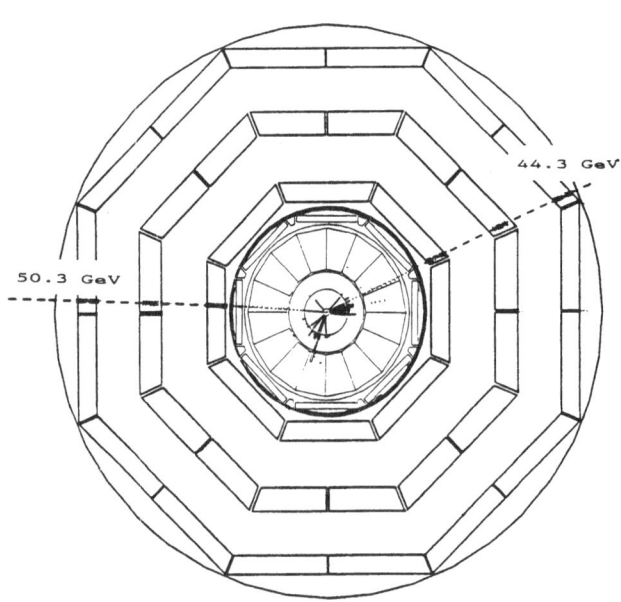

Fig. 38.
End view of L3 with Monte Carlo event ee (160 GeV) → Z(93 GeV) + Higgs (50 GeV) with Z → μμ and H → bb jets.

350

The detector is well suited to detect this reaction in ee and is μμ channel and in fact was designed for this. Figure 39 shows how the signal in the missing mass spectrum may be observed despite the small cross section. The measured resolution of 136 μm will make the signal even clearer than shown here for the design value of 250 μm.

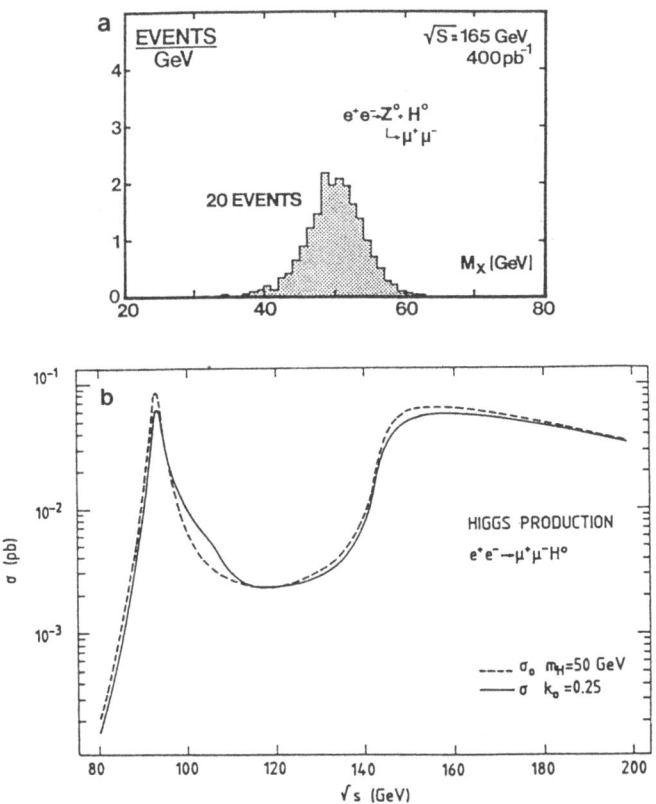

Fig. 39   a) Peak in the missing mass spectrum calculated for a 50 GeV Higgs and proposed L3 resolution. The signal is inversely proportional to the dimuon resolution.
b) Cross section for Higgs production.

Multi TeV colliders like the Eloisatron, and the SCC will need: **Precision Muon Detectors in the TeV Region** [17], especially if the Higgs sector is not clarified at LEP 2. As in the case of weak interactions at $\sqrt{s}$ = 100 GeV, one may expect at a few TeV the mass-generating mechanism to reveal his true nature. Whereas detection of e^+e^- and hadrons may have rate, multiplicity and identification problems at the luminosities envisioned, the muon pairs do not have this problem. Provided a good resolution can be achieved:

$$\Delta m/m \simeq 1\text{-}2\%,$$

the above questions can be addressed in a distinct manner.

Figure 40 gives the side view of a detector based on L3 experience which is able to achieve this goal. With a magnet assembly three times heavier than L3 generating 7.5 kgauss, the analyzing power $B\ell^2$ is increased by a factor 6.

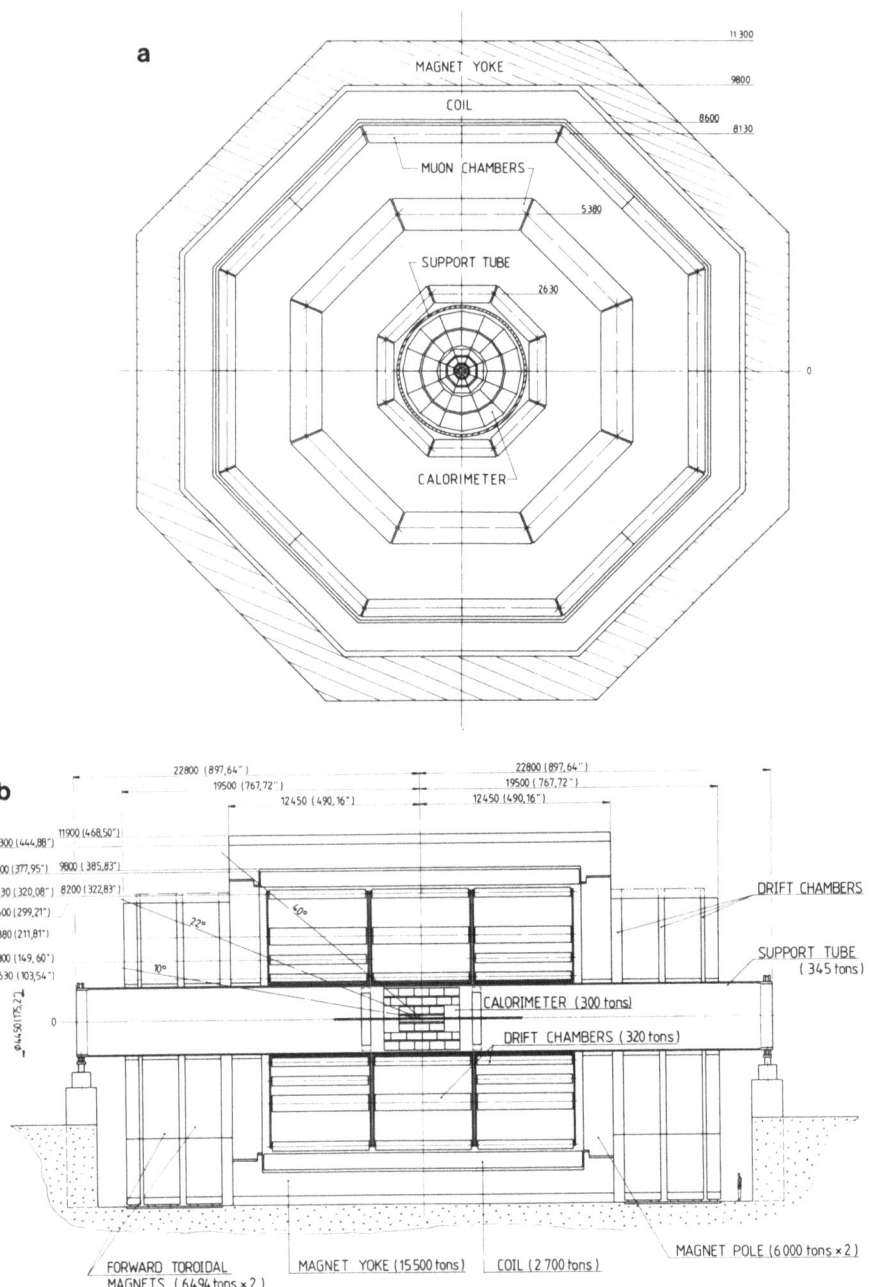

Fig. 40   a) End view of L3+1 detector.
b) Side view. Note the muon chamber array and the toroids in the forward region.

Using 32,64,32 wires in the three multisampling drift chambers and 150μm rotation we obtain $\Delta p/p = 2\%$, hence

$$\Delta m/m = 1,4\% \quad \text{at 1 TeV.}$$

This can be reached (see Figure 41) **provided** that all systematic errors of these 5.5 m structures can be kept **< 20μm**. This critical requirement is common to all precision detectors in the TeV region and it will be addressed in a special LAA sponsored R + D program [18].

If nature chooses the mass of Higgs to be 1 TeV, the dominating decay signal $H \to Z_0 \, Z_0 \to \mu\mu + x$ is given in Figure 41a on top of background [19] from semileptonic Top decay for the high resolution detector.

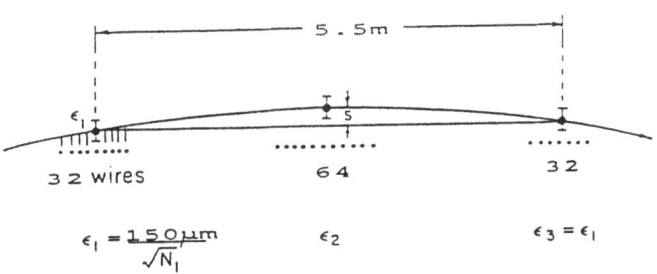

$$\text{chamber resolutions:} \quad = \sqrt{0.5\epsilon_1^2 + \epsilon_2^2} = 27.3 \ \mu m.$$

$$\text{alignment to } 20 \ \mu m: \quad \Delta S = \sqrt{27.3^2 + 20^2} \ \mu m = 33.8 \ \mu m.$$

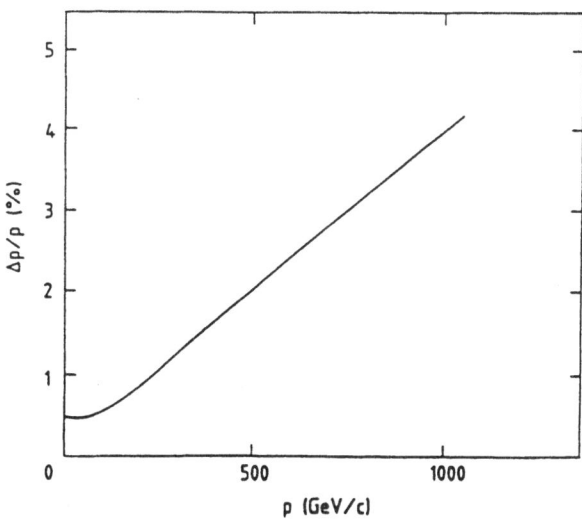

Fig. 41   Principle of muon trajectory measurement in the TeV region. Compare to Fig.25. Estimate of monentum resolution.

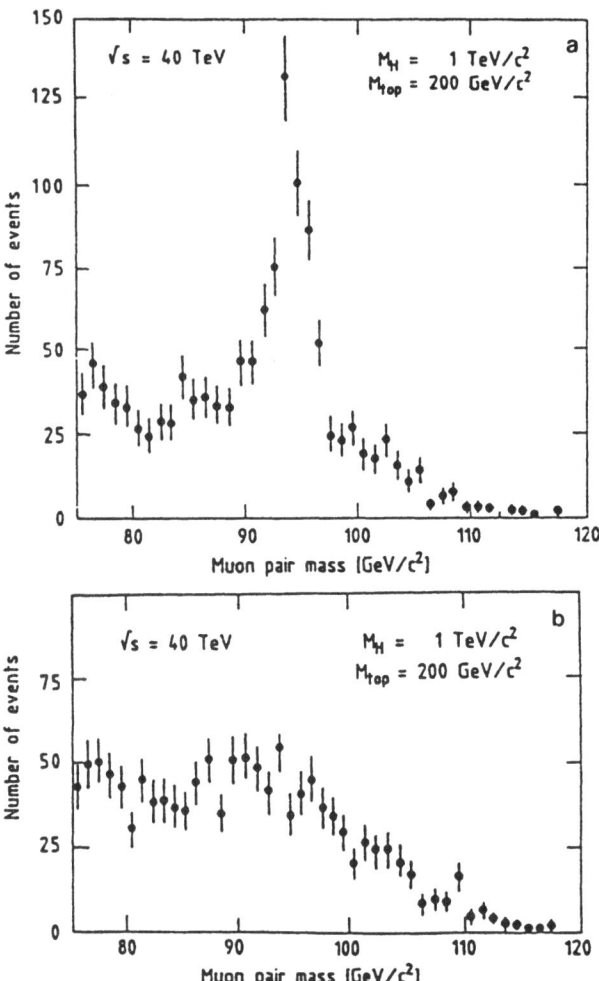

Fig. 42 Signal from pp → Hx, H → ZZ, Z → μμ, and background
from pp → TT → Wb → μμx,(Monte Carlo).
a) for L3+1) detector
b) for an iron toroid detector.

Figure 42b shows the same reaction as observed with a 12% resolution
toroid or other present detectors. The message again is clear:

Resolution is imperative!

In summary, we observed that most discoveries in the past were made
by clean γ, e, or μ detection in p-N collision with good resolution.
Presently, 1% resolution for 50 GeV has been demonstrated. The absence of
theoretical guidance beyond standard models forces to look with high
resolution and clear signals for new insights. Given accelerators like
Eloisatron and SSC we describe how muon detection to **1-2% accuracy** can be
realized, as to explore the new realm of physics.

ACKNOWLEDGEMENTS

Foremost, I am indebted to Professor Ting for the work described, and I thank my colleagues from the L3 collaboration in particular, those involved in the construction and evaluation of the muon detector:
B.Adeva, M. Aguilar-Benitez, D. Antreasyan, J. Berdugo, G.J. Bobbink, J. Branson, C. Burgos, J. Burger, D. Campana, F. Carbonara, M. Cerrada, Y.H Chang, M.L. Chen, G. Chiefari, E. Drago, P. Duinker, I. Duran, F. Erne, R. Fabbretti, M. Fabre, S.S. Gau, M. Gettner, E. Gonzalez, H. van der Graaf, F.G. Hartjes, V. Innocente, A. Konig, J.M. Legoff, S. Lanzano, L. Li, C. Mana, L. Martinez-Laso, G.G.G. Massaro, P. McBride, L. Merola, M. Napolitano, J. Onvlee, G. Paternoster, S. Patricelli, Y. Peng, O. Prokofiev, J.M. Qian, P. Razis, L. Romero, J.A. Rubio, H. Rykaczewski, J. Salicio, A. Schetkovsky, P. Schmitt, C. Sciacca, P. Seiler, V. Suvorov, K. Strauch, C. Timmermans, S. Volkov, J.H. Wang, Q.F. Wang, T. Wenaus, M. White, R. Wilhelm, C. Willmott, B. Wyslouch, S.C. Yeh.

Professor A.Zichichi and Dr.L.Chifarelli I thank for their kind hospitality.

References

[1] M.Conversi, T.Massam, Th.Müller, A.Zichichi, Nuovo Cim XL A, 2, 690, (1965).
[2] J.G.Asbury et al., Phys.Rev.Lett.18, (1967), 56.
    H.Alvensleben et al., Phys.Rev.Lett.25, (1970), 1373.
[3] J.J.Aubert et al., Phys.Rev.Lett.33 (1974), 1404.
[4] "A Proposal to Search for Leptonic Quarks and Heavy Leptons produced by Adone" A.Zichichi et al., Bologna-CERN-Frascati Coll. INFNAE-67/3, 20 March 1967.
[5] M.L.Perl et al., Phys.Rev.Lett.35, (1975), 1489.
[6] D.P.Barber et al., Phys.Rev.Lett.45, (1980), 1904.
[7] S.W.Herb et al., Phys.Rev.Lett.39, (1977), 252.
[8] D.Antreasyan et al., Phys.Rev.Lett 45, (1980), 863.
[9] G.Arnison et al., Phys.Lett. B 122 (1983), 103, M.Banner et al., Phys.Lett. 122 (1983) 476.
[10] "The L3 Technical Proposal" S.C.C.Ting et al, CERN, LEPC(1983).
[11] A.H. Walenta, IEEE NS-26,(1979), 73 and J.Fehlmann, PhD Thesis, ETH (1988).
[12] J.A.Bakken et al., NIM A 270 (1988), 397 and references therein.
[13] A.Ariefiev et al, NIM A 245 (1986), 71 and H.S.Chen et al., L3 DOC 5-62 (to be published).
[14] U.Becker et al., NIM 180 (1981), 61.
     P.Duinker et al, NIM 201 (1982), 357.
[15] U.Becker et al., NIM 196 (1982), 381.
     H.Groenstege et al., NIKHEF int.Report 1985.

[16] D.Antreasyan et al., NIM A252 (1986) 304 U.Becker et al,
NIM 225(1984) 456.
U.Becker et al., NIM 225 (1984) 456.
[17] U.Becker et al., NIM A 253 (1986) 15 and NIM A263 (1988)14.
[18] A.Zichichi, LAA Project, CERN 1986.
[19] G.Herten, unpublished.

21

PERSPECTIVES FOR A NEW DETECTOR AT A FUTURE SUPERCOLLIDER: THE LAA PROJECT

A. Zichichi

CERN
Geneva, Switzerland

The following Physicists, Engineers and Technicians represent the core of the LAA Project:

A. Ali, G. Anzivino, M. Arneodo, F. Arzarello, G. Bari, M. Basile, R. Battiston, U. Becker, J. Berbiers, F. Bergsma, R. Bertin, R.K. Bock, R. Bouclier, G. Bruni, L. Caputi, G. Cara Romeo, R. Casaccia, G. Charpak, M. Chiarini, N.H. Christ, L. Cifarelli, F. Cindolo, E. Colavita, A. Contin, I. Crotty, G. D'Ali, C. D'Ambrosio, S. D'Auria, M. Dardo, S. De Pasquale, R. De Salvo, C. Del Papa, R. Dobinson, J. Dupont, J. Dupraz, T. Ekelöf, J.P. Fabre, P. Ford, F. Frasconi, J. Gaudaen, P. Giusti, K. Goebel, C. Grinnel, B. Guerard, T. Gys, E. Heijne, S. Hellman, M. Hourican, G. Iacobucci, P. Jarron, P. Jenni, L. Jones, W. Krisher, I. Laakso, J.C. Labbé, H. Larsen, G. Laurenti, T.D. Lee, H. Leutz, L. Lone, G. Maccarrone, T. Massam, K.H. Meier, G. Million, R. Nania, V. O'Shea, A. Oliva, H.P. Paar, P. Pelfer, C. Peroni, E. Perotto, V. Peskov, D. Piedigrossi, S. Qian, J.C. Santiard, G. Sartorelli, F. Sauli, E. Schenvit, J. Schipper, H. Schönbacher, D. Scigocki, P. Sharp, G. Simonet, P. Sonderegger, L. Sportelli, M. Suffert, G.C. Susinno, S. Tailhardat, A.E. Terraneo, L. Votano, T. Weidberg, R. Wigmans, C.H. Yeh, T. Ypsilantis, A. Zichichi and K. Zographos.

1 Introduction

The present status of accelerators and detectors is as follows (see Fig. 1).

No one knows, at present, how to design an (e^+e^-) collider in the TeV energy range, while a conceptual design of ELOISATRON [1] (the 100 TeV (pp) collider) has shown that there are no basic difficulties for the machine to be designed in all details.

The situation is reversed for the case of detectors.

If a multi-TeV (e^+e^-) collider would be operative, we could immediately design an experimental set-up. The reason being that, due to the photon and Z° boson propagator dumping, the cross-section becomes very small at these energies.

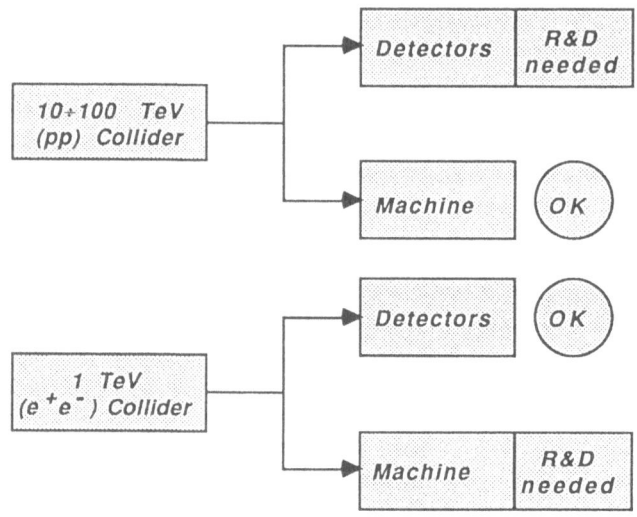

Figure 1. The present situation of Accelerators and Detectors.

But, if a multi-TeV hadron collider would be operating, no physicist in the planet would know how to perform a meaningful experiment, apart from the trivial "full screened" systems. The basic reasons being the high rate of events and the strong radiation doses which the detectors are exposed to.

As discussed later on, for the first time in the field of High Energy Physics there are basic reasons to justify a jump in energy as big as possible. The multi-TeV energy domain is—at present—exclusive privilege of proton machines, in spite of the gluon and quark structure of the colliding particles. This is why the choice emerging from Fig. 1 is R&D for detectors, in proton colliders. A new multi-TeV Collider opens up a new series of problems. These problems are of great general interest for all those who want to do physics in the next generation of Accelerators. The 100 TeV level is off-limits for (e^+e^-) colliders.

The urgency of undertaking a series of studies where new ideas and new instruments could be investigated, has brought us to the present status of the CERN LAA Project [2–6].

The LAA Project consists of ten sub-projects which are open to all physicists and engineers who are interested in participating. The ultimate goal is to prove, on the basis of prototypes, the feasibility of essential components for a detector to operate in a future multi-TeV hadron Collider. Special attention is paid to radiation hardness, rate capability, momentum resolution and hermeticity of such a detector assembly.

In what follows, the motivations, the requirements, the choices, and the final achievements of the LAA Project are reported.

2　The Motivations

Is this large R&D programme justified in terms of Physics goals?

Back in 1979 it was clear to some of us that we were entering a new *era* in physics. In spite of the great development of our physics, little was done to promote the necessary support required by these *new frontiers of advanced research*. *Why?* Probably because the physics community was not convinced that this new *era* was really there.

This new *era* indicated that a big jump was needed in the *energy level*.

The strong feeling that we were entering a new *era* in physics initiated the ELOISA-TRON Project, and LAA is part of this programme.

New ideas, new concepts and new phenomena make physics of just twenty years ago seem as old as millenia. Einstein's quadrimensional space-time seemed to be a conquest beyond which no one would be able to go. This, however, seems to be an incredibly narrow outlook for two reasons, both fundamental: the number of dimensions and the property of those dimensions.

No one had thought, before the sixties, that there could exist space-time dimensions with *fermionic* properties. Those of Einstein are *bosonic*. This is how the new concept of *superspace* was born, and with it, *superparticles* and *supermatter*.

The world in which we live and the matter which we are made of, could have their roots in a *bosonic* superspace with ten dimensions, plus the 32 *fermionic* ones. And this is not all.

The concept of "point" that has held its position for centuries and centuries, falls by the wayside. In its place is the "superstring" [7]: a unidimensional entity with a pointless structure in a 42-dimensional superspace.

In this extraordinary progress of our knowledge, the winning parameter has so far been and will certainly remain the **energy**. The final goal is the unification of all the fundamental forces of Nature. Figure 2 shows the unification scale, and where the ELOISATRON hadron collider stands. The straight lines and the unification point are just to show the conceptual game. For example, the point at 10^{15} GeV is more like an ellipse. And the straight lines should not be taken as if no basic problems were there.

There are four basic theoretical problems to be solved in the multi-TeV (1–100 TeV) energy domain:

- The **family** problem.

- The **hierarchy** problem.

- The **proliferation** problem.

- The **compositness** problem.

The questions which arise when translating these theoretical problems into physically observable phenomena are the following:

i. Do **new**, heavier, **quarks** and **leptons** exist?

ii. Are there **other intermediate vector bosons?**

iii. How many **Higgs bosons** exist?

iv. Do **supersymmetric partners** exist?

v. Are **quarks** and **leptons composite?**

vi. Would some **unexpected exotic process** occur?

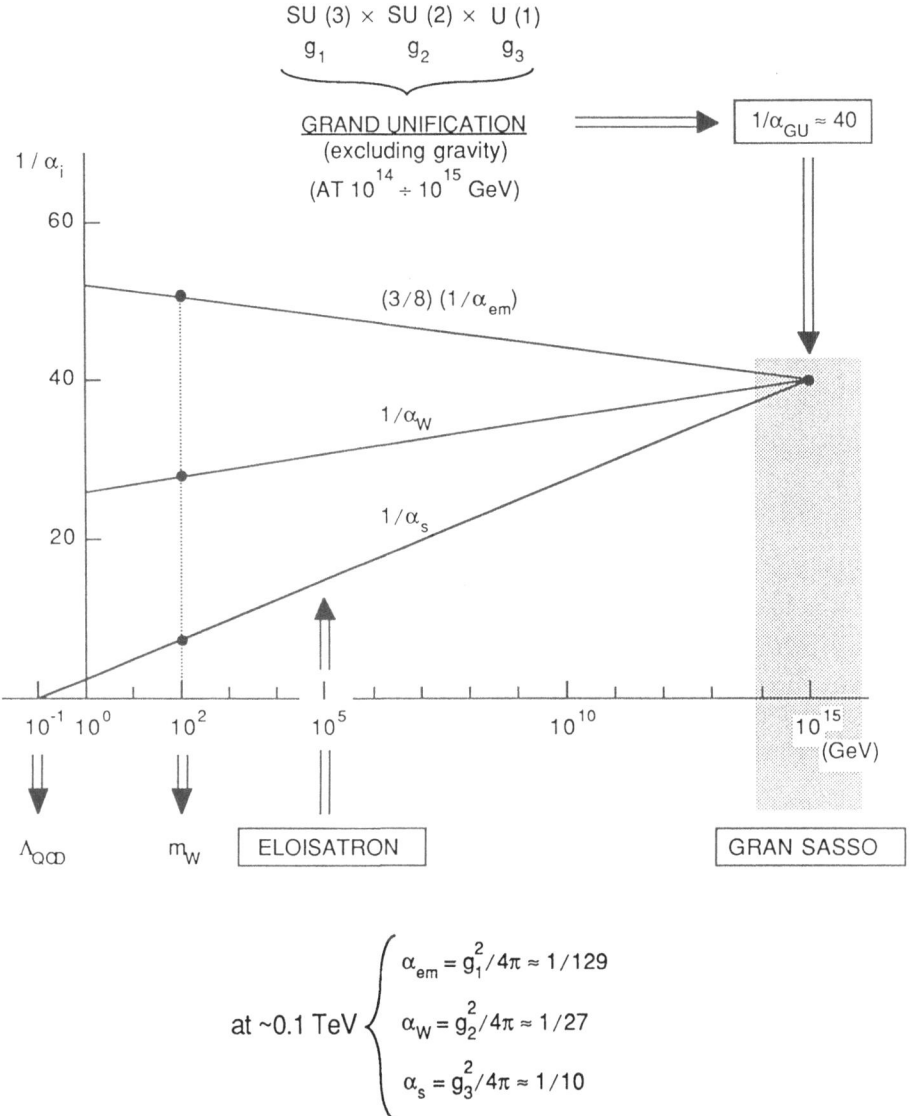

Figure 2. All the fundamental forces of nature should be generated by a unique force. The convergence of the three fundamental constants g_1, g_2, g_3, is the basis for GUT.

In order to go from theory to down-to-earth physics the detectors should be able to observe and measure in the multi-TeV range:

 i. electrons and photons;

 ii. muons and other lepton-like long-lived particles;

 iii. neutrinos and other non-interacting particles (i.e. missing objects);

 iv. leading protons (for hermeticity and new physics);

 v. hadrons and jets, with and without leptons inside.

In addition to all these new effects, there is an impressive series of "expected" phenomena to be studied. The multi-TeV range is overcrowded of short- and long-lived hadrons.

3 The Basic Data

Let me discuss the basic data for a new detector at a future SUPER-COLLIDER.

The first and main requirement is that a luminosity as high as possible (10^{33}–10^{34}–10^{35} cm^{-2}s^{-1}) should be aimed at. This imposes severe conditions on new detectors. Among them, a vital one is radiation hardness.

A basic feature of new physics is to produce undetectable events. Hermeticity will be essential for the discovery of new phenomena. Therefore the average number of events per bunch crossing $\langle n \rangle$ must be ONE if the missing energy is to be used as a signature in event selection and analysis[1].

The limiting luminosity is:

$$\mathcal{L}_{pp} = \frac{\langle n \rangle}{\Delta t_b \times \sigma_{pp}} \tag{1}$$

where: $\langle n \rangle$ is the average number of events per bunch crossing, Δt_b time between bunch crossings and σ_{pp} total (pp) cross-section.

Figure 3 shows the detection limit for rare events (fixed at 10 events per year), and the total minimum bias rate, as a function of luminosity. Note that a total running time of 10^7 seconds per year, and a total (pp) cross—section of 100 mb are assumed. The "magic" limit of observability, at the 10^{-40} cm^2 level in the cross–section for new physics, is reached if a luminosity at the level of $10^{34} cm^{-1}$ s^{-1} can be achieved.

At present, the following machine parameters:

$$\Delta t_b \sim 100 \text{ ns}, \quad \text{and} \tag{2}$$
$$\mathcal{L}_{pp} \sim 10^{32} \text{ cm}^{-2}\text{s}^{-1}, \tag{3}$$

are well within reach from a technological point of view.

On the other hand, the total (pp) cross-section is expected to be:

$$\sigma_{pp} \cong 100 \text{ mb} = 10^{-25} \text{cm}^2. \tag{4}$$

---

[1]It must be remembered, however, that $\langle n \rangle = 1$ means that in 37% of the cases the number of events per crossing is 0, in 37% of the cases it is equal to 1, and in 26% of the cases it is greater than 1.

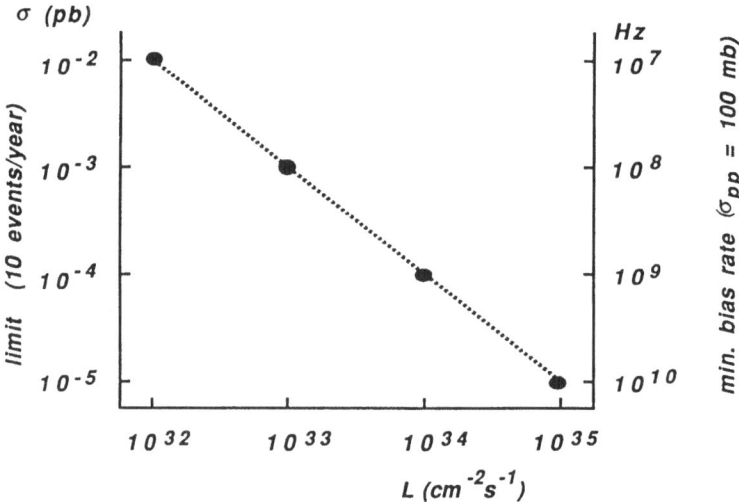

Figure 3. Discovery limit for rare events and total rate as a function of luminosity.

These three values together produce $\langle n \rangle \sim 1$. Figure 4 summarize the present status, including the radiation level.

But what is wanted is $\langle n \rangle = 1$ at higher luminosities. And therefore the corresponding Δt_b reaches prohibitive figures:

$$
\begin{array}{ccccccc}
\mathcal{L} & = & 10^{33} & \to & 10^{34} & \to & 10^{35} \quad \mathrm{cm^{-2}\,s^{-1}} \\
& & \downarrow & & \downarrow & & \downarrow \\
\Delta t_b & = & 10 & \to & 1 & \to & 0.1 \quad \mathrm{ns}
\end{array} \tag{5}
$$

High values of luminosity, such as 10^{34}–10^{35} cm^{-2}s^{-1}, must be the aim to be reached in future supercolliders, and detectors must be ready to cope with this high luminosity, in order to study the existence of rare phenomena. Conclusion: machine builders [1] and physicists have to solve very difficult problems. This is a great challenge for all of us.

At present, no detector is able to make full use of a (pp) collider working with $\mathcal{L} > 10^{32}$ cm^{-2}s^{-1}, thus a strong R&D effort on **Detectors** must be pursued. **This is the aim of the LAA Project**.

4 The Major Requirements

In this section, the major requirements on multi-TeV detectors are outlined, and the R&D necessary to obtain the necessary performances are indicated.

Vertex detector technologies

Requirements:

- Unambiguous and precise determination of primary and secondary vertices;

- Tracking back to the vertex, measuring the charged multiplicities and defining the topology of the event;

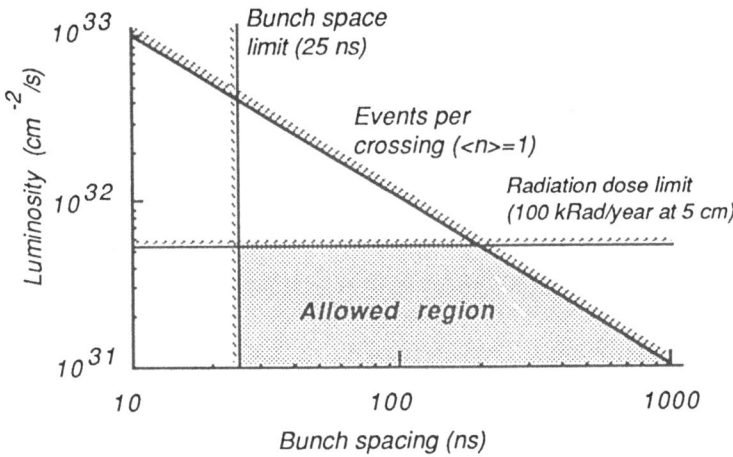

Figure 4. Luminosity as a function of bunch spacing. The limits correspond to the three extreme values: bunch spacing = 25ns, average events per crossing = 1, radiation dose = 100 kRad/year at 5 cm from the beam. The grey region is allowed.

- Very high spatial accuracy, double-track resolution and redundancy, because of:

    - large multiplicities (~ 100),
    - high particle densities in jets,
    - high momenta.

- Particle identification.

Superconducting, very high field magnets should be used in association with very high precision vertex detectors.

The above requirements imply R&D on:

- Multidrift gas proportional tubes;

- High resolution scintillating fibres;

- *GaAs* and Silicon microstrips;

- Superconducting magnets (~ 10 T);

- Current-carrying beam pipe [9]: iron-free space with toroidal field (~ 10 T).

Calorimetry

Requirements:

- Large coverage and hermeticity:

    - full detection of showering particles and jets;

- study of missing p_T (neutrinos, photinos and other "inos").

- High granularity:

  - for electron identification inside jets,

  - increasing when going to more forward regions.

- High rate capability:

  - to cope with the high luminosity and short time interval between bunch crossings (≤ 100 ns).

- Optimum energy resolution. It depends on:

  - response to electromagnetic and hadron showers (compensation),

  - choice of absorbers and sampling medium,

  - longitudinal segmentation and depth,

  - control of systematics, which become crucial at very high energies.

The above requirements imply R&D on:

- Lead/scintillating fibres;

- BaF_2 scintillators and photosensitive wire chambers;

- Liquid Xenon calorimeters with electrical and optical read-out.

Muon detection

Requirements:

- Good rejection against hadron decays and hadron punch-through.

- Maximum angular coverage for detector hermeticity.

- Good momentum analysis over very large volumes, i.e. high resolution and precision alignment.

- Easy and economic construction.

- Minimum number of electronic channels.

The above requirements imply R&D on:

- Very large area detectors:

  - Limited streamer tubes;

  - Large toroidal and solenoidal magnets (air/iron);

  - Very large, high-precision drift chambers.

- Alignment.

Leading particle detectors

Requirements:

- Large coverage down to the smallest polar angles, needed for:

    - exploring the leading effect and the forward physics domain;
    - studying the longitudinal momentum balance;
    - tagging multiple interactions in the same crossing.

- Capability of facing serious background problems.

- Very high precision in measuring the space points (some μm).

- Stability in position (survey) and response (calibration) over large distances and long periods of time.

Particularly relevant for new heavy flavour research and other expected and unexpected phenomena.

The above requirements imply R&D on:

- *GaAs* microstrip detectors;

- Silicon microstrip detectors;

- Compact calorimeters;

- Ring Imaging Cherenkov counters (RICH);

- Completely new detectors.

Data acquisition and analysis

This is the possible scenario for the trigger and data acquisition system:

- First Level trigger:

    - Reduction in rate from 10^7 to $\sim 10^5$ Hz;
    - Decision time: some hundreds of nanoseconds;
    - Simple cuts on energy and p_T;
    - Each detector element provides an independent trigger;
    - Pipelined with the bunch crossing frequency to avoid completely the problem of dead time;
    - Massive use of custom-programmable processors, realized in close connection with industry.

- Second Level trigger:

    - Reduction in rate from 10^5 to $\sim 10^3$ Hz;
    - Decision time: $\sim 10\,\mu$s;
    - Refines the first level trigger results using the digitized signals;
    - Massive use of custom-programmable processors.

- Third Level trigger:

  - Most important part of the whole trigger and data acquisition system;
  - Reduction in rate from $\sim 10^3$ to 1 Hz;
  - Capable of handling event sizes of about 1 Mb;
  - Analysis of the full event in a multiprocessor stack of about 1000 CPUs linked by fast data buses;
  - Flexible to explore new energy domains;
  - Implemented in strong connection with off-line (by using the same analysis programmes) to allow a fast implementation of algorithms, coping with new and interesting event topologies.

The above requirements imply R&D on:

- Microelectronics;

- Radiation hardness on the relevant components of the detector;

- Real-time data processing with ultra-high event rates;

- Dedicated Supercomputers.

Theory (QCD lattice calculations for dynamics) and Monte Carlo Simulations

Requirements: Event simulation must be based on QCD calculations.
This implies R&D on:

- Dedicated Supercomputers for lattice QCD calculations.
  This may eventually lead to a quantitative understanding of non-perturbative dynamical aspects like the structure and fragmentation functions.

- QCD perturbative models.
  The background to exotic and new physics must be known very precisely. Hence, high order QCD radiative effects must be calculated and included in realistic general purpose Monte Carlo simulation programs.

5 The Choice

The previous analysis in terms of physics goals and machine parameters have brought us to the following choices for the LAA Project: ten basic components (see Fig. 5).

1. High Precision Tracking.
   Here three parts are needed. The closest to the vertex uses gaseous detectors consisting of multidrift tubes. The surrounding one, uses the technology of scintillating fibres. The third one, along the beam, is based on microstrips made of a new type of detector material: Gallium Arsenide ($GaAs$).

2. Calorimetry.

Here we follow two lines. One is for the "fixed-target" mode of operation, and it is based on BaF_2 scintillators coupled to photosensitive wire chambers. The other, for a "collider" mode is based on the so-called "Spaghetti Calorimeter" (Lead-plastic fibre calorimeter). Of course, no one can exclude that the first approach becomes so successful to be extended also to the collider mode.

3. Large Area Devices.

Here there are two basic parts. The first studies the problem of constructing large area devices sensitive to charged particles and with high precision. The other part of the R&D refers to the problem of positioning and monitoring the alignment of these large area devices. There is, in fact, no point in constructing a large area device if we do not know how to position it with great accuracy.

4. Leading Particle Detection.

This is a crucial component of the LAA project. All, in fact, started with the discovery of the "leading" effect at the ISR [10,15]. To work near the beam of a high energy collider presents a series of very difficult, and therefore extremely exciting, problems to be solved. From fast removal of the detectors to high precision positioning, to high radiation resistance.

5. Subnuclear Multichannel Integrated Detector Technologies.

The very high number of channels needed in the detectors for high luminosity colliders, and, consequently, the problem of compactification of the very many elements, the electric power consumption, etc., lead naturally to the development of integrated electronics. Moreover, the problem of producing radiation resistant electronic components, must be solved if these have to be operated very close to the detector. Therefore, this component of the LAA Project consists of two parts, one dedicated to the design of general-purpose integrated circuits with standard technologies (silicon), and the other which will concentrate mostly on new, radiation-hard, technologies.

6. Data Acquisition and Analysis.

Here, the objective is the real-time processing of the detector signals. Solutions have to be found to the triggering and data compaction problems, filling the gap which exists today between the custom-made electronics and the programmable devices. Specific solutions have to be found for each of the trigger levels, with particular emphasis on the communication between the various components.

7. Supercomputers and Monte Carlo Simulations.

A full understanding of the theoretical predictions on high energy hadron collisions would be absolutely necessary, in order both to design the detector, and to analyze in a meaningful way the data. QCD non-perturbative calculations (QCD lattice) have never been applied to the dynamics of the hadronic interactions. Powerful supercomputers are needed to perform the calculations in a reasonable time. Moreover, the theoretical understanding of the algorithms is not yet complete. Therefore, a strong effort is needed in this direction. Meanwhile, a unification of the various Monte Carlo programs, based on QCD perturbative models, is necessary. The goal is to produce a single, QCD "Super-Monte Carlo" program, which incorporates all the higher order perturbative information, thus eliminating the uncertainties and contradictions still present today in this field.

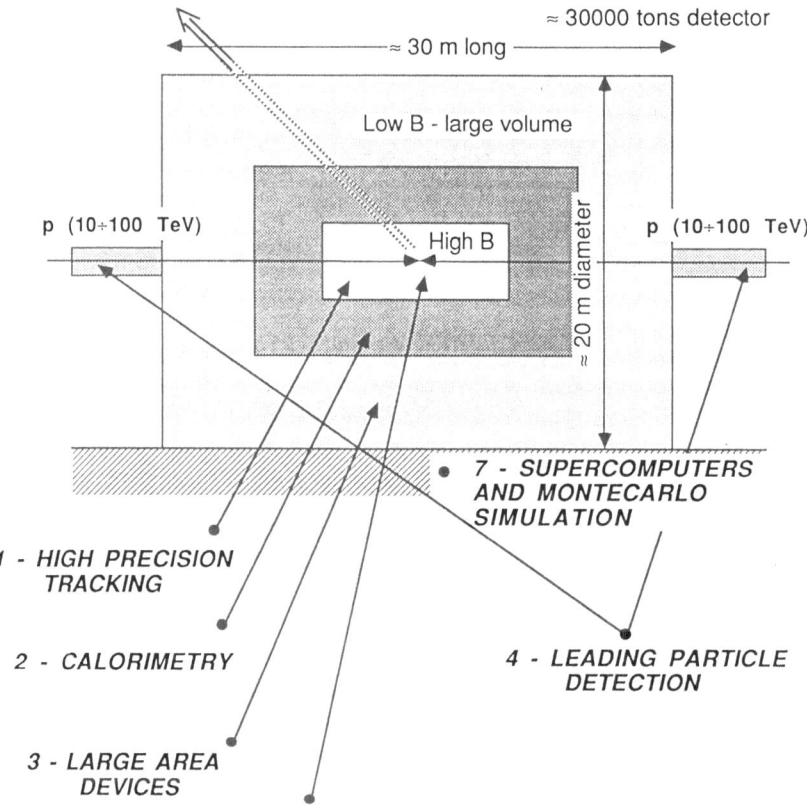

- 5 - SUBNUCLEAR MULTICHANNEL INTEGRATED DETECTOR TECNOLOGIES
- 6 - DATA ACQUISITION AND ANALYSIS

≈ 30000 tons detector

≈ 30 m long

Low B - large volume

p (10÷100 TeV)

High B

≈ 20 m diameter

p (10÷100 TeV)

- 7 - SUPERCOMPUTERS AND MONTECARLO SIMULATION

1 - HIGH PRECISION TRACKING

2 - CALORIMETRY

4 - LEADING PARTICLE DETECTION

3 - LARGE AREA DEVICES

- 8 - VERY HIGH MAGNETIC FIELDS
- 9 - SUPRACONDUCTIVITY AT HIGH TEMPERATURE

- 10 - RADIATION HARDNESS

Figure 5. The ten components of the LAA Project.

8. Very High Magnetic Fields.

Here the aim is to design a magnet to produce an extremely high magnetic field in the interaction region. Coupled to very precise tracking detectors, this magnetic field will allow the measurement of the momentum and of the charge sign of the particles produced in multi-TeV interactions. The higher is the magnetic field, the more compact the full detector will be.

9. Superconductivity at High Temperature.

The recent discovery of superconductivity at high temperatures opens up a very large range of possibilities in High Energy Physics. Here, most of the problems to be solved concern the technology of materials: for example, how to produce reliable cables, capable of sufficiently high current density to be of practical use. The LAA Project will follow the developments in this field through a series of contacts with the various research centres in the World.

10. Radiation Hardness.

Here, all the studies on the radiation hardness of the various materials and electronics, developed by the different LAA components, are co-ordinated.

6 Summary of the Results after one Year of Activity

The present status of the LAA Project, after one year of activity, is fully reported in ref. [6]. Here I will limit myself to a very telegraphic summary.

- A new plastic scintillator, PMP, has been found: it will allow to produce 50 μm diameter scintillating fibres.

- A new wire, radiation resistant at the MRad level has been found.

- A new substance, EF, has been found which is better in time resolution and mechanical operation than TMAE. This will allow the construction of the so much needed Solid Scintillator Proportional Counters (SSPC).

- It has been proved that $GaAs$ competes very well with the so far dominating Si.

- It has been discovered that intrinsic fluctuations in Pb nuclei are lower than in U. This means that future calorimeters will be free from the complex of Uranium.

- A new microelectronic circuit—4 channels amplifier and multiplexer—has been built and shown to work in spite of its 100 μm pitch.

7 Conclusions

There is no doubt that the feasibility of future experiments will totally rely upon intensive R&D studies on detectors carried out by a large community of researchers. The first results of the LAA Project make us confident that the ambitious goal of operating a detector at the next supercollider, which we hope will be the 10% ELOISATRON is not out of reach.

References

[1] K. Johnsen, in *The ELOISATRON Project*, June 1988.

[2] A. Zichichi, *Report on the LAA Project*, Volume 1, 15 December 1986.

[3] A. Zichichi, *Report on the LAA Project*, Volume 2, 25 June 1987.

[4] A. Zichichi, *Report on the LAA Project*, Volume 3, 15 June 1988.

[5] A. Zichichi, *Report on the LAA Project*, Volume 4, CERN-LAA/88-1, 25 July 1988.

[6] A. Zichichi, *Report on the LAA Project*, Volume 5, CERN-LAA/88-2, 19 September 1988.

[7] J.H. Schwarz, Proceedings of the XXIV Course of the Ettore Majorana International School of Subnuclear Physics: *Old and new Forces of Nature*, (Plenum Press Inc., New York- London), Erice, 1985;
M. Green, Proceedings of the XXVI Course of the Ettore Majorana International School of Subnuclear Physics: *The SuperWorld-III* (Plenum Press Inc., New York-London), Erice, 1988.

[8] M. Duff, Proceedings of the XXV Course of the Ettore Majorana International School of Subnuclear Physics: *The SuperWorld-II* (Plenum Press Inc., New York-London), Erice, 1987;
M. Duff, Proceedings of the XXV Course of the Ettore Majorana International School of Subnuclear Physics: *The SuperWorld-III* (Plenum Press Inc., New York-London), Erice, 1988.

[9] J.C. Sens, private communication.

[10] M. Basile et al., Nuovo Cimento **63A** (1981) 230.

[11] M. Basile et al., Nuovo Cimento **66A** (1981) 129,
M. Basile et al., Nuovo Cimento Letters **32** (1981) 321.

[12] K. Alpgard et al., UA5 Collaboration, Phys. Lett. **B121** (1983) 209.

[13] K. Alpgard et al., UA5 Collaboration, Phys. Lett. **B123** (1983) 381.

[14] M. Basile et al., Nuovo Cimento Letters **41** (1984) 298.

[15] M. Basile et al., Nuovo Cimento Letters **38** (1983) 359,
see also:
A. Zichichi, Proceedings of the XXI Course of the Ettore Majorana International School of Subnuclear Physics: *How far are we from the Gauge Forces* (Plenum Press Inc., New York-London), Erice, 1983.

22

EXPERIMENTATION AT SUPERCOLLIDERS

W. Bartel

Deutsches Elektronen Synchrotron DESY
Notkestrasse 85 · D 2000 Hamburg 52
Germany

Abstract

Experimental techniques, which may be used to exploit the physics potential of supercolliders are reviewed. The most stringent requirements have to be met at proton-proton colliders, while ep and $e^+ e^-$ experiments are less demanding. Technical problems which have to be solved are related the high density of final state particles, the short duty cycle and the high radiation level. The high standard physics background to new physics events has to be suppressed by novel trigger schemes. Thus R&D work is required in several areas ranging from detector development, digital and analog electronics design to data-acquisition and computing.

1 Introduction

In order to fully exploit the physics potential of TeV collisions, experiments have to be designed which are far more sophisticated and more complex than todays current experiments. The increasing demands on experimental technology are not restricted to one area, R&D is necessary in various fields like detector technique, read out electronics, data-acquisition and computing.

The most hostile environment will be experienced at pp colliders and if an experiment can be designed to survive at such a machine, all the ingredients will be available to construct experiments for ep or $e^+ e^-$ colliders. Therefore the discussion on experimenting at supercolliders will be focused on pp machines and I will only briefly refer to differences which arise when other particle species are collided.

Experiments at future TeV hadron colliders have been studied on several occasions. There have been SSC [1] and LHC [2] workshops, which are well documented and which were helpful in preparing this note. The HERA experiments H1 [3] and ZEUS [4] will be the first large scale experiments to operate with a TeV proton beam at a bunch separation of about 100 nsec and experience gained at HERA will provide useful input to the design of experiments at supercolliders.

In order to study interactions between proton constituents, luminosities of the order of 10^{33} cm^{-2} sec^{-1} are required. Though the pointlike constituent cross-sections are small, the total cross-section for proton-proton interactions is large and the high flux of secondary particles puts severe limitations on any detector. Implications on radiation hardness of detector elements and electronics will be discussed. Due to the high density of secondary particles fine grained detectors are required with an emphasis on calorimetry and lepton identification. Special attention has to be payed already in the design phase of a detector to triggering and data acquisition, because the short time between successive interactions (order of 25 ns) demands that the electronics be capable of making fast decisions at high data rates. Under these aspects possible detector concepts for a supercollider are discussed in defining the physics requirements and evaluating the technical limitations.

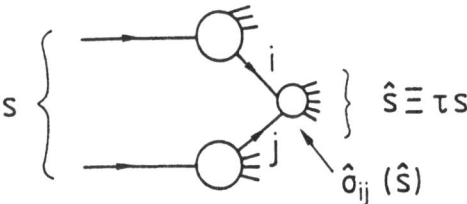

Figure 1. Parton-parton interaction in proton-proton scattering.

2 Cross-sections and Luminosity

New physics is expected to emerge from the study of parton-parton collisions i.e. pointlike qq, $q\bar{q}$, gg and qg interactions. The cross-sections for two parton processes decrease like

$$\sigma_{ij} \propto \frac{1}{\hat{s}}$$

where σ_{ij} is the cross-section for scattering parton i in one proton off parton j in the other one. $\sqrt{\hat{s}}$ is the parton-parton center of mass energy. The diagram of Fig.1 schematically shows the kinematics for a parton process in a proton-proton collision at a total center of mass energy squared s and a parton energy $\hat{s} = \tau \cdot s$. Lab. cross-sections are calculated by folding parton cross-sections with the corresponding parton distributions inside the proton.

$$\frac{d\sigma_{ij}}{\hat{s}} = \int_0^1 \int_0^1 f_i(x_i, \hat{s}) f_j(x_j, \hat{s}) \cdot \sigma_{ij}(\hat{s}) dx_i dx_j$$

This integral is interpreted as the product of a parton luminosity L_{ij} and the parton cross-section

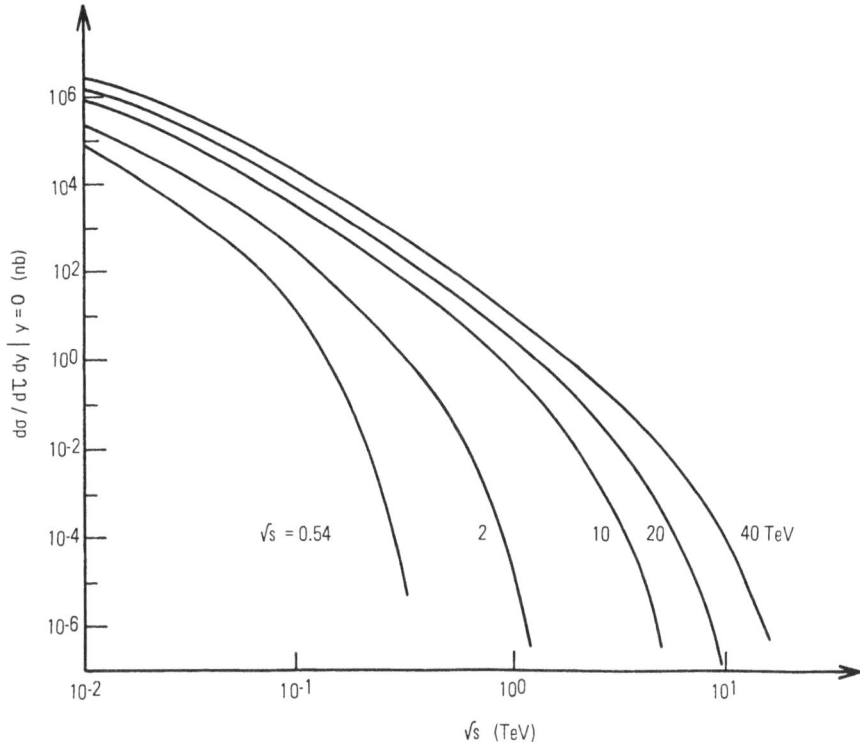

Figure 2. Glue-glue pseudo cross-section at different center of mass energies.

$$\frac{d\sigma_{ij}}{\tau} = \tau \cdot \frac{dL_{ij}(s,\tau)}{d\tau} \cdot \sigma_{ij}(\hat{s})$$

For glue-glue hard scattering the pseudo cross-section $d\sigma/d\tau$ in the central region at rapidity y=0 is displayed in Fig.2, which is taken from a report by Ali et al. [5].

To give an example of cross-sections and rates consider a hadron collider with E_{cm}=20 TeV. For observing qq collisions with a reasonable rate the parton-parton center of mass energy will be about 25% of the total energy typically $\sqrt{\hat{s}} \sim 5$ TeV with cross-sections between $\sigma \sim 10^{-1}$ and 10^{-3} nb, corresponding to rates between 10^{-1} and 10^{-3} events per second at a luminosity of L=10^{33} cm^{-2} sec^{-1}. The total cross-section for proton-proton scattering is , however, of the order of 100 mb, where for the present investigation only the inelastic part of about 60 mb counts, because the secondaries of elastic and diffractive processes normally disappear down the beam pipe. Thus the rate of interactions producing secondary particles in an area which will be covered by detector elements is $6 \cdot 10^{7}$ sec^{-1} and therefore between 8 and 10 orders of magnitude higher than the rate for high p_t parton-parton processes. Therefore efficient trigger schemes have to be developed to suppress the high background from standard "soft" hadron physics.

The experimental difficulties are enhanced by the fact that the beams are bunched and instantaneous rates are thus higher than the average rates quoted above. There are several reasons why the beams have to be bunched. Three of them are listed below:

- The synchrotron radiation losses have to be compensated by an accelerating RF system. (For a hadron collider in the LEP tunnel the losses amount to 13 keV per turn at a beam energy of 8.2 TeV.)

- A machine with DC beams like the ISR would require a finite crossing angle in order to keep the length of the luminous region small and two independent magnet systems would be necessary. That essentially doubles the price of the machine.

- The luminosity increases linearly with the number of stored particles, while the stored energy increases quadratically. Thus the amount of stored energy may become prohibitive for a DC machine in the TeV energy range.

For projected hadron colliders the intervals between successive crossings range between 10 and 25 ns. For a rate of $6 \cdot 10^7$ interactions per second there will be on the average $<n> = 1.5$ interactions per crossing at a bunch spacing of 25 ns. This implies that for every high p_t trigger a detector will register on the average secondaries from 2.5 interactions, because it is technically impossible to resolve a time structure inside one beam beam crossing. Especially in calorimeters the information from ovelapping events will be superimposed without the possibility of later separation. It has been suggested [1] that it may be possible to distinguish different events in one bunch crossing via their vertices in the luminous region. This is a viable method for assigning charged particles to a particular primary interaction. Neutrals, however, especially when they have large transverse momenta with respect to the event axis, cannot be so assigned and will mix events.

Hence quantities like total energy or missing transverse energy cannot be attributed to a particular event. These quantities are measured as averages over two to three events, and only a decrease in luminosity from $L=10^{33}$ cm^{-2} sec^{-1} by about a factor of 10 to $L=10^{32}$ cm^{-2} sec^{-1} at 20 TeV center of mass energy would solve the problem of overlapping events.

High luminosity is essential to investigate rare processes, and instead of compromising on the luminosity, it is better to sacrifice some knowledge of the event structure. In view of these alternatives it seems reasonable to design two types of interaction regions at a supercollider. Those with high luminosity would serve detectors which are specially designed for high rates and multiple event overlay, and those with low luminosity would serve general purpose detectors for detailed studies of individual events.

The physics scopes of these two alternative solutions are complementary:

- At high luminosity one may look for event classes with a unique signature e.g. four final state muons from heavy Higgs decays. The information one gets on individual events is diluted, but rare events are in general overlayed with standard event types which have a well known pattern and which may be subtracted on a statistical basis.

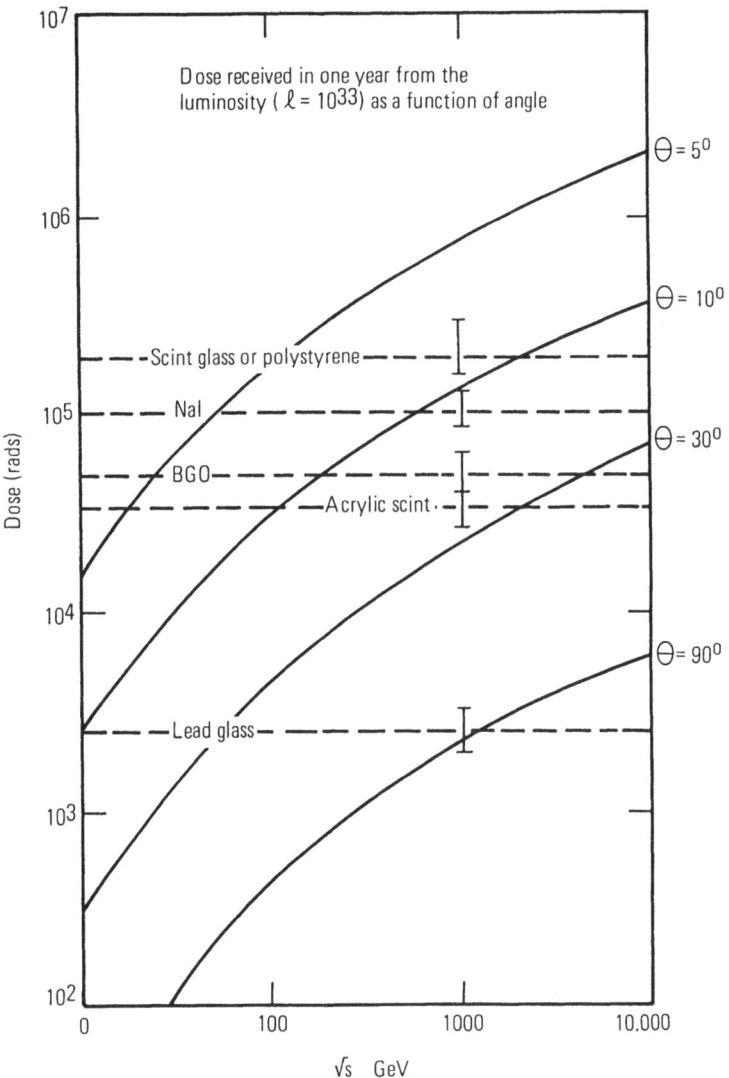

Figure 3. Expected dose rates received by detector components at various angles Θ with respect to the beam line. Indicated are also doses at which commonly used detector components start to fail.

- If there is only one interaction per trigger registered in a detector, the full power of a fine grained detector will help to study details of individual events and thus also be sensitive to new physics.

Both approaches to TeV physics have their virtues and both methods enable new physics to be discovered. Which of the techniques is more effective depends on the type of new events which show up and on the cross sections involved. A priori neither of the two alternatives should be rejected and space should be found for two types of experiments, using the different philosophies.

Various detectors could be designed which are intermediate between the two alternatives. There are, however, good arguments in favour of clear cut solutions one way or the other. For the start up of a new machine, an experiment designed for high luminosity may have some advantage, not because the luminosity will be high, but because the experiment is more robust and less sensitive to background.

At a supercollider with L=10^{33} cm^{-2} sec^{-1} an experiment will be exposed to dose rates between about 10^4 rad per year at $\Theta = 90°$ and some 10^6 rad per year at $\Theta = 5°$ with respect to the beam direction. These numbers are shown in graphical form in Fig.3, where the expected dose rates are plotted which detector components will receive at different angles Θ as a function of the center of mass energy. In such an environment acrylic scintillators or BGO crystals would not survive the first year of operation. Also electronics attached to the detector components would suffer from radiation damage. Present integrated circuits tend to fail after an exposure to 1 to 2 $\cdot 10^5$ rad. In this field R & D is badly needed, especially because current experiments would also profit from more resistant electronics components.

At ep colliders cross-sections are smaller by a factor of α and the time interval between crossings has to be larger, because for a head on collision geometry tens of meters are needed to separate the two beams. For a spacial separation of 30 m between successive beam-beam interactions the corresponding bunch separation would be 100 nsec as at HERA. In some cases one may be forced to lengthen the time between crossings even more because the revolution frequencies and the frequencies of the accelerating RF of the hadron part and the electron part of the accelerator complex do not match e.g. at CERN. A major problem for ep machines is the design of the interaction regions with head on collisions, since any deflection of the electron beam is accompanied by the emission of synchrotron radiation and it is difficult to protect the super conducting magnets of the proton machine against the radiated power.

The considerations of the last chapter are not relevant for e$^+$ e$^-$ colliders. Cross sections are smaller, jets and other secondaries from annihilation events are more uniformly emitted into the whole solid angle and the time between collisions will be larger. Detector requirements are more relaxed with the exception that a high level of synchrotron radiation and beam strahlung can make life difficult.

3   Physics Requirements on Detectors

In standard hard scattering processes at the TeV scale, particle jets from energetic partons will be the principle items to be observed in a detector. Jet multiplicities, angular distributions, energy distributions of jets and jet masses will be studied as well as particle distributions inside jets. In particular an investigation of jet-jet mass spectra may give hints cocerning the existence of new particles. New physics may also show up in missing transverse energy distributions if non-interacting particles have left the interaction region. Often leptonic and semileptonic weak decays give a good signature for new particles so that lepton identification will play an essential role.

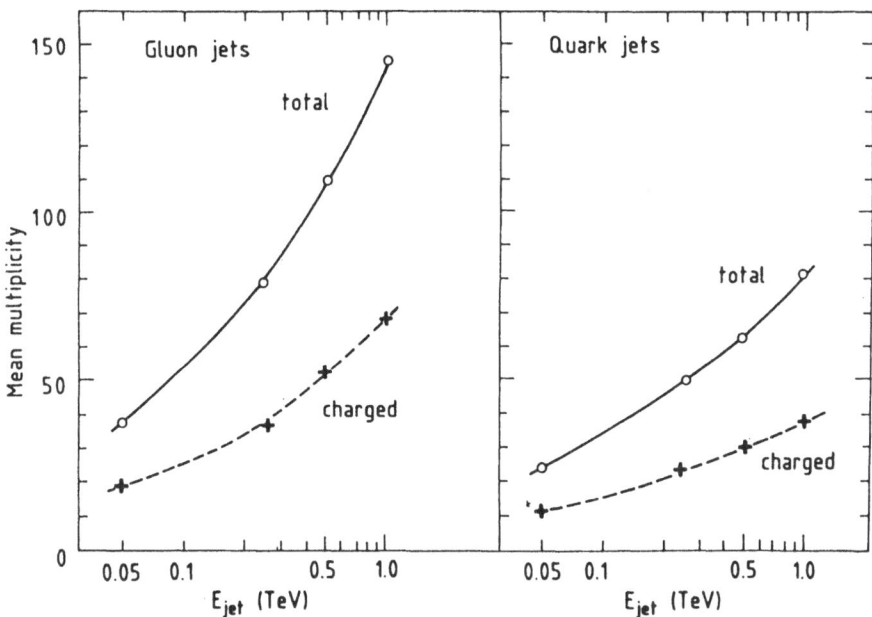

Figure 4. Particle multiplicities in 1 TeV gluon and quark jets.

3.1 TeV Jets

Before designing a detector for a TeV collider, event topologies have to be studied. From the point of view of detector design it is sufficient to consider the most difficult case with narrow jets in the final state, because a detector which is good for jets will also be adequate for final states where particles are more uniformly distributed in solid angle.

As a reference 1 TeV jets will be used, which are characterised through the plots of Figs. 4 to 6. At this level differences between quark and gluon jets, heavy quarks or light quarks will be neglected and only average values are considered. For detector design the following basic quantities are relevant:

- The average number of charged particles is $< n_{ch} > = 55$ (Fig.4).
- The total multiplicity of an event is $< n > = 110$ (Fig.4).
- On the average the angular separation between neighbouring charged particles is $< \delta\Theta > = 30$ mrad.
- 50% of the particles are contained in a cone with a half opening angle of $\delta\Theta = 20°$ (Fig.5).
- The energy flux is more collimated than the particle flux. 50% of the energy flow is contained in a $5°$ cone (Fig.5).
- In spite of the strong collimation of the energy flux, a cone of $\Theta = 60°$ is required to catch 90% of the jet energy.
- 15% of the particles have an energy $E \geq 80$ GeV (Fig.6).

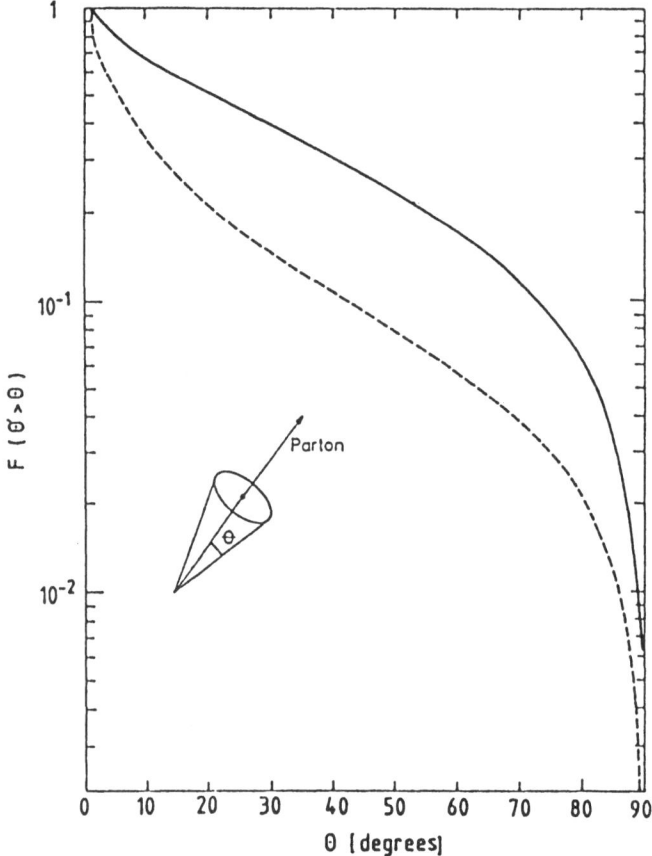

Figure 5. Energy fraction (dotted line) and particle fraction (solid line) outside a cone with half opening angle Θ.

- 50% of the jet energy is carried by particles with E$\geq$ 80 GeV.
- 10% of the particles have an energy E$\geq$ 300 GeV (Fig.6).
- 10% of the jet energy is carried by particles with E$\geq$ 300 GeV.

A few comments on jet properties will be made here as far as they have an impact on the design of detectors at a supercollider. Fine grained total absorbtion calorimeters are adequate instruments for the reconstruction of high energy jets, because calorimeters are sensitive to charged and neutral particles, their energy resolution improves with increasing energy and their performance is insensitive to high local particle fluxes. Tracking devices backed up by electromagnetic calorimeters, which is a typical LEP set up, are not sufficient to reconstruct jets at hadron supercolliders, because the momentum resolution for charged particles recorded in a tracking device inside a magnetic field deteriorates with increasing momentum, the high density of particles leads to overlapping tracks which cannot be resolved and tracks cannot be uniquely matched with energy depositions in the calorimeter so that the overall jet energy resolution is degraded. Thus calorimeters will play an essential role at supercolliders.

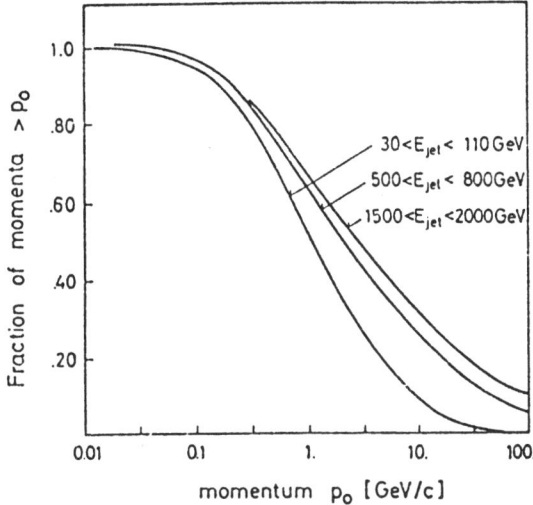

Figure 6. Fraction of particles with momenta $\geq p_0$.

3.2   Lepton Identification

The reconstruction of leptons is an important tool in the search for new physics. In addition muon identification is essential to back up calorimetry, because muons do not deposit their energy in hadron calorimeters. Their momentum has to be measured separately and added to the total jet energy later. Furthermore neutrinos and other non-interacting particles are recognised by missing p_t measurements. This is a technique which relies on the full reconstruction of the energy flow of interacting particles including muons.

The inclusive muon spectrum coming from heavy flavour decays is shown in Fig.7. The muons have high momenta and may be recognised as penetrating tracks. Their momenta could be measured in a magnetised iron spectrometer, behind a hadron calorimeter. For isolated muons track linking to the interaction point will be possible in most cases with the help of a central tracking device. For muons associated with the core of a jet the linking procedure may fail due to higher local particle densities.

Electron identification is more difficult than muon identification. Electrons can be recognised by transition radiation devices, by tracing shower profiles in an electromagnetic calorimeter or by comparing a magnetic momentum measurement with the energy deposit in a shower counter or by a combination of these methods. Electron identification is severely affected by converted photons unless care is taken to minimize the amount of material between the interaction point and the electron identifier. Therefore electrons have to be identified in detector components close to the interaction point, where overlap problems severely reduce the detection efficiency even for isolated electrons. In the core of a jet, electron identification will not be possible.

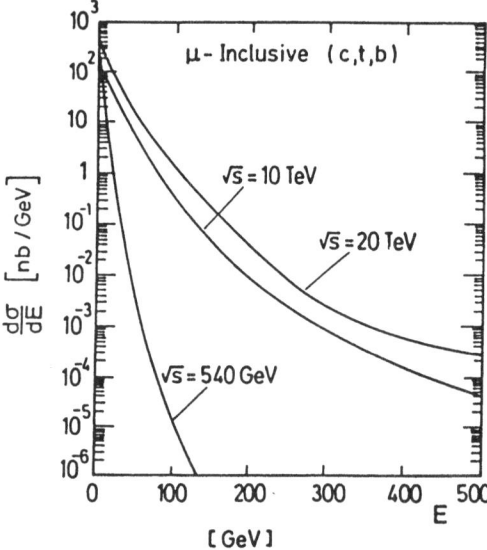

Figure 7. Inclusive μ spectrum from heavy flavour decays.

4 Detector Performance

In this chapter a detector design will be outlined and the performance of the relevant components will be discussed. Two alternative solutions one for high, the other for low luminosity interaction regions will be described.

4.1 Detection Principle

In a typical experimental set up at a collider sketched in Fig.8, the interaction point is surrounded by a tracking section, preferable in a magnetic field. That part of the tracking device which is closest to the luminous region needs a particularly high spatial resolution and good multi-track resolution. It is referred to as a vertex detector. The tracking section may also contain a transition radiation device for particle identification. The tracking area is surrounded by a total absorption calorimeter, the innermost layer of which consists of a high Z material electromagnetic shower counter. The outermost part of the detector is made of magnetised iron slabs interleaved with tracking chambers to identify muons and to measure their momenta. At the same time the muon spectrometer may serve as a backup calorimeter, in which showers leaking out of the main calorimeter are registered.

4.2 Calorimetry

Calorimetry will be the essential part of any detector [6]. Because missing transverse momentum measurements will be used as a tool to trace non interacting

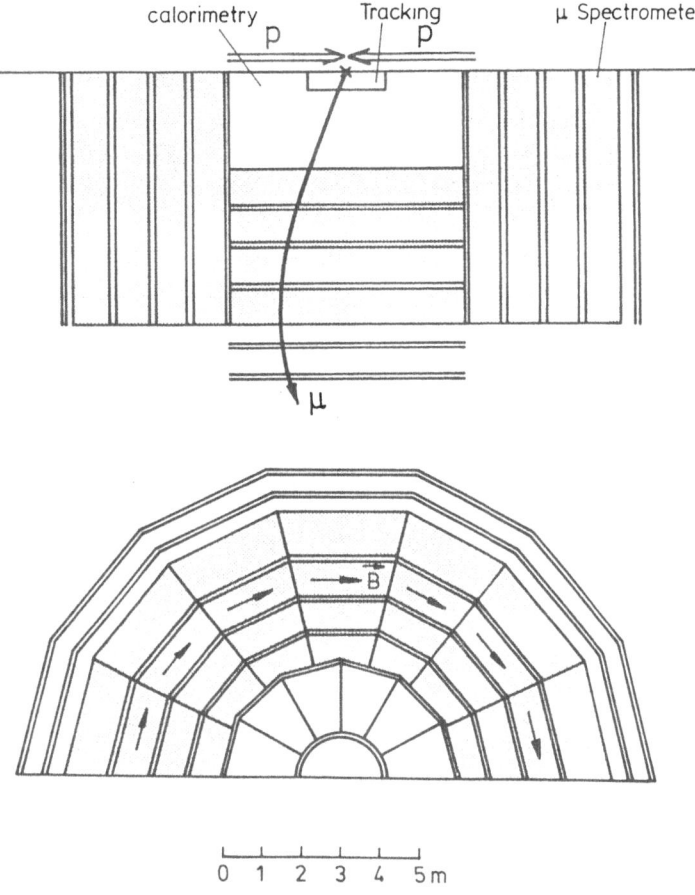

Figure 8. Typical detector arrangement at a supercollider.

particles, a hermetic calorimenter is obligatory with as small a beam hole as possible.

The granularity of a calorimeter should match the size of a typical jet core and allow a determination of the jet direction with an accuracy which it not worse than the smearing of the parton direction due to fragmentation effects. Besides by the size of incident particle jets the spatial resolution of a calorimeter is limited by the finit size of the showers themselves. The performance of the calorimeter does not improve if the segmentation is smaller than the lateral extension of an average hadronic or electromagnetic shower. Therefore a careful optimisation of calorimeter cell sizes is necessary.

Electromagnetic and hadronic showers are different in shape. While the transverse size of a hadronic shower grows linearly within the first 1.5 absorption lengths and remains constant thereafter, the width of an electromagnetic shower keeps growing up to the end of the shower as indicated in Fig.9. Also transverse and longitudinal dimensions of electromagnetic and hadronic showers are widely different. The characteristic length for electromagnetic showers is the radiation length

X_o, while hadronic showers are measured in terms of the nuclear absorption length λ. For typical detector materials the two quantities may differ by as much as a factor of 10.

The mean radius of an electromagnetic shower is given by the Moliere radius of the detector material.[1] Approximately 95% of the shower energy is deposited inside a cylinder of $2 \times R_m$. A typical lead scintillator sandwich counter has a Moliere radius of $R_m \approx 2.5$ cm.

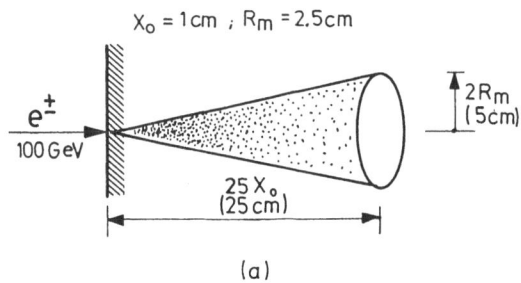

(a)

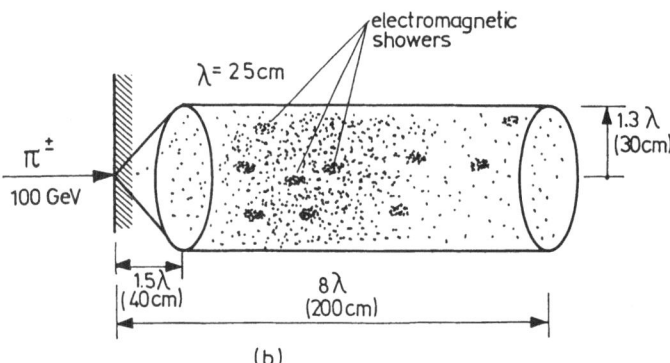

(b)

Figure 9. Shapes and particle densities in a) electromagnetic and b) hadronic showers.

The lateral extension of a hadronic shower is typically one nuclear absorption length and 95% of the energy is contained in a cylinder with a radius of about 1.3 λ around the incident particle direction. For modern hadron calorimeters the effective nuclear absorption length is $\lambda_{eff} \approx 25$ cm for an Uranium/scintillator sandwich (ZEUS [4]) and $\lambda_{eff} \approx 24$ cm for a steel/liquid Argon calorimeter (H1 [3]). To obtain a good energy resolution it is important to keep the leakage through the back of the calorimeter small. The required depth to contain 99% of the

---

[1] Moliere radius $R_m = 21/E_c$; E_c = critical energy in MeV, R_m in units of radiation length.

hadronic energy increases only logarithmically with energy and a depth of about 9λ is sufficient to contain TeV showers.

As a consequence of the growth of the shower width with depth, the granularity of calorimeters may decrease with depth without loss of information. If a calorimeter starts with a cell size of 2.0×2.0 cm the core (50% energy) of a 1 TeV jet with a full opening angle of 10° would cover about 20 calorimeter cells in a calorimeter starting 0.5 m from the interaction point. With this segmentation the direction of isolated particles will be measured with an accuracy of $\delta\Theta \leq 10$ mrad. This suffficient to even study details inside jets. At hadron colliders the number of secondary particles per pseudo-rapidity interval $\Delta\eta$ is constant and in order to economise on the number of electronic channels, one may increase the cell size in the central part and have smaller cells in the foreward and backward cones of the detector.

Besides granularity, energy resolution is the main characteristic of a hadron calorimeter. The energy resolution can be parametrised by

$$\frac{\sigma E}{E} = \frac{C_1}{\sqrt{E}} \bigoplus C_2 \cdot (e/h - 1) \bigoplus C_3.$$

C_1, C_2 and C_3 are constants which depend on the calorimeter design, the ratio e/h describes the difference in the response function of the calorimeter to electromagnetically showering particles and hadrons. Is this ratio one (compensating calorimeter) the response of the calorimeter is the same for electrons and hadrons of the same energy. The symbol $\bigoplus$ means add in quadrature.

The term C_1 arises from shower and sampling fluctuations. In modern experiments the value of C_1 for hadrons lies between 33% according to test beam measurements of the ZEUS Uranium scintillator stacks and 53% for the fine grained non-compensating liquid Argon hadron calorimeter of the H1 collaboration. For electromagnetic showers C_1 is of the order of 10%. At high energy the terms C_2 and C_3 dominate. The term proportional to C_2 vanishes for compensating calorimeters and as C_2 can be big, typically 8% for $e/h=1.4$, it is advisable to use compensating calorimeters. The term C_3 arises from calibration errors and other experimental imperfections. An effort should be made to keep this term small. It seems possible to keep the constant term as small as 1 to 2% for large calorimeters over long periods of time.

A good measure with which to describe the performance of a calorimeter is the jet-jet mass resolution. It turns out that the error introduced by fragmentation is about 1% at 1 TeV jet-jet mass and 0.5% at 4 TeV. The experimentally achievable mass resolution has been estimated by Monte Carlo calculations with realistic jets including jet fluctuations and simple models for a calorimeter [7]. Typical results of such calculations are shown in Fig.10. It turns out that the mass resolution is insensitive to the solid angle subtended by one calorimeter cell. It does not deteriorate if the size is increased by a factor of 4 w.r.t. the calorimeter discussed above. The resolution is, however, sensitive to overlapping events [7].

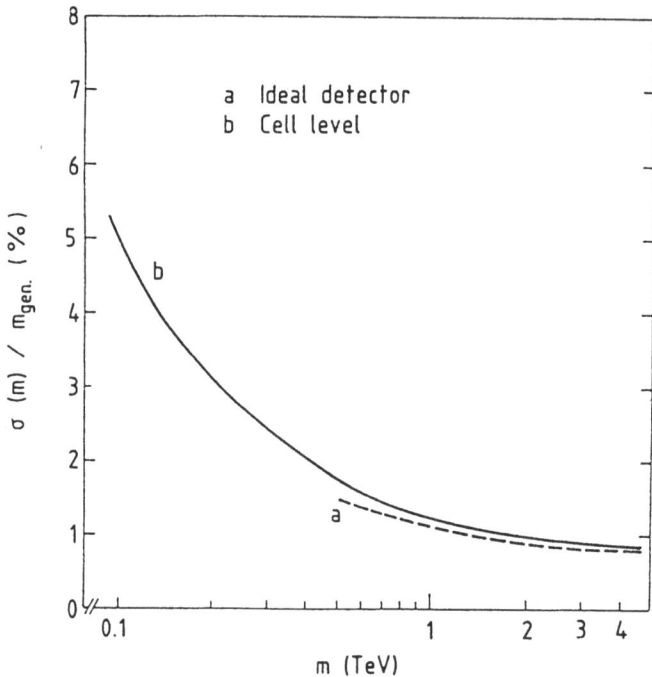

Figure 10. Jet-jet mass resolution.

4.3 Tracking

The installation of tracking systems at TeV hadron colliders has a considerably smaller impact on the physics performance of an experiment than at e^+e^- colliders like LEP for, obvious reasons:

- The momentum resolution deteriorates with increasing momentum and the tracking volume required to obtain a useful resolution quickly becomes prohibitively large.
- Tracking devices are sensitive to overlapping events and general machine background.
- Insufficient double-track resolution is a problem.
- Tracking devices, in particular if they have high spatial resolution, tend to fail at high rates.

Despite the fact that tracking systems do not play the key role in an experimental set up, a compact tracker inside a solenoidal magnetic field of 0.5 to 1 Tesla with tracking radii between 0.5 and 1 m is a good investment. Such a device will support and improve the physics capability of a detector in many respects.

- Electron identification cannot be done without at least some information on the track entering the electromagnetic part of the calorimeter.

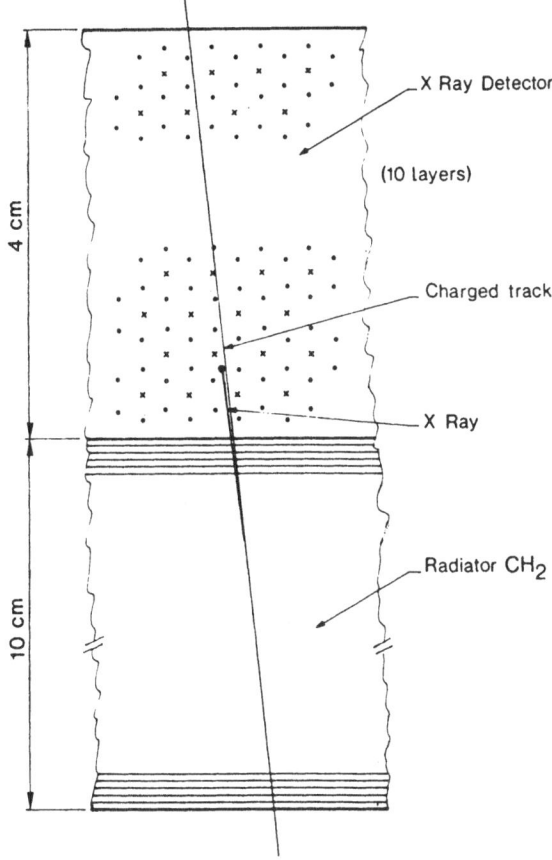

Figure 11. Typical cell of a transition radiation detector.

- A knowledge of the event vertex improves the jet-jet mass resolution and counting event vertices is the simplest way of determining the number of overlapping events in a trigger.
- A vertex detector could find secondary vertices from long lived particles.
- In cases where a muon track identified in the outside muon spectrometer can be linked to a track coming from the main event vertex the background to the prompt muon signal is significantly reduced.
- A very important aspect of tracking is that it gives redundancy and it can be of great help in the interpretation of some rare events.

A special case of a tracking device is a transition radiation detector, sketched in Fig. 11. The transition radiation, generated in a foil or fibre radiator, is registered in X-ray sensitive chambers, which are at the same time part of a general tracking system. Such a transition radiation detector is used to support electron identification. At present very little experience exists with transition radiation devices in an environment with a high density of tracks. At HERA they will be installed by H1 and ZEUS and valuable information on their operation will be gained with the start up of HERA.

4.4 Electron Identification

Electron identification will only be possible for isolated tracks. The simplest method, which can be employed at a supercollider is based on longitudinal and lateral shower sampling in an electromagnetic calorimeter. To make this method work, the incoming charged track has to be recognised in order to distinguish electrons from photons. Furthermore two close tracks have to be recognised to enable the identification of converted photons and single electrons. By this technique, which might even work in a high luminosity environment rejection rates of a few 10^4 can be achieved.

Methods in which the result of a magnetic momentum measurement is compared with a calorimetric energy measurement will not be applicable above about 100 GeV. The observation of transition radiation for electron identification is in principle a powerful tool, but it requires space and considerable R & D work has to be done before one could envisage employing this method at a TeV collider. The estimated rejection rate of a transition radiation device is of the order of 10^3. Therefore such an instrument is a valuable tool to support other methods.

4.5 Muon Identification

Muons are detected as penetrating particles behind a total absorption hadron calorimeter in a magnetised iron spectrometer. Such a spectrometer, which is sketched in Fig. 12, will need a total of about 4 m magnetised iron to suppress punch through and to achieve the necessary momentum resolution of at least 30% to enable charge determination. The iron has to be segmented and a muon track should be sampled about every meter. With a spatial resolution of 300 μm in the sampling chambers the momentum resolution will be 10% up to 1 TeV and it will reach 30% at 3 TeV as shown in Fig.13. With sampling chambers which have a resolution of 200 μm, the 30% limit could be pushed to 10 TeV.

The muon spectrometer is by far the largest component of any collider experiment and it will essentially determine the size and the weight of the experiment. Therefore, in order to keep the experiment down to a manageable size and weight, the tracking system and the hadron calorimeter have to be as compact as possible. Some space can be saved if the first part of the muon calorimeter has a fine longitudinal segmentation and is used to back up the central calorimeter.

For a muon spectrometer of the size discussed here, the alignment of the tracking chambers which has to be better than their spatial resolution, will be difficult and will put stringent demands on the machanical design of an experiment. With the experiment installed in the interaction region the calibration may be monitored by isolated muons from heavy flavour decays.

5 Detector Arrangements

The detector components which have been described in the previous sections, have to be assembled to form a detector. Arguments have been given that two types of experiments should be designed. One for high luminosity and one general purpose detector for low luminosity. The two alternatives are sketched in Fig.14.

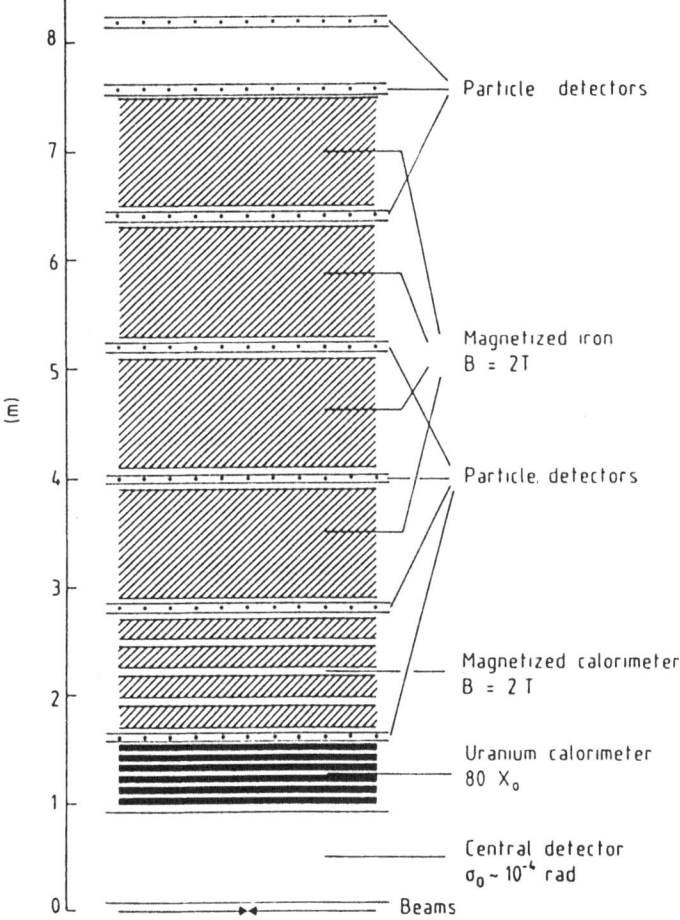

Figure 12. Schematic of a μ spectrometer at a collider.

5.1 High Luminosity Detector

A detector, which is capable of operating at high luminosity with multiple event overlaps, has to be robust with respect to background and it has to have a highly selective trigger system. Because not all details of a particular event will be measured, the tracking system can be reduced to the bare minimum of a vertex detector inside a weak (~ 0.7 Tesla) axial magnetic field in order to suppress low energy background and to recognise low energy tracks. To support electron identification a device should be installed to tag converted photons. One possibility would be to use silicon pads or strips which are sensitive to one minimum ionising particle. This is a technique which has been successfully employed by UA2 at the CERN collider.

Due to the compactness of the central detector, the fine grained calorimeter starts at a small radius and the size of the muon spectrometer is correspondingly reduced. The lower part of Fig.14 shows the arrangement of components around the interaction region for a high luminosity detector. The total weight of the experiment can be kept below about 20,000 tonnes.

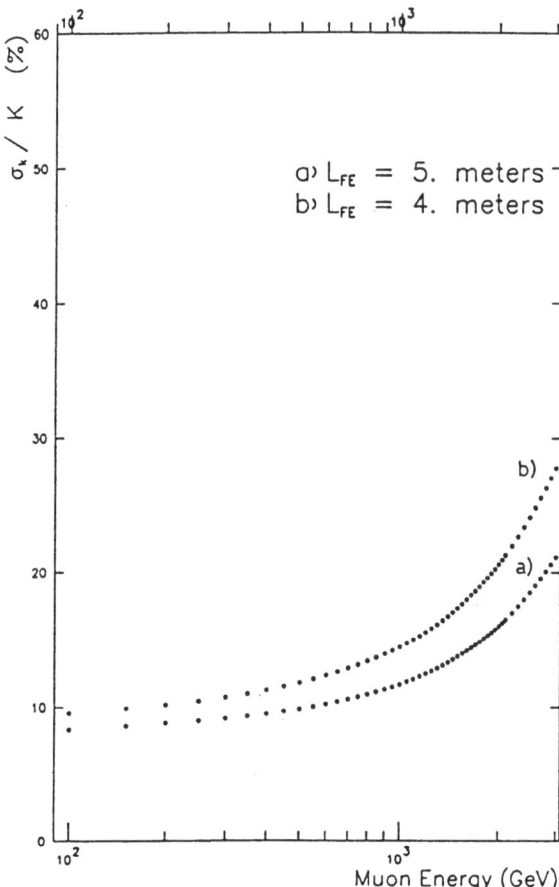

Figure 13. Momentum resolution of a μ spectrometer as a function of μ momentum for two thicknesses of the iron absorber.

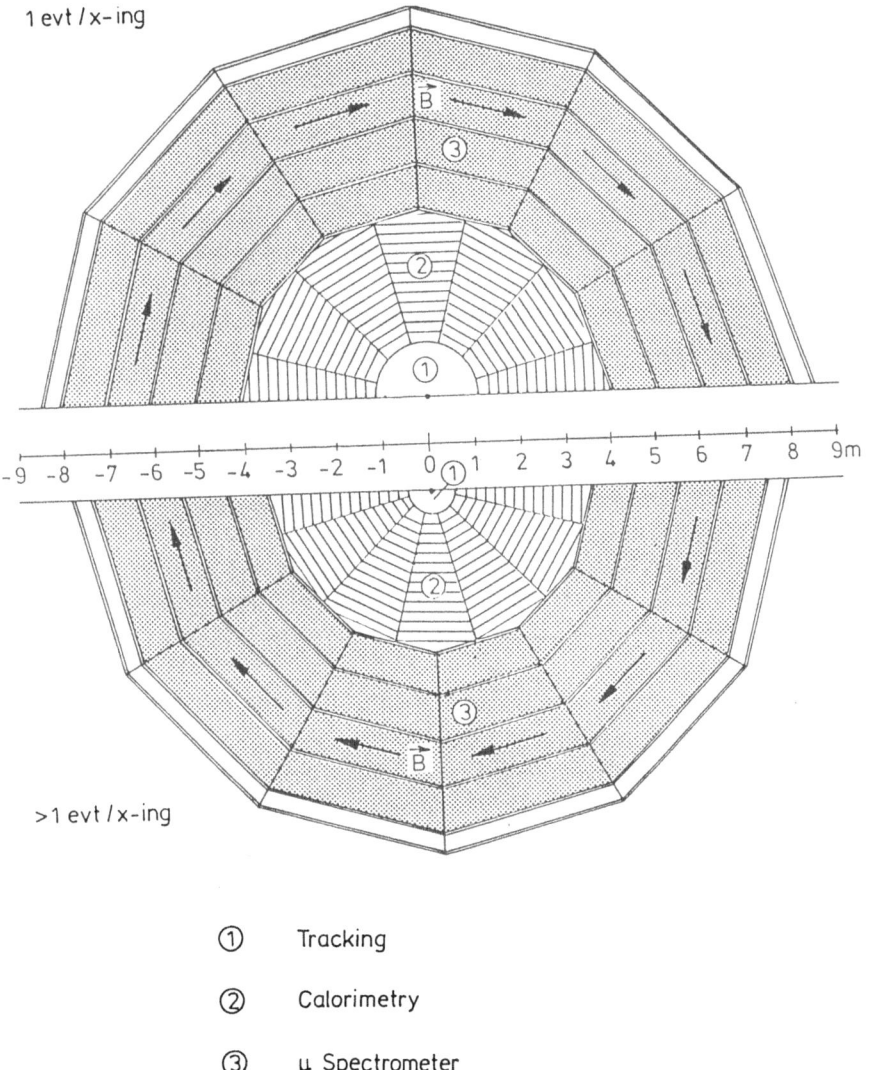

1 evt / x-ing

>1 evt / x-ing

①   Tracking

②   Calorimetry

③   μ Spectrometer

Figure 14. Transverse view of a high and low luminosity detector.

5.2   General Purpose Detector

A detector for operation at low luminosity with only one event per trigger will typically consist of a sophisticated tracking system inside a solenoidal magnetic field, a fine grained calorimeter followed by a muon spectrometer of magnetised iron. A typical set up is sketched in the upper half of Fig.14. It will weigh between 25.000 and 30,000 tonnes. Because of its high weight it has to be built in small units which can be moved independently of one another.

The trigger for such an experiment has to be very flexible in order to adjust to varying conditions and physics demands.

5.3   Trigger and Data Acquisition

Due to the mismatch between the high background from "soft" proton-proton collisions and the rate at which large p_t parton collisions occur any trigger has to be very selective. The data logging rate for off line analysis is limited to a few Hertz not only by the transfer speed of data busses but also by the amount of data which can be processed within a reasonable time.

The time required for complex trigger decisions is of the order of msec while the bunch crossing time measures tens of nsec. Therefore decisions are taken in steps rejecting at each level events which do not meet the requirements of a valid physice event. First level trigger decisions are processed by hard wired logic circuits within a time span of less than 1 μsec using only those signals from the detector which arrive less than typically 100 nsec. For second level trigger decisions digital information is used which is fed to specialised trigger processors which need a few hundred microseconds for their analysis. The third level trigger decision is then taken by by processor arrays making use of all detector information. At this level some secondary quantities are derived during the event validation like charged tracks, jets, clusters in calorimeters or energy momentum balance which are also used in off line analysis steps. During the time of the trigger decisions all detector information is stored in analog and digital pipe lines. These pipe lines represent a major fraction of the cost for electronics.

6   Conclusions and Outlook

In spite of the technical difficulties one will meet at the next generation hadron colliders, it is not unrealistic to consider building a detector for a 50 TeV hadron machine based on current technology with an R&D program similar to the effort which went into the construction of LEP and HERA experiments. For more ambitious physics programs beyond 100 TeV a major breakthrough, in particular on the electronics sector, has to be achieved before the physics potential of such a machine can be exploited.

Extrapolating from current experience to 100 or 200 TeV colliders involves large uncertainties. The total cross-section may go up by 50% and so will the multiple interaction rate, while the average parton-parton cross-section, taken at

$\sqrt{\hat{s}} = 25$ TeV, will decrease by about a factor of 25. This decrease will, however, be compensated by the higher parton luminosity. In order to exploite the new energy regime a luminosity between 10^{33} and 10^{34} cm^{-2} sec^{-1} is necessary.

An extensive detector R&D program has to accompany the construction of a supercollider. In particular an effort should be made to improve muon spectrometers, because more efficient muon detection may drastically reduce the size and cost of an experiment. Calorimetry has also to be improved. A technique is needed for building fine grained compensating calorimeters avoiding dead zones which can be read out at repetition rates of more than 100 MHz. Tracking systems have to be designed with improved spatial resolution and double track resolution. Current tracking chambers are sensitive to high rates and an improvement is necessary before they can be built into a supercollider detector. The development of radiation resistant electronics is essential.

The demands on read out electronics and data acquisition at a supercollider are far more stringent than can be fulfilled with current technology. The high data rate and the vast amount of information from an experiment require new computing concepts. In addition collaborations at a supercollider will have members from many laboratories from all over the world, who all participate in the data analysis. Remote computing has therefore to be further developed.

Acknowledgement. I greatfully acknowledge the hospitality extended to me during my stays at Erice. I would like to thank Prof. A. Zichichi who initiated the workshop and the staff at the Ettore Majorana for their excellent support. I am graetful to A. Ali for organising an interesting meeting with stimulating discussions.

References

[1] Proc. 1984 Summer Study on the Design and Utilisation of the Superconducting Super Collider
Snowmass, Colo. eds. R.Donaldson and J.Morfin (API,NY,1985)

[2] Proc. ECFA-CERN Workshop on a Large Hadron Collider in the LEP Tunnel, Lausanne and CERN, 1984, ed. M.Jacob, CERN 84-10 (1984)
Proc. of the Workshop on Physics at Future Accelerators, La Thuile and CERN, CERN 87-07 (1987)

[3] Letter of Intent for an Experiment at HERA, H1 Collaboration DESY (1985), Internal Document H1
Technical Proposal for the H1 Detector, H1 Collaboration DESY (1986), Internal Document H1
Technical Progress Report H1 Collaboration, H1 Collaboration DESY (1987), Internal Document H1

[4] Letter of Intent ZEUS a Detector for HERA, ZEUS Collaboration DESY (1985), Internal Document ZEUS

Technical Proposal for the ZEUS Detector, ZEUS Collaboration DESY (1986), Internal Document ZEUS

Technical Progress Report H1 Collaboration, ZEUS Collaboration DESY (1987), Internal Document ZEUS

[5] A. Ali et al. in Proc. ECFA-CERN Workshop on a Large Hadron Collider in the LEP Tunnel, Lausanne and CERN, 1984, CERN 84-10 (1984) Vol.2

[6] R. Wigmans. Report on calorimetry, This Proceedings

[7] T. Akesson et al. in Proc. ECFA-CERN Workshop on a Large Hadron Collider in the LEP Tunnel, Lausanne and CERN, 1984, CERN 84-10 (1984) Vol.1

23

CALORIMETRY IN THE SUPERCOLLIDER ERA

Richard Wigmans

CERN, Genève, Switzerland

ABSTRACT

The required precision, the event rates to be handled, and the radiation levels envisaged for experiments at proposed high-luminosity pp Supercolliders like the Eloisatron are a formidable challenge to detector technology. High quality calorimetry will be the key to success for such experiments. We review the requirements needed for calorimeters operating at a Supercollider, and the present state of the art concerning these devices. In the framework of the LAA project, an attempt is being made to develop a new type of calorimeter that could meet the Supercollider challenge. The goals and achievements of this project are discussed in some detail.

1. INTRODUCTION

It has become increasingly clear that the success of experiments at (proton-proton) supercolliders will depend to a large extent on the quality of the available calorimetry[1] . Calorimeters or total absorption detectors are devices in which the particle to be measured interacts and deposits all its energy in the form of a shower of decreasingly lower-energy particles. The detector is made such that a certain (usually small and hopefully constant) fraction of the initial particle energy is transformed into a measurable signal (light, electrical charge).

Large calorimeters play already a key role in almost any big experiment currently running or being prepared at high-energy colliding beam machines, in fixed-target neutrino or heavy-ion scattering, and in nucleon decay experiments.

Some major discoveries in particles physics, *e.g.* the discovery of the intermediate vector bosons W and Z, would have been impossible without calorimeters.

The reasons why calorimeters have become so important that experimental particle physicists are nowadays frequently spending more than half of their detector budget on them, can be divided into two classes. Firstly, there are reasons related to the calorimeter *properties*:

a) Calorimeters are sensitive to both charged and neutral particles.

b) Owing to differences in the characteristic shower patterns some crucial particle identification is possible (hadron/electron/muon/neutrino separation).

c) Since calorimetry is based on statistical processes, the measurement accuracy *improves* with increasing energy, in contrast to other detectors, *e.g.* magnetic spectrometers.

d) The calorimeter dimensions needed to contain showers increase only slightly with the energy (log E), which means that even at the highest energies envisaged one can work with a rather compact instrument (cost!).

e) Calorimeters can be fast; response times much better than 100 ns are entirely feasible, which is important if the detect[Bor has to operate in a high rate environment.

f) They don't require a magnetic field for energy measurements.

g) They can be segmented to a very high degree, which allows a precise measurement of the direction of the incoming particles.

Secondly, there are reasons related to the *physics* that one wants to study. Here, the emphasis has clearly shifted from a precise measurement of the 4-vectors of all individual reaction products to more global event characteristics, indicative for interesting processes at the constituent level. These event characteristics include missing (transverse) energy, total transverse energy, jet production, *etc.* Calorimeters are extremely well suited for this purpose. They can make sense out of the sometimes very discouraging forest of tracks that (electronic) bubble chambers would yield in the TeV era (fig.1). Moreover, they can make this sense very fast, within 100 ns. This triggering capability is perhaps the most crucial advantage offered by calorimeters.

The cross sections at a multi-TeV pp supercollider are such that new physics may be expected in at maximum 10^{-8} of all collisions, probably even several orders of magnitude lower[2] . That is why the design luminosity of these machines

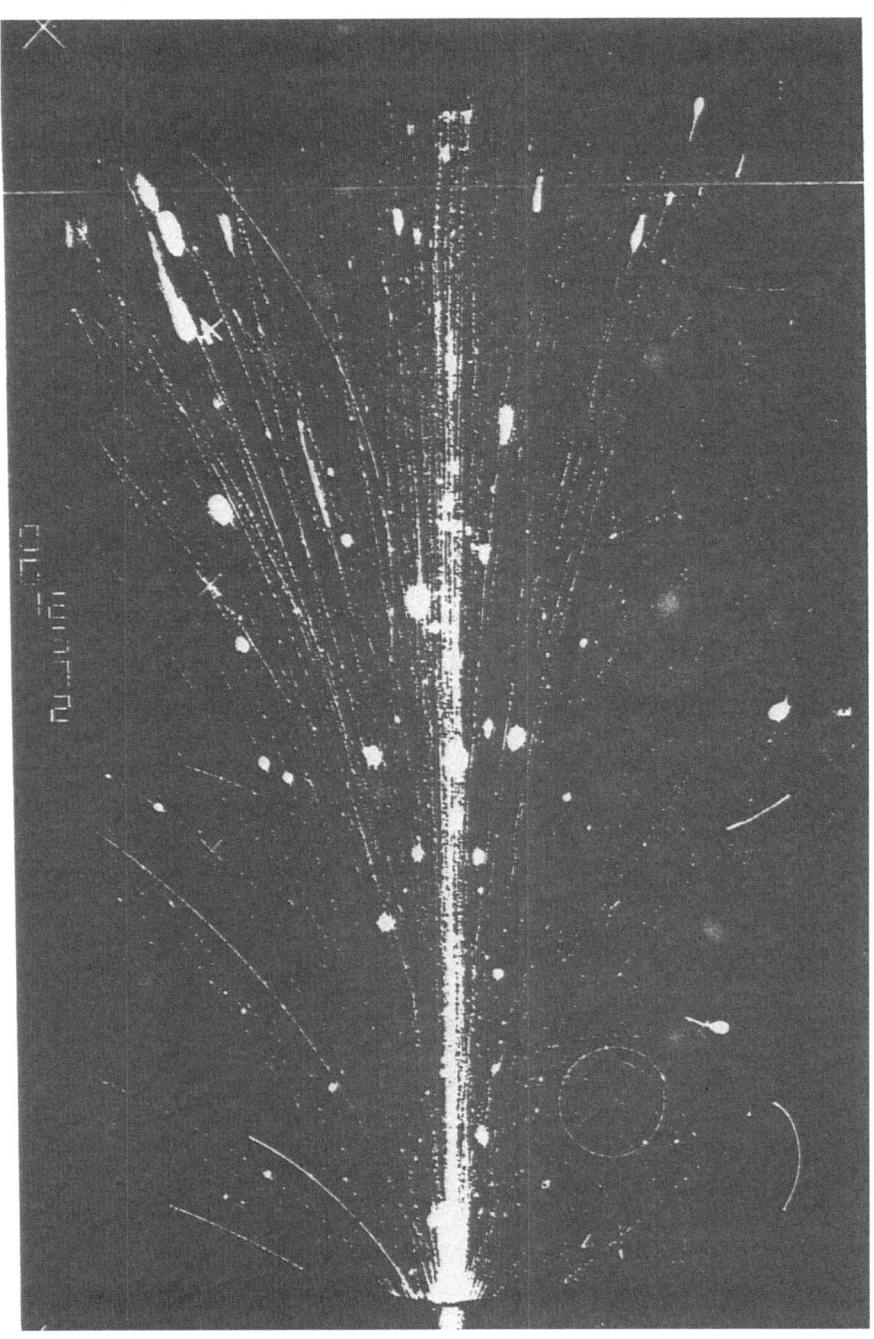

Figure 1

Streamer chamber picture of a collision between a 3.2 TeV ^{16}O ion and a tungsten nucleus at rest. The picture does *not* show the few hundred neutral particles also produced in this interaction. Data from NA35.

must be so high. Calorimeters are unique in their capability of reducing the primary event rate of $\sim 100\text{MHz}$ down to a level that can be handled by the readout and data recording electronics ($\sim 1\text{Hz}$), while retaining all those events that might contain signatures of new physics.

Being so crucial for future experimental successes, the LAA project naturally contains an important component aiming for the best possible calorimetry in the supercollider era.

2. REQUIREMENTS FOR A CALORIMETER AT A SUPERCOLLIDER

The calorimeter serving experiments at a supercollider will have to fulfill the following tasks:

1) Provide the triggers that select the potentially interesting events. Since these represent only a tiny fraction of the total number of interactions, extraordinary requirements on trigger selectivity, efficiency and reliability have to be met.

2) Provide the experimental information needed for the analysis of hadron production in the pp collisions (jet physics).

3) Provide crucial information on lepton production.

In order to be able to properly fulfill these tasks, supercollider calorimeters should respond to showering particles with a signal that is fast, independent of the type of particle or its impact point, and narrowly distributed as a Gaussian around a mean value that is proportional to the particle energy. It should be possible to distinguish electromagnetically interacting particles (e, γ) from hadrons with a high degree of reliability by means of the shower profile, and the noise should be sufficiently low to measure muons.

The performance of supercollider calorimeters will, therefore, crucially depend on (in random order of importance):

a) The hermeticity

b) The granularity

c) The energy resolution

d) The e/h signal ratio

e) The uniformity of the response

f) The response time

g) The noise level

h) The signal stability

Because of the harsh environment in which the calorimeters will have to operate, their components and in particular the active material and the electronics mounted inside the sensitive volume, should be able to sustain high radiation levels. Although the calorimeters should be sufficiently large to make the effects of shower leakage negligible, they should also be as compact as possible in order to limit the size of costly downstream equipment (magnet, muon detector).

The latter requirement calls for the use of high-density shower absorption material. The desire to do excellent electron identification naturally favours high-Z absorbers. The ratio between the nuclear interaction length and the radiation length, the relevant parameter in this respect, is for materials like uranium and lead a factor three larger than for copper or iron.

3. THE STATE OF THE ART

In the previous section the three main calorimeter tasks were mentioned: Triggering, lepton identification and jet analysis. The powerful calorimeter capabilities can be illustrated with the discovery of the W particle, where the first two aspects were employed. The calorimeter trigger selected from many millions of interactions the few ones that had a large missing transverse energy, combined with an energetic electron in the opposite hemisphere; these turned out to be events in which the goldplated intermediate vector bosons were produced.

In experiments at a future supercollider one will be particularly interested in events where such bosons are produced, since much of the new physics looked for is expected to be accessible through such events. For example, the decay mode $H^0 \rightarrow WW$ is among the most promising ones for detecting the Higgs particle. An efficient and reliable W trigger is therefore very important. It would be very profitable if one could use jets for this purpose, since the branching ratio $W \rightarrow q\bar{q}$ is an order of magnitude larger than for the process $W \rightarrow e\nu$, used for the discovery of the W particle. The Higgs search would thus be two orders of magnitude more sensitive when jets are employed than with leptons alone, if the latter option can be used at all because of the confusion created by 2 escaping neutrinos.

An interesting study on the production of jets was done by the UA2 Collaboration[3] . Figure 2 shows the invariant mass of jet-jet systems produced in $p\bar{p}$ collisions at 630 GeV. A clear enhancement over the QCD background is observed in the $W - Z$ mass region. This figure shows that a better mass resolution, i.e. a better energy resolution for the jet detection, would be very important, since it would increase the signal/background ratio for a given mass bin, and resolve the W and Z peaks.

Other new physics expected in the TeV region will be accessible through the missing (transverse) energy signature, e.g. the production of supersymmetric particles. Apart from having a good energy resolution, the calorimeter should also be as hermetic as possible for these studies, since any crack will increase the number of fake triggers. The energy resolution, hermeticity and rate capability of the calorimeter are therefore intimately related to prime physics goals at the supercollider.

The hadron calorimeters employed in experiments that are either running or in an advanced stage of preparation at present, are certainly not meeting the

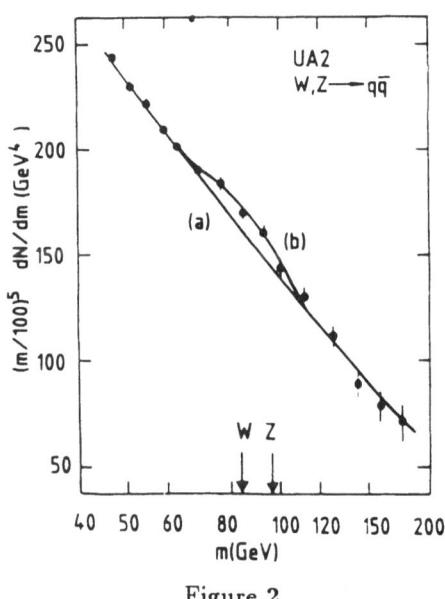

Figure 2

Jet-jet invariant mass distribution, measured in $p\bar{p}$ collisions at $\sqrt{s}$ = 630GeV. Data from ref. 3.

supercollider requirements for these important issues. The calorimeters with a readout medium based on electron drifting (*e.g.* gaseous media in wire chambers or liquids in ionization chambers) are definitely too slow, with their time constant of typically $1\mu s$ compared to the 10 ns repetition rate (for a luminosity of 10^{33}). The hermeticity is severely jeopardized in cryogenic systems (liquid argon), while the energy resolution is by no means optimal in most systems.

By far the best energy resolution for hadron detection has been achieved in so-called *compensating* calorimeters, *i.e.* calorimeters that respond with equal signals to electromagnetic and hadronic showers at the same energy. In this way, the hadron signal becomes independent on the fraction of the initial particle energy that is spent on π^0 production. In non-compensating calorimeters, the hadron signal *is* dependent on this fraction, and the large fluctuations in it have several highly undesirable consequences[4,5,6] : The calorimeter is *not* linear, *i.e.* the signal is not proportional to the deposited energy, the signal distribution at a given energy is *not* Gaussian, and the energy resolution σ/E does *not* improve as $1/\sqrt{E}$ at increasing energy, but more slowly and levels off at values for σ/E of typically $5 - 10\%$. There is general agreement that calorimeters at supercollider experiments should be compensating[5] .

So far, compensation has only been experimentally demonstrated in calorimeters with plastic-scintillator readout. The HELIOS Collaboration at CERN are operating a uranium/plastic-scintillator detector with an energy resolution $\sigma/E = 35\%/\sqrt{E}$, and have obtained resolutions better than 2% at the highest energies available nowadays[7] (see fig.3). The ZEUS Collaboration at DESY are building a uranium detector with the same (sandwich) structure, which will operate as from 1990 at HERA. None of the non-compensating hadron calorimeters used by other experiments have given an energy resolution better than $55\%/\sqrt{E}$, and frequently considerably worse than that, especially at high energies.

Recently, a detailed analysis of the working of hadron calorimeters showed that compensation is by no means a property unique to uranium absorber[4] . It was predicted that other absorber materials, and in particular lead, when used in combination with hydrogen-rich active media such as plastic, could give compensation as well, provided that absorber and readout material were used in the right proportion. This prediction was shortly afterwards experimentally confirmed for a lead/plastic-scintillator calorimeter, which yielded $\sigma/E = 43\%/\sqrt{E}$ (predicted was 42%), and $\epsilon/h = 1.05 \pm 0.04$[8] .

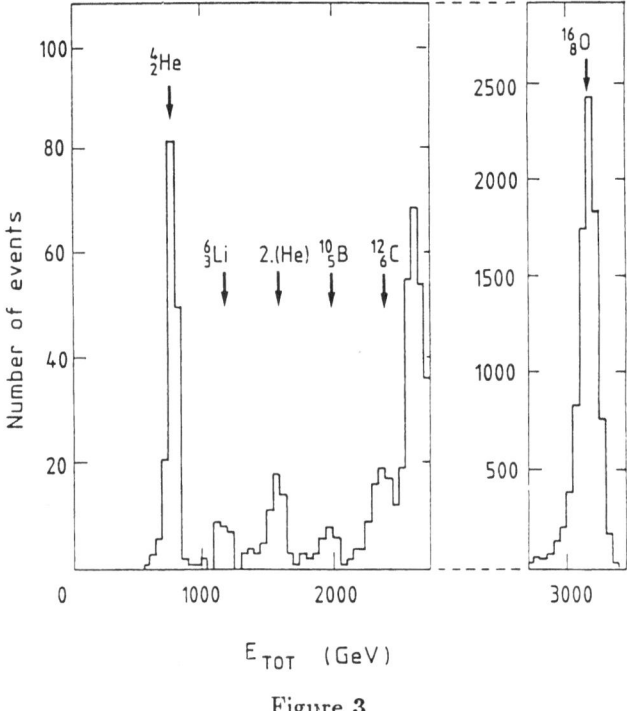

Figure **3**

The HELIOS calorimeter as a high resolution spectrometer. Total energy measured in the calorimeter for minimum bias events, showing the composition of the CERN heavy-ion beam. Data taken from ref. 7.

4. THE SPAGHETTI CALORIMETER PROJECT AT LAA

The recent breakthrough in our understanding of the working of hadron calorimeters, marked by the fact that the predicted excellent performance of a very unconventional calorimeter (thick lead plates, thin scintillator) was experimentally confirmed in great detail, makes it possible to define a dedicated $R\&D$ effort to optimize the technique.

The fact that compensation can be achieved with lead absorber is in itself a major achievement. In contrast to the cumbersome uranium, lead is available in unlimited quantities, easy to handle and to machine, not radioactive and, last but not least, cheap. It is interesting to note that because of the fact that a much larger fraction of the detector has to consist of absorber in the case of lead, the *effective* nuclear interaction lengths of *compensating* lead/scintillator and uranium/scintillator calorimeters are about equal, ~ 20cm. Therefore, in spite of its much larger density uranium does not allow for constructing more compact compensating detectors than does lead.

Apart from the fact that it makes compensating calorimeters possible, plastic scintillator has one additional major advantage: it is very fast. Signal rise times of the order of 1 ns are feasible, and therefore it is one of the very few techniques with which the envisaged supercollider event rates can possibly be handled.

The large hadron calorimeters based on plastic-scintillator readout which are either operating or under construction today (HELIOS, CDF, UA2, ZEUS), all have the classical sandwich structure, *i.e.* alternating plates of absorber and plastic perpendicular to the direction of the incoming particles. In order to be able to read the signals, the light has to make a turn of 90 degrees, which is achieved with wave length shifting plastic plates that run perpendicular to the other ones. The scintillation light is absorbed in these plates, and reemitted at a longer wave length. This inefficient process has the disadvantage of introducing regions without absorber, which usually span $5 - 10\%$ of the total detector volume. Another source of inhomogeneity is caused by light attenuation in the plastic, which may be quite important since the attenuation lengths are typically considerably less than 1 meter.

The main disadvantage of the sandwich structure is, however, the impossibility to achieve a fine granularity. Although some experiments have demonstrated a lot of ingenuity to limit the effects of the mentioned disadvantages, the results are far from ideal.

In the LAA project, we are therefore using a different structure, based on scintillating plastic fibers. This eliminates all the mentioned problems. The fibers are embedded in a lead matrix and are running (roughly) in the direction of the incoming particles. This technique was already successfully employed on a small scale, for electromagnetic shower detection in various experiments (Omega, NA38)[9].

Apart from the ultrafast timing and the compensation capability common to all scintillator structures, the potential advantages of fiber calorimeters at a supercollider include:

a) The possibility of making a completely hermetic detector (no cracks).

b) An arbitrarily fine granularity.

c) An excellent energy resolution, due to the very frequent shower sampling. We expect $15\%/\sqrt{E}$ for electromagnetic showers and $30\%/\sqrt{E}$ for jets.

d) A large signal/noise ratio, due to the high light yield (no 90° wave length shifting).

e) The possibility of using very stable light detecting devices that can operate in a magnetic field, thanks to the same high light yield.

f) A very good homogeneity. Thanks to the long attenuation length in fibers (at least several meters), the signal practically does not depend on the position where the particles interacted.

g) Easy calibration and signal stability monitoring, a crucial point at very high energies.

A detailed·discussion on all these points can be found in the proposal[10] . In the first phase of the project, which is running since October 1987, a prototype calorimeter will be built, with a simple structure where all the fibers are running parallel to one another. This prototype will be used to extensively check the optimistic expectations concerning its performance. In order to be able to do proper tests, it will be made sufficiently large to contain the highest energy hadron showers available at CERN (450 GeV protons) at a level of 99%, such as to make the effects of shower leakage small compared to the expected energy resolution. This leads to a detector of about 15 tons.

If this phase turns out to be successful, a more sophisticated structure, optimized for operating in an experiment at a storage ring, is envisaged. The quasi-pointing geometry of the fibers will then allow obtaining additional shower information (see ref. 10), which will definitely lead to improved electron identification possibilities.

5. ACHIEVEMENTS SO FAR

Although the project was started less than a year ago, some major progress towards achieving its goals has already been made.

Compensation, energy resolution

In collaboration with some physicists from the ZEUS experiment, we have studied in more detail the properties of compensating lead/plastic-scintillator calorimetry, using the detector that was built to verify the original predictions[4.8] This detector consists of 10 mm thick lead plates interleaved with 2.5 mm plastic-scintillator plates (the ratio 4 : 1 in these thicknesses is the secret of compensation).

The energy resolution of any hadron calorimeter consists of a component due to *sampling* fluctuations, arising from the fact that only a small fraction of the particle energy is deposited in the active layers, and an *intrinsic* component, due to the fact that the fraction of the particle energy transformed into ionization (the "visible" energy) fluctuates from event to event. The effects of these fluctuations on the energy resolution of the calorimeter add in quadrature:

$$\sigma_{tot}^2 = \sigma_{samp}^2 + \sigma_{intr}^2$$

The new measurements were intended to separate these two contributions, to be able to make realistic predictions for the energy resolution of the fiber calorimeter. Since the showers are sampled much more frequently in the fiber case, σ_{samp} will be considerably smaller than in the rather crude-sampling sandwich detector, which will lead to a further improvement of the already impressive energy resolution obtained with the latter.

The contributions of sampling and intrinsic fluctuations were separated in the standard way, by blocking the light of half the scintillator plates. In doing so, the contribution of sampling fluctuations in the quadratic sum is doubled, and by comparing the total resolutions measured with and without blocking, one may disentangle σ_{samp} and σ_{intr}.

We found that the energy resolution of the sandwich calorimeter was completely dominated by sampling fluctuations. When half of the plates were read out, the resolution for hadron detection went up from $42.8\%/\sqrt{E}$ to $59.2\%/\sqrt{E}$, *i.e.* by almost a factor $\sqrt{2}$. The *intrinsic* energy resolution, *i.e.* the limit for an infinitely fine sampling device, was found to be $12.7^{+4.4}_{-7.5}\%/\sqrt{E}$, averaged over the energy range 10 - 75 GeV. This is significantly better than the intrinsic limit for a compensating uranium/scintillator detector, which was measured to be $22\%/\sqrt{E}$ with the same method[11] . This can be explained as follows (see ref.4).

The intrinsic resolution is largely dominated by fluctuations in the amount of energy used to release nucleons from atomic nuclei in the nuclear spallation reactions (binding energy losses). Since most of these nucleons are neutrons in the case of high-Z target material, there is a correlation between these invisible nuclear binding energy losses and the kinetic energy carried away by neutrons. Efficient neutron detection, a crucial ingredient for compensation, therefore re-

duces the effect of fluctuations in the nuclear binding energy losses on the energy resolution. The extent to which this mechanism works depends then on the degree of correlation between the nuclear binding energy losses and the kinetic neutron energy. This correlation is considerably better in lead than in uranium, since in the latter case many of the neutrons come from fission processes. These fission neutrons are not correlated at all to the nuclear binding energy losses.

The results of our measurements are extremely encouraging, since they mean that the envisaged energy resolution for jet detection of $30\%/\sqrt{E}$, which in itself would already be a world record, is probably a conservative estimate (see fig.4).

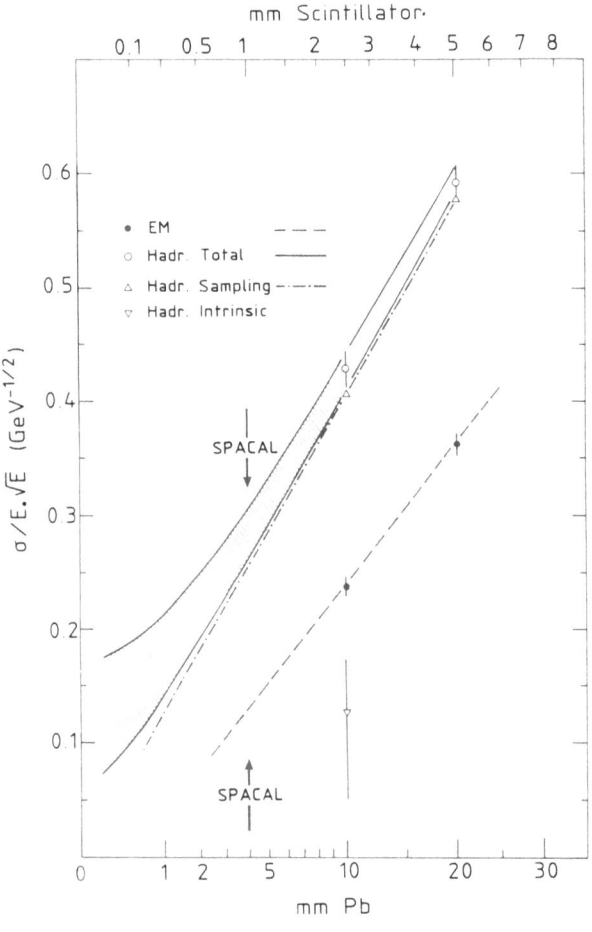

Figure 4

The energy resolution for electron and hadron detection with a compensating lead/plastic-scintillator calorimeter, as a function of the thickness of the readout layers. The experimental measurement results are included. The arrow denotes the working point of the fiber calorimeter under construction.

In a second series of measurements, half of the scintillator plates were physically *removed*. In doing so, the ratio lead/scintillator was changed by a factor 2 (8 : 1 instead of 4 : 1). We found that the e/h signal ratio was only marginally affected. This is also an important and encouraging result, since the detector envisaged for the second phase of the project will have a varying fraction of scintillator as a function of depth (pointing fibers). The measurements showed that this will not significantly affect the compensation.

Calorimeter design

Much thought has been given to the calorimeter design. We have chosen a modular structure, with each module having its own readout attached to it. The main advantage of this scheme is a very flexible setup, where modules can be rapidly exchanged, and many different things can be measured.

Previous fibre calorimeters were all built by machining grooves in lead sheets, and glueing fibres into these grooves. Especially in view of the future, more complicated projective structures, we investigated several alternative methods, among which a casting technique gave remarkably good results. In this method, a structure is set up consisting of very thin (50μm) stainless steel capillary tubes, held together by spacers. Liquid lead, or rather an alloy that contains 4% Sb for improving the mechanical properties, is poured around these tubes, yielding a Swiss cheese-like lead structure of any wanted shape. The fibres are slided into the holes afterwards. This method allows for a modular structure of the detector in a natural way.

The shape of the modules was chosen to be hexagonal. This shape matches the cylinder symmetric shower development better than rectangular shapes, it makes possible a very stable detector structure with minimal support, and avoids edge effects (fig. 5).

The size of the modules, or the final granularity of the detector, was determined by many considerations, like the typical lateral dimensions of hadronic showers and the segmentation requirements for a calorimeter at a supercollider; moreover, the useful surface of standard photomultiplier tubes, the weight of the modules, and the possibility of calibrating them with a radioactive source were important criteria. We finally chose for a design that contains 1141 fibres, weighs 86 kg, and has a lateral cross section of 48.7cm^2 (about a factor 10 smaller than

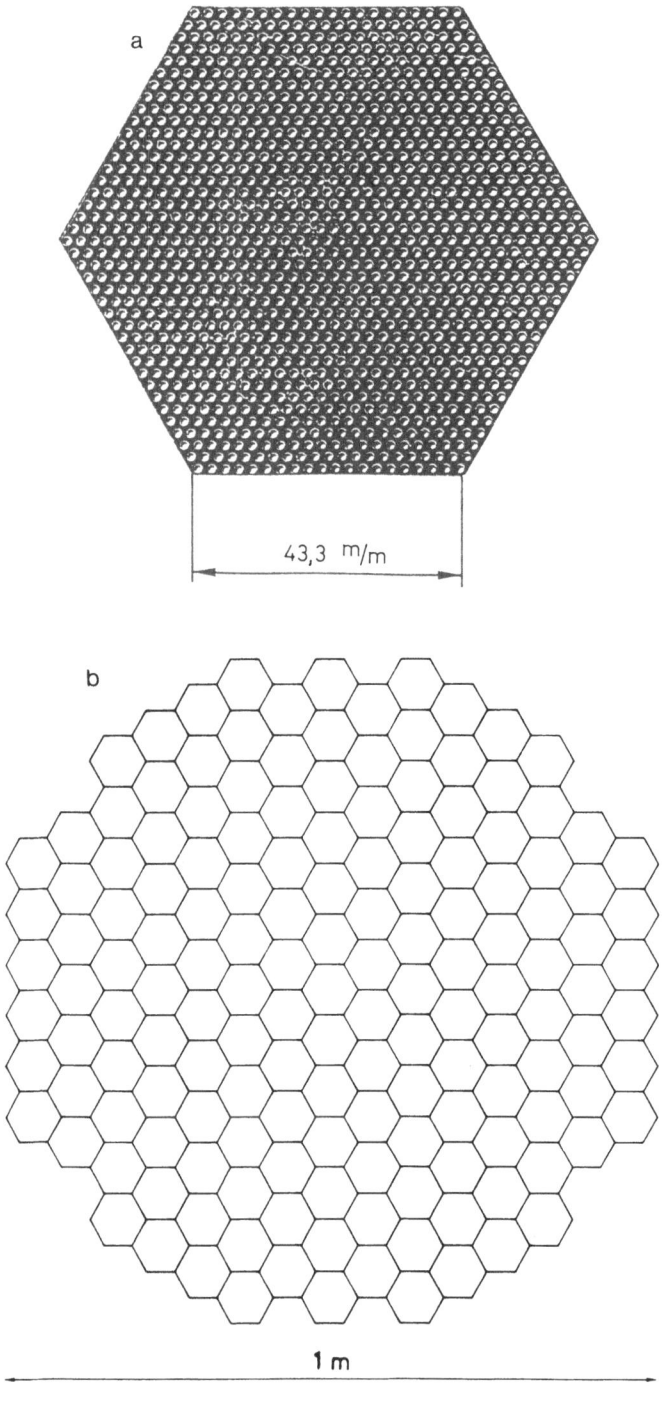

a

43,3 $^m/m$

b

1 m

155 MODULES

Figure 5

The structure of one calorimeter module (a), and the layout of the whole calorimeter (b).

the cell size of the HELIOS and ZEUS hadron calorimeters). In total 155 modules are foreseen to be built (fig. 5b). The effective radiation and nuclear interaction lengths are 7.5 mm and 21.0 cm, respectively, almost equal to the figures of the HELIOS and ZEUS uranium detectors.

After many tests the casting technology, which is far from trivial, is now well under control. We have successfully built at CERN a first module with the nominal lateral size, and 20 cm (27 radiation lengths) deep. This modulle will primarily be used to study the optimization of the detector in terms of readout (what is the best way to couple the fibres to the PM?), light tightness, *etc.*

Apart from that, several other modules, among them a full size (2 m deep) one, are being constructed by industry. They are expected to arrive at CERN by the end of July and will be tested in particle beams during this summer.

Fiber studies

The quality of the scintillating fibres will be crucial for the success of this project. Therefore, we have put much effort in studying the relevant aspects of the fibre quality, which concern the signal uniformity, the rate capability and the radiation sensitivity of the detector.

Signal uniformity implies that the calorimeter response should *not* depend on *which* fibres are hit by the shower particles, and *where* they are hit. Since we are aiming for calorimetry with 1% precision, the instrumental effects coming from non-uniformities should be kept below this level. That this is a very ambitious goal may be illustrated by the experience from ZEUS (and also HE-LIOS) who suffer from non-uniformities that are considerably larger (fig. 6). Non-uniformities of the type shown in this figure will not only seriously affect the em performance, but will also make the energy resolution for jet detection considerably worse than for single hadrons.

In the fibre calorimeter, two types of non-uniformities have to be worried about:

a) Fibre-to-fibre response fluctuations. These will mainly affect the em energy resolution, since the average number of fibres contributing to the signal from a hadronic shower is so large (typically more than 1000) that the effects are completely washed out.

b) Light attenuation in the fibres. Because of the depth profile of the light production, this phenomenon will mainly affect the hadronic energy resolution.

Fibre-to-fibre response fluctuations may be caused by variations in the fibre diameter, by differences in the light yield or light transmission, and by non-uniformities in the quantum efficiency of the light detecting element. Simulations with the em shower programme EGS4 showed that in order to limit the systematic

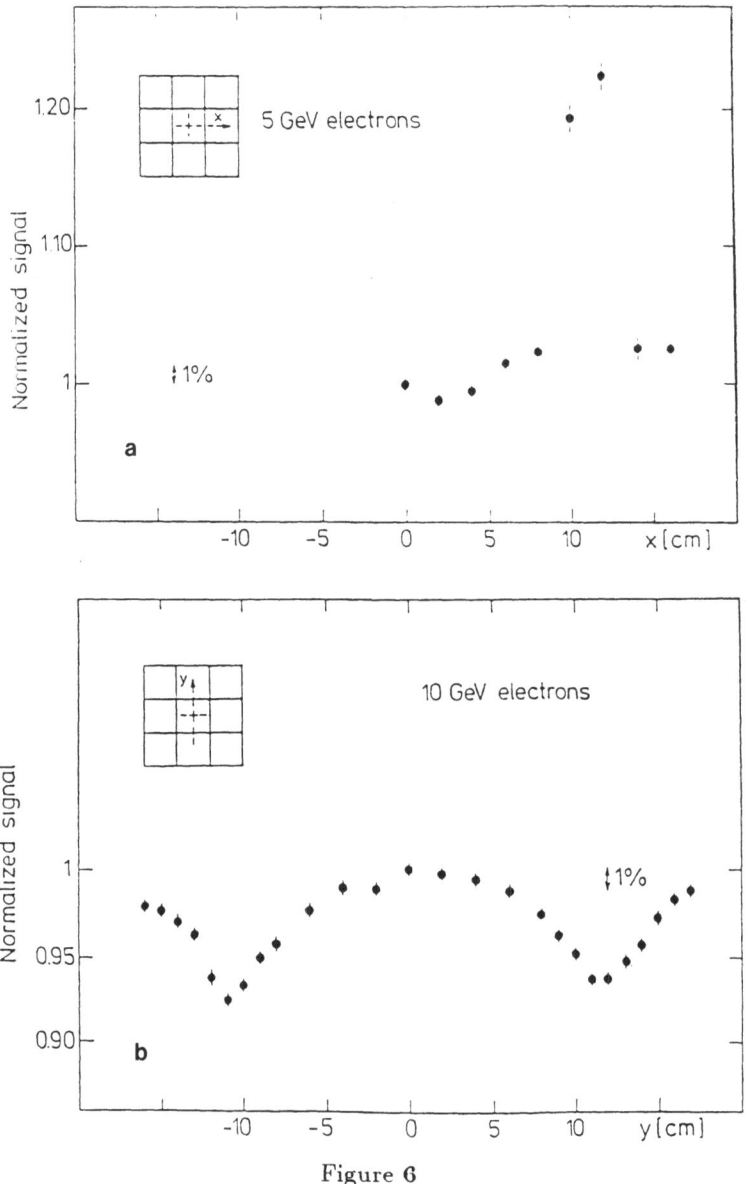

Figure 6

Results of a horizontal (a) or vertical (b) scan with an electron beam across a prototype module of the ZEUS hadron calorimeter. Data taken from ref. 8.

error on the energy measurement of em showers to 1%, fibre-to-fibre response fluctuations should be limited to 6%, which means that the average number of fibres effectively contributing to the em shower signal amounts to $\sim$ 35 in our detector.

We have built a test setup that will allow us to study the response of individual fibres in our prototype detector. The variations in the fibre diameter have already been measured, and were found to be negligible ($\sigma_{diam} \sim 1\%$).

In order to limit the contribution of light attenuation to the energy resolution of hadron detection to the required level, we calculated that the attenuation length should be larger than 6 m for single hadrons and 3 m for jets. The fibres used by the UA2 Collaboration for their tracking detector did certainly not meet these requirements.

Measurements of the light spectra revealed that the light attenuation is strongly wave length dependent. Due to the Stokes shift of the dyes, light at short wave lengths (400 - 420 nm) is much more attenuated than the light beyond 450 nm. Therefore, one can considerably *increase* the light attenuation length by filtering the outcoming light, at the expense of a somewhat reduced signal (fig. 7). However, since photon statistics is by no means a limiting factor for the detector performance, this technique will be applied.

Another factor that may limit the the attenuation length of the fibres, and determine the level of fibre-to-fibre fluctuations, concerns the optical quality of the core-cladding interface. This is apparently one of the most difficult aspects of fibre production, and not every company masters this technology equally well.

Apart from light filtering at the photomultiplier side, we also use a mirror at the open end of the fibres. This gives a further improvement of the uniformity as a function of depth. The fibre tests show that the calorimeter response can be made constant to within 6% over the full depth of 2 m (fig. 8), which is in fact better than needed and hoped for.

A very nice surprise is the extreme fastness of the fibre signals (fig. 9). The decay time of the light production by the dyes is only 2 ns, and the whole signal does not take more than $\sim$ 7 ns. This is extremely important for experiments at high luminosity machines. The measurements were done with a 2 m long fibre, with a mirror at the open end. When the fibre was excited, the direct and reflected light signals were clearly recorded separately, even if the fibre was excited as close as 23 cm from the mirror. These measurements illustrate that

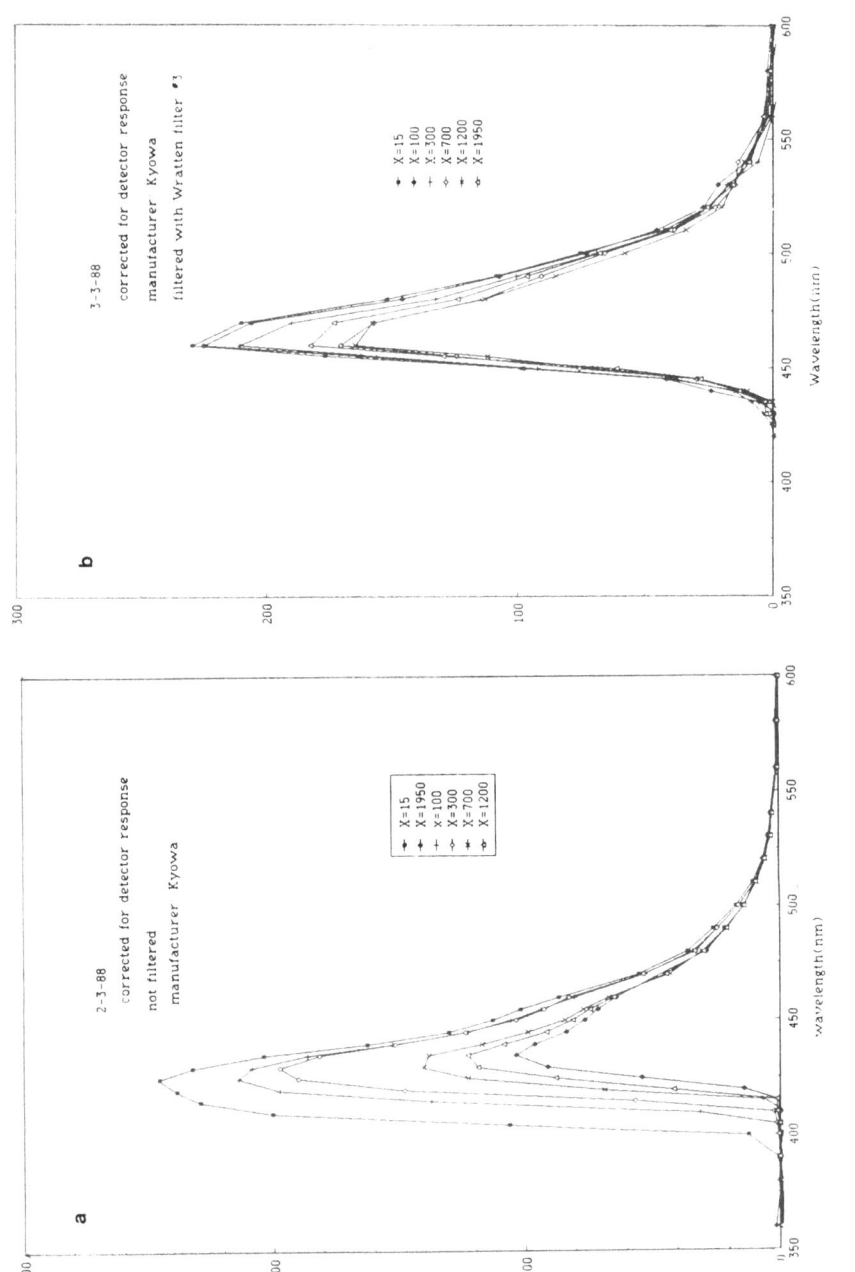

Figure 7. Wavelength spectra for SCSN-38 scintillating plastic fibres, measured at various distances from the point where the fibre was excited (a). Idem, using a yellow filter (b).

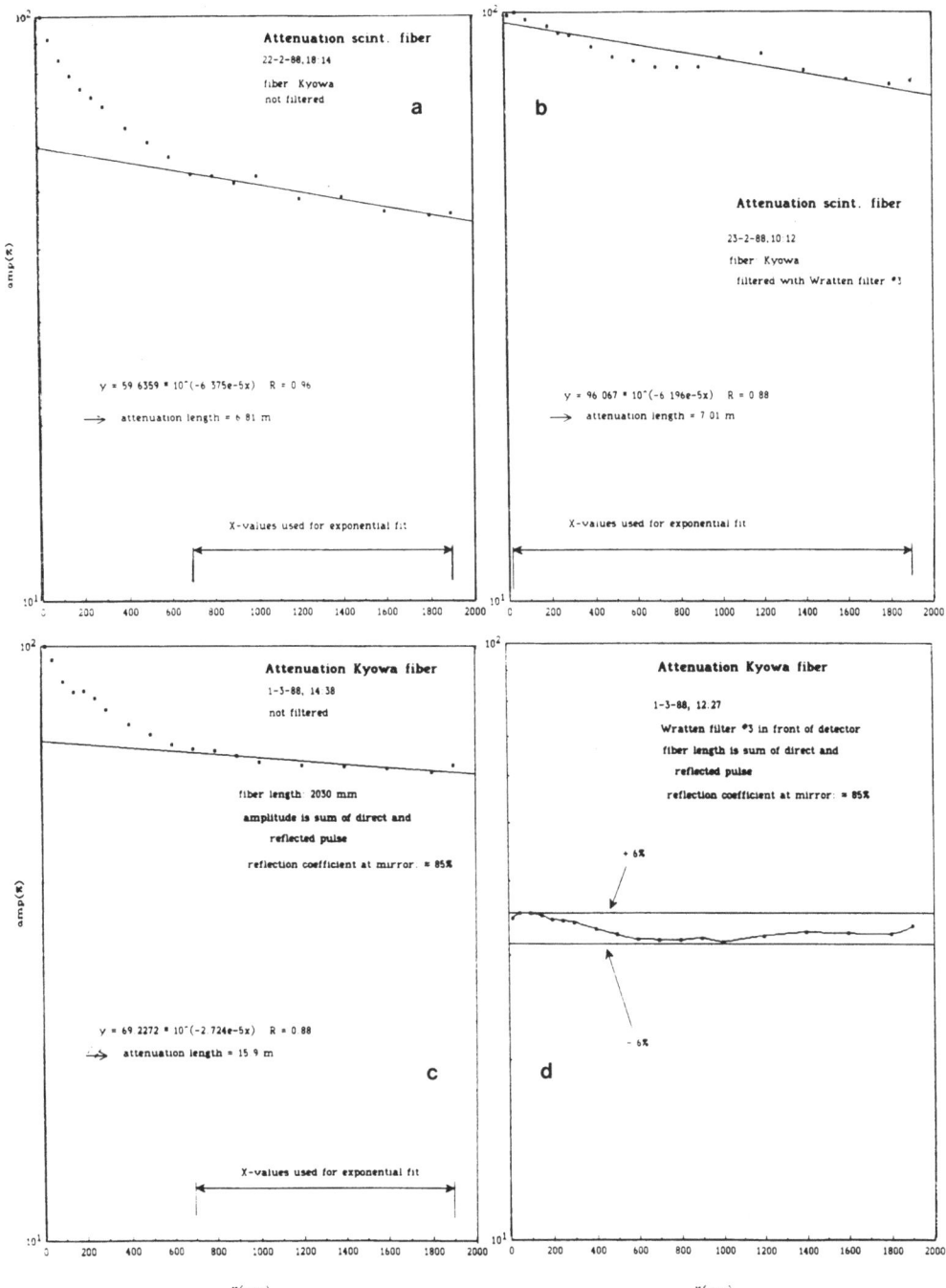

Figure 8

Effects of filter and mirror on the light attenuation curves of SCSN-38 scintillating plastic fibres. No filter/mirror (a). Effect of filter (b). Effect of mirror (c). Effect of filter plus mirror (d).

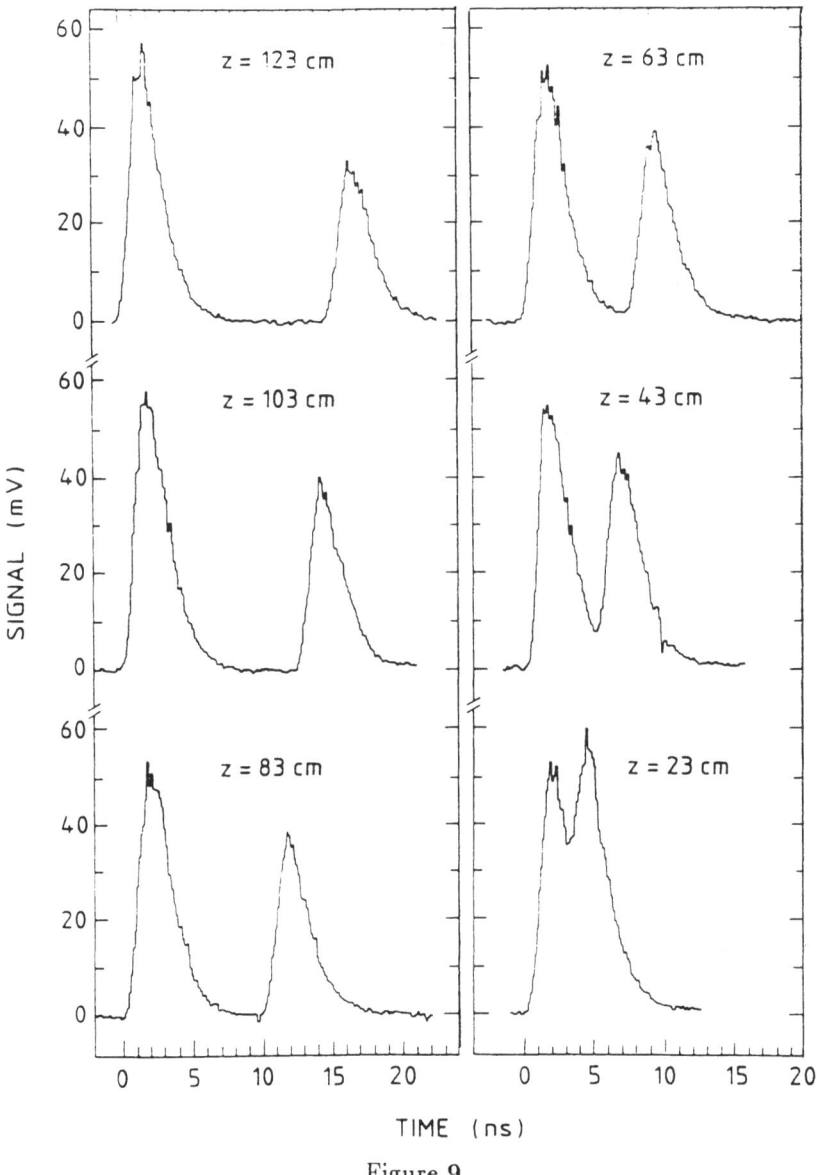

Figure 9

Time structure of the signal from SCSN-38 scintillating plastic fibres, for excitation at various distances z from the mirror at the open fibre end.

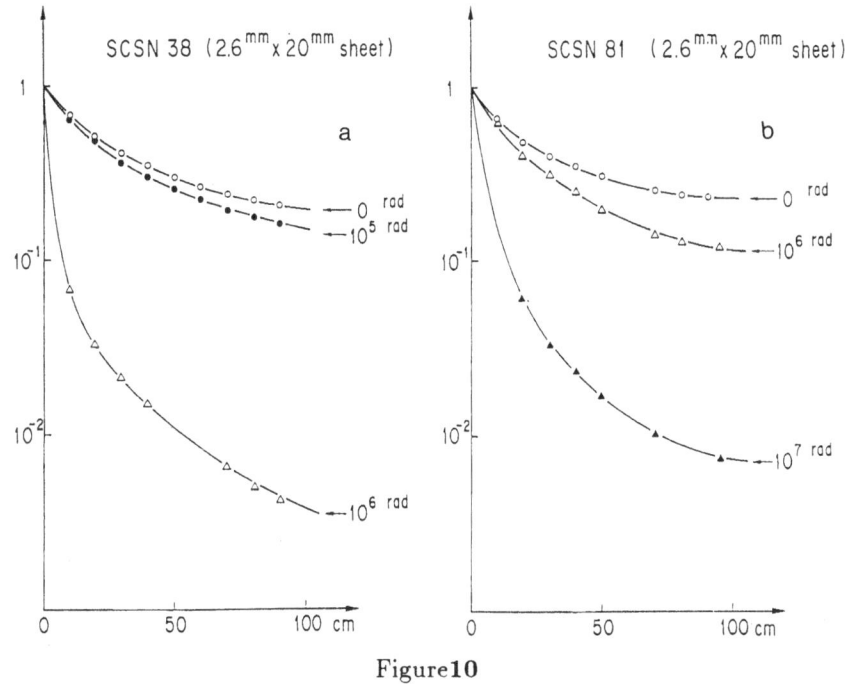

Figure10

Light attenuation curves for two types of scintillating plastic fibres, before and after exposure to ionizing radiation at various levels.

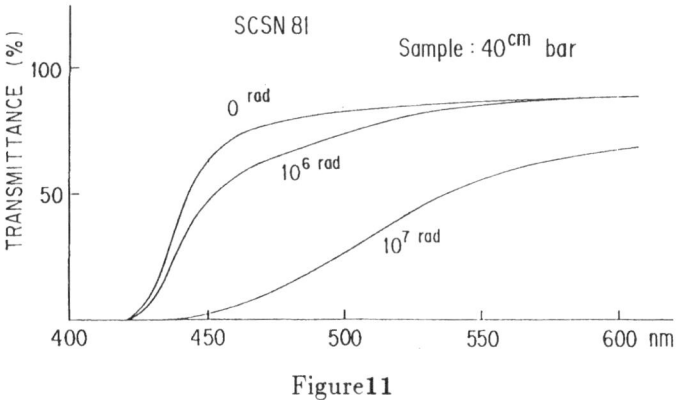

Figure11

The light transmission as a function of wave length for two different types of scintillating plastic fibres, before and after exposure to ionizing radiation at various levels.

the time structure of the calorimeter response may contain important depth information, which may be used *e.g.* for discrimating between em and hadronic showers without the need of splitting the detector into separate parts.

A crucial issue for experiments at a high-luminosity supercollider is the radiation sensitivity of the detector elements. Plastic scintillators do not have the best reputation in this respect. However, with the mechanisms of radiation damage becoming better understood, dedicated efforts to improve the radiation stability of scintillators are undertaken, sometimes with remarkably good results. Kyowa Gas Inc. have recently developed a fibre that is claimed to be an order of magnitude more resistant than the SCSN-38 ones that we are using in our prototype modules (fig. 10). Radiation stability up to a level of 1 MRad, which seems to be achievable for modern plastic scintillators, is considered acceptable for detectors that have to operate at the SSC[13] .

Moreover, our modules are designed in such a way that replacement of damaged fibres is a relatively easy job, which could if necessary well be done for part of the detector in between runs.

And finally, the light filtering applied to improve the signal uniformity is also likely to improve the radiation resistance, since the effect of ionizing radiation on the light transmission is wave length dependent (fig. 11). Summarizing, radiation damage seems to be a manageable problem for this type of detector.

6. CONCLUSIONS

The quality of the available calorimetry is of crucial importance for the success of experiments at future high-luminosity *pp* supercolliders. The spaghetti calorimeter investigated in the framework of the LAA project promises substantial improvements with respect to present-day technology in terms of energy resolution, hermeticity, and signal speed, which are all directly linked to the quality of the accessible physics.

REFERENCES

1. Proceedings of the Workshop on Experiments, Detectors and Experimental Areas for the Supercollider, Berkeley (1987).

2. U. Amaldi, Physics and Detectors at the Large Hadron Collider and the CERN Linear Collider, Proceedings of the Workshop on Physics at Future Accelerators, La Thuile and Geneva (1987).

3. R. Ansari *et al.* (UA2 Collaboration), Phys. Lett. **186** (1987) 452).

4. R. Wigmans, Nucl. Instr. and Meth. **A259** (1987) 389.

5. R. Wigmans, Calorimetry at the SSC, Proceedings of the Workshop on Experiments, Detectors and Experimental Areas for the Supercollider, Berkeley (1987) 608.

6. R. Wigmans, Nucl. Instr. and Meth. **A265** (1988) 273.

7. T. Akesson *et al.*, Performance of the Uranium/Plastic Scintillator Calorimeter for the HELIOS Experiment at CERN, Nucl. Instr. and Meth. **A262** (1987) 243.

8. E. Bernardi *et al.*, Performance of a Compensating Lead-Scintillator Calorimeter, preprint DESY 87-041 (1987).

9. P. Sonderegger, Nucl. Instr. and Meth. **A257** (1987) 523.

10. P. Jenni *et al.*, The High Resolution Spaghetti Hadron Calorimeter, in Report on the LAA Project, presented by A. Zichichi, vol. 2, 25 June 1987, p. 603; preprint NIKHEF-H/87-7.

11. C.W. Fabjan and T. Ludlam, Ann. Rev. Nucl. Part. Sci. **32** (1982) 335.

12. J.M. Gaillard *et al.* (UA2 Collaboration), Proc. of the DPF Conf., Eugene, Oregon (1985) 912.

13. Radiation Levels in the SSC Interaction Regions, Task Force Report of the SSC Central Design Group, D.E. Groom, editor, Berkeley, June 1988.

24

FAST INTELLIGENT SYSTEMS AT THE ELOISATRON

R.K. Bock

CERN
Geneva, Switzerland

1   The Data Acquisition Environment

For the high data rates expected at future multi-TeV hadronic colliders like the ELOISATRON, it will be of utmost importance to take decisions in real time on *partial data* and as fast as possible. At a first level and shortest timescale, some customized electronics will reduce the rates, but at the next level, some semi-custom devices will have to take decisions with more flexibility and with some algorithmic component, i.e. *intelligent decisions*. The target detector we will have to consider is one which

- is *complex*, i.e. contains multiple detector elements including tracking and calorimetry for electromagnetic and hadronic particles, possibly track identification devices, at various production angles;

- is made of elements of *fine grain*, i.e. records interactions in considerable detail, producing a very large total amount of information;

- can operate in an environment of *high rates*, i.e. is capable of recording events at a hadronic collider with a luminosity of the order of 10^{33} cm^{-2}sec^{-1}.

With an assumed total inelastic cross section of around 100 mbarn, the assumed luminosity translates into a primary event rate of 10^8/sec. We will assume that custom-made fast (analogue) electronics will reduce this at least by a factor 1000. Further we ignore the event separation problems arising from the recording of multiple events. Event overlap in recording becomes likely as the average time between events shortens with increasing luminosity, and as event separation in time by bunching the colliding beams becomes ineffective.

We assume, then, that at a rate of 10^5/sec events are presented to some *second level trigger* processors, most likely in a pipeline just after analogue-to-digital conversion (see Fig. 1). We postulate, that at a rate of one event every 10 μsec or faster, these processors will have to decide with some degree of intelligence about the suitability of further processing.

Presently, no digital systems with computer-like characteristics exist to take on such a task. Existing trigger systems rely on full custom analogue or digital electronics to reduce event rates way below the 10^5/sec mark, and anything operating faster than, say, the msec level has litte or no programmability. Computer-like systems used for triggering today are microprocessors, operating on full events and with several milliseconds of response time. Their parallelism is thus implemented at the event level, and we consider such system as *third level* triggers. We should note that substantial improvements in the computing capacity of the third level will also have to be an objective for the ELOISATRON experiment planning.

It is our specific interest to find suitable ways to bridge the existing *serious time gap* between custom-made electronics, analogue or digital, and commercially available programmable systems.

2 Data Compression

The fine granularity of a detector translates into a large number of "channels" and a large per-event data volume. Our assumption is that 10^6 bytes is a lower limit for the raw information content of an event, after removal of trivial zero channels. Both for taking intelligent decisions and for economizing on transmission bandwidth, it is mandatory to *correlate* raw signals in order to extract *physics meaning* from them. In as much as the extracted information like "electromagnetic shower", "local track segment", or "parton jet", contains sufficient information for further processing, albeit in need of subsequent fine calibration, it may serve as a suitable local description of event features in the given detector element and may thus replace the original raw information, which is much more voluminous (like the amount of light recorded in calorimeter cells or the drift times on wires etc.). A *triggering* task executed in a processor may thus also become a *data compression* task, and this compression of data volume may even turn out to be a major asset of embedded intelligence. Today, the common use of digital signal processors (e.g. for extracting multi-hit information on single drift wires) could be viewed as a forerunner of the systematic use of image processors extracting features like track segments from two-dimensional information such as a group of tracking signals.

3 Significant Trigger Tasks

Any novel technique to be introduced in a future detector will be expected to demonstrate its feasibility in a real-life test. Full scale trigger tests are, of course detector dependent and thus not possible before accelerator and detector parts exist. We thus have to define for ourselves a number of *significant* triggering tasks, that can be realistically simulated by existing hardware in order to achieve the same purpose. A number of assumptions and educated guesses can not be avoided. Most prominent amongst the hypotheses we have to make (apart from the rate assumptions made above), are the *physics* objectives at a multi-TeV collider and the global *detector* characteristics we have to deal with.

As for physics, we can rely on a wealth of material collected at various workshops in the past. From the meetings at Snowmass (1986 proceedings edited by R. Donaldsen and J. Marx), La Thuile (1987 proceedings edited by J. Mulvey, CERN 87-07), and

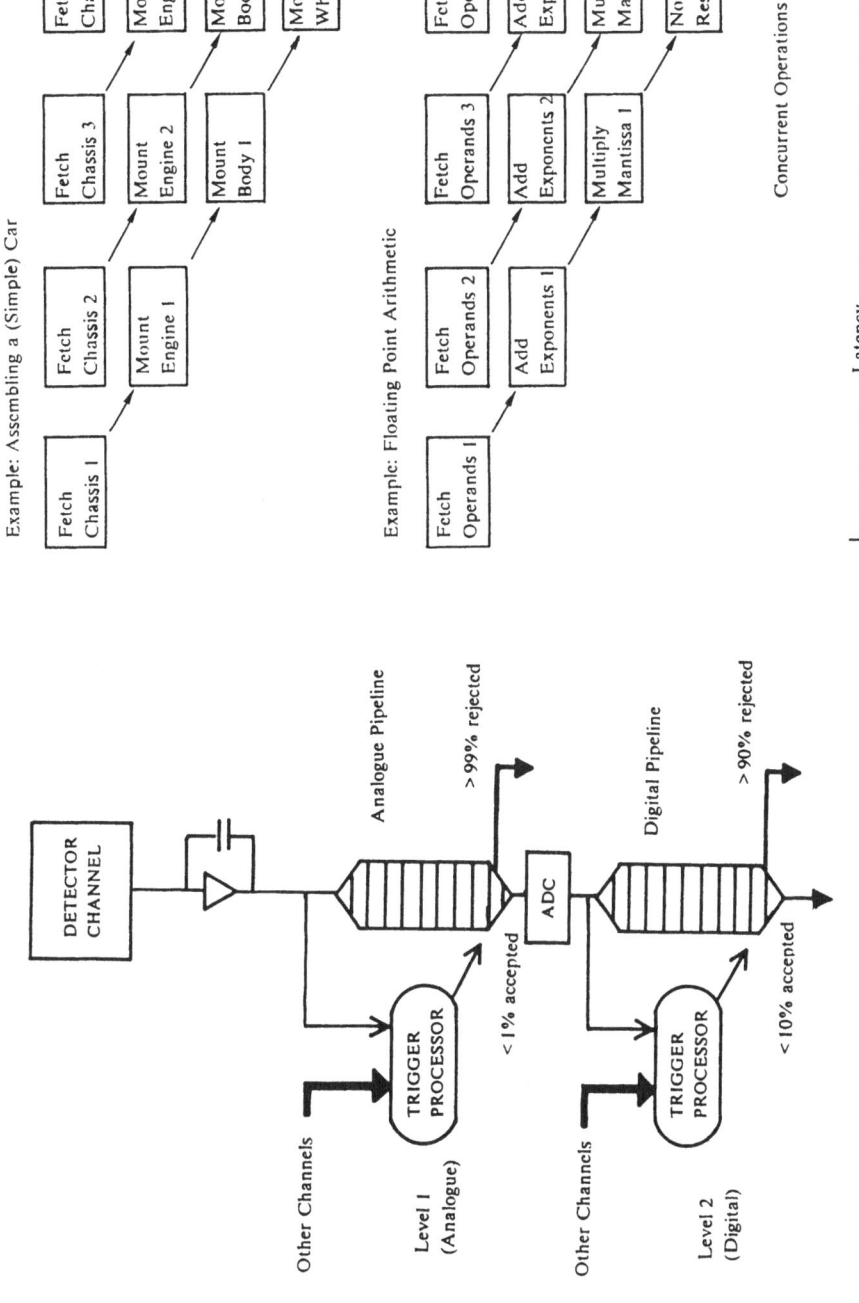

Example: Assembling a (Simple) Car

Fetch Chassis 1	Fetch Chassis 2	Fetch Chassis 3	Fetch Chassis 4
	Mount Engine 1	Mount Engine 2	Mount Engine 3
		Mount Body 1	Mount Body 2
			Mount Wheels 1

Example: Floating Point Arithmetic

Fetch Operands 1	Fetch Operands 2	Fetch Operands 3	Fetch Operands 4
	Add Exponents 1	Add Exponents 2	Add Exponents 3
		Multiply Mantissa 1	Multiply Mantissa 2
			Normalize Result 1

Concurrent Operations

Latency

Figure 2. Principle of Pipelined Operations

DETECTOR CHANNEL

Analogue Pipeline
> 99% rejected

ADC

Digital Pipeline
> 90% rejected

Other Channels

Level 1 (Analogue)
TRIGGER PROCESSOR
< 1% accepted

Other Channels

Level 2 (Digital)
TRIGGER PROCESSOR
< 10% accepted

To Event Builder and Level 3

Figure 1. Schematic Data Flow for Data Acquisition

419

from the present discussions at Erice, we can extract as main physics objectives and their signatures the following.

- The Higgs boson: Depending on its mass which is subject to speculation, and possibly on the mass of the top quark, the Higgs will manifest itself in heavy quarks or W/Z. The most relevant signature in either case is a high-p_t lepton, possibly a high-p_t jet with lepton content.

- Heavy Sequential Quarks and Leptons: The expected signature consists of high-p_t jets, constrained by effective mass, and associated to high-p_t neutrinos.

- Additional Gauge Bosons: Their decay is expected to be analogous to the Z and W, i.e. into lepton pairs of high p_t. (e/μ and ν for the W', $e^+e^-/\mu^+\mu^-$ for the Z'). The dominant quark decay is considered difficult to observe, as for the known vector bosons.

- Supersymmetric Particles: The expected signatures from squark decay into photino and quarks consist of quark jets and missing p_t (photino). Electroweak sparticles leave a charged lepton and a photino signature.

- Leptoquarks: The expected signature would be jets associated to high-p_t leptons, charged or neutral.

These signatures can be abstracted to very few significant elements: The recognition of high-p_t charged leptons or jets, or of escaping transverse energy, possibly with mass constraints on them. Most high-p_t phenomena occur over a very central rapidity range, typically at angles larger than 40° with respect to the beam direction. "High p_t" stands for at least 30 GeV/c. Charged leptons are single charged particles with a signature in an electromagnetic calorimeter or muon chamber (we ignore τ leptons), possibly supplemented by track identification information. Missing transverse energy is obtained by summing energy vectors over all cells of a hermetically closed hadronic calorimeter. Jets are defined by local clusters of energy (e.g. peaks), again in a hadronic calorimeter. Supporting information on jets from charged tracks is not considered relevant at this level.

Beyond these elements, the usual "unexpected" physics must be catered for. At the trigger level, this can only mean that a maximum degree of flexibility of concepts and modularity of construction must be aimed at. Triggers must be expressed in terms of basic physics concepts, and operated by parameters and logical connections both easy to change.

As for data compression, the degree to which this must be a facility that can be "turned off" to learn more about unforeseen details, will have to be learned in the field. A detector which has the (idealized) facility to record physics-related quantities like complete shower information, jet and neutrino energy vectors, certified muons, etc., would seem an ideally flexible tool even for the unexpected physics, if the correlations and constraints on any of these quantities remain fully programmable.

4 Parallel Systems in General

No single computer system can be expected to handle the processing load arising from the task outlined above. This is only partly a problem of computing capacity, but one of making the large amounts of raw data available to the processor. Maximum parallelism of operation is therefore mandatory, and some structure has to connect the individual processing elements, providing the necessary synchronization.

In this paragraph, some general principles of parallel computing operations are recalled; the applications to high-energy physics problems and our specific triggering or data compression tasks will be discussed later.

Historical Note

The potential limits of the classical digital computer with one instruction and data memory and a single sequential processing unit were foreseen by J. von Neumann even before practical computers had been built, and many early publications dealt with the aspects of parallelism. S.H. Unger published a paper in 1958 (A Computer Oriented Towards Spatial Problems, Proceedings of the IRE, October 1958) which is the direct basis for a system available commercially today (the DAP from ICL, now from AMT). Different forms of parallelism have also been introduced early into the more complex and performing general computer systems: Concurrency between input/output and processing, separate units to deal simultaneously with mantissae and exponents in floating point operations, multiple registers to allow overlapping access to data, operation pipelines to achieve concurrency for different types of instructions and, of course, multiple CPU's run by a single operating system.

All designs of this type have as objective a simple *enhancement* of performance of a perfectly general system. No work is expected from the user of such a system to adapt his problem formulation to the architectural peculiarities, the parallelism remains *invisible*.

More recently, systems have appeared on the market that contain parallelism which *does* require architecture-specific code to make use of the system's potential. Vector processors belong into this category and so do various other parallel architectures, some built for specific applications like the processing of two-dimensional images, others being in a way solutions to problems yet to be defined.

In the following, the basic concepts which allow classifying parallel systems are introduced; the taxonomy of systems is, of course, not unique and simple, most systems being hybrids of one sort or another.

4.1   Pipelining Architectures

A frequent type of parallelism is to execute sequential functions concurrently on different data, like automobiles are assembled on an assembley line. A given data unit undergoes various successive processing steps; when processing step 1 on data unit 1 is terminated the data unit 1 undergoes step 2, whilst unit 2 is subjected to step 1 (see Fig. 2). A pipeline may result in remarkable throughput due to the possible high *functional* parallelism, but is also chracterized by its "latency", the time that elapses between the beginning of the first and the end of the last processing step, for a given data unit.

SIMD (Single-Instruction-Multiple-Data) Architectures

Multiple processors can be built to operate in full lockstep on different data (Fig. 3). Controlled from a single program, a SIMD architecture can achieve extraordinary peformance on problem formulations that keep all processors busy a good

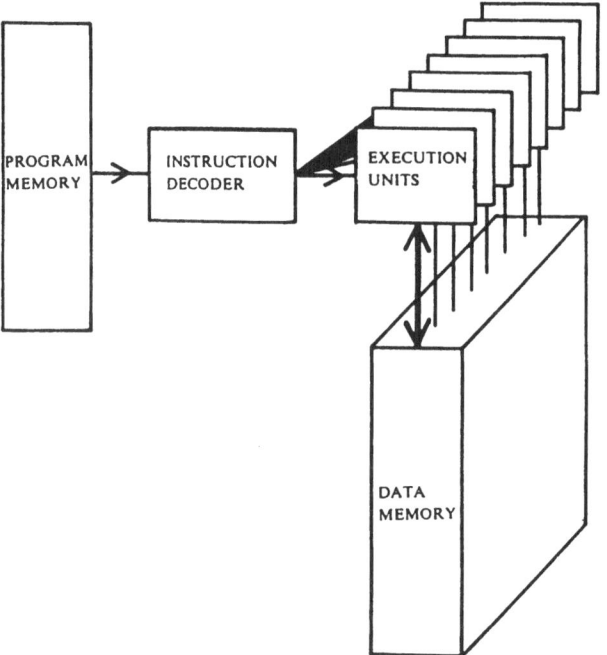

Figure 3. Schematic Architecture of a SIMD Computer

fraction of the available time. Depending on the available memory and the mechanism of flowing data into and from it, vectorial problems may be the most suitable for such architectures, and vector processors do, of course, exploit the SIMD principle. An example of a commercially available pure SIMD architecture is AMT's DAP 500: 1024 processing elements operate under a single master control unit, each unit performs single bit operations and has 32 Kbit of memory associated.

MIMD (Multiple-Instruction-Multiple-Data) Architectures

If multiple processors work independently on different data, general problems can take more easily advantage of the concurrency, at some level of "data granularity". Existing examples are multiple CPU's in large systems (the data concurrency unit is the job, routing is done by the operating system), or high-energy physics event processing farms (emulators or the ACP, where the data concurrency unit is the event, and special dispatching software has been written into a host computer).

MIMD machines destined for finer data grain have also been marketed, with a variety of network structures interconnecting the nodes. The Intel iPSC/2 (Hypercube) systems have an "n-space" connectivity minimizing general node-to-node transmission time, INMOS transputers have four fast links each. This allows connecting them into complex networks, although a simple two-dimensional grid layout is naturally favoured.

Examples of Implementing Parallelism

Image processors typically are hybrids between SIMD machines and pipelines, often custom-configured and thus allowing the application to determine the precise ar-

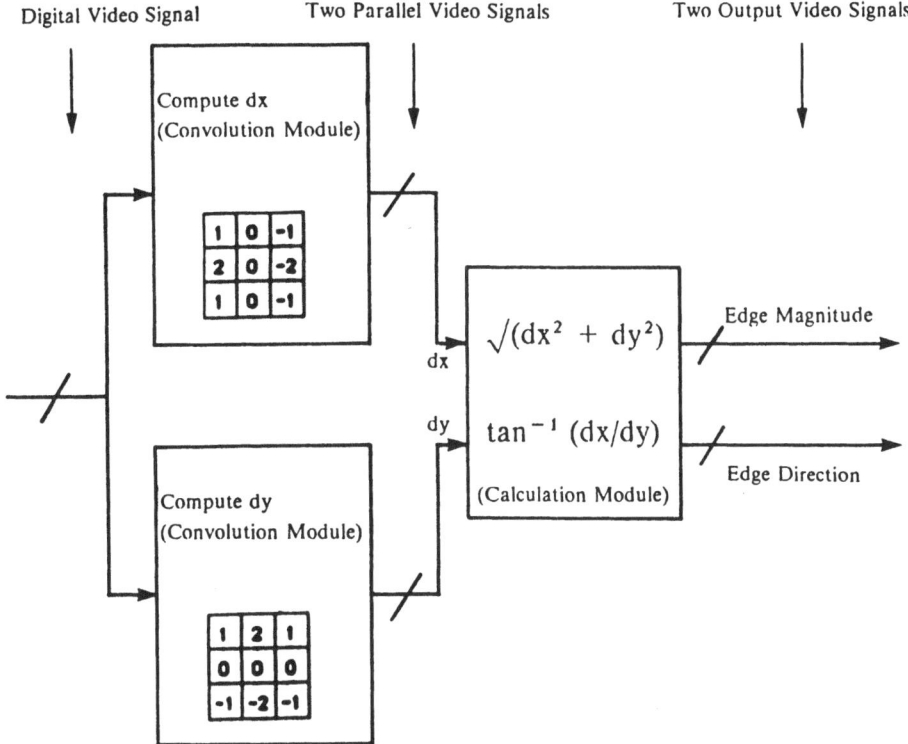

Figure 4. Implementation of an Edge Detection Algorithm (Sobel Operator).

chitecture. Operating on a single system clock and processing two-dimensional arrays of typically 8-bit information (pixels), the processing units have limited instruction sets and programmability tailored to image operations like convolutions, simple transforms or thresholding. The processing proceeds at "video rate", i.e. at least 30 images of 512×512 pixels per second, equivalent to 10 MHz.

Fig. 4 shows an example of a simple image processing system which implements an edge finder. Two identical 3×3 convolvers and an arithmetic merger module are a realization of the "Sobel operator", a transformation defining a zone boundary (edge) with its magnitude and direction. Modules of this type are available commercially, e.g. from Datacube Inc.

The rhythmic operation characteristic for image proceesing systems makes them part of a larger category of systems referred to as *systolic arrays* (systoles are the heart's rhythmic contractions). Such systems have been conceived for many other parallel tasks like matrix operations, and again are typically SIMD architectures with pipelining added at some higher level.

5 Parallel Architectures in High-Energy Physics

Physicists are near-insatiable in their requirements of computing power, where data analysis is concerned. More or less ingenious systems of the MIMD type have therefore been introduced for off-line data analysis in the past, the ACP system originated at FNAL and the emulator farms in the US and Europe are well known. Higher accelerator energies and increased event rates will undoubtedly accentuate the need for such systems in the future. It seems likely that the associated logistics for the expected large data volumes will push them in the direction of maximal use in real time.

The large number of microprocessors that are being used in the data acquisition systems of large experiments, do, of course, constitute another class of very successful MIMD systems operating in real time. More often than not their task is one of data flow control, many have data compaction tasks, in some cases a (third-level) trigger function is executed.

The Fast Triggering Problem

For genuine fast triggering, however, as was outlined above, fully programmable MIMD systems using the event as data concurrency unit are thinkable only at the last ("third") level. They require event builder functions to be performed before any computation can be launched, and the necessary compacting and pipelining or buffering operations have to be implemented further upstream. We believe that the second level triggering needs a different approach: the triggering functions need very deep embedding with those of detector readout, buffering, and data compaction. We think that a combination of special-purpose processors inspired by the image processing approach (and possibly following commercially available architectures very closely) with standard digital buffering techniques will have to be grafted locally onto groups of detector readout modules. Numerous such systems will have to operate independently on sub-event data units. The communication between systems will have to be as between nodes in MIMD architectures, to achieve the final trigger. Only the retained fraction of events will then require full event building, and may even consist largely of processed (compacted) data only. The resulting event and data rate must, of course, be compatible with the next-level processing, which is likely to be of pure MIMD architecture and with full programmability.

Ongoing Work in LAA

In the framework of the LAA detector development project at CERN (see A. Zichichi, ICFA Instrumentation Bulletin 3 (1987) p. 17), we have started defining significant tasks and promising candidate architectures. Among the parallel systems available commercially, we concentrate on SIMD and image processing systems.

We have benchmarked, by simulation, a DAP 500 system with 32×32 processors, on various characteristic tasks (parameterized pixel transformations, histogramming, Hough transforms, peak finders for showers), and found, after some invested programming effort, a remarkable performance in execution (≥ 30 times Vax 8700 for the array). Also clearly apparent were difficulties to achieve the required rates in particular in data transmission and reordering into the DAP's fast memory.

We are also building up pilot systems based on three further manufacturer's hardware: a Datacube system for pure image processing, an NCR 45CG72 GAPP hybrid SIMD/image processing system (in collaboration with Laben, Milan), and an ASP (Associative String Processor) system (in collaboration with Brunel University, Uxbridge).

No practical or simulated experience exists for an application as we envisage. From specifications, the expected results should indicate that these systems are suitable from their architecture to take decisions on limited-size arrays of detector data, but

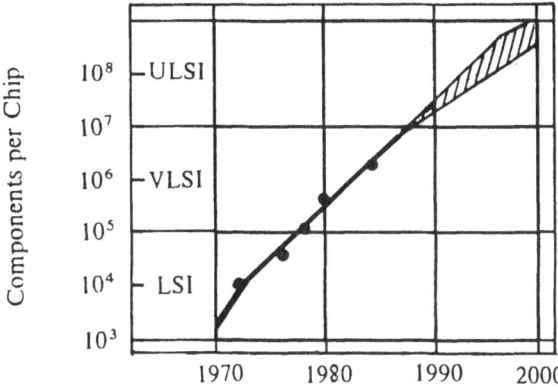

Figure 5. Evolution of Component Density on Chips since 1970.

not or only marginally within the necessary response time. We expect therefore, that both for data transfer logistics and performance, full integration of the systems or at least of systems memory into the data flow of the detector readout will be necessary. The likely conclusion is that *existing* commercial parallel *systems* turn out *inadequate* without major modifications.

Technology Outlook

We therefore consider it our prime task to investigate in which direction improvements in the data flow can be expected or achieved, without deviating too far from marketed architectures.

As for performance, the increase in *component density* as observed over the last 20 years may be extrapolated towards "ultra-large scale integration"; the present yearly increase is about a factor 1.5 (see Fig. 5). Manufacturers' chips announced today use 1μm geometry and contain $> 10^6$ transistors/chip; the next steps to submicron geometries are already under way. These numbers will directly translate into performance. Some manufacturers are also integrating more layers onto their chips; Mitsubishi has prototyped a three-layer chip with image processing functions, outperforming present-day components by orders of magnitude.

As for data *transmission speed*, the necessary steps into the future bandwidths of $>$ Gbyte/sec are also being taken. Optical components like photodiode arrays may even become directly part of some of our light-producing detectors (scintillators in calorimeters, fibres). Internal (on-chip) speed can take advantage of the increased component density and the absence of constraints imposed by interfacing standards. External "buses" are likely to move increasingly towards high-performance point-to-point links (Meeting on "Superbus" at CERN, July 15, 1988), with or without optical transmission.

Most importantly, perhaps, the possibilities of *integration* of components seem to be improving rapidly. Industry does talk about "wafer-scale integration", implying fault-tolerant complex systems in which a faulty component does not cause the system to fail. In designing modern VLSI systems, increasing use is made of high-level ("behavioural") languages, of full simulation, and of automatic (computer-supported) layout. It is conceivable to accumulate libraries of modularly designed components,

425

described in a standard language (e.g. VHDL), and translatable into hardware of a given technology. The integration of processing functions with digitization and buffering components, in a multi-chip system, may thus become a reality.

6   Summary

The *triggering problem* for future accelerators will be a *critical* one, in particular at the level where simple custom-made electronics becomes too complex and commercial computer systems do not offer enough performance. Unless decisions of some intelligence can be taken at a sustained rate of few microseconds per event, physics and/or accelerator time will be wasted.

Classical computer architectures and off-the-shelf solutions to computer parallelism will be inadequate to solve this problem. The *ingredients* for a future solution, however, do exist. Some of the existing parallel architectures, especially image processing devices, hold sufficient promise for being able to run the significant physics feature extraction tasks locally in a detector. Such tasks will be relevant both for triggering and data compaction.

With the foreseeable *advances of technology* and, in particular, with the trends towards standardizing the system design tools at the level of high level hardware description languages, we are hopeful that the *integration* of suitable processing architectures into future readout and buffering electronics can become reality. This technology transfer from industry, at the hardware design level, may be one of the most challenging tasks facing embedded systems of the future.

Several pilot projects are being built up as part of the LAA detector development project at CERN. It is their goal to evaluate commmercially available parallel architectures for their suitability to act as trigger processor elements, and to acquire the skills for their full integration as active elements in future acquisition electronics. To succeed on both accounts seems a necessary condition to channel the data rates at future colliders.

25

REVIEW OF EXPERIMENTAL RESULTS AND PERSPECTIVES FROM UA2 AT THE CERN p̄p COLLIDER

The UA2 Collaboration

presented by Barbara De Lotto
I.N.F.N. Pavia

1. INTRODUCTION

The success of the Sp̄pS Collider program at CERN [1] in the years 1981-1985 has given a contribution of fundamental importance to the field of particle physics.

Encouraged by such achievements and motivated by equally exciting physics topics still unrevealed and possibly within reach (the top quark and supersymmetry searches are two prominent examples) CERN has undertaken an important upgrading of the p̄p Collider complex [2].

In order to fully exploit the physics potentials of the upgraded Collider, a substantial upgrade of the UA2 detector [3] was approved. This paper will review, in Section 2, the motivations and the main aspects of the upgraded UA2 detector (for convenience, we will refer to it simply as UA2′). The Physics that UA2′ intends to address is discussed in Section 3 (topics related of the Standard Model) and in Section 4 (minimal extension to the Standard Model and search for exotics); some of the results obtained so far at the CERN Collider are reviewed, and the physics perspectives for the near future are also discussed. Finally, the experience gained with the new detector during the first run in November-December 1987 is reported in Section 5.

2. THE UPGRADING OF UA2

The main motivations for the upgrade of the UA2 detector are i) a better electron identification in the central region and ii) a better "neutrino" identification through an improved resolution of the missing transverse momentum measurement.

A detailed description of the upgraded UA2 detector can be found elsewhere [4]. Only the general features are outlined here, together with preliminary indications of the performance during the first data-taking period.

2.1 Electron identification

The major difficulty in improving the electron identification in the central region of UA2 was the limited space available, defined by the Central Calorimeter edges [3] and almost entirely occupied by multi-wire proportional and drift chambers for track finding and vertex reconstruction.

It was only with the development of a scintillating fibre detector [5], providing 18 points in three projections in a radial space of less then 4 cm, that enough room could be made for the introduction of transition radiation detectors which, together with a silicon hodoscope, constitue the main tools of the improved electron identification in UA2'. The resulting rejection power against fake electrons is expected to increase by a factor of 25-30 with respect to UA2, as measured with electron and pion beams. Fig. 1 outlines the different detector configurations in the central region of UA2 (1a) and UA2' (1b). The latter consists of:

i. A cylindrical drift chamber of the "jet type" (Jet Vertex Detector, JVD) to measure tracks close to the vertex, which surrounds a beryllium beam pipe. From test beam results the resolution of the JVD is expected to be about 0.15 mm in $R \times \phi$ and about 1 cm in z from charge division (strips on the outer layer provide a z resolution of 1 mm) . With the present level of calibration the resolution from Collider data is 0.35 mm and 3 cm, respectively.

ii. A Silicon Hodoscope [6] made of 3024 40 × 8 mm pads, supported by a cylinder of 140 mm radius. The main functions of the silicon hodoscope are the rejection of gamma-conversion by dE/dx measurement and the removal of "ghost" tracks. The behaviour of the silicon detector during the 1987 run is shown in Fig. 2, where the pulse height distribution of pads traversed by reconstructed tracks from a minimum-bias trigger is compared to test beam data.

iii. A Scintillating Fibre Detector (SFD) [5] provides track segments and the start of electromagnetic showers in front of the Central Calorimeter. It consists of about 60000 scintillating fibres, 1 mm in diameter, arranged in 24 coaxial cylindrical layers forming 8 stereo triplets. A lead converter, 1.5 radiation length thick, is inserted before the last two stereo triplets, all along the angular range covered by the central calorimeter. The fibres are read by means of 32 read-out chains, each consisting of three Image Intensifiers, one CCD and a FASTBUS digitizer, providing the necessary amplification, image demagnification, multiplexing and data compaction. The performance of the SFD during the first Collider run is illustrated in Fig. 3.

iv. A Transition Radiation Detector (TRD) fills the gap between the Silicon detector and the SFD. It consists of two coaxial cylindrical chambers, each made of a stack of polypropylene radiators followed by a xenon chamber, which has a drift region followed by an amplification region. The X-ray photons produced by electrons in the converter are rapidly absorbed by the xenon in the early part of the chamber and generate large signals at later times then the ones produced in the amplification

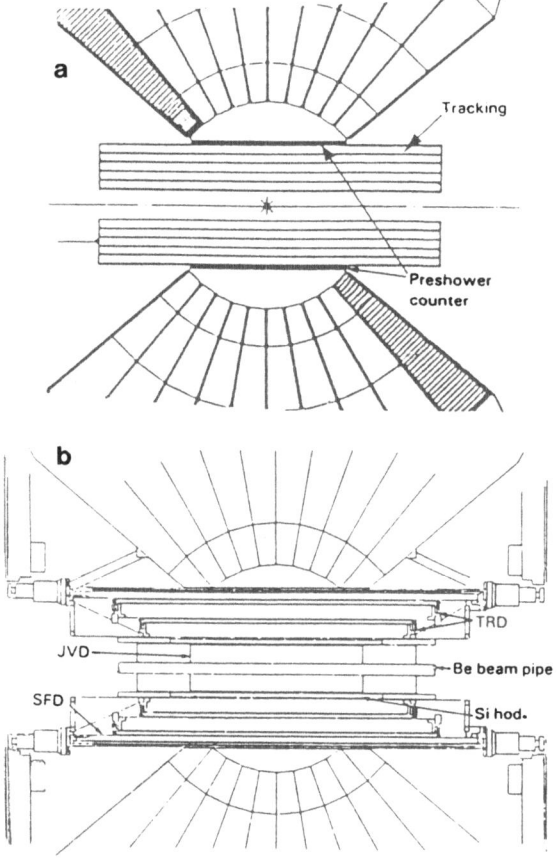

Fig. 1. Longitdinal view of the central detectors of UA2 (a) and UA2' (b). Note the smaller beryllium beam pipe and the shortened calorimeter edges of UA2'.

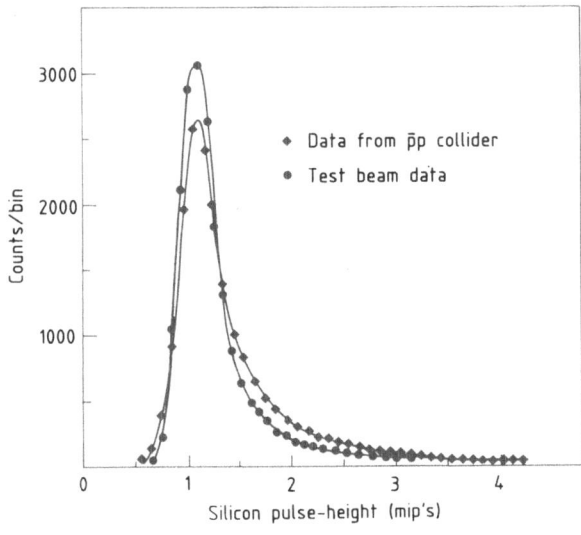

Fig. 2. Pulse height distribution of the silicon detector signals from test beam and Collider data .

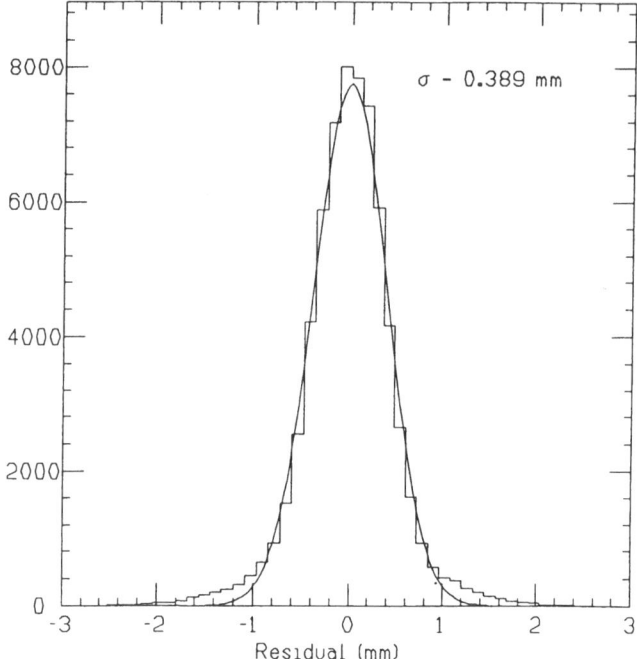

Fig. 3. Distribution of the residuals of the 18 tracking layers of the SFD with respect to the fitted track.

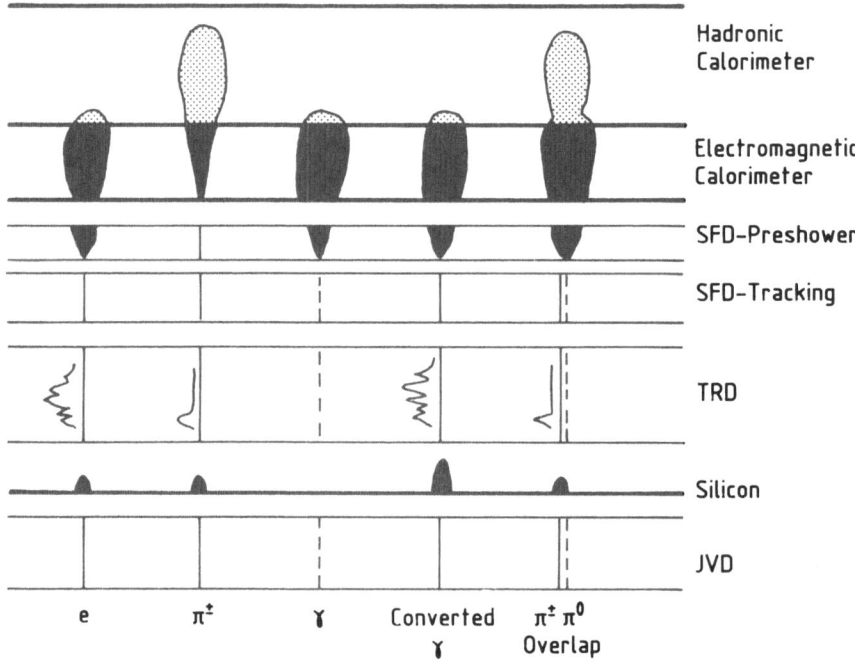

Fig. 4. Principle of electron idendification in the UA2′ central region.

region by any ionising particle. Signals are digitized by 100 Mhz FADCs [7]. The outer cathode of each chamber, made of helicoidal strips, is also connected to 15 Mhz FADCs enabling the charge induced by the sense wire avalanches to be measured. The different signatures of electrons and other particles in the UA2′ central detectors are schematically summarized in Fig. 4.

Tracking and electron identification in the forward regions are provided by a set of proportional tube chambers (ECPT) located in front of the new End-Cap calorimeters. The ECPT detector is described in ref. [4], which also gives details of a fast time-of-flight system aimed at measuring the vertex position with good precision.

2.2 "Neutrino" Identification

Non-interacting particles, such as neutrinos, among the final state products can in principle be detected by measuring the total transverse momentum vector associated with the visible particles. The missing transverse momentum vector is defined as $\vec{p_T}^{miss} = -\Sigma \vec{p_T}^i$, where $\vec{p_T}^i$ is a vector with magnitude given by the energy of the ith calorimeter cell and directed from the event vertex to the cell centre. The sum extends over all calorimeter cells. The quality of the $\vec{p_T}^{miss}$ measurement is a function of the angular coverage of the calorimeter. The UA2 apparatus had full calorimetry in the central region ($40° < \theta < 140°$) and only electromagnetic calorimeters between 20° and 40°, while there was no detection of particles below 20°. Such incomplete coverage was detrimental to the calculation of the missing transverse momentum.

In UA2′ the central calorimeter (CC) is unchanged, while the forward regions have been equipped with new End Cap Calorimeters (EC) extending to 5° from the beam axis. The EC being of the same kind as the CC, uniform hadronic measurement is guaranteed from $5° < \theta < 175°$ over the full azimuth.

Preliminary results of the quality of the $\vec{p_T}^{miss}$ measurement in UA2′ during the 1987 run are summarized in Fig. 5. Since the two components of the $\vec{p_T}^{miss}$ vector are gaussian (Fig. 5a) with the same $\sigma = \sigma_x = \sigma_y$, one can show that $dn/d(p_T^{miss})^2 \sim \exp[-(p_T/\Delta^2)]$, where $\Delta = \sqrt{2} \sigma$ depends on ΣE_T. Fig. 5b shows the behaviour of Δ as a function of ΣE_T, which can be parametrised by $\Delta = 1.06 (\Sigma E_T)^{04}$ GeV (solid line in Fig. 5b). Also shown for comparison is the equivalent parametrisation for UA1. The UA2′ performance is likely to improve with final energy calibration.

3. THE PHYSICS - STANDARD MODEL

3.1 W AND Z PHYSICS

The accurate measurement of the intermediate vector boson masses is of primary importance to test the Standard Model. While m_Z will be measured very accurately in the near future at SLC and LEP [we expect $\Delta (m_Z) \sim 50$ MeV], the W sector will remain the domain of $\bar{p}p$ Colliders until LEP Phase II.

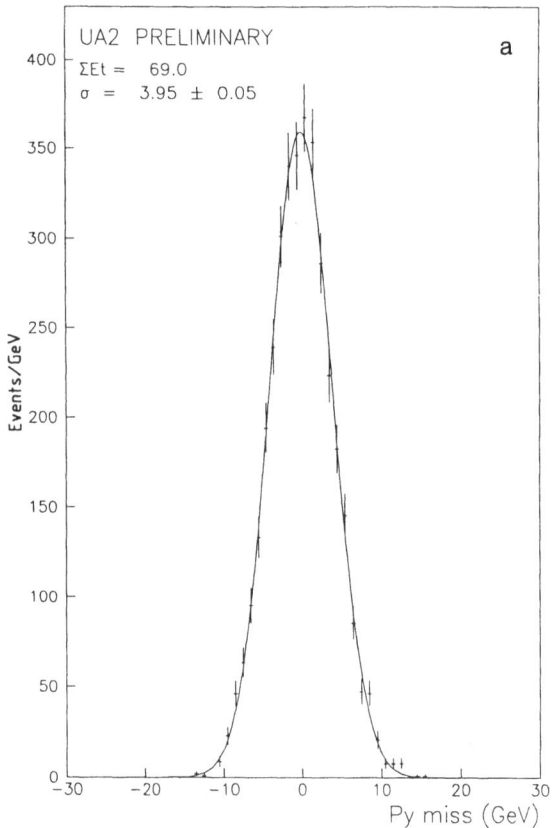

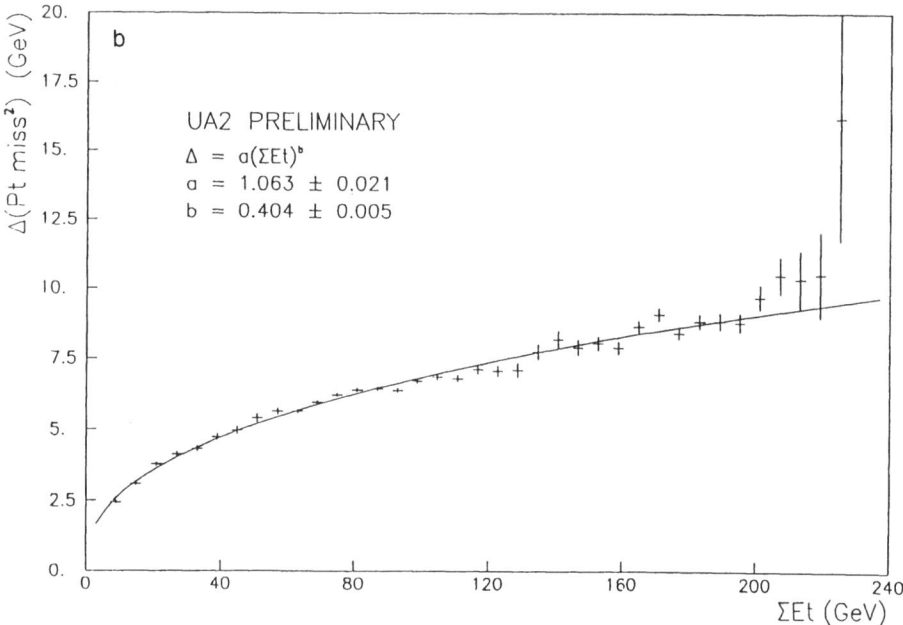

Fig. 5. a. One component of the $\vec{p}_T^{\,miss}$ vector at a given ΣE_T. The $\vec{p}_T^{\,miss}$ components show gaussian behaviour for all values of ΣE_T.

b. Resolution function of the p measurement in UA2'.

3.1.1 Measurement of the W and Z masses

The published results from UA2 [8] are based on 251 $W \to e\nu$ and 39 $Z \to e^+e^-$ decays, corresponding to an integrated luminosity of 910 nb^{-1}. The mass values of W and Z measured by UA2 are:

i) $m_W = 80.2 \pm 0.6(\text{stat}) \pm 0.5(\text{syst}_1) \pm 1.3(\text{syst}_2)$ GeV

ii) $m_Z = 91.5 \pm 1.2(\text{stat}) \pm 1.7(\text{syst})$ GeV

as obtained by fitting i) the transverse mass distribution of the e-ν system, $m_T^{e\nu}$, in events consistent with the reaction $\bar{p}p \to W + X \to e\nu + X$, ii) the mass distribution of the e^+e^- system in events consistent with the reaction $\bar{p}p \to Z + X \to e^+e^- + X$. The quoted systematic errors reflect uncertainties from the fit (syst$_1$) and from the knowledge of the calorimeter energy scale (syst$_2$).

For a total of 10 pb^{-1} of integrated luminosity, ~ 3000 $W \to e\nu$ and ~ 350 $Z \to e^+e^-$ are expected in the upgraded UA2 detector. After applying fiducial cuts to select events with good energy measurement, the previous numbers will reduce to 2000-2500 and 200-250, respectively. The expected improvement in the measurement of m_W and m_Z is based on the following considerations :

i. With 10 pb^{-1} the expected statistical error is ~ 220 MeV for the W and ~ 300 Mev for the Z.

ii. The systematic error from the fit of the W mass depends on theoretical and experimental uncertainties on the width of the W, Γ_W, the production transverse momentum of the W, p_T^W, and the evaluation of p_T^ν through the measurement of p_T^{miss}. The uncertainty on Γ_W can be reduced by combining the measurements of the ratio $R_{exp} = \sigma_W^\ell/\sigma_Z^\ell$, at the CERN $\bar{p}p$ Collider, and of Γ_Z at LEP/SLC, where a precision of 30 MeV is expected. The relation $\Gamma_Z/\Gamma_W = R_{exp}\, \sigma_Z/\sigma_W\, \Gamma_Z^\ell/\Gamma_W^\ell$ will give then Γ_W. The error from p_T^W would decrease by using p_T^Z from $Z \to e^+e^-$ to scale the theoretical calculation of p_T^W. Finally, the uncertainty on p_T^ν can be reduced by extracting m_Z with the same method used for m_W. It has been shown [8] that with high statistics the fit to the electron p_T distribution, sensitive to p_T^W but insensitive to p_T^ν, gives a more precise result than the fit to $m_T^{e\nu}$, once the method just described to extract p_T^W is used. In this case, the expected systematic error from the fit of m_W is of order 200 MeV.

iii. The uncertainty on the knowledge of the energy scale (syst$_2$) is expected to improve from the 1.6% of UA2 to 1.0%, thanks to accurate beam calibrations and an improved calibration system of the UA2' calorimeters.

In conclusion, one expects

$$\Delta m_W = \pm 0.22 \pm 0.20 \pm 0.80 \text{ GeV}$$
$$\Delta m_Z = \pm 0.30 \pm 0.90 \text{ GeV}.$$

The error on the energy scale, by far the dominant one, cancels to a large extent in the ratio m_W/m_Z :

433

$$\Delta (m_W/m_Z) = \pm 0.003 \pm 0.002$$

Assuming then that m_Z will be measured with an accuracy of ± 50 MeV at e^+e^- machines, the combination of $\bar{p}p$ Collider and LEP/SLC measurements will give

$$\Delta m_W = \pm 350 \text{ MeV}.$$

3.1.2 Standard Model Parameters

The measurements of m_W and m_Z lead to further checks of the predictions of the Standard Model, once properly renormalized and radiatively corrected quantities are used. In the scheme where $\sin^2\theta_W$ is defined [9] as

$$\sin^2\theta_W = 1 - (m_W/m_Z)^2 \qquad (1)$$

the SM predicts

$$m_W^2 = A^2/(1 - \Delta r) \sin^2\theta_W \qquad (2)$$
$$m_Z^2 = 4 A^2/(1 - \Delta r) \sin^2 2\theta_W \qquad (2')$$

where $A = (\pi\alpha/\sqrt{2} \, G_F)^{1/2} = (37.2810 \pm 0.0003)$ GeV using the experimental measurements of the Fermi coupling constant and of the fine structure constant. The value of Δr, which accounts for one-loop radiative corrections at the W mass [10], is not constrained by experimental data. Recent calculations [11] based on a value for the top mass of 45 GeV and for the Higgs of 100 GeV give $\Delta r = 0.0711 \pm 0.0013$.

The Standard Model can be tested by extracting i) $\sin^2\theta_W$, ii) ρ, iii) Δr from the measured values of m_W and m_Z as follows :

i. Equation (1) provides a direct measurement of $\sin^2\theta_W$ independent of other experiments and of theoretical uncertainties. UA2 finds

$$\sin^2\theta_W = 0.232 \pm 0.025(\text{stat}) \pm 0.010(\text{syst})$$

where the systematic error is mostly related to the m_W fit, the mass scale uncertainty cancelling out in the ratio m_W/m_Z. In UA2′ using (1) and for 10 pb^{-1}, we expect

$$\Delta(\sin^2\theta_W) = \pm 0.006(\text{stat}) \pm 0.004(\text{syst}).$$

A more precise value of the weak mixing angle can be extracted by a simultaneous fit of the measured m_W and m_Z to the theoretical equations (2) and (2′). The UA2 result is

$$\sin^2\theta_W = 0.232 \pm 0.003(\text{stat}) \pm 0.008(\text{syst})$$

where now the systematic error is dominated by the error on the mass scale, and the measurement relies on other experimental results and theoretical calculations [A and Δr in (2) and (2′), respectively]. The expected precision of this measurement in UA2′ is :

$$\Delta(\sin^2\theta_W) = \pm\, 0.001(\text{stat}) \pm 0.005(\text{syst})$$

ii. Any departure from the Minimal Standard Model would result in modifications to the formalism presented above. In particular values of $\sin^2\theta_W$ from (1),(2) and from low-energy neutrino experiments would in general differ. Although all existing measurements are in agreement with the Standard Model, they can be used to place limits on possible deviations from the Minimal Standard Model. In particular, the quantity

$$\rho = m_W^2/m_Z^2 \cos^2\theta_W$$

is expected to be 1 in the Minimal Standard Model. UA2 measures

$$\rho = 1.001 \pm 0.028(\text{stat}) \pm 0.006\ (\text{syst})$$

while the sensitivity of UA2′ is expected to be

$$\Delta\rho = \pm\, 0.008(\text{stat}) \pm 0.009(\text{syst}).$$

iii. The radiative correction parameter Δr can also be extracted from the measured values of m_W and m_Z. Eliminating $\sin^2\theta_W$ from (2) and (2′) the UA2 result is

$$\Delta r = 0.068 \pm 0.087(\text{stat}) \pm .030(\text{syst})$$

while using $\sin^2\theta_W$ from low-energy experiments we get

$$\Delta r = 0.068 \pm 0.022(\text{stat}) \pm 0.032(\text{syst}).$$

This result can be visualised in the m_W-m_Z plane (Fig. 6). The present measurements, although in good agreement with the Standard Model, are not precise enough to test the presence of Δr. The expected sensitivity at the upgraded CERN p$\bar{\text{p}}$ Collider is also summarized in Fig. 6, from which we can conclude that UA2′ will test radiative corrections, once a precise measurement of m_Z is available from SLC or LEP. The importance of a precise measurement of Δr resides in the fact that Δr would deviate from the computed value [11] if a new fermion family existed with a large mass splitting, or if there were additional gauge bosons, or again if the top quark were very heavy.

3.2 SEARCH FOR THE TOP QUARK

At $\bar{\text{p}}$p Colliders the main mechanisms of top production are:

$$\begin{aligned}\bar{\text{p}}\text{p} &\to \text{W} + \text{X} \\ &\to \text{t}\bar{\text{b}}\end{aligned} \qquad (1)$$

if kinematically allowed, and QCD production

$$\bar{\text{p}}\text{p} \to \text{t}\bar{\text{t}} + \text{X} \qquad (2)$$

via q-$\bar{\text{q}}$ annihilation or gluon-gluon fusion. At $\sqrt{s}$ = 630 GeV reaction (1) is the dominant one for 40 < m_{top} < 80 GeV. The production cross section can be precisely estimated through the relation

$$(W \rightarrow t\bar{b}) = 3\ \sigma(W \rightarrow e\nu)\ \phi(m_t)$$

where $\sigma(W \rightarrow e\nu)$ has been measured at the CERN Collider, $\phi(m_t)$ is a known phase space suppression term and the factor of 3 is for colour. Outside the above mass range, top production is dominated by direct $t\bar{t}$ production (2). A recent complete calculation of the next-to-leading order QCD corrections to the total cross-section for heavy flavour production [12] and a set of proton structure functions including up-to-date experimental results [13] has allowed the authors of ref. [14] to give reliable predictions for top production.

Top decay proceeds as

$$t \rightarrow (Wb) \rightarrow 3\ \text{jets} \quad \text{or} \quad t \rightarrow (Wb) \rightarrow \ell\nu + \text{jet}.$$

The final state will therefore consist of > 4 jets or a charged lepton accompanied by two or more jets in events with missing transverse energy. Given the high rate of multi-jet events from QCD processes, the interesting final states at $\bar{\text{p}}$p Colliders are those connected to the semi-leptonic decay modes which, at the price of a branching ratio of 1/9 for each charged lepton, give a much cleaner signature.

The analysis requires very strict selection criteria in order to improve the signal to background ratio. With the standard UA2 electron cuts [8] about 10 top events are expected over a background of 300 in the existing UA2 data sample. Strong requirements of electron isolation, both in tracking and preshower detectors and in the calorimeter, are necessary to improve the signal-to-background ratio. In this case, for an expectation of 3-4 events UA2 finds 30 candidates in the 910 nb^{-1} data sample, entirely consistent with background, both in rate and event topology.

The greatly improved electron identification power of UA2' should allow the detection of the top at the CERN $\bar{\text{p}}$p Collider if its mass is smaller than the W mass. For 10 pb^{-1} we estimate that about 70 events containing a t $\rightarrow$ beν are expected, over a background of about 10 events, if m_t = 40 GeV, where the contributions from W $\rightarrow$ t$\bar{b}$ and t$\bar{t}$ are about equal. For m_t = 60 GeV the contribution from t$\bar{t}$ is about 25% of the total and the predicted signal is 20 events with a background of 4 events. For top masses approaching m_W the b jets are softer and therefore it becomes more and more difficult to distinguish W $\rightarrow$ t$\bar{b}$ from W + jets production. On the other hand, for the same kinematical reasons the t$\bar{t}$ channel will result in a final state containing a $\ell\nu$ and a two-jet pair with invariant masses close to m_W. Finally for m_t > m_W the most promising channel is t$\bar{t}$ $\rightarrow$ W$^+$W$^-$b$\bar{b}$ followed by a semileptonic decay of one W. For m_t = 100 GeV, for example, about 10 events are expected before any selection cut necessary to reduce the background of W + two-jet production. We can consider, therefore, m_t = 100 GeV as an (optimistic) upper limit of the sensitivity of top search in UA2'.

3.3 QCD MEASUREMENTS

Among the many tests of QCD already made by UA2, only two topics will be briefly mentioned here.

3.3.1 Multi-jet studies

The study of final states containing hadron jets of high transverse energy has been one of the main activities of UA1 and UA2 at the $Sp\bar{p}S$ Collider. The properties of the two and three-jet configurations have been successfully compared with QCD to leading and next-to-leading order in α_S. To higher order in α_S QCD predicts multi-gluon bremsstrahlung giving rise to final states containing many jets. The study of such configurations would provide additional tests of perturbative QCD. In particular, four-jet events have been the subject of a recent analysis by UA2, which has also compared such configurations with the hypothesis of double parton interactions. Restricted to the central region, $|\eta| < 1$, and for $p_T^{jet} > 10$ GeV, the UA2 four-jet analysis finds no evidence of multi-parton scattering [15].

The study of these and more complicated many-jet final states will obviously benefit from the increased acceptance of UA2', $|\eta| < 3$, from the refined jet triggers, which would allow a reduced p_T threshold, and from the increased integrated luminosity.

3.3.2 Measurement of α_S

As mentioned in the previous section, QCD predicts the occurrence of gluon bremsstrahlung as a first order perturbative correction in the strong coupling constant α_S to parton-parton scattering. Final states containing three jets should be observed at a rate dependent on the value of α_S. Therefore, the yield of three-jet relative to two-jet events gives a measure of $\alpha_S k_3/k_2$. Unfortunately the contributions k_3 and k_2 from higher order corrections in α_S to the two and three-jet cross-sections have not yet been calculated and only the quantity $\alpha_S k_3/k_2$ can be extracted with this method. UA2 finds $\alpha_S k_3/k_2 = 0.23 \pm 0.01 \pm 0.04$ [16].

A similar approach consists of using the yield R of the W + jet relative to W production. By fitting the prediction of a QCD Monte Carlo [17] to the measured value of R, one can extract the value $\alpha_S k_1/k_0$, where the k factors take into account the corrections of higher order diagrams to the W + jet and W cross-sections, respectively. The advantage of this method is given by a recent evaluation of k_1 and k_0 [18] which allows one to extract an absolute value for α_S. A preliminary analysis of the UA2 data [19] gives

$$\alpha_S (m_W) = 0.13 \pm 0.03 (\text{stat}) \pm 0.03 (\text{syst})$$

where the systematic error contains theoretical and experimental uncertainties.

Some of these uncertainties, such as the one related to the knowledge of the energy scale or the effect of the underlying event, will improve in UA2'. Taking into account statistical error corresponding to an integrated luminosity of 10 pb^{-1}, the overall error on the determination of α_S in UA2' with this method is expected to be of the order of 0.015-0.020.

4. EXTENSIONS OF THE MINIMAL STANDARD MODEL

4.1 New electroweak gauge bosons

Additional vector bosons arise from any extension of the minimal SU(2)xU(1) group of the Standard Model, composite models, or models derived from Superstring theories.

At $\bar{p}p$ Colliders they would be produced through $q\bar{q}$ annihilation and they would be detected via their leptonic decay, the decay into quarks being dominated by the two-jet background from QCD processes. The sensitivity of experiments is a function of their masses, $m_{W'}$ and $m_{Z'}$, their coupling to quarks and their branching ratio to leptons.

A search in UA2 excludes at a 90% confidence level the existence of a W' of $m_{W'} < 209$ GeV and a Z' of $m_{Z'} < 180$ GeV, in the mass region above $m_{W'}$ and $m_{Z'}$ and for standard coupling to quarks and electrons. Details of the analysis, variation of the limits for different couplings and for masses around and below $m_{W'}$ and $m_{Z'}$ can be found in Ref [20].

The sensitivity of UA2' at ACOL will be about 300 GeV for $m_{W'}$ and 250 GeV for $m_{Z'}$, for an integrated luminosity of 10 pb^{-1}.

5. SEARCH FOR EXOTICS

5.1 Supersymmetric particles

Among the many mechanisms proposed to overcome the theoretical difficulties associated with the Higgs sector of the Standard electroweak theory, Supersymmetry seems to be the most far-reaching, since it provides a natural framework for spontaneously broken gauge theories involving elementary scalars. At $\bar{p}p$ Colliders the highest cross sections, for any given mass of the supersymmetric particles, would be for the strongly interacting ones, the squark $\tilde{q}$ and the gluino $\tilde{g}$, the dominant production channels being $\bar{p}p \to \tilde{g}\tilde{g}$, $\bar{p}p \to \tilde{g}\tilde{q}$, $\bar{p}p \to \tilde{q}\tilde{q}$. The subsequent decay depends on the relative values of the squark mass, $m_{\tilde{q}}$, with respect to the gluino mass, $m_{\tilde{g}}$: i) $\tilde{q} \to q\tilde{g}$, $\tilde{g} \to q\bar{q}\tilde{\gamma}$ if $m_{\tilde{q}} > m_{\tilde{g}}$, or ii) $\tilde{g} \to \tilde{q}q$, $\tilde{q} \to q\tilde{\gamma}$ if $m_{\tilde{q}} < m_{\tilde{g}}$. Since in most models the photino is the lightest non-interacting SUSY particle, the final states would consist in many (2 to 6) jets and p_T^{miss}. At the Sp$\bar{p}$S Collider the UA1 detector was better configured than UA2 for the study of such processes, owing to its larger η-coverage.

The better granularity of UA2 allowed us to study the case of an unstable photino which, through the decay $\tilde{\gamma} \to \gamma\tilde{H}$, would produce a final state with two photons and jets [20]. The absence of events with a two-photon pair with invariant mass larger than 10 GeV and at least two jets with $p_T > 10$ GeV translates into the limits of Fig. 7. UA2 excludes gluinos in the mass range 15-50 GeV and squarks in the range 9-46 GeV.

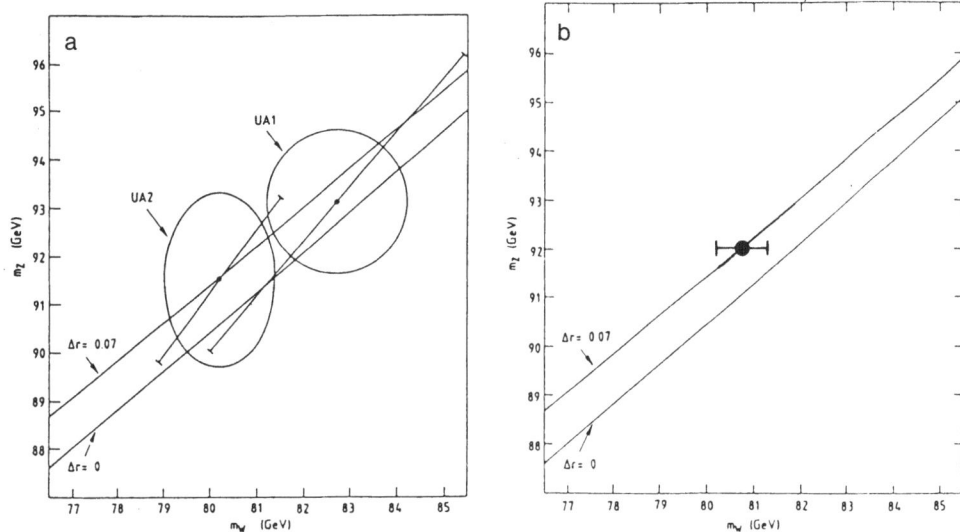

Fig. 6. Present m_W versus m_W measurements from UA1 and UA2 (black points). The ellipses represent the one-sigma statistical errors and the bars centred on the points represent the systematic uncertainty on the mass scale. The expected accuracy with 10 pb^{-1} in UA2′, following a precision measurement of m_Z at SLC/LEP, is shown in Fig. 6 b). The position of the point in the m_W-m_Z plane is arbitrary.

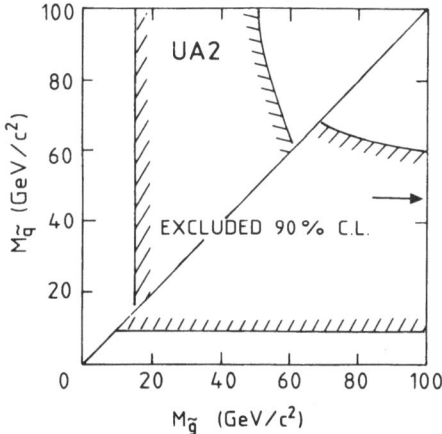

Fig. 7. Limits on the $\tilde{g}$ versus $\tilde{q}$ masses obtained from an analysis of photon pairs accompanied by jets are shown as 90% confidence level contours, in the case of an unstable photino.

UA2' will take advantage of the improved p_T^{miss} measurement and of the increased acceptance. Assuming a selection efficiency of 10% the sensitivity to scalar quarks and gluinos for L = 10 pb^{-1} should reach mass values of order 100 GeV.

6. THE FIRST RUN OF UA2'

Many of the solutions adopted in upgrading the original UA2 detector have been constrained by the tight schedule imposed by the improvement program of the p̄p complex and the choice of being ready with the full configuration for the starting of the improved machine. All the detectors of the substantially modified apparatus have been successfully operational through the whole p̄p running period in November and December 1987 and preliminary results on their performance have been reported in Section 2. Unfortunately, technical problems both in the AA/AC and in the SPS resulted in a poor performance of the Collider, with a peak luminosity never exceeding 3×10^{29} cm^{-2}sec^{-1}, the same as the old machine, and a tiny integrated luminosity of 46 nb^{-1}, only 5% of that accumulated so far by UA2. None of the problems is of fundamental nature and they are expected to be overcome for the next running periods, when luminosities close to the design value of 4×10^{30} cm^{-2}sec^{-1} should be reached (*).

One major item in the UA2 upgrading, not mentioned so far, is indeed related to the capability of the experiment to face the expected high luminosity and the large data volume generated by the sophisticated UA2' detectors, i.e. the trigger and the data acquisition (DAQ) systems [21]. Two additional levels of software triggers were added to the hardware trigger of UA2. The DAQ system itself had to be upgraded to provide two stages of event processing and buffering. At the design luminosity of the upgraded p̄p Collider the expected rate from the first level trigger will be of order 100 Hz and the amount of data to be collected per event will average 80-100 kBytes. Several special runs were made to simulate running conditions at 4×10^{30}cm^{-2}sec^{-1} and test the performance of the trigger and DAQ systems. No particular problems were detected. The full detector is ready now to start the '88 data taking.

REFERENCES

1. The Staff of the CERN Proton-Antiproton Project, Phys. Lett. 107B (1981) 306.

2. E. Jones, Proc of the 6th Topical Workshop on Proton-Antiproton Collider Physics, Aachen, 1986, World Scientific, Singapore (1987), p.691.

3. UA2 Collaboration, "Proposal to Improve the Performance of the UA2 Detector" CERN/SPSC 84-30 SPSC/P93 Add.2 (1984);
   UA2 Collaboration, "Proposal to Improve the Performance of the UA2 Central Detector" CERN/SPSC 84-95 SPSC/P93 Add.3 (1984).

(*) At the end of September '88, a peak luminosity of 1.9×10^{30}cm^{-1}sec^{-2} was reached; this performance is likely to be improved during the Collider run.

4. UA2 Collaboration, C. Booth, Proc. 6th Topical Workshop on Proton-Antiproton Collider Physics, Aachen 1986, World Scientific, Singapore (1987) 381.

5. R.E. Ansorge et al., Nucl. Inst. Meth. 265A (1988) 33.

6. K. Borer et al., Nucl. Inst. Meth. 265A (1988) 33.

7. F. Bourgeois, Nucl. Inst. Meth. 219 (1984) 153.

8. UA2 Collaboration, R. Ansari et al., Phys. Lett. 186B (1987) 440;
UA2 Collaboration, R. Ansari et al., Phys. Lett. 194B (1987) 158. See references therein for previous UA2 publications.

9. W.J. Marciano and A. Sirlin, Phys Rev. D29 (1984) 945.

10. W.J. Marciano, Phys Rev. D20 (1979) 274;
A. Sirlin, Phys Rev. D22 (1980) 971 ;
F. Antonelli et al., Phys. Lett. 91B (1980) 90 ;
M. Veltman, Phys. Lett. 91B (1980) 95.

11. F. Jegerlehner, Bielefeld preprint BI-TP 1986/8 (1986).

12. P. Nason et al., FNAL preprint Pub-87/222-T (9187).

13. M. Diemoz et al., CERN preprint CERN-TH.4751/87 (1987).

14. G. Altarelli et al., CERN preprint CERN-TH.4978/88 (1988).

15. UA2 Collaboration, see L. Mapelli in Proc. XVIII Int. Symp. on Multiparticle Dynamics, Tashkent (1987).

16. UA2 Collaboration, J.A. Appel et al., Z. Phys. C30 (1986) 341.

17. S.D. Ellis et al., Phys. Lett. 154b (1985) 435.

18. A.C. Bawa and W.J. Stirling, Durham preprint DPT/87/42 (1987).

19. UA2 Collaboration, V. Ruhlmann, Ph.D. Thesis, CEN Saclay (1988);
UA2 Collaboration, K. Jacobs, Ph.D. Thesis, Heidelberg (1988).

20. UA2 Collaboration, R. Ansari et al. Phys. Lett. 195B(1987) 613.

21. G. Blaylock et al., The UA2 data-acquisition system, to appear in Proc. Adriatic Conf. on Impact of Digital Microelectronics and Microprocessors on Particle Physics, ICTP, Treste, Italy.

26

ELECTROWEAK SYMMETRY BREAKING STUDIES AT THE pp COLLIDERS OF THE 1990'S AND BEYOND

Michael S. Chanowitz

Lawrence Berkeley Laboratory
University of California
Berkeley, CA 94720

ABSTRACT

Within the conventional framework of a spontaneously broken gauge theory, general principles establish that the electroweak symmetry is broken by a new force that may be weak with associated new quanta below 1 TeV or strong with quanta above 1 TeV. The SSC parameters, $\sqrt{s} = 40$ TeV and $\mathcal{L} = 10^{33}\mathrm{cm}^{-2}\mathrm{s}^{-1}$, define a minimal facility with assured capability to observe the signals of symmetry breaking by a strong force above 1 TeV. Foreseeable luminosity upgrades would not be able to compensate a much lower collider energy for these physics signals. If the strong WW scattering signal were seen at the SSC in the 1990's it would provide a clear imperative for a collider with the physics reach of the ELOISATRON to begin detailed studies of the new force and quanta early in the next century.

1. SYMMETRY BREAKING AND THE TEV SCALE

In the energy landscape reaching up to the Planck scale, one landmark stands out clearly at the TeV scale. General arguments tell us that the origin of W and Z masses, the mechanism of electroweak symmetry breaking, waits to be discovered at the TeV scale or below. It is therefore very exciting that we may soon be able to explore this region with multi-TeV pp collisions.

One purpose of this talk is to review the reasons that a pp collider with the design energy and luminosity of the SSC, $\sqrt{s} = 40$ TeV and $\mathcal{L} = 10^{33}\mathrm{cm}^{-2}\mathrm{s}^{-1}$, is a *minimal* facility with an assured capability to point the way to the symmetry breaking mechanism. In particular, such a pp collider will teach us whether electroweak symmetry breaking is due to new, weakly interacting quanta below 1 TeV or to new, strongly interacting quanta above 1 TeV. In the latter case the signals at the SSC are observable but small, and an even higher energy facility would be needed to study the new phenomena in detail. This might be a 200 TeV pp collider like the ELOISATRON that is the subject of this workshop and/or a multi TeV e^+e^- collider of energy 10 TeV or more. Though it may seem reckless to imagine such incredible devices when 16 and 40 TeV pp colliders are still just gleams in their designers' eyes, it is nevertheless clear that symmetry breaking physics at or above 1 TeV would impel us irresistibly in this direction.

In this talk I will briefly review the general analysis which identifies the weak boson pair channel (WW, WZ, ZZ) between 1 and 2 TeV as the critical physics signal that we must be able to detect. If we are able to detect this signal, then we will learn from its presence or its absence. If present, we learn that symmetry breaking is due to new strongly interacting physics at or above 1 TeV, and we will probably see the first signs of the new quanta in the 1-2 TeV signal. If absent, we learn that symmetry breaking is due to weakly interacting quanta below 1 TeV, and we will be strongly motivated to redouble the search below 1 TeV. Therefore we win whether the signal occurs or not, in contrast to the more typical situation in which a negative result leaves all possibilities open.

The unified theory of weak and electromagnetic interactions[1] implies the existence of a new (fifth) force and a world of new particles, the "symmetry breaking sector". This force induces spontaneous breaking of the electroweak symmetry, so that the W and Z become massive while the photon remains massless. At the same time the W and Z acquire longitudinal polarization states, which are essentially particles (Goldstone bosons) from the symmetry breaking sector. In this sense, three new particles from the symmetry breaking sector have already been discovered: the longitudinal modes $W_L^{\pm}$ and Z_L. They are the pions of this new world of particles.

By measuring W_L and Z_L scattering amplitudes we therefore measure the strength of the symmetry breaking force. This is done at colliders by means of the process shown in figure (1). The W's are initially virtual (off-shell) and must rescatter to become real (on-shell). The number of W_L and Z_L pairs produced by this mechanism thus measures the strength of the unknown force.[2]

Model-independent arguments, based only on symmetry and unitarity,[2,3] specify the two possibilities: a weak force with associated particles below 1 TeV (Higgs bosons) or a strong force with a rich spectrum of new particles (probably not simply Higgs bosons) above 1 TeV. The latter case could be a confining gauge theory, as in technicolor,[4] or it could be something completely unanticipated.

Deciding between these alternatives is one of the fundamental problems of particle physics, and one which we have the best chance of solving in the immediate future. The first stage in a systematic approach would be to build a collider capable of detecting the W_L pairs that would arise from figure (1) if the new force is strong $i.e.$, strong WW scattering. Such a device could tell us definitively whether the new physics lies below or above 1 TeV. If there is strong WW scattering the lightest resonances would probably appear at $\sim 2 \pm 1$ TeV, $i.e.$, at the onset of the strong scattering domain. The first stage collider might be able to detect the lightest resonances but it would not be capable of detailed studies of the spectrum that would exist above 1 TeV. It would be as if we knew hadrons existed but had not yet built the Bevatron or the PS to study them. A new higher energy facility would be imperative.

The strong WW scattering signal would, if it exists, emerge between 1 and 2 TeV in the WW invariant mass. Observation of this signal places great demands on the collider luminosity and, especially, on the collider energy. There are two reasons for this extreme sensitivity to the energy: one has to do simply with the size of the signal and the other has to do with the effect of the background on the significance of the signal.

Like the analogous photon-photon scattering process at e^+e^- colliders, WW

fusion[5] places great demands on beam energy since the final state fermions "want" to keep a large fraction of their initial energy. The analogue of the Weiszacker-Williams luminosity function for longitudinal W's is[6,2]

$$\frac{\partial \mathcal{L}}{\partial x} = \frac{\alpha^2}{16\pi \sin^4 \theta_W} \frac{1}{x} \left[(1+x) \ln \frac{1}{x} + 2x - 2 \right] \tag{1.1}$$

where $x = s_{WW}/s_{qq}$. Here $\partial \mathcal{L}/\partial x$ is the effective distribution function to find incident W_L beams in colliding quark (or antiquark) beams. The steep dependence on energy is clear in table 1. We see that for fixed s_{WW}, doubling the initial quark energy from $\sqrt{s_{qq}} = 2\sqrt{s_{WW}}$ to $4\sqrt{s_{WW}}$ causes the luminosity to rise by a factor of 20. The consequences for the WW fusion event rate as a function of collider energy are evident in the results presented in Section 4.

The second aspect of the importance of collider energy has to do with the signal:background ratio. For $\sqrt{s_{WW}} \cong 1 - 2$ TeV, the quark distribution functions at LHC and SSC energies tend to force us toward the threshold region in eq. (1.1). In the case of the WW fusion signal we are dealing with a four body final state, $qq \to qq\, WW$, so the dependence of the phase space and the cross section is a very steep function of energy. On the other hand the dominant backgrounds such as $\bar{q}q \to WW$ have two body final states and therefore depend much less sensitively on energy. Consequently as we raise the collider energy not only the signal increases but also the signal:background ratio. This means, as discussed more extensively in Section 4, that to gain statistical parity a lower energy collider must be upgraded in luminosity by a ratio that is larger than just the ratio of signal cross sections. In evaluating the trade-off between energy and luminosity, it is also important to keep in mind that the largest gauge boson pair decay modes which are observable at $\mathcal{L} = 10^{33}$cm^{-2}s^{-1} are probably not observable in the environment of $\mathcal{L} \gg 10^{33}$ (see ref. (7) and Section 4). The task of preparing for physics at $\mathcal{L} = 10^{33}$ is already a major challenge.

Unlike other physics goals of multi-TeV colliders which may or may not exist in nature, we know quite generally that the effects of the symmetry breaking sector must emerge in the $W_L W_L$, $W_L Z_L$, and $Z_L Z_L$ channels at or below ~ 2 TeV — provided only that we accept the conventional framework of a spontaneously broken gauge theory. The basic physical point is that W_L and Z_L are essentially (as made explicit by the equivalence theorem discussed in Section 2) the "pions" of the symmetry breaking sector, and like the pions of hadron physics they obey low energy theorems characteristic of the scattering of Goldstone bosons.[2] If there are no other light (compared to 1 TeV) particles in the symmetry breaking sector, the low energy scattering amplitudes depend only on the known parameters G_F and $\rho = (M_W/M_Z \cos\theta_W)^2$ and not at all on the unknown physics of the symmetry breaking sector,[3] denoted by its lagrangian $\mathcal{L}_{SB}$. For example, one of the low energy amplitudes is

$$\mathcal{M}(W_L^+ W_L^- \to Z_L Z_L) = \sqrt{2}\, G_F \frac{s}{\rho}. \tag{1.2}$$

Substituting

$$\sqrt{2}\, G_F = \frac{1}{v^2} \tag{1.3}$$

where $v = 246$ GeV is the vacuum expectation value of the standard Higgs boson model[1] and using the experimental fact that $\rho \cong 1$, eq. (1.2) can be rewritten as

$$\mathcal{M}(W_L^+ W_L^- \to Z_L Z_L) = \frac{s}{v^2}. \tag{1.4}$$

Equation (1.4) has a more than coincidental resemblance to the pion low energy theorem,[8]

$$\mathcal{M}(\pi^+\pi^- \to \pi^0\pi^0) = \frac{s}{F_\pi^2}.$$ (1.5)

However, as shown in Section 2, eqs. (1.2) and (1.4) are more properly called "intermediate energy theorems" and are only relevant if there is a strongly coupled symmetry breaking sector above 1 TeV.

A priori we know neither the mass scale M_{SB} nor the coupling strength λ_{SB} of the symmetry breaking sector lagrangian $\mathcal{L}_{SB}$. The low energy theorem provides the correlation between the mass scale M_{SB} and the interaction strength λ_{SB}. Unitarity requires that the amplitude cannot be proportional to s for arbitrarily large s, and the most likely scenario (Section 3) is that the growth in s is cut off at the mass scale M_{SB}. This observation allows us to identify the strong coupling regime with the mass domain $M_{SB} \gtrsim 1$ TeV (see Section 3).

The low energy theorems are the key to a general "no-lose" strategy because they provide an estimate of the scale of $\sigma(W_L W_L)$ if $\mathcal{L}_{SB}$ is strongly coupled (I cannot bring myself to use the more accurate term "intermediate energy theorems"). A collider at which we can detect the strong WW scattering signal is sure to lead us to the symmetry breaking physics, whether strong WW scattering occurs or not.

For the SSC the crucial observation is that strong W_L, Z_L scattering is observable at the SSC by virtue of increased yields of gauge boson pairs produced with the WW fusion mechanism shown in figure (1). At the design energy and luminosity these extra gauge boson pairs will be observable above the background sources of gauge boson pairs that are present whether $\mathcal{L}_{SB}$ is a strong or weak coupling theory. The conclusion is the

No-Lose Corollary:

Either there are light ($\ll 1$ TeV) particles from $\mathcal{L}_{SB}$ that can be produced and studied directly

and/or

Excess WW, WZ, ZZ production is observable, signaling strongly-coupled $\mathcal{L}_{SB}$ with $M_{SB} \gtrsim 1$ TeV.

For the strong coupling case, if (as in hadron physics) resonances occur when the partial wave amplitudes are $O(1)$, then probably $M_{SB} \lesssim O(2)$ TeV and the low-lying spectrum of $\mathcal{L}_{SB}$ is (just) visible. However, the 1–2 TeV strong coupling signal would be observable even if $M_{SB} >> 2$ TeV and the new particles were too heavy to produce. With $\sqrt{s} = 40$ TeV and $\mathcal{L} = 10^{33} \text{cm}^{-2}\text{s}^{-1}$ the SSC would be a minimal pp collider at which the no-lose strategy could be pursued (see Section 4).

In the weak coupling, sub-TeV case there is one known exception that should be mentioned: if $\mathcal{L}_{SB}$ is given by the minimal Higgs model and if the mass happens to lie in the interval $2m_t < m_H < 2M_W$, then although 10^6 Higgs would be produced in an SSC year, it is not now known how to detect them in their $H \to \bar{t}t$ decay mode.[9] For now the only known way of discovering the Higgs boson in this mass range is to build a $\sqrt{s} > 330$ GeV e^+e^- collider with a luminosity of $\gtrsim 10^{32}\text{cm}^{-2}\text{s}^{-1}$. Even in this rather unlikely scenario (becoming more unlikely as the lower bound on m_t rises toward M_W), the SSC would contribute to our understanding of symmetry breaking by verifying the absence of the excess gauge boson pairs associated with new strong interactions. (Strong coupling models might have light scalars that would approximately mimic a light Higgs boson.) In general the absence of additional gauge boson pairs from WW fusion would be our cue for a redoubled search of the sub-TeV mass scale.

I want to make a few comments to put the above statements in perspective. TeV scale symmetry breaking physics puts the greatest demands on energy and luminosity. Some of the other physics that has been contemplated is less demanding and less sensitive to the difference between 15 and 40 TeV. However, symmetry breaking physics has a special claim on our attention, since while other possible new physics (e.g., compositeness, Z', $\cdots$) may or may not occur in nature, we know for sure that the electroweak symmetry is broken. The no–lose corollary says that a facility able to "see" the 1–2 TeV signal of strongly–coupled symmetry breaking gives us valuable information about the mass scale M_{SB} of the symmetry breaking physics whether the new particles occur below, within, or above the 1–2 TeV region. As will become clear from the results presented below, a pp collider with $\sqrt{s} = 40$ TeV and $\mathcal{L} = 10^{33}\text{cm}^{-2}\text{s}^{-1}$ is *just* able to observe the 1–2 TeV strong–interaction signal but is not sufficient to study it in detail. In this sense it is an optimal probe, since we would not want to consider a more ambitious facility without first knowing that there is TeV–scale physics requiring detailed study. Conversely, and very much to the point of this meeting, if the strong scattering signal is seen it is clear that we will badly need some more ambitious facility.

Though the problem of observing the strong interaction symmetry breaking signal at TeV e^+e^- colliders has not been as extensively studied, several authors have had a first look at the signal and/or backgrounds.[10] The conclusion from Bento and Llewellyn–Smith is that the signal of the conservative model discussed below would not be visible at an e^+e^- collider with $\sqrt{s}, \mathcal{L} = 2$ TeV, $10^{33}\text{cm}^{-2}\text{s}^{-1}$ but requires a collider in the range $\sqrt{s}, \mathcal{L} \sim 3 - 5$ TeV, $(1 - 2\frac{1}{2})10^{33}\text{cm}^{-2}\text{s}^{-1}$

Section 2 reviews relevant aspects of spontaneous symmetry breaking, presents the equivalence theorem which makes precise the connection between the longitudinal modes and the underlying Goldstone bosons, and sketches a proof of the low energy theorems. Section 3 is concerned with the implications of unitarity and a simple strong interaction model of WW scattering based on the low energy theorems. Section 4 surveys the experimental signals for a strongly interacting symmetry breaking sector, considering the LHC, the SSC, and the ELOISATRON. Understanding of the signals and backgrounds is far from complete. Much more work is needed to insure effective utilization of the multi-TeV pp colliders.

2. SPONTANEOUS SYMMETRY BREAKING AND LOW ENERGY THEOREMS

In order to implement spontaneous symmetry breaking, the lagrangian of the symmetry breaking sector, $\mathcal{L}_{SB}$, must possess a global symmetry group G — analogous to the flavor symmetry of QCD — which breaks by asymmetry of the vacuum to a smaller group H,

$$G \to H. \qquad (2.1)$$

Gauge invariance requires that G include the electroweak $SU(2)_L \times U(1)_Y$ and that H include the unbroken electromagnetic $U(1)_{EM}$. For each broken generator of G there is a massless Goldstone boson in the spectrum of $\mathcal{L}_{SB}$. Three of these couple to the weak currents and are denoted $w^{\pm}$, z. Others, if any, are denoted by $\{\phi_i\}$. Including the electroweak gauge interactions, the Goldstone triplet $w^{\pm}$, z become longitudinal gauge boson modes $W_L^{\pm}$, Z_L, and the $\{\phi_i\}$ acquire small masses $O(gM_{SB})$, becoming "pseudo-Goldstone" bosons.

As an example, for two flavor QCD with massless quarks the global symmetry G is $SU(2)_L \times SU(2)_R$. After spontaneous symmetry breaking the surviving invariance group is $H = SU(2)_{L+R}$ which is just the isospin group. There are three broken generators, corresponding to the axial generators of $SU(2)_{L-R}$, so that three massless Goldstone bosons emerge, $\pi^{\pm}$ and π^0. If there were no other symmetry breaking physics, $\mathcal{L}_{SB}$, $\pi^{\pm}$ and π^0 would indeed become longitudinal modes of $W^{\pm}$ and Z, which would however have masses of ~ 30 MeV (see $e.g.$ ref. (10a)).

The statement that the longitudinal modes $W_L^{\pm}$, Z_L are identified with the Goldstone bosons $w^{\pm}$, z is given a precise meaning by the equivalence theorem,

$$M\left(W_L(p_1)W_L(p_2)\ldots\right) = M\left(w(p_1)w(p_2)\ldots\right)_R + O\left(\frac{M_W}{E_i}\right). \qquad (2.2)$$

In eq. (2.2) the left side is a gauge-invariant S-matrix element involving longitudinal modes while the right side is the corresponding Goldstone boson amplitude in an R or renormalizeable gauge. As indicated in eq. (2.2) the equivalence holds at energies large compared to the W and Z masses. We can use the equivalence theorem to translate statements about Goldstone boson scattering amplitudes into statements about scattering of longitudinally polarized W's and Z's.

The equivalence theorem was proved in tree approximation in reference (11) and to all orders in both gauge and symmetry breaking interactions in reference (2) (see also reference (12)). The validity of the theorem to all orders in λ_{SB} is crucial since we wish to apply it when $\mathcal{L}_{SB}$ is strongly interacting and perturbation theory in λ_{SB} fails. Intuitively the theorem is a plausible consequence of the Higgs mechanism that transmutes the Goldstone bosons w and z into the longitudinal gauge boson modes W_L and Z_L. This is seen explicitly by the gauge transformation from a renormalizable gauge — in which the Goldstone boson fields appear in the lagrangian — to the unitary gauge in which the Goldstone fields do not appear. Nevertheless the proof to all orders is lengthy and complicated, making use of the BRS identities which embody the full content of gauge invariance in nonabelian gauge theories. Here I will only state the theorem and illustrate it with a simple example.

In addition to being useful in the derivation of the $W_L W_L$ low energy theorems, eq. (2.2) greatly simplifies perturbative calculations for heavy — and therefore strongly coupled — Higgs systems (see references (13, 14, 15, 2)). For instance, to correctly evaluate heavy Higgs boson production and decay by WW fusion in unitary gauge requires evaluation of many diagrams with "bad" high energy behavior that cancel to give the final result.[16-18] But to leading order in the strong coupling $\lambda = m_H^2/2v^2$ it suffices using eq. (2.2) to compute just a few simple diagrams using the interactions of the Higgs scalar potential. The result embodies the cancellations of many diagrams in unitary gauge and trivially has the correct high energy behavior. On the other hand, unitary gauge calculations of Higgs boson production in the s–channel pole approximation[19] have bad high energy behavior and overestimate the yield for heavy Higgs bosons, $m_H \geq 800$ GeV.

As a simple example, consider the decay of a heavy Higgs boson to $W_L^+ W_L^-$. In unitary gauge the $HW_L^+ W_L^-$ amplitude is

$$\mathcal{M}(H \to W_L^+ W_L^-) = g M_W \epsilon_L(p_1) \cdot \epsilon_L(p_2). \tag{2.3}$$

For $m_H >> M_W$ we neglect terms of order M_W/m_H, so that $\epsilon_L^\mu(p_i) \cong p_i/M_W$ and similarly from $m_H^2 = (p_1 + p_2)^2 \cong 2 p_1 \cdot p_2$ we find

$$\mathcal{M}(H \to W_L^+ W_L^-) = g \frac{m_H^2}{2 M_W} + O\left(\frac{M_W}{m_H}\right). \tag{2.4}$$

In a renormalizable gauge the corresponding amplitude can be read off (taking care with factors of 2) from the Hww vertex in the Higgs potential

$$\mathcal{M}(H \to w^+ w^-) = 2\lambda v. \tag{2.5}$$

Using $M_W = \frac{1}{2} g v$ and $m_H^2 = 2\lambda v^2$ it is easy to see that eqs. (2.4) and (2.5) are indeed equal up to M_W/m_H corrections.

The accuracy of the equivalence theorem can be judged in figure (2) taken from Bento and Llewellyn Smith.[10] The scattering cross sections for $W_L^+ W_L^- \to W_L^+ W_L^-$ and $W_L^+ W_L^- \to Z_L Z_L$ are computed in the standard model with $m_H = 1$ TeV. The exact calculations are compared with the result obtained from the equivalence theorem. For $\sqrt{s} \gtrsim 800$ GeV the agreement is very good and above 1 TeV the two calculations are indistinguishable on the figure.

As an immediate application, consider the case[2] in which the global symmetry G includes $SU(2)_L \times SU(2)_R$ and H includes an $SU(2)_{L+R}$. For such theories, which includes the case of the standard Higgs model, $\rho = 1$ up to electroweak corrections and we may immediately apply the pion low energy theorems that were derived from current algebra for just this case. For pions we have[8]

$$M(\pi^+ \pi^- \to \pi^0 \pi^0) = \frac{s}{F_\pi^2} \qquad s \ll 1 \text{ GeV}^2 \tag{2.6}$$

and for $\mathcal{L}_{SB}$ with no particles other than W, Z that are light compared to M_{SB} we would have

$$M(w^+ w^- \to zz) = \frac{s}{v^2} \qquad s \ll M_{SB}^2 \tag{2.7}$$

where $v = 0.246$ TeV. Using the equivalence theorem this becomes a statement about the scattering of W_L and Z_L in an intermediate energy domain:[2]

$$M(W_L^+ W_L^- \to Z_L Z_L) = \frac{s}{v^2} \qquad M_W^2 \ll s \ll M_{SB}^2. \tag{2.8}$$

It is important to stress that the pion low energy theorems were derived *before* the discovery of QCD. This was possible because they depend only on the symmetry and not on dynamics. Furthermore they are valid to all orders in the strong interactions. The low energy theorems and other current algebra results were important steps toward the discovery of QCD because they provided a reliable method to study the symmetries of a strongly coupled theory that could not be studied using perturbation theory.

The assumptions used above, $G \supset SU(2)_L \times SU(2)_R$ and $H \supset SU(2)_{L+R}$, are sufficient to guarantee $\rho = 1$ to all orders in λ_{SB} but they are not necessary conditions for $\rho = 1$. We are therefore motivated to derive the low energy theorems for all candidate groups G and H and for all values of ρ. The problem we face is equivalent to that of obtaining the pion–pion scattering low energy theorems in the absence of isospin symmetry. The low energy theorems are derived by three different methods:[3] a perturbative power counting analysis, nonlinear chiral lagrangians, and current algebra. I will sketch the current algebra derivation below. Along with the low energy theorems for general values of ρ, the derivation establishes a limited converse to the result quoted above: we find that if $\rho = 1$ then the Goldstone boson sector consisting of $w^\pm$, z possesses an effective $SU(2)_{L+R}$ symmetry ("custodial" $SU(2)$) in the low energy domain $s \ll M_{SB}^2$.

Briefly the derivation is as follows. The global symmetry G must be at least as large as the gauge group, $G \supset SU(2)_L \times U(1)_Y$, so in particular we have the $SU(2)_L$ charge algebra

$$[L_a, L_b] = i\varepsilon_{abc}L_c \qquad (2.9)$$

where the corresponding local currents L_a^μ can generally be expanded in terms of the Goldstone triplet $w^\pm$, z as

$$L_a^\mu = \frac{1}{2}r_a \varepsilon_{abc} w_b \partial^\mu w_c - \frac{1}{2}f_a \partial^\mu w_a + \ldots \qquad (2.10)$$

with terms involving heavy fields omitted. Since $H \supset U(1)_{EM}$ we have $f_1 = f_2$ and $r_1 = r_2$. The f_a are analogues of the PCAC constant and determine the gauge boson masses,

$$M_W = \frac{1}{2}g f_1, \qquad (2.11)$$

$$\rho = (f_1/f_3)^2. \qquad (2.12)$$

Corrections are suppressed by inverse powers of order M_{SB} or, because of quantum corrections, by inverse powers of $4\pi f_a$.

It is straightforward to show that the $SU(2)_L$ algebra (2.9) requires

$$r_1 = \frac{1}{\sqrt{\rho}}, \qquad (2.13)$$

$$r_3 = 2 - \frac{1}{\rho}, \qquad (2.14)$$

so that the parameters r_a and f_a in eq. (2.10) are completely determined in terms of G_F and ρ. In particular, $\rho = 1$ implies $f_1 = f_2 = f_3$ and $r_1 = r_2 = r_3 = 1$ which means that the Goldstone boson contributions to L_a^μ are the difference of $SU(2)$ vector and axial vector currents. The existence of this vector $SU(2)$ triplet of currents establishes the limited converse alluded to above.

The rest of the derivation is much like the usual current algebra derivation[8] except that we do not assume an $SU(2)_{L+R}$ isospin invariance. Consequently pole terms which are forbidden by G-parity in the pion case are not forbidden here. Assuming that $w^{\pm}$, z saturate these pole terms we find Goldstone boson low energy theorems such as

$$M(w^+ w^- \to zz) = \frac{s}{f_1^2} \frac{1}{\rho} \qquad s \ll M_{SB}^2 \qquad (2.15)$$

which using (2.11) reduces to (2.7) for the case $\rho = 1$. By the equivalence theorem we have then

$$M(W_L^+ W_L^- \to Z_L Z_L) = \frac{s}{v^2} \frac{1}{\rho} \qquad M_W^2 \ll s \ll M_{SB}^2 \qquad (2.16)$$

with $v = f_1 \cong 2M_W/g$. The other two independent amplitudes are

$$M(W_L^+ W_L^- \to W_L^+ W_L^-) = -\frac{u}{v^2}\left(4 - \frac{3}{\rho}\right), \qquad (2.17)$$

$$M(Z_L Z_L \to Z_L Z_L) = 0, \qquad (2.18)$$

and by crossing we have also

$$M(W_L^{\pm} Z_L \to W_L^{\pm} Z_L) = \frac{t}{v^2}\frac{1}{\rho}, \qquad (2.19)$$

$$M(W_L^+ W_L^+ \to W_L^+ W_L^+) = M(W_L^- W_L^- \to W_L^- W_L^-) = -\frac{s}{v^2}\left(4 - \frac{3}{\rho}\right). \qquad (2.20)$$

Like (2.16) eqs. (2.17-2.20) are valid in the intermediate domain $M_W^2 \ll E_i^2 \ll M_{SB}^2, (4\pi v)^2$.

It is also instructive to consider the perturbative power counting analysis.[3] It does not use the equivalence theorem since it is carried out in unitary gauge. This derivation shows directly that the low energy theorems follow from the form of the $SU(2)_L \times U(1)_Y$ gauge interactions. It reveals an amusing coincidence of the threshold behavior determined by the low energy theorems and the famous "bad" high energy behavior that a massive Yang-Mills theory would have if the masses were not "softly" generated by spontaneous symmetry breaking.

Consider first the minimal $SU(2)_L \times U(1)_Y$ Higgs model. Though we are ultimately interested in working to order g^2 in the electroweak gauge coupling and to all orders in the Higgs coupling λ_{SB}, we begin by examining $W_L^+ W_L^- \to Z_L Z_L$ scattering to tree approximation in both couplings. The tree approximation amplitude in unitary gauge can be decomposed into the sum of a gauge sector term and a symmetry breaking sector term,

$$\mathcal{M}(W_L^+ W_L^- \to Z_L Z_L) = \mathcal{M}_{gauge} + \mathcal{M}_{SB}. \qquad (2.21)$$

We will evaluate the amplitude for $s \gg M_W^2$. The first term, $\mathcal{M}_{gauge}$, is given by the sum of t and u channel W exchanges and by the four point contact interaction. Independent of the nature of the symmetry breaking sector it is a universal function of M_W and ρ,

$$\mathcal{M}_{gauge} = \frac{g^2 s}{4\rho M_W^2}. \qquad (2.22)$$

The second term, $\mathcal{M}_{SB}$, is in tree approximation just given by s-channel Higgs exchange,

$$\mathcal{M}_{SB} = -\frac{g^2 s}{4M_W^2} \frac{s}{s - m_H^2}. \qquad (2.23)$$

$\mathcal{M}_{SB}$ has the famous "bad" high energy behavior that is cancelled at infinite s by $\mathcal{M}_{SB}$ (since $\rho = 1$ in the minimal Higgs model). However for $s \ll m_H^2$, $\mathcal{M}_{SB}$ is negligible, so for the low energy domain $M_S^2 \ll s \ll m_H^2$ we have

$$\mathcal{M}(W_L^+ W_L^- \rightarrow Z_L Z_L) \cong \mathcal{M}_{gauge}$$

$$\cong \frac{g^2 s}{4\rho M_W^2}. \qquad (2.24)$$

Using $M_W = gv/2$ this agrees precisely with eq. (2.16) obtained from the current algebra derivation. So the form of the low energy amplitude is determined by the "bad" ultraviolet behavior of the gauge sector.

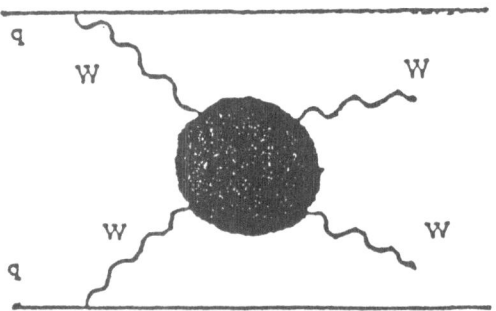

Figure 1. Production of WW pairs by WW fusion.

Though to this point we have only obtained the theorems in tree approximation for the minimal Higgs model, we can extend the derivation to all orders in λ_{SB} for any symmetry breaking sector with no light particles. By power counting one can show that the only strong corrections to the low energy amplitudes are absorbed as renormalizations of M_W and ρ. All other quantum corrections due to the symmetry breaking sector are screened by an extra power of the electroweak coupling constant, $\alpha_W/\pi = g^2/4\pi^2$, or they are suppressed by powers of s/M_{SB}^2, where M_{SB} is the characteristic scale of the spectrum of the symmetry breaking sector. For details see reference (3).

If other light particles are present, such as pseudo-Goldstone bosons, they may or may not cause the low energy theorems to be modified. QCD is an example of a theory in which additional light particles do not cause modifications. That is, the $\pi\pi$ low energy theorems obtained from $SU(2)_L \times SU(2)_R$ symmetry are not affected by the existence of K and η Goldstone bosons if $SU(2)_L \times SU(2)_R$ is embedded in a spontaneously broken $SU(3)_L \times SU(3)_R$.

3. UNITARITY AND A CONSERVATIVE STRONG INTERACTION MODEL

Perhaps the most interesting application of the low energy theorems is to use them to estimate the generic signal we should expect if $\mathcal{L}_{SB}$ is a strongly coupled sector at the TeV scale or above. The problem we face is like the one that physicists of the 1930's would have faced if they knew nothing of nuclei, baryons or other hadrons, but had discovered the pion, measured the PCAC constant F_π, and recognized (!?) the pion as an almost Goldstone boson. They would have then been able to derive the pion-pion low energy theorems, such as eq. (2.6), and the problem would be to use this information to reconstruct the scale of hadron physics. Though it would take a skilled writer of science fiction to make this a plausible plot line for the 1930's, it is precisely the situation we are in today if $\mathcal{L}_{SB}$ is strongly coupled: our pions are the longitudinal modes of W and Z, our "PCAC" constant is $v = 0.25$ TeV, and we have the low energy theorems eqs. (2.16-2.20).

The central ingredient in our considerations is unitarity. The linear growth in s, t, u of the amplitudes (2.16-2.20) cannot continue indefinitely or unitarity would be violated. For instance the $W_L W_L \to Z_L Z_L$ amplitude (2.16) is pure s-wave. If we adopt the low energy amplitude (2.16) as a model of the absolute value of the scattering amplitude, then the $J = 0$ partial wave amplitude is[2]

$$\left| a_0(W_L^+ W_L^- \to Z_L Z_L) \right| = \frac{s}{16\pi v^2} \tag{3.1}$$

where here and hereafter we set $\rho = 1$. Unitarity requires $|a_0| \le 1$ so we see that the growth of a_0 must be cut off at a scale Λ with

$$\Lambda \le 4\sqrt{\pi} v = 1.75 \text{ TeV.} \tag{3.2}$$

At the cutoff $\sqrt{s} = \Lambda$ the order of magnitude of the amplitude is

$$|a_0(\Lambda)| = \frac{\Lambda^2}{16\pi v^2}. \tag{3.3}$$

For $\Lambda \lesssim \frac{1}{2}$ TeV we have $|a_0(\Lambda)| \ll 1$ indicating weakly interacting symmetry breaking dynamics $\mathcal{L}_{SB}$, while for $\Lambda \gtrsim 1$ TeV we have $|a_0(\Lambda)| \cong O(1)$, the hallmark of a strong interaction theory. The most likely dynamics is that the cutoff scale Λ is of the order of M_{SB}, the typical mass scale of the new quanta. Then for $\Lambda \cong O(M_{SB})$ eq. (3.3) generalizes the Higgs model relationship

$$\frac{\lambda}{8\pi} = \frac{m_H^2}{16\pi v^2} \tag{3.4}$$

between the mass scale of the new quanta and the strength of the new interactions: weak coupling for $M_{SB} \ll 1$ TeV and strong coupling for $M_{SB} \gtrsim O(1)$ TeV.

A weak coupling example is provided by the standard Higgs model with a light Higgs boson, $m_H \ll 1$ TeV, which can be treated perturbatively. Then $a_0(W_L W_L \to Z_L Z_L)$ is given in tree approximation by (where I neglect M_W^2/s)

$$a_0 = \frac{-s}{16\pi v^2} \frac{m_H^2}{s - m_H^2}. \tag{3.5}$$

For $s \ll m_H^2$ this agrees with the low energy theorem (3.1) while for $s \gg m_H^2$ it saturates at the constant value $m_H^2/16\pi v^2$. Comparing with (3.3) we see that m_H indeed provides the scale for Λ in the standard Higgs model.

A strong interaction example is provided by hadron physics. For the $J = I = 0$ partial wave, the low energy theorem gives

$$a_{00}(\pi^+\pi^- \to \pi^0\pi^0) = \frac{s}{16\pi F_\pi^2} \tag{3.6}$$

with $F_\pi = 92$ MeV. Equation (3.6) would saturate unitarity at $4\sqrt{\pi}F_\pi = 650$ MeV which is indeed the order of the hadron mass scale. The a_{11} and a_{02} amplitudes saturate at 1100 and 1600 MeV.

The two generic possibilities are illustrated in figure (3). For weak coupling the partial wave amplitudes saturate at values small compared to 1 giving rise to narrow resonances at masses well below 1 TeV. For strong coupling they saturate the unitarity limit with broad resonances in the TeV range.

These results suggest a general experimental strategy to search for the symmetry breaking sector. $W_L W_L$ fusion at a pp collider probes the interaction of the symmetry breaking sector as shown in figure (1). Since the initial $W_L W_L$ pair are off mass shell, they must rescatter to appear in the final state as real on-shell particles. Therefore this process measures the strength of the $W_L W_L$ interaction which we know from the equivalence theorem is essentially the strength of the interaction of the symmetry breaking sector $\mathcal{L}_{SB}$. Counting powers the $W_L W_L$ fusion amplitude is $O(g^2 \lambda_{SB})$. It must be compared to the $O(g^2)$ background due to $\bar{q}q \to WW$. The gauge bosons produced in this way are predominantly transversely polarized, but they cannot be efficiently separated from the longitudinal pairs at the necessary statistical level. Therefore the $W_L W_L$ fusion signal can be visible above the $\bar{q}q \to WW$ background only if $\lambda_{SB} = O(1)$, that is, if $\mathcal{L}_{SB}$ is strongly interacting.

We will see in detail in Section 4 that for the ZZ, WZ, and W^+W^- channels, the signal can only emerge over the $\bar{q}q$ annihilation backgrounds for $M_{WW} > 1$ TeV. This in turn requires a pp collider of at least the SSC design parameters, $\sqrt{s} = 40$ TeV and $\mathcal{L} = 10^{33}\text{cm}^{-2}\text{s}^{-1}$. It is unlikely that lower energy can be sufficiently compensated by higher luminosity for this particular physics, as discussed in Section 4.

The strategy then is to look for an excess of gauge boson pair events with M_{WW} between 1 and 2 TeV. If there is no excess then we learn that $\mathcal{L}_{SB}$ is weakly interacting and that the new physics, probably Higgs bosons, lies below 1 TeV. If an excess is found we learn that $\mathcal{L}_{SB}$ is strongly interacting and that the new quanta lie above 1 TeV.

Since the low energy theorems for pion scattering apply approximately for $\sqrt{s} \lesssim 300$ MeV, we guess that the $W_L W_L$ theorems may apply out to $\sqrt{s} \lesssim 300$ MeV $\times v/F_\pi \cong 1$ TeV. In the next section we consider the experimental implications of a crude but conservative model[2] of strong $W_L W_L$ scattering based on the low energy theorems (2.16-2.20). As in eq. (3.1) we use the low energy theorems as the model for the absolute value of the relevant partial wave amplitudes which are $a_{JI} = a_{00}, a_{11}, a_{02}$ where I denotes the effective low energy custodial $SU(2)$. Above the energy at which partial wave unitarity is saturated, the amplitudes are set equal to one. For a_{00} saturation occurs at $\sqrt{16\pi v^2} = 1.7$ TeV, eq. (3.2), so in that case the model is

$$|a_{00}(s)| = \frac{s}{16\pi v^2}\theta(16\pi v^2 - s) + 1 \cdot \theta(s - 16\pi v^2). \tag{3.7}$$

The detailed form is not critical, the essential point being to extrapolate smoothly from the known behavior at low energy into the domain above 1 TeV where the amplitude becomes strong, of order 1, without assuming enhanced resonant behavior.

The model is conservative in that it assumes no resonant behavior. For instance, the ZZ continuum signal for $M_{ZZ} > 1$ TeV is only half the corresponding signal from the 1 TeV Higgs boson. Comparatively little is gained at the SSC from the high energy region where $|a_J| = 1$ because of the rapidly falling luminosities at the relevant collider energies. The model is also conservative in that only the leading partial waves, $I, J = (0,0), (1,1), (2,0)$, are kept while higher partial waves are set to zero.

Figure (4) from reference (20) shows how well the model fits $\pi\pi$ scattering data. The "dictionary" is that ~ 650 MeV corresponds to the unitarity limit at 1.75 TeV. Our model corresponds to curve a in the plots of $|T_0^0|$, $|T_1^1|$, and Re T_0^2 (since T_0^2 is exotic in QCD, the imaginary part is negligible below 1 GeV). The model describes the trend of the $0, 0$ channel well. In the $1, 1$ channel it underestimates the data by a large factor, as a result of the rho meson. In the $2, 0$ channel there is considerable disagreement between different experiments; the model interpolates the data out to ~ 600 MeV beyond which it overestimates the data. At the SSC this is not as serious as it might seem since $\sqrt{s}_{\pi\pi} > 600$ MeV corresponds to $\sqrt{s}_{WW} > 1.7$ TeV where the WW luminosity is small.

In the next section we will consider the experimental signals (and backgrounds) of a strongly interacting symmetry breaking sector for pp collider energies corresponding to the LHC, the SSC, and the ELOISATRON.

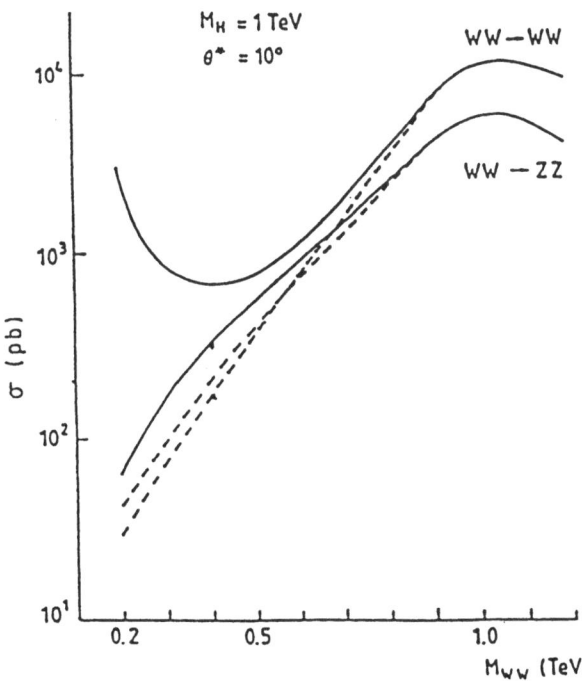

Figure 2. Comparison of WW cross sections in unitary gauge and with use of the equivalence theorem (from Bento and Llewellyn Smith, ref. (10)).

4. EXPERIMENTAL SIGNALS FOR STRONG INTERACTION MODELS

In this section we consider the experimental signals of a strongly interacting symmetry breaking sector. An important general lesson emerges from technicolor models:[21] the Higgs mechanism does not require the existence of a physical Higgs boson. As shown in Sections 2 and 3 the generic signal of a strongly interacting $\mathcal{L}_{SB}$ is strong $W_L W_L$ scattering between 1 and 2 TeV, giving rise by the mechanism of figure (1) to production of longitudinally polarized gauge boson pairs in all channels: $ZZ, W^{\pm}Z, W^+W^-, W^+W^+, W^-W^-$. (The longitudinal polariztion of the W's provides little leverage against the predominantly transversely polarized W's in the $\bar{q}q \to WW$ background; I do not assume polarization information in the results given here.) For a collider powerful enough to observe this signal we have a No-Lose Corollary: if the signal is not present we learn that the symmetry breaking sector is below 1 TeV, probably in the form of one or more Higgs bosons.

In this section we will review some of the possible experimental signals. We will see that a collider with the SSC energy and luminosity is a minimal configuration for the 1-2 TeV signals. A pp collider with half the energy would not suffice, nor does it seem practicable for this purpose to compensate the lower energy with higher luminosity. (The necessary luminosity upgrade would be of order 150 to 600, i.e., $\mathcal{L} = 1.5 \cdot 10^{35}$ to $6 \cdot 10^{35} \mathrm{cm}^{-2}\mathrm{s}^{-1}$, as discussed below.) The section begins with some general remarks about the relationship of collider energy and luminosity to the "reach" for TeV-scale symmetry breaking signals. Next is a survey of the cross sections and discovery strategies for the pp colliders projected for the 1990's. Finally I present estimates of the cross sections to be expected at a 200 TeV pp collider. The WW fusion mechanism of figure (1) has an important consequence for the design of such a pp collider (or for $\gtrsim 10$ TeV e^+e^- collider): like the more familiar $\gamma\gamma$ cross section it increases with energy, rather than falling like $1/s$ as is typical of annihilation and hard scattering processes. As a result, $\mathcal{L} = 10^{33}\mathrm{cm}^{-2}\mathrm{s}^{-1}$ would suffice at these 21'st century "hypercolliders" to ensure large yields of TeV-scale quanta from a strongly interacting symmetry breaking sector. (Of course higher luminosity will be welcome if 21'st century experimenters are able to cope with it.)

4.1 Energy, Luminosity, and Physics Reach

The four-body phase space of the WW fusion process, $qq \to qqWW$, increases steeply with energy near threshold, which is in fact the region inhabited by the LHC and SSC for $\sqrt{s_{WW}} \gtrsim 1$ TeV. The result is that the cross sections reviewed below in Section 4.2 are an order of magnitude larger for the SSC than the LHC. The background process $\bar{q}q \to WW$ is two-body and therefore does not rise as steeply with energy in this region. This fact cannot be ignored in appraising the relationship of energy and luminosity.

To what extent can lower energy be compensated by higher luminosity to obtain equivalent reach for TeV symmetry breaking physics? An answer requires consideration of two factors:

1) what luminosity upgrade would be needed if all the same measurements could be carried out at $\mathcal{L} = 10^{33}\mathrm{cm}^{-2}\mathrm{s}^{-1}$ and at $\mathcal{L} \gg 10^{33}$?

2) what is the factor resulting from decay modes which are detectable at $\mathcal{L} = 10^{33}$ but not at much higher luminosity?

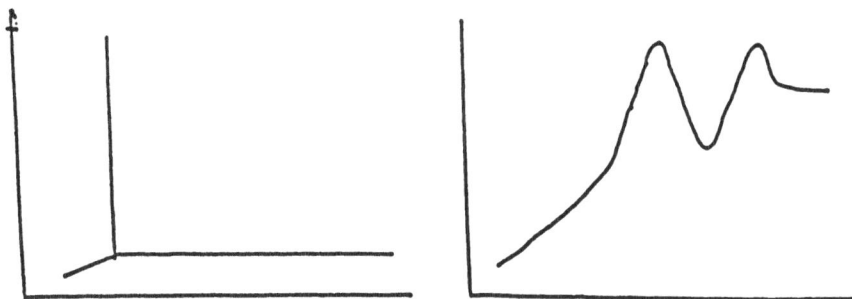

Figure 3. Typical behavior of partial wave amplitudes for $W_L W_L$ scattering. The plot on the left represents a weak coupling model with a narrow (Higgs) resonance below 1 TeV. The plot on the right shows strong coupling behavior, with saturation of unitarity and broad resonances in the 1-2 TeV region.

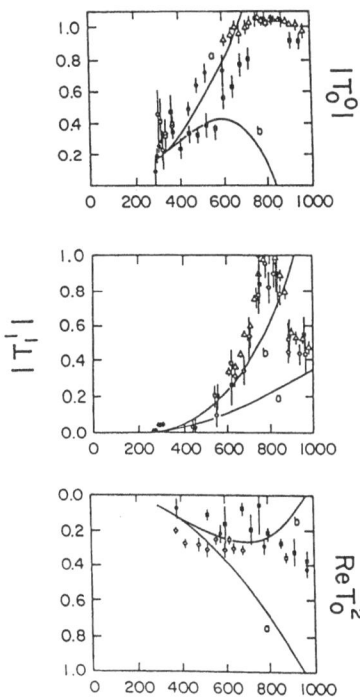

Figure 4. Data for $\pi\pi$ partial wave amplitudes compared with extrapolated low energy theorems (curves a) as in equation 3.12 (from ref. 20).

We consider these two factors in turn.

Suppose first that it is actually possible to make the same experimental observations at luminosities far above 10^{33}cm^{-2}s^{-1} as at 10^{33}. Let us first compare two colliders with different energies, $E_2 > E_1$, and the same luminosity, say $\mathcal{L}_1 = \mathcal{L}_2 = 10^{33}$cm^{-2}s^{-1}. The physics reach is determined not simply by the number of events in the signal S_i but by its statistical significance, proportional to $S_i/B_i^{1/2}$ where B_i is the number of background events. The steeper energy dependence of the signals S_i than the backgrounds B_i implies greater statistical significance at the higher energy collider,

$$\frac{S_2}{B_2^{1/2}} > \frac{S_1}{B_1^{1/2}}. \tag{4.1}$$

Next suppose that we upgrade the luminosity of the lower energy collider from $\mathcal{L}_1 = \mathcal{L}_2$ to $\mathcal{L}_1'$ by a large factor $\mathcal{R}$,

$$\mathcal{R} \equiv \frac{\mathcal{L}_1'}{\mathcal{L}_2} \gg 1 \tag{4.2}$$

so that the two machines achieve statistical parity,

$$\frac{S_2}{B_2^{1/2}} = \frac{S_1'}{B_1'^{1/2}} = \frac{\mathcal{R}S_1}{(\mathcal{R}B_1)^{1/2}}. \tag{4.3}$$

We see then that the upgrade factor $\mathcal{R}$ must satisfy

$$\mathcal{R} = \left(\frac{S_2}{S_1}\right)^2 \left(\frac{B_1}{B_2}\right). \tag{4.4}$$

Here S_i and B_i are always the number of signal and background events *before* upgrade, *i.e.*, for $\mathcal{L}_1 = \mathcal{L}_2$. Therefore the ratios S_2/S_1 and B_2/B_1 are equal to the ratios of the signal and background cross sections at the two energies.

Physics "reach" is by definition the largest mass signal that could be observed with some minimum level of significance, *e.g.*, 3σ. For any such signal eq. (4.3) is then the criterion for two colliders to have equal reach and eq. (4.4) gives the luminosity upgrade needed for the lower energy collider. We can use the signal and background cross sections presented below to evaluate the upgrade factor R for specific TeV-scale phenomena. For example, considering the ZZ cross sections in tables 2-4 for the strong scattering model and the $\bar{q}q$ annihilation backgrounds, the upgrade factors are

$$\left(\frac{S^2}{B}\right)_{16\text{ TeV}} : \left(\frac{S^2}{B}\right)_{40\text{ TeV}} : \left(\frac{S^2}{B}\right)_{200\text{ TeV}} \cong 1 : 25 : 500. \tag{4.5}$$

Assuming the same detection capability, a 16 TeV collider with $\mathcal{L} = 2.5 \cdot 10^{34}$cm^{-2}s^{-1} would be equivalent to a 40 TeV collider with $\mathcal{L} = 10^{33}$. A 40 TeV collider would need $\mathcal{L} = 2 \cdot 10^{34}$ to be equivalent to a 200 TeV collider with $\mathcal{L} = 10^{33}$.

Under the experimental conditions likely for the 1990's the advantage of 40 TeV over 16 TeV is much greater than these idealized estimates suggest. In Section 4.2 we consider the particular final states that are likely to be detectable at $\mathcal{L} = 10^{33}$cm^{-2}s^{-1}. These include most certainly the leptonic final states $ZZ \to \ell^+\ell^- + \ell^+\ell^-$, $ZZ \to \ell^+\ell^- + \bar{\nu}\nu$, $WW \to \ell\nu + \ell\nu$ (especially like-charged) and perhaps also the mixed lepton-hadron final states $WW \to \ell\nu + \bar{q}q$. Allowing $\ell = e$ or μ but not τ and $\bar{q}q = u\bar{d}$ or $c\bar{s}$ but not $t\bar{b}$, the branching ratios for these four decays are respectively

$4.4 \cdot 10^{-3}$, $2.6 \cdot 10^{-2}$, $2.8 \cdot 10^{-2}$, and $1.7 \cdot 10^{-1}$. Because of unavoidable effects of event pile-up it was concluded in the high luminosity option study for the LHC[7] that only the first ("gold-plated") decays, $ZZ \rightarrow \ell^+\ell^- + e^+e^-$, could be detected for $\mathcal{L} \gtrsim 5 \cdot 10^{34}$. Furthermore, as is carefully noted in reference (7), it is not clear that detectors capable of electron identification could survive radiation damage at $\mathcal{L} \geq 5 \cdot 10^{34}$. If not then only $ZZ \rightarrow \mu^+\mu^- + \mu^+\mu^-$ would be detectable, reducing the branching ratio by another factor four to $1.1 \cdot 10^{-3}$. While one cannot give a precise number without detailed simulations to assess the relative efficiencies of the different decay channels, it is clear from the branching ratios into leptonic final states that a 40 TeV collider gains another order of magnitude advantage from the "silver-plated" decays that include neutrinos in the final state. The advantage would be much larger still if the mixed lepton-hadron decays prove to be detectable. (The latter were also judged to be undetectable at high luminosity in reference (7).)

Assuming electron identification it was concluded in reference 7 that the LHC Higgs boson reach could be raised from $m_H = 600$ GeV at $\mathcal{L} = 10^{33}$ using $ZZ \rightarrow \bar{\ell}\ell + \bar{\nu}\nu$ to $m_H = 800$ GeV at $\mathcal{L} = 5 \cdot 10^{34}$ with $ZZ \rightarrow \ell^+\ell^- + \ell^+\ell^-$ assuming $\ell = e$ or μ. Since the 1-2 TeV continuum signal of the strong scattering model is much smaller than even the 1 TeV Higgs boson signal, the no-lose criterion is not met even assuming electron identification. (*E.g.*, the strong scattering model for ZZ would then predict 11 signal events in $ZZ \rightarrow e^+e^-/\mu^+\mu^- + e^+e^-/\mu^+\mu^-$ for 10^7 sec. The background would be ~ 30 events assuming $\bar{q}q \rightarrow ZZ$ is augmented 30% by $gg \rightarrow ZZ$.)

Combining the effect of detectability with the R factor, eq. (4.5), the effective luminosity for the strong scattering model signals of a 40 TeV collider is at least two orders of magnitude larger than that of a 16 TeV collider and is nearly three orders of magnitude larger if electrons cannot be identified for $\mathcal{L} = 10^{33}$.

4.2 Experimental Signals at 16 and 40 TeV

We consider what might actually be observed at the LHC and SSC if $\mathcal{L}_{SB}$ is a strongly interacting theory. The generic strong interaction signal is longitudinally polarized W, Z pairs produced by WW fusion, figure (1). The W polarization does not provide much leverage against the predominantly transverse $\bar{q}q \rightarrow WW$ background, and I will *not* assume polarization information in the results given here. Since the irreducible background is $\mathcal{M}(\bar{q}q \rightarrow WW) = O(g^2)$ while the signal is $\mathcal{M}(qq \rightarrow qqWW) = O(g^2\lambda_{SB})$, we expect a discernible signal if and only if $\mathcal{L}_{SB}$ is strongly interacting, $\lambda_{SB} = O(1)$. The signal may occur in W^+W^- and ZZ, as for the standard Higgs boson, but also more generally in $W^{\pm}Z$ and even W^+W^+ and W^-W^- ("I" = 1 and 2 channels) as in eqs. (2.16 - 2.20).

To compute the expected yields we must convolute the quark luminosity distribution functions with the luminosity distribution to find longitudinally polarized gauge bosons in the incident q's or $\bar{q}$'s, convoluted finally with the $2 \rightarrow 2$ scattering cross section of the longitudinally polarized gauge bosons:

$$\sigma(pp \rightarrow Z_L Z_L + \cdots) = \int_\tau \left.\frac{\partial \mathcal{L}}{\partial \tau}\right|_{qq/pp} \cdot \int_x \sum_{V_L} \left.\frac{\partial \mathcal{L}}{\partial x}\right|_{V_L V_L/qq} \cdot \sigma(V_L V_L \rightarrow Z_L Z_L) \quad (4.6)$$

where $\tau = s_{qq}/s_{pp}$ and $x = s_{ZZ}/s_{qq}$. In the effective W approximation the luminosity function for $W_L^+ W_L^-$ pairs is given in eq. (1.1). The x dependence, tabulated in table 1, shows the strong dependence on the available phase space that will be familiar to practitioners of two photon physics at e^+e^- colliders.

459

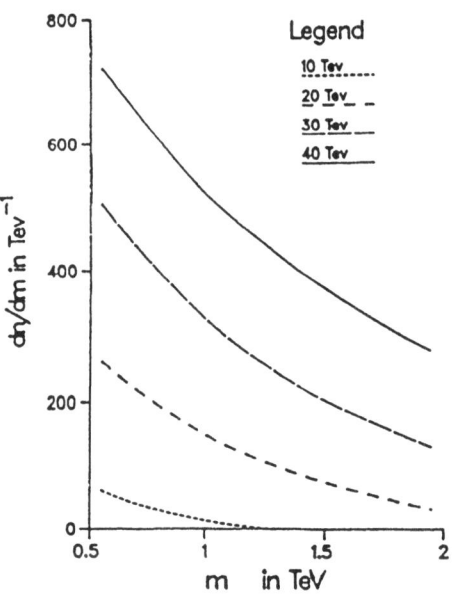

Figure 5. Differential yield per 10 fb^{-1} with respect to m_{ZZ} for the model of equation 3.12, at pp colliders of 10, 20, 30, and 40 TeV (from ref. (2)). Rapidity $|y_Z| < 1.5$ is required.

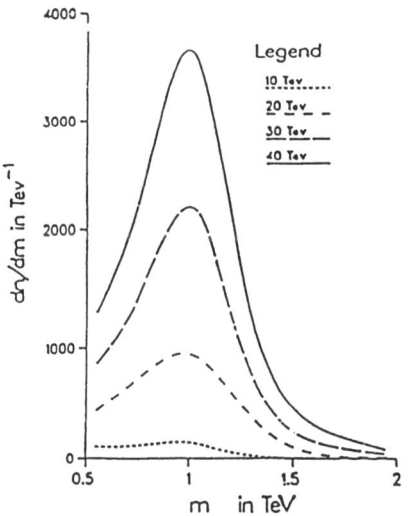

Figure 6. Differential yield per 10 fb^{-1} with respect to m_{ZZ} for production of a 1 TeV Higgs boson at pp colliders of 10, 20, 30, and 40 TeV (from ref. (2)). Rapidity $|y_Z| < 1.5$ is required.

The effective W approximation has been compared with analytical[22] and numerical evaluations[23,24,17] of Higgs boson production. The most definitive results are probably the analytical calculations of reference (22). They show good agreement for $WW \to H$ for $m_H \geq 500$ GeV, with errors $\lesssim O(10\%)$ and decreasing with increasing m_H, while for the relatively less important process $ZZ \to H$ the errors are roughly twice as large.

I will consider three examples of strong interaction symmetry breaking signals: the 1 TeV Higgs boson, the strong scattering model described in Section 3, and the techni-rho meson resonance expected in technicolor models. Total cross sections (without Z or W decay branching ratios) are shown in figures (5-7) from reference (2). Figures (5) and (6) show the ZZ signals for the strong scattering model of Section 3 and for the 1 TeV Higgs boson respectively, assuming integrated luminosity of 10^4pb^{-1} ($10^{33} \text{cm}^{-2}\text{s}^{-1}$ for 10^7s) at pp colliders with $\sqrt{s} = 10, 20, 30, 40$ TeV. A rapidity cut $|y_Z| < 1.5$ has been imposed in the figures to reduce the $\bar{q}q \to ZZ$ background which is strongly forward while the signal is relatively isotropic. Figure (7) shows the two signals for the 40 TeV collider as increments to the $\bar{q}q \to ZZ$ background, again with $|y_Z| < 1.5$. From figure (7) we see the necessity to detect gauge boson pairs at large invariant mass, $m_{VV} \gtrsim O(1)$ TeV, where the signal can emerge from the rapidly falling background.

Yields are presented in tables 2 and 3 for $\sqrt{s} = 16$ and 40 TeV with the rapidity cut $|y_V| < 1.5$ and invariant mass cut $m_{VV} > 1.0$ TeV. The values are taken from reference (2) except that an error in the Higgs boson yield has been corrected (reducing those yields by $\sim 20\%$ relative to reference (2)).

We must include the decay branching ratios to particular final states in order to translate the tables into estimates of the observable yields. The relevant branching ratios for leptonic channels are

$$B(ZZ \to e^+e^-/\mu^+\mu^- + e^+e^-/\mu^+\mu^-) \cong 0.0044 \tag{4.7}$$

$$B(ZZ \to e^+e^-/\mu^+\mu^- + \nu\bar{\nu}) \cong 0.026 \tag{4.8}$$

$$B(WW \to (e/\mu)\bar{\nu} + (\bar{e}/\bar{\mu})\nu) \cong 0.028 \tag{4.9}$$

For the "mixed" lepton-hadron channel we have the larger rate

$$B(WW \to u\bar{d}/c\bar{s} + e\bar{\nu}/\mu\bar{\nu}) \cong 0.17. \tag{4.10}$$

However the latter has a formidable QCD background from $W + 2$ jets, two orders of magnitude larger than the signals even if excellent jet-jet mass resolution is assumed. This background is not shown in the tables.

The cleanest and rarest channel is $H \to ZZ \to \ell^+\ell^-\ell^+\ell^-$ with $\ell = e$ or μ. The background is from $\bar{q} \to ZZ$ and $gg \to ZZ$, the latter[25] proceeding by a quark loop. Recent Monte Carlo simulations suggest that the Higgs boson can be detected in this channel for $m_H \leq 300$ GeV at the LHC[26] and for $m_H \leq 600$ GeV at the SSC.[27] (Here and elsewhere unless otherwise stated both SSC and LHC are assumed to operate at $\mathcal{L} = 10^{33} \text{cm}^{-2}\text{sec}^{-1}$.) At these values of m_H the Higgs boson appears as a recognizable peak above the continuum background. For $m_H = 1$ TeV an essentially optimal set of cuts is $y_Z < 1.5$ and $m_{ZZ} > 1$ TeV, as in the tables. We then obtain 4 signal events over a background (augmented by 50% for $gg \to ZZ$) of $1\frac{1}{2}$ events for an integrated luminosity of 10^4pb^{-1} at the SSC. (Yields are always quoted per 10^4pb^{-1}

461

Table 1. Effective $W_L^+W_L^-$ luminosity as a function of $\sqrt{s_{qq}}/\sqrt{s_{WW}}$.

$\dfrac{1}{\sqrt{x}} = \sqrt{\dfrac{s_{qq}}{s_{WW}}}$	1	2	3	4	5	10
$\dfrac{16\pi}{\alpha_W^2}\dfrac{\partial \mathcal{L}}{\partial x}$	0	0.9	6	17	36	270

Table 2. Yields at a 16 TeV pp collider from WW fusion and $\bar{q}q$ annihilation in events per 10^4pb^{-1}. Cuts imposed are $|y_W| < 1.5$ and $s_{WW} > 1$ TeV^2. No gauge boson decay branching ratios are included.

$\sqrt{s} = \mathbf{16}$ **TeV**	1 TeV Higgs	Strong Scattering Model	$\bar{q}q$ annihilation
ZZ	110	50	100
W^+W^-	220	60	440
$W^+Z + W^-Z$		60	180
$W^+W^+ + W^-W^-$		70	

Table 3. As in Table 2, for a 40 TeV pp collider.

$\sqrt{s} = \mathbf{40}$ **TeV**	1 TeV Higgs	Strong Scattering Model	$\bar{q}q$ annihilation
ZZ	890	490	370
W^+W^-	1800	630	1600
$W^+Z + W^-Z$		670	580
$W^+W^+ + W^-W^-$		710	

Table 4. As in Table 2, for a 200 TeV pp collider.

$\sqrt{s} = \mathbf{200}$ **TeV**	1 TeV Higgs	Strong Scattering Model	$\bar{q}q$ annihilation
ZZ	9800	6100	1800
W^+W^-	20000	8000	8000
$W^+Z + W^-Z$		9500	2500
$W^+W^+ + W^-W^-$		8900	

corresponding to $\mathcal{L} = 10^{33} \mathrm{cm}^{-2}\mathrm{sec}^{-1}$ for 10^7sec) At the LHC the corresponding signal is a factor ten times smaller while the background is about four times smaller. Such a signal at the SSC would not be statistically significant, but augmented by additional years of running and/or results from several experiments it would become significant. It would also be a valuable confirmation of larger signals detected in other channels.

Of course to detect such a structureless signal (for $m_H = 1$ TeV the width is $\Gamma_H = 0.5$ TeV) it is necessary to know the magnitude of the background, requiring a variety of calibrations in *situ* to confirm knowledge of the relevant distribution functions and couplings. After such calibration studies are completed the ZZ background should be known to within 30% uncertainty,[27] sufficient accuracy given the expected 3:1 signal to background ratio.

Ascending the ladder of rates while descending on the scale of "purity" of signal, we come next to the decay $H \to ZZ \to \ell\bar{\ell}\bar{\nu}\nu$.[2,28] Monte Carlo simulations have suggested a reach in this channel to $m_H \lesssim 600$ GeV at the LHC[26] and to at least 800 GeV at the SSC[27] (A Monte Carlo study for $m_H > 800$ GeV at the SSC has not yet been done). The two studies are not directly comparable because while both considered the $\bar{q}q \to ZZ$ and $gg \to ZZ$ backgrounds, only the SSC study considered the background from $Z + $ jet where the jet generates large missing energy, faking a $Z \to \bar{\nu}\nu$ decay. This background is sharply reduced by cutting on visible hadronic transverse energy on the side opposite to the observed $Z \to \ell^+\ell^-$. A very hermetic detector was found to be critical. The analysis may be improved in the future by adding cuts involving event topology and the rapidity distribution of the visible hadronic clusters.

The 1 TeV Higgs boson is probably also observable at the SSC in this mode. Requiring the observed $Z \to \ell^+\ell^-$ to satisfy $p_T > 0.45$ TeV and $|y| < 1.5$, the estimated yield is a signal of 27 events over a $(\bar{q}q$ or $gg) \to ZZ$ background of ~ 12 events.

The mixed decay mode, $H \to WW \to \bar{\ell}\nu\bar{q}q$ with $\ell = e$ or μ and $\bar{q}q = u\bar{d}$ or $c\bar{s}$, has a large branching ratio, $2 \times \frac{1}{6} \times \frac{1}{2} = 1/6$. However the QCD background from Wjj is two orders of magnitude bigger than the signal even if we optimistically assume a 5% measurement of the dijet mass.[28] Two approaches have been taken to attempt to winnow the signal from this enormous background. One method is to cut on the p_T of the jets, using the tendency of the longitudinally polarized W's from the signal to decay into jets transverse to the W line of flight, unlike the QCD dijets and the transversely polarized W's that both tend to produce jets along the line of flight. Applying this approach to the SSC for $m_H = 800$ GeV a parton level calculation results in a signal of 500 events over an equal background.[29] Smaller but still encouraging yields have been reported based on Monte Carlo studies.[30] The prospects to follow this strategy in the real world are more difficult to assess than for the purely leptonic decays, being more sensitive to detailed aspects of jet physics and detector performance.

The second approach to the mixed modes borrows a trick from photon–photon scattering experiments at e^+e^- colliders where detecting an $e^\pm$ near the forward direction is a powerful way to isolate a clean sample of two photon events. The analogous idea[19] is to tag the forward jets that occur in WW fusion, $qq \to qqWW$, with transverse momentum of order M_W. Of course tagging suffers its own QCD background, due to processes with a real $W \to \ell\bar{\nu}$ plus a dijet to fake the second

$W \rightarrow \bar{q}q$ and one or two jets near the forward direction that fake the tagged quark or quarks. The necessary background calculation has not yet been performed. A recent calculation assuming 100% efficient tagging for a 700 GeV Higgs boson resulted in 20 events for an LHC year and 160 events for an SSC year[31] (a "year" is always 10^7 seconds). The QCD backgrounds after tagging were estimated at 12 and 140 events respectively, however the authors remark that their background estimates will probably prove to be small by a factor of 2. Highly segmented forward calorimeters would be essential.

An additional serious background would occur if $m_t > M_W$ so that the top quark decays by $t \rightarrow Wb$.[32] For 100 GeV $\leq m_t \leq$ 200 GeV the contribution to the W^+W^- continuum from $\bar{t}t \rightarrow W^+W^-\bar{b}b$ would be two orders of magnitude larger than $\bar{q}q \rightarrow W^+W^-$, eliminating any hope of detecting $H \rightarrow W^+W^-$. The leptonic decay signals from $H \rightarrow ZZ$ would not be affected.

The experimental signals for the strong scattering model described in Section 3 have not yet been studied at the level of Monte Carlo simulations. Since the continuum signal is shifted to larger values of M_{ZZ} relative to the 1 TeV Higgs boson, it puts even greater stress on the beam energy. The most promising leptonic signal for the ZZ channel is $ZZ \rightarrow \ell^+\ell^-\bar{\nu}\nu$ with $\ell = e$ or μ. Requiring the observed Z to satisfy $|y| < 1.5$ and $p_T > 0.45$ TeV, the yield at $\sqrt{s} = 40$ TeV is 15 events over a background of $\sim$ 12 including $gg \rightarrow ZZ$. The corresponding signal at $\sqrt{s} = 16$ TeV is 1 event over a background of 4 or 5. The possible background from $Z+$ jet has not yet been examined but is probably less pernicious than for the 800 GeV Higgs boson considered in reference (27).

Unlike the standard Higgs boson signal, strong interaction continuum signals also occur in the $W^\pm Z$ and like-charge WW channels. WZ can be detected in clean leptonic decays $WZ \rightarrow \ell\nu\ell\ell$, $\ell = e$ or μ, with a $\sim$ 1% branching ratio, augmented to $\sim$ 1.5% if $W \rightarrow \tau\nu$ is also feasible. Assuming only e's and μ's the expected signal is $\sim$ 7-1/2 events over a $\bar{q}q \rightarrow WZ$ background of $\sim$ 3 after cuts of $|y_{W,Z}| < 1.5$ and $M_{WZ} > 1$ TeV. As discussed below, more significant WZ signals would occur if resonances are present.

The like charge $W^+W^+ + W^-W^-$ signal[2,33] has no $\bar{q}q$ or gg annihilation backgrounds but does have a background comparable to the signal from e.g. $uu \rightarrow ddW^+W^+$ via single gluon exchange.[34] The charge can only be detected in leptonic decays so that M_{WW} cannot be measured. An estimate of the strong scattering model signal (defined by $y_\ell < 3$ and $p_T(\ell) > 50$ GeV) yielded[33] $\sim$ 40 events at $\sqrt{s} = 40$ TeV and $\sim$ 4 events at $\sqrt{s} = 15$ TeV. Work is in progress to develop cuts to reduce the gluon exchange background.

If either of the methods to detect the mixed decay modes $WW \rightarrow \ell\bar{\nu}\bar{q}q$ prove practicable, they can also be applied to the WW continuum signal. Including W^+W^-, W^+W^+, and W^-W^-, the net WW continuum signal with $|y_W| < 1.5$ and $M_{WW} > 1$ TeV is $\sim$ 2/3 of the $H \rightarrow W^+W^-$ signal for $m_H = 1$ TeV.

Though we have pessimistically concentrated here on continuum signals of strong WW scattering, where there are strong interactions there are probably also resonances. As a concrete example consider the techni-rho meson of $N_c = 4$ technicolor, with a mass of 1.8 TeV and width of 260 GeV. In pp colliders both $\bar{q}q \rightarrow \rho_T$[35] and $WW \rightarrow \rho_T^2$ contribute comparably to the production cross section. In analogy to the hadronic ρ we expect the charged ρ_T to decay predominantly to WZ, $\rho_T^\pm \to W^\pm Z$, which can be detected in the leptonic decay mode $WZ \to \ell\bar{\nu}\ell\ell$ that occurs with a 1% branching ratio for $\ell = e$ or μ. Then at the SSC with 10^4pb^{-1} the signal in the central region, $|y_W|$ and $|y_Z| < 1.5$, is 12 events over a $\bar{q}q \to WZ$ background of only 1 event, while the LHC signal is 1 event over a somewhat smaller background.[2]

4.3 Prospects for the ELOISATRON

I have argued in Section 4.2 that the SSC parameters, $\sqrt{s} = 40$ TeV and $\mathcal{L} = 10^{33}\text{cm}^{-2}\text{s}^{-1}$, provide an assured capability to see the signals of a strongly interacting symmetry breaking sector above 1 TeV. The SSC occupies strategic ground in that it is close to being the minimal machine about which this statement can be made. We have seen in Section 4.2 that the numbers of events are only a few ten's per leptonic channel and a few hundred's for the mixed lepton-hadron channel if the latter proves to be detectable. If such signals are observed there will be an irresistable motivation to find a way to a higher energy facility where the new quanta could be produced in sufficient numbers to begin detailed studies. This higher energy facility could be the 200 TeV pp collider that is the focus of this workshop or it might be an e^+e^- collider. (In the latter case the center of mass energy would clearly need to be much greater than the 2 TeV being considered for CLIC.) Here I consider only the pp collider option.

The technological, financial, social, and political implications of such a device are enormous and to me at least they are essentially imponderable. It is however straightforward to estimate the physics signals at such a collider if the $SU(2)_L \times U(1)_Y$ breaking condensate arises from a strong fifth force with associated quanta above 1 TeV. To begin we can simply apply the models considered in Section 4.2 to the ELOISATRON energy. The result is table 3, where we see that the signals of the 1 TeV Higgs boson and the strong scattering model are an order of magnitude larger than at the SSC and two orders larger than at the LHC. The effective luminosity that accounts for both signal and background was given in eq. (4.5). All three tables assume the same integrated luminosity, 10^4pb^{-1}. Like the photon-photon scattering cross section, these WW fusion cross sections grow with energy. We are not therefore faced with the same imperative to raise the luminosity that is implied by annihilation and hard scattering subprocess cross sections that scale like $1/\hat{s}$ (though the maximum useable luminosity would obviously be desireable).

Table 4 underestimates the yield of quanta from a strong $\mathcal{L}_{SB}$ by at least one order of magnitude and perhaps more. For the signals of table 4 we are using the ELOISATRON as a source of colliding longitudinally polarized W beams, the precise analogue of a pion-pion scattering facility in hadron physics. The strong scattering model extrapolates the low energy theorems, eqs. (2.16 - 2.20), as in eq. (3.16) (see reference (2) for the formulation in the other partial waves). It therefore represents only the three partial waves that dominate at threshold, $I, J = (0,0), (1,1), (2,0)$. At ELOISATRON energies many partial waves contribute, up to $\ell = pR$, as in hadron scattering where here R is the W_L radius. A geometrical cross section may be a more appropriate estimate,

$$\hat{\sigma}(W_L W_L)_{\gg 1 \text{ TeV}} \cong \pi R^2 \qquad (4.11)$$

where R is the W_L radius. For comparison the strong scattering model gives

$$\hat{\sigma}(W_L W_L)_{\text{1-2 TeV}} \cong \frac{1}{2v^2} \cong O(1nb). \tag{4.12}$$

To compare eqs. (4.11) and (4.12) we need an estimate for the radius R. Hadron radii might be characterized by their "pion clouds" extending to $R \cong m_\pi^{-1}$ which in the present context becomes $R \cong M_W^{-1}$. Then eq. (4.11) gives ~ 100 nb, two order of magnitude bigger than eq. (4.12). A more conservative estimate in a technicolor model would use the confinement radius, $R \cong \Lambda_{TC}^{-1} \cong v^{-1}$, in which case eq. (4.11) is $\pi/v^2 \cong O(10)$nb. Table 4 is probably conservative by at least one order of magnitude. At ELOISATRON energies inelasticity will begin to set in and $n > 2$ particle final states will begin to emerge.

If the new quanta carry QCD color or have colored constituents then they can be directly produced by gluon-gluon fusion. In that case we would expect still larger cross sections.

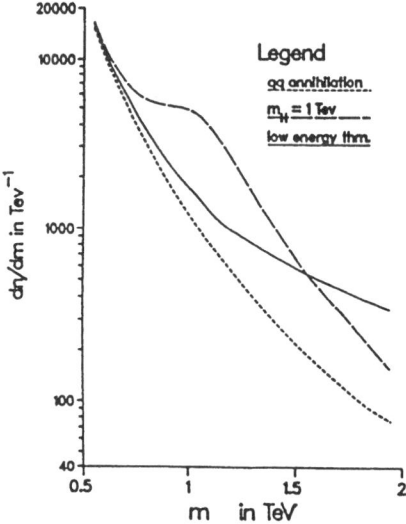

Figure 7. Differential yield per 10 fb^{-1} with respect to m_{ZZ} of the 1 TeV Higgs boson and the model of equation 3.12, both shown as increments on the $\bar{q}q$ annihilation background with $\sqrt{s} = 40$ TeV (from ref. (2)). Rapidity $|y_Z| < 1.5$ is required.

5. CONCLUSION

There is always an element of risk that accompanies investment in a major new research facility. Even if the instrument performs as designed, the fact that we are exploring unknown phenomena means we cannot predict whether the discoveries will justify the investment. The historical record is reassuring: scientifically our investments up to now have proved well founded, and from scientific success other benefits are sure to come, both indirect and direct. Nevertheless as the scale of investment grows, the risk also increases. Though it will always continue to be true that some of the most important discoveries are completely unanticipated, it also becomes important to know in advance of phenomena that the new facility is sure to elucidate.

Such concerns apply to the LHC and SSC projected for the 1990's and would apply with even greater force to the ELOISATRON in the next century. Two points from these lectures are relevant to these concerns:

- *If the signals of a strongly interacting symmetry breaking sector above 1 TeV are found during the 1990's, the physics case for a facility with the capability of the ELOISATRON will be as clear as was the case for the particle accelerators built in the 50's and 60's to study the emerging physics of hadrons.* The overriding concern of particle physics will then be to solve the technical, economic and political problems that will allow us to study in detail the new TeV-scale force and particles. Without such detailed studies high energy physics will be at an impasse. Theoretical efforts to skip directly to the Planck scale will prove to be as premature as was the effort to unify gravity and electromagnetism without incorporating the weak and strong nuclear forces. And nonaccelerator experimental studies, while valuable as exploratory probes, can never replace the controlled accelerator laboratory setting for the detailed studies that would be needed.

- *If the $SU(2)_L \times U(1)_Y$ symmetry is broken by a new strong force and quanta above 1 TeV, the only likely prospect to discover that new force in the 1990's lies with the SSC.* The SSC is sure to establish if the force is weak or strong, to determine if the new quanta are lighter or heavier than 1 TeV, and probably to discover the new quanta provided the lightest of them are not much heavier than the $O(2)$ TeV scale of the unitarity limit, eq. (3.2). Only by measuring the symmetry breaking force directly in $W_L W_L$ scattering between 1 and 2 TeV will we know for sure if it is strong. Other predictions, such as the pseudoGoldstone bosons of some technicolor theories, are model dependent and would not have unambiguous interpretations if found. Even with the luminosity upgrades that have been considered, a 15 TeV pp collider has little prospect to detect the 1-2 TeV $W_L W_L$ strong scattering signal. Though capable of important discoveries it cannot replace the SSC in this respect.

In this sense the SSC and the ELOISATRON are strategically linked. If a strongly coupled symmetry breaking sector exists above 1 TeV, we will need to build a facility with the physics reach of the ELOISATRON to study its properties. But only by constructing the SSC are we likely to establish in this century whether a strongly coupled symmetry breaking sector exists in nature.

REFERENCES

1. S.L. Glashow, *Nucl. Phys.* **22**, 579 (1961); S. Weinberg, *Phys. Rev. Lett.* **19**, 1264 (1967); A. Salam, in Proc. 8th Nobel Symp., ed. N. Svartholm (Almqvist and Wiksells, Stockholm, 1968) p. 367.
2. M.S. Chanowitz and M.K. Gaillard, *Nucl. Phys.* **B261**, 379 (1985).
3. M.S. Chanowitz, M. Golden, H. Georgi, *Phys. Rev. Lett.* **57**, 2344 (1986); *Phys. Rev.* **D36**, 1490 (1987).
4. S. Weinberg, *Phys. Rev.* **D13**, 974 (1976); **D20**, 1277 (1979); L. Susskind, *Phys. Rev.* **D20**, 2619 (1979).
5. R.N. Cahn and S. Dawson, *Phys. Lett.* **136B**, 196 (1984); E **138B**: 464 (1984).

6. M.S. Chanowitz and M.K. Gaillard, *Phys. Lett.* **142B**, 85 (1984); S. Dawson, *Nucl. Phys.* **B29**, 42 (1985); G. Kane, W. Repko, W. Rolnick, *Phys. Lett.* **148B**, 367 (1984).

7. The Feasibility of Experiments at High Luminosity at the LHC, ed. J. Mulvey, CERN 88-02 (1988).

8. M. Chanowitz and R. Cahn, *Phys. Rev. Lett.* **56**, 1327 (1986).

9. B. Cox and F. Gilman, p. 87, *Proc. 1984 Summer Study on Design and Utilization of the SSC*, eds. R. Donaldson and J. Morfin, American Physical Society.

10. M. Bento and C.H. Llewellyn Smith, *Nucl. Phys.* **B289**, 36 (1987); G. Altarelli, B. Mele, F. Pitolli, *Nucl. Phys.* **B287**, 205 (1987); J. Gunion, A. Tofighi-Niaki, *Phys. Rev.* **D36**, 2671 (1987).

10a. M. Chanowitz, *Ann. Rev. Nucl. Part. Sci.* **38**, 323 (1988).

11. M. Cornwall, D. Levin, G. Tiktopoulos, *Phys. Rev.* **D10**, 1145 (1974).

12. G. Gounaris, R. Kogerler, H. Neufeld, *Phys. Rev.* **D34**, 3257 (1986).

13. C. Vayonakis, *Lett. Nuovo Cim.* **17**, 383 (1976).

14. B.W. Lee, C. Quigg, H. Thacker, *Phys. Rev.* **D16**, 1519 (1977).

15. M. Chanowitz, M. Furman, I. Hinchliffe, *Phys. Rev. Lett.* **78B**, 285 (1978); *Nucl. Phys.* **B153**, 402 (1979).

16. M. Duncan, G. Kane, W. Repko, *Nucl. Phys.* **B272**, 571 (1986).

17. D. Dicus, R. Vega, *Phys. Rev. Lett.* **57**, 1110 (1986).

18. J. Gunion, J. Kalinowski, A. Tofighi-Niakis, *Phys. Rev. Lett.* **57**, 2351 (1986).

19. R.N. Cahn *et al.*, *Phys. Rev.* **D35**, 1626 (1987).

20. J. Donoghue, C. Ramirez, G. Valencia, *Phys. Rev.* **D38**, 2195 (1988).

21. See. ref. (16).

22. G. Altarelli *et al.*, ref. (10).

23. R. Cahn, *Nucl. Phys.* **B255**, 341 (1985).

24. See ref. (18).

25. H. Georgi *et al.*, *Phys. Rev. Lett.* **40**, 692 (1978).

26. D. Froidevaux *et al.*, p. 61, *Proc. Workshop on Physics at Future Accelerators*, ed. J. Mulvey, CERN 87-07 (1987).

27. R. Cahn *et al.*, p. 20, *Proc. Workshop on Experiments, Detectors, and Experimental Areas for the SSC*, July 7-17, 1987, Berkeley, eds. R. Donaldson and M. Gilchriese (World Scientific, Singapore, 1988).

28. W. Stirling *et al.*, *Phys. Lett.* **163B**, 261 (1985); J. Gunion *et al.*, *Phys. Lett.* **163B**, 389 (1985).

29. J. Gunion and M. Soldate, *Phys. Rev.* **D34**, 826 (1986).

30. A. Savoy-Navarro, p. 68, ref. (27).

31. R. Kleiss, W. Stirling, *Phys. Lett.* **200B**, 193 (1988); W.J. Stirling, these proceedings.

32. G. Herten, p. 103, ref. (27).

33. M. Chanowitz, M. Golden, *Phys. Rev. Lett.* **61**, 1053 (1988). Due to a programming error the gluon exchange background was underestimated in this paper—see ref. (34) and Chanowitz and Golden, in preparation.

34. D. Dicus and R. Vega, U.C. Davis preprint (1988).

35. E. Eichten *et al.*, *Rev. Mod. Phys.* **56**, 579 (1984).

27

NON-PERTURBATIVE ASPECTS OF THE HIGGS SECTOR IN THE STANDARD ELECTROWEAK THEORY

István Montvay

Deutsches Elektronen-Synchrotron DESY
D-2000 Hamburg, FRG

ABSTRACT

Non-perturbative aspects of quantum field theories with elementary scalars are reviewed. Such theories play an important rôle in connection with the Higgs mechanism of mass generation in the standard model. Recent non-perturbative results in pure ϕ^4 models imply an upper limit $m_H \leq 630\ GeV$ on the mass of the Higgs boson. The effect of the SU(2) gauge coupling on the pure ϕ^4 sector is discussed. It is pointed out that the influence of the Yukawa-couplings of the Higgs scalar to heavy fermions may be important, because a non-trivial infrared fixed point structure can arise. The problem of chiral fermion gauge theories is summarized. In these theories the chiral fermions always appear in mirror pairs if the mirror Yukawa-couplings are attracted by the trivial infrared fixed point. This provides a strong motivation of the experimental search for mirror fermions and for their indirect effects in low energy phenomenology. In case if mirror fermions do exist in nature, future high energy colliders have a very important rôle in the exploration of their properties.

INTRODUCTION

The SU(2) $\otimes$ U(1) electroweak interaction of elementary particles is weak, therefore the question naturally arises, why is it necessary to study the electroweak theory non-perturbatively? The answer to this question has several parts:

1. A mathematical aspect: quantum field theories cannot be defined by perturbation theory, unless the perturbative expansion is convergent. The perturbation series, however, can only be expected to be asymptotic.

2. A general physical aspect: spontaneous symmetry breaking providing the vacuum expectation value of the Higgs scalar field is a genuinely non-perturbative phenomenon. There is a lot of accumulated knowledge about such phenomena in statistical physics which may be very useful.

3. A special physical aspect: even if the SU(2) $\otimes$ U(1) gauge couplings are weak the quartic scalar self-coupling can in principle be strong, implying a non-perturbative Higgs sector.

Before going into the details of recent non-perturbative investigations of the Higgs sector let us summarize some general concepts defining the theoretical framework. It is natural to assume that quantum field theory defined on a flat space-time is valid only up to the Planck-scale ($1.2 \, 10^{19} \, GeV$) where the effects of quantum gravity become important. Beyond this natural physical cut-off some new framework is necessary describing the physics at still higher energies. This more general framework may be the "theory of everything", it may involve "superstrings", "wormholes" or "chaos with random dynamics" etc. From the point of view of quantum field theory the most important aspect is that it has to specify the free parameters at the natural cut-off scale.

Within quantum field theory the simplest possibility is that the minimal $SU(3) \otimes SU(2) \otimes U(1)$ model is valid up to the Planck scale. This might, however, be impossible either because this assumption is mathematically inconsistent or because it is in conflict with some presently unknown experimental result. Another possibility is that the $SU(3) \otimes SU(2) \otimes U(1)$ model is embedded at some scale intermediate between $100 \, GeV$ and $10^{19} \, GeV$ in some larger quantum field theory involving e. g. supersymmetry, grand unification etc. In a renormalizable quantum field theory the cut-off dependence is weak if the cut-off scale (Λ) is much higher that the scale of physical masses (m). In most cases the dependence behaves as some positive integer power of the ratio m/Λ. Since the electroweak scale is so much smaller than the Planck-scale ($m/\Lambda \simeq 10^{-17}$) the cut-off dependence can be neglected for every practical purpose and a simple hypercubical lattice can be used for the cut-off. One has to keep in mind, however, that the replacement of the continuous space-time by a discrete lattice is only a mathematical tool and is only valid as long as the details of the discretization (lattice type, form of lattice action etc.) do not matter.

Every quantum field theory has a number of free parameters which are primarily defined by the independent bare parameters in the lattice action. Since these are free parameters, the determination of there values lies beyond the scope of quantum field theory. Every point in the space of bare parameters is equally possible. There is no inherent concept of "naturalness" for the choice of bare parameters. In this sense there is *parameter democracy*. In other words, there is nothing wrong in fine tuning the bare parameters of a given quantum field theory. Just the contrary, some bare (mass) parameters have to be tuned always to a high accuracy, because $m/\Lambda \simeq 10^{-17} << 1$. The consequence of the smallness of physical masses in lattice units (= cut-off units) is that the physically interesting points are very close to the *critical hypersurface* in bare parameter space, where the masses in lattice units are zero. Therefore, parameter democracy holds only in the immediate vicinity of this critical hypersurface. Aspects like "naturalness" or "simplicity" may play a rôle in the choice between different possible quantum field theories. For instance, theories with a smaller number of free parameters can generally be considered simpler and more natural. Since symmetries usually reduce the number of free parameters considerably, they are the most common source of "naturalness". From this point of view the standard $SU(3) \otimes SU(2) \otimes U(1)$ model is only moderately natural because of the large number (about 20) of its free parameters.

NON-PERTURBATIVE RESULTS IN PURE ϕ^4 MODELS

The simplest working laboratories for renormalizable quantum field theories with a possibility of spontaneous symmetry breaking are the ϕ^4 models with a quartic self-interaction. For an N-component scalar field $\phi_n(x)$, $n = 1, \ldots, N$ the interaction term in the action can be written as

$$\sum_{n_1 n_2 n_3 n_4} \lambda_{n_1 n_2 n_3 n_4} \phi_{n_1}(x) \phi_{n_2}(x) \phi_{n_3}(x) \phi_{n_4}(x) \tag{1}$$

The simplest case is the single component model $N = 1$. The complex scalar doublet in the standard model corresponds to $N = 4$. Another simple limit is $N \to \infty$, where the $1/N$-expansion offers a possible non-perturbative framework.

A basic property of pure ϕ^4 models is the *triviality of the continuum limit* [1]. This means that in the infinite cut-off limit $\Lambda/m \to \infty$ mathematical consistency requires a vanishing renormalized quartic self-coupling: $\lambda_r \to 0$. This is at the first sight a rather surprising result, which shows in a dramatic way the importance of the mathematical aspect of non-perturbative investigations mentioned in the introduction. The point is that in a perturbation theory framework the renormalized coupling is *assumed* to be a free parameter. Questions about the mathematical consistency of this assumption are usually difficult to formulate. Even if an apparent inconsistency appears it can be interpreted as a limit of applicability of perturbation theory.

The simplest possibility for a non-perturbative definition of quantum field theories is *lattice regularization*. The euclidean space-time continuum (with imaginary time) is approximated by a finite hypercubical lattice with sites x. The N-component scalar field ϕ_{nx} lives on the sites of the lattice. In the case of O(N)-symmetry the lattice action S contains two independent relevant parameters. The field normalization is a further irrelevant parameter which is, however, sometimes useful to keep, therefore S can be written as

$$S = \sum_x \left\{ \mu \phi_{nx} \phi_{nx} + \lambda (\phi_{nx} \phi_{nx})^2 - \kappa \sum_\nu \phi_{nx+\hat\nu} \phi_{nx} \right\} \qquad (2)$$

Here an automatic summation over the O(N)-index n is understood, but not over the lattice sites x. One out of the three parameters can be fixed by an appropriate choice of the field normalization. For instance, in lattice bare perturbation theory $\kappa = \frac{1}{2}$ is convenient, whereas in numerical studies (especially for large λ) $\mu = 1 - 2\lambda$ is a correct choice. In this case the two free parameters are: the quartic self-coupling λ and the *hopping parameter* κ which stands in front of the quadratic coupling of nearest neighbours ϕ_{nx} with $\phi_{nx+\hat\nu}$, $\nu = \pm 1, \pm 2, \pm 3, \pm 4$. With this choice of parameters the lattice action (2) goes over into

$$S = \sum_x \left\{ \phi_{nx} \phi_{nx} + \lambda (\phi_{nx} \phi_{nx} - 1)^2 - \kappa \sum_\nu \phi_{nx+\hat\nu} \phi_{nx} \right\} \qquad (3)$$

The limit of infinitely strong bare self-coupling $\lambda \to \infty$ is particularly interesting. In this case the length of the field is frozen to

$$\phi_{nx} \phi_{nx} = 1 \qquad (4)$$

This looks like losing one of the degrees of freedom. For instance, in the standard Higgs sector with $N = 4$, where the field components are usually denoted by σ_x and π_{rx}, $r = 1, 2, 3$, the above constraint implies

$$\sigma_x^2 = 1 - \pi_{rx} \pi_{rx} \qquad (5)$$

In this case the $\lambda = \infty$ limit is usually called the *non-linear σ-model*, in contrast to the linear σ-model for $\lambda < \infty$ [2]. According to Eq. (5) at $\lambda = \infty$ σ_x is a function of the π-field and therefore one is tempted to assume that the σ-field, which is the physical Higgs-field in the standard Higgs sector, is removed from the physical spectrum. This is, however, not correct. The physical spectrum at $\lambda = \infty$ is not essentially different from the physical spectrum at finite λ. A state with the same quantum numbers as the σ-field remains in the physical spectrum, as one can see in a *numerical simulation* [3]. In other words, the σ-field cannot be removed from the spectrum by taking the infinite bare coupling limit. The state with the quantum numbers of σ can be considered to be a bound state of two (or more) π's. Therefore, the non-linear σ-model with the action

$$S = -\kappa \sum_{x,\nu} \phi_{nx+\hat{\nu}}\phi_{nx} \qquad (6)$$

is equivalent to the linear σ-model defined by Eq. (3). Here the restriction to the simplest minimal actions (3,6) is important. For instance, the non-linear sigma model as a low-energy approximation of QCD contains many higher dymensional couplings, which do influence the physical content in an essential way.

This simple example shows, how difficult the correct identification of the spectrum of a quantum field theory can be, due to the appearance of all sorts of non-perturbative bound states. Another important point is, that perturbative renormalizability is not necessary. In perturbation theory the non-linear σ-model is non-renormalizable, whereas the $\lambda < \infty$ case is renormalizable. If one defines these models non-perturbatively, the $\lambda \to \infty$ limit is smooth and finite. The $\lambda = \infty$ model is *in the same universality class* as $\lambda < \infty$. Instead of perturbative renormalizability one is interested in the existence of critical points with zero mass in lattice units ($m/\Lambda = 0$). If such critical points do exist, one can perform a continuum limit by going to one of them. In the vicinity of critical points there is a quasi-continuum situation where the cut-off dependence is negligible and the model can be used for the description of a physical theory.

The critical line for the O(N)-symmetric ϕ^4 model (3) is qualitatively shown in Fig. 1 (see [4-7] and references therein). Below the critical line there is the symmetric phase with vanishing vacuum expectation value. Above the critical line the symmetry is spontaneously broken by the non-zero vacuum expectation value of the scalar field. In both phases near the critical line there is the *scaling region* where the mass in lattice units m/Λ is small and the cut-off effects are negligible. An important information about the model is contained in the *curves of constant physics*. These are the curves in the (κ, λ)-plane where the renormalized coupling λ_r is constant. Their qualitative behaviour is shown in Fig. 2 (for more details see the papers in Ref. [4,5]. As it is also shown by the figure, the renormalized coupling is zero on the critical line in accordance with the triviality of the continuum limit. The consequence of the triviality of the continuum limit is that for increasing ratio of the cut-off to the renormalized mass Λ/m_r there is an upper limit for λ_r. In Fig. 3 this is represented by the excluded area. The mapping from the bare parameters to the renormalized parameters is such that no points in the excluded area can be reached by any choice of the bare parameters.

The numerical determination of the upper limit for the renormalized coupling in the phase with spontaneously broken symmetry involves some interesting technical problems. Without going into details here let us mention the problem of infrared singularities due to the presence of Goldstone-bosons in the O(N)-symmetric ϕ^4 models with $N \geq 2$. As it is well known, the zero mass Goldstone-bosons are the consequence of spontaneous symmetry breaking. The long range correlations due to the zero mass Goldstone-bosons (and the corresponding infrared singularities) imply strong finite volume effects in the numerical simulations. These can, however, be used to extract the required infinite volume information from the study of finite volume systems [8,9] In a similar way, the study of finite volume effects in the single component ϕ^4 models ($N = 1$) can also be used to obtain interesting physical information, for instance about low energy scattering and vacuum tunneling [10,11].

The cut-off dependent upper limit on the renormalized coupling in the broken phase of the O(4)-symmetric ϕ^4 model implies an upper limit on the ratio of the Higgs-boson mass to the W-mass in the standard model. The renormalized quartic coupling is usually defined as

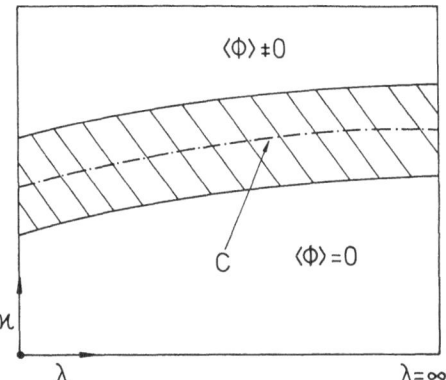

Fig. 1.   The qualitative behaviour of the critical line(C) in the (λ, κ)-plane of the O(N)-symmetric ϕ^4 model. The hatched area on both sides of the critical line is the scaling region where the cut-off is much higher than the physical scale.

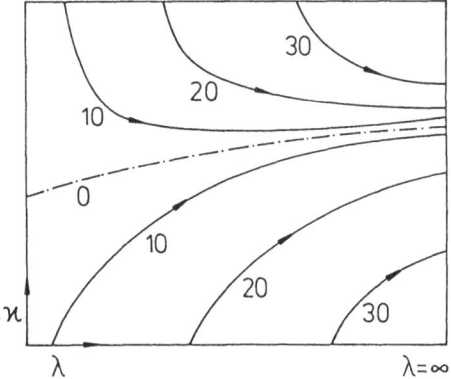

Fig. 2.   The curves of constant physics (i. e. constant renormalized coupling) in the (κ, λ)-plane for the O(N)-symmetric ϕ^4 model. The dashed-dotted line is the critical line where the mass in lattice units is zero. The arrows point in the direction of increasing cut-off (decreasing lattice spacing).

$$\lambda_r = \frac{3m_r^2}{v_r^2} \tag{7}$$

where v_r denotes the renormalized vacuum expectation value. For small SU(2) gauge coupling (g) the physical Higgs-boson mass (m_H) is equal to a good approximation to the renormalized mass in the pure ϕ^4 model ($m_H = m_r$), whereas the W-mass is given by the Dashen-Neuberger formula [12]

$$m_W = \frac{1}{2} g v_r \tag{8}$$

This relation is a consequence of the O(4) Ward-identities and together with Eq. (7) implies

$$\frac{m_H^2}{m_W^2} = \frac{4\lambda_r}{3g^2} \tag{9}$$

From this relation and the upper limit on λ_r one obtains [5-7]

$$m_H < 630 \ GeV \tag{10}$$

This is at the lowest cut-off where the lattice regularization is applicable, actually at $\Lambda > 2m_r$. Taking the natural value of the cut-off $\Lambda = 1.2 \cdot 10^{19} \ GeV$ the result is

$$m_H < 145 \ GeV \tag{11}$$

These are surprisingly low limits which imply the absence of a strongly interacting Higgs sector, because the tree-level unitarity limit signalizing strong interaction is at $m_H \simeq 1 \ TeV$ [13] In the case of the very low cut-off corresponding to Eq. (10) the upper limit depends a little on the way how the discretization was done (lattice structure, lattice action) [14], but the corresponding change in Eq. (11) is rather small. For $\Lambda/m_r \to \infty$ the asymptotic behaviour of the upper limit is given by the perturbative renormalization group equations (see Sec. 4):

$$\lambda_r < \frac{4\pi^2}{\ln(\Lambda/m_r)} \tag{12}$$

This is why the value of the upper limit in Eq. (11) is so similar to the limits obtained by the requirement that the renormalized coupling be small up to a scale close to the Planck-scale ("perturbative grand unification") [15].

THE INCLUSION OF THE GAUGE COUPLINGS

The weak SU(2) $\otimes$ U(1) gauge couplings are usually assumed to be small perturbations on the pure ϕ^4 model. This assumption is plausible, but an exact proof is not easy in the case if the bare quartic scalar coupling is large. For a not too large number of fermions the SU(2)- and U(1)-couplings behave differently, because SU(2) is asymptotically free and U(1) is not. Most of the non-perturbative work was up to now done on the SU(2)-coupling, therefore let us concentrate here on it (for studies of the phase structure of the full SU(2) $\otimes$ U(1) Higgs model see [16]).

The phase structure of the standard SU(2) Higgs model was extensively studied in a large number of papers (for references see the reviews [17]). The bare parameter space and the established phase structure is shown in Fig. 4. On the surface shown in the figure there is a phase transition which is most probably of first order everywhere except for some parts of the boundary. Above the phase transition surface there is the *Higgs phase* where the W-boson gets a mass due to

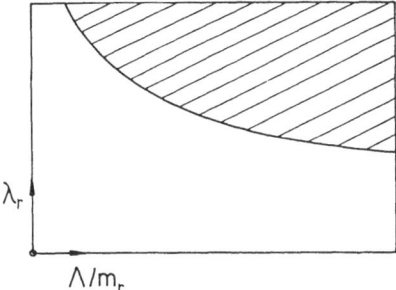

Fig. 3. The cut-off dependent upper limit on the renormalized coupling in the $(\Lambda/m_r, \lambda_r)$-plane (m_r is the renormalized mass, λ_r the renormalized coupling). The hatched area is excluded.

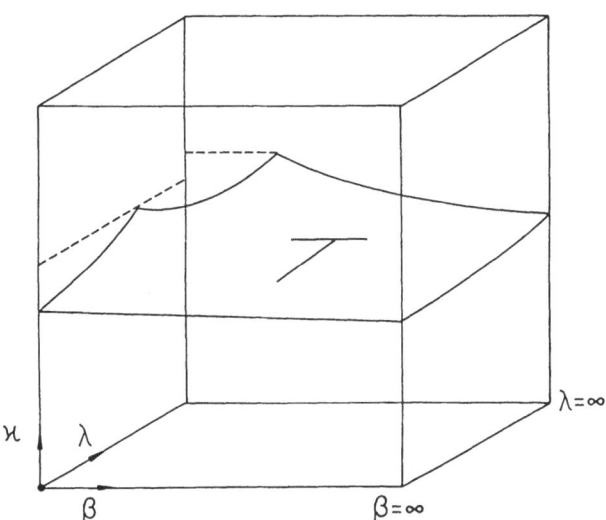

Fig. 4. The bare parameter space of the standard SU(2) Higgs model. λ and κ are the bare parameters of the scalar field and $\beta \equiv 4/g^2$ stands for the SU(2) gauge coupling. Above the phase transition surface (T) there is the Higgs phase, below it the confining phase.

475

the Higgs mechanism. In the *confining phase* below this surface the model describes a QCD-like theory with scalar "quarks". The usual assumption is that the standard electroweak model is in the Higgs phase. The confining phase is relevant in the formulation of the *strongly interacting electroweak model* [18]. In this phase the broken symmetry is "restored", therefore the phase transition is usually called *symmetry restoring phase transition* [19].

The effect of the small SU(2) gauge coupling on the ϕ^4 model can be described in the framework of the weak gauge coupling expansion [20], where the Green's functions of the Higgs model are expressed by power series in the gauge coupling. The coefficients of the series depend on the Green's functions of the pure ϕ^4 model. With reasonable assumptions one can show that the higher terms of the expansion give only small corrections, but at the same time one can also see why the smallness of the corrections is not completely trivial. Namely, in the higher loop contributions an integration over all momenta has to be performed, and at the momenta near the cut-off scale the scalar self-interaction is roughly equal to the bare quartic coupling which can be large. A direct control over the effect of the small gauge coupling can be achieved in numerical simulations which are possible at small gauge couplings in the Higgs phase [21]. The results are consistent with the weak gauge coupling expansion. The direct simulation with a weak gauge coupling has the advantage that the finite mass of the gauge W-boson acts as an infrared regulator and therefore there is no problem with infrared divergencies. The upper limit on the Higgs-boson mass can also be obtained in this way and the results [22,23] are consistent with the upper limit obtained in the ϕ^4 model (see previous section).

Besides the gauge couplings in the standard model there are also the Yukawa-couplings of the fermions which can influence the Higgs sector. In particular, heavy fermions imply strong Yukawa-couplings which can have an important effect because they can change the renormalization group behaviour qualitatively. These questions will be discussed in the next section.

INFRARED AND ULTRAVIOLET FIXED POINTS

Pure Scalar ϕ^4 Model

The absence of strong interactions in the ϕ^4 model can be qualitatively understood from the infrared behaviour of the renormalization group equations (RGE's). For this purpose the convenient form of the RGE's involves the renormalized coupling (λ_r) as a function of the cut-off (actually, the ratio of the cut-off to the physical mass Λ/m) [24]. Using the natural variable

$$\tau \equiv \ln \frac{\Lambda}{m} \tag{13}$$

the RGE can be written as

$$\frac{d\lambda_r(\tau)}{d\tau} = -\beta_r(\lambda_r, \tau) = -\frac{1}{16\pi^2} \, 4\lambda_r^2 + \cdots \tag{14}$$

Here β_r is an appropriate Callan-Symanzik β-function, which in the scaling region depends only on λ_r. (The τ-dependence implies scale breaking.) The last equality in (14) shows the universal one-loop contribution, which dominates for small λ_r. This equation determines the change of the renormalized coupling for fixed bare coupling λ, that is along vertical lines in Fig. 1. A numerical check of the RGE behaviour in the single component ϕ^4 model at infinite bare quartic coupling (Ising limit) was performed in Ref. [25], and a good agreement with the three-loop β-function was found. Near the critical point λ_r is small, therefore the one-loop term dominates and drives the solution for $\tau \to \infty$ (i. e. for $m/\Lambda \to 0$) to the fixed point

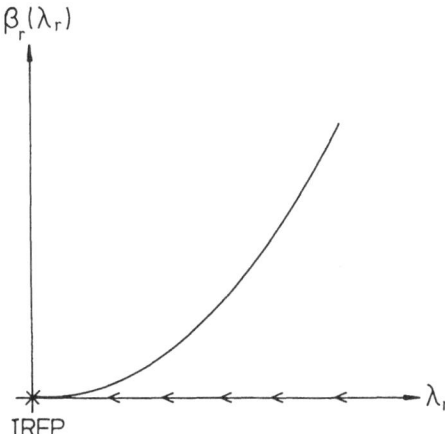

Fig. 5. The qualitative behaviour of the Callan-Symanzik β-function in the ϕ^4 model. The β-function has no other zeros besides the IRFP at $\lambda_r = 0$. The curves of constant physics look in this case like Fig. 2.

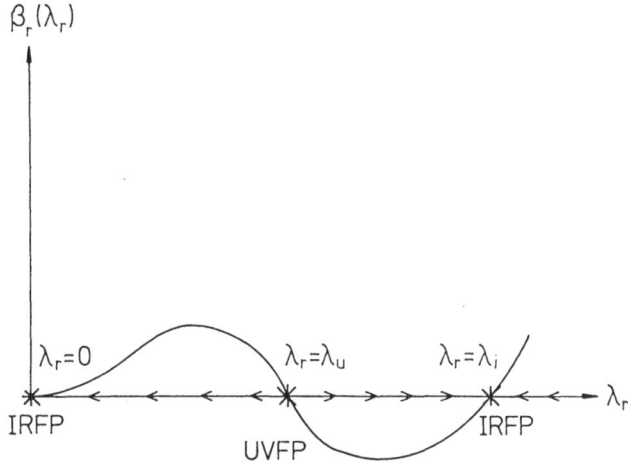

Fig. 6. An illustrative Callan-Symanzik β-function. It has two infrared fixed points (IRFP) at $\lambda_r = 0$ and at $\lambda_r = \lambda_i$ and an ultraviolet fixed point (UVFP) at $\lambda_r = \lambda_u$.

477

of the renormalization group equation at $\lambda_r = 0$. The asymptotic behaviour near $\lambda_r = 0$ is, in accordance with Eq. (12),

$$\lambda_r \simeq \frac{4\pi^2}{\tau} \tag{15}$$

Since $\tau \to \infty$ is the limit when the physical mass in cut-off units tends to zero, $\lambda_r = 0$ is called an *infrared fixed point* (IRFP) of the renormalization group equation. The qualitative consequence of the IRFP at $\lambda_r = 0$ is that once the cut-off is large compared to the physical mass, the renormalized coupling is small because the solution of Eq. (14) is attracted to the IRFP.

The situation in a U(1) gauge theory like QED can be similar to ϕ^4, because the leading term of the Callan-Symanzik β-function is similar to Eq. (14). This would imply the triviality of the continuum limit of QED and, consequently, a cut-off dependent upper limit on the fine structure constant. However, if the cut-off is at the Planck-scale this upper limit is much higher than the physical value 1/137. In the continuum formulation of perturbation theory the triviality of the continuum limit is signalized by inconsistencies which were discovered long ago by Landau ("Landau-ghosts") [26].

The triviality of the continuum limit follows from Eq. (14) only if the leading term in Eq. (14) is at least qualitatively correct also for large renormalized couplings (see Fig. 5). In the ϕ^4 model this is most probably true but in QED there are arguments that the behaviour of the β-function is different at large couplings [27,28]. In order to illustrate how a non-trivial continuum limit can arise let us consider a more complicated β-function depicted in Fig. 6. It starts at $\lambda_r = 0$ as a parabola (see Eq. (14)), but for larger λ_r it has two other zeros. The second zero at $\lambda_r = \lambda_i$ is another IRFP which is attractive for increasing τ. The intermediate zero at $\lambda_r = \lambda_u$ is a repulsive fixed point of the RGE (14). One can consider another RGE which describes the change of the bare coupling as a function of the cut-off for fixed renormalized coupling:

$$\frac{d\lambda(\tau)}{d\tau} = \beta(\lambda, \tau) = \frac{1}{16\pi^2} 4\lambda^2 + \cdots \tag{16}$$

The functional form of the β-function in this equation is similar but not exactly equal to β_r in Eq. (14). The leading term for small coupling is the same and the qualitative behaviour for large couplings is also given by Fig. 6. Let us assume that the zeros of $\beta(\lambda)$ are at the same place as those of $\beta_r(\lambda_r)$, for instance $\beta(\lambda = \lambda_u) = 0$. (This can always be achieved by an appropriate redefinition of the renormalized coupling.) At $\lambda = \lambda_u$ the differential equation (16) has an attractive fixed point, which is called *ultraviolet fixed point* (UVFP) because for $\tau \to \infty$ the cut-off is infinitely large compared to the physical mass. (Note that an equation similar to (16) also describes the change of the renormalized coupling in the continuum theory as a function of μ/m, where μ is the renormalization scale.)

Combining the two equations (14) and (16) one can construct the curves of constant renormalized coupling (λ_r) in the bare parameter space. Staying for simplicity in the symmetric phase, one obtains Fig. 7. (In the broken phase a qualitatively similar picture is repeated, only upside down.) Since the UVFP corresponds to a singular point of the theory, there may be all sorts of singularities near to it. For instance, the critical line may have a cusp or some other type of singularity at $\lambda = \lambda_u$.

The consequence of Fig. 7 is that at the UVFP it is possible to define a non-trivial continuum limit. Near this point the renormalized coupling can have any value between the two neighbouring IRFP's, namely

$$0 \leq \lambda_r \leq \lambda_i \tag{17}$$

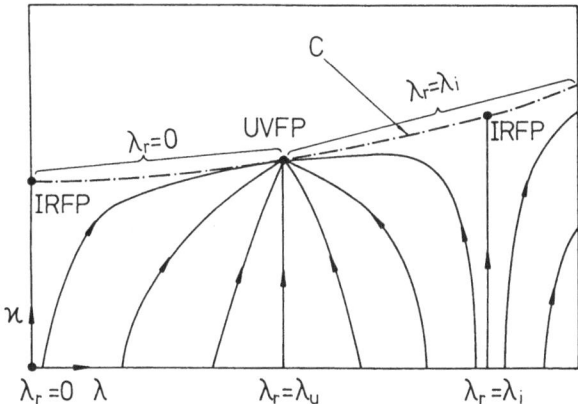

Fig. 7. The qualitative behaviour of the curves of constant physics for the β-function shown in Fig. 6. C is the critical line, and here only the symmetric phase is shown for simplicity.

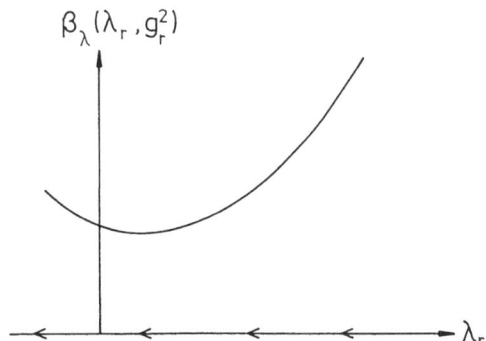

Fig. 8. The qualitative behaviour of the Callan-Symanzik β-function for λ_r ($g_r = fixed$) in the standard SU(2) Higgs model.

In QCD the existence of the continuum limit is guaranteed by an UVFP, which is at zero coupling corresponding to asymptotic freedom. The triviality of the continuum limit in the ϕ^4 model is equivalent to the absence of an UVFP in the β-function. As mentioned above, in QED there might be a non-trivial UVFP, therefore Fig. 6 can be qualitatively correct (perhaps without the second IRFP).

The importance of the IRFP's can also be inferred from Fig. 7. First, they imply limits for the renormalized coupling in the continuum limit (see Eq. (17)). Second, as it is shown by the figure, on the critical line outside the UVFP the value of the renormalized coupling is given by the IRFP's. This means that defining a continuum limit at some critical point different from the UVFP, the value of the renormalized coupling always tends to an IRFP. Therefore, continuum theories with IRFP couplings are special points in the space of all theories (in this context see Ref. [29], where the notion of IRFP's was approached more generally).

Inclusion of Gauge Couplings

After this general discussion of infrared and ultraviolet fixed points let us return to the renormalization group behaviour in the standard SU(2) Higgs model. In the following let us always consider the version of RGE's corresponding to Eq. (14), which gives the change of the renormalized couplings for fixed bare couplings. If the SU(2) gauge coupling is added to the ϕ^4 model there are two renormalized couplings: the quartic coupling λ_r and the gauge coupling g_r. Along the lines $\lambda = const.$; $\beta \equiv 4/g^2 = const.$ in Fig. 4 the change of λ_r and g_r^2 as a function of $\tau \equiv \ln(\Lambda/m)$ is determined by

$$\frac{d\lambda_r(\tau)}{d\tau} = -\beta_\lambda(\lambda_r, g_r^2, \tau) = -\frac{1}{16\pi^2}\left(4\lambda_r^2 - 9\lambda_r g_r^2 + \frac{27}{4}g_r^4\right) + \cdots$$

$$\frac{dg_r^2(\tau)}{d\tau} = -\beta_g(\lambda_r, g_r^2, \tau) = \frac{1}{16\pi^2}\frac{43}{3}g_r^4 + \cdots \qquad (18)$$

Here the one-loop terms are explicitly given. The SU(2) gauge coupling has an UVFP at $g_r^2 = 0$, corresponding to asymptotic freedom. The general behaviour of β_λ for fixed g_r is shown by Fig. 8. Since this function has no zeros, nothing can stop the decrease of λ_r which, therefore, goes to negative values and makes the theory unstable. The consequence is the first order Weinberg-Linde phase transition [19] at the surface separating the two phases (see Fig. 4).

Inclusion of Yukawa-couplings

The inclusion of light fermions with small Yukawa-couplings to the scalar field does not influence the Higgs sector in an essential way. The only noticeable difference is due to the change in the β-functions of the gauge couplings. Heavy fermions and the corresponding strong Yukawa-couplings are, however, important. In particular, heavy quarks establish a strong coupling between QCD and the electroweak sector and induce a qualitatively new renormalization group behaviour. In order to illustrate this let us consider here the RGE's for a colour triplet weak doublet quark. The SU(2) gauge coupling is not essential here, therefore let us only consider the SU(3) colour coupling (g_r), the Yukawa-coupling (G_r) and the scalar quartic coupling:

$$\frac{dg_r^2(\tau)}{d\tau} = -\beta_g(g_r^2, G_r^2, \lambda_r, \tau) = \frac{1}{16\pi^2}\frac{58}{3}g_r^4 + \cdots$$

$$\frac{dG_r^2(\tau)}{d\tau} = -\beta_G(g_r^2, G_r^2, \lambda_r, \tau) = -\frac{1}{16\pi^2}(24G_r^4 - 16G_r^2 g_r^2) + \cdots$$

$$\frac{d\lambda_r(\tau)}{d\tau} = -\beta_\lambda(g_r^2, G_r^2, \lambda_r, \tau) = -\frac{1}{16\pi^2}(4\lambda_r^2 + 48\lambda_r G_r^2 - 288 G_r^4) + \cdots \qquad (19)$$

For $\tau \to \infty$ the colour gauge coupling grows, the higher corrections to the one-loop β-function become important and at some point perturbation theory breaks down. In the perturbative region, where one-loop gives a good approximation, the right hand sides of the last two equations have zeros (see Fig. 9). Therefore, as long as the colour gauge coupling is small and slowly varying, the other two couplings are "dragged" with it, close to the values where β_G and β_λ vanish. This situation can be called *quasi-IRFP*. Since in this case the quartic and Yukawa-couplings can be predicted, there is a large number of papers in the literature exploiting this possibility (see for instance [30]). Of course, once the colour coupling gets large, the perturbative β-functions are not applicable, the quasi-IRFP becomes irrelevant and the question of the continuum limit at $\tau \to \infty$ remains open. In an imaginary world with only heavy quarks, where the colour coupling remains perturbative at the quark mass scale and therefore all the couplings can stay perturbative, one can show that the quasi-IRFP does not imply a non-trivial continuum limit of the quartic and Yukawa-couplings [31]. This means that for infinite cut-off the model tends to pure QCD with heavy quarks and a non-interacting Higgs-boson. The additional couplings go to zero for $\tau \to \infty$, although very slowly. For fixed g_r we have [31]:

$$G_r(\tau) \to \tau^{-\frac{5}{58}} \qquad \lambda_r(\tau) \to \tau^{-\frac{10}{29}} \qquad (20)$$

Such a slow change has practically no consequences. For instance, for the Yukawa-coupling the asymptotic change between 100 and 10^{19} GeV is only a decrease by a factor of about 1.4. In other words, in the case of a quasi-IRFP the upper limits on the quartic and Yukawa-couplings are slowly changing with the cut-off. The value of the upper limit for the Higgs-boson mass is in this case somewhat higher than in Eq. (11). For the natural cut-off at the Planck-scale both this limit and the upper limits for heavy quark masses are in the range of $200 - 300$ GeV, depending on the details of the model [30].

Is there a Non-trivial Infrared Fixed Point?

Since models with a true IRFP are special, it is conceivable that for some yet unknown reason they play a important rôle in the understanding of the standard model. The above example of a quasi-IRFP fulfils almost all requirements. The only missing ingredient is the infrared attractive zero of the β-function for the gauge coupling. In the one-loop approximation this is impossible, therefore one has to go to higher loops or to a non-perturbative approach. In fact, as it was pointed out by Banks and Zaks [32], in the 2-loop approximation some non-abelian gauge theories with an intermediate number of flavours have an IRFP. The 2-loop β-function of the gauge coupling can be written in models with fermions as

$$\beta_r(g_r^2) = -2\beta_0 \frac{g_r^4}{16\pi^2} - 2\beta_1 \frac{g_r^6}{(16\pi^2)^2} \qquad (21)$$

The constants $\beta_{0,1}$ depend on the group and on the number of fermions. In the case of SU(2) and a number N_f of fermions in the fundamental representation we have

$$\beta_0 = \frac{22}{3} - \frac{2}{3} N_f \qquad \beta_1 = \frac{136}{3} - \frac{49}{6} N_f \qquad (22)$$

The same for SU(3) is:

$$\beta_0 = 11 - \frac{2}{3} N_f \qquad \beta_1 = 102 - \frac{38}{3} N_f \qquad (23)$$

This shows that in the case of SU(2) for N_f between 6 and 10, and in the case of SU(3) for N_f between 9 and 16 the 2-loop β-function has an IRFP zero, as indicated

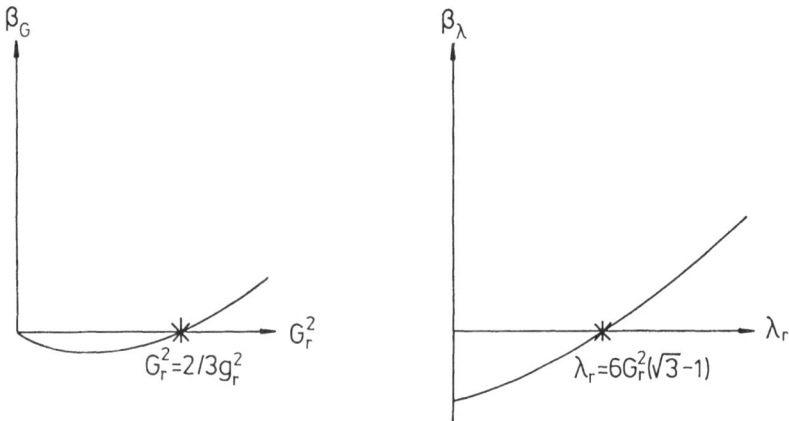

Fig. 9. The qualitative behaviour of the Callan-Symanzik β-functions for the Yukawa-coupling (β_G) and for the quartic coupling (β_λ) at small couplings according to Eq. (19).

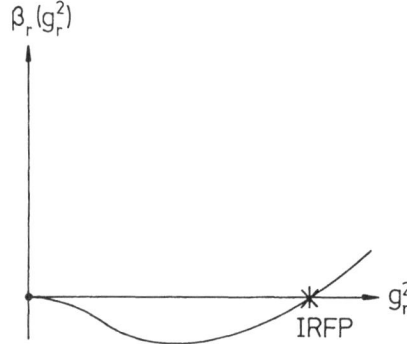

Fig. 10. The 2-loop β-function of the gauge coupling for intermediate number of fermion flavours.

by Fig. 10. Taking as an example 6 standard families, or which is from this point of view the same, 3 mirror pairs of standard families (see next section), we have $N_f = 12$ both for SU(3) and SU(2) and the 2-loop β-function predicts a non-trivial IRFP for the SU(3) colour coupling at

$$\frac{g_r^2}{16\pi^2} = \frac{3}{50} \tag{24}$$

This implies a non-trivial IRFP also for the quartic and Yukawa-couplings (see above), whereas the IRFP for the SU(2) and U(1) couplings is at zero. (The absolute value of the 2-loop β-function for the SU(2) coupling is, however, very small.)

Therefore, the 2-loop beta functions predict the possibility of non-trivial IRFP's. The question is, of course, whether the IRFP remains there after taking into account the higher loop corrections and the mass- (threshold-) effects. Also the question of the phase structure of such models has to be clarified, for instance, whether there is a second order phase transition allowing a non-trivial continuum limit or not.

CHIRAL GAUGE THEORIES AND MIRROR FERMIONS

Chiral Gauge Theories with Mirror Fermions on the Lattice

The electroweak interactions in the standard model are described by a *chiral gauge theory* where left- and right-handed components of the fermion fields are transforming differently under the SU(2) $\otimes$ U(1) gauge symmetry. This implies that left-right symmetry is broken at low energy. Nevertheless, it can be restored at high energy above the scale of spontaneous symmetry breaking. There are two different ways how this can happen:

1. by enlarging the gauge group to a left-right symmetric one, for instance to SU(2)$_L$ $\otimes$ SU(2)$_R$ $\otimes$ U(1) [33];

2. by doubling the fermion spectrum with *mirror fermions*. The mirror fermions are defined in chiral gauge theories by interchanging the transformation properties of the L- and R-handed field components with respect to the gauge group. The idea of mirror fermions is as old as the idea of parity breaking. In fact, the possibility of the existence of "elementary particles exhibiting opposite asymmetry" was discussed already in the classical paper by Lee and Yang [34]. Mirror fermions also occur naturally in connection with many interesting modern theoretical ideas. To mention a few typical examples, mirror fermions were introduced in order to cancel anomalies [35], they occur in grand unified theories with large orthogonal groups [36], in Kaluza-Klein theories [37], in extended supersymmetry [38] and also in superstring inspired models [39].

Since spontaneous symmetry breaking is a non-perturbative phenomenon, for the study of the left-right symmetry restoration a non-perturbative framework (such as lattice regularization) is needed. Chiral gauge theories with mirror fermions were intorduced in lattice regularization in Ref. [40,31]. Let us denote the "normal" fermions with $V - A$ coupling to the W-boson by ψ and the mirror partners with $V + A$ coupling by χ. The generic form of the fermion mass matrix in the broken symmetry phase on the (ψ, χ)-basis is

$$\mu = \begin{pmatrix} G_\psi v & \mu_{\psi\chi} \\ \mu_{\psi\chi} & G_\chi v \end{pmatrix} \tag{25}$$

Here v is the vacuum expectation value of the scalar doublet field, G_ψ, respectively G_χ, are the Yukawa-couplings of the ψ- and χ-fermions and $\mu_{\psi\chi}$ is a chiral invariant mass parameter. The physical states are mixtures of ψ and χ characterized by some mixing angle α. The corresponding physical fermion masses $\mu_{1,2}$ are, in general, different and are usually of the order of $\max(|\mu_{\psi\chi}|, |v|)$. In the specific case of $\mu_{\psi\chi}^2 = G_\psi G_\chi v^2$, however, one of the mass eigenvalues is zero. The very small fermion masses in the standard model (for neutrinos, electron, u- and d-quarks etc.) could be due to such a cancellation mechanism. Although this looks at the first sight as an ugly fine tuning of parameters, it is conceivable that there is some dynamical or symmetry principle in a more general theoretical framework determining the parameters of quantum field theory which implies such a relation. In this respect it is worth to emphasize that in the usual perturbative setup of the Higgs sector the smallness of the fermion masses compared to the vacuum expectation value is due to the fine tuning of the corresponding Yukawa-couplings to values very close to zero. (The smallness of the Yukawa-couplings does not necessarily have to do with local chiral symmetry.)

As it will be discussed in some detail below, models with three mirror pairs of fermion families can be constructed which are consistent with all the presently known phenomenology (see for example Ref. [41]). It is a very interesting experimental question whether this way of parity symmetry restoration is realized in nature. Mirror fermions should be searched for directly in high energy production experiments or indirectly by looking for their effects in low energy phenomenology. Of course, as long as mirror fermions are not found experimentally, we have to ask the exciting theoretical question, whether the mirror partners can be removed from the physical spectrum of a quantum field theory?

Can the Mirror Fermions be Removed?

In a lattice regularized theory with fermions the mirror partners are always present due to the Nielsen-Ninomiya theorem [42], provided some rather plausible assumptions are fulfilled. This is the *fermion doubling* phenomenon on the lattice. Nevertheless, in vectorlike theories as QCD the superfluous additional states can be kept at the cut-off scale and hence are removed from the physical spectrum in the large cut-off ("continuum") limit [43]. The question is whether the mirror doublers can also be kept at the scale of the cut-off in chiral gauge theories?

For the correct formulation of the question it is important to note that the mirror partners *can* be removed from the physical spectrum in scalar-fermion theories without gauge fields. For instance, in the broken phase this can be achieved by an appropriate choice of scalar field vacuum expectation values and Yukawa-couplings. In a model with SU(2) gauge symmetry the simples possibility is to introduce a second scalar doublet which has a vacuum expectation value of order one in lattice units. It can be shown that by tuning the parameters it is possible to arrange in this case that one of the masses of a mirror fermion pair remains at the cut-off scale.

The difficulty comes in the physically interesting case with W- and Z-gauge fields, because then the scale of the scalar doublet vacuum expectation values is fixed by the W- and Z-masses. Let us first assume that in the large cut-off limit the Yukawa-couplings $G_{\psi,\chi}$ are attracted by the trivial IRFP at vanishing (renormalized) couplings, as it is suggested by the perturbative β-functions. In this case, similarly to the triviality upper limit for the Higgs-boson mass, there are cut-off dependent upper limits for the masses of the mirror fermion partners which follow from the requirement of mathematical consistency of the quantum field theory. As a consequence, the mirror partners cannot have much higher masses than the vacuum expectation value.

The only viable alternative to the physical existence of mirror fermions seems to be that there is a non-trivial UVFP in the Yukawa-couplings of fermions $G_{\psi,\chi}$.

In this case, as it was generally discussed in the previous section, the upper limit on the renormalized Yukawa-couplings of the mirror fermion partners can be very large (in principle also infinite, see Eq. (17)), and the mirror families can be moved in the continuum limit to very high (maybe also infinite) masses. The existence of a non-trivial UVFP may also be connected to the existence of chirally asymmetric phases in quantum gauge field theories with mirror fermions [44].

At present it is not known whether a non-trivial UVFP in the Yukawa-couplings of mirror fermion pairs does exist or not. Nevertheless, its existence or non-existence is a genuine property of the quantum field theory, because it is generally assumed that the physical content of a quantum field theory is independent of the regularization scheme. Therefore, the possibility of removing the mirror fermion partners from the physical spectrum is a general quantum field theory problem and not only a question in lattice regularization.

Phenomenology of Mirror Fermions

As it was stated before, it is possible to construct extensions of the minimal standard model which contain mirror fermions at the 100 GeV scale and at the same time are consistent with all presently known phenomenology. Needless to say that if such mirror fermions do indeed exist in the 100 GeV range, then future high energy colliders are very important for the exploration of their properties.

Let us first consider the constraints on mirror fermion models imposed by phenomenology. The most important consequence of the mirror fermions at low energies is that the weak currents, instead of being pure $V - A$, have a generic form

$$(V - A)\cos\alpha + (V + A)\sin\alpha \qquad (26)$$

Here α is a mixing angle in the (ψ, χ)-basis. In a model with three mirror pairs of standard fermion families an important constraint is the absence of lepton number violations and the absence of flavour changing neutral currents. These constraints can be satisfied, for instance, with fermion mixing schemes having a one-to-one correspondence between fermions and mirror fermions (such mixing schemes can be called *monogamous*) [41]. In such a scheme the mixing in the heavy mirror quark sector is given by the same Kobayashi-Maskawa matrix as in the light quark sector (including the top quark).

The phenomenological upper bounds on the mixing angles $\alpha_{e,\mu,u,...}$ were derived in a more general framework recently by Langacker and London [45]. (For earlier works see also the references in [46].) In the best cases, namely for the first fermion family and for the muon, the bounds are typically $|\sin\alpha| \leq 0.2$. The lower bounds on mirror fermion masses are similar to the bounds on the masses of a fourth heavy fermion family, typically $m \geq 30 \ GeV$ for leptons and $m \geq 50 \ GeV$ for quarks.

The important couplings for the production and decay of mirror fermions are the off-diagonal couplings to real or virtual W- and Z-bosons (see Fig. 11). In the monogamous mixing scheme these couplings are always between corresponding pairs, that is between electrons and mirror electrons, u-quarks and mirror u-quarks etc. They are always proportional to the corresponding $\sin\alpha$. Depending on the values of the masses and mixing angles, the dominant decay pattern of the mirror fermions can be either a direct decay to the light partner, or first a decay to some lighter mirror fermion. In the monogamous scheme these latter are suppressed by the fact that the mass splittings between the mirror families are relatively small, in fact similar to the mass splittings among the light families. Therefore, the decay signature of a heavy mirror fermion is quite spectacular: the mirror leptons can decay to 3 leptons or to a lepton plus 2 jets, the mirror quarks to 3 jets or to a jet plus a lepton pair. At very high energy hadron colliders, like the Eloisatron, the

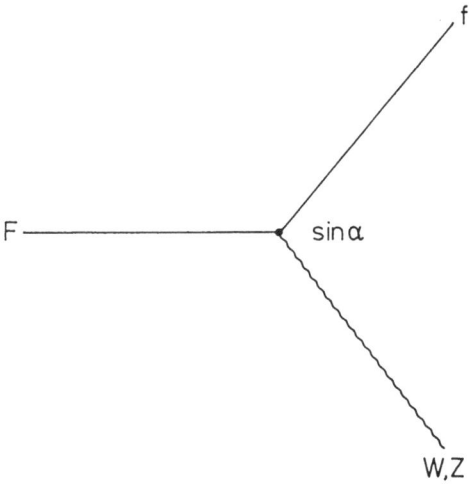

Fig. 11. The off-diagonal couplings of the mirror fermions (F) to the corresponding light fermions (f).

mirror fermions can probably be pair-produced. This would allow a detailed study of their spectrum and of many of their decay modes.

REFERENCES

[1] K. G. Wilson, Phys. Rev. *B4* (1971) 3184;
   K. G. Wilson, J. Kogut, Phys. Rep. *12C* (1974) 75;
   R. Schrader, Phys. Rev. *B14* (1976) 172;
   M. Aizenmann, Phys. Rev. Lett. *47* (1981) 1;
   Commun. Math. Phys. *86* (1982) 1;
   J. Fröhlich, Nucl. Phys. *B200 [FS4]* (1982) 281;
   C. Aragao de Carvalho, C. S. Caracciolo, J. Fröhlich, Nucl. Phys. *B215 [FS7]* (1983) 209;
   J. Fröhlich, in *Progress in gauge field theory*, Cargese lecture 1983, ed. G. 't Hooft et al., Plenum Press 1984; for further references see this review.
[2] M. Gell-Mann, M. Lévy, Nuovo Cimento, *16* (1960) 705;
   B. W. Lee, *Chiral dynamics*, Gordon and Breach 1972, New York
[3] I. Montvay, Phys. Lett. *150B* (1985) 441
[4] M. Lüscher, P. Weisz, Nucl. Phys. *B290 [FS20]* (1987) 25;
   Nucl. Phys. *B295 [FS21]* (1988) 65
[5] M. Lüscher, P. Weisz, Phys. Lett. *212B* (1988) 472;
   DESY preprint 88-146 (1988)
[6] A. Hasenfratz, K. Jansen, C. B. Lang, T. Neuhaus, H. Yoneyama, Phys. Lett. *199B* (1987) 531;
   A. Hasenfratz, K. Jansen, J. Jersák, C. B. Lang, T. Neuhaus, H. Yoneyama, Jülich preprint HLRZ-88-02 (1988)
[7] J. Kuti, L. Lin, Y. Shen, Phys. Rev. Lett. *61* (1988) 678;
   San Diego preprint UCSD/PTH 88-05 (1988)
[8] H. Neuberger, Phys. Rev. Lett. *60* (1988) 889;
   Nucl. Phys. *B300 [FS22]* (1988) 180
[9] H. Leutwyler, Nucl. Phys. Proc. Suppl. *B4* (1988) 248;
   J. Gasser, H. Leutwyler, Nucl. Phys. *B307* (1988) 763
[10] M. Lüscher, Comm. Math. Phys. *105* (1986) 153
[11] I. Montvay, P. Weisz, Nucl. Phys. *B290 [FS20]* (1987) 327;
   K. Jansen, J. Jersák, I. Montvay, G. Münster, T. Trappenberg, U. Wolff, Phys. Lett. *213B* (1988) 203;
   K. Jansen, I. Montvay, G. Münster, T. Trappenberg, U. Wolff, DESY preprint (1988)
[12] R. Dashen, H. Neuberger, Phys. Rev. Lett. *50* (1983) 1897
[13] B. W. Lee, C. Quigg, H. B. Thacker, Phys. Rev. *D16* (1977) 1519
[14] G. Bhanot, K. Bitar, Phys. Rev. Lett. *61* (1988) 798
[15] L. Maiani, G. Parisi, R. Petronzio, Nucl. Phys. *B136* (1978) 115;
   N. Cabibbo, L. Maiani, G. Parisi, R. Petronzio, Nucl. Phys. *B158* (1979) 295
[16] R. E. Shrock, Phys. Lett. *162B* (1985) 165; Nucl. Phys. *B267* (1986) 301;
   Phys. Rev. Lett. *56* (1986) 2124; Nucl. Phys. *B278* (1986) 380
[17] J. Jersák, in *Lattice gauge theory - a challenge to large scale computing*, (Wuppertal, 1985), edited by B. Bunk, K. H. Mütter, K. Schilling, Plenum 1986, p. 133;
   I. Montvay, in *Proceedings of the International Europhysics Conference on High Energy Physics* (Uppsala, 1987), edited by O. Botner, Vol. I. p. 298;
   R. E. Shrock, Nucl. Phys. Proc. Suppl. *B4* (1988) 373;
   H. G. Evertz, M. Marcu, DESY preprint 88-133 (1988), to appear in *Proceedings of the 12th Johns Hopkins Workshop on Current Problems in Particle Theory* (Baltimore, 1988)
[18] M. Claudson, E. Farhi, R. Jaffe, Phys. Rev. *D34* (1986) 873
[19] D. A. Kirzhnitz, A. D. Linde, Phys. Lett. *42B* (1972) 471;
   S. Weinberg, Phys. Rev. *D9* (1974) 3357;
   S. Weinberg, Phys. Rev. Lett. *36* (1976) 294;
   A. D. Linde, JETP *23* (1976) 64

[20] I. Montvay, Phys. Lett. *172B* (1986) 71; Nucl. Phys. *B293* (1987) 479
[21] W. Langguth, I. Montvay, P. Weisz, Nucl. Phys. *B277* (1986) 11
[22] W. Langguth, I. Montvay, Z. Phys. *C36* (1987) 725
[23] A. Hasenfratz, T. Neuhaus, Nucl. Phys. *B297* (1988) 205
[24] E. Brezin, J. C. Le Guillou, J. Zinn-Justin, in *Phase transitions and critical phenomena*, eds. C. Domb, M. S. Green (Academic Press, London, 1976) vol. 6, p. 125
[25] I. Montvay, G. Münster, U. Wolff, Nucl. Phys. *B305 [FS23]* (1988) 143
[26] L. D. Landau, in *Niels Bohr and the Development of Physics*, edited by W. Pauli (Mc Graw Hill, New York, 1955)
[27] K. Johnson, M. Baker, R. Willey, Phys. Rev. *136B* (1964) 111;
S. Adler, Phys. Rev. *D5* (1972) 3021;
P. Fomin, V. Gusinyn, V. Miransky, Yu. Sitenko, Riv. Nuovo Cimento, *6* (1983) 1;
V. Miransky, Nuovo Cimento, *90A* (1985) 149;
C. N. Leung, S. T. Love, W. A. Bardeen, Nucl. Phys. *B273* (1986) 649
[28] J. Kogut, E. Dagotto, A. Kocić, Phys. Rev. Lett. *60* (1988) 772
[29] W. Zimmermann, Commun. Math. Phys. *97* (1985) 211
[30] B. Pendleton, G. Ross, Phys. Lett. *98B* (1981) 291;
C. T. Hill, Phys. Rev. *D24* (1981) 691;
S. Dimopoulos, S. Theodorakis, Phys. Lett. *154B* (1985) 153
[31] I. Montvay, Nucl. Phys. Proc. Suppl. *B4* (1988) 443
[32] T. Banks, A. Zaks, Nucl. Phys. *B196* (1982) 189
[33] S. Weinberg, Phys. Rev. Lett. *29* (1972) 388;
J. C. Pati, A. Salam, Phys. Rev. *D10* (1974) 275;
G. Senjanovic, R. N. Mohapatra, Phys. Rev. *D12* (1975) 1502
[34] T. D. Lee, C. N. Yang, Phys. Rev. *104* (1956) 254
[35] D. J. Gross, R. Jackiw, Phys. Rev. *D6* (1972) 477;
P. Fayet, in *Proceedings of the 17th Rencontre de Moriond on elementary particles*, ed. J. Tran Thanh Van, Gif-sur-Yvette (Editions Frontiers, 1982), p. 483
[36] G. Senjanovic, F. Wilczek, A. Zee, Phys. Lett. *141B* (1984) 389
[37] E. Witten, Nucl. Phys. *B186* (1981) 412
[38] F. de Aguila, M. Dugan, B. Grinstein, L. Hall, G. G. Ross, P. West, Nucl. Phys. *B250* (1985) 225
[39] A. L. Kagan, C. H. Albright, Phys. Rev. *D38* (1988) 917
[40] I. Montvay, Phys. Lett. *199B* (1987) 89
[41] I. Montvay, Phys. Lett. *205* (1988) 315
[42] H. B. Nielsen, M. Ninomiya, Nucl. Phys. *B185* (1981) 20;
errata: Nucl. Phys. *B195* (1982) 541
[43] M. Bochicchio, L. Maiani, G. Martinelli, G. Rossi, M. Testa, Nucl. Phys. *B262* (1985) 331
[44] E. Eichten, J. Preskill, Nucl. Phys. *B268* (1986) 179
[45] P. Langacker, D. London, Phys. Rev. *D38* (1988) 886
[46] J. Maalampi, M. Roos, Helsinki preprint HU-TFT-88-17 (1988)

28

GLUINO SIGNATURES AND DISCOVERY WHEN R-PARITY IS BROKEN

P. Binétruy

LAPP

Annecy-le-Vieux, France

J.F. Gunion

Department of Physics

UC Davis, USA

1 Introduction

The signature for gluino pair production at a hadron collider that has received the greatest attention is large missing energy. For this signature to be appropriate two requirements must be met: i) there must be a significant branching ratio for a gluino to decay directly to the lightest supersymmetric particle (LSP); and ii) the LSP must be stable, thereby appearing as missing energy in the detector. Initial studies of gluino discovery have presumed both that the LSP is indeed stable and that the branching ratio for the direct-LSP decay

$$\tilde{g} \to q\bar{q}\tilde{\chi}_1^0 \tag{1}$$

is 100%.[1] (For a convenient summary, see ref. [1].) In ref. [2] it was shown that there can be substantial suppression in the branching ratio for the direct-LSP decay once either the parameter μ, responsible for mixing between the two Higgs doublet fields of the minimal supersymmetric model, is large or the gluino mass $m_{\tilde{g}}$ is large. Indeed, once $m_{\tilde{g}} \gtrsim 500$ GeV the branching ratio for the decay of eq. (1) is never larger than 14% and can be even smaller. A first estimate of the consequences of such a branching ratio for gluino discovery was made in ref. [3] with a more detailed effort in progress [4]. Certainly, the reduced branching ratio for the decay (1) makes it desirable to identify additional final states in gluino pair production that can be used for gluino discovery. A particularly useful and universal signature is that which emerges when neither gluino decays directly to $\tilde{\chi}_1^0$, but each decays first to a more massive neutralino or chargino that in turn decays to a charged lepton, a neutrino and a lighter chargino or neutralino (perhaps $\tilde{\chi}_1^0$). An example is

$$\tilde{g} \to q\bar{q}\tilde{\chi}_1^{\pm}(\to l^{\pm}\nu\tilde{\chi}_1^0) \ . \tag{2}$$

---

[1] Here we have assumed that the LSP is the lightest neutralino, denoted by $\tilde{\chi}_1^0$, of the supersymmetric model. This is generally the case.

When both gluinos decay in this manner it is possible to obtain a significant number of events in which there are two energetic charged leptons of the same sign:

$$\tilde{g}\tilde{g} \to l^+ l'^+ (\text{or } l^- l'^-) + 4 \; jets + E_T^{miss}. \tag{3}$$

The E_T^{miss} arises in part from the neutrinos produced and in part from the soft $\tilde{\chi}_1^0$'s that appear at the end of the decay chain. The 4 jets produced by the initial $\tilde{g}$ decays are always energetic enough to be isolated when $m_{\tilde{g}}$ is large. Additional softer jets can also be present for the more complicated decay chains. In ref. [5] it was shown that this signature for gluino pair production is particularly useful and fairly universal at large $m_{\tilde{g}}$. In the minimal model without R-parity breaking, the net branching ratio for this final state, with $l, l' = e, \mu$, is typically of order 1% for large $m_{\tilde{g}}$. This signature is especially suitable for gluino discovery since its viability depends upon both the Majorana nature of the gluino, which allows a gluino to decay with equal probability to leptons of either sign, and upon the strong production cross section for gluino pairs necessary to yield an adequate event rate for such a small effective branching ratio. The discovery limits at the LHC ($\sqrt{s} = 17$ TeV), SSC ($\sqrt{s} = 40$ TeV), and ELOISATRON ($\sqrt{s} = 200$ TeV), using this mode in the unbroken R-parity case are reviewed in a companion contribution to the present paper, and are found to be comparable to or better than those so far claimed for the classic missing energy techniques.

If R-parity is broken, then the classic missing energy signature for gluino pair production is completely absent, and one must depend entirely upon signatures that do not rely on missing energy from the $\tilde{\chi}_1^0$. The appropriate signature(s) depend upon the precise way in which R-parity is broken. Here we examine those signatures that are most closely related to the like-sign dilepton signature discussed above for the case of unbroken R-parity. We show that they should provide a very viable way of searching for heavy gluinos in the case of broken R-parity.

2   R-parity Breaking and LSP Decay

In the context of superstrings and grandunification, there are three basic types of superpotential Yukawa couplings that can arise which would lead to the breaking of R-parity [6]. They are:

$$[\hat{L}_i \hat{L}_j] \hat{e}_k^c, \tag{4}$$

$$[\hat{Q}_i \hat{L}_j] \hat{d}_k^c, \tag{5}$$

and

$$\hat{u}_i^c \hat{d}_j^c \hat{d}_k^c. \tag{6}$$

In the above the hatted fields are the superfields, the i, j, k are family indices (these Yukawa couplings need not be diagonal in family index), $\hat{L}$ denotes the $(\hat{e}, \hat{\nu})$ lepton doublet superfield, $\hat{Q}$ denotes the $(\hat{u}, \hat{d})$ quark doublet superfield, $\hat{e}^c$ denotes the SU(2) singlet conjugate field of $\hat{e}$, $\hat{d}^c$ denotes the SU(2) singlet conjugate field of $\hat{d}$, and the bracket construct in eqs. (4) and (5) denotes the SU(2) singlet (i.e. antisymmetric) combination of the superfields in the two doublets. Thus $i = j$ is not possible in the case of (4). In addition, in the case of $i = 1, j = 2, k = 2$ the coupling (4) can allow the e_2 and $\bar{e}_2$ (i.e. μ and $\bar{\mu}$) masses to induce a mass term for the electron neutrino ν_1. A Yukawa coupling of the type (4) that is not so constrained corresponds to

$i = 1$, $j = 2$, $k = 1$, which we shall take as our prototype example. In terms of more intuitive notation this particular choice for (4) becomes:

$$(\hat{\nu}_e\hat{\mu} - \hat{\nu}_\mu\hat{e})\hat{e}^c. \tag{7}$$

For (5) and (6) we shall assume that family diagonal contributions are dominant. Thus, in the case of (5) the first family superpotential term takes the form:

$$(\hat{u}\hat{e} - \hat{d}\hat{\nu}_e)\hat{d}^c. \tag{8}$$

Finally, for (6) we have in this alternative notation

$$\hat{u}^c\hat{d}^c\hat{d}^c \tag{9}$$

for the first family coupling. In eqs. (7)–(9) the c superscript refers to charge conjugation: the scalar component of $\hat{e}$ is $\tilde{e}_L$, whereas the scalar component of $\hat{e}^c$ is $\tilde{e}_R^*$ (the L and R subscripts refer to left and right handedness, and the $\tilde{l}_{L,R}$ fields annihilate negatively charged sleptons while $\tilde{l}_{L,R}^*$ fields annihilate positively charged sleptons). We note that a Yukawa of the form (6) may not be present when either (4) or (5) is; otherwise one would have rapid proton decay. We shall, for simplicity, analyze the three scenarios that arise when we take, in turn, each of the above three Yukawas to be non-zero with the other two absent.

At the component field level, each of the Yukawa's of eqs. (7)–(9) yields a variety of couplings that can lead to LSP decay. From (7) we obtain couplings yielding

$$
\begin{array}{llll}
\tilde{e}_R & \rightarrow & \nu_e\mu^- & \qquad \tilde{e}_R & \rightarrow & \nu_\mu e^- \\
\tilde{\mu}_L & \rightarrow & \bar{\nu}_e e^- & \qquad \tilde{e}_L & \rightarrow & \bar{\nu}_\mu e^- \\
\tilde{\nu}_e & \rightarrow & e^-\mu^+ & \qquad \tilde{\nu}_\mu & \rightarrow & e^+e^-\,,
\end{array}
\tag{10}
$$

and appropriate charge conjugates. In the above, the fields with a tilde are the slepton and sneutrino fields. For the Yukawa of eq. (8) we obtain component level couplings of the type

$$
\begin{array}{llll}
\tilde{u}_L & \rightarrow & e^+d & \qquad \tilde{d}_R & \rightarrow & \nu_e d \\
\tilde{d}_R & \rightarrow & e^-u & \qquad \tilde{e}_L & \rightarrow & \bar{u}d \\
\tilde{d}_L & \rightarrow & \bar{\nu}_e d\,, & \qquad \tilde{\nu}_e & \rightarrow & d\bar{d}
\end{array}
\tag{11}
$$

and appropriate conjugates. Finally, in the case of Yukawa (9) we find

$$
\begin{array}{lll}
\tilde{u}_R & \rightarrow & dd \\
\tilde{d}_R & \rightarrow & d\bar{u}\,,
\end{array}
\tag{12}
$$

and conjugates thereof.

Such Yukawa couplings can induce decays of the LSP as outlined in the following. We begin by emphasizing that the LSP eigenstate, $\tilde{\chi}_1^0$, is a mixture of wino ($\widetilde{W}_3$), bino ($\tilde{B}$), and higgsino fields. At large $m_{\tilde{g}}$ one typically finds [7] that $\tilde{\chi}_1^0$ is largely $\tilde{B}$ when $|\mu| \gtrsim m_{\tilde{g}}/4$, while for small $|\mu|$ it is primarily a superposition of higgsino fields with a small $\tilde{B}$ and $\widetilde{W}_3$ admixture. In no case is it even approximately pure photino ($\tilde{\gamma}$). Decays of $\tilde{\chi}_1^0$ are induced through its $\widetilde{W}_3$ and $\tilde{B}$ components, which have direct gauge couplings to $f\tilde{f}$ ($f = q, f = l$, or $f = \nu$) deriving from the standard supersymmetric Lagrangian. (Couplings of the higgsino components to such $f\tilde{f}$ channels are suppressed by factors of m_f/m_W just like Higgs couplings to $f\bar{f}$.) The decay of $\tilde{\chi}_1^0$ proceeds via virtual $\tilde{f}$ exchanges through chains of the type (for example)

$$\tilde{\chi}_1^0 \rightarrow f\tilde{f}^*, \quad \tilde{f}^* \rightarrow f''\bar{f}', \tag{13}$$

where the transition from the virtual $\tilde{f}^*$ to final SM fermions is induced by the R-parity breaking Lagrangian. Clearly, the SM particles that emerge from $\tilde{\chi}_1^0$ decay depend upon which one of the R-parity breaking Yukawas we take to be non-zero, and these, in turn, determine the appropriate signatures for gluino pair production.

3 Gluino Production Signatures

We proceed to analyze the consequences of the three different types of R-parity breaking Yukawa couplings for gluino discovery. We discuss the different cases in order of complexity.

Yukawa case (9)

The simplest case to analyze is that corresponding to a Yukawa of the form (9). The $\tilde{\chi}_1^0$ decays are

$$\tilde{\chi}_1^0 \rightarrow udd, \quad \tilde{\chi}_1^0 \rightarrow \bar{u}\bar{d}\bar{d}. \tag{14}$$

Thus, the experimental signature of $\tilde{\chi}_1^0$ decay would consist of three jets and no missing energy. If $2nd$ and $3rd$ family analogues of (9) are present then the decays of the $2nd$ and $3rd$ family quarks would yield additional soft jets and/or associated soft leptons. In order to simplify the discussion and language, we adopt the specific 1st family case. It is then apparent that the dilepton signature of eq. (3) for gluino pair production discussed in the introduction is essentially unaltered, except that the missing energy component is reduced in comparison to the case of a stable $\tilde{\chi}_1^0$, being largely replaced by additional soft jets. In other words we still look for final states that are generated by gluino decays that do not directly produce $\tilde{\chi}_1^0$, but rather produce a more massive neutralino or chargino which, in turn, decays to a lighter neutralino or chargino plus a charged lepton and a neutrino, as in the example of eq. (2). For more details see the companion paper, ref. [8], and ref. [5]. The effective branching ratio for such final states and our ability to isolate the gluino events from backgrounds would not be affected by the $\tilde{\chi}_1^0$ decay being visible instead of invisible. Of course, for this like-sign dilepton signature the only evidence of R-parity breaking would be in finding less E_T^{miss} than anticipated for a stable $\tilde{\chi}_1^0$. However, the visibility of the $\tilde{\chi}_1^0$ decays makes it possible to consider other final state signatures that might allow us to confirm a signal found in the dilepton mode, and verify that R-parity is indeed broken.

An example of such a signature is that obtained when one of the produced gluinos decays directly to $\tilde{\chi}_1^0$ as in eq. (1), while the other gluino decays indirectly (as for the dilepton signature) to a channel containing a single lepton ($l = e^{\pm}$ or $l = \mu^{\pm}$), a single neutrino, two or more jets, and the $\tilde{\chi}_1^0$. A typical branching ratio for this latter decay for any one type of lepton is $\sim 3.6\%$, yielding a combined branching ratio for all four cases of $\sim 14.5\%$. If the maximal 14% branching ratio for the direct decay (1) is used, we obtain an effective branching ratio for

$$\tilde{g}\tilde{g} \rightarrow l^{\pm} + 7 \; jets + E_T^{miss} \tag{15}$$

of $2 \times (.145) \times (.14)$ or $\sim 4\%$. (The factor of 2 corresponds to the fact that either of the two gluinos can be chosen as the directly decaying one.) In the above, we have presumed that the 3 jets coming from the decay of the directly produced $\tilde{\chi}_1^0$ would be sufficiently energetic to be isolated, whereas those of the indirectly produced $\tilde{\chi}_1^0$ would be too soft to distinguish from the mini-jets present in all events. This then implies a total of 5 jets observable from the gluino decaying directly to the $\tilde{\chi}_1^0$, and 2 jets observable from the indirectly decaying gluino. The amount of E_T^{miss} in the final state of eq. (15) would obviously be smaller on average than that for the dilepton final state of eq. (3). The primary background is probably that from $W +$ jets production. One could reasonably expect to be able to isolate the signal from such backgrounds using the total mass of the final state and the isolation of the single lepton from the energetic jets. However, detailed Monte Carlo work would be necessary to be certain.

Yukawa case (8)

The $\tilde{\chi}_1^0$ decay channels and their relative strength depend upon several factors. First, there is the question of whether the couplings of this type are equally strong for all three families. We shall presume that this is the case as an illustrative example. Second, there is the question of the exact composition of the $\tilde{\chi}_1^0$ and the relative magnitude of the various squark, slepton and sneutrino masses. These two factors determine the relative weighting between decays mediated via $\tilde{d}$, $\tilde{u}$, $\tilde{e}$ and $\tilde{\nu}_c$. The possible decay sequences are (see eq. (11))

$$
\begin{aligned}
\tilde{\chi}_1^0 &\rightarrow \nu_e(\tilde{\nu}_e^* \rightarrow d\bar{d}) \\
\tilde{\chi}_1^0 &\rightarrow e^-(\tilde{e}_L^* \rightarrow u\bar{d}) \\
\tilde{\chi}_1^0 &\rightarrow u(\tilde{u}_L^* \rightarrow e^-\bar{d}) \\
\tilde{\chi}_1^0 &\rightarrow d(\tilde{d}_L^* \rightarrow \nu_e\bar{d}) \\
\tilde{\chi}_1^0 &\rightarrow \bar{d}(\tilde{d}_R \rightarrow \nu_e d, e^-u),
\end{aligned}
\tag{16}
$$

and their charge conjugates. (Here the stars on the squark, sneutrino and slepton fields indicate the charge conjugate fields of the unstarred fields.) Those mediated by $\tilde{d}$ give (generally different numbers of) ν_e and e final states, those mediated by $\tilde{u}$ or $\tilde{e}$ give only e final states, while those mediated by $\tilde{\nu}_e$ give only ν_e final states. Of course, regardless of the relative weighting of these different virtual exchange diagrams, e^+ and e^- are produced with equal probability. For simplicity, we assume that, for the first family, e^+, e^-, ν_e and $\bar{\nu}_e$ are all equally probable, with analogous statements for the 2nd and 3rd family leptons. For these simplistic assumptions, the decays

$$
\tilde{\chi}_1^0 \rightarrow l + 2 \ jets,
\tag{17}
$$

for $l = e^+, e^-, \mu^+, \mu^-$ each has a probability of 1/12.

What is the most appropriate signature for such a scenario? Since the $\tilde{\chi}_1^0$ decay yields a single charged lepton a significant fraction of the time, it is desirable to use this as a means of distinguishing gluino pair production from other processes. Several different decay sequences for the two gluinos allow this. First there is the case where both gluinos decay directly to the $\tilde{\chi}_1^0$ as in eq. (1). However, as we have discussed, the double direct-LSP decay mode has a branching ratio of at most 2% at high $m_{\tilde{g}}$. Thus, if we require (say) e^+e^+ via two such decays the effective branching ratio is only $0.02(1/12)(1/12)$, with 8 such like-sign dilepton modes (for $l = e$ or μ) yielding a net branching ratio for like-sign dilepton final states via double direct-LSP decay of only 1.1×10^{-3}. Thus we must allow at least one of the produced gluinos to decay indirectly.

Let us consider the case where one gluino decays indirectly to a final state containing a single lepton and neutrino plus jets, while the other gluino decays via the direct-LSP mode. The directly decaying gluino produces a final state of 4 jets and a

e^+ with probability $(0.14) \times (1/12) \sim 0.012$. The indirectly decaying gluino produces a final state containing two energetic jets, an energetic e^+, and the $\tilde{\chi}_1^0$ with probability ~ 0.036. (This is the number which yields the net $0.01 = 8(0.036)^2$, 1% branching ratio for the originally discussed dilepton mode of eq. (3). We remind the reader that the factor of 8 here and below corresponds to the 8 like-sign dilepton possibilities for $l = e$ or μ.) Ignoring the decay products of this second $\tilde{\chi}_1^0$, which would be much softer than those final state objects already enumerated, we would have a net probability for

$$\tilde{g}\tilde{g} \to l^+ l'^+ (\text{or } l^- l'^-) + 6 \text{ jets} + E_T^{miss} + \text{soft leptons and jets}, \qquad (18)$$

with l, $l' = e$, μ, of $2 \times (8) \times (0.036) \times (0.012) \sim 0.0069$, or, roughly, 0.7%.

In addition, we still have the original dilepton signal with effective branching ratio of 1% coming from the case in which both gluinos decay indirectly yielding the final state of eq. (3) (with the replacement of some of the E_T^{miss} by soft jets and leptons from the two $\tilde{\chi}_1^0$ decays). Thus, it seems safe to conclude that the $\tilde{\chi}_1^0$ decay only serves to improve the already significant discovery potential deriving from final states containing two energetic like-sign dileptons and four or more energetic jets. The evidence for R-parity breaking would derive from the increased number of events, and, possibly, the somewhat more energetic nature of the primary leptons in those events where one gluino has decayed directly to the $\tilde{\chi}_1^0$ or the presence of additional soft leptons in those events where both gluinos decay indirectly.

Yukawa case (7)

Finally, let us turn to the most complex of the three cases, corresponding to the Yukawa coupling of eq. (7). As stated earlier we will presume that Yukawa couplings analogous to that of eq. (7) involving the $3rd$ family are absent, and we consider only the explicit coupling given. Here, we obviously have the potential of a plethora of energetic leptons in the final state. Once again, the exact weighting of the different types of final states in $\tilde{\chi}_1^0$ decay that can result from the couplings of eq. (10) depends upon the composition of the $\tilde{\chi}_1^0$, as it influences the relative coupling to different virtual sleptons and sneutrinos, and upon the relative size of the masses of these latter particles. The possible final decay channels for the $\tilde{\chi}_1^0$ are:

$$\begin{array}{cc} e^+ e^- \nu_\mu & e^- e^+ \bar{\nu}_\mu \\ e^+ \mu^- \nu_e & e^- \mu^+ \bar{\nu}_e \end{array} \qquad (19)$$

with a channel containing $\mu^- \mu^+$ being notably absent. In order to obtain a representative idea of the importance of different signatures, we will imagine that the relative weightings of these final states are such that $e^+ e^-$, $e^+ \mu^-$, and $e^- \mu^+$ lepton pairs emerge with equal probability of $1/3$ in the $\tilde{\chi}_1^0$ decays.

As in the previous case, we can consider three options for the decays of the two gluinos produced in a hadron-hadron collision: a) both decay directly to the $\tilde{\chi}_1^0$ plus two jets, as in eq. (1); b) one decays directly, and the other decays indirectly to a state containing an energetic lepton, a neutrino, and at least two energetic jets in addition to the $\tilde{\chi}_1^0$, as in eq. (2); and c) both decay indirectly to channels of the latter type and we imagine triggering on events where the two primary energetic leptons have the same sign.

As always, the net branching ratio for (a) is at most $\sim 2\%$ before accounting for the $\tilde{\chi}_1^0$ decay branching ratios to any given channel. However, the final state, which consists of 4 energetic jets, 2 pairs of oppositely charged leptons, and some missing energy, would have a large reconstructed total mass and large lepton pair invariant masses and would be sufficiently dramatic that relatively few events of this type would be sufficient to signal the gluino pair production process. For instance, events with $e^+e^+\mu^-\mu^-$ or $e^-e^-\mu^+\mu^+$ (these choices clearly eliminate random QED backgrounds) would have a net effective branching ratio (under our simplistic equal weighting assumption) of $0.02 \times (2) \times (1/3)^2$ or about 0.5%. Referring to the companion contribution, ref. [8], and adjusting the event numbers of Table 1 for the 50% smaller branching ratio of the present case, we see that for a gluino mass of 2 TeV there would be 12 such events at the SSC and 1900 events of this type at the ELOISATRON, in one year's running at standard $10^{33} cm^{-2} sec^{-1}$ luminosity. The only conceivable standard model background would derive from $t\bar{t}$ production, in which

$$t \to e^+\nu_e b(\to c\mu^-\bar{\nu}_\mu), \quad \bar{t} \to \mu^-\bar{\nu}_\mu\bar{b}(\to \bar{c}e^+\nu_e), \tag{20}$$

and the $\mu \leftrightarrow e$ flip thereof. (We presume that we may ignore leptons emerging from the tertiary c decay as being much softer and less isolated than those emerging directly from the primary t decay, or from the secondary b decay.) Using branching ratios of 1/9 for the virtual decay $\wp \to e^+\nu_e$ and 1/8 for $b \to c\mu^-\bar{\nu}_\mu$, etc., the net branching for the $t\bar{t}$ final state to appear in either of the two channels is $2 \times (1/9)^2 \times (1/8)^2$ or $\sim 0.04\%$. Based on the studies of ref. [5], such an effective branching ratio, combined with minimal requirements on the lepton and jet energies and on the total reconstructed mass of the event would be sufficient to eliminate the $t\bar{t}$ background, while allowing near 100% acceptance of the signal events at large $m_{\tilde{g}}$. Of course, one could also imagine using all of the two-charged-lepton-pair modes, yielding a factor of 4 higher signal event rate, and, if necessary, using $\mu^-\mu^-\mu^+\mu^+$ events, which do not arise from the signal, to normalize any residue from the $t\bar{t}$ background.

Turning next to possibility (b), we can eliminate random QED backgrounds by focusing on events containing $\mu^+e^-e^-$, $e^+\mu^-\mu^-$, $\mu^-e^+e^+$ or $e^-\mu^+\mu^+$. In addition, the final state will have 4 or more energetic jets, some missing energy, and softer charged leptons coming from the decay of the indirectly produced $\tilde{\chi}_1^0$. Each of the above four final states has an effective branching ratio of $2 \times (0.036) \times (0.14) \times (1/3)$ (assuming the maximal 14% for the direct-LSP decay of the one gluino), for a combined effective branching ratio of $\sim 1.4\%$. This gives us three times the number of signal events quoted for the previous case (a). Again, $t\bar{t}$ production produces the only conceivable background. For instance, the $\mu^+e^-e^-$ final state can be obtained via $t\bar{t}$ production followed by

$$t \to q\bar{q}b(\to e^-c\bar{\nu}_e), \quad \bar{t} \to e^-\bar{\nu}_e\bar{b}(\to \mu^+\nu_\mu\bar{c}), \tag{21}$$

or

$$t \to \mu^+\nu_\mu b(\to e^-c\bar{\nu}_e), \quad \bar{t} \to e^-\bar{\nu}_e\bar{b}(\to jets). \tag{22}$$

The combined branching ratio for these latter two chains is $\sim 0.2\%$, a probably adequate suppression when combined with cuts on other kinematic variables.

Finally, just as for the previous two Yukawa coupling cases, the double indirect decay is essentially unchanged from the analysis of ref. [5] as summarized in ref. [8]. After triggering on the energetic like-sign dileptons, the soft charged lepton pairs emerging from the soft $\tilde{\chi}_1^0$ decays might be visible and provide evidence of R-parity breaking. But direct observation of the previously discussed modes would be the best means of knowing that R-parity was broken.

4 Conclusions

In this report we have outlined the consequences of R-parity breaking for the detection of heavy gluinos. (Corresponding results for light gluinos as well as further details for heavy gluinos will be reported elsewhere [9]. We have considered three representative possibilities for the superpotential Yukawa coupling responsible for R-parity breaking and discussed the phenomenological consequences of each for gluino pair production at a hadron collider. In general, we have found that like-sign dilepton or multi-lepton signatures are the key to detecting gluino pair production and that a detailed survey of a variety of possible final states will enable one to distinguish between the case where R-parity is broken and that where it is not. The fact that each of the various classes of final state has a small effective branching ratio (typically $\lesssim 1\%$), combined with the fact that for heavy gluinos the Standard Model backgrounds are small, implies that our ability to discover and study gluinos is event rate limited. These abilities would clearly be enhanced by the high event rates predicted at the energy of the ELOISATRON.

Acknowledgements

We would like to acknowledge the warm hospitality and support of the Ettore Majorana Center. Work supported, in part, by the Department of Energy.

References

[1] F. Pauss, R. Batley, M.A. Marquina, and A. Nandi, CERN-88-02 (1988).

[2] M. Barnett, J.F. Gunion, and H.E. Haber, Phys. Rev. **D37** (1988) 1892.

[3] H. Baer et al., Proceedings of the 1987 Berkeley Workshop on "Experiments, Detectors and Experimental Areas for the Supercollider", edited by R. Donaldson and M. Gilchriese.

[4] F. Paige, et al., Proceedings of the 1988 Snowmass Workshop.

[5] M. Barnett, J.F. Gunion, and H.E. Haber, preprint UCD-88-30, to appear in Proceedings of the 1988 Snowmass Workshop, and in preparation.

[6] L. Hall and M. Suzuki, Nucl. Phys. **B231** (1984) 419;
M.C. Bento, L. Hall, and G.G. Ross, Nucl. Phys. **B292** (1987) 400;
S. Dimopoulos and L. Hall, preprint LBL-24738 (1988).

[7] See J.F Gunion and H.E. Haber, Nucl. Phys. **B307** (1988) 445, for example.

[8] J.F. Gunion, "A Test Case ... ', preprint UCD-88-31, these Proceedings.

[9] P. Binétruy and J.F. Gunion, in preparation.

29

TESTING THE $WW\gamma$ COUPLING AT FUTURE MULTI-TeV pp-COLLIDERS

U. Baur

CERN, Geneva, Switzerland

ABSTRACT

The structure of the $WW\gamma$ vertex can be measured in $W^{\pm}\gamma$ production at future multi-TeV pp-colliders. Deviations from the gauge theory structure can be parametrized in terms of 4 form factors whose high-energy behavior is severly limited by unitarity. Signatures of these anomalous couplings are described and it is found that multi-TeV hadron colliders can improve the measurement of the $WW\gamma$ vertex by up to two orders of magnitude beyond what is expected from $e^{+}e^{-} \to W^{+}W^{-}$ at LEP200.

1. INTRODUCTION

One of the main tasks of future colliders will be to test the self interactions of the electroweak vector bosons. Within the standard model of electroweak interactions (S.M.) the vector boson self interactions are uniquely given by the gauge theory structure of the S.M. So far this crucial part of the model has not been tested experimentally and it is conceivable that anomalous contributions to the $WW\gamma$ and WWZ vertices exist, $e.g.$ due to novel interactions affecting the weak boson sector.

Such deviations from the gauge theory prediction, if large enough, will produce observable signals at future accelerators. A suitable process to study $WW\gamma$ and WWZ couplings at LEP200 is W pair production [1]. At the tree level $e^{+}e^{-} \to W^{+}W^{-}$ proceeds via photon-, Z- and neutrino exchange (see Fig. 1). The interference between these graphs and in particular the large gauge cancellations between them make this reaction sensitive to deviations of the $WW\gamma$ and WWZ vertices from the S.M. However, since the center of mass energy at LEP200 ($\sqrt{s} \approx 190$ GeV) is only slightly above the W pair threshold, threshold effects dominate and limit the sensitivity to anomalous couplings. In particular, CP violating terms turn out to be only marginally observable at LEP200. Furthermore, large cancellations among different anomalous contributions to the helicity amplitudes may take place, unless longitudinal or transverse beam polarization is available [2]. As a result, experiments at LEP200 can measure anomalous $WW\gamma$ and WWZ couplings only with an accuracy of $\pm 0.1 \ldots \pm 0.7$.

These limits may be improved significantly at future multi-TeV pp-colliders such as the LHC ($\sqrt{s} = 17$ TeV), the SSC ($\sqrt{s} = 40$ TeV) or the ELOISATRON ($\sqrt{s} = 100 - 200$ TeV), due to the large center of mass energy available at these machines. Suitable processes for testing the three vector boson vertices at hadron colliders are $W^{\pm}\gamma$ and $W^{\pm}Z$ production. At tree level these reactions proceed via $q\bar{q}$ annihilation with t-channel quark and s-channel W exchange (see Fig. 2). $W\gamma$ production is therefore only sensitive to anomalous $WW\gamma$, and WZ production only to anomalous WWZ couplings. Here we concentrate on $W^{\pm}\gamma$ production, and study the most general $WW\gamma$ vertex which is accessible in the annihilation process $q\bar{q} \to W\gamma$ of effectively massless quarks (Section 2). Anomalous three vector boson couplings result in contributions to the helicity amplitudes which grow with energy and, eventually, violate unitarity if constant couplings are assumed. This means that any anomalous $WW\gamma$ coupling has to be introduced as a form factor which vanishes at high energies. As we shall discuss in Section 2, partial wave unitarity applied to $f\bar{f} \to W\gamma$ and WW results in upper bounds for the low energy values of the form factors [3]. Since multi-TeV hadron colliders cover a wide range of center of mass energies, this form factor behavior should not be neglected in a realistic study.

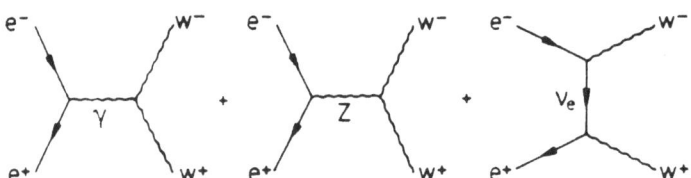

Fig. 1. Feynman diagrams for the process $e^+e^- \to W^+W^-$.

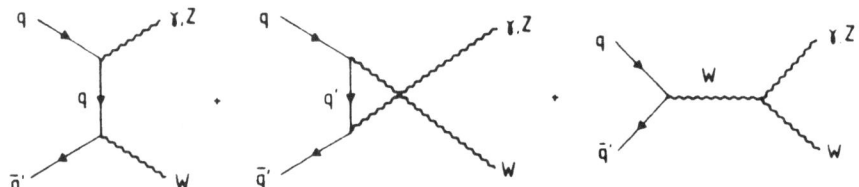

Fig. 2. Feynman graphs for the subprocesses $q\bar{q} \to W\gamma$ and $q\bar{q} \to WZ$.

Our analysis [4] of the $W\gamma$ production process at hadron colliders is based on the calculation of helicity amplitudes for the complete process

$$q\bar{q} \to W^{\pm}\gamma , \qquad W^{\pm} \to \ell^{\pm}\nu \qquad (1.1)$$

with $\ell = e, \mu$. In order to obtain realistic estimates for the sensitivity to anomalous couplings expected at multi-TeV pp-colliders we have imposed quite stringent cuts. These cuts are described in Section 3 where we also briefly discuss the background from W jet production, with the jet misidentified as a photon.

Already at present, severe constraints on anomalous $WW\gamma$ couplings exist, either derived from low energy experiments [5-10], or from unitarity considerations [3,6]. In order to avoid confusion we shall first discuss the expected signals in $pp \to W^{\pm}\gamma X$, without taking into account any low energy bounds on anomalous couplings (Section 4). However, tree level unitarity bounds will always be respected. Only afterwards shall we compare the sensitivity, expected from multi-TeV pp-colliders, with the low energy bounds (Section 5).

2. THREE-VECTOR-BOSON COUPLINGS AND UNITARITY CONSTRAINTS

The most general $WW\gamma$ vertex, which can contribute to $W\gamma$ production in the annihilation of effectively massless fermions, can conveniently be parametrized by the following effective Lagrangian [1,11]

$$
\mathcal{L}_{WW\gamma} = -ie\left\{ \left(W_{\mu\nu}^\dagger W^\mu A^\nu - W_\mu^\dagger A_\nu W^{\mu\nu} \right) + \kappa W_\mu^\dagger W_\nu F^{\mu\nu} + \frac{\lambda}{M_W^2} W_{\lambda\mu}^\dagger W_\nu^\mu F^{\nu\lambda} \right.
$$
$$
\left. + \tilde{\kappa} W_\mu^\dagger W_\nu \tilde{F}^{\mu\nu} + \frac{\tilde{\lambda}}{M_W^2} W_{\lambda\mu}^\dagger W_\nu^\mu \tilde{F}^{\nu\lambda} \right\} .
\tag{2.1}
$$

Here A^μ and W^μ are the photon and W^- fields, respectively, $W_{\mu\nu} = \partial_\mu W_\nu - \partial_\nu W_\mu$, $F_{\mu\nu} = \partial_\mu A_\nu - \partial_\nu A_\mu$, and $\tilde{F}_{\mu\nu} = \frac{1}{2}\epsilon_{\mu\nu\rho\sigma} F^{\rho\sigma}$. e is the charge of the proton.

The first term in Eq. (2.1) arises from minimal coupling of the photon to the $W^\pm$ fields and is completely fixed by the charge of the W boson for onshell photons. κ and λ do not violate any discrete symmetries while $\tilde{\kappa}$ and $\tilde{\lambda}$ are P odd and CP violating. Within the S.M., at tree level,

$$
\kappa = 1 , \quad \lambda = 0 ,
$$
$$
\tilde{\kappa} = 0 , \quad \tilde{\lambda} = 0 .
\tag{2.2}
$$

The $\kappa(\tilde{\kappa})$ and $\lambda(\tilde{\lambda})$ terms are related to the magnetic (electric) dipole moment $\mu_W(d_W)$ and the electric (magnetic) quadrupole moment $Q_W(\tilde{Q}_W)$ of the W^+

$$
\mu_W = \frac{e}{2M_W} \left(1 + \kappa + \lambda \right) ,
\tag{2.3a}
$$

$$
Q_W = -\frac{e}{M_W^2} (\kappa - \lambda) ,
\tag{2.3b}
$$

$$
d_W = \frac{e}{2M_W} (\tilde{\kappa} + \tilde{\lambda}) ,
\tag{2.3c}
$$

$$
\tilde{Q}_W = -\frac{e}{M_W^2} (\tilde{\kappa} - \tilde{\lambda}) .
\tag{2.3d}
$$

It is well known that tree level unitarity, e.g. for the process $e^+e^- \to W^+W^-$, uniquely restricts the $WW\gamma$ couplings to their (S.M.) gauge theory values at asymptotically high energies [12]. This implies that any deviation $a = \kappa - 1, \ldots, \tilde{\lambda}$ from the S.M. expectation has to be described by a form factor $a(\hat{s}, q_W^2, q_\gamma^2)$ which vanishes when either $\hat{s}$, q_W^2 or q_γ^2 becomes large. For deviations of the three vector boson couplings from the gauge theory value, which are produced by some novel interactions operative at a scale Λ, one should expect that the form factors stay essentially constant for center of mass energies $\sqrt{\hat{s}} < \Lambda$ and start decreasing only when the scale Λ is reached or surpassed, very much like the well-known nucleon form factors. The precise $\hat{s}$-dependence is unknown, of course. For our subsequent analysis of $W\gamma$ production we use form factors

$$
a(\hat{s}, q_W^2 = M_W^2, q_\gamma^2 = 0) = \frac{a_0}{(1 + \frac{\hat{s}}{\Lambda^2})^n} ,
\tag{2.4}
$$

usually with $n = 2$. We will comment later to what extent our results depend on this choice.

The low energy values a_0 of these form factors are constrained by partial wave unitarity applied to fermion-pair annihilation processes into W^+W^- and $W^\pm\gamma$ [3]. For $\Lambda \gg M_W$ and $M_W = 81$ GeV, $\sin^2\theta_W = 0.23$ and $\alpha = \alpha(M_W) = 1/128$ the following upper limits are obtained from unitarity:

$$|\kappa_0 - 1| \leq \frac{n^n}{(n-1)^{n-1}} \frac{1.86 \text{ TeV}^2}{\Lambda^2} \, ,$$

$$|\tilde{\kappa}_0| \leq \frac{(2n)^n}{(2n-1)^{n-1/2}} \frac{11.5 \text{ TeV}}{\Lambda} \, , \quad (2.5)$$

$$|\lambda_0|, |\tilde{\lambda}_0| \leq \frac{n^n}{(n-1)^{n-1}} \frac{0.99 \text{ TeV}^2}{\Lambda^2} \, .$$

From Eq. (2.5) we see that $n \geq 1$ for κ, λ and $\tilde{\lambda}$, and $n \geq \frac{1}{2}$ for $\tilde{\kappa}$ is sufficient to guarantee that unitarity is not violated. When the bounds in Eq. (2.5) are observed tree level unitarity is satisfied throughout the entire $\hat{s}$ range, provided one anomalous coupling is nonvanishing at a time. For the more likely case that several anomalous couplings contribute, the bounds differ slightly and can be deduced from the results of Ref. 3.

3. $W\gamma$ PRODUCTION: SIGNAL AND BACKGROUND

The effective Lagrangian (2.1) may now be used to obtain cross-section formulas for $W\gamma$ production from $q\bar{q}$ annihilation, including the effect of any possible deviation of the $WW\gamma$ vertex from the S.M. prediction. For our later phenomenological discussion it is convenient to have available a rough approximation for the amplitudes and to display the contributions of anomalous couplings to the $q\bar{q} \rightarrow W\gamma$ helicity amplitudes in the high energy limit ($\hat{s} \gg M_W^2$). Neglecting the masses of the incoming quarks, the helicity of the quark (antiquark) is fixed to be $-\frac{1}{2}$ $(+\frac{1}{2})$ by the $V-A$ structure of the $Wq\bar{q}$ coupling. This means that the anomalous contributions to the $W\gamma$ production amplitudes only depend on the W and photon helicities λ_W and λ_γ. Denoting these contributions by $\Delta\mathcal{M}_{\lambda_W \lambda_\gamma}$ one finds

$$\Delta\mathcal{M}_{0\pm} = \frac{e^2}{\sin\theta_W} \frac{\sqrt{\hat{s}}}{2M_W} \left[\kappa - 1 + \lambda \mp i(\tilde{\kappa} + \tilde{\lambda}) \right] \tfrac{1}{2}(1 \mp \cos\Theta)$$

$$\Delta\mathcal{M}_{\pm\pm} = \frac{e^2}{\sin\theta_W} \frac{\hat{s}}{2M_W^2} \left[\lambda \mp i\tilde{\lambda} \right] \frac{1}{\sqrt{2}} \sin\Theta \, , \quad (3.1)$$

where Θ denotes the scattering angle of the photon with respect to the quark direction and $\sqrt{\hat{s}}$ is the invariant mass of the $W\gamma$ pair.

The helicity combinations in Eq. (3.1) are the only ones which get affected by anomalous couplings. The fact that only the above four helicity combinations of the $J = 1$ partial wave can be reached by s-channel W exchange explains why four free parameters suffice to parametrize the effects of the most general $WW\gamma$ vertex in $W\gamma$ production.

While the S.M. contribution to the $q\bar{q} \rightarrow W\gamma$ scattering amplitudes is bounded from above for fixed scattering angle Θ, the anomalous contributions (3.1) rise without limit as $\hat{s}$ increases, eventually violating unitarity. This is the reason, of course, why the anomalous couplings have to show a form factor behaviour at high energies. On the other hand, when the amplitudes (3.1) approach the unitarity limit, they will completely dominate the much smaller S.M. contribution to the cross-section. Hence, at large $W\gamma$ invariant masses, Eq. (3.1) suffices to understand differential distributions of the photon and the W decay products.

The signal we are investigating consists of an isolated, high p_T photon and an onshell W which may decay either leptonically or hadronically. It is obvious, that the hadronic decay modes of the W will be swamped by the QCD prompt photon background [14] very much like W-pair production with one W decaying hadronically is essentially unobservable due to the W jet jet background [15]. By considering leptonic decay modes of the W only, roughly 20% of all W decays are still observable (we neglect the τ decay mode of the W in the following). We include the leptonic branching fraction

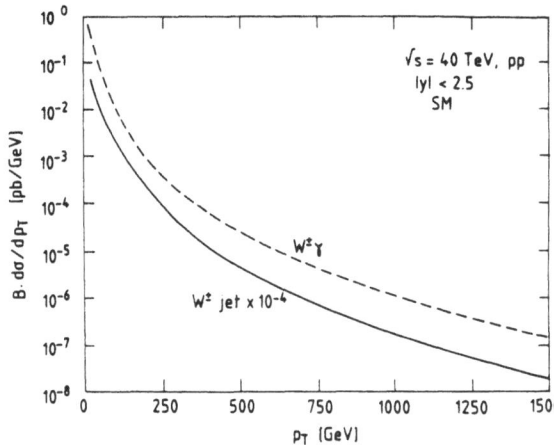

Fig 3. Photon p_T spectrum for the signal $pp \rightarrow W^{\pm}\gamma X$ (dashed line) and the background $pp \rightarrow W^{\pm}$ jet X (solid line). The background is multiplied by a factor 10^{-4} to represent the misidentification probability of a jet as a photon. Rapidity cuts $|y| < 2.5$ are imposed on the W, the photon and the jet.

$$B = \mathrm{Br}\,(W \rightarrow e\nu,\ \mu\nu) = 0.186 \qquad (3.2)$$

(corresponding to a top-quark mass of 50 GeV) in all subsequent figures.

The experimental signal thus consists of an isolated, high p_T, electromagnetic cluster and an isolated charged lepton in association with missing p_T. We expect isolation cuts to effectively remove any background from heavy flavor production, e.g. $gg \rightarrow t\bar{t}\gamma$ with one top decaying semileptonically. On the other hand W jet production with the jet misidentified as a photon may cause a much more serious background. Such misidentification may arise due to fluctuations in the calorimetry or because the jet hadronizes, with a leading π^0 carrying away most of its energy. At SSC energies the probability for either fluctuations was estimated to be below 10^{-4} for jets of $p_T > 200$ GeV and $2\text{-}5 \times 10^{-4}$ for $p_T > 100$ GeV from calorimetric information alone [16]. Since the thus misidentified photon will generally have a smaller p_T than its parent jet, and since no tracking information was included in the analysis of Ref. 16, a misidentification probability below 10^{-4} for $p_{T_\gamma} > 100$ GeV appears to be a reasonable estimate. No calculation of the photon jet misidentification probability exists so far for LHC or ELOISATRON energies. Generally one would expect, however, that, for a fixed jet p_T cut, the misidentification probability becomes smaller (larger) for energies below (above) the SSC value of $\sqrt{s} = 40$ TeV.

In Fig. 3 we compare the p_T distribution of the signal ($pp \rightarrow W^+\gamma X,\ W^-\gamma X$) with the W jet background at $\sqrt{s} = 40$ TeV in the S.M. Rapidity cuts $|y| < 2.5$ are imposed on W, photon, and jets in order to roughly simulate detector response. Since, in the case of misidentification, a jet of given p_T will always be viewed as a photon of lower p_T, a 10^{-4} misidentification probability will produce an even lower background than suggested by the figure. We conclude that a misidentification probability below 10^{-4} is sufficient to remove the W jet background. (In Fig. 3 and all subsequent figures we use the distribution functions of Duke and Owens, set I [17].)

The misidentification problem will become more severe for small p_{T_γ} and it is not clear at present, down to which value the background rejection will be sufficient to obtain a clean signal. Hence we have imposed an, admittedly conservative, cut $p_{T_\gamma} > 100$ GeV in all our simulations. In addition we impose a rapidity cut $|y_\gamma| < 2.5$. The p_T cut removes both the collinear and the infrared singularity of the $q\bar{q} \rightarrow W\gamma$

subprocess. In order to simulate the performance of realistic detectors, for the decay products of the W we require $|y_{e,\mu}| < 3$, $p_{Te,\mu} > 30$ GeV, and $p_{T,\text{miss}} > 30$ GeV. For W identification we further require the transverse mass M_{trans} reconstructed from the charged lepton and $p_{T,\text{miss}}$ to essentially agree with the W mass, $|M_{\text{trans}} - M_W| < 20$ GeV.

It is clear that the cuts that we impose are somewhat arbitrary without simulating a definite detector. Within our cuts and with an expected integrated luminosity of 10^4pb^{-1} after one or two years of running, the LHC (SSC) will provide about 500 (1200) clean $W^\pm\gamma$ events [4] in a $p_{T\gamma}$ range which is particularly sensitive to anomalies in the $WW\gamma$ vertex, as we shall now see. At ELOISATRON energies, the rate would be about a factor 2-4 larger than at the SSC.

4. SIGNALS FOR ANOMALOUS $WW\gamma$ COUPLINGS

The prime advantage of high energy pp colliders is the very high center of mass energy $\sqrt{\hat{s}}$ that is available in parton collisions. In the $q\bar{q} \to W\gamma$ subprocess, the effects of anomalies in the $WW\gamma$ vertex are greatly enhanced at large energies, due to the $\hat{\gamma} = \sqrt{\hat{s}}/2M_W$ factors in the anomalous contributions to the amplitudes (3.1). Compared to $W\gamma$ production at the Tevatron [18,19] or W^+W^- production at LEP200 [1,2] this enhancement factor will particularly favor the observability of anomalous values of λ or $\tilde{\lambda}$, which are enhanced by $\hat{\gamma}^2$ in the amplitude whereas terms containing κ and $\tilde{\kappa}$ grow only linearly with $\hat{\gamma}$. As discussed in Section 2, this enhancement is limited by partial wave unitarity. As a rule of thumb, anomalous $WW\gamma$ photon couplings cannot enhance the $q\bar{q} \to W\gamma$ cross-section by more than a factor 10^3 ($\approx \sin^2\theta_W/\alpha^2$) over the S.M. cross-section at a fixed center of mass energy $\sqrt{\hat{s}}$, after cuts have been imposed to eliminate partial waves with high angular momentum, $i.e.$ photons collinear to the initial state fermions. Unitarity of the $q\bar{q} \to W\gamma$ scattering amplitude will be insured in the following by using the form factors (2.4), usually with $n = 2$ and $\Lambda = 2$ TeV, for anomalous couplings.

A typical signal for anomalous couplings will be a broad increase in the W-photon invariant mass spectrum at large values of $m_{W\gamma} = \sqrt{\hat{s}}$, possibly cut off above $m_{W\gamma} = \Lambda$, the scale of the form factor. In models of composite weak bosons [20] one would expect additional resonance structure around $m_{W\gamma} \approx \Lambda$ due to the production and subsequent radiative decay of excited W's [21]. In the following we shall not consider such resonance structure; however, we shall allow for its existence above the form factor enhancement that we study by choosing the low energy values of the anomalous couplings well below their unitarity limits (2.5).

The resulting effect on $B \cdot d\sigma/dm_{W\gamma}$ at SSC energies is shown in Fig. 4 for a representative choice of anomalous couplings: $\kappa_0 = 1.5$, $\lambda_0 = 0.1$, $\tilde{\kappa}_0 = 0.8$, and $\tilde{\lambda}_0 = 0.05$, respectively, in the form factor (2.4) with $\Lambda = 2$ TeV and $n = 2$ ("dipole form factor"). Only one $WW\gamma$ coupling at a time is chosen different from the S.M. prediction. For comparison the S.M. curve is included as a solid line.

At hadron colliders the $W\gamma$ invariant mass cannot unambiguously be determined, of course, due to the nonobservation of the neutrino arising from W-decay. Identifying the transverse momentum of the neutrino with the missing transverse momentum of a given $W\gamma$ event, the unobservable longitudinal neutrino momentum can be reconstructed, albeit with a twofold ambiguity, by imposing the constraint that the neutrino and charged lepton four-momenta combine to form the W rest mass [18,22]. What is shown in Fig. 4 is actually the "reconstructed" $W\gamma$ invariant mass spectrum obtained by using the two solutions for the neutrino longitudinal momentum for each event. Due to the transverse mass cut described in the last section, the reconstruction of $m_{W\gamma}$ is excellent and the "true" and "reconstructed" $W\gamma$ invariant mass distributions are almost identical.

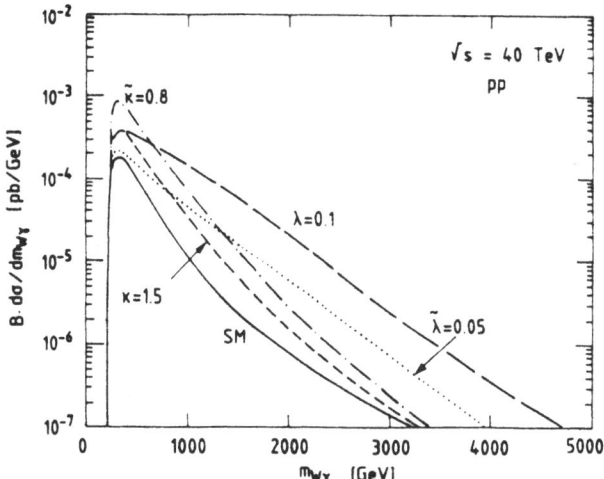

Fig 4 . W photon invariant mass spectrum for the process $pp \to W^+\gamma X$ at the SSC. Cuts are specified in the text. The curves are for the S.M. (solid line), $\kappa = 1.5$ (dashed line), $\lambda = 0.1$ (long-dashed line), $\tilde{\kappa} = 0.8$ (dot-dashed line) and $\tilde{\lambda} = 0.05$ (dotted line). Dipole form factors ($n = 2$ in Eq. (2.4)) with $\Lambda = 2$ TeV are chosen for all anomalous couplings.

In Fig. 4 the anomalous coupling curves are clearly distinguishable from the S.M. prediction, when remembering the expected luminosity of 10^4pb^{-1} year^{-1} at the SSC: a differential cross-section of 10^{-7}pb/GeV corresponds to one event per year per TeV interval.

A pronounced feature of $W\gamma$ production in $q\bar{q}$ annihilation is the S.M. prediction of radiation zeros in all contributing helicity amplitudes at one value of the photon scattering angle and hence in the differential cross-section [23]. For $u\bar{d} \to W^+\gamma$ this radiation zero occurs at $\cos\theta^* = -\frac{1}{3}$, where θ^* is the scattering angle of the photon relative to the u-quark direction, measured in the $W\gamma$ rest frame. Since the $W\gamma$ rest frame can be reconstructed, θ^* could be determined as well, if one knew which of the two incoming protons to associate with the u- and $\bar{d}$-quarks, respectively. In practice, hence, only $|\cos\theta^*|$ is measurable and the radiation zero is considerably washed out.

In order to eliminate the strong peaking of the differential cross-section at $\cos\theta^* = \pm 1$ which arises from the collinear singularity, we study the rapidity distribution $d\sigma/d|y^*|$ of the photon instead, with

$$y^* = \frac{1}{2}\ln\frac{1+\cos\theta^*}{1-\cos\theta^*} . \qquad (4.1)$$

In the $|y^*|$ distribution the radiation zero is signalled by a dip at $|y^*| = 0$, which remains even when taking into account the finite acceptance of detectors, as simulated by our cuts.

In the presence of any anomalous contribution to the $WW\gamma$ vertex the radiation zero will be eliminated. This is obvious from Eq. (3.1): none of the anomalous contributions to the scattering amplitudes vanishes at $\cos\theta^* = -\frac{1}{3}$. As a result the dip in the rapidity spectrum of the photon at $|y^*| = 0$ will be filled at least partially. This effect is shown in Fig. 5 for the same parameters as in Fig. 4.

Figure 5 clearly demonstrates that anomalous couplings mainly affect the region of small center of mass rapidities. This is due to the fact that anomalous couplings only contribute via the s-channel W-exchange graph of Fig. 2, and hence only to the

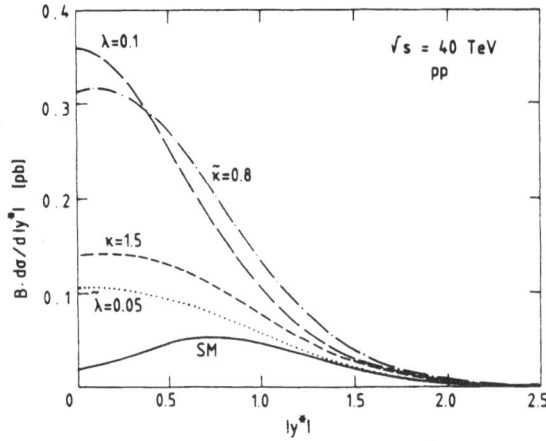

Fig 5 . Rapidity spectrum of the photon in $pp \rightarrow W^+\gamma X$ in the $W\gamma$ rest frame at the SSC. Anomalous couplings, form factors and cuts are the same as in Fig. 4.

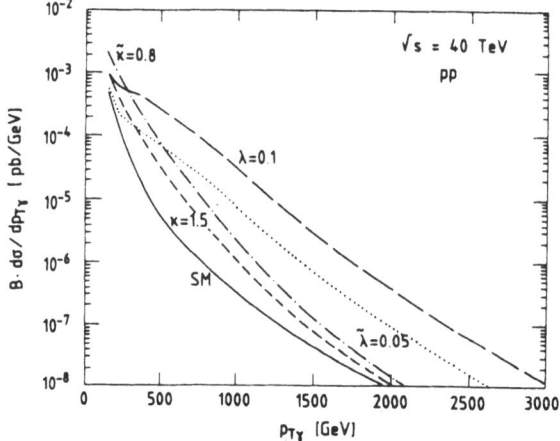

Fig 6 . Transverse momentum spectrum of the photon in $pp \rightarrow W^+\gamma X$ at SSC energies. Parameters and cuts are chosen as in Fig. 4.

$J = 1$ partial wave, when fermion masses are neglected. The anomalous contributions are, therefore, almost isotropic in the center of mass frame, while the u- and t-channel graphs of Fig. 2 result in a strong enhancement of the high rapidity region. As a very fortunate result of this behaviour the finite acceptance cuts will largely eliminate the well known fermion exchange contributions to the cross-section and reduce possible signals of new physics, $i.e.$ the effects of anomalous $WW\gamma$ interactions, by a much smaller amount.

The population of the small rapidity region, induced by anomalous couplings, considerably increases the average photon transverse momentum of events produced at a fixed value of the W photon invariant mass, as compared to the S.M. The p_T distribution of the photon, $d\sigma/dp_{T\gamma}$, should hence be particularly sensitive to anomalous couplings. This fact is clearly visible in Fig. 6 where the $p_{T\gamma}$ spectrum is plotted for SSC energies.

As we have demonstrated so far, the transverse momentum and rapidity distribu-

TABLE 1

Sensitivities achievable at the 99.99% CL for the four anomalous $WW\gamma$ couplings at the SSC ($\sqrt{s} = 40$ TeV, $\int \mathcal{L} \, dt = 10^4 \text{pb}^{-1}$) [4]. Dipole form factors with a cutoff scale Λ are assumed. (See text for details.) The bounds are derived from the $W^+\gamma$ channel only. The value for $\Delta\kappa_0 = \kappa_0 - 1$ at $\Lambda = 10$ TeV is larger than what is allowed by unitarity (see Eq. (2.5)) and, therefore, has been included in parentheses.

Λ	0.5 TeV	1 TeV	2 TeV	5 TeV	10 TeV
$\Delta\kappa_0$	± 0.46	± 0.26	± 0.20	± 0.14	(± 0.13)
λ_0	± 0.095	± 0.039	± 0.014	± 0.0066	± 0.0046
$\bar{\kappa}_0$	± 0.43	± 0.26	± 0.20	± 0.13	± 0.12
$\bar{\lambda}_0$	± 0.094	± 0.039	± 0.013	± 0.0065	± 0.0047

tions of the produced photon, and the $W\gamma$ invariant mass spectrum are very sensitive indicators of anomalous couplings. This statement can be quantified by deriving the minimal values which would give rise to detectable effects at multi-TeV pp-colliders. Table 1 shows the results of such a calculation for the SSC [4]. Since the anomalous couplings are actually form factors, the answer depends to some extent on their functional behaviour. As before a dipole form factor ($n = 2$ in Eq. (2.4)) is chosen, and the form factor dependence is explored by varying the cutoff scale Λ. The following sensitivity criterion is used in Table 1: an anomalous coupling is called detectable if it produces a deviation from the S.M. at the 99.99% confidence level (CL) in either the y^*, the $m_{W\gamma}$ or the $p_{T\gamma}$ distribution calculated with the cuts specified in Section 3. The confidence levels are calculated by splitting the $m_{W\gamma}$ and p_T spectra into 6 or 7 and the y^* distribution into 12 bins with, typically, more than 8 events. In each bin the Poisson statistics is approximated by a Gaussian distribution. In order to achieve a sizable counting rate in each bin, within the S.M., all events with $p_T > 400$ GeV, or $m_{W\gamma} > 1.7$ TeV, are collected into a single bin. This procedure guarantees that in our calculation a high confidence level cannot arise from a single event at very high p_T, where the S.M. predicts, e.g. less than 10^{-2} events. The sensitivity estimates that we derive will thus be conservative. Furthermore we allow for a factor 2 normalization uncertainty of the S.M. prediction.

Results similar to the ones listed in Table 1 can be also derived for the LHC and the ELOISATRON. Typically, at the LHC the minimal observable couplings are a factor 1.5-2 weaker than the values given in Table 1 [4]. At ELOISATRON energies one expects a sensitivity which is about a factor 2-3 better than for the SSC.

The limits presented in Table 1 directly reflect the different powers of $\hat{\gamma} = \sqrt{\hat{s}}/2M_W$ multiplying the various anomalous contributions to the amplitudes (3.1). Since at high energies ($\hat{s} \gg M_W^2$) terms proportional to λ and $\bar{\lambda}$ grow much faster than κ or $\bar{\kappa}$ terms, the sensitivity limits achievable for λ and $\bar{\lambda}$ are far better than the ones for κ and $\bar{\kappa}$.

Table 1 has been derived using the $W^+\gamma$ signal only. The corresponding sensitivity bounds derived from the $W^-\gamma$ signal are up to 20% (40%) weaker for κ and $\bar{\kappa}$ (λ and $\bar{\lambda}$) because of the smaller counting rate in this channel. Since at pp-colliders the $\bar{u}d$ luminosity decreases more rapidly with $m_{W\gamma}$ than the $\bar{d}u$ luminosity, the effect is particularly large at $\Lambda = 5$ and 10 TeV and for λ and $\bar{\lambda}$.

Effectively the sensitivity to anomalous couplings in $pp \rightarrow W^\pm\gamma X$ stems from regions of phase space where the anomalous contributions to the cross-section are considerably larger than the S.M. expectation. As a result interference effects between the S.M. amplitude and the anomalous amplitude contributions (3.1) play a minor role, and an excess in counting rate, beyond the S.M. expectation, essentially scales like the sum of squares of anomalous couplings. This fact has a number of important consequences:

i) The sensitivity bounds of Table 1 depend only marginally on the phases of the various anomalous couplings. Hence the limits are rather insensitive to the sign of the anomalous coupling. Furthermore, the limits for λ and $\tilde{\lambda}$ (κ and $\tilde{\kappa}$) are very similar because terms in the helicity amplitudes containing λ and $\tilde{\lambda}$ (κ and $\tilde{\kappa}$) differ, for $\hat{s} \gg M_W^2$, only by a factor $\pm i$. The marginal dependence of the differential cross-sections on the phases implies of course that it will also be difficult to measure phases of anomalous couplings, once they are found.

ii) Contributions from different anomalous couplings cannot cancel at a significant level. Hence the sensitivities of Table 1, though derived by assuming only one anomalous coupling at a time to be different from the S.M., represent model independent upper bounds on individual anomalous couplings that can be set by experiments.

iii) Increasing statistics by a factor 2 will reduce the statistical error by a factor $\sqrt{2}$ and hence improve the sensitivity to any of the anomalous couplings by, roughly, a factor $\sqrt[4]{2}$, *i.e.* by less than 20%. An even smaller improvement is achieved by combining the $W^+\gamma$ and $W^-\gamma$ spectra. Incidentally, this is comparable to the estimated error of the bounds in Table 1 [4].

The improvement in sensitivity with increasing Λ in Table 1 has a simple explanation. It is due to the additional events at large W photon invariant masses which were cut off by the form factor when the scale Λ had a smaller value. Similarly the less drastic cutoff for $n = 1$ instead of $n = 2$ in the form factor allows for additional high $m_{W\gamma}$ events and therefore leads to an increased sensitivity to the low energy values a_0 of the form factors. For $\Lambda = 1$ TeV, for example, the SSC is sensitive to $\lambda_0 = 0.016$ (at the 99.99% CL) if $n = 1$ and to $\lambda_0 = 0.039$ if $n = 2$, which can be understood by noting that these two values of the low energy couplings give roughly the same value for the form factor in the critical energy region around $\sqrt{\hat{s}} = \Lambda$. It should be stressed that the difference between the $n = 1$ and $n = 2$ case becomes less pronounced for larger values of Λ, because the decrease in parton luminosity in the $\sqrt{\hat{s}} = \Lambda$ region shifts importance to the low energy region where differences in form factors are small. The sensitivities in Table 1 can thus be taken as representative for a much wider class of form factors.

So far we have demonstrated that anomalous $WW\gamma$ couplings produce clearly visible deviations from the S.M. at multi-TeV pp-colliders. Thus the question arises to what extent experiments can distinguish these couplings. Inspection of the helicity amplitudes (3.1) shows that λ and $\tilde{\lambda}$ lead to an excess production of transverse W's while a deviation of κ and $\tilde{\kappa}$ from its S.M. value 1 will enhance longitudinal W production [24]. The effect of these different W polarizations shows clearly in the angular distribution of the charged leptons arising from W decay. Figure 7 shows $B \cdot d\sigma(pp \to W^+\gamma X)/d\cos\bar{\theta}$ at SSC energies for the S.M. and the anomalous couplings and form factors as in Fig. 4. Here $\bar{\theta}$ is the e^+/μ^+ polar angle in the W^+ rest frame with respect to the W^+ direction in the $W\gamma$ rest frame. The cuts imposed on $p_{Te,\mu}$ and $p_{T,\mathrm{miss}}$ (see Section 3) force the cross-section to vanish at $\cos\bar{\theta} = \pm 1$. Due to the $V - A$ structure of the $We\nu$ coupling, transversely polarized W^+'s of helicity $\lambda_W = \pm 1$ produce a $(1 \pm \cos\bar{\theta})^2$ distribution of the charged decay lepton, while longitudinal W's result in a $\sin^2\bar{\theta}$ distribution. In the S.M. the dominant $W^\pm$ helicity is $\lambda_W = \pm 1$ [24], implying that the charged lepton will predominantly be emitted in the direction of the parent W. The S.M. $\cos\bar{\theta}$ distribution in Fig. 7 clearly exhibits the pronounced dominance of left-polarized W^+'s in the standard model, as well as

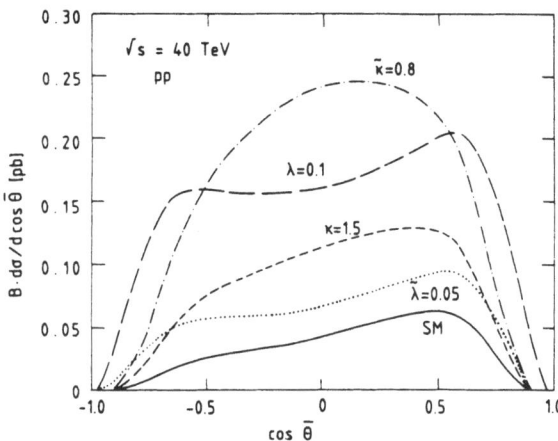

Fig. 7 . Polar angle distribution of the charged decay lepton from W^+ decay in $pp \rightarrow W^+ \gamma X$ at the SSC. See text for definition of $\bar{\theta}$. Cuts and anomalous couplings are the same as in Fig. 4.

of longitudinal W's for large anomalous values of κ [24], and of equal numbers of left- and right-handed W's for anomalous values of λ and $\bar{\lambda}$. The W decay distribution will thus allow a separation of $\kappa(\bar{\kappa})$ from $\lambda(\bar{\lambda})$. A separation of $\kappa(\lambda)$ from $\bar{\kappa}(\bar{\lambda})$ requires the measurement of interference terms between anomalous and S.M. contributions to individual scattering amplitudes or between helicity amplitudes corresponding to the same photon but different W polarizations. Since these interference terms can easily change sign as a function of center of mass energy and/or scattering angle, their measurement will most likely require the investigation of correlations of observables, which may well prove prohibitive in view of the limited statistics of clean $W\gamma$ events, at least for the LHC and the SSC.

5. DISCUSSION

We have described, in the last section, the signatures which anomalies in the $WW\gamma$ vertex will produce in $W\gamma$ production at future hadron colliders, and, for the SSC, we have determined how large deviations from the S.M. have to be in order to yield visible effects. Naturally the question arises how the sensitivity of multi-TeV hadron colliders compares with existing low energy limits on anomalous couplings and with the sensitivity to non-gauge theory terms in the three boson vertex reachable at LEP by investigating the process $e^+e^- \rightarrow W^+W^-$. It must be realized, of course, that these three approaches to determining anomalous couplings measure the form factors corresponding to the couplings in quite different regions of momentum space: timelike momenta in the vector boson production processes and spacelike momenta in the low energy bounds, which are derived by performing loop calculations involving the $WW\gamma$ vertex. In this sense the three approaches are complementary. Since we have parametrized the form factors κ, λ, etc. in terms of the small momentum transfer values κ_0, λ_0, etc. which necessarily are the same in all three approaches, low energy bounds and bounds from $e^+e^- \rightarrow W^+W^-$ must nevertheless be incorporated by the time the hadron collider experiments are performed. Of course these bounds will not rule out the possible existence of resonances in the $W\gamma$ channel, which also can be parametrized via these form factors.

Low energy bounds on κ and λ have been derived by calculating loop contributions to $(g-2)_\mu$ [5-7], the photon propagator as measured at PETRA [7], and the W/Z mass ratio [6,7]. The W/Z mass ratio produces the most stringent, but also the most controversial bounds [6,7]. Since any non-gauge model of the W and the Z

has to explain their small mass (as compared to the compositeness scale Λ) and their mass ratio, probably by some as yet unknown symmetry between the various $WW\gamma$ and WWZ couplings, bounds on individual anomalous couplings derived from the W and Z masses should not be used to predict null results in high energy scattering experiments. The most stringent bound is then obtained from $(g-2)_\mu$ [5-7], namely

$$\left| (\kappa - 1) \ln \frac{\Lambda^2}{M_W^2} + \lambda \right| < 4.4 \ . \tag{5.1}$$

For cutoff scales larger than a few TeV, this bound gets surpassed by the unitarity bounds (2.5). From Table 1 and the discussion in Section 4 it is obvious that experiments at future pp-colliders will considerably improve the present low energy bounds on κ and λ. This statement is not true for the CP violating coupling $\tilde{\kappa}$, which produces a potentially large contribution to the electric dipole moment of the neutron. The present experimental limit on the neutron electric dipole moment limits $|\tilde{\kappa}|$ to be less than $O(10^{-3})$ [9] which clearly excludes any observable effect in $W\gamma$ production, as shown by a comparison with Table 1. Not even the ELOISATRON will be able to improve this number. Curiously no such bound exists for the other CP violating coupling $\tilde{\lambda}$. Contributions of $\tilde{\lambda}$ to the neutron electric dipole moment are suppressed by at least another factor of $(M_N/M_W)^2$ (M_N being the neutron mass) [10], as compared to the $\tilde{\kappa}$ bound, and hence experiments at the LHC, the SSC, or the ELOISATRON will explore a range of three orders of magnitude in $\tilde{\lambda}$.

As we have mentioned already in the introduction, experiments at LEP200, studying W^+W^- production in e^+e^- annihilation at 190 GeV, can measure κ, λ, $\tilde{\kappa}$, and $\tilde{\lambda}$ with an accuracy of $\pm 0.1 \ldots \pm 0.7$, depending on the values of the interfering WWZ couplings [2]. While it will be difficult for future hadron colliders to improve the measurement of κ in W photon production, the determination of λ and $\tilde{\lambda}$ can be improved by up to two orders of magnitude.

In this report, we have focused on anomalous $WW\gamma$ couplings and their signals in $W^{\pm}\gamma$ production at future multi-TeV pp-colliders. Using a similar strategy, signals for anomalous WWZ couplings in $W^{\pm}Z$ production have been analyzed in Ref. 25. The expected sensitivities to WWZ anomalous couplings were found to be somewhat weaker (by at most a factor 2) than the corresponding sensitivities to anomalous $WW\gamma$ couplings in $W\gamma$ production [25]. The larger sensitivity in $W\gamma$ production is partially due to the larger counting rate at high p_T (the limiting factor in WZ production is the small branching fraction of the Z into electron or muon pairs). Partially it arises because the S.M. prediction of radiation zeros produces additional structure in $W\gamma$ differential cross-sections which is not available in the WZ channel.

It is quite instructive to relate the accuracy of the various methods for determining non-gauge theory $WW\gamma$ interactions with the magnitude of the scale of new physics, Λ, which can be probed. Assuming that possible deviations are due to some novel strong interactions in the electroweak boson sector, the anomalous couplings may be expected close to their unitarity limits (2.5). Saturating these unitarity limits and assuming dipole form factors for all the anomalous couplings, the low energy bounds probe scales up to $\Lambda \approx 2$ TeV at most, LEP200 is sensitive up to $\Lambda \approx$ 4-5 TeV while the SSC (LHC) probes scales up to 32 TeV (21 TeV) in $W\gamma$ production [4]. The ELOISATRON would even be sensitive to Λ-values of $O(100 \text{ TeV})$. Thus W photon production at future multi-TeV pp-colliders provides a tool to investigate the properties of the weak boson sector well above the few TeV region where the production of new particles may yield a more direct signature.

ACKNOWLEDGEMENTS

I would like to thank Dieter Zeppenfeld for a very enjoyable collaboration, and Ahmed Ali for the invitation to Erice. I am also grateful to Alan Martin and the members of the High Energy Physics Group at the University of Durham, where this report was completed, for their warm hospitality.

REFERENCES

1. K. Hagiwara *et al.*, <u>Nucl. Phys.</u> B282:253 (1987), and references therein.

2. D. Zeppenfeld, <u>Phys. Lett.</u> 183B:380 (1987).

3. U. Baur and D. Zeppenfeld, <u>Phys. Lett.</u> 201B:383 (1988).

4. U. Baur and D. Zeppenfeld, <u>Nucl. Phys.</u> B308:127 (1988).

5. F. Herzog, <u>Phys. Lett.</u> 148B:355 (1984);
   J. C. Wallet, <u>Phys. Rev.</u> D32:813 (1985);
   A. Grau and J. A. Grifols, <u>Phys. Lett.</u> 154B:283 (1987).

6. M. Suzuki, <u>Phys. Lett.</u> 153B:289 (1985).

7. J. J. van der Bij, <u>Phys. Rev.</u> D35:1088 (1987).

8. J. A. Grifols, S. Peris, and J. Solà, <u>Int. J. Mod. Phys.</u> A3:255 (1988).

9. W. J. Marciano and A. Queijeiro, <u>Phys. Rev.</u> D33:3449 (1986).

10. F. Hoogeveen, preprint MPI-PAE/PTh 25/87.

11. K. Gaemers and G. Gounaris, <u>Z. Phys.</u> C1:259 (1979).

12. J. M. Cornwall, D. N. Levin, and G. Tiktopoulos, <u>Phys. Rev. Lett.</u> 30:1268 (1973); <u>Phys. Rev.</u> D10:1145 (1974);
    C. H. Llewellyn Smith, <u>Phys. Lett.</u> 46B:233 (1973);
    S. D. Joglekar, <u>Ann. Phys.</u> 83:427 (1974).

13. Particle Data Group, M. Aguilar-Benitez *et al.*, <u>Phys. Lett.</u> 204B:1 (1988).

14. F. Halzen and D. M. Scott, <u>Phys. Rev.</u> D18:3378 (1978).

15. W. J. Stirling, R. Kleiss, and S. D. Ellis, <u>Phys. Lett.</u> 163B:261 (1985);
    J. F. Gunion, Z. Kunszt, and M. Soldate, <u>Phys. Lett.</u> 163B:389 (1985).

16. Y. Morita, in "Proceedings of the Summer Study on the Physics of the Superconducting Supercollider", Snowmass, Colorado, 1986, edited by R. Donaldson and J. Marx (Division of Particles and Fields of the APS, New York, 1987), p. 194.

17. D. Duke, and J. Owens, <u>Phys. Rev.</u> D30:49 (1984).

18. J. Cortes, K. Hagiwara, and F. Herzog, <u>Nucl. Phys.</u> B278:26 (1986).

19. J. C. Wallet, <u>Z. Phys.</u> C30:575 (1986).

20. For a review see *e.g.* W. Buchmüller, <u>Acta Phys. Austr. Suppl.</u> 27:517 (1985) and references therein.

21. U. Baur, D. Schildknecht, and K. H. G. Schwarzer, <u>Phys. Rev.</u> D35:297 (1987).

22. J. Stroughair and C. Bilchak, <u>Z. Phys.</u> C26:415 (1984);
    J. Gunion, Z. Kunszt, and M. Soldate, <u>Phys. Lett.</u> 163B:389 (1985);
    J. Gunion and M. Soldate, <u>Phys. Rev.</u> D34:826 (1986);
    W. Stirling *et al.*, <u>Phys. Lett.</u> 163B:261 (1985).

23. Zhu Dongpei, <u>Phys. Rev.</u> D22:2266 (1980);
    C. J. Goebel, F. Halzen, and J. P. Leveille, <u>Phys. Rev.</u> D23:2682 (1981);
    S. J. Brodsky and R. W. Brown, <u>Phys. Rev. Lett.</u> 49:966 (1982);
    R. W. Brown, K. L. Kowalski, and S. J. Brodsky, <u>Phys. Rev.</u> D28:624 (1983);
    M. A. Samuel, <u>Phys. Rev.</u> D27:2724 (1983).

24. C. L. Bilchak, R. W. Brown, and J. D. Stroughair, <u>Phys. Rev.</u> D29:375 (1984).

25. D. Zeppenfeld and S. Willenbrock, <u>Phys. Rev.</u> D37:1775 (1988).

30

ASPECTS OF THE GLUONIC PRODUCTION OF
WEAKLY INTERACTING PARTICLES

J. J. van der Bij

CERN, Geneva, Switzerland

Abstract

In the first part of this paper we study an approximate formula for the production of an arbitrary number of light Higgses at hadron colliders. The formula indicates that the relevant scale for strong interactions in the Higgs sector may be 3 TeV rather than 1 TeV. In the second part we give a simple relation for determining the relative importance of gluon fusion versus quark-antiquark annihilation in the production of vectorboson pairs. This relation is a good approximation away from threshold.

I. Multi Higgs production via gluon fusion

The main 'missing link' in the standard model of electroweak interactions is the lack of direct evidence for the Higgs sector. High energy pp-colliders like the LHC (16 TeV), SSC (40 TeV) or Eloisatron (100 TeV) have as one of their main objectives the search for a Higgs particle. However finding the Higgs is by itself not enough to establish the nature of spontaneous symmetry breaking. In the standard model there are also Higgs self couplings with a strength, that is prescribed by the requirement of renormalisability of the model. In more complicated models different couplings could exist, possibly giving strong interactions in the Higgs sector. Present limits on such couplings, coming from the study of radiative corrections, are very weak [ref.1].

In order to settle this question one has to study the production of more than one Higgs particle. The three-Higgs self-coupling appears first in two-Higgs production, where a virtual Higgs can decay into two real ones. The four-Higgs self-coupling only appears in processes with three or more outgoing Higgses. In hadron collisions there are two main processes contributing to multi-Higgs production, WW and gluon-gluon fusion. We limit ourselves to gluon-gluon fusion. The production of Higgses via gluon-fusion proceeds via an intermediary heavy quark loop. Complete results are only known for the case of two outgoing Higgses [ref.2,3]. The cross sections are very small, unless there is an extra generation of heavy quarks. Some results relevant for a fourth generation are given in fig.(1). In this graph also the result for $m_q \rightarrow \infty$ is given. This limit is qualitatively reasonable for $m_H < m_q$.

Extending these calculations to the case of more than two Higgses requires the evaluation of pentagon, hexagon etc. graphs which is beyond present capabilities. However the large quark mass limit can be extended to multi Higgs production. The large quark mass limit can be calculated by considering the Higgs field as a slowly varying background field. Due to the presence of this background field the fermion *propagator is modified*. The fermion loop contributon to the gluon propagator leads to the following effective interaction between gluons and Higgs particles [ref.4] :

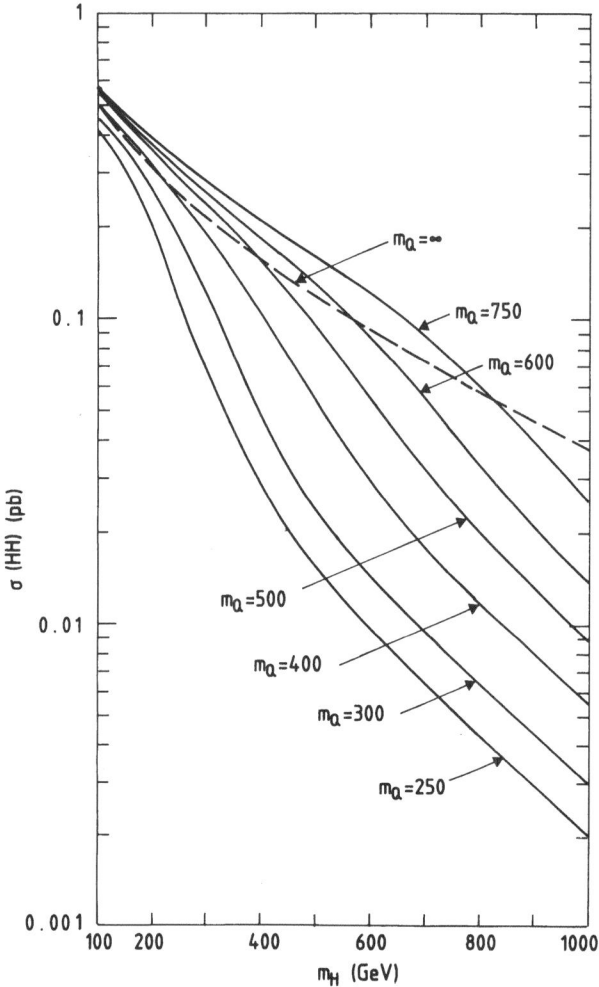

Fig. 1. The cross-section for $gg \to HH$ for pp collisions at $\sqrt{s} = 40 TeV$ for $m_q=$ 250,300,400,500,600 and 750 GeV. The large mass approximation is shown as a dashed curve. Only one heavy quark is assumed. For a generation of heavy quarks the curves should be multiplied by four.

$$\mathcal{L}_{eff} = \frac{\alpha_s}{12\pi} F_{\mu\nu}^2 log(1 + H/v) \tag{1}$$

$F_{\mu\nu}$ is the gluon field strength and $v = \frac{2M_W}{g} \approx 250 GeV$ is the vacuum expectation of the Higgs field. From this effective interaction one derives a completely analytic form for the multi Higgs cross-section in the case of a light Higgs particle, strictly speaking only for $m_H = 0$. If the Higgs particle is light one can ignore the Higgs self couplings and therefore the matrix element for nH production is direcly found by expanding the logarithm in formula(1). As a result one finds the following cross-section at the parton level :

$$\sigma(gg \rightarrow nH) = \frac{(2\pi)^{4-3n}}{N^2 - 1} \frac{\alpha_s^2 \hat{s}}{144\pi^2} \frac{[(n-1)!]^2}{v^{2n}n!} \times phasespace(n) \tag{2}$$

where N is the number of colors. Next one uses the formula for massless n-body phase space:

$$phasespace(n) = \int \prod_i \frac{d^3 p_i}{2E_i} \delta_4(\sum p_i - P) = (\frac{\pi}{2})^{n-1} \frac{n-1}{[(n-1)!]^2} \hat{s}^{n-2} \tag{3}$$

This gives :

$$\sigma(gg \rightarrow nH) = \frac{\alpha_s^2}{N^2 - 1} \frac{2\pi}{9\hat{s}} (\frac{\hat{s}}{16\pi^2 v^2})^n \frac{n-1}{n!} \tag{4}$$

Summing the series :

$$\sigma(gg \rightarrow\geq 2H) = \frac{\alpha_s^2}{N^2 - 1} \frac{2\pi}{144\pi^2 v^2} F(\hat{s}/16\pi^2 v^2) \tag{5}$$

where :

$$F(x) = \frac{(x-1)e^x + 1}{x} \tag{6}$$

Folding in the gluon structure functions gives finally :

$$\frac{d\sigma}{dM} = \frac{4\pi}{9M^3} \frac{\alpha_s^2}{N^2 - 1} G(M^2/16\pi^2 v^2)\mathcal{L}_{gg}(\tau) \tag{7}$$

where M is the invariant mass of the produced multi-Higgs system and $G(x) = xF(x)$. It is to be noticed that the scale at which this cross-section blows up exponentially is given by $4\pi v \sim 3TeV$. Therefore maybe this scale is the relevant scale for which the Higgs interactions become strong, rather than the canonical 1 TeV.

Clearly the above formulae have only a limited validity. The used approximation breaks down for $m_q \sim 1.3 TeV$ [ref.5], the use of massless phase space is not justified for a large number of Higgses and the approximation $m_q = \infty$ is incorrect for $\hat{s} > m_q$. Still for a light Higgs the formula should give a reasonable qualitative picture for multi-Higgs production at parton energies below 3 TeV. Above this energy the modifications will be severe. Therefore one cannot use the above to give predictions for the SSC or Eloisatron. The predictions, with one heavy quark, for the Tevatron(2 TeV) $\sim 3.7~10^{-3}pb$, UNK(6 TeV) $\sim 2.8~10^{-2}pb$ and LHC(16 TeV) $\sim 0.18pb$ should be roughly correct and can be used as an upper estimate of light Higgs production. For a complete degenerate generation these numbers should be multiplied by four.

II. Vector boson pair production via gluon fusion

The study of the production of vectorboson pairs at high energy hadron colliders is of importance for a variety of physics reasons, the main one being the search for a Higgs particle. When it is heavy enough the Higgs decays virtually always into a WW or ZZ pair. If the Higgs is below threshold for W boson pair production it will mainly decay into quark pairs for which the QCD background is overwhelming. However in this case one can consider rare decay modes of the Higgs, proceeding via loop graphs, like the decays into photon-photon or photon Z-boson.

In order to see if the production of vectorboson pairs can really be used to get information about the existence of the Higgs particle one needs also to calculate the background coming from known processes. For hadron colliders with an energy of O(1 TeV) the background comes essentially only from the direct quark-antiquark annihalation into vectorbosons (see fig. 2a). However when one goes to hadron colliders in the energy range of O(10 TeV) also gluon initiated processes (see fig. 2b) become important. The reason for this is that by going to higher energies one enters the low x region, where the gluon structure function is steeply rising. As a consequence processes proceeding via quark loop graphs, which are nominally of higher order in α_s can become important.

The best studied example is the production of photon-pairs [ref.6-9]. At the SSC the background to the two-photon signal for an intermediate mass Higgs is for two-thirds due to the $gg \rightarrow \gamma\gamma$ process. A prediction for the invariant mass spectrum of the photon pairs at LHC, SSC and Eloisatron is given in fig.(3). In this figure we ignored the existence of the top quark by assuming it to be heavy. If the top quark is light the $gg \rightarrow \gamma\gamma$ cross section has to be multiplied by $(15/11)^2$. Also the full m_{top} dependence can be taken into account relatively easy [ref.9].

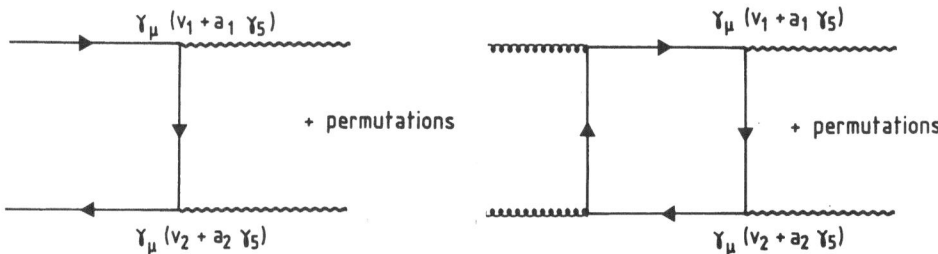

Fig. 2.   Feynman graphs contributing to vector boson pair production. Light quarks add coherently in the amplitude of the gluon-gluon process.

The extension of the above calculations to the production of other pairs of vectorbosons is technically rather complicated, in particular if one wants to keep track of the exact m_{top} dependence. Fully analytic results are at present known only for the γZ case [ref.7,10]. For the ZZ case a numerical calculation exists [ref.11]. However these exact calculations do not give much insight into the relative importance of the gluon-gluon process versus the quark-antiquark process. For instance it is not obvious how the different processes scale by going to different energy machines. In the following we will establish some rough scaling relations by considering the limit $m_Z, m_W \rightarrow 0$. The top quark will be considered to be either light $m_{top} = 0$ or very heavy $m_{top} = \infty$, the true result being somewhere in between.

In the limit $m_Z, m_W = 0$ the production of arbitrary vectorboson pairs can be directly related to the production of photon pairs. The production of vector boson pairs is described by the graphs in fig.(2), the relevant coupling constants are given in table 1.

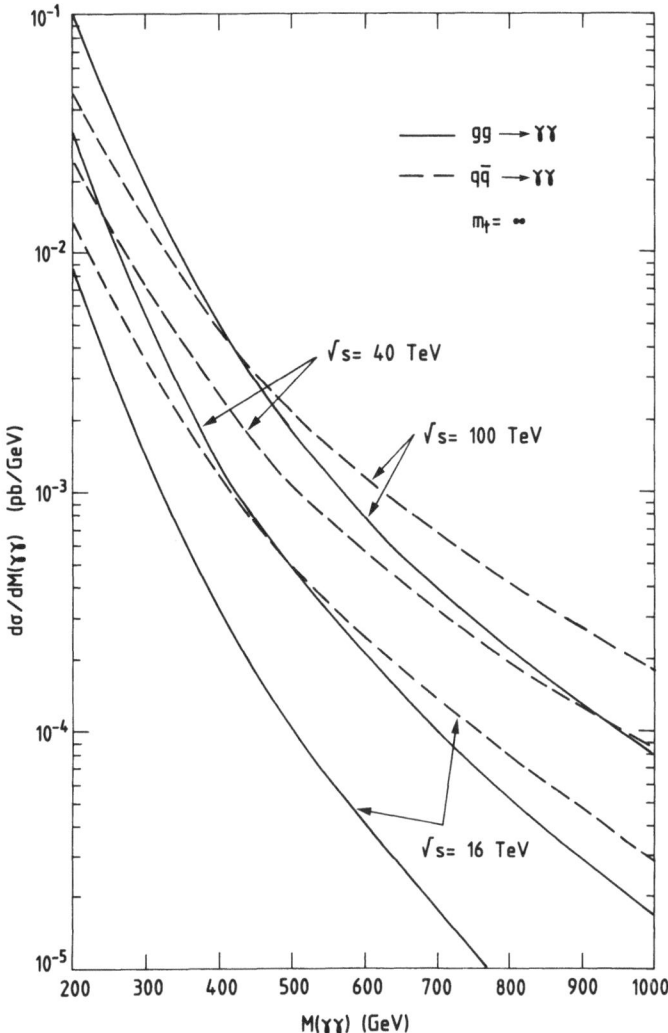

Fig. 3. Two-photon production at hadron colliders. A y-cut $|y_\gamma| \leq 2.5$ and a p_T-cut of $40 GeV$ on the photons have been applied. Duke-Owens structure functions have been used with an effective scale of $Q^2 = \hat{s}/4$.

Table 1. Coupling constants of quarks to vector bosons.

a,v	v_γ	a_γ	v_Z	a_Z	v_W	a_W
up	2/3	0	0.230	0.594	0.737	0.737
down	-1/3	0	-0.412	-0.594	0.737	0.737

The parton level cross section for quark-antiquark annihilation (fig. 2a) is proportional to the following combination of coupling constants :

$$(v_1^2 + a_1^2)(v_2^2 + a_2^2) + 4v_1 v_2 a_1 a_2 \tag{8}$$

For the gluon-gluon process (fig. 2b) the relevant combination of coupling constants is :

$$\left(\sum_{quarks} (v_1 v_2 + a_1 a_2) \right)^2 \tag{9}$$

These combinations are listed in table 2. The numbers in table 2 contain an additional factor 1/2 in the $\gamma\gamma$ and ZZ channels to account for the identical particle phase space.

Table 2. Combination of coupling constants relevant for the production of vectorboson pairs in $\bar{q}q$ annihilation and gluon fusion. Both the cases $m_t = 0$ and $m_t = \infty$ are given.

	$\gamma\gamma$	γZ	ZZ	WW
$\bar{u}u$	0.099	0.180	0.120	2.36
$\bar{d}d$	0.0062	0.058	0.256	2.36
gg(5 quarks)	0.747	0.516	2.83	29.5
gg(6 quarks)	1.39	0.760	3.88	42.5

One can use the numbers from table 2 directly to give an estimate for arbitrary vectorboson pair production in terms of the known two-photon production rate. For instance one has the following relation :

$$R_1(\gamma Z) = \frac{\bar{q}q \to \gamma Z}{\bar{q}q \to \gamma\gamma} = \frac{0.180(dL/d\tau)_{\bar{u}u} + 0.058(dL/d\tau)_{\bar{d}d}}{0.099(dL/d\tau)_{\bar{u}u} + 0.0062(dL/d\tau)_{\bar{d}d}} \approx 2.26 \tag{10}$$

where $(dL/d\tau)_{\bar{u}u}$ and $(dL/d\tau)_{\bar{d}d}$ are the luminosities of up-type respectively down-type quarks. The final number follows from the fact that these luminosities are approximately equal for the invariant mass range of interest, since the preponderance of valence up-quarks over valence down-quarks is compensated by the presence of the strange quark sea. Similar results are derived for the ratio:

$$R_2(X) = \frac{gg \to X}{gg \to \gamma\gamma} \tag{11}$$

While the ratio's R_1 and R_2 are strictly valid only in the limit $m_W, m_Z = 0$, the combination $R_3 = R_2/R_1$ should be fairly reliable not too close to threshold. Table 3 shows the various ratio's R_1, R_2 and R_3.

Table 3.Approximate ratio's $R_1 = \bar{q}q \to X/\bar{q}q \to \gamma\gamma, R_2 = gg \to X/gg \to \gamma\gamma$ and $R_3 = R_2/R_1$. The left part is for 5 light quarks, the right part for 6 quarks.

	R_1	R_2	R_3		R_1	R_2	R_3
γZ	2.26	0.691	0.306		2.26	0.547	0.242
ZZ	3.57	3.79	1.06		3.57	2.79	0.782
WW	44.9	39.5	0.880		44.9	30.6	0.682

One can use the given values of R_3 to give an estimate of the relative importance of the gluon fusion process, using the known results for the $\gamma\gamma$ case in fig.(4). For instance at the Eloisatron at an invariant mass of 1 TeV where $gg \to \gamma\gamma$ is about 40% of $\bar{q}q \to \gamma\gamma$, one expects :

$$\frac{gg \to WW}{\bar{q}q \to WW} = 0.88 \times 40\% = 35\% \tag{12}$$

assuming only 5 quarks are important.

In order to get some idea about the validity of the approximation we give in table 4 the exact ratio's for the γZ case at the SSC. We see that R_1 and R_2 are somewhat larger than expected, but the ratio R_3 approximates the limit rather well for large values of $M(\gamma Z)$.

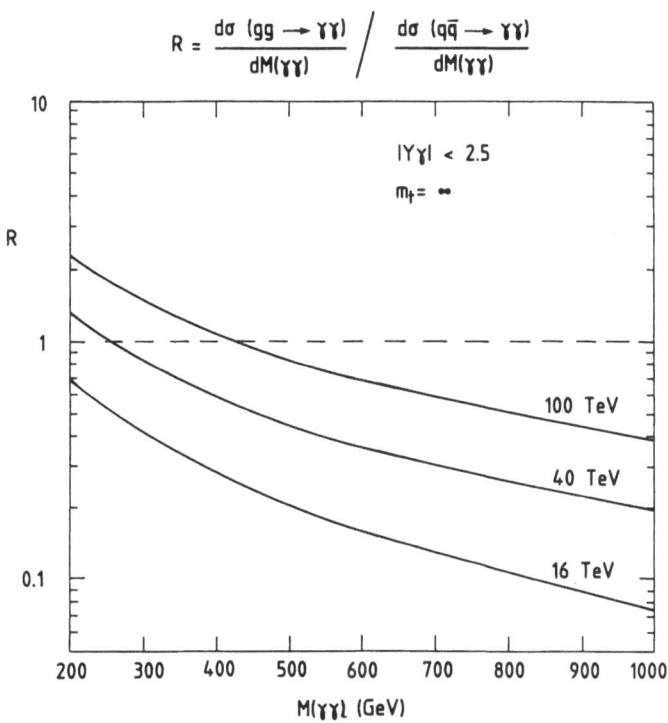

Fig. 4.   The ratio $gg \to \gamma\gamma/\bar{q}q \to \gamma\gamma$ at different hadron colliders. The same parameters as in fig.(3) have been used.

Table 4. The exact ratio's R_1, R_2, R_3 for γZ production in the case of 5 light quarks.

$M(\gamma Z)$	200	300	400	500	600
R_1	3.7	3.4	3.1	2.9	2.7
R_2	0.55	0.71	0.75	0.76	0.76
R_3	0.15	0.21	0.24	0.26	0.28

Acknowledgement. This work was performed in collaboration with E. W. N. Glover.

References

1. J.J. van der Bij, Nucl. Phys. B267(1986),557.

2. E.W.N. Glover and J.J.van der Bij, CERN-Th.4934/87, Nucl. Phys. B(in press).

3. D. A. Dicus and S. S. D. Willenbrock, Phys. Lett. 203B(1988),457.

4. A. I. Vainshtein, M. B. Voloshin, V. I. Zakharov and M. A. Shifman, Sov. J. Nucl. Phys. 30,711(1979).

5. F. Hoogeveen, Nucl. Phys. B259(1985),19.

6. B. Combridge, Nucl. Phys. B174(1980),243.

7. Ll. Amettler, E. Gava, N. Paver and D. Treleani, Phys. Rev. D32(1980),1699.

8. R. K. Ellis, I. Hinchliffe, M. Soldate and J. J. van der Bij, Nucl. Phys. B297(1988),221.

9. D. A. Dicus and S. S. D. Willenbrock, Phys. Rev. D37(1988),1801.

10. E. W. N. Glover and J. J. van der Bij, Phys. Lett. 206B(1988),701.

11. D. A. Dicus, Chung Kao and W. W. Repko, Phys. Rev. D36(1987),1570.

31

EXCITED QUARKS AND LEPTONS*

M. Spira and P.M. Zerwas

Institut f. Theor. Physik, RWTH Aachen, D-5100 Aachen, FRG

ABSTRACT

The existence of excited states is a natural consequence of composite models for quarks and leptons. Production rates and signatures are discussed for a 200 TeV proton-proton collider. Such particles could be discovered with masses up to about 50 and 25 TeV, respectively.

1 THE PHYSICAL SET-UP

The proliferation of quarks and leptons is often taken as evidence for possible substructures of these particles. Their masses and mixing angles are considered as a result of the interactions between the more elementary preonic constituents. The analysis of static properties, such as $(g-2)$ of the electron and muon, and the estimate of quark and lepton radii from high-energy e^+e^- and quark-quark collisions limit the distance Λ^{-1} at which these preons are bound, to 10^{-16} cm and less. Experiments at the CERN $p\bar{p}$ collider and the Tevatron, at HERA and LEP will raise the compositeness scale Λ up to a level of order 1 to 10 TeV.

One of the classical signatures of compositeness on the quark and lepton level would be the observation of excited states towering over the leptonic and quark ground states [1–11]

$$l, l^*, l^{**}, \ldots$$
$$q, q^*, q^{**}, \ldots$$

The masses of excited quark and lepton states are expected to be of order Λ so that we will assume a lower limit of ~ 5 TeV for $m_* \sim \Lambda$ in the subsequent analysis. This limit can be considered as a natural range for possible deviations from the Standard Model as currently treated at low energies. Because a detailed approach to preon dynamics is lacking, we will focus on a few representative examples illuminating the gross features of excited states that could eventually be observed in a 200 TeV proton-proton collider.

---

*Supported in part by the W. German Bundesministerium für Forschung und Technologie.

Spin and isospin of the excited fermions will be set to 1/2 to limit the number of parameters in this study. The assignment of left- and right-handed components to isodoublets, e.g. for the first generation

$$
\begin{bmatrix} \nu_e \\ e^- \end{bmatrix}_L \qquad e_R^- \qquad \begin{bmatrix} \nu_e^* \\ e^{*-} \end{bmatrix}_L \qquad \begin{bmatrix} \nu_e^* \\ e^{*-} \end{bmatrix}_R
$$

$$
\begin{bmatrix} u \\ d \end{bmatrix}_L \qquad \begin{matrix} u_R \\ d_R \end{matrix} \qquad \begin{bmatrix} u^* \\ d^* \end{bmatrix}_L \qquad \begin{bmatrix} u^* \\ d^* \end{bmatrix}_R
$$

allows for non-zero masses prior to SU(2)×U(1) symmetry breaking and it protects $(g-2)_l$ quadratically in the mass ratio $(m_l/m_*)^2$ [2].

The coupling to gluons, $\gamma, W^\pm$ and Z gauge particles is vector-like:

$$
\mathcal{L}_{eff}^a = \bar{f}^* \gamma_\mu [g_s \frac{\vec{\lambda}}{2} \vec{G}_\mu + g \frac{\vec{\tau}}{2} \vec{W}_\mu + g' \frac{Y}{2} B_\mu] f^* \tag{1}
$$

where $f^* = \nu_e^*, \ldots, d^*$ are the excited states. The weak hypercharge Y of the excited states is -1 and $1/3$ in the lepton and quark sector, respectively; g_s and $g(g') = e/\sin\theta_w (\cos\theta_w)$ are the strong and electroweak gauge couplings. Each of the vertices will in general be modified by form factors and anomalous magnetic-moment couplings in a way similiar to the light fermion-sector [12,13].

Transitions between light (left-handed) and excited (right-handed) fermions can be mediated by gauge particles. They can be described in an SU(3)×SU(2)×U(1) invariant form by the magnetic-moment type interaction

$$
\mathcal{L}_{eff}^b = \frac{1}{2\Lambda} \bar{f}_R^* \sigma_{\mu\nu} [g_s \frac{\vec{\lambda}}{2} \vec{G}_{\mu\nu} + g \frac{\vec{\tau}}{2} \vec{W}_{\mu\nu} + g' \frac{Y}{2} B_{\mu\nu}] f_L + h.c. \tag{2}
$$

where $\vec{G}_{\mu\nu}$ is the gluon field strength, etc. Form factors, not necessarily normalized to 1, will soften the vertices.

Contact interactions, as a result of preon exchange mechanisms, may lead to the excitation of quarks and leptons as well (Fig.1c): $f + f \rightarrow f + f^*$ and $f^* + f^*$. In parallel to the light-fermion case, this energy density will be parametrized in the following (simplified) current×current form [5,8]

$$
\mathcal{L}_{eff}^c = \frac{g_*^2}{\Lambda^2} \frac{1}{2} j_\mu j_\mu \tag{3}
$$

$$
\begin{aligned}
j_\mu = & \ \eta_L \bar{f}_L \gamma_\mu f_L + \eta_L' \bar{f}_L^* \gamma_\mu f_L^* + \eta_L'' \bar{f}_L^* \gamma_\mu f_L + h.c. \\
& + (L \rightarrow R)
\end{aligned}
$$

As usual, g_*^2 is chosen equal to 4π, and the η factors of the left-handed currents equal to 1, while the right-handed currents will be neglected for the sake of simplicity. Such

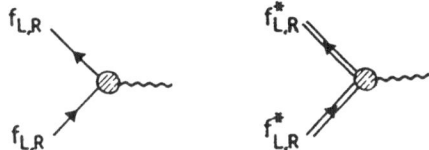

Figure 1.a. *Gauge interactions of light and excited fermions.*

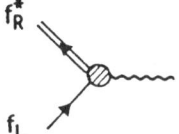

Figure 1.b. *Transitions via gauge boson emission.*

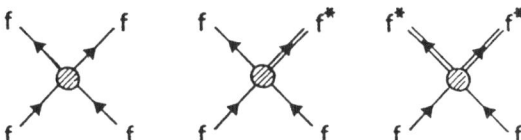

Figure 1. c. *Contact interactions.*

a form can be derived for quark and lepton energies below the compositeness scale. We shall, however, assume its approximate validity up to the range $\sqrt{\hat{s}} \leq O(\Lambda)$ for the parton energies, after excited boson formation is seperated off [10]. Beyond this range the pointlike hard coupling Eq.(3) must be dissolved and the rise of the cross sections with energy will be damped. Nevertheless, this ansatz should be sufficient to provide a rough estimate of how preon interchange will affect the widths of excited fermionic states and their production cross sections.

2 THE WIDTHS OF EXCITED QUARKS AND LEPTONS

Heavy excited fermions will decay into light fermions plus gauge bosons, but also, through preon-pair creation, into bunches of quarks and leptons.

The partial widths for the various electroweak decay channels follow from

$$\Gamma(f^* \to f\gamma) \;=\; \frac{\alpha e_f^2}{4}\frac{m_*^3}{\Lambda^2} \tag{4}$$

$$\Gamma(f^* \to fZ) \;=\; \frac{\alpha[I_3^f - e_f \sin^2\theta_w]^2}{\sin^2 2\theta_w}\frac{m_*^3}{\Lambda^2} \tag{5}$$

$$\Gamma(f^* \to fW) \;=\; \frac{\alpha}{8\sin^2\theta_w}\frac{m_*^3}{\Lambda^2} \tag{6}$$

and for gluon decays of excited quarks,

$$\Gamma(q^* \to qg) = \frac{\alpha_s}{3}\frac{m_*^3}{\Lambda^2} \tag{7}$$

The widths for decays into gauge particles are fairly small; taking for illustration $m_* = \Lambda$, one finds the values listed in Table 1a, i.e. approximately 70 GeV for e^* and 400 GeV for q^* if the masses are set to $m_* = 10$ TeV. The relative gauge branching ratios $B_G = \Gamma(f^* \to fV)/\sum_V \Gamma(f^* \to fV)$ are collected in Table 1b.

The states, however, will presumably be significantly widened by decays which are mediated by the novel strong preon interactions. This is apparent from the estimates of the decay widths for contact interactions built up from left-handed currents:

$$\Gamma(f^* \to f + f'\bar{f'}) = \frac{m_*}{96\pi} \left(\frac{m_*}{\Lambda}\right)^4 N_c' S' \tag{8}$$

$N_c' = 3$ or 1 is the color factor due to the f' pair production; for $f = f'$ an additional statistics factor $S' = 4/3$ or 2 is to be added for quarks or leptons, respectively. For 3 generations and $m_* = \Lambda = 10$ TeV, Eq.(8) leads to f^* widths of order 1 TeV, i.e. 10% of the mass of the excited states. This is still in the same ball-park as gauge decay widths of excited quarks; however excited lepton states can be substantially wider than predicted by the electroweak decay channels – a natural consequence of

Table 1. *a. Gauge widths*   *b. Relative branching ratios for excited fermions*

	$\sum_V \Gamma(f^* \to fV)/m_*$ [$m^* = \Lambda$]
ν^*	$6.5 \cdot 10^{-3}$
e^*	$6.5 \cdot 10^{-3}$
u^*	$3.9 \cdot 10^{-2}$
d^*	$3.9 \cdot 10^{-2}$

e^*	B_G	ν^*	B_G	u^*	B_G	d^*	B_G
$e\gamma$	0.28	νZ	0.39	ug	0.85	dg	0.85
eZ	0.11	eW	0.61	$u\gamma$	0.02	$d\gamma$	0.005
νW	0.61			uZ	0.03	dZ	0.05
				dW	0.10	uW	0.10

the *strong* interactions on the subconstituent level. More details on the total widths and branching ratios are summarized in Table 2. Excellent signatures are predicted by the large percentage of decays with leptons in the final state. [Note that decays mediated by contact interactions are more strongly suppressed than gauge decays if Λ rises above m_*, a consequence of the different powers with which the compositeness scale enters the decay widths.]

Table 2. *Decay widths of excited fermions mediated by gauge (G) and contact interactions (CT); for $m_* = \Lambda$.*

	Γ_{tot}/m_*	Γ_G/Γ_{tot}	Γ_{CT}/Γ_{tot}	lept. decays/all
ν^*	$8.9 \cdot 10^{-2}$	0.07	0.93	100%
e^*	$8.9 \cdot 10^{-2}$	0.07	0.93	100%
u^*	$1.2 \cdot 10^{-1}$	0.32	0.68	16.3%
d^*	$1.2 \cdot 10^{-1}$	0.32	0.68	16.3%

3 PRODUCTION IN pp COLLISIONS

Excited quarks can be produced in pp collisions through a variety of mechanisms. The most obvious reaction is $q^*\bar{q}^*$ pair creation in quark-antiquark annihilation and gluon-gluon fusion [11]. This cross section can be reliably predicted, yet it turns out to be so small that it would be very difficult to discriminate the signal against the ordinary QCD background. Much more promising are the gluonic excitation of quarks $g + q \to q^*$, and the excitation through preon interactions $qq \to qq^*$ and q^*q^*. Through preon interactions excited leptons, too, could eventually be produced at observable rates, $q\bar{q} \to ee^*$ and e^*e^*. The signatures of excited quarks are bumps in the invariant energies of jets, jet + gauge boson and jet + lepton pair combinations. Excited leptons would reveal themselves in bumps of leptons, leptons + gauge particles or leptons + quark jets. The first indication for the production of novel excited fermions could be the copious production of leptons, at large rates unexpected in the framework of the Standard Model.

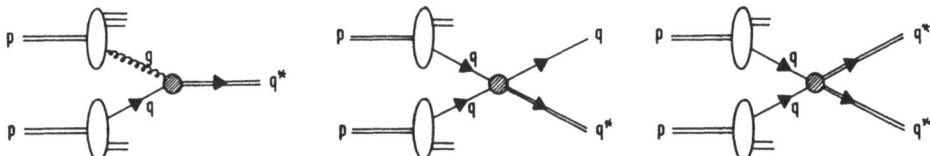

Figure 2. *Representative examples of various mechanisms for q^* production in pp collisions.*

3.1 q^*: QCD gauge interactions

The cross section for the gluonic excitation of quarks

$$g + q \to q^* \quad \begin{array}{l} {}^{G}\!\!\!\nearrow \; jj,\, j\gamma,\, jZ,\, jW \\ {}_{CT}\!\!\!\searrow \; jjj,\, jll \end{array}$$

in proton-proton colliders (Fig.2) is given by

$$\sigma = \frac{\alpha_s \pi^2}{3\Lambda^2} \tau \frac{d\mathcal{L}^{gq}}{d\tau} \tag{9}$$

where the coupling of the quark, the excited quark and the gluon is of the magnetic-moment type defined in Eq.(2). $d\mathcal{L}^{gq}/d\tau$ is the quark-gluon luminosity for the proton-proton beams [11] at $\tau = m_*^2/s$. The production cross section is shown by the full line in Fig.3a for $\sqrt{s} = 200$ TeV. For the sake of simplicity Λ is identified with m_*; other choices can easily be realized by scaling the curve appropriately. Given an integrated luminosity of 10 fb^{-1}/a, a mass range of 30–40 TeV can be reached in this channel, based on 100 to 1000 events/a.

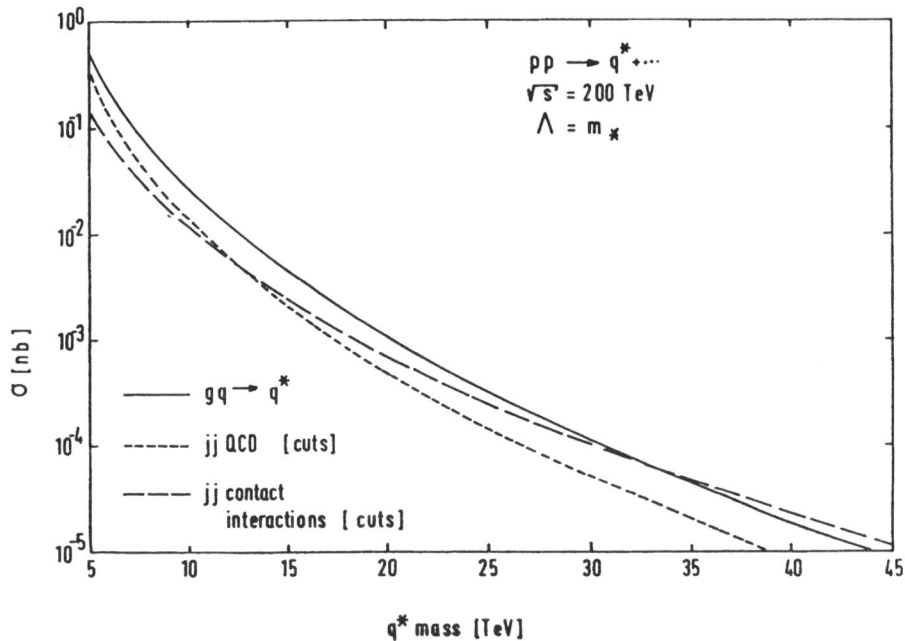

Figure 3.*a*. *Cross section for the production of excited quarks in $g + q \to q^*$ at a pp collider of $\sqrt{s} = 200$ TeV (full line); the 2-jet backgrounds from QCD and contact interactions are shown by the broken lines.*

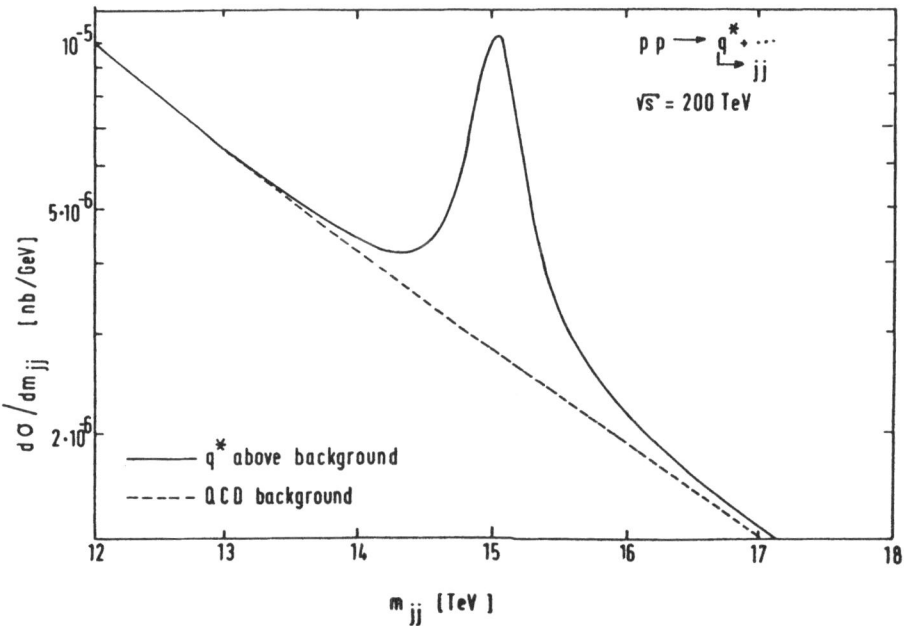

Figure 3. *b*. *Peak in the jet-jet invariant mass [gauge interactions only] due to q^* events above the QCD background.*

If the gauge interactions are dominating, the signals for singly produced excited quarks are large transverse momentum jj, $j\gamma$, jZ or jW pairs peaking at the mass of the resonance. The jj mass distribution above the QCD background is shown in Fig.3b for a resonance mass of 15 TeV. Because the final states of the signal consist of large $p_\perp$ jets with large angles θ_{jj} between the jets of each pair, we introduced the following cuts to reduce the background ($\eta =$ pseudorapidity):

$$\theta_{jj} > 30° \qquad\qquad\qquad p_\perp > 100\,GeV$$

$$\eta < 2.5 \qquad\qquad\qquad E^{tr} > \begin{cases} m_*/2 & \text{for } gq \to q^* \\ m_* & \text{other reactions} \end{cases}$$

The mass resolution is determined by the decay width Γ of the resonance and the experimental jet resolution, which is taken to be $\Delta E/E = 0.35/\sqrt{E} + 0.02$.

If however the contact interactions overwhelm the gauge interactions in the decay process and in the background, the picture changes quite dramatically. The peak becomes broader, yet in addition to jjj final states one expects a copious production of semi-leptonic jll final states that could easily reveal the existence of a new state. The main source of background 3-jet final states are quark-quark scattering processes with additional gluon bremsstrahlung or gluon induced 3-quark final states where the quark interactions are dominated by contact terms. Characteristic examples are sketched below (Fig.4).

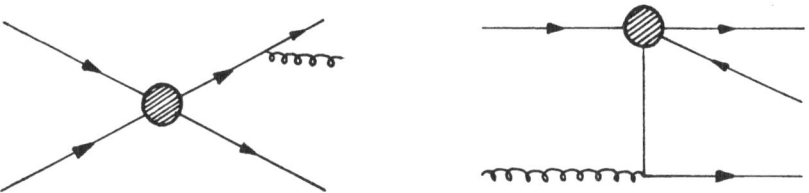

Figure 4. *3-jet final states from quark-quark contact interactions with additional gluon bremsstrahlung or gluon induced.*

The signal is compared with the background cross section in Fig.3. Since the background cross section is smaller than the q^* production cross section, it is clear that order 100 to 1000 events are sufficient to extract the q^* signal.

3.2 q^*: contact interactions

Excited quarks can also be created directly in quark-quark collisions (Fig.2)

$$qq \to qq^* \begin{cases} \xrightarrow{\text{G}} & j + jj,\, j + j\gamma,\, j + jZ,\, j + jW \\ \xrightarrow{\text{CT}} & j + jjj,\, j + jll \end{cases}$$

525

A set of representative cross sections has been derived for left-handed currents in the contact interactions:

$$\sigma(qq' \to qq'^*) = \frac{\pi}{\hat{s}} \left(\frac{\hat{s}}{\Lambda^2} \right)^2 \left[1 - \frac{m_*^2}{\hat{s}} \right] \tag{10}$$

$$\sigma(qq' \to qq'^*) : \sigma(qq \to qq^*) = 1 : \frac{8}{3}$$

and

$$\sigma(q\bar{q} \to q'\bar{q}'^*) = \frac{\pi}{4\hat{s}} \left(\frac{\hat{s}}{\Lambda^2} \right)^2 \left[1 + \frac{v}{3} \right] \left[1 - \frac{m_*^2}{\hat{s}} \right]^2 \left[1 + \frac{m_*^2}{\hat{s}} \right] \tag{11}$$

$$\sigma(q\bar{q} \to q'\bar{q}'^*) : \sigma(q\bar{q}' \to q\bar{q}'^*) : \sigma(q\bar{q} \to q\bar{q}^*) = 1 : 1 : \frac{8}{3}$$

[v is the velocity of the excited quark in the cm frame of the subprocess.] The validity of these expressions is limited to parton energies below Λ. Above this range, the growth $\propto \hat{s}$ will be damped. However, because the quark-quark luminosities fall rapidly with τ, the damping does not affect the overall proton-proton cross section very much. The production cross section for $\sqrt{s} = 200$ TeV is displayed in Fig.5.

Background processes to gauge decays of excited quarks are (i) 3-jet QCD reactions, and (ii) 3-jet events due to qq contact interactions with additional gluon bremsstrahlung or gluon induced as exemplified in Fig.4. These background cross sections are compared with the cross section for the qq^* signal in Fig.5. It is evident that the signal is not submerged by the background. Assuming instead the contact interactions to dominate the decay process, multijet clusters and jet + lepton pair final states would signal the production of excited quarks, with rates far above the standard model background. A detection of these states up to masses of ~ 50 TeV appears feasible.

The production of excited quark pairs

$$qq \to q^*q^* \qquad q^* \begin{array}{c} \overset{G}{\nearrow} \quad jj, j\gamma, jZ, jW \\ \underset{CT}{\searrow} \quad jjj, jll \end{array}$$

requires a larger energy in the subprocess than qq^* production, and these processes are suppressed correspondingly. For left-handed couplings of the contact interactions, the collider cross section shown in Fig.5 for $\sqrt{s} = 200$ TeV has been derived from

$$\sigma(qq' \to q^*q'^*) = \frac{\pi v}{\hat{s}} \left(\frac{\hat{s}}{\Lambda^2} \right)^2 \left[1 - \frac{2m_*^2}{\hat{s}} \right] \tag{12}$$

$$\sigma(qq' \to q^*q'^*) : \sigma(qq \to q^*q^*) = 1 : \frac{1}{3}$$

and

$$\sigma(q\bar{q}' \to q^*\bar{q}'^*) = \frac{\pi v}{3\hat{s}} \left(\frac{\hat{s}}{\Lambda^2} \right)^2 \left[1 - \frac{m_*^2}{\hat{s}} \right] \tag{13}$$

$$\sigma(q\bar{q}' \to q^*\bar{q}'^*) : \sigma(q\bar{q} \to q'^*\bar{q}'^*) : \sigma(q\bar{q} \to q^*\bar{q}^*) = 1 : 1 : \frac{8}{3}$$

526

The final states consist of jet bunches $jjj + jjj$ and jets accompanied by gauge bosons or lepton pairs $jll + jll$ [with small background from Standard Model processes]. Excited quarks up to masses of ~35 TeV are accessible in this channel.

3.3 e^*: contact interactions

The possibility to create leptons copiously through contact interactions in proton-proton collisions is one of the most exciting phenomena expected to occur in composite models in which quarks and leptons share common preons. In such a case, also excited leptons could be produced copiously in hadron colliders,

$$q\bar{q} \to ee^* \qquad\qquad e^* \xrightarrow{\text{G}} e\gamma, eZ, \nu W$$
$$q\bar{q} \to e^*e^* \qquad\qquad \xrightarrow{\text{CT}} eee, ejj$$

The cross sections, based on the interaction density Eq.(3) for L couplings, follow from

$$\sigma(q\bar{q} \to ee) = \frac{\pi}{9\hat{s}}\left(\frac{\hat{s}}{\Lambda^2}\right)^2$$

$$\sigma(q\bar{q} \to ee^*) = \frac{\pi}{6\hat{s}}\left(\frac{\hat{s}}{\Lambda^2}\right)^2\left[1+\frac{v}{3}\right]\left[1-\frac{m_*^2}{\hat{s}}\right]^2\left[1+\frac{m_*^2}{\hat{s}}\right] \qquad (14)$$

$$\sigma(q\bar{q} \to e^*e^*) = \frac{\pi v}{12\hat{s}}\left(\frac{\hat{s}}{\Lambda^2}\right)^2\left[1+\frac{v^2}{3}\right]$$

In a similar way muonic states can be produced. The cross sections are displayed for a 200 TeV collider in Fig.6. Background reactions to these channels are very rare in the

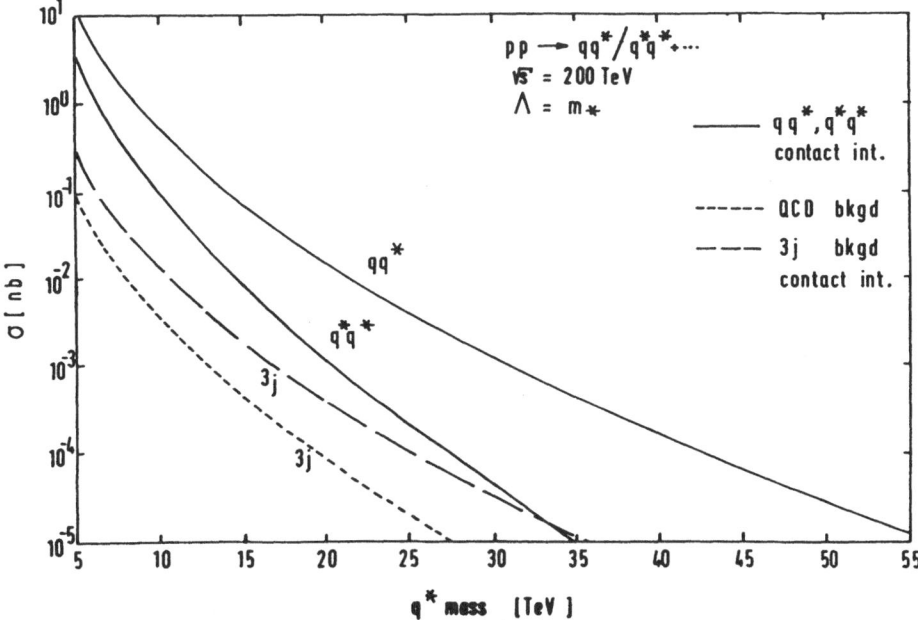

Figure 5. *Cross sections for qq^* and q^*q^* production due to contact interactions for $\Lambda = m_*$ (full lines); the 3-jet backgrounds from QCD and contact interactions are represented by the broken lines.*

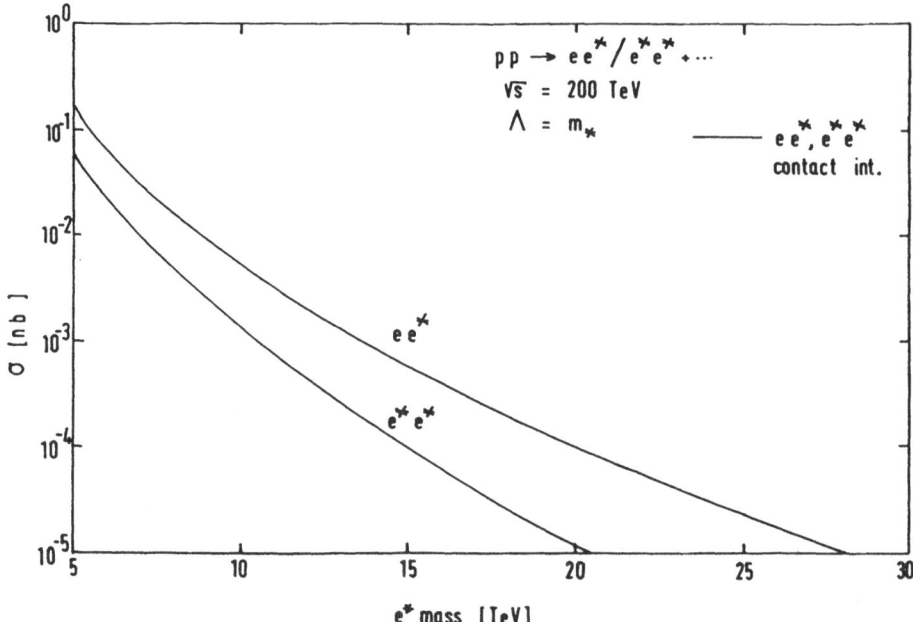

Figure 6. *Cross sections for ee* and e*e* production due to contact interactions for* $\Lambda = m_*$.

Standard Model. For the set of parameters adopted throughout this analysis, excited leptons with masses up to 25 TeV could be accessible. Pure leptonic decay channels would provide very clear signatures for the experimental identification of these novel states.

4 CONCLUSIONS

The discovery limits for excited quarks and leptons in a 200 TeV proton-proton collider are summarized in Table 3. They are compared with the corresponding limits of other colliders at lower energies that are presently under discussion. The parameters on which these numbers are based, have been defined in the interaction densities Eqs.(1–3). The compositeness scale is identified with the masses of the excited fermions. It appears that even if the parameters are varied within large margins, a 200 TeV collider will safely reach the multi-TeV range on the subconstituent level: a new layer of fundamental constituents of matter could eventually be revealed in this facility.

Table 3. *Discovery limits for excited quarks and leptons at pp colliders presently under discussion.*

	LHC/16 TeV	SSC/40 TeV	Eloisatron/200 TeV
q^*	8 TeV	14 TeV	50 TeV
e^*	4 TeV	7 TeV	25 TeV

ACKNOWLEDGEMENTS

We are grateful to Prof. A. Ali and Prof. A. Zichichi for the invitation to the Eloisatron Workshop in Erice where this work was initiated.

REFERENCES

[1] F. Low, Phys. Rev. Lett. 15 (1965) 900.

[2] F.M. Renard, Phys. Lett. 139B (1984) 449.

[3] N. Cabibbo, L. Maiani and Y. Srivastava, Phys. Lett. 139B (1984) 459.

[4] A. DeRujula, L. Maiani and R. Petronzio, Phys. Lett. 140B (1984) 253.

[5] J. Kühn and P. Zerwas, Phys. Lett. 147B (1984) 189.

[6] J. Kühn, H.D. Tholl and P. Zerwas, Phys. Lett. 158B (1985) 270.

[7] K. Hagiwara, S. Komamiya and D. Zeppenfeld, Z. Phys. C29 (1985) 115.

[8] R. Kleiss and P. Zerwas, Proceedings, Workshop "Physics at Future Accelerators", La Thuile/CERN 1987.

[9] U. Baur, I. Hinchliffe and D. Zeppenfeld, Proceedings, Workshop "From Colliders to Supercolliders", Madison (Wisconsin) 1987.

[10] I. Bars and I. Hinchliffe, Phys. Rev. D33 (1986) 704.

[11] E. Eichten, I. Hinchliffe, K. Lane and C. Quigg, Rev. Mod. Phys. 56 (1984) 579.

[12] M.S. Chanowitz and S.D. Drell, Phys. Rev. Lett. 30 (1973) 807.

[13] G.B. West and P. Zerwas, Phys. Rev. D10 (1974) 2130.

32

PROBING NEW CONTACT INTERACTIONS AT FUTURE *pp*, *ep* and *e⁺e⁻* COLLIDERS

Fridger Schrempp

Deutsches Elektronen Synchrotron DESY

Notkestraße 85, D-2000 Hamburg 52

Abstract

A comparative discussion of the 'reach' Λ^* for possible new contact interactions at future pp, $e^{\pm}p$ and e^+e^- colliders is presented. First of all, some introductory background is provided concerning the generality of the concept of contact interactions and the interpretation of the associated scale Λ^*. Useful analytical insight comes, moreover, from a simple scaling law and a more specialized approximate formula for the energy and luminosity dependence of Λ^*. In a second part, 'state of the art' numerical simulations for extracting Λ^* are discussed. For e^+e^- collisions, an analysis is reviewed which extends from PETRA/ PEP energies, via the Z^0 peak, up to the TeV regime of a possible linear collider (CLIC). For pp and $\bar{p}p$ collisions, a new detailed analysis was performed, extending from $S\bar{p}pS$ energies, via the LHC and SSC regime, up to a possible 200 TeV pp collider (ELOISATRON). The importance of longitudinal beam polarization for filtering out contact interactions is illustrated for the case of HERA and LEP.

1 Introduction

A major objective of experiments at future high energy accelerators will be to search for new particles and interactions announcing the onset of physics beyond the Standard Model.

Although the Standard Model so far perfectly manages to accomodate all existing experimental data, there are a number of well known theoretical arguments, calling for some sort of modification of the present theory, not far above the Fermi scale of weak interactions, $\Lambda_F \simeq 250$ GeV. While in the light of such arguments the appearance of new particles and interactions seems unavoidable, their specific properties entirely depend on the preferred scheme, such as the supersymmetry-superstring pathway[1] or, perhaps, a world with composite quarks and leptons and/or Higgs bosons [2].

The discovery of dramatic formation peaks or the crossing of new open production thresholds at one of the forthcoming high energy colliders would certainly provide the most unambigous signals of new particles. Yet, the mass range accessible in this case *is strictly limited by the* available beam energy.

A considerably larger mass range may be explored, however, through the study of effects from *virtual* new particle exchanges. For masses much larger than the beam energy such indirect signatures may be conveniently and model independently investigated by adding general contact interaction terms to the Standard Model lagrangian, quite analogously to Fermi's early description of the weak interactions. The potential of such a strategy was particularly emphasized by Eichten et. al. [3,4,5] and Ref. [6] in the context of quark and lepton substructure. It is important to note, however, that (experimental) limits on the strength of contact interactions provide information on possible new physics in a far more general sense (c.f. Sect. 2.1).

The purpose of this report is to present a comparative discussion of the 'reach' for new contact interactions at the various forthcoming pp, ep and e^+e^- colliders.

In Sect. 2.1 I shall provide some introductory background concerning the generality of the concept of contact interactions and the interpretation of the associated scale Λ^* . The four-fermion contact interaction terms as relevant for the various types of colliders are then introduced in Sect. 2.2. Useful analytical insight comes from a simple scaling law and a more specialized approximate formula for the energy and luminosity dependence of the 'reach' Λ^* (Sect. 2.3). In Sects. 3 I proceed to discuss selected results from 'state of the art' simulations for the various colliders. I start in Sect. 3.1 with reviewing an analysis of contact interactions at future e^+e^- colliders, which covers the entire energy range from PETRA/PEP up to a possible 2 TeV linear collider (CLIC) under realistic assumptions about expected detector performances and machine luminosities. An important issue in the discussion about future machines concerns the question of beam polarization. In Sect. 3.2 I shall illustrate the impact of longitudinal polarization on contact interactions for the case of HERA and LEP. Finally, Sect. 3.3 deals with a new detailed study of general contact terms for pp and $\bar{p}p$ colliders. It covers the entire energy range from the $S\bar{p}pS$ up to a possible 200 TeV pp collider (ELOISATRON). The bounds on Λ^* are extracted from the two-jet angular distribution $\frac{1}{\sigma_{2j}}d\sigma_{2j}/d\chi$ at large two-jet masses which enjoys a number of advantages over previous methods and has been proven at the $S\bar{p}pS$ to be relatively free of systematic errors.

Finally, the results for Λ^* to be expected at the various future colliders are compared to each other in the Conclusions.

2 Reach for Contact Interactions

2.1 Setting the Stage

Including contact interactions both in real experiments and in simulations for future colliders has by now become a standard matter. Yet, since in the literature such a procedure is often *identified* with ideas of quark and lepton substructure, it may be worth spending a few words to point out the actual generality of this (effective lagrangian) method.

The basic assumption is that the Standard Model is, in fact, part of a larger theory, involving more (heavier) fields besides the ones we know already[1]. At this point it is irrelevant whether the full theory e.g. involves a grand unification of gauge forces, supersymmetry or describes the interactions of quark and lepton constituents. What matters is that at energies small as compared to some characteristic scale $\Lambda > m_W$ the

---

[1]or more generally, field degrees of freedom which are unobservable at present energies

$SU(3)_c \times SU(2)_L \times U(1)_Y$ symmetric Standard Model results to good approximation, after the unobservable fields (particles) have been integrated out. In this case, the total lagrangian, valid up to energies $\sqrt{s} < (<)\Lambda$, may be written in terms of presently *relevant* fields only (quarks, leptons, gluons, W's, γ, Higgs), in form of an expansion in powers of $1/\Lambda$

$$\mathcal{L}_{eff} = \mathcal{L}_{SM} + \frac{1}{\Lambda}\mathcal{L}_5 + \frac{1}{\Lambda^2}\mathcal{L}_6 + \ldots \tag{1}$$

Here, the leading term of mass dimension 4, $\mathcal{L}_{SM}$, is to be identified with the (renormalizable) Standard Model lagrangian. The additional terms $\mathcal{L}_5, \mathcal{L}_6, \ldots$ generally involve all possible $SU(3)_c \times SU(2)_L \times U(1)_Y$ *symmetric* interaction terms of dimensions 5, 6, ... *They are the (non- renormalizable) remnants from the integration over the (presently) unobservable fields of the full theory* and if nonzero, provide small but crucial corrections for $\sqrt{s} \ll \Lambda$. A detailed classification of the terms allowed in $\mathcal{L}_5$ and $\mathcal{L}_6$ may be found in Refs.[7].

The new terms contribute to a variety of processes. The strength of each term is characterized by a dimensionless effective coupling parameter which is unknown at this level. A prediction of such couplings would require the detailed knowledge of the underlying theory. The most stringent experimental limits on the allowed strengths of such additional interaction terms come from rare decays and from high-energy scattering experiments, the subject of this paper.

Henceforth, I shall restrict the discussion to contact terms of the four-fermion type as contained in $\mathcal{L}_6$. They are relevant for a prominent class of $f_1 f_2 \to f_3 f_4$ (sub) processes to be studied at pp, $\bar{p}p$, $e^{\pm}p$ and $e^+ e^-$ colliders.

These contact terms may be visualized to arise predominantly from the exchange of virtual, new bosons X which, on the scale $\sqrt{s}$, are too heavy to propagate. Integrating these particles out amounts on the tree level to 'collapsing' the respective exchange diagrams into contact terms as depicted in Fig. 1.

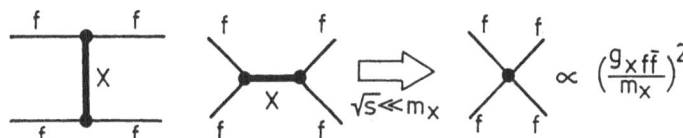

Figure 1. Heavy boson exchanges giving rise to four-fermion contact terms of effective strength $\propto (g_{Xf\bar{f}}/m_X)^2$.

By comparing the result with the expected effective strength $(g_{eff}/\Lambda)^2$ of a four-fermion contact term, one infers (c.f. Fig. 1)

$$g_{eff} = g_{Xf\bar{f}} \qquad \text{and} \qquad \Lambda \propto m_X. \tag{2}$$

Of course, only the ratio g_{eff}/Λ can be determined experimentally. In the context of quark and lepton compositeness it has become customary to quote, nevertheless, experimental bounds for Λ by adopting the rather *ad hoc* normalization

$$g_{eff}^2/4\pi \equiv 1, \tag{3}$$

suggested only by a non-compulsive analogy to hadron physics. Since, in this paper I want to take a more general attitude as to the nature of the new physics involved, it is sensible to consider instead the scale

$$\Lambda^\star = \Lambda/\sqrt{g_{eff}^2/4\pi}, \tag{4}$$

which is both directly measurable and such that

$$\Lambda^\star = \Lambda \sim m_X \qquad \text{if} \qquad g_{eff}^2 \simeq 4\pi. \tag{5}$$

However, if the contact term is e.g. due to the exchange of a distant Z' vector boson from some underlying GUT theory, say, one expects

$$g_{eff} \sim g_{weak} \simeq 0.64 \qquad \text{and then} \qquad \Lambda^\star \simeq 5.5\Lambda \gg m_X \tag{6}$$

Let me illustrate the preceeding discussion with a very instructive example which I shall come back to also in Sect. 2.3. We consider $\sqrt{s} \ll m_{W,Z}$ such that the *weak* neutral current interactions due to Z exchange degenerate into the well known four-fermion *contact interactions*

$$\delta \mathcal{L}_{eff}^{NC} = -\frac{g_{weak}^2}{2m_W^2} J_{NC}^\nu J_{NC,\nu} \tag{7}$$

providing a small perturbation to pure QED processes. For example, by picking out of the product of weak neutral currents (7) the contact terms contributing to $e^+ e^- \to \mu^+ \mu^-$, one finds with $s^2 \equiv \sin^2 \theta_W$

$$\delta \mathcal{L}_{eff}^{\bar{e}e\bar{\mu}\mu} = \frac{(2-4s^2)^2}{8\Lambda_F^2} [\eta_{LL}(\bar{\mu}_L \gamma^\nu \mu_L)(\bar{e}_L \gamma_\nu e_L) + \eta_{LR}(\bar{\mu}_L \gamma^\nu \mu_L)(\bar{e}_R \gamma_\nu e_R) +$$
$$+ \eta_{LR}(\bar{\mu}_R \gamma^\nu \mu_R)(\bar{e}_L \gamma_\nu e_L) + \eta_{RR}(\bar{\mu}_R \gamma^\nu \mu_R)(\bar{e}_R \gamma_\nu e_R)] \tag{8}$$

where

$$\Lambda_F = \frac{2m_W}{g_{weak}} \simeq 246 \text{ GeV}, \tag{9}$$

and

$$\eta_{LL} \equiv -1; \; \eta_{LR} = \frac{s^2}{\frac{1}{2} - s^2}; \; \eta_{RR} = -\left(\frac{s^2}{\frac{1}{2} - s^2}\right)^2, \tag{10}$$

i.e. the η's factorize, $\eta_{LL} \eta_{RR} = \eta_{LR}^2$, and are defined such as to satisfy

$$\max(|\eta|) = 1 \text{ and } |\eta| \le 1. \tag{11}$$

By comparing Eq. (8) with the generic ansatz for contact terms (see Sect. 2.2) with strength $(g_{eff}/\Lambda)^2$ and identifying $g_{eff} = g_{weak} \simeq 0.64$, we find

$$\Lambda_{e\mu} = \frac{4\cos\theta_W}{2-4s^2} m_Z \simeq 0.29 \; TeV \tag{12}$$

$$\Lambda^\star_{e\mu} = \frac{4\sqrt{\pi}\Lambda_F}{2-4s^2} \simeq 1.6 \; TeV. \tag{13}$$

Thus, in this example, $\Lambda^\star$ is directly related to the scale of electroweak symmetry breaking, Λ_F, defined in analogy to the scale of chiral flavor symmetry breaking, $f_\pi \simeq 95$ MeV, appearing in an effective lagrangian description of low energy hadron physics. In fact, $\Lambda^\star$ precisely corresponds to the energy, where (partial wave) unitarity would be violated if there was no physical Higgs boson in the Standard Model[8]! On the other hand, the scale Λ being much smaller than $\Lambda^\star$ due to the smallness of g_{weak}, is directly related to the mass of the 'new particle' Z^0 (as seen from low-energy QED).

After these attempts to illuminate contact interactions in general and the meaning of the associated scales Λ and $\Lambda^\star$, let me now introduce the form of the contact interactions as relevant for future collider experiments.

2.2 Four-Fermion Contact Terms

Let me start by listing the various types of colliders together with the $f_1 f_2 \to f_3 f_4$ (sub) processes/contact terms of interest

Table 1

collider	four-fermion (sub) processes to be probed		
e^+e^-	$e^+e^- \to l^+l^-$	$e^+e^- \to \bar{q}q$	—
$e^\pm p$	—	$e^\pm q \to e^\pm q$	—
$pp, \bar{p}p$	—	$\bar{q}q \to \bar{l}l$	$qq \to qq,\ \bar{q}q \to \bar{q}q$

We note the *unique*, complementary ability of e^+e^- and pp or $\bar{p}p$ colliders to provide information on possible four-fermion contact terms involving *only* leptons and *only* quarks, respectively. $e^\pm p$ colliders, instead, specialize on probing mixed $e-q$ interaction terms. These can be related via crossing or T invariance to mixed $e-q$ interaction terms at e^+e^- and pp or $\bar{p}p$ colliders. Clearly, the various scales Λ_{ll}, Λ_{lq} and Λ_{qq} may, *a priori*, be drastically different and, thus, the complementary information from different types of future colliders is vital for a *systematic, model independent* search programme.

Unfortunately, the most general ansatz for the contact terms involves (too) many unknown parameters. Therefore, a traditional restriction has been to include in experimental analyses or in simulations only those $SU(3)_c \times SU(2)_L \times U(1)_Y$ invariant four-fermion contact terms which conserve both helicity and flavor and correspond to color and isospin singlet exchanges,[2].

$$\delta\mathcal{L}_{eff}^{4-f} = \frac{4\pi}{\Lambda^{\star\,2}} \sum_{i,j} \frac{1}{1+\delta_{ij}} [(\bar{f}_L^i \gamma^\mu f_L^i)(\bar{f}_L^j \gamma_\mu f_L^j)\eta_{LL}^{ij} + (\bar{f}_R^i \gamma^\mu f_R^i)(\bar{f}_R^j \gamma_\mu f_R^j)\eta_{RR}^{ij} +$$
$$+ (\bar{f}_R^i \gamma^\mu f_R^i)(\bar{f}_L^j \gamma_\mu f_L^j)\eta_{RL}^{ij} + (\bar{f}_L^i \gamma^\mu f_L^i)(\bar{f}_R^j \gamma_\mu f_R^j)\eta_{LR}^{ij}]. \tag{14}$$

Here, i and j run through 'up' and 'down' quarks and leptons. The ignorance about the helicity structure in Eq. (14) is reflected in the unknown weights η with $|\eta| \leq 1$ and the largest $|\eta| \equiv 1$. $SU(2)_L$ invariance requires for 'up' and 'down' flavors

$$\eta_{LL}^{ij} = \eta_{LL};\ \eta_{RL}^{ij} = \eta_{RL}^i;\ \eta_{LR}^{ji} = \eta_{LR}^i. \tag{15}$$

Moreover,

---

[2]It is, indeed, hard to imagine that these terms -if present- are abnormally suppressed by (approximate) symmetries. In contrast, e.g. helicity changing contact terms are expected to involve extra suppression factors $\propto (m_{fermion}/\Lambda)^n$ due to an approximate chiral symmetry usually associated with light fermions

$$\eta^{ii}_{RL} = \eta^{ii}_{LR}. \tag{16}$$

Hence, at this level, one encounters, in principle,

- η_{LL}, η_{LR} and η_{RR} in case of $e^+e^- \to e^+e^-$,

- η_{LL}, η^i_{LR}, η^i_{RL} and η^i_{RR}; $i = u, d$ in case of $e^\pm q \to e^\pm q$,

- η_{LL}, η^i_{LR} and $\eta^{ij}_{RR} = \eta^{ji}_{RR}$; $i = u, d$ in case of $qq \to qq$ or $\bar{q}q \to \bar{q}q$.

It has become customary, to concentrate on standardized helicity configurations as summarized in Table 2. As to the four-quark contact terms, to my knowledge, only the first configuration, LL_+, in Table 2 has actually been taken into account in numerical analyses up to now.

Table 2. Standard helicity configurations for the contact terms

configuration	η_{LL}	η_{RR}	η_{RL}	η_{LR}
$LL_\pm$	± 1	0	0	0
$RR_\pm$	0	± 1	0	0
$RL_\pm$	0	0	± 1	0
$LR_\pm$	0	0	0	± 1
$VV_\pm$	± 1	± 1	± 1	± 1
$AA_\pm$	± 1	± 1	∓ 1	∓ 1
$WCS_\pm$	± 1	∓ 1	0	0

In the bottom line of Table 2, I have included a peculiar choice of η values, corresponding to the so-called 'worst-case-study' (WCS) configuration[9,6,10,11,13]. It is motivated from the observation that often the dominant sensitivity to contact terms arises from their *interference* with the exchanged, *massless* vector bosons (γ, gluons) of the Standard Model. The WCS configuration in Table 2 just corresponds to combinations of η's such as to precisely cancel this interference term in the *unpolarized* cross section. As we shall see, however, the WCS configuration is only 'worst' in certain observables and processes, which illustrates already the importance of a systematic investigation at different colliders using a variety of observables (polarization!). It should also be mentioned that the WCS configuration is perhaps not naturally expected to occur in *elastic* (sub) processes, since there for exchanges of new, heavy vector particles, η_{RR} and η_{LL} represent sums of couplings squared and, therefore, cannot have different signs.

In a new analysis for future pp colliders in Sect. 3.3, I shall also consider the effects from contact terms due to the most general helicity conserving, $SU(3)_c \times SU(2)_L \times U(1)_Y$ invariant contact interaction among up and down quarks[4,5]. This amounts to including four additional terms in Eq. (14)

$$
\begin{aligned}
\delta \mathcal{L}'_{eff} = \ & \delta \mathcal{L}^{4-f}_{eff} + \\
& + \frac{4\pi}{\Lambda^{\star 2}} \{ \eta^1_{LL}(\bar{q}_L \gamma^\mu \frac{\tau^a}{2} q_L)(\bar{q}_L \gamma_\mu \frac{\tau^a}{2} q_L) + \eta'^{ud}_{RR}(\bar{u}_R \gamma^\mu d_R)(\bar{d}_R \gamma_\mu u_R) + \\
& + \eta^{8u}_{LR}(\bar{q}_L \gamma^\mu \frac{\lambda^A}{2} q_L)(\bar{u}_R \gamma_\mu \frac{\lambda^A}{2} u_R) + \eta^{8d}_{LR}(\bar{q}_L \gamma^\mu \frac{\lambda^A}{2} q_L)(\bar{d}_L \gamma_\mu \frac{\lambda^A}{2} d_R) \}
\end{aligned} \tag{17}
$$

with

$$q_L = (u,d)_L^t, \tag{18}$$

and the three $SU(2)_L$ and eight $SU(3)_c$ generators $\tau^a/2$ and $\lambda^A/2$, respectively. Thus, there are 10 independent four-quark contact terms of strengths

$$\eta_{LL}, \; \eta_{LL}^1, \; \eta_{LR}^u, \; \eta_{LR}^d, \; \eta_{LR}^{8u}, \; \eta_{LR}^{8d}, \; \eta_{RR}^{uu}, \; \eta_{RR}^{ud}, \; \eta_{RR}^{\prime ud}, \; \eta_{RR}^{dd}, \tag{19}$$

the contributions of which to the various $qq \to qq$ cross sections is found after a lengthy calculation[12]

$$\frac{d\hat{\sigma}}{d\hat{t}}(ij \to i'j') = \frac{\pi}{\hat{s}^2}|A(ij \to i'j')|^2, \tag{20}$$

with $(i = u, d)$

$$
\begin{aligned}
|A(ii \to ii)|^2 &= \alpha_s(Q^2)^2 \{ \frac{4}{9}(\frac{\hat{s}^2 + \hat{u}^2}{\hat{t}^2} + \frac{\hat{s}^2 + \hat{t}^2}{\hat{u}^2} - \frac{2}{3}\frac{\hat{s}^2}{\hat{t}\hat{u}}) + \\
&+ \frac{8}{9}(\frac{\hat{s}^2}{\hat{t}} + \frac{\hat{s}^2}{\hat{u}})(\mu_{LL} + \mu_{RR}^{ii}) + \frac{8}{9}(\frac{\hat{u}^2}{\hat{t}} + \frac{\hat{t}^2}{\hat{u}})\mu_{LR}^{8i} + \\
&+ \frac{8}{3}\hat{s}^2(\mu_{LL}^2 + (\mu_{RR}^{ii})^2) + (\hat{u}^2 + \hat{t}^2)(\frac{4}{9}(\mu_{LR}^{8i})^2 + 2(\mu_{LR}^i)^2)\},
\end{aligned} \tag{21}
$$

and

$$
\begin{aligned}
|A(ud \to ud)|^2 &= \alpha_s(Q^2)^2 \{ \frac{4}{9}\frac{\hat{s}^2 + \hat{u}^2}{\hat{t}^2} - \frac{8}{9}\frac{\hat{s}^2}{\hat{t}}(\mu_{LL}^1 + \mu_{RR}^{\prime ud}) + \frac{4}{9}\frac{\hat{u}^2}{\hat{t}}(\mu_{LR}^{8u} - \mu_{LR}^{8d}) + \\
&+ \hat{s}^2[\mu_{LL}^2 + \frac{8}{3}(\mu_{LL}^1)^2 - \frac{8}{3}\mu_{LL}\mu_{LL}^1 + (\mu_{RR}^{ud})^2 + (\mu_{RR}^{\prime ud})^2 - \frac{2}{3}\mu_{RR}^{ud}\mu_{RR}^{\prime ud}] + \\
&+ \hat{u}^2[(\mu_{LR}^u)^2 + (\mu_{LR}^d)^2 + \frac{2}{9}((\mu_{LR}^{8u})^2 + (\mu_{LR}^{8d})^2)]\},
\end{aligned} \tag{22}
$$

with the abreviations

$$\mu = \frac{\eta}{\alpha_s(Q^2)\Lambda^{*\,2}}. \tag{23}$$

All remaining cross sections are obtained from Eqs. (21, 22) by $\hat{s} \leftrightarrow \hat{u}$ and $\hat{s} \leftrightarrow \hat{t}$ crossing. First of all, let me point out that the contributions from the square of the contact terms to the $uu \to uu$ and $ud \to ud$ cross sections as quoted for the special case LL in Refs. [3,4,5], are actually incorrect[3]. By using the wrong form, one obtains significantly lower values of Λ^* in pp collisions.

The generalized ansatz, Eqs. (21, 22), exhibits the following features in comparison with the usual LL form.

- Apparently, seven out of the ten helicity couplings occur in interference terms which will to some extent enhance the sensitivity to these parameters. The cross section for $ud \to ud$ (and $u\bar{d} \to u\bar{d}, u\bar{u} \to d\bar{d}, \ldots$) now may have an interference term, too.

---

[3]I thank Estia Eichten for a communication on this point.

- For the case of all η's contributing equally with equal signs $(VV_{\pm})$, one expects a significant improvement[6] in the 'reach' (c.f. Sect. 3.3)

$$\Lambda^{\star}{}_{VV} \approx 1.5 \Lambda^{\star}{}_{LL}, \tag{24}$$

as compared to previous studies of pp and $\bar{p}p$ collisions, where only the $LL_{\pm}$ configurations were considered.

2.3 Analytical Insight

Usually, in real experiments or simulations, the 'reach' $\Lambda^{\star}$ for contact interactions is obtained via numerical fits. They certainly provide the most precise results. However, notably in case of simulations for future colliders, the operation of which will start years from now, it seems appropriate to trade some precision for analytical insight into the crucial dependences of $\Lambda^{\star}$ on energy $\sqrt{s}$, integrated luminosity $\int L dt$ and the process in question.

Therefore, in this Section, I shall discuss a simple scaling law and semi quantitative analytical expressions[6,9,10] for $\Lambda^{\star}$. Such results are particularly useful for quickly estimating and comparing the potential 'sensitivity' of future colliders to new physics. A comparison with 'state of the art' numerical simulations is deferred to Sect. 3.

Let me begin with the scaling law.

In our problem, there are, in general, four relevant *dimensionful* quantities

$$\Lambda^{\star} , \sqrt{s}, \int L dt \text{ and } \Lambda_{QCD}. \tag{25}$$

The QCD scale Λ_{QCD} enters, in principle, to characterize the effects from QCD scaling violations. However, even in case of hadron colliders where the cross sections are

$$\hat{\sigma} \propto \alpha_s (Q^2/\Lambda_{QCD}^2)^2, \tag{26}$$

and hadronic structure functions enter, the QCD scaling violations turn out to largely cancel in expressions for the bounds on $\Lambda^{\star}$, if extracted from suitable observables.

A bound on $\Lambda^{\star}$ is defined via a 'confidence' criterion, like a χ^2. The χ^2-function is a complicated expression, involving summation over (real or simulated) data, errors and theory depending on $\Lambda^{\star}$. Nevertheless, the χ^2 is *dimensionless* and a typical 95% confidence level criterion

$$\Delta\chi^2 = 4, \tag{27}$$

then should depend only on dimensionless *ratios* of our dimensionful quantities (25), while any additional dimensionless (kinematical) variables are summed (integrated) over. Thus, ignoring effects from Λ_{QCD}, there are only two independent ratios, $\Lambda^{\star}/\sqrt{s}$ and $\int L dt/s$, say, and the criterion (27) implicitly defines

$$\frac{\Lambda^{\star}{}_{95\%CL}}{\sqrt{s}} = f(\frac{\int L dt}{s}). \tag{28}$$

Clearly, the functional dependence in this simple scaling law depends on the details of the specific criterion applied (like kinematical cuts, systematical errors, etc.), the observable from which $\Lambda^{\star}$ is extracted and the process under consideration[4].

---

[4]More involved criteria or extraction of $\Lambda^{\star}$ from certain observables (c.f. Ref. [4]), in fact, invalidate the simple scaling (28), typically through an additional dependence on $\sqrt{s}/\Lambda_{QCD}$.

538

In Sect. 3.3, we shall confront Eq. (28) to realistic numerical simulations, e.g. the complex case of pp collisions and illustrate its validity and usefulness.

At this point, we may draw two conclusions:

- Future colliders satisfying $\int L dt/s \approx const.$ (which is advisable for reasons of rate), give

$$\Lambda^\star \propto \sqrt{s}. \tag{29}$$

- Knowing the function $f(\int L dt/s)$ in Eq. (28) for a range of $\int L dt$ and/or s values either from present experiments or existing simulations, allows an immediate evaluation of $\Lambda^\star$ for any different s but fixed $\int L dt/s$.

What can we say about the functional form entering Eq. (28)?

Let us concentrate on the c.m. angular distribution $d\hat\sigma/dz$, $z = cos\theta$, of a four-fermion (sub) process with total c.m. energy $\sqrt{\hat{s}}$. Of course, for $e^+ e^- \to \bar f f$, $\hat{s} = s$, $\hat\sigma = \sigma$.

After adding the contact terms to the Standard Model (SM) amplitudes, the cross section may be written as

$$\frac{d\hat\sigma}{dz_{|\Lambda^\star}} = \frac{d\hat\sigma}{dz_{|SM}} + 2\sqrt{\frac{d\hat\sigma}{dz_{|SM}}}\frac{\sqrt{\hat{s}}}{\Lambda^{\star\,2}}\{\eta h_1(z) + h_2(z)(\frac{\eta\sqrt{\hat{s}}}{\Lambda^\star})^2\} \tag{30}$$

The first term in the bracket is due to an *interference* of the contact term with the SM amplitude and often dominates (typically in $e^+ e^-$ collisions). The second term corresponds to the square of the contact term. It gains increasing importance for very high energy pp or $\bar p p$ collisions. The parameter η is to characterize the helicity configuration of the contact term, $|\eta| \leq 1$, as in Eq. (14). Moreover, one typically finds

$$\int_{-1}^{+1} dz h_1^2(z) = O(1). \tag{31}$$

For $e^+ e^- \to \bar f f$, the situation is then quite simple.

Ignoring systematical errors and keeping only the interference term in (30), we compare by means of the χ^2-function the relative deviation

$$\Delta \equiv \left(\frac{\frac{d\sigma}{dz}_{|\Lambda^\star}}{\frac{d\sigma}{dz_{|SM}}} - 1\right) \simeq 2\frac{h_1(z)}{\sqrt{\frac{d\sigma}{dz_{|SM}}}}\eta\frac{\sqrt{s}}{\Lambda^{\star\,2}}, \tag{32}$$

with the statistical errors

$$\epsilon_\Delta = \frac{1}{\sqrt{dN(z)}} = \frac{1}{\sqrt{\frac{d\sigma}{dz_{|SM}}\Delta z \int L dt}}, \tag{33}$$

where $dN(z)$ is the number of events falling between z and $z + \Delta z$.

The 95% CL bound on $\Lambda^\star$ is obtained from

$$\Delta\chi^2 = \sum_{z_i} \frac{\Delta^2}{\epsilon_\Delta^2} \qquad (=4)$$

$$= \underbrace{[\sum_{z_i} \Delta z h_1^2(z_i)]}_{\approx 1 \text{from } (31)} \frac{4\eta^2 s \int L dt}{\Lambda^{\star\,4}}. \tag{34}$$

Solving for $\Lambda^\star / \sqrt{s}$, one finds the simple formula[6,9,10,11]

$$\frac{\Lambda^\star}{\sqrt{s}} \simeq (4\eta^2/\Delta\chi^2)^{1/4} \left(\frac{\int L dt}{s}\right)^{1/4}, \tag{35}$$

of course in agreement with the more general scaling form (28).

We conclude from Eq. (35)

- Since $\Lambda^\star$ increases only with the fourth root of $\int L dt$, 'patience' does not really pay off!

- The numerical coefficient $(4\eta^2/\Delta\chi^2)^{1/4}$ is typically $O(1)$ and - due to the fourth root involved - only weakly dependent on the χ^2 criterion, a restricted angular acceptance or the details of the Standard Model amplitudes.

For a first, pedagogical application of formula (35), let us return to our illustration in terms of the four-fermion contact term limit of the weak interactions from Sect. 2.1.

We may use Eq. (35) to immediately quote the luminosity needed in an e^+e^- machine to observe effects from the *weak* scale $\Lambda^\star_{weak} \simeq 1.6$ TeV, Eq. (13), in $e^+e^- \to \mu^+\mu^-$, say. Eq. (31) reads in this case

$$\int_{-1}^{+1} dz h_1^2(z) = \frac{2\pi}{(2 - 4\sin^2\theta_W)^4} \int_{-1}^{+1} dz \frac{z^2}{1 + z^2} = \frac{\pi(4 - \pi)}{(2 - 4\sin^2\theta_W)^4} \simeq 2.0, \tag{36}$$

such that from Eq. (35)

$$\left(\int L dt\right)_{95\%CL} \simeq \left(\frac{\pi}{4 - \pi}\right) \frac{(4\Lambda_F)^4}{s}. \tag{37}$$

In Fig. 2, the luminosity needed according to Eq. (37) is displayed as a function of $\sqrt{s}$ together with the luminosity taken by HRS/ PEP[14], JADE/ PETRA [15] and AMY/ TRISTAN [16]. While at the lower energies rather high statistics experiments are needed to spot the *weak* contact terms, statistics is no problem for TRISTAN due to the higher energy available.

For ep and pp or $\bar{p}p$ collisions it becomes an increasingly complex task to derive useful approximations for the functional form appearing in the scaling law (28). The complications come, of course, from the nucleon structure functions and for pp and $\bar{p}p$ collisions, in particular, from the gg and qg subprocesses which dominate over the four-quark subprocesses in a wide range of phase space.

Nevertheless, let me make a few 'analytical' remarks on the complex case of pp and $\bar{p}p$ collisions.

As will be argued in more detail in Sect. 3.3, a very suitable observable for extracting bounds on $\Lambda^\star$ is the 2-jet angular distribution, normalized to the integrated

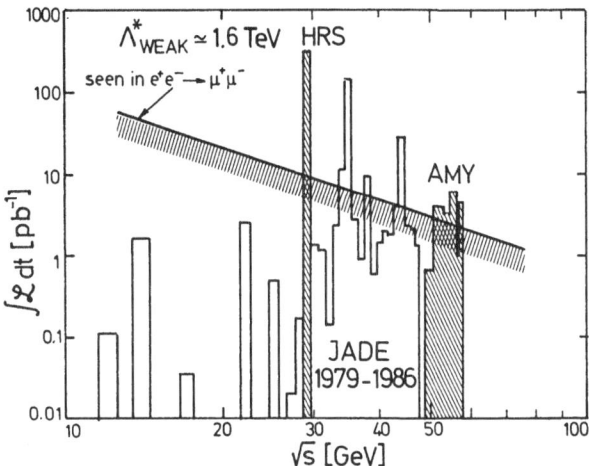

Figure 2. Spotting effects from *weak* contact terms in $e^+e^- \to \mu^+\mu^-$. The necessary luminosity according to Eq. (37) is displayed along with the one taken by representative PEP, PETRA and TRISTAN experiments.

2-jet cross section[17,18]. In order to enhance the signal from possible contact interactions relative to the qg and gg subprocesses, one has to consider sufficiently large 2-jet masses $m_{2-jet} \equiv \sqrt{\hat{s}}$

$$\sqrt{\tau} = m_{2-jet}/\sqrt{s} \geq 0.25, \text{ say}, \tag{38}$$

as apparent from Fig. 3.

In this domain of $\hat{s}$ the square of the contact term becomes increasingly important. By repeating the same steps as above, however, taking into account in Eq. (30) the square of the contact term only, one arrives at a similar formula as (35) with the power 1/4 replaced by 1/8.

In conclusion, we may expect in pp or $\bar{p}p$ collisions a behaviour like in (35), but with a power around 1/6, say.

3 Selected Results From Simulations

3.1 e^+e^- Collisions - A Clean Reference Case

A systematic (numerical) study of the 'reach' Λ^* for possible new contact interactions at e^+e^- colliders has been performed in Ref. [10,11,13]. Let me summarize some essential aspects and also compare the numerical results for Λ^* with the simple analytical formula (35), as derived in Sect. 2.3. The reactions under consideration were $e^+e^- \to e^+e^-, \mu^+\mu^-$. The analysis is distinguished by the following features

- A large energy range was considered, from PETRA/ PEP energies at the lower end, via the Z^0 peak (LEP/ SLC) and LEP II, up to the TeV regime of a potential future linear e^+e^- collider (CLIC).

- Realistic assumptions on the expected luminosities and detector performances were made.

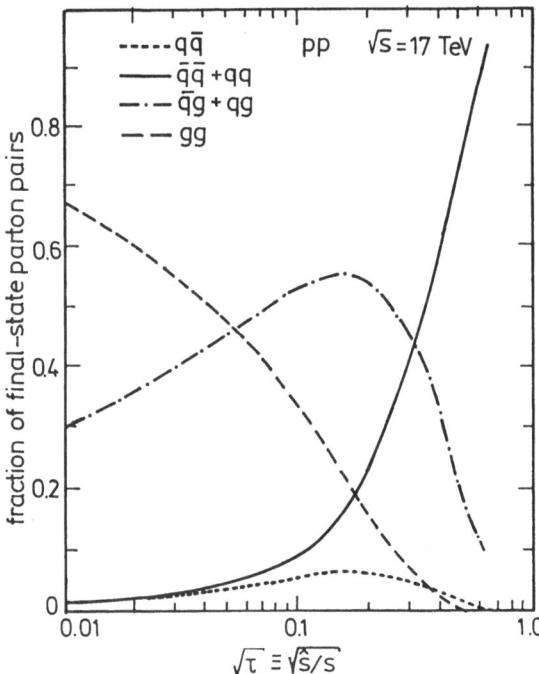

Figure 3. Fraction of final-state parton pairs versus $\sqrt{\tau}$ in case of the LHC, from Ref. [18].

- Besides the expected statistical errors, a systematic error of 3% coming from the error on the luminosity measurement was taken into account.

- The bounds for Λ^* resulting from simulated 'data' were shown to agree with the actual experimental limits from PETRA and PEP within the range allowed by the various experimental measurements.

- The impact of *longitudinal* and transverse beam polarization was studied in detail (c.f. Sect. 3.2).

- The question of radiative corrections was considered.

Fig. 4 illustrates the quality of expected data and the effects of contact interactions in representative helicity configurations (c.f. Table 2) for LEP II with $\int L dt = 500\ pb^{-1}$ and $\Lambda^* = 5\ TeV$. Plotted is the deviation

$$\Delta \equiv \left(\frac{d\sigma}{dz_{|\Lambda^*}} \Big/ \frac{d\sigma}{dz_{|SM}} - 1 \right),\tag{39}$$

versus $z = cos\theta$.

The analytic formula for Λ^*, Eq. (35), suggests to define a *reduced* 'reach' Λ_{red}, by dividing out the machine dependent 'patience' factor $(\int L dt/s)^{1/4}$,

$$\Lambda_{red} = \Lambda^*{}_{95\%CL} \left[\frac{s}{(190\ GeV)^2} \frac{500\ pb^{-1}}{\int L dt} \right]^{1/4} \equiv \Lambda^*{}_{LEP\ II}.\tag{40}$$

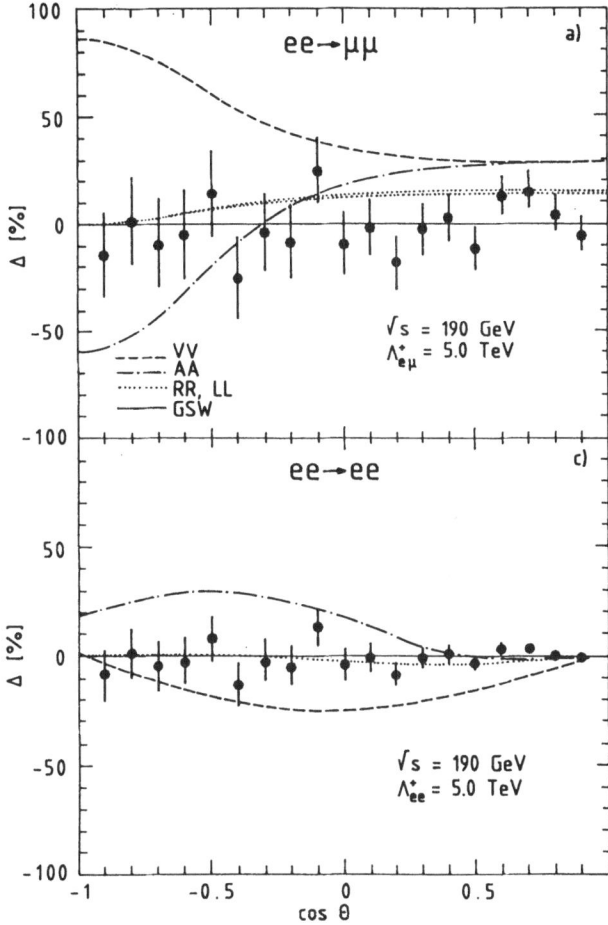

Figure 4. Quality of expected data (LEP II) and the effect of contact terms on $d\sigma/dcos\theta$ for $e^+e^- \rightarrow \mu^+\mu^-$ (a), and $e^+e^- \rightarrow e^+e^-$ (c). Δ is the relative deviation from the Standard Model cross section.

In Eq. (40), I have normalized to the conditions expected at LEP II within a few years of operation. Moreover, it so happens that $\Lambda_{red} = \Lambda^*$ if $\int Ldt \approx$ 65 pb^{-1}, 100 pb^{-1} and 15 fb^{-1} for PETRA (47 GeV), LEP and CLIC (1 TeV), respectively, which allows a uniform comparison of the 'reach' for these e^+e^- colliders.

Fig. 5 shows the results for Λ_{red} from the numerical simulation as a function of $\sqrt{s}$ from 30 GeV to 5 TeV. The solid lines correspond to the two extreme helicity configurations VV and WCS, as well as to LL. Indeed, apart from the Z^0 pole, Λ_{red} shows to very good approximation a $\sqrt{s}$ behaviour, as expected from Eqs. (35, 40).

Energy is, therefore, the determining factor for the 'reach'. On the Z^0 pole, the Standard Model amplitude is imaginary and, thus, cannot interfere with the real contact term. Since in e^+e^- collisions the interference term is crucial, the sensitivity is very bad right on the Z^0!(see also the inserts in Fig. 5). The general importance of the interference term in e^+e^- collisions is also made obvious by the strong suppression of the WCS configuration (c.f. Sect. 2.3).

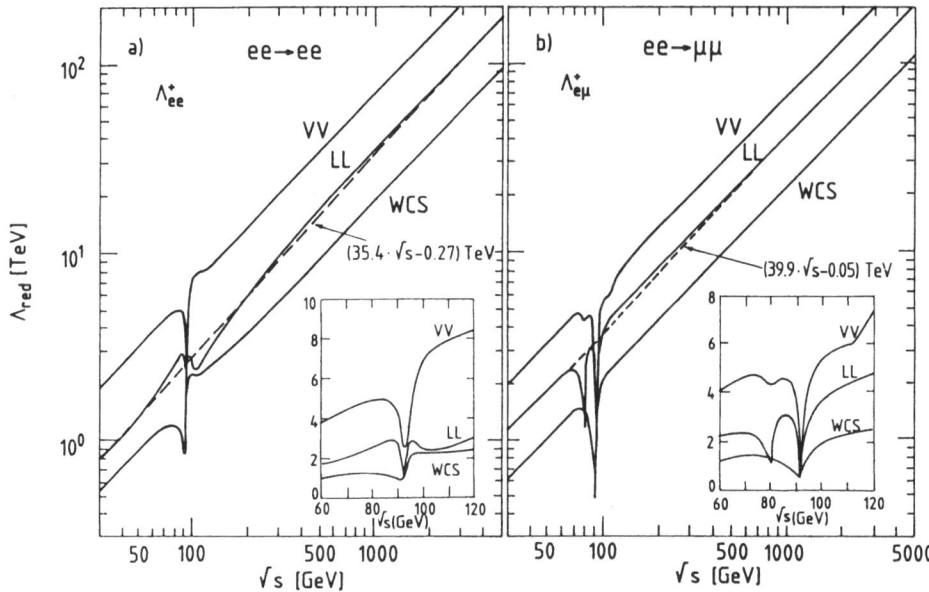

Figure 5. The reduced 'reach', Λ_{red} of Eq. (40) for e^+e^- colliders as a function of $\sqrt{s}$ behaving as expected $\propto \sqrt{s}$ (dashed line). The most favorable (VV), the most insensitive (WCS) and the LL helicity configurations are displayed. The inserts magnify the 'insensitive' region around the Z^0 peak.

3.2 Impact of Longitudinal Polarization at HERA and LEP

The interest in polarized beams for the study of contact terms arises mainly for the following reasons

- By means of longitudinal beam polarization one may hope to increase the sensitivity to particular helicity couplings, where unpolarized data are comparatively insensitive (e.g. WCS, LL, RR in e^+e^-).

- Once a signal has been found in unpolarized data, longitudinal polarization will be crucial in disentangling the various helicity couplings.

The impact of polarization on contact interactions has been extensively studied for $e^{\pm}p$ collisions (HERA, LHC-ep) in Refs. [19,20,21] and for e^+e^- collisions in Ref. [10,11].

One may hope that at a later stage of operation longitudinally polarized e^- and e^+ beams will be made available at HERA, such that the differential cross sections

$$d^2\sigma(e_{L,R}^{\mp}p \to e_{L,R}^{\mp}X)/dxdQ^2 \qquad (41)$$

may be studied experimentally.

In Fig. 6 the bounds for Λ^*_{eq} as extracted from simulated 'data' [20] for these four cross sections are displayed for HERA with $\sqrt{s} = 314$ GeV, $\int Ldt = 100\ pb^{-1}$ and $P = 80\%$. The ability of polarized cross-section measurements to filter out a particular chirality component in the contact terms is nicely apparent in this type of plot. The LL (RR) couplings are most sensitive to e_L^- (e_R^-) beams and least sensitive to e_L^+ (e_R^+) beams, to which the RL(LR) couplings, in turn, are most sensitive.

544

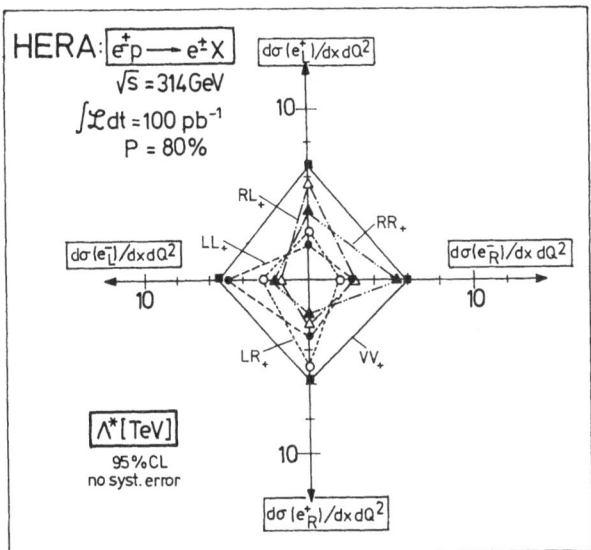

Figure 6. Λ^* extracted from simulated HERA 'data'. The ability of polarized cross-section measurements to filter out a particular chirality component in the contact terms is nicely apparent.

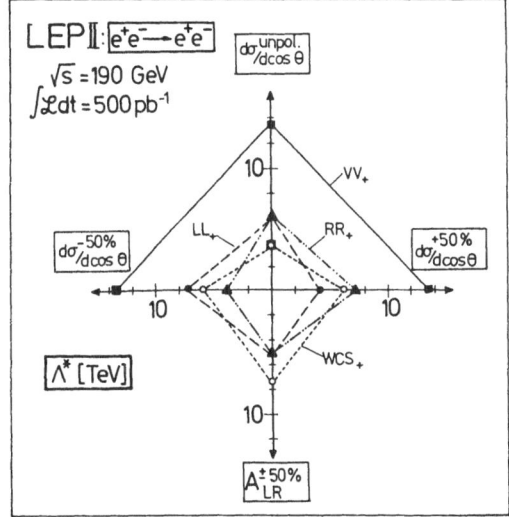

Figure 7. Impact of longitudinal beam polarization at LEP II. While the sensitivity to the WCS (VV) helicity configuration was worst (best) in $d\sigma^{unpol.}/dcos\theta$, it is best (worst) in the left-right asymmetry A_{LR}!

Fig. 7 illustrates an analogous pattern for the case of $e^+e^- \rightarrow e^+e^-$ (LEP II). Here, the bounds for $\Lambda^*_{\bar{e}e}$ corresponding to LL_+, RR_+, VV_+, and WCS_+ couplings from the unpolarized $d\sigma/dcos\theta$ are compared to those from $d\sigma^{\pm 50\%}/dcos\theta$ and from the left-right asymmetry $A_{LR}^{\pm 50\%}$ with +50% and -50% longitudinal beam polarization, respectively. While the sensitivity to WCS (VV) was worst (best) in $d\sigma^{unpol.}/dcos\theta$, it is best (worst) in A_{LR}!

These two examples clearly underlign the virtues of longitudinal beam polarization in the context of contact interactions.

3.3 A New Analysis for pp and $\bar{p}p$ Collisions

Several years ago, a detailed, well known study of the 'reach' Λ^* at future pp and $\bar{p}p$ colliders in the multi-TeV range has been performed[4] (EHLQ). The values for Λ^*_{qq} were extracted from the inclusive jet production cross section, $d^2\sigma/dp_\perp dy_{|y=0}$, at high $p_\perp$, e.g. with the result

$$\Lambda^*_{qq} = O(20 \; TeV) \tag{42}$$

for the SSC(40 TeV) with $\int Ldt = 10^4 \; pb^{-1}$. Only the $LL_\pm$ helicity couplings were considered in this work.

Experimentally, the inclusive jet cross section at high $p_\perp$ is known to be subject to considerable normalization uncertainties and other systematical errors. Given the huge range of extrapolation, the inherent theoretical uncertainties are, indeed, surprisingly small, as pointed out in[4]. Nevertheless, the analysis of $d^2\sigma/dp_\perp dy_{|y=0}$ requires the detailed knowledge of *both* the x and Q^2 dependences of the nucleon structure functions, and the predictions do vary from one parametrization to another in the $p_\perp/\sqrt{s}$ range of interest. In view of these reasons a relatively 'coarse' confidence criterion for extracting Λ^* was applied in the EHLQ analysis, leading, in turn, to somewhat low values of Λ^*.

On the other hand, actual *experimental* studies of the normalized two-jet angular distribution for high 2-jet masses at the $S\bar{p}pS$ indicate[17,22] that this observable is best suited for a reliable extraction of bounds on Λ^*. Experimentally, the systematical errors are fairly small, essentially, because the c.m. scattering angle depends primarily on the measurement of jet directions which can be well determined in a large calorimetric detector. The (normalized) two-jet angular distribution has the advantage that in the limit of vanishing QCD scaling violations it is virtually *independent* of the structure functions, due to an approximate factorization into an effective structure function times an angular distribution corresponding to a single effective subprocess.[23]

Let me summarize next the results from a new analysis of the 'reach' Λ^*, I have performed for pp (and $\bar{p}p$) colliders. For the reasons mentioned above, it is based on (simulated) data for the normalized 2-jet angular distribution

$$\frac{1}{\sigma_{2-jet}(\chi < 19)} \frac{d\sigma^{2-jet}}{d\chi} \quad \text{for} \quad \sqrt{\tau_0} \le \sqrt{\tau} \equiv \frac{m_{2-jet}}{\sqrt{s}} \le \sqrt{\tau_1}. \tag{43}$$

Here, χ is the popular angular variable[23,17]

$$\chi = \frac{1 + \cos\theta}{1 - \cos\theta}, \tag{44}$$

in terms of which the prominent forward (backward) peaks from the exchange of vector particles

$$\frac{d\sigma}{d\cos\theta} \propto \frac{1}{(1-\cos\theta)^2},\tag{45}$$

are transformed into

$$\frac{d\sigma}{d\chi} \simeq const.,\tag{46}$$

such that Eq. (46) represents a good approximation to the angular distribution from the lowest order subprocesses.

Let me point out some important features of the present analysis

1. A large energy range was considered, from $S\bar{p}pS$ energies, via the LHC and SSC regimes, up to a potential 200 TeV super collider (ELOISATRON), to which this workshop is devoted.

2. The kinematical cuts are directly adapted from the experience gained by actual $S\bar{p}pS$ experiments. In particular[17]

    (a) $1 \leq \chi \leq 19$ is used which corresponds to $0 \leq \cos\theta_{2-jet} \leq 0.9$.

    (b) The cross section (43) is integrated over scaled two-jet masses in the range $\sqrt{\tau_0} \leq \sqrt{\tau} \leq \sqrt{\tau_1} = 0.45$ with

    $$\sqrt{\tau_0} = 0.15 \text{ to } 0.35,$$

    depending on $\int Ldt$, such that at least $O(50)$ events are left.

3. A systematic error of 15% is included throughout.

4. EHLQ structure functions, set 1 with $\Lambda_{QCD} = 200$ MeV, are used and $Q^2 = p_\perp^2$ is taken to incorporate QCD scaling violations in the structure functions and via $\alpha_s(Q^2)$.

5. The present analysis is very close in spirit to the one performed[10] for the $e^+e^- \to \bar{f}f$ angular distribution, thus allowing for a direct comparison of the 'reach' at e^+e^- and pp colliders. In particular, the same 95 % CL criterion is used.

6. The most general and correct (c.f. Sect. 2.2) ansatz (17 , 21, 22) for the four-quark contact terms was incorporated.

Fig. 8 illustrates simulated 'data' for the normalized 2-jet angular distribution (43) as a function of χ according to the Standard Model at the $S\bar{p}pS$ and the SSC (40 TeV) with $\int Ldt = 10^4\ pb^{-1}$. The error bars shown include a 15% systematical error besides the statistical one[5]. The 2-jet masses are integrated over in the range corresponding to $0.35 \leq \sqrt{\tau} \leq 0.45$. Also shown in Fig. 8 are the effects of LL_- contact terms (here corresponding to $\Delta\chi^2 \simeq 11$). At 95% CL ($\Delta\chi^2 = 4$) I find

---

[5]While my computer program was checked against a number of different distributions (including absolute normalization) from different authors, there remains a 'mystifying' disagreement with [17] *both* as to the size of the error bars (computed here from the SM cross sections and given $\int Ldt$) and the effect of the contact terms. Amazingly, the end result (47) *agrees!* The normalized χ distribution also agrees with the one shown in[17].

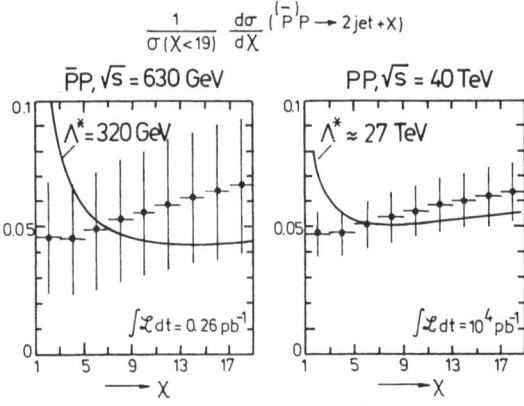

$$\frac{1}{\sigma(X<19)} \frac{d\sigma}{dX} (\overset{(-)}{P}P \to 2\,jet + X)$$

Figure 8. Simulated 'data' according to the Standard Model and effects of contact terms at the $S\bar{p}pS$ and the SSC.

$$\Lambda^* \simeq 392 \text{ GeV} \tag{47}$$

for the $S\bar{p}pS$ with $\sqrt{s} = 630$ GeV and $\int L\,dt = 260\ nb^{-1}$, in good agreement with actual experimental results[17,22].

Fig. 9 displays the dependence of $\sigma_{2-jet}(\chi < 19)$ and Λ^* as extracted from (43) on different parametrizations for the nucleon structure functions[24]. While $\sigma_{2-jet}(\chi < 19)$ varies by as much as a factor of 1.5 both for the $S\bar{p}pS$ and the SSC (40 TeV), Λ^* is very insensitive, indeed, which underligns the reliability of the method used.

The next issue concerns the choice of the starting value $\sqrt{\tau_0}$ in the integration of (43) over the scaled 2-jet masses.

If $\sqrt{\tau_0}$ is decreased, the statistical errors decrease due to a strong rise of σ_{2-jet}, while the systematical errors, of course, remain unaffected. The importance of the contact terms, however, decreases due to the increasing importance of the qg subprocess. If $\sqrt{\tau_0}$ is increased, the importance of the contact terms in the 4-quark cross sections increases, but, eventually, one runs out of statistics!

This is illustrated in Fig. 10a, where $\Lambda^*_{95\% CL}$ has been determined as a function of $\sqrt{\tau_0}$ for the SSC and the ELOISATRON with $\int L\,dt = 10^4\ pb^{-1}$ and $10^3\ pb^{-1}$. The larger luminosities, of course, admit larger $\sqrt{\tau_0}$ values.

Fig. 10b demonstrates that to good approximation the $\sqrt{\tau_0}$ dependence of Λ^* *factorizes* and indicates already that the scaling law (28) holds well for pp collisions. Fig. 10b contains the same points as Fig. 10a. A good parametrization is

$$\frac{\Lambda^*}{\sqrt{s}} \simeq g(\sqrt{\tau_0}, \sqrt{\tau_1}) \left(\frac{\int L\,dt}{s}\right)^{0.155} \tag{48}$$

with $\sqrt{\tau_0}$ optimally chosen in the range 0.15 to 0.35, depending on $\int L\,dt$.

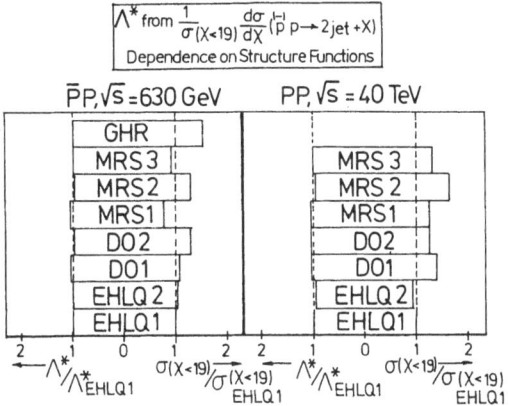

Figure 9. Dependence of $\sigma_{2-jet}(\chi < 19)$ and Λ^* on different parametrizations of the nucleon structure functions[24]. Unlike σ_{2-jet}, Λ^* is quite insensitive which underligns the reliability of the method used to extract Λ^*.

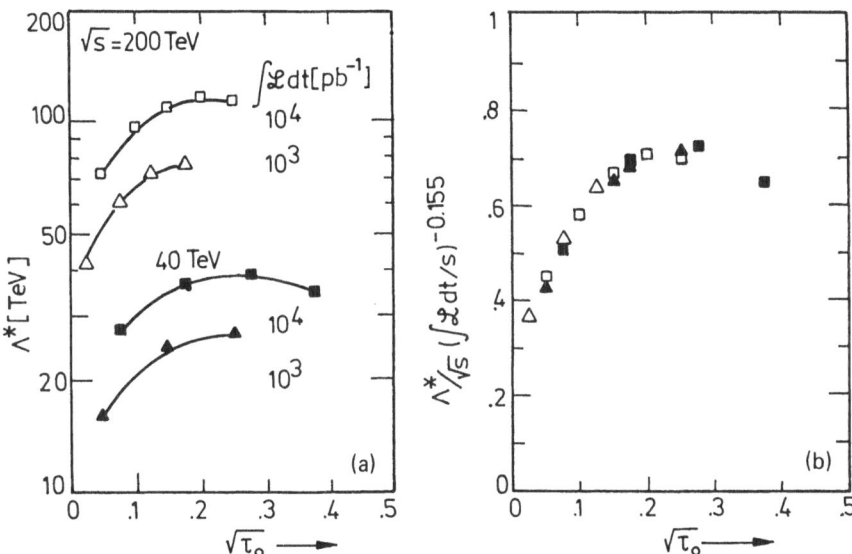

Figure 10. (a) Λ^* as a function of $\sqrt{\tau_0}$; (b) the same points replotted, demonstrating a factorization of the τ_0 dependence.

Figs. 11, 12 contain the main results of this analysis.

Fig. 11a shows the 95% CL bounds on Λ^* for pp colliders as a function of $\sqrt{s}$ for a large range of $\int L dt$. The LL_- helicity configuration is displayed. For the ELOISATRON one finds, for instance,

$$\Lambda^*_{qq}(200\ TeV) \simeq O(100\ TeV), \tag{49}$$

for $\int L dt = 10^4\ pb^{-1}$.

Fig. 11b contains the same points as Fig. 11a, however, this time $\Lambda^* / \sqrt{s}$ is plotted versus $\int L dt/s$.

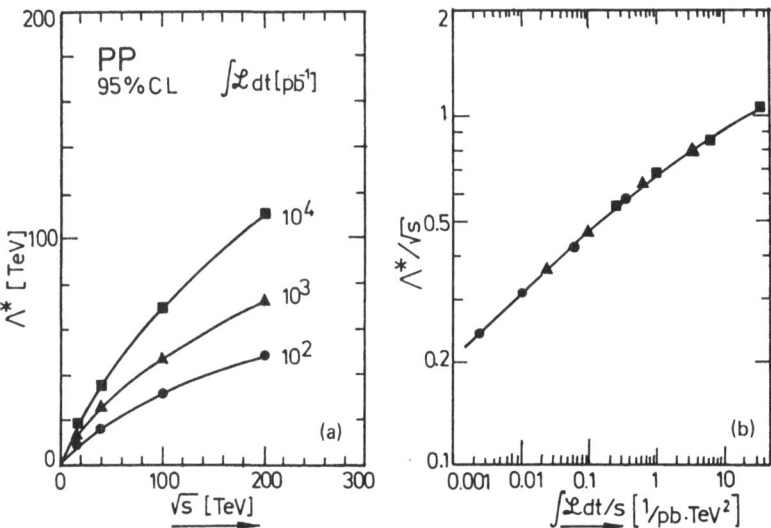

Figure 11. (a) Λ^* at 95% CL versus $\sqrt{s}$ for LL_- helicity couplings and pp colliders with different values of $\int L dt$; (b) same points replottet, demonstrating the validity of the scaling law (28).

All points fall nicely on a *universal* curve and the very good validity of the scaling law (28) becomes apparent! The functional form is approximately a power law[25]

$$\frac{\Lambda^*}{\sqrt{s}} \propto \left(\frac{\int L dt}{s} \right)^\gamma, \tag{50}$$

where

$$\gamma \simeq 1/6, \tag{51}$$

ranging as expected from Sect. 2.3, in between $\gamma = 1/4$ (interference term only) and $\gamma = 1/8$ (square of contact term only). The flattening of the universal curve in Fig. 11b for the larger $\int L dt/s$ values, i.e. larger event numbers, is an effect due to the systematical error which begins to win over the decreasing statistical one there. Nevertheless, the scaling law holds irrespective of systematical errors. There is also no trace of QCD scaling violations visible from Fig. 11b.

The usefulness of the scaling law should be obvious, given the substantial computing effort necessary in pp collisions to extract values of Λ^* . Amazingly, while the scaling curve in Fig. 11b was made up from pp points in the range $\sqrt{s} \geq 17$ TeV and $\int L dt \geq 10^2\ pb^{-1}$, even the $\bar{p}p$ result (47) for the $S\bar{p}pS$ with $\sqrt{s} = 0.63$ TeV and $\int L dt = 0.26\ pb^{-1}$, fits right onto the same curve!

Let me point out again, however, that the validity of the scaling law refers to a given 'confidence' criterion and a particular (suitable) observable from which Λ^* is extracted. For instance, it does not hold as well for the Λ^* values from the EHLQ analysis[4].

Finally, in Fig. 12, Λ^* corresponding to different helicity configurations is displayed versus $\sqrt{s}$ for pp collisions. At high energies, the most sensitive coupling is VV_- i.e. all 10 η's in Eq. (19) being equal to -1, and the least sensitive one is LL_+. Indeed, one finds roughly

$$\Lambda^*_{\ VV_-} \simeq 1.4\Lambda^*_{\ LL_-}, \tag{52}$$

as expected. The WCS configuration, not shown for clarity in Fig. 12, is only somewhat suppressed at lower energies, where the interference term is more important.

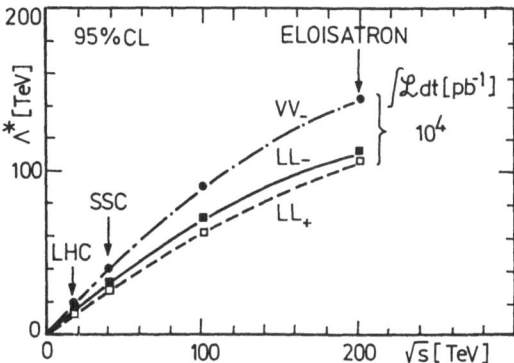

Figure 12. Λ^* at 95% CL versus $\sqrt{s}$ for various helicity couplings and pp colliders with $\int L dt = 10^4\ pb^{-1}$.

4   Conclusions

As we have discussed, the study of contact interactions at future colliders represents a *general* tool to hunt for traces of new physics (new heavy bosons,...) much above the available machine energy.

Fig. 13 contains a summary of the 'reach' Λ^* , as resulting from the various numerical analyses discussed in Sect. 3. The lengths of the white, black and white, and black bars denote the ranges of Λ^* values resulting for the standard helicity couplings of e^+e^-, eq and qq type contact terms, respectively.

For instance, the 'reach' at the ELOISATRON may be expected to correspond to the one at CLIC. HERA ranges between LEP/SLC and LEP II. As explained, the 'reach' right on the Z^0 peak is exceptionally low.

The entries for PETRA, TRISTAN, $S\bar{p}pS$ and the TEVATRON correspond to actual experimental results[17,26] and fit well onto the general trend as a function of energy.

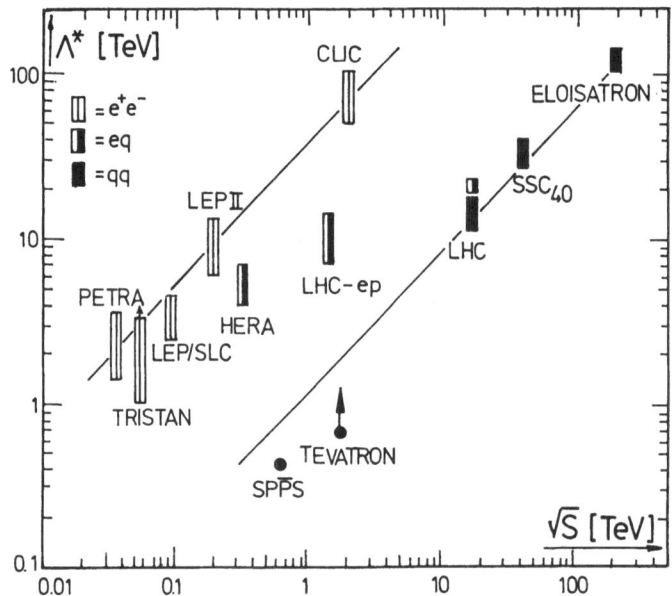

Figure 13. Summary of the 'reach' Λ^* at operating[17,26] and future pp, ep and e^+e^- machines. The two parallel lines are to guide the eye.

I wish to thank A. Ali for the pleasant and lively working atmosphere at the Erice Center, where this ELOISATRON workshop took place, G. Ingelman for providing me with a program for the nucleon structure functions, B. Naroska for experimental advice and, last not least, my wife Barbara for numerous helpful discussions and suggestions.

The support from the Deutsche Forschungsgemeinschaft until the end of 1988 is also acknowledged.

References

[1] for reviews see e.g
J. Ellis, Proc. 1985 Int. Symp. on Lepton and Photon Int.'s at High Energies, Kyoto/Japan;
D.V. Nanopoulos, Rivista del Nuovo Cimento 8 (1985) 1.

[2] for reviews see e.g.
W. Buchmüller, Schladming Lectures 1985, Acta Physica Austriaca, Suppl. XXVII (1985) 517;
F. Schrempp, MPI-Munich Report MPI-PAE/PTh 32/86, 1986 ('Habilitationsschrift').

[3] E. Eichten, K. Lane and M.E. Peskin, Phys. Rev. Lett. 50 (1983) 811.

[4] E. Eichten, I. Hinchliffe, K. Lane and C. Quigg, Rev. Mod. Phys. 56 (1984) 579;
erratum, Fermilab-Pub-86-75 (1986).

[5] E. Eichten, Fermilab-Conf-85/178-T (1986).

[6] B. Schrempp, in "New Aspects of High-Energy Proton-Proton Collisions" (Ed. A. Ali, Plenum), 1989, p. 143.

[7] W. Buchmüller and D. Wyler, Nucl. Phys. B268 (1986) 621;
C.N. Leung, S.T. Love and S. Rao, Z. Phys. C31 (1986) 433.

[8] M. Chanowitz and M.K. Gaillard, Nucl. Phys. B261 (1985) 379
M. Chanowitz, M. Golden and H. Georgi, Phys. Rev. Lett. 57 (1986) 2344

[9] F. Schrempp, Proc. XXIII Int. Conf. on High Energy Physics, Berkeley/USA, 1986, Vol II, p. 1243.

[10] B. Schrempp, F. Schrempp, N. Wermes and D. Zeppenfeld, Nucl. Phys. B296 (1988) 1.

[11] U. Baur, A. Blondel, D. Bloch, D. Dominici, H. Fesefeld, K. Hamacher, L. Levinson, M. Lindner, L. Lyons, P. Mättig, P. Mery, E. Milotti, M. Perottet, F. Renard, D. Schildknecht, B. Schrempp, F. Schrempp, K.H. Schwarzer, D. Treille, N. Wermes and D. Zeppenfeld, Proc. ECFA Workshop on LEP200, Aachen, 1986, ECFA 87-108, 1987, Vol II, p. 414.

[12] B. Schrempp, unpublished.

[13] N. Wermes (B. Schrempp, F. Schrempp and D. Zeppenfeld), Proc. Workshop on Physics at Future Accelerators, La Thuile, CERN 87-07, 1987, Vol II, p. 305.

[14] HRS Collab., P. Baringer et al., Phys. Lett. B206 (1988) 551.

[15] B. Naroska, private communication.

[16] AMY Collab., A. Bacala et al., Phys. Lett. B218 (1989) 112

[17] UA1 Coll., G. Arnison et al., Phys. Lett B177 (1986) 244.

[18] A.K. Nandi, Proc. Workshop on Physics at Future Accelerators, La Thuile, CERN 87-07, 1987, Vol II, p. 270.

[19] R. Rückl, Phys. Lett. 129B (1983) 363; Nucl. Phys. B234 (1984) 91;
F. Cornet, Proc. 25th Int. Winter Meeting on Fundamental Physics, Sevilla/Spain, 1987, and DESY-Report 87-131 (1987).

[20] H.-U. Martyn, Proc. HERA Workshop, DESY, Hamburg/FRG, 1987, (Ed. R.D. Peccei), Vol II, p. 801.

[21] F. Cornet and R. Rückl, Proc. Workshop on Physics at Future Accelerators, La Thuile, CERN 87-07, 1987, Vol II, p. 287.

[22] UA2 Collab., P. Bagnaia et al., Phys. Lett. B138 (1984) 430.

[23] B.L. Combridge and C.J. Maxwell, Nucl. Phys. B239 (1984) 429.

[24] EHLQ1,2: Ref.[4];

DO1,2: D. Duke and J.F. Owens, Phys. Rev. D30 (1984) 49;

GHR: M. Glück, E. Hoffmann and E. Reya, Z. Phys. C13 (1982) 119;

MRS1,2,3: A.D. Martin, R.D. Roberts and J.Stirling, private communication by G. Ingelman.

[25] C.H. Llewellyn Smith, Proc. Second Hellenic Summer School, Corfu/Greece, 1985.

[26] PLUTO Collab., Ch. Berger et al., Z. Phys. C27 (1985) 341;

JADE Collab., W. Bartel et al., Z. Phys. C30 (1986) 371;

TASSO Collab., W. Braunschweig et al., Z. Phys. C40 (1988) 163;

HRS Collab., M. Derrick et al., Phys. Lett. B166 (1986);

MAC Collab., E. Fernandez et al., Phys. Rev. D35 (1987) 10;

M.J. Shochet, Proc. XXIV Int. Conf. on High Energy Physics, Munich/FRG, 1988 (Eds. R. Kotthaus and J.H. Kühn) p. 18.

33

MONTE CARLO SIMULATION OF GENERAL HARD PROCESSES WITH QCD COHERENCE

G. Marchesini

Dipartimento di Fisica, Università di Parma
INFN, Gruppo Collegato di Parma, Italy

In this talk I review some recent progress, obtained in collaboration with Bryan Webber, in the simulation of QCD hard process. In particular I shall concentrate on HERWIG, a QCD Monte Carlo program we have recently constructed[1] for the simulation of general hard processes. The most important new feature of HERWIG is that the interference of QCD radiation is fully included to leading order. Thus HERWIG correctly describes the coherence of the soft radiation emitted both from incoming and outgoing quarks and gluons.

In this talk I shall describe the following topics: 1) General structure of HER-WIG. 2) Schematic description of the new theoretical results which are the basis of HERWIG. 3) A recent test[2] of the coherence of the radiation from initial and final partons. This consists in the study of the transverse energy associated to hard jet production in $p\bar{p}$ collisions. 4) Description of coherence in processes involving heavy flavours, and the recent implementation[3] in HERWIG of these results.

1. General strucure of HERWIG

HERWIG simulates many processes in the Standard Model, ranging from e^+e^- annihilation and deep inelastic lepton scattering to Drell-Yan processes and hadronic jet production. It thus extends our earlier work[4,5], which was limited to processes in which QCD partons are involved only in the final state, such as e^+e^- annihilation.

The simulation of all these processes in one package realizes one of the most characteristic features of perturbative QCD: the factorization of hard processes.

The advantages of using a single package consists in a unified treatment of: i) tuning of parameters; ii) inclusion of new hard processes; iii) improvements of QCD results.

There are three basic elements in the construction of HERWIG which correspond to three different operative stages:

1) *Perturbative QCD Stage.* Here QCD hard cascades lead to the emission of partons. For the mass of gluons an infrared cut off Q_g is needed with the constraint $\alpha(Q_g)/\pi \lesssim 1$. Essentially this is the only arbitrary parameter entering in this Stage and is the main indication of our ignorance of the confinement of QCD partons. We take $Q_g \simeq (0.5\text{-}1.)$ GeV.

2) *Colour Singlet Hadronization Stage.* The hadronization model used in HER-WIG is essentially that developed earlier for e^+e^- annihilation[5], with some minor improvements. After the perturbative parton branching process, all outgoing gluons are split non-perturbatively into quark-antiquark or diquark-antidiquark pairs. The perturbative parton branching is such that a colour line can be followed, in the planar approximation, from each quark to an antiquark or diquark with which it can form a colour-singlet cluster. These clusters are 'pre-confined' – they have a distribution of mass and spatial size that peaks at low values, falls rapidly for large cluster masses and sizes, and is asymptotically independent of the hard subprocess scale[6]. The average mass of these colour singlet clusters is controlled by the infrared cutoff Q_g and is actually about $3Q_g$. Such hadronization model of course has not a theoretical basis. However the property of preconfinement[6] suggests that a different hadronization model which conserves the colour singlet structure should not lead to significantly different results. Actually this can be checked in e^+e^- annihilation process: Monte Carlo codes[1,4,5,7] with the same level of perturbative QCD accuracy and with different colour singlet hadronization models give in general similar results.

3) *Minimum Bias Stage.* In hadron-hadron collisions there are usually two 'beam clusters' containing the spectators from the incoming hadrons. In HERWIG these are used to generate the underlying soft event. The model used for minimum-bias soft hadronic collisions is based on that of the UA5 Collaboration[8] and may be regarded simply as a parametrization of the CERN $p\bar{p}$ minimum-bias data.

In conclusion we have: the beam cluster model for the *Minimum Bias* events is arbitrary and is just a parametrization of the CERN $p\bar{p}$ minimum-bias data. *Colour Singlet Hadronization Model* is also arbitrary but it is strongly constrained by the usually small phase space available (preconfinement) and by the Particle Data Table. So that the main results are expected to be largely independent of the details. The *Perturbative QCD Stage* is instead on solid theoretical grounds. Let us then schematically illustrate its basic features and the new results discussed in Ref. [1,9] which made possible this Stage.

2. Coherence of radiation from initial and final partons

Let me start by recalling that the possibility of using Monte Carlo methods to sum classes of asymptotically dominant Feynman diagrams was one of the important results of the first studies of leading collinear singularities. From this analysis one not only obtains the well-known scaling violations and factorization properties of parton distributions, but also recovers a parton model type of description of QCD emission.

In general, however, infrared contributions are asymptotically as important[9] as the collinear ones, and therefore they have to be included in the Monte Carlo simulation. This was shown to be possible[9] in the case of e^+e^- annihilation and it is due to the fact that the principal interference effect is fully destructive (to leading order) and simply reduces the available phase space region. This property was subsequently used[4,5,7] to construct Monte Carlo simulations in e^+e^- annihilation.

The structure of coherence in general hard processes is considerably more subtle than that in e^+e^- annihilation. However, also for these processes involving both outgoing and incoming partons, interference can be described[1,9,10] by a Markov process. In Ref. [1] we have been able to extended the QCD Monte Carlo simulation to general hard processes.

Let me schematically explain the features of coherence in these cases. Consider hadron-hadron collisions in which high transverse momentum jets are emitted, a process which exhibits all the QCD complications of both incoming and outgoing partons. To leading twist order[11] the parton distributions may be factorized into the following basic components:

1) The elementary hard scattering cross section for two incoming and two outgoing partons. One can show[12] that, because of interference, the radiation emitted from the four partons in the QCD hard $2 \to 2$ vertex is confined to cones determined by the colour structure of the hard subprocess. Subsequent parton branching takes place within these angular cones.

2) The Sudakov form factors for both timelike and spacelike branching vertices, which take into account the contributions of diagrams involving virtual partons. The relevant expressions[9,11] are known in both the spacelike and the timelike cases.

3) QCD parton cascades for timelike outgoing partons emitted both from the hard vertex and from the spacelike branching of incoming partons. Each timelike parton cascade is similar to those in e^+e^- annihilation and then they are correctly described by a QCD branching in the reduced region of phase space in which the angles decrease as one moves along the cascade from the first emission vertices to the final partons. In HERWIG there is a further improvement on this stage since we take into account also the azimuthal correlation due both to interference[1,9] and gluon polarization[13].

4) The spacelike branching of the incoming partons. This constitutes the main novel development in our work and therefore let me sketch the main points.

Let me start by recalling the results[11] of the early studies of the leading collinear contributions. In this approximation the structure function of the incoming hadron is given by the Altarelli-Parisi equations, which correspond to the sum of the ladder type of Feynman diagrams in which the incoming spacelike parton p undergoes successive branching with the emission of timelike partons q_i. The process is then described by the successive branchings:

$$p_i \rightarrow p_{i-1} + q_i, \tag{1}$$

where p_i and p_{i-1} are spacelike. In order to obtain the leading collinear contribution, these diagrams are usually evaluated in the phase space region in which the virtual masses of the successive spacelike partons p_i in the cascade are ordered:

$$Q_s^2 < \left|p^2\right| = \left|p_n^2\right| < \cdots < \left|p_i^2\right| < \cdots < \left|p_1^2\right| < \left|p_0^2\right| < Q^2. \tag{2}$$

The Altarelli-Parisi equations themselves have the structure of a Markov process for the spacelike cascade in (1). It proves most convenient to generate[14] the parton cascade 'backwards', that is, starting from the hard vertex scale Q and terminating at a lower spacelike scale Q_s. The Markov process in the phase space (2) is the basis of the presently available[15] Monte Carlo simulations of initial-state radiation.

Soft gluon coherence in spacelike branching however modify the phase space in Eq. (2). As expected[9] the reduction of phase space for spacelike branching is similar, but not identical, to that for timelike branching. In the latter the branching angles decrease as one goes from the hard vertex to the final emitted partons, while in spacelike branching the angles increase going from the initial parton p to the parton p_0 which undergoes the hard collision. However, there is a subtle difference in the detailed identification of the ordered angles due to the asymmetry of the parton virtualities: in each branching two partons are spacelike and one timelike. Furthermore, in spacelike branching the energy ordering is typically opposite to the virtuality ordering.

The main result of the infrared analysis[1,10] is that, for small-angle, the spacelike branching in Eq. (1) the phase space is reduced by interference to the region

$$E_i \theta_{pq_i} < E_{i-1} \theta_{pq_{i-1}} \tag{3}$$

where E_i is the energy of the i-th spacelike parton p_i and θ_{pq_i} is the angle between the incoming parton p and the emitted parton q_i.

In conclusion, in the case of branching of an incoming parton, destructive interference can be taken into account by a simple modification of the Altarelli-Parisi equations, as follows. Instead of using $|p_i^2|$ as evolution variable, one should use an angular type of variable proportional to the quantity in Eq. (3). The actual variable used in HERWIG for coherent spacelike branching is:

$$Q_i = E_i\sqrt{\xi_i} \quad ; \quad \xi_i = p \cdot q_i / E\omega_i, \tag{4}$$

where E, E_i and ω_i are the energies of partons p, p_i and q_i, respectively. Then the new ordering condition is

$$Q_{i+1} < Q_i, \tag{5}$$

which reduces to (3) in the small-angle region.

Note that $|p_i^2|$–ordering is equivalent to Q_i–ordering only when all $\omega_i \to 0$. However in the complementary region with $E_i \to 0$ Q_i–ordering is only a small portion of the phase space of $|p_i^2|$–ordering. The phase space reduction from $|p_i^2|$- to Q_i–ordering is the result of destructive interference.

An important consequence of this is that Q_i–ordering leads to the correct inclusive cancellation of infrared singularities: the leading singularity of the Sudakov form factor is cancelled when the distribution is integrated over this region. If one neglecs the reduction of phase space due to interference and considers the incoming parton cascade in the $|p_i^2|$–ordering phase space, the infrared singularities generated by the soft gluon emission would not be compensated by the infrared singularities in the Sudakov form factor[1,9] and one would need to arbitrarely use its squared. This would distort for instance all the parton p_t distributions.

3 Transverse Energy in Hadronic Jet Production

I come now to describe a study of some physical consequence of the coherence of QCD emission from initial and final partons. I shall briefly review the results of Ref. [2] in which we have studied the *pedestal transverse energy* $\left\langle \omega_T^{ped} \right\rangle$ for a jet of given tranverse energy. This quantity, introduced and measured in Ref. [16], is sensitive to the structure of the QCD radiation emitted outside the jet cone where interference takes place. Let me schematically recall its definition. For events with a jet of a given transverse energy E_T^{jet}, consider the distribution in transverse energy ω_T integrated over the azimuthal angle ϕ on the same side of the jet axis ($|\Delta\phi| \leq \pi/2$). This distribution is a function of $\Delta\eta$, the difference in (pseudo-) rapidity ($\eta = -\log\tan\frac{1}{2}\theta$) with respect to the jet axis. One then defines

$$\omega_T^{ped} = \tfrac{1}{2}(\omega_T^L + \omega_T^R) \tag{6}$$

where ω_T^L and ω_T^R are the transverse energies in the rapidity intervals $-2 < \Delta\eta < -1$ and $1 < \Delta\eta < 2$ respectively. The quantity $\left\langle \omega_T^{ped} \right\rangle$ is thus the average transverse energy per unit of rapidity measured in the pedestal, 1.5 units of rapidity away from the jet axis. The integration over $|\Delta\phi| < \pi/2$ avoids contributions to $\left\langle \omega_T^{ped} \right\rangle$ from the recoiling jet.

This quantity has been measured by the UA1 collaboration[16] at the CERN $p\bar{p}$ Collider for $\sqrt{s} = 630$ GeV and has the following features: It increases from $\left\langle \omega_T^{ped} \right\rangle \simeq 2$ GeV to about 4 GeV as E_T^{jet} increases to 10 GeV, then remains around this value for E_T^{jet} up to the maximum measured value of 60 GeV. An analogous trend is observed in a similar quantity measured by the UA2 collaboration[17].

The quantity $\left\langle \omega_T^{ped} \right\rangle$ takes contributions from the radiation emitted both from incoming and outgoing partons and is then sensitive to their interference. However in hadron collisions the radiation outside the jet cone also has a component originating from the low-p_t interaction involving the spectator partons: the *soft underlying event*. Therefore $\left\langle \omega_T^{ped} \right\rangle$ simultaneously tests perturbative QCD and the soft underlying dynamics.

In Ref. [2] we have studied this quantity by using HERWIG. Here I report only the main features of this study but see the original paper for a detailed description of the results. Particularly intreseting is the region of E_T^{jet} above 10 GeV, where genuine jet QCD production takes place. Taking advantage of the factorized structure of the simulation (and of perturbative QCD) we have been able to study independently the contributions to $\left\langle \omega_T^{ped} \right\rangle$ from perturbative QCD and from the soft underlying event. In particular we have performed three independent analyses: i) a *parton* analysis based directly on the parton momenta generated in the perturbative phase of the simulation; ii) a *hard* analysis using only the particles from the hard process, i.e., those resulting from the hadronization of the partons; and iii) a *full* analysis using all final-state particles including those from the soft underlying event.

We have found that the results of HERWIG are in good agreement with the data. Roughly speaking, the total pedestal height of about 4 GeV is composed of: i) a lowest-order perturbative part of about 0.8 GeV; ii) higher-order corrections (including rescaling the argument of α_s), about 0.8 GeV; iii) hadronization contribution, about 0.4 GeV; iv) soft underlying event, about 2 GeV.

The model of the underlying event incorporated in the Monte Carlo program assumes that it is essentially just a minimum-bias soft event superimposed on the hard process. Since we can account for the data with this hypothesis, we do not yet see a need for more complicated models such as that of Ref. [18] in which a multiple interaction model for the soft underlying event is proposed.

In order to be able to perform a clear analysis of the radiation in this phase space region one should be able to disentangle the perturbative QCD from the underlying event contribution. In Ref. [1] we have proposed quantities which are most sensitive either to the hard perturbative or to the soft underlying components. By studying these quantities one would then be able not only to perform a direct analysis of features related to perturbative QCD, but also to obtain, at the same time, phenomenological constraints on models for the soft underlying event.

The definition of these quantities is suggested by observing that $\left\langle \omega_T^{ped} \right\rangle$ is an infrared finite quantity and therefore its leading hard contribution can be computed in perturbation theory from the $(2 \rightarrow 3)$ matrix elements. Therefore one has that to this order ω_T^L and ω_T^R, defined in(6), cannot both be different from zero. Therefore we introduced the quantity:

$$\omega_T^{dif} = \tfrac{1}{2}|\omega_T^L - \omega_T^R| \tag{7}$$

which, to leading order, coincides with ω_T^{ped}.

On the other hand, the soft underlying contribution to $\langle \omega_T^{dif} \rangle$ should be small, for the following reason. The underlying event is expected to be similar to a minimum-bias soft collision. It is well established[19] that such collisions show strong positive long-range rapidity correlations. Thus the underlying contributions to ω_T^L and ω_T^R should be strongly correlated and they should cancel in the difference.

Conversely, the quantity

$$\omega_T^{min} = \min\left(\omega_T^L, \omega_T^R\right) = \omega_T^{ped} - \omega_T^{dif} \tag{8}$$

vanishes in lowest-order perturbation theory, while the correlations in the underlying event mean that its contributions to ω_T^{min} and ω_T^{ped} should be comparable. Therefore $\langle \omega_T^{min} \rangle$ should be much more sensitive to the soft underlying contribution than $\langle \omega_T^{dif} \rangle$.

Thus measurements of the two quantities $\langle \omega_T^{dif} \rangle$ and $\langle \omega_T^{min} \rangle$ would be helpful in disentangling the hard perturbative and the soft underlying components respectively. Since only the sum of the two contributions has been measured so far, we have had to rely on the Monte Carlo simulation to test the method. Our study shows that it is quite efficient at CERN Collider energies: $\langle \omega_T^{dif} \rangle$ is dominated by the hard component and so one can test perturbative QCD more directly with this quantity. The opposite happens for $\langle \omega_T^{min} \rangle = \langle \omega_T^{ped} \rangle - \langle \omega_T^{dif} \rangle$, which can thus be used to test different hypothesis about the soft underlying event. The only essential property that the underlying event must have in order to be enhanced in $\langle \omega_T^{min} \rangle$ is a strong long-range rapidity correlation.

We have studied the energy dependence of these quantities and in particular we have presented predictions at the energy of the FNAL Tevatron Collider. The Monte Carlo simulation shows, as expected, that the enhancement of the hard component in $\langle \omega_T^{dif} \rangle$ is greater at higher energies, while the dominance of the underlying event in $\langle \omega_T^{min} \rangle$ becomes less strong.

The quantity $\langle \omega_T^{ped} \rangle$ is defined to suit the characteristics of the UA1 detector and may not be the best quantity to measure with other detectors. We should emphasise therefore that the proposed method for separating the hard QCD and soft underlying components is simply based on the fact that the contribution of the former is asymmetrical while that of the latter is strongly correlated. Hence one should be able to base the method on almost any pair of quantities involving the sum and difference of contributions in disjoint, comparable phase space regions.

4 Simulation of QCD Coherence in Heavy Quark Production and Decay

In perturbative QCD, one of the main feature of heavy flavour production processes is that the heavy quark masses largely affect the phase space region in which destructive interference takes place. In Ref. [3] we extended the analysis of destructive interference of soft QCD radiation to the QCD diagrams with heavy quark and showed that also in this case the leading contributions can be approximately summed by a Markov process. This is the basis for the extension we have done of

HERWIG to simulate hard physical processes involving also heavy flavours. I shall review the results in Ref. [3] which are necessary for a Monte Carlo simulation, namely i) the destructive interference in the QCD branching of heavy quark; and ii) Coherence and Colour structure in heavy quark production.

4.a Soft gluon radiation from heavy quarks

To leading infrared order we find that heavy quarks undergo a QCD branching cascade similar to the one for light quarks. However the phase space available for the branching is now even more reduced by the heavy quark masses. The phase space available to the soft gluon radiated by a heavy quark of momentum p_1 and velocity v_1 is approximately reduced to:

$$v_2 \cos\theta_{12} < \cos\theta_1 < \cos\theta_1^{min} \simeq v_1 \tag{9}$$

where v_2 is the velocity of parton p_2, which is colour connected to p_1. For $v_1 < 1$ we have $\theta_1^{min} \neq 0$ which corresponds to the fact that the collinear singularity is now screened.

This result can be obtained, as in the massless case, by analyzing the gauge invariant distribution of soft gluon emission from two colour charges of momentum p_i and p_j. In the soft limit the spin of emitting partons is not relevant and one finds the well known eikonal distribution:

$$W_{ij} = -\frac{\omega^2}{2}(\frac{p_i}{k \cdot p_i} - \frac{p_j}{k \cdot p_j})^2 = \frac{\xi_{ij}}{\xi_i \xi_j} - \frac{1}{2}\left(\frac{1}{\gamma_i \xi_i}\right)^2 - \frac{1}{2}\left(\frac{1}{\gamma_j \xi_j}\right)^2 \quad , \tag{10}$$

where ω is the energy of soft gluon k and as usual

$$\xi_{ij} = p_i \cdot p_j / E_i E_j = 1 - v_i v_j \cos\theta_{ij},$$

$$\xi_i = 1 - v_i \cos\theta_i \quad , \quad \gamma_i = E_i/m_i = 1/\sqrt{1 - v_i^2}, \tag{11}$$

v_i is the velocity of parton i, and θ_i is the angle between the directions of motion of the soft gluon and parton i. Note that the main effect of the second term in W_{ij} is to suppress the radiation in the region $\xi_i \lesssim \gamma_i^{-2}$ (that is, at small angles with respect to parton i), introducing an effective angular cutoff $\cos\theta_i < v_i$. This cutoff becomes sharper as $v_i \to 1$. Similarly, the third term introduces the approximate cutoff $\cos\theta_j < v_j$. On the other hand, the interference between radiation from i and j tends to suppress values of ξ_i and ξ_j above ξ_{ij}. Again, these become sharp cutoffs (after azimuthal averaging) when $v_{i,j} \to 1$. Altogether, then, the dominant phase space region for soft gluon radiation from partons i and j is $R^i_{ij} \cap R^j_{ij}$ where

$$R^i_{ij} : \quad v_j \cos\theta_{ij} < \cos\theta_i < v_i,$$

$$R^j_{ij} : \quad v_i \cos\theta_{ij} < \cos\theta_j < v_j. \tag{12}$$

Notice that these two angular regions coincide in any frame in which i and j are back-to-back ($\theta_{ij} = \pi$). If either hard parton is at rest, the corresponding phase space region vanishes.

In order to describe the soft gluon radiation as a Markov process we need to approximate the distribution in (10) as the sum of the two contributions for the emission from partons i and j. To this end, as in the massless case, we subdivide W_{ij} into two terms;

$$W_{ij} = W_{ij}^i + W_{ij}^j \tag{13}$$

where W_{ij}^i and W_{ij}^j approximately describe the radiation emitted from partons i and j respectively. We require that in the massless limit W_{ij}^i and W_{ij}^j contain the leading collinear singularity as $\theta_i \to 0$ and $\theta_j \to 0$ respectively. We write then:

$$W_{ij}^i = \frac{1}{2\xi_i} \left[1 + \frac{\xi_{ij} - \xi_i}{\xi_j} - \frac{(1 - v_i^2)}{\xi_i} \right], \tag{14}$$

and similarly for W_{ij}^j. As one can easily check, W_{ij}^i is positive definite inside the region R_{ij}^i in (12) and here can be used as distribution for the radiation emitted from parton i interfering with parton j. Outside the region R_{ij}^i the radiation function W_{ij}^i is not in general positive definite (destructive interference) and can not be interpreted any more as a probability distribution for emission from parton i. However if we consider, as in the massless case, the average of W_{ij}^i over the azimuthal angle around the direction of the emitting parton i we find that the contribution outside R_{ij}^i is generally small and to a good approximation can be neglected. Actually it vanishes in the massless case. One can then approximate the radiation distribution by:

$$W_{ij} \simeq W_{ij}^{(a)} \equiv W_{ij}^i \Theta(R_{ij}^i) + W_{ij}^j \Theta(R_{ij}^j) \tag{15}$$

where $\Theta(R_{ij}^i) = 1$ in region R_{ij}^i and 0 outside. The first term $W_{ij}^i \Theta(R_{ij}^i)$ can now be used as a probability distribution to generate, in a factorized form, the soft gluon radiation emitted by parton i. Similarly for the second term. As in the massless case this can now be generalized to describe the full QCD cascade for the heavy quark. Apart for the approximation in (15), this expression is accurate to the leading infrared order.

In HERWIG we use also a simpler approximate expression for the radiation distribution. Taking the azimuthal average around i inside R_{ij}^i, the factor in the square bracket of (14) is approximately equal to $2v_i$. One can then approximate W_{ij} by:

$$W_{ij} \simeq W_{ij}^{(b)} \equiv \frac{v_i}{\xi_i} \Theta(R_{ij}^i) + \frac{v_j}{\xi_j} \Theta(R_{ij}^j) \tag{16}$$

The approximations $W_{ij}^{(a)}$ and $W_{ij}^{(b)}$ are quite close[3] to the exact distribution apart for a small region around the direction in which the soft gluon is emitted parallel to the heavy quarks. This is due to the sharp cut off of phase space for $\cos \theta_i > v_i$ introduced in R_{ij}^i. Of course one could improve the approximation, but, taking into account the expected smearing of the pattern in the hadronization process, both approximations look satisfactory.

4.b Coherence and Colour structure in heavy quark production

I recall now the result of the analysis of the coherence and the corresponding colour connection structure of the elementary subprocess governing heavy flavour production. Also for this problem we have shown that it is possible to formulate a Markov process which takes into account the leading infrared sructure of soft gluon radiation and the colour connection structure of all partons emitted in the QCD cascades.

Consider the production of heavy quark Q in collisions between hadrons that do not contain Q or $\bar{Q}$ as constituents. This is governed by the light-quark and gluon fusion processes:

$$
\begin{aligned}
F & \quad q(p_1) + \bar{q}(p_2) \to Q(p_3) + \bar{Q}(p_4), \\
G & \quad g(p_1) + g(p_2) \to Q(p_3) + \bar{Q}(p_4).
\end{aligned}
\tag{17}
$$

Consider also the subprocess of Higgs production:

$$
H \quad g(p_1) + g(p_2) \to H \to Q(p_3) + \bar{Q}(p_4).
\tag{18}
$$

We performed this study by following the same procedure we applied in the massless case[1,12]. For each elementary $2 \to 2$ hard distribution governing the physical processes, we first identify the contributions corresponding to a definite colour flow in the planar limit, i.e. to leading order in the inverse of the number of colours N. By approximating the soft gluon distribution of a given $2 \to 3$ subprocess (*Radiation Pattern*) we are able to identify the contribution of the radiation emitted from each hard parton, its corresponding angular phase space and its colour connection. In this way one has the necessary elements to set up the Monte Carlo simulation with colour coherence and definite colour flow structures.

We considered the squared amplitudes[20] for the subprocesses $2 \to 3$, which correspond to subprocesses (17), (18) with an additional soft gluon of momentum k. In the soft $k \to 0$ limit we found that, to a good approximation, the corresponding *Radiation Patterns* $W^F(k)$, $W^G(k)$ and $W^H(k)$ have the same structure we have found for the massless case:

$$
W^{F,G,H}(k) \simeq \sum_{i \neq j} W_{ij}^{iF,G,H}(k)
\tag{19}
$$

where in the approximation (15) we have

$$
W_{ij}^{iF,G,H}(k) = C_i h_{ij}^{F,G,H}(s',t',u') W_{ij}^i \Theta(R_{ij}^i)
\tag{20}
$$

and in the approximation (16) we have

$$
W_{ij}^{iF,G,H}(k) = C_i h_{ij}^{F,G,H}(s',t',u') \frac{v_i}{\xi_i} \Theta(R_{ij}^i)
\tag{21}
$$

For a quark $C_i = 2C_F$ and for a gluon $C_i = C_A$. This shows that, due to soft gluon interference and to the heavy quark mass screening, the radiation emitted by parton i, interferring with parton j, is essentially confined within the region R^i_{ij}.

The functions $h^{F,G,H}_{ij}$ depend on $p_i \cdot p_j$, are positive and provide the probabilities for the emission around the hard parton direction within R^i_{ij}. For subprocesses F and H there is a single colour flow diagram and then we have:

$$h^F_{13} = h^F_{31} = h^F_{24} = h^F_{42} = H^F$$
$$h^H_{12} = h^H_{21} = h^H_{34} = h^H_{43} = H^H$$
(22)

where H^F and H^H are the colour and spin averaged squared amplitudes for the corresponding $2 \to 2$ subprocesses. In the case of subprocess G we have two colour flow diagrams and then the exact $2 \to 2$ squared amplitude H^G contains a term suppressed by $1/N^2$ which corresponds to the interference of diagrams with different colour flows. The planar part of h^G_{ij} is just the squared amplitudes in which partons i and j are colour connected. As in Ref. [1], we have distributed the non planar parts of H^G in such a way that h^G_{ij} are uniquely identified by requiring: i) each h^G_{ij} should remain positive definite in order to be interpreted as a probability distribution; ii) each h^G_{ij} should have the same pole structure and crossing symmetry as its planar part.

Eq. (19) can be used as in the massless case as the bases of the Monte Carlo simulation for heavy flavour production, in which multiple soft gluon emission correctly takes into account both the coherence of soft radiation and the colour connection structure of the finally emitted partons.

5 Conclusions

The most important new theoretical property included in HERWIG is the coherence of soft gluon radiation emitted not only from the outgoing QCD partons, including heavy quarks, but also from incoming partons. Let me recall again that the correct inclusion of colour interference is essential in order to sum all the asymptotically dominant terms of Feynman diagrams and consequently to ensure the correct cancellation of infrared singularities in inclusive quantities. Coherence effects have been seen in e^+e^- annihilation. The next step consits in searching evidence for coherence in hard reactions with incoming hadrons. In hadron-hadron reactions the effect of interference among the radiation emitted from initial and final partons is masked by the presence of the low-p_t interaction involving the spectators (soft underlying event). We have initiated this analysis by studying the pedestal transverse energy and found that QCD branching with coherence in HERWIG, together with a mere superposition of typical minimum bias events, provides a good descriptions of the CERN data. We have also suggested a method to define physical quantities in hadron-hadron reactions which should differentiate the perturbative QCD hard component from the soft underlying event and should then help in the future analysis of coherence in initial state radiation.

All the new results here reviewed have been obtained in collaboration with B.

Webber. It is a pleasure to acknowledge many helpful discussions with M. Ciafaloni, R.K. Ellis and A.H. Mueller. This research was supported in part by the Italian Ministero della Pubblica Istruzione.

References

[1]   G. Marchesini and B.R. Webber, *Monte Carlo Simulation of General Hard Processes with Coherent QCD Radiation*, to be published in Nucl. Phys. B.

[2]   G. Marchesini and B.R. Webber, *Associated Transverse Energy in Hadronic Jet Production*, to be published in Phys. Rev. D.

[3]   B.R. Webber, *Monte Carlo simulation with QCD coherence*, Talk at 7th Topical Workshop on Proton Antiproton Collider Physics, Fermilab, June 1988; G. Marchesini and B.R. Webber, in preparation.

[4]   G. Marchesini and B.R. Webber, Nucl. Phys. B238 (1984) 1.

[5]   B.R. Webber, Nucl. Phys. B238 (1984) 492.

[6]   D. Amati and G. Veneziano, Phys. Lett. 83B (1979) 87; see also A. Bassetto, M. Ciafaloni and G. Marchesini, Phys. Lett. 83B (1979) 207; G. Marchesini, L. Trentadue and G. Veneziano, Nucl. Phys. B181 (1981) 335.

[7]   T.D. Gottschalk and D.A. Morris, Nucl. Phys. B288 (1987) 729; M. Bengtsson and T. Sjöstrand, ibid. B289 (1987) 810.

[8]   UA5 Collaboration: G.J. Alner et al., Nucl. Phys. B291 (1987) 445

[9]   For a review see: L.A. Gribov, E.M. Levin and M.G. Ryskin, Phys. Rep. 100 (1983) 1 and A. Bassetto, M. Ciafaloni and G. Marchesini, Phys. Rep. 100 (1983) 201. See also S. Catani, M. Ciafaloni and G. Marchesini, Nucl. Phys. B264 (1986) 588; B.R. Webber, Ann. Rev. Nucl. Part. Sci. 36 (1986) 253; Yu.L. Dokshitzer, V.A. Khoze, S.I. Troyan and A.H. Mueller, Rev. Mod. Phys. 60 (1988) 377.

[10] M. Ciafaloni, Nucl. Phys. B296 (1987) 249.

[11] W. Marciano and H. Pagels, Phys. Rep. 36 (1978) 138; Yu.L. Dokshitzer, D.I. Dyakanov and S.I. Troyan, Phys. Rep. 58 (1980) 270; A.H. Mueller, Phys. Rep. 73 (1981) 237; G. Altarelli, Phys. Rep. 81 (1982) 1.

[12] R.K. Ellis, G. Marchesini and B.R. Webber, Nucl. Phys. B286 (1987) 643.

[13] B.R. Webber, Phys. Lett. 193B (1987) 91.

[14] T. Sjöstrand, Phys. Lett. 157B (1985) 321; T.D. Gottschalk, Nucl. Phys. B277 (1986) 700; B.R. Webber, Ann. Rev. Nucl. Part. Sci. 36 (1986) 253.

[15] COJETS: R. Odorico, Computer Phys. Comm. 32 (1984) 139; PYTHIA: H.-U. Bengtsson and G. Ingelman, Computer Phys. Comm. 34 (1985) 251; FIELDAJET: R.D. Field, Nucl. Phys. B264 (1986) 687; ISAJET: F. Paige and S.D. Protopopescu, BNL-38034 (1986); JETSET: T. Sjöstrand and M. Bengtsson, Computer Phys. Comm. 43 (1987) 367; EUROJET: A. Ali, B. van Eijk and E. Pietarinen, to be published; B. van Eijk, Amsterdam thesis (1987).

[16] UA1 Collaboration: C. Albajar et al., CERN-EP/88-29.

[17] UA2 Collaboration: J.A. Appel et al., Phys. Lett. 165B (1985) 441.

[18] T. Sjöstrand and M. van Zijl, Phys. Rev. D 36 (1987) 2019.

[19] UA5 Collaboration: G.J. Alner et al., Nucl. Phys. B154 (1987) 247.

[20] R.K. Ellis and J.C. Sexton, Nucl. Phys. B282 (1987) 642.

34

SEMIHARD INTERACTIONS AND INELASTIC CROSS SECTION IN HIGH ENERGY HADRONIC COLLISIONS

D. Treleani[*]

Dipartimento di Fisica Teorica dell'Università di Trieste, Italy
INFN, Sezione di Trieste, Italy

ABSTRACT

The contribution of semihard interactions to the inelastic hadronic cross section is discussed in the framework of the QCD parton model. The role played by multiple parton collisions is emphasized.

Introduction

In hadronic interactions with C.M. energies of several TeV a large fraction of all events will be characterized by a momentum transfer greater than a few GeV[1].

Such a situation is already present at CERN Collider energy, where, looking at jets, one finds a jet cross section that becomes a large fraction of the total inelastic one when going to the minijet region[2].

One can then naively expect to be able to describe a sizeable fraction, or even the total inelastic cross section just using the perturbative QCD parton model that is the standard tool for large p_t physics.

One will however notice that in the perturbative QCD parton model there is no intrinsic scale, in such a way that all dynamical quantities that are constructed in its framework get their dimensions by the kinematical variables.

For example the inclusive cross section to produce large p_t jets is given by

$$\sigma_{incl} = \int_{p_t^{min}} G_A(x) G_B(x') \hat{\sigma}(xx') dx dx'$$

with $G(x)$ the parton distributions, $\hat{\sigma}$ the parton parton cross section and p_t^{min} the cut off distinguishing a large p_t event from a low p_t one.

Then σ_{incl} is a quantity that, as a first approximation, is of size $(1/p_t^{min})^2$.

The consequence is that, for p_t^{min} small, σ_{incl} will become very large, consistently with the expectation that in high energy regimes a large fraction of all events will be related to short distance dynamics.

The problem is, however, that the total inelastic cross section is characterized by an intrinsic scale.

The regime of interest (that will then be presently discussed) is therefore that one where the inclusive cross section for production of large p_t jets becomes comparable to the total inelastic one. This is the region where the perturbative QCD parton model (at least in its standard formulation) will show to be inadequate to describe most of the features of inelastic events.

A better understanding of the situation will be gained when taking into account multiple parton collisions: A natural scale, of the order of the hadronic geometrical size, will then enter into the picture (because of the multiparton distributions[3]) and a connection with the total inelastic cross section will then be possible[4].

An heuristic case, showing all the main features of interest (actually the self shadowing of the inclusive cross sections in high energy hadron nucleus collisions) will be discussed in the next paragraph. Semihard interactions in high energy hadronic collisions will then be the following argument and a few final remarks will be presented in the conclusions.

Self shadowing for inelastic events in hadron nucleus collisions

Considering the ideal case of pure imaginary hadron-nucleon amplitude the elastic scattering amplitude for a collision of a hadron with a nucleus of atomic weight A is given by the optical expression:

$$f^A(\mathbf{q}) = \frac{ik}{2\pi}\int d^2b\, e^{i\mathbf{q}\mathbf{b}}(1 - exp[-\frac{1}{2}\sigma_T T(b)]$$

with k incoming hadron energy in the laboratory system, $\mathbf{q}$ momentum transfer, σ_T total hadron nucleon cross section, $T(b)$ nuclear thickness function defined in terms of the nuclear density $\rho(b,z)$ as $T(b) = \int_{-\infty}^{+\infty}\rho(b,z)dz$ and $\int T(b)d^2b = A$.

One can then write the total hadron-nucleus cross section using the optical theorem:

$$\sigma_T^A = \frac{4\pi}{k}Im f^A(0) = 2\int d^2b(1 - exp[-\frac{1}{2}\sigma_T T(b)])$$

and the elastic hadron nucleus cross section:

$$\sigma_{el}^A = \int |f^A(\mathbf{q})|^2\, d\Omega = \int d^2b(1 - exp[-\frac{1}{2}\sigma_T T(b)])^2$$

one can look at the difference $\sigma_T^A - \sigma_{el}^A$:

$$\sigma_T^A - \sigma_{el}^A = \int d^2b(1 - exp[-\sigma_T T(b)]) = \int d^2b\sum_{n=1}^{\infty}\frac{1}{n!}[\sigma_T T(b)]^n exp[-\sigma_T T(b)]$$

All inelastic events are therefore distributed according to a poissonian probability distribution of multiple collisions with average number of collisions given by $\sigma_T T(b)$ (and therefore different at each impact parameter).

One may now decide to select some particular events according to some selection criterium C. Then $\sigma_T = \sigma_C + \sigma_N$ on a nucleon (with $\sigma_C =$ cross section for events of kind C and $\sigma_N =$ cross section for all other events).

One may ask which will be the cross section for selecting events of kind C when interacting with a nuclear target.

A simple answer[5] will be obtained if the criterium C is such that any superposition of events of kind C and kind N will also be of kind C. Then one will write:

$$\sigma_T^A - \sigma_{el}^A = \int d^2 b \sum_{n=1}^{\infty} [\sum_{k=0}^{n} \binom{n}{k} \sigma_C^k \sigma_N^{n-k}] \frac{1}{n!} T(b)^n \times exp[-\sigma_T T(b)]$$

and

$$\sigma_C^A = \int d^2 b \sum_{n=1}^{\infty} [\sum_{k=1}^{n} \binom{n}{k} \sigma_C^k \sigma_N^{n-k}] \frac{1}{n!} T(b)^n \times exp[-\sigma_T T(b)]$$

One has however

$$\sum_{k=1}^{n} = \sum_{k=0}^{n} - (k = 0) = \sigma_T^n - \sigma_N^n$$

so that

$$\sigma_C^A = \int d^2 b (exp[-\sigma_T T(b)] - exp[-\sigma_N T(b)]) \times exp[-\sigma_T T(b)]$$

in such a way that one can write:

$$\sigma_C^A = \int d^2 b (1 - exp[-\sigma_C T(b)]) = \int d^2 b \sum_{n=1}^{\infty} \frac{1}{n!} [\sigma_C T(b)]^n exp[-\sigma_C T(b)]$$

Therefore also σ_C^A is obtained from a sum on multiple collisions distributed according to a poissonian with average number at each impact parameter b given by $\sigma_C T(b)$.

Notice that

$$P_n(b) \equiv \frac{1}{n!} (\sigma_C T(b))^n exp[-\sigma_C T(b)]$$

is the probability for having n events each of them of kind C. It is a quantity normalized to 1:

$$\sum_{n=0}^{\infty} P_n(b) = 1$$

Suppose that one wants to sum over all the configurations in which the interaction of kind C appears at least n times. One has then:

$$\sum_{k=n}^{\infty} \binom{k}{n} \frac{1}{k!} (\sigma_C T(b))^n \exp[-\sigma_C T(b)] = \frac{1}{n!} [\sigma_C T(b)]^n$$

that is not a normalized quantity.

Suppose now that one makes an experiment on a nucleus and he wants to measure the inclusive cross section for events of kind C. Then he will look in the final state if C is present at least once. The inclusive cross section for C production on a nuclear target will then be given by:

$$(\sigma_C^A)_{inclusive} = \int d^2 b \, \sigma_C T(b)$$

since n=1 in this case. Therefore

$$(\sigma_C^A)_{inclusive} = A \sigma_C$$

to be compared with

$$\sigma_C^A = \int d^2 b (1 - \exp[-\sigma_C T(b)]) = \int d^2 b \sum_{n=1}^{\infty} \frac{1}{n!} [\sigma_C T(b)]^n \exp[-\sigma_C T(b)]$$

Therefore while σ_C^A depends on the whole series of multiple scatterings where C is produced, the inclusive cross section $(\sigma_C^A)_{inclusive}$ is given by the single scattering expression only.

When computing the inclusive cross section all initial and final state interactions will cancel[6]

Self shadowing in semihard hadronic interactions

The perturbative QCD parton model expression for the inclusive cross section to produce large p_t partons is not a normalizable quantity because of the divergence of the parton-parton cross section when the C.M. scattering angle goes to zero or to π and because of the divergence of the parton distributions for $\hat{s} \to 0$. Because of these divergent behaviours a merging of large p_t inelastic events with low p_t inelastic events is, at present, not possible. More precisely large p_t physics needs to be defined with a cut off (p_t^{min}) that distinguishes low p_t events from large p_t ones. All quantities computed in the QCD parton model will then be cut off dependent and when the cut off is moved towards small values they will show a very strong dependence, actually they will grow as an inverse power of the cut off.

This behaviour is, on the other hand, related to the nature of the inclusive cross section itself, that, as noticed in the previous paragraph, being related to the multiplicity, is not normalized a priori.

The situation, in this respect, will improve looking, rather at the hard cross section, namely at the contribution to the total inelastic cross section coming from hard partonic collisions. In the case of hadron-nucleus interactions, discussed previously, this amounts to focus on σ_C^A rather than on $(\sigma_C^A)_{inclusive}$.

While $(\sigma_C^A)_{inclusive}$ is related to the single scattering term, and therefore is greatly simplified by the cancellation of all initial and final state interactions, σ_C^A is rather related to the whole multiple scattering series and therefore needs a lot of more information to be constructed. σ_C^A is however constructed with normalized quantities, or equivalently $\sigma_C^A \leq \sigma_{inel}^A$.

Analogously one may expect that constructing the hard cross section σ_H one will get a physical quantity with a smooth behaviour as a function of the cut off because of the bound $\sigma_H \leq \sigma_{inel}$ making therefore a step towards the merging of large p_t physics with low p_t physics.

The problem with σ_H is that it will depend on the whole multiple scattering series and therefore on an infinite set of non perturbative inputs.

In order to construct σ_H it will be convenient to write explicitly the dependence of the hard parton cross section on the transverse partonic coordinates.

The single parton distribution $G(x)$ is constructed integrating the parton wave function squared on the transverse parton momentum a. One can then define:

$$G(x) \equiv \int \mid \psi(x,\mathbf{a}) \mid^2 d^2 a = \int \mid \tilde{\psi}(x,\mathbf{b}) \mid^2 d^2 b$$

and

$$\Gamma(x,\mathbf{b}) \equiv \mid \tilde{\psi}(x,\mathbf{b}) \mid^2$$

with **b** the parton transverse coordinate, so that the inclusive cross section for 2 large p_t parton production becomes:

$$\frac{d\sigma}{dx\,dx'\,dcos\theta} = \int \Gamma_A(x,\mathbf{b})d^2 b\,\Gamma(x',\mathbf{b}')d^2 b' \frac{d\hat{\sigma}}{dx\,dx'\,dcos\theta}$$

One can then replace the integration on **b'** with an integration on the impact parameter β between the two colliding hadrons:

$$\frac{d\sigma}{dx\,dx'\,dcos\theta} = \int d^2\beta \int \Gamma_A(x,\mathbf{b})\Gamma_B(x',\mathbf{b}-\beta)d^2 b \frac{d\hat{\sigma}}{dx\,dx'\,dcos\theta}$$

In the case of a double collision one can also work out a similar expression[4]:

$$\frac{d\sigma}{dx_1\,dx_1'\,dcos\theta_1\,dx_2\,dx_2'\,dcos\theta_2} = \int \Gamma_A(x_1,x_2;\mathbf{b}_1,\mathbf{b}_2)\Gamma_A(x_1',x_2';\mathbf{b}_1-\beta,\mathbf{b}_2-\beta)\times$$

$$\times \frac{d\hat{\sigma}}{dx_1\,dx_1'\,dcos\theta_1}\frac{d\hat{\sigma}}{dx_2\,dx_2'\,dcos\theta_2}d^2 b_1 d^2 b_2 d^2\beta$$

where the quantities $\Gamma_A(x_1,x_2;\mathbf{b}_1,\mathbf{b}_2)$ are the double parton distributions depending explicitly on the parton fractional momenta x_1, x_2 and transverse coordinates $\mathbf{b}_1$, $\mathbf{b}_2$. Being the multiple parton distributions non perturbative quantities presently unknown one is forced to make assumptions to proceed further.

The assumption will be that the soft parton distribution, both as a function of the fractional momentum x and as a function of the transverse coordinate **b**, is a Poissonian, so that the probability density of having n partons with fractional momenta $x_1,...,x_n$ and with transverse coordinates $\mathbf{b}_1,...,\mathbf{b}_n$, is given by:

$$\frac{1}{n!}\Gamma(x_1,\mathbf{b}_1)...\Gamma(x_n,\mathbf{b}_n)exp(-\int_\lambda \Gamma(x,\mathbf{b})dx\,db)$$

(where λ is the cut-off that regularizes the integral at small x).

The average number of partons with fractional momentum x is then given by:

$$G(x) = \int \Gamma(x, \mathbf{b}) d\mathbf{b}$$

and the average number of pairs of partons with fractional momenta x_1, x_2 and transverse coordinates $\mathbf{b_1}, \mathbf{b_2}$ is

$$\Gamma(x_1, x_2; \mathbf{b_1}, \mathbf{b_2}) \equiv \frac{1}{2} \Gamma(x_1, \mathbf{b_1}) \Gamma(x_2, \mathbf{b_2})$$

One can now look at the integrated cross section introducing p_t^{min} as cut off. Defining then

$$\hat{\sigma}(x_1, x_1') \equiv \int_{p_t > p_t^{min}} \frac{d\hat{\sigma}}{dx_1 dx_1' dcos\theta_1} dcos\theta_1$$

the integrated cross section for double collision becomes:

$$\sigma_{Double}(p_t^{min}) = \frac{1}{2} \int d^2\beta \int_{p_t^{min}} \Gamma_A(x_1, \mathbf{b}) \Gamma_A(x_2, \mathbf{b} - \beta) \hat{\sigma}(x_1 x_1') dx_1 dx_1' d^2 b$$

One can work out analogously the inclusive cross section for n parallel hard collisions getting:

$$\sigma_{2n} = \frac{1}{n!} \int d^2\beta [\int_{p_t^{min}} \Gamma_A(x_1, \mathbf{b}) \Gamma_B(x_1', \mathbf{b} - \beta) \hat{\sigma}(x_1 x_1') dx_1 dx_1' d^2 b]^n$$

that is analogous to the expression

$$\frac{1}{n!} [\sigma_C T(b)]^n$$

obtained for the inclusive cross section to produce C n times in the case of hadron nucleus reactions. One may then define

$$\langle n(\beta) \rangle \equiv \int_{p_t^{min}} \Gamma_A(x_1, \mathbf{b}) \Gamma_B(x_1', \mathbf{b} - \beta) \sigma(x_1 x_1') dx_1 dx_1' d^2 b$$

so that

$$\sigma_{2n} = \frac{1}{n!} \int d^2\beta \langle n(\beta) \rangle^n$$

notice that $\langle n(\beta) \rangle$ is a dimensionless quantity ($\Gamma(x, \mathbf{b})$ has dimensions length^{-2}). Similarly to the hadron-nucleus case one will then interpret $\langle n(\beta) \rangle$ as the average number of parton-parton collisions at a given impact parameter β so that the hard cross section will be given by:

$$\sigma_H = \int d^2\beta (1 - exp[-\langle n(\beta) \rangle])$$

One can get this expression for the hard cross section σ_H introducing the probability for a parton with fractional momentum x and transverse coordinate $\mathbf{b}$ to have an hard interaction with the hadron B whose C.M. transverse coordinate is β:

$$P_B(x, \mathbf{b} - \beta) \equiv \int_{p_t^{min}} \Gamma_B(x', \mathbf{b} - \beta)\hat{\sigma}(x, x')dx'$$

actually $\Gamma_B(x', \mathbf{b} - \beta)$ is the average number density of partons per unit area in the transverse plane and $\hat{\sigma}(x, x')$ represents the transverse size of one parton (for the interaction on interest) so that P_B is the ratio of area occupied by partons to the available area. The hard cross section is then the cross section counting all the events in which there is at least one hard partonic interaction:

$$\sigma_H = \sum_{n=1}^{\infty} \int d^2\beta \frac{1}{n_1} \Gamma_A(x_1, \mathbf{b_1})...\Gamma_A(x_n, \mathbf{b_n}) exp[-\int_\lambda \Gamma_A(x, \mathbf{b})dx d^2 b] \times$$

$$\times [1 - \Pi_{i=1}^n(1 - P_B(x_i, \mathbf{b}_i - \beta))]dx_1...dx_n d^2 b_1...d^2 b_n$$

One is, in fact, summing over all partonic configurations of hadron A (each one with its own probability) and for each configuration with n partons one is looking for the probability of having at least one hard interaction.

The sum can be evaluated:

$$\sigma_H = \int d^2\beta \left(1 - e^{-\int_\lambda \Gamma_A(x, \mathbf{b})dx d^2 b} - e^{-\int_\lambda \Gamma_A(x, \mathbf{b})dx d^2 b} \times \right.$$

$$\left. \times \left[exp(\int_\lambda \Gamma_A(x, \mathbf{b})P_B(x, \mathbf{b} - \beta)dx \quad d^2 b) - 1 \right] \right.$$

giving

$$\sigma_H = \int d^2\beta \left(1 - exp[-\int_\lambda \Gamma_A(x, \mathbf{b})P_B(x, \mathbf{b} - \beta)dx \quad d^2 b] \right)$$

and being

$$\langle n(\beta) \rangle = \int_\lambda \Gamma_A(x, \mathbf{b})P_B(x, \mathbf{b} - \beta)dx \quad d^2 b$$

one gets

$$\sigma_H = \int d^2\beta (1 - exp[-\langle n(\beta) \rangle])$$

Oviously σ_H can be expressed in terms of multiple scatterings:

$$\sigma_H = \sum_{n=1}^{\infty} \int d^2\beta \frac{1}{n!} \langle b(\beta) \rangle^n exp(-\langle n(\beta) \rangle)$$

and observing only one pair of large p_t partons in each event one will write the inclusive cross section as:

$$\sigma_{inclusive} = \sum_{n=1}^{\infty} \int d^2\beta \frac{n}{n!} \langle b(\beta) \rangle^n exp(-\langle n(\beta) \rangle)$$

575

Summing over n one gets back the QCD parton model expression:

$$\sigma_{inclusive} = \int d^2\beta \langle n(\beta) \rangle = \int_{p_t^{min}} G_A(x) G(x') \hat{\sigma}(xx') dx dx'$$

One can now study the dependence of the hard cross section on the cut off p_t^{min}.

One limiting case is that one of $p_t^{min} \to \infty$. One has then; expanding the exponential:

$$\sigma_H = \int d^2\beta (1 - exp[-\langle n(\beta) \rangle]) \simeq \int d^2\beta \langle n(\beta) \rangle = \sigma_{inclusive}$$

so that there is no distinction between σ_H and $\sigma_{inclusive}$.

The case of interest is however that one of $p_t^{min} \to 0$.

In this case $\sigma_{inclusive}$ is badly divergent:

$$\sigma_{inclusive} = \int d^2\beta \langle n(\beta) \rangle \to \infty$$

however $\langle n(\beta) \rangle$ cannot be always large, in fact, if β is larger than some typical hadronic scale R, for any value of the cut off, $\langle n(\beta) \rangle = 0$ since the overlap function

$$\int \Gamma_A(x, \mathbf{b}) \Gamma_B(x', \mathbf{b} - \beta) d^2 b$$

will be equal to zero when the two interacting hadrons are too far. One will then have $exp(-\langle n(\beta) \rangle) \simeq 0$ for $\beta < R$ and $exp(-\langle n(\beta) \rangle) \simeq 1$ for $\beta > R$ so that:

$$\sigma_H \simeq \int d^2\beta (1 - \theta(R - \beta)) = \pi R^2$$

When the cut off p_t^{min} is small, while the inclusive cross section becomes very large, the hard cross section does not, rather it reaches some limiting value that is provided by the transverse hadronic dimension.

Some other infrared safe cross section

The picture that comes out is then that one of a poissonian distribution of multiple scatterings of partons with average number varying with the impact parameter. When the cut off (distinguishing hard collisions from soft ones) is large (more precisely $\sigma_{2j}(p_t^{min}) << \sigma_{inel}$) one has only one hard collision and the single scattering picture given by the QCD parton model is fine.

When the cut off is small $(\sigma_{2j}(p_t^{min}) \simeq \sigma_{inel})$ the inclusive cross section is still given by the QCD parton model expression but this one does not represent properly the final state.

In the final state there will be several pairs of partons and the inclusive cross section is becoming a very large cross section because it counts the multiplicity of the pairs. The hard cross section is, on the contrary, limited because it counts only the inelastic hadronic hard events and not the multiplicity of the pairs of partons.

Given this picture of the final state one can construct also other cross sections that are not divergent when the cut off p_t^{min} goes to zero.

One can, for example, in each event, measure only the parton pair with the largest transverse momentum. Since one is not counting the multiplicity one will have also in this case a finite quantity for $p_t^{min} \to 0$.

The cross section corresponding to the above operational definition is easily written down:

$$\frac{d\sigma}{dp_t} = \int d^2\beta \frac{d\langle n(\beta)\rangle}{dp_t} exp[-\langle n(\beta)\rangle_{p_t}]$$

where $\langle n(\beta)\rangle_{p_t}$ is the average number expression computed with p_t as cut off, and

$$\frac{d\langle n(\beta)\rangle}{dp_t} = \int \Gamma_A(x,\mathbf{b})\Gamma_B(x',\mathbf{b}-\beta)\frac{d\hat{\sigma}}{dxdx'dcos\theta} \times$$

$$\times \delta(p_t - \frac{1}{2}sen\theta\sqrt{\hat{x}x's})dxdx'dcos\theta d^2b$$

This expression for $\frac{d\sigma}{dp_t}$ corresponds to a single scattering term in a multiple scattering expansion. One may, in fact, consider p_t as the cut off separating large p_t events from soft ones. Therefore with that cut off (that will change event per event) one always has only one hard collision.

The exponential factor $exp[-\langle n(\beta)\rangle_{p_t}]$ will give the probability that there is no other pair of partons with a larger p_t.

Because of the dependence of the exponential on p_t this expression is well behaved for $p_t \to 0$. Actually:

$$\lim_{p_t \to 0} \frac{d\sigma}{dp_t} = 0$$

meaning that when p_t is small there will almost always be a pair of partons in the inelastic event with a larger p_t.

One may also notice that

$$\frac{d\sigma_{QCD}}{dp_t} = \int d^2\beta \frac{d\langle n(\beta)\rangle}{dp_t}$$

(where $\frac{d\sigma_{QCD}}{dp_t}$ is the single scattering expression given in the QCD parton model), and that

$$\int_{p_t^{min}}^{\infty} \frac{d\sigma}{dp_t}dp_t = \int d^2\beta(1 - exp[-\langle n(\beta)\rangle_{p_t}]) = \sigma_H$$

Final remarks

The main observation, in the present note, is that in semihard interactions (namely in the regime where the momentum transfer at partonic level is such to allow a perturbative approach but the corresponding large p_t parton production cross section is of the order of the total inelastic one) the perturbative QCD parton model expression for large p_t parton production, while giving correctly the inclusive cross section, fails to describe correctly the structure of the final partonic state.

The simplest consistent description one may then conceive (and that has been considered here) is that one of poissonian distribution of multiple hard parton collisions with average number depending on the impact parameter.

Such a description of the interaction, while being consistent with the known cancellation rules involving the inclusive cross section[6,7], allows to connect measurable physical quantities to (presently) unknown non perturbative properties of the hadronic bound state that are independent on the parton structure functions as measured in deep inelastic scattering experiments.

A better understanding of the final partonic state will also allow the definition of physical observables that are "infrared safe", in the sense that are smooth functions of the cut off distinguishing large p_t from low p_t physics.

This statement does not mean that the infrared problem in QCD has found a solution. One has still a whole set of divergent quantities to deal with: all average quantities, like the average transverse energy generated by the hard interaction will still be given by the standard QCD parton model expression[8] and therefore will be badly divergent as a function of the cut off.

A solution to these problems can be obtained only when also the parton distributions themselves are unitarized[9].

Still the simple exponentiation discussed here can be justified by the following observation:

When looking at the values of the fractional momentum x where the unitarization of the parton distributions is of relevance, one gets[10] (for values of Q^2 of interest) $x < 10^{-4}$, while one can get integrated jet cross sections comparable with the total inelastic one significantly before (one can in fact notice that the minijet region at CERN Collider corresponds to values of $x \geq 10^{-2}$).

As a consequence one will start having multiple hard parallel parton scattering also at kinematical regimes where the inclusive jet cross section will be safely given by the perturbative QCD expression, without taking into account unitarity corrections to the parton distributions.

REFERENCES

1) G. Pancheri and Y.N. Srivastava, Phys. Lett. **159B** (1985) 69, Phys. Lett. **182B** (1986) 199; T.K. Gaisser and F. Halzen, Phys. Rev. Lett. **54** (1985) 1754; T. Sjostrand and M. van Zijl, Phys. Rev. **D36** (1987) 2019.

2) G. Arnison et al. Phys. Lett **172B** (1986) 461; F. Ceradini (UA1 Collaboration) in Proc. of the International Europhysics Conf. on High Energy Physics, Eds. L. Nitti and G. Preparata (Laterza, Bari, 1985) 363.

3) N. Paver and D. Treleani, Nuovo Cimento **70A**(1982) 215; Z. Phys. **C28** (1985) 187.

4) Ll. Ametller and D. Treleani, Int. J. Mod. Phys. **A3** (1988) 521.

5) R. Blankenbecler, A. Capella, C. Pajares, J. Tran Tranh Van and A.V. Ramallo, Phys. Lett. **107B** (1981) 106.

6) V.A. Abramovskii, V.N. Gribov and O.V. Kancheli, Yad. Fiz. **18** (1973) 595, (Sov. J. Nucl. Phys. **18** (1974) 308).

7) C.E. De Tar, S.D. Ellis and P.V. Landshoff, Nucl. Phys. **B87** (1975) 176.

8) K. Kajantie, P.V. Landshoff and J. Lindfors, Phys. Rev. Lett. **59** (1987) 2527.

9) L.V. Gribov, E.M. Levin and M.G. Ryskin, Phys. Reports **100** (1983) 1.

10) A.H. Mueller, Talk given at the Division of Particles and Fields Meeting, Eugene, August 12-15, 1985; J.C. Collins, Supercollider Physics, Proceedings of the Oregon workshop on super high energy physics (1985) 62.

35

THE POMERON IN QCD AND EXPERIMENT

Jochen Bartels

II.Institut für Theoretische Physik
Universität Hamburg, W-Germany

Abstract

An overview is given of the Pomeron in experiment and its interpretation within QCD. The discussion also includes the small-x limit in deep-inelastic scattering.

I Introduction

In this talk I will attempt to give an overview of the present situation of the Pomeron. This includes the main experimental features and, in particular, the question how well the Pomeron is understood within QCD. As it is well-known, the theoretical explanation of the Pomeron is an old problem in particle physics, and despite numerous and elaborate efforts in the past, we have, in my view, not yet reached a satisfactory solution in the framework of the Standard Model. On the other hand, hadron colliders which are presently working or being planned for the future (CERN-collider, Tevatron, HERA, SSC, LHC, Eloisatron,...) are or will be full of events where the majority of particles is produced close to the forward direction. In would also be hard to argue that we can reach an understanding of the dynamics of quark and color confinement without eventually being able to explain diffraction scattering. Therefore, interest in the Pomeron problem remains, and an orientation of where we are standing now may be of some use.

First a few introductory remarks. The name "Pomeron Physics" refers to phenomena seen in the small-t region of hadron-hadron scattering at high energies. Quantities of primary interest are: the energy dependence of the total cross section, energy and t-dependence of the elastic cross section (in the range $-t \ll s$, the energy dependence of the real part of the elastic forward scattering amplitude F(s,t). Given a successful theory for these quantities, standard Regge theory then allows to describe also inelastic processes in multiregge limits and inclusive particle spectra. New and recent interest in the Pomeron is also connected with the small-x behavior of deep-inelastic structure functions. Here one also probes the Pomeron, although in a different kinematic limit. As I will discuss below, the advantage of this limit is that one is somewhat closer to the region of validity of perturbation theory. In the language of QCD, in both contexts one probes the content of very slow quarks and gluons inside the hadron, but in slightly different spatial regions. Hadron hadron scattering at small angles probes the slow partons at large impact parameters, whereas deep-inelastic scattering in looks at small transverse distances. Only if, at fixed Q^2, x is made very small, one also enters the region of larger and larger impact parameters. In the first case, one cannot ignore confinement (which makes the problem so hard), whereas in the second case one at least begins with perturbation theory and then tries to move deeper and deeper into the nonperturbative regime. That is why it may be helpful to have both approaches described in one paper. To give them a name, I will call them the "soft" and the "hard" Pomeron, resp.

The talk will be organized in three parts. In the first part I summarize a few gross features of the experimental situation in hadron-hadron scattering ("soft" Pomeron). Then I discuss a few models and how they relate to experimental data. The main conclusions of this section are: (i) at present day energies the total cross section shows a rather clear lns^2-behavior, but there is little doubt that we are still far away from truly asymptotic energies. One should therefore not rule out the possibility that some new physics will leave its signature also in the total cross section (e.g.in the TeV-region).(ii) The recent UA4 measurement of the real part of the forward scattering amplitude is of extreme interest, since it contains information on the behavior of the total cross section at higher energies. It should, therefore, be confirmed by another independent measurement.

In the second part I will review what has been done in order to derive the Pomeron from QCD. Starting point are extensive perturbative calculations (most economically done with the help of dispersion relations). They seem to result in a complete (nonlocal) reggeon field theory. When trying to solve this field theory one faces a principal difficulty: in the infrared region the Feynman diagrams from which the reggeon field theory is derived become sensitive to confinement dynamics. Suggestions have been made how to overcome this difficulty and how to solve the reggeon field theory. The most promising candidate seems to be critical reggeon field theory.

The third part reviews the situation of the small-x limit of deep-inelastic structure functions. It is well-known that the Standard QCD-description comes into conflict with s-channel unitarity when $x^{-1} = s/Q^2$ is getting too large. By taking into account much larger classes of Feynman diagrams (so-called fan diagrams, the building units of which are the standard QCD-ladders) it is possible to extend the range of validity of perturbative QCD up to larger values of $1/x$. Existing data, however, do not yet allow to test this "improved" QCD description. HERA will be the first machine where measurements down to $x = 10^{-4}$ will be possible and where these theoretical ideas can be tested. It will also be discussed how this "hard" Pomeron appearing in the small-x region comes closer and closer to the reggeon field theory description of the "soft" Pomeron. Finally, a few words will be said about the concept of the recently proposed "Pomeron structure function".

II   The Experimental Situation

In this part I first review the most important experimental facts of the Pomeron in hadron-hadron scattering and confront them with a few theoretical models.

I begin with the total cross section in $p\bar{p}$-scattering (Fig.1a)[1]. The arrow should remind us that we are still waiting for the Tevatron measurement of the total cross section. The rise of σ_{tot} is well-described by the classical fit of Amaldi et al.[2] (Fig.2a):

$$\sigma_{\pm} = C_1 E^{-\nu_1} \mp C_2 E^{-\nu_2} + B_1 + B_2 (ln\, s)^{\gamma} \tag{1}$$

with

$$C_1 = 41.9,\ C_2 = 29.2,\ B_1 = 27.0,\ B_2 = 0.17,\ \nu_1 = 0.37,\ \nu_2 = 0.55,\ \gamma = 2.1 \tag{2}$$

(all constants B and C are in mb, energy in Gev/c). A more recent fit is due to Bourrely and Martin [3], (Fig.3). For the scattering amplitude they make the ansatz:

$$F = is\frac{A + B[ln\, s/s_0 - i\pi/2]^2}{1 + C[ln\, s/s_0 - i\pi/2]^2}\ . \tag{3}$$

The constants A,B,C are fitted to the data, and two different scenarios are used: in (I) C=0 (which implies $\sigma_{tot} = const * lns^2$); in (II) C is allowed to be different from zero (which implies asymptotically that σ_{tot} approaches a constant value). The results of the fits are:

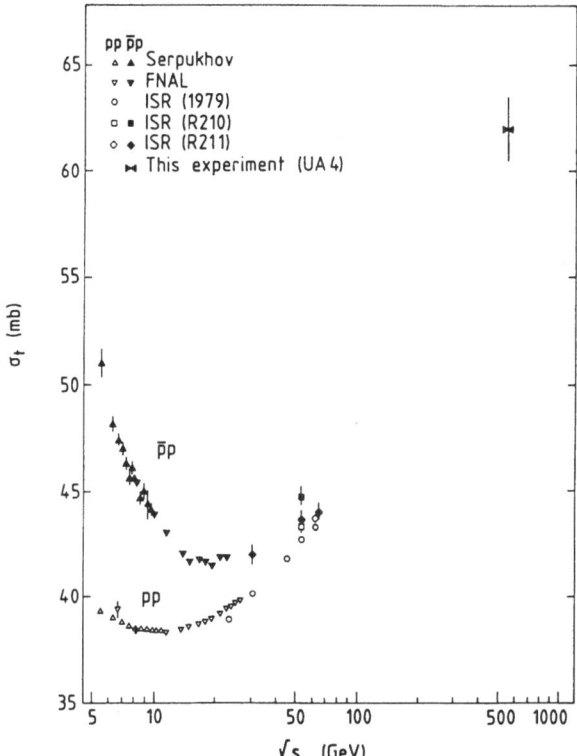

Figure 1a. The total cross section for pp and for $p\bar{p}$-scattering (from [1]). The arrow marks the Tevatron point which has not been published yet.

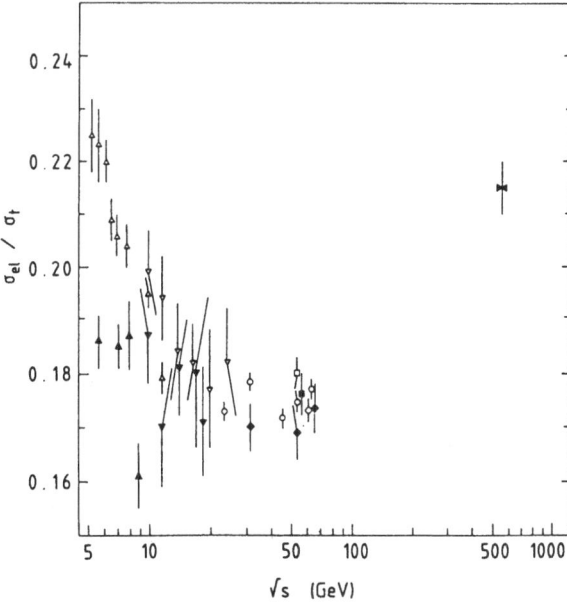

Figure 1b. The ratio σ_{el}/σ_{tot} (from [1]).

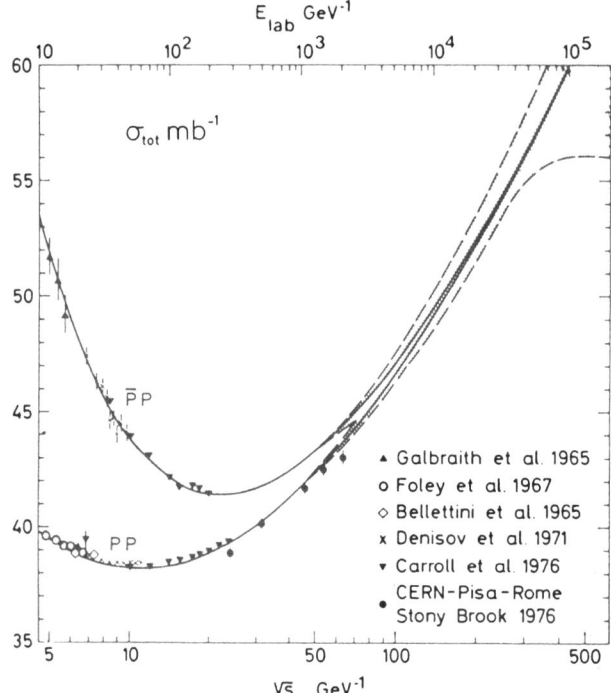

Figure 2a. Amaldi's fit to σ_{tot} for pp and $p\bar{p}$-scattering (from [2]).

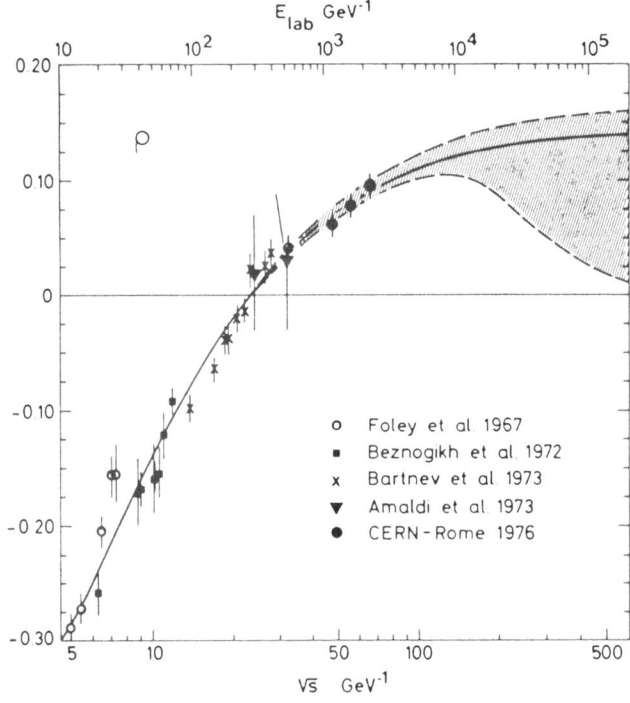

Figure 2b. Amaldi's fit to $\rho = ReF(s,0)/ImF(s,0)$ (from [2]).

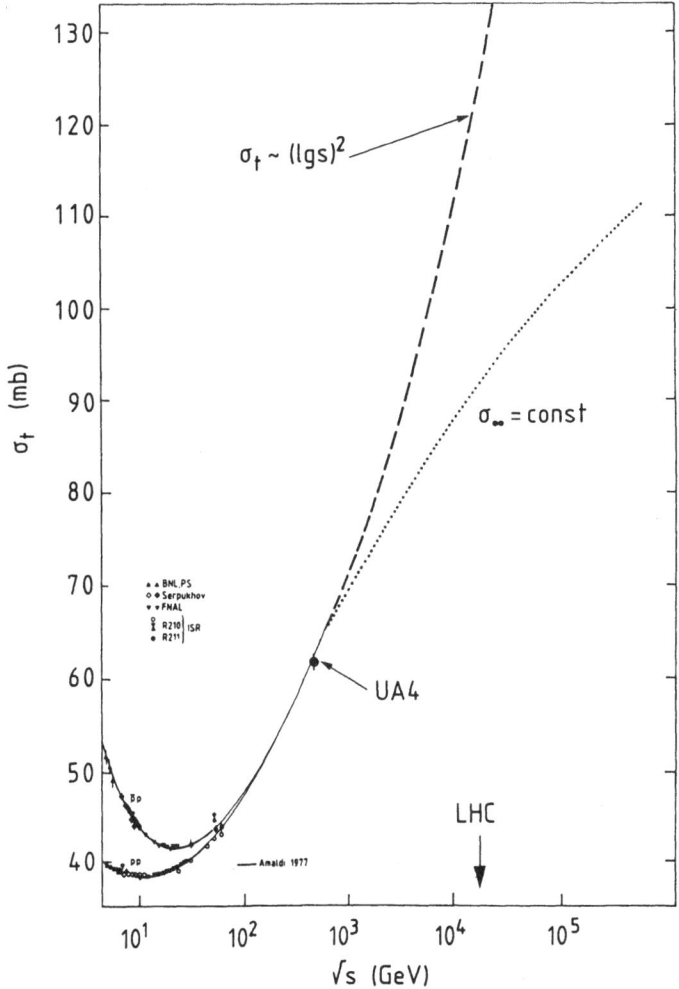

Figure 3. Fit of Bourrely and Martin (taken from [3]).

$$
\begin{array}{ccc}
 & I & II \\
A & 41.95 & 41.824 \\
B & 0.43 & 0.815 \\
C & 0 & 0.006 \\
s_0 & 243.6 & 277.7
\end{array}
\tag{4}
$$

One notices that in scenario II the value for C is very small: the plateau σ_{tot} = constant is reached only for extremely high energies. Hence these fits support the conclusion that, for present energies, σ_{tot} is best described by a lns^2-behavior, i.e.it goes as the Froissart bound (although it is far from saturating it).(See, however, also [4] where slightly different conclusions are drawn).

The next quantity of interest is the elastic cross section. As shown in Fig.1b [1], the ratio σ_{el}/σ_{tot} rises from 0.175 at ISR to 0.215 at the collider. For the differential cross section $d\sigma_{el}/dt$ a compilation of data [5] is shown in Fig.4: in the $p\bar{p}$-data a shoulder is seen at $/t/$=0.9 $(GeV)^2$. With a little imagination one also sees some structure at $/t/$=1.6$(GeV)^2$. A very interesting topic is the measurement of $\rho = ReF(s,0)/ImF(s,0)$. The UA4 measurement [6] gives (Fig. 5)

$$
\rho = 0.24 \pm 0.04 \ . \tag{5}
$$

This has to be confronted with the extrapolations from data at lower energies (Fig. 2b). They all lie in the range ρ=0.10 to 0.15 (e.g. ρ=0.119 for scenario II and ρ=0.154 for scenario I of Bourelly and Martin [3]). The measured value is thus off by almost a factor of two from the expected value. If it is correct, it implies that, via the use of dispersion relations, at $s = 0(TeV^2)$ the total cross section should rise faster than seen presently. Fig.6 shows a few "predictions", taken from [7]. Since such a growth cannot continue forever and eventually has to flatten off again, one expects to see some "structure", e.g. a broad enhancement. It is this "predictive" power of ρ which makes it such an interesting quantity. It would, therefore, be very desirable to have an independent experimental confirmation of this unexpected high value of ρ.

In a more detailed survey one would now go on and discuss other quantities which allow more specific tests of theoretical ideas, for example: σ_{inel}, the rise of the plateau of inclusive cross sections, multiplicities, KNO-scaling,.... For this rather general review, however, I have to restrict myself to the very few quantities discussed above.

In the next step one might ask what kind of model agrees with these data. Here is a list of a few:

1) Critical reggeon field theory [8]
2) Flavoured reggeon field theory [9]
3) Eikonal Model [10]
4) Models containing an Odderon [11]
5) Model based upon the analogy Pomeron-photon[12]
6) QCD inspired three gloun exchange [13]
7) Dual Topological Unitarization [14]

Since I am unable to present a critical discussion of all these models, I make the subjective selection and say a few words only about 1) and 3). I do this for the following reason. Model 1) is, from the theoretical point of view, the most sophisticated one, and as I will argue below, the efforts of analyzing the high energy behavior of QCD point in the direction as if critical reggeon could be an effective field theory which described the high energy limit of QCD. On the other hand, the asymptotic predictions of critical reggeon field theory do not match with presently observed behavior. Some of the other models, on the other hand, do provide successful descriptions of present data, and model 3) seems to do particularly well. Moreover, it has some basis in quantum field theory (massive QED).

I begin with critical reggeon field theory. It has been set up as a solution to the t-channel partial wave unitarity equations and has its roots in general S-matrix theory rather than any specific quantum field theory model. It has also been shown to satisfy constraints that follow

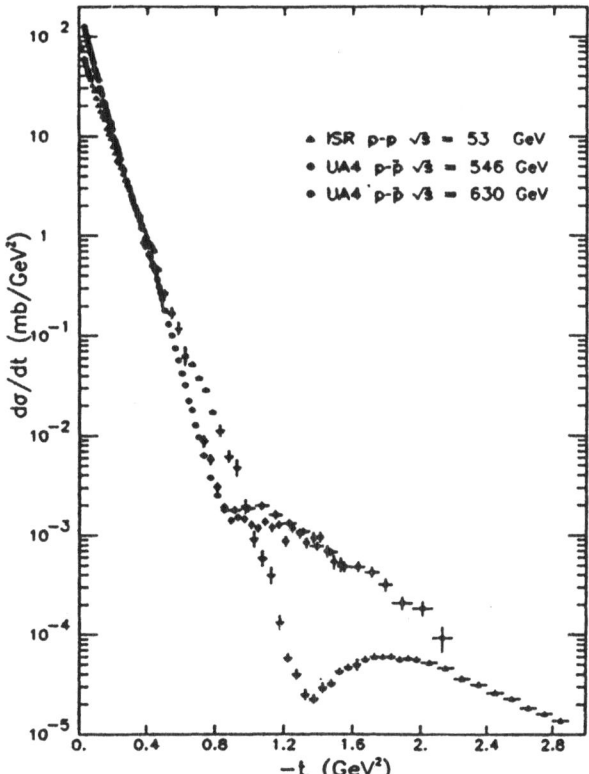

Figure 4. Compilation of $d\sigma_{el}/dt$ for pp at ISR and for $p\bar{p}$ at the collider (from [5]).

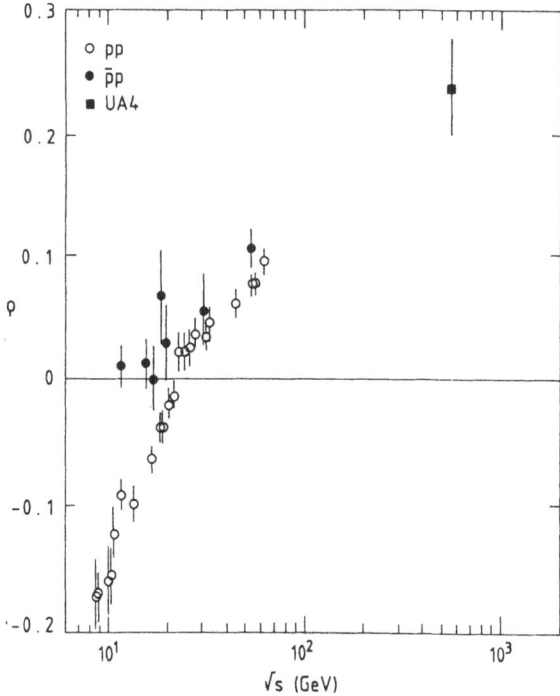

Figure 5. Data points for the ρ-parameter (from [6]).

from s-channel unitarity and that have been problematic for other models. All this makes it, from an esthetic point of view, a very attractive candidate for a self consistent theory for asymptotic hadron-hadron scattering. It contains, however,a large number of parameters (in particular, the couplings of reggeons to external hadrons) which have to be computed from the underlying theory, QCD. This defines the task of "deriving critical reggeon field theory from QCD", which I will discuss below in more detail. At this place let me list the most important asymptotic predictions which follow from critical reggeon field theory without any further reference to any underlying theory. For the total cross section one finds:

$$\sigma_{tot} = const(ln\ s/s_o)^{-\gamma}[1 + ln(s/s_0)]^{-\gamma}[c_0 + C_1 ln\ ln(s/s_0) + \ldots] \tag{6}$$

with the approximate numerical values $-\gamma = 0.2, \lambda = 1$. The elastic scattering amplitude obeys the scaling law:

$$F_{el}(s,t) = i\beta_A(t)\beta_B(t)ln(s/s_o)^{-\gamma}f[t(ln\ s/so)^z] \tag{7}$$

where f denotes a universal function which is independent of the scattering particles A and B, and z=1.1 approximately. Integrating $|F_{el}(s,t)|^2$ over t leads to:

$$\frac{\sigma_{el}}{\sigma_{tot}} = const \cdot (lns/s_0)^{-2\gamma-z} \ , \tag{8}$$

$$-2\gamma - z \approx -3/4 < 0 \ , \tag{9}$$

i.e., the ratio σ_{el}/σ_{tot} is predicted to decrease with energy. All these formulae are asymptotic and have corrections which vanish as s goes to infinity. All predictions have been compared with present data: the total cross section in [15], the elastic cross section in [16]. None of them fits well, but the strongest argument against this theory is presently the prediction (8) which is clearly violated. The conclusion then is that critical reggeon field theory, at least in its asymptotic limit, is not applicable at present energies. However, for the obvious possibility that we may still be too far away from truly asymptotic energies, it seems to me that we cannot dismiss this theory altogether: it still may turn out to be the right theory at much higher energies.

Next we turn to a successful theory, model 3). As it was said before, this form of high-energy behavior has been derived from massive QED, by iterating in the s-channel the so-called tower diagrams [17]. In this picture the hadron looks like a black disc with the radius increasing as lns. I again list the main predictions. For the scattering amplitude one has:

$$F(s,t) = \frac{is}{2\pi} \int d^2b e^{i\mathbf{q}\mathbf{b}}[1 - e^{\Omega(s,b)}] \tag{10}$$

$$\Omega(s,b) = F(b^2) \left[\frac{s^c}{(lns)^{c'}} + \frac{u^c}{(lnu)^{c'}} \right] + R_o(s,b) \tag{11}$$

where $q^2 = -t$, $c = 0.167$, and $c' = 0.748$, and $R_0(s,b)$ stands for secondary Regge pole exchange which becomes negligible when s goes to infinity. From these expressions on obtains the predictions:

$$\sigma_{tot} \sim const \cdot ln\ s^2 \tag{12}$$

$$\frac{\sigma_{el}}{\sigma_{tot}} \to 0.5 \tag{13}$$

A numerical evaluation of (10) for $d\sigma_{el}/dt$ gives impressive agreement for both pp-scattering at ISR and $p\bar{p}$-scattering at the collider. The authors also present their predictions for much higher energies. If one defines asymptopia to be at those energies where observables reach their limiting values (e.g. ρ or σ_{el}/σ_{tot}), then one has to conclude that we are still far below [18] (Fig.7). Very similar conclusions could also be drawn from Model 4 [19].

Attempting to draw simple conclusions from this brief overview, I would like to stress the following three points: (i) Up to collider energies the behavior of σ_{tot} shows lns^2 behavior ,and the eikonal model 3) gives a good description. The asymptotic formulae of critical reggeon field theory cannot be used. (ii) A crucial question then seems to be: do we observe already the truly asymptotic behavior? The simplest possibility is that the behavior seen

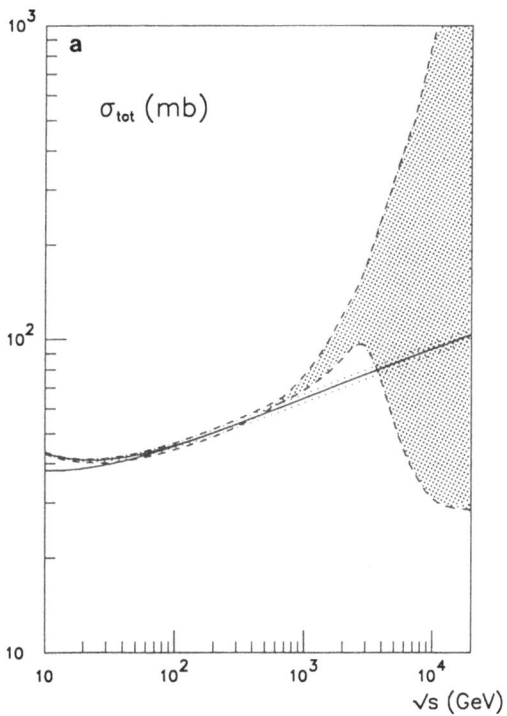

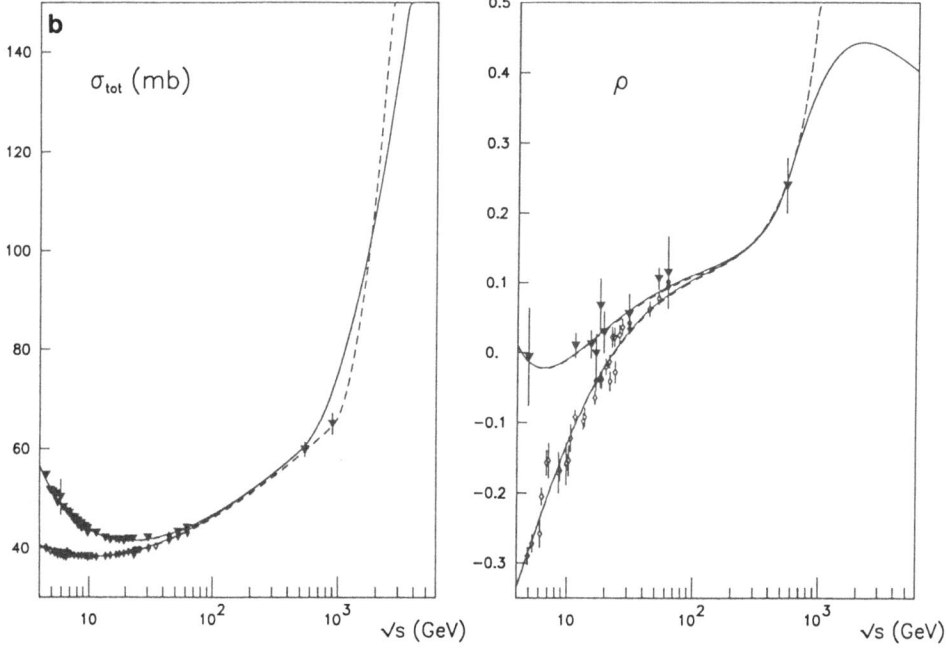

Figure 6. "Predictions" for σ_{tot} and ρ (using the UA4 measurement of ρ and dispersion relations). The models are described in [7], and the curves are taken from the same reference. (a) No energy threshold is assumed, and the dashed area shows the allowed variation of the constant in front of the $ln^2 s$ term. (b) An energy threshold is assumed at $\sqrt{s_0} = 500 GeV$.

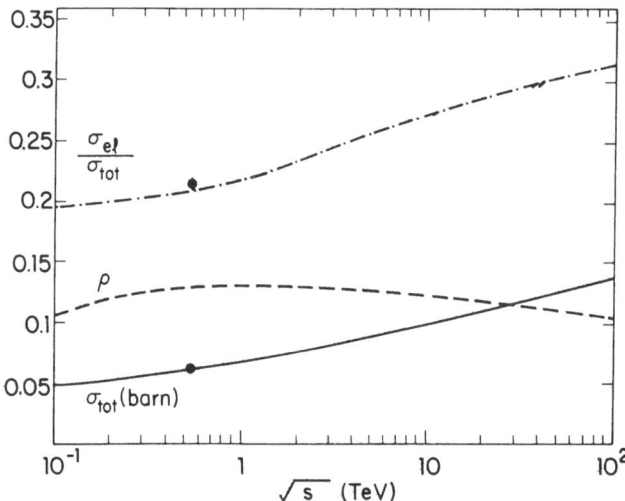

Figure 7. Extrapolations for σ_{el}/σ_{tot}, ρ, and σ_{tot} to higher energies (from [18]).

Figure 8. QCD ladder diagrams in the Regge limit (leading -lns approximation): The wavy lines denote reggeized gluons, and the open circles stand for momentum dependent reggeon interaction vertices.

now can smoothly be extrapolated up to infinite energies. But then still the comparison of present values for, e.g. ρ and σ_{el}/σ_{tot}, with the limiting values predicted by the successful models indicates that we are far below asymtopia. From the theoretical point of view there remains the task of obtaining the $\ln s^2$ behavior from QCD. Alternatively, the behavior seen at present energies may not yet be the asymptotic behavior, and some changes will be observed at higher energies. Theoretical explanations for such a change could be: the increasing importance of hard QCD processes and their contribution to σ_{tot}, heavy flavour production thresholds, new physics in the TeV-region which not only affects the electroweak sector but also leaves some signature in the strong sector,.... In any of these cases, of course, the question of the asymptotic behavior of hadron-hadron scattering will remain open until we can reach much higher energies. (iii) An independent measurement of the ρ-value at the collider or at the Tevatron would be very important. Because of its predictive power it sheds some light on the behavior of σ_{tot} at higher energies.

Let me conclude this section by repeating an argument [20] which, in my opinion, is important because it is related to one of the striking features of hadron-hadron scattering at present energies, the validity of the additive quark model. Within the QCD-inspired quark parton model, the validity of the quark counting rules is explained by assuming that inside the fast hadron with transverse radius R_h each valence quark is surrounded by a cloud of (virtual) sea quarks and glouns. The radius of this cloud, r_q, is smaller than the hadron radius R_h, and most of the time the clouds of the two or three valence quarks do not overlap. Therefore the total cross section is to a good approximation given by the sum over valence quark - valence quark interaction cross sections. As a function of energy, r_q^2 grows as $4*\alpha'_{pomeron} * \ln s/s_0$, and using the estimates of [20], r_q is expected to reach R_h at $\sqrt{s} = O(10^6 GeV)$. Once the clouds start to overlap, the additive quark model should break down and one expects to see a change in the hadronic cross sections. This argument supports the view that what we are seeing at present energies is not yet the true asymptotic behavior in soft hadron scattering.

III The Soft Pomeron in QCD

In this section I will try to give an overview on our present understanding of the high energy behavior of QCD. There is no doubt that the Pomeron is an intrinsically nonperturbative object, i.e. it cannot be calculated from perturbation theory alone. Nevertheless, existing calculations start from the analysis of the perturbative (mainly gluonic) content of the Pomeron and then try to move deeper and deeper into the nonperturbative region. It will become clear that this program is not completed, and more theoretical efforts will be needed in the future.//

This overview can naturally be organized into three steps. In the first step I want to describe what may be called the "leading-lns" Pomeron in QCD [21]-[23]. It consists of the sum of ladder diagrams (Fig.8) with reggeized gluons along the side lines. The rungs of the ladders are due to gluons being produced multiperipherally. They can also be viewed as reggeon interaction vertices: two reggeons $\rightarrow$ two reggeons. They have a nontrivial momentum dependence which is related to the reggeization of the gluon. This set of QCD diagrams represents the leading-lns approximation to the scattering of colorless hadrons in the limit $s \rightarrow \infty$, t small and fixed: for each power of the strong coupling constant α_s, these diagrams yield the highest power of lns. As an example, Fig.9 shows the scattering of two heavy flavor mesons [24]: the momentum scale of α_s is then given by the mass of the heavy quark. (These QCD ladder diagrams with reggeized gluons play a role analogous to that of the tower diagrams in QED [17]: there is, however, the fundamental difference between the photon and the nonabelian vector particles. The latter reggeize whereas the former does not. This makes the analysis of the high energy behavior of QED from the start different from that of nonabelian gauge theories.) The QCD ladder diagrams when coupled to colorless external bound states are all infrared finite and ultraviolet finite; the strong coupling constant remains fixed in the leading-lns approximation.

There are a few remarkable properties of these diagrams that should be mentioned. First of all one would like to know what kind of high energy behavior comes out if the ladder diagrams are summed up to all orders. The answer is [23]:

$$F(s,t) = i\, const \cdot s^{\alpha_c} \;,\;\; \sigma_{tot} = const' \cdot s^{2\alpha_c - 1} \qquad (14)$$

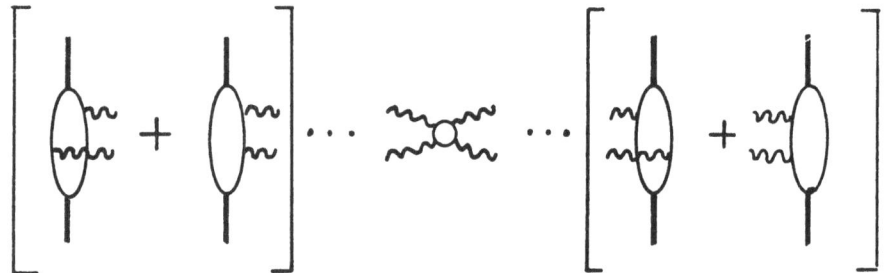

Fig.9 . Meson-meson scattering in the leading-lns approximation (with heavy quarks).

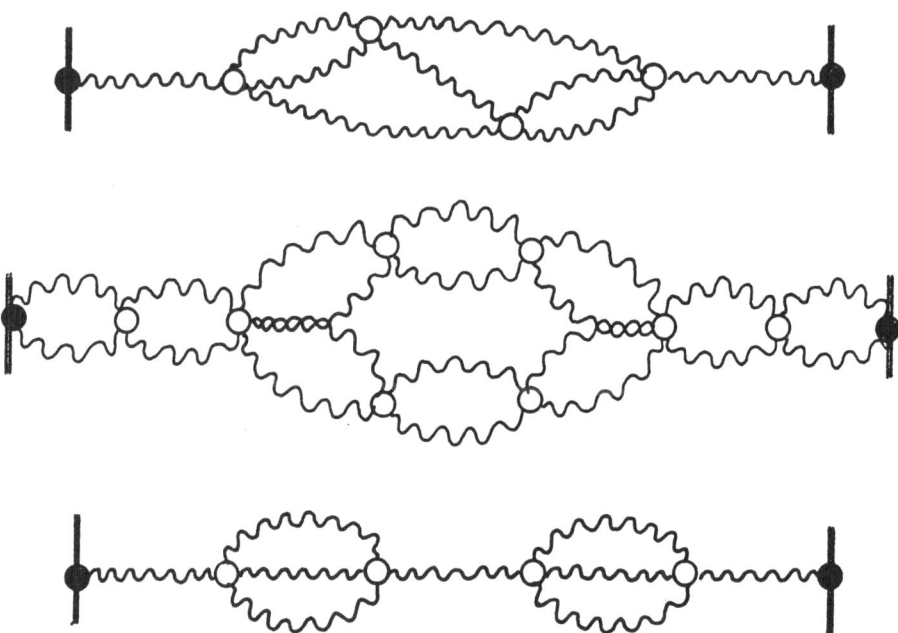

Fig.10. A few reggeon diagrams which emerge from the s-channel unitarization descibed in [27].The notation is the same as in Fig. 8.

with

$$\alpha_c = 1 + \frac{3g^2 ln2}{\pi^2} \quad (\alpha_s = \frac{g^2}{4\pi}) \tag{15}$$

That is one finds a fixed-cut singularity in the angular momentum plane to the right of j=1, and the total cross section violates the Froissart bound. This clearly shows that these diagrams provide an invalid approximation if s becomes too large. In order to have a better understanding of what goes wrong it is useful to take a closer look of how this bad high energy behavior is generated. As it has been said before, at each order of α_s all internal momentum integrations (integration is always over two-dimensional transverse momentum.Its conjugate variables can be understood as the transverse distance between wee partons inside the scattering hadron) are perfectly convergent both in the infrared and ultraviolet region, and the average value of momentum is of the order of the scale in α_s (in our example, the mass of the heavy quark). However, if one starts to sum over all orders in α_s and includes higher and higher powers of the coupling constant, and if one asks which values of internal momentum are responsible for building up the leading singularity in the j-plane, then, according to the analysis of [23], one finds:

$$< ln^2 \frac{k_t^2}{m^2} > \sim const \cdot n \tag{16}$$

where n is the order of α_s. Hence the leading-s behavior depends largely upon the behavior of the diagrams in both the infrared and the ultraviolet regions. But there the leading-lns approximation is not quite correct: for large momenta, the dependence of the strong coupling α_s upon momentum ought to be taken into account, and in the infrared region where the strong coupling is becoming large perturbation theory is expected to break down. One therefore has to conclude that these ladder diagrams, as they come out from the leading-lns analysis, are valid only as long as the internal momenta are neither to small nor too large, i.e. one is not allowed to go to infinite order in α_s, and s has to be such that $\alpha_s * lns < 1$.

It is also instructive to see to what extent a modification of the ladder diagrams in the ultraviolet and/or the infrared region is able to cure the bad high energy behavior. This has been studied by L.N.Lipatov in [25]. For large momentum he replaces the fixed coupling constant by the momentum dependent one, and in the low-momentum region where the coupling becomes strong he imposes as certain boundary condition onto the behavior of the perturbative amplitude which represents confinement dynamics. As a result of this, the leading fixed-cut singularity is resolved into a string of moving poles, but the right-most one is still to the right of j=1. (The slope of the moving Regge poles depends upon the assumed boundary condition, i.e upon the unknown strong interaction dynamics). Thus one of the undesirable features has disappeared, namely the fixed-cut nature of the leading j-plane singularity. The other one, however,the violation of the Froissart bound is still present. These modifications alone are therefore not yet sufficient to provide an acceptable high energy description.

In the next step one therefore has to ask which further modifications are necessary. The answer is: more diagrams have to be included. The reason is that the ladder diagrams with two reggeized gluons are lacking s-channel unitarity. An iterative way of restoring s-channel unitarity while preserving at each step partial wave unitarity in the t-channel has been outlined (and partly been carried out [26]) in [27]. The central idea to this program is the use of dispersion relations for multiparticle amplitudes rather than doing perturbation theory order by order. In order to apply this powerful technique to QCD, the gluons have first to made massive (by introducing Higg's scalars and invoking the Higg's mechanism). Then the gluon reggeizes like an ordinary massive vector particle, and extended use can be made of the known analytic structure of multiparticle amplitudes in multiregge limits. (Note that this technique could, most likely, not be used in massive QED: since the photon does not reggeize, scattering amplitudes have a much more complicated analytic structure than in nonabelian theories where the vector mesons do reggeize). Once all these calculations have been done, the gluon mass has to be removed (see below).

The outcome of this unitarization scheme (still with the gluon mass being nonzero) is a complete reggeon field theory: the reggeized gluons are the "elementary fields", all n-reggeon → m-reggeon couplings are present (obeying the rules of signature conservation), and all

591

these vertices are momentum dependent, i.e. they are nonlocal. A few diagrams are shown in Fig.10. In elastic scattering amplitudes with color zero exchange channel (which is, of course, what we are interested in) only even numbers of gluon lines appear in the t-channel, and the Pomeron appears as a bound state (glueball). For full QCD this description is still incomplete: fermions have to be included by introducing flavor nonsinglet exchange channels (reggeization of fermions in nonabelian gauge theories has been studied in [28]).

As it has been said before, after all these diagrams have been included and the elements of the reggeon field theory have been defined, it still remains to remove the gluon mass. The whole program described above relies upon the assumption that QCD can be reached by taking a certain limit in the parameter space of a spontaneously broken SU(3) gauge theory. There are arguments which strongly support this assumption: (i) in the framework of perturbation theory, one can check whether the two approaches give the same answer: either calculating order by order in QCD or applying the dispersion relation technique to a spontaneously broken gauge theory and then removing the scalar fields. This has been done for the two-gluon ladder diagrams in SU(2) gauge theory. Further support comes from the study of the small-x limit of hadronic structure functions in QCD (see below): this analysis seems to lead to the same diagrams as the dispersion relation-based analysis of the massive theory.(ii) Beyond perturbation theory, an analysis of the phase structure of lattice gauge theories [29] shows that, as long as the Higg's scalars are in the fundamental representation of the color group, the Higg's phase and the confinement phase are analytically connected. There exists therefore a path in the parameter space which leads from the phase with massive reggeizing vector particles to that of confinement. (iii) Finally, one may ask directly how the massive reggeon diagrams behave when the mass of the vector particle is taken to zero. Most remarkably, one sees striking differences between color zero and color-nonzero exchange channels. In the former case, there is an enormous cancellation of infrared divergencies (which, however, is not expected to be complete in more complicated reggeon diagrams). In channels with color-nonzero exchange, on the other hand, the infrared logarithms always seem to pile up in such a way that these amplitudes are suppressed: this is, of course, consistent with the expectation that open color should not be produced at high energies. Taking together all these bits of evidence, it seems very probable that this scheme of isolating QCD diagrams which in the high energy limit satisfy both t-channel and s-channel unitarity might really work.

In the third step, which to me seems to be the most difficult one, on has to solve this reggeon field theory. In doing so, one has to keep in mind the restrictions which I have mentioned in the context of the two-gluon ladders. That means that, in addition to the difficulty that the obtained reggeon field theory is nonlocal, comes with arbitrary high order interaction vertices and that we are asking for bound states rather than corrections to the elementary reggeon field, both the infrared and the ultraviolet part of the momentum integrals of the underlying Feynman amplitudes must be handled with great care. In the ultraviolet part it may be enough to simply put in by hand the momentum dependence of the strong coupling, but in the infrared region confinement dynamics must be brought in. A signal for this are the infrared logarithms, which are expected to appear in more complicated reggeon diagrams when the gluon mass is removed. All experience from statistical mechanics and the use of the renormalization group suggests that when trying to solve this theory one should look for a phase transition: higher order interactions and nonlocalities in the interaction vertices could then be dismissed as irrelevant operators. This is why critical reggeon field theory appears to be the strongest candidate for a solution.

As the simplest attempt one could think of the following scheme of arranging the summation of all reggeon diagrams [25]. Define the two-gluon ladders - with ultraviolet and infrared modifications being built in following the ideas of [25] - to be the "bare Pomeron". It is a moving pole with intercept above one, and its slope depends upon the (unknown) confinement properties of the low-momentum region. The vertex: two reggeized gluons → four reggeized gluons then generates a triple Pomeron vertex which leads to self interactions of the bare Pomeron and brings the intercept down by some amount. By gradually including more and more of the higher order reggeon interaction vertices one then might hope to lower the (renormalized) intercept down to one (one could, of course, also end up below one). In carrying out these steps it might be helpful to make use of the conformal invariance of the interaction vertices [25]. A somewhat unsatisfactory aspect of this scheme is the lack of an argument why one should end up with the critical theory. In any case, before this program can be carried out, more information on the higher order reggeon interaction vertices is needed. Work on this line is in progress.

A much more sophisticated and ambitious program for deriving critical reggeon field theory from QCD has been formulated by A.White [30]. Strictly speaking, his ideas do not apply to QCD with the usual number of color triplet quarks, but they require the existence of two additional quarks in color sextet representations. They are expected to bind together into three color sextet pions which serve as the standard Higg's scalars in the electroweak sector. As an important consequence of this, the longitudinal components of the W's and the Z vector bosons couple to the glouns. At sufficiently high energies - most likely in the TeV region - the production of W's and Z's then start to contribute to the Pomeron (which at lower energies consisted mainly of gluons), and one expects to see some signal of this in hadron-hadron scattering. The true high energy behavior, therefore, can set in only at energies sufficiently above the onset of these effects. The main dynamical idea for obtaining for this theory the critical Pomeron is a subtle mapping of the phase of the underlying QCD onto different phases of reggeon field theory. Initially QCD (augmented by the additional quark sextets and provided with an ultraviolet cutoff) is taken to be in the spontaneously broken phase (Higg's phase), and its high energy behavior is described by the sort of reggeon field theory which I have described above. At this stage no critical behavior for the Pomeron is expected, and the total cross section might even be falling with energy. It is then only after a very careful removal of gluon mass and ultraviolet cutoff (which actually has to proceed in several steps) that the associated reggeon field theory becomes critical. It would be very helpful to have more of the details of this very interesting proposal written up, such that one could take a closer look.

What can be said to summarize this review on the theoretical work on the high energy behavior of QCD: on the whole it seems to me as if the perturbative content of the Pomeron is under control, although quite a few details of the reggeon field theory still have to be worked out. The solution of this reggeon field theory seems to me to be the major task which has to be worked on. Needless to say, that more understanding of the strong interaction confinement dynamics will be of extreme importance.

IV   The Hard Pomeron in QCD

In this section I want to review another kinematical limit which also probes the Pomeron, although from a slightly different direction. One of the motivations for including this into my review is the consistency which seems to emerge when both limits are studied within QCD: the Regge limit and the small-x limit. But there is also some interest in the small-x behavior of structure functions by itself, in particular with the HERA machine coming up soon.

I begin with a few basic features of the standard QCD description of deep-inelastic scattering [31]. The Q^2-evolution of a structure function $D_a^b(x, Q^2)$ is usually described in terms of the Altarelli-Parisi equations (the structure function $D_a^b(x, Q^2)$ expresses the probability of finding, at scale Q^2, parton b with momentum fraction x inside parton a. Parton stands for quark or gluon. The scaling variable x is defined as usual: $x = -q^2/2pq$. For small x the approximate relation holds: $1/x = s/Q^2$, where s is the energy of the system: photon + hadron). This description is completely equivalent to representing $D_a^b(x, Q^2)$ as the sum of flavor singlet QCD ladder diagrams (Fig.11). In an obvious manner, they describe the cascade of partonic decay processes, beginning with parton a at low momentum scale Q_0^2 and momentum fraction 1, and ending with the slower parton with momentum fraction x and scale Q^2. At their lower end, these diagrams have still to be linked to the valence quarks inside the hadron, and the x-distribution at the low-momentum scale Q_0^2 has to be put in by hand. The ladders then describe how this x-distribution changes as a function of Q^2. A characteristic feature of these diagrams is the ordering of the momenta:when moving along the ladder from the bottom to the top, the momentum fraction decreases from 1 down to x, whereas the virtuality of the parton (or, in the language of Sudakov variables, the square of its transverse momentum) increases from the initial low scale up to Q^2. In the region of small x, gluons generally dominate over quark contributions:in the following I will restrict myself to gluonic ladders (it should be clear that these ladders are not identical of those of the previous section).

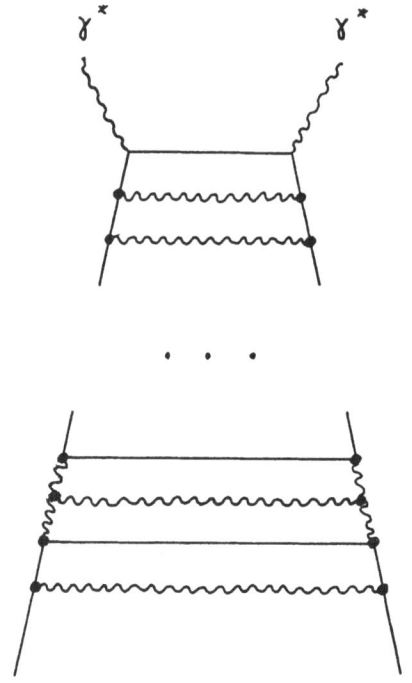

Fig.11. QCD diagrams in the small-x limit. Straight lines and wavy lines are quarks and gluons, resp.

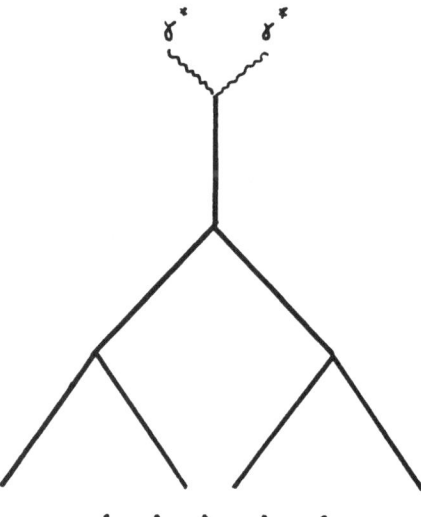

Fig.12. A few "fan diagrams". The black lines denote ladders as shown in Fig.11 (but with gluons only).

These QCD ladders, together with a suitable ansatz for the x-distribution at some low-momentum scale, provide a good description for hadronic structure functions in deep-inelastic scattering and in other hard processes, provided the variable x is not taken to be too small (there are also words to be said about the limit $x \to 1$, but this is not my topic). In the limit of very small x the gluon ladders have the following behavior [32]:

$$F(\neq 6^2) \approx \frac{\exp[\sqrt{2(\xi - \xi_0)y}]}{\sqrt{2\pi}[2(\xi - \xi_0)]^{1/4}} \qquad (17)$$

where

$$\xi = \quad ln\, lnQ^2/\Lambda^2 \ , \quad y = \frac{8N}{\beta_0}ln\frac{1}{x}$$
$$\beta_0 = \quad \frac{11}{3}N - \frac{2}{3}n_F \ , \quad N = 3 \text{ for } QCD \ . \qquad (18)$$

This growth is too strong and violates unitarity: if one uses the photon-hadron analogy (vector dominance), hadronic unitarity requires:

$$\sigma_{tot}^{\gamma} = \frac{4\pi^2 \alpha_{em}}{Q^2} F(x, Q^2) < 2\pi R_h^2(s) \ , \qquad (19)$$

with

$$R_h(s) \sim ln s \qquad (20)$$

From (17) it follows, however, that F grows faster than any power of lns, and eqs.(19), (20) are violated. To put this differently [32],the QCD ladders when coupled to a fast constituent quark describe the cascade of decays into slower and slower partons. The photon then "sees" the quark as being surrounded by a cloud of gluons and sea quarks, and as long as the clouds of the different constituent quarks are not too dense, it interacts with one at a time. This situation changes if the density increases. When the probability of an interaction between partons of different clouds becomes of order unity, the single ladder approximation no longer holds, and the interaction of several ladders has to be taken into account. To make this more quantitative, interpret $F(x, Q^2)/R_h^2$ as the density of gluons with scale Q^2 and momentum fraction x, and take for the interaction cross section of partons at scale Q^2 $\alpha_s(Q^2)/Q^2$. Then the probability for partons to interact is:

$$P(x, Q^2) \approx \frac{\alpha_s(Q^2)}{Q^2} \cdot \frac{F(x, Q^2)}{R_h^2} \ , \qquad (21)$$

This probability function should be less than one in order to justify the one-ladder approximation. With (17), however, P reaches unity at some small value x, and the approximation breaks down.

It is therefore necessary to "improve" this QCD description in the small-x region by including more diagrams. This has been worked out by the Leningrad group [32]. As it has been said above, these new contributions represent interactions between ladders evolving from different constituent quarks at low momentum. After the interaction the ladders combine and continue their further evolution in one common ladder (Fig.12). These diagrams have suggestively been named "fan diagrams". Their sum is given as a solution of an integral equation and has been discussed in [32] (see also [33]). It allows, at fixed Q^2, to extend the QCD description of structure functions further down in x, but at low x there is again a limit to its validity. In order to cross this limit two modifications have to be made. The first one concerns the QCD ladder itself: so far the QCD gluon ladders of this section differ from those of the previous one in that the gluon has not been reggeized. These effects have now to be included, and from now on the QCD ladders consist of the same diagrams as those of the Regge limit. Because of the different kinematic limits and choices of variables (x and Q^2 versus s and t), however, the analytic expressions one is dealing with are different. (As a consequence of the reggeization of the gluon lines in the QCD ladders, the small-x behavior in (17) becomes even stronger [34], and the conflict with unitarity more acute). The second modification to be applied to the fan diagrams is the removal of a restriction:

when moving from the top to the bottom, in the fan diagrams the number of QCD ladders never decreases. This restriction is now lifted, as a result of which the set of diagrams to be summed takes the form of a field theory (Fig. 13). The propagator of the "bare field is derived from the QCD ladders [32]. Since the ladder diagrams are the same as those of the Regge limit, we clearly see a convergence appearing. It is only because of the different limits and variables, that the analytic expressions obtained from ladder graphs are not the same and the rules of the field theories are different. In fact, because of this difference it has not yet been possible to find a method of carrying out calculations in this new form of a "reggeon field theory" [32]. Nevertheless, the mere fact that the perturbative analysis of both the Regge limit and the small-x limit in deep-inelastic scattering in QCD lead to the same set of diagrams presents a very important consistency check and clearly supports that we are going in the right direction. Once it has been established that both limits require the same set of diagrams, it seems inevitable that the difficulties mentioned in the context of the Regge limit will eventually also show up in the small-x limit. In particular, at some very small value x the ordering of momenta inside the QCD ladders which so far has protected us from the infrared problems becomes invalid and the difficulties of dealing with this region will appear. As a consequence, even if one could handle the new "reggeon field theory" which appears in the small-x limit, also this description would not yet provide the ultimate theory for $x \to 0$.

Next one has to answer the important question for what numerical values of x and Q^2 these various versions of QCD approximations should be used. Let me first repeat the estimate given by the inventors of this theory [32] (Fig.14). Starting from the ξ-axis below and moving upwards in y, one first is in a region where the standard ladders of the beginning of this section provide a good description. The line "1" marks the boundary of this region: above this line the first step of improvement, the fan diagrams, have to be used. The second curve "2" shows where this step of approximation fails: above this line one would have to solve the new "reggeon field theory", and eventually the infrared problems mark the end of any perturbative treatment. Line "1" is tangent to curve "2" and the point ξ_0. A numerical evaluation of these curves is shown in Fig.15: the results depend upon the QCD scale Λ and the momentum scale Q_0^2, where the QCD evolution is assumed to start. The results indicate a little inconsistency at the left end: for realistic values of Λ and Q_0^2, the boundary curves "1" and "2" intersect rather than being tangent. This implies that the expressions for the two curves presumably have to be modified for low ξ-values. Leaving this question aside and taking the third figure to be valid for larger values in Q^2, the conclusion would be that as long as x is not smaller than 10^{-2} the effects of the fan diagrams will not be seen.

Another numerical estimate has been carried out by Kwiecinski [36]. He has investigated at which values x the first fan diagram becomes nonnegligeable compared to the standard ladders. For Q^2 between 10 and 10^4 GeV^2, the first fan diagram is less than approximately 5% of the standard ladder, if x is bigger than 10^{-2}. The 10% level is reached at about $x \approx 10^{-5}$. He has also calculated the probability function $P(x, Q^2)$ (eq.(4.5)): it is well below one in all these regions. This supports the conclusions stated before: in order to test the "improvement" due to fan diagrams, x has to be as small as 10^{-4}. Experimental data so far lie in the region $x > 10^{-2}$. HERA will be the first machine which will allow to reach the interesting $x \approx 10^{-4}$.

A somewhat different analysis of Glueck et al.[37] leads to a similar conclusion. The standard QCD-ladder analysis, as it has been said before, depends upon the x-dependence which is used as an input at low momentum. In particular, the small-x behavior at higher Q^2 values will be sensitive to this input. The numerical analysis of [37] however shows that these effects will also not be seen unless x becomes smaller than 10^{-2} (see also [38]). All this makes it extremely desirable to have data for x-values as small as 10^{-4} or even less.

To conclude this part of the discussion, I would like to stress three points.(i) From the theoretical viewpoint I find it very satisfactory that both the Regge limit and the small-x limit consistently lead to the same sort of Feynman diagrams. In both cases, however, the extreme asymptotic limits ($s \to \infty$ and $x \to 0$) require nonperturbative components and are still waiting for a complete solution. (ii) The ideas of Gribov et al.[32] have not been tested yet, since existing data have not yet reached the interesting region of small x. For practical purposes it would be desirable to have manageable expressions for the sum of the fan diagrams (see, for example, [33]). (iii) HERA will be the first machine which probes the

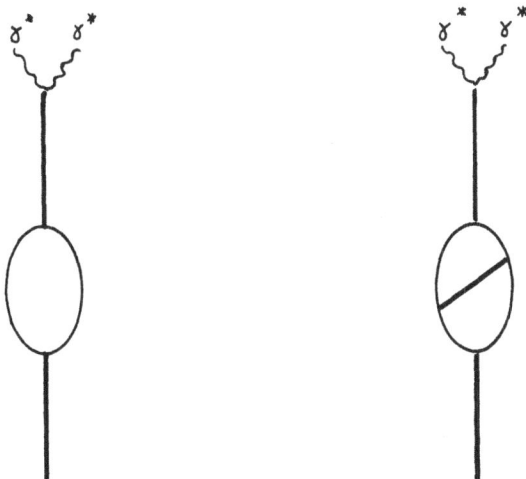

Fig.13. More diagrams have to be included if one goes to even smaller x-values.The ladders are now the same as in Fig. 8.

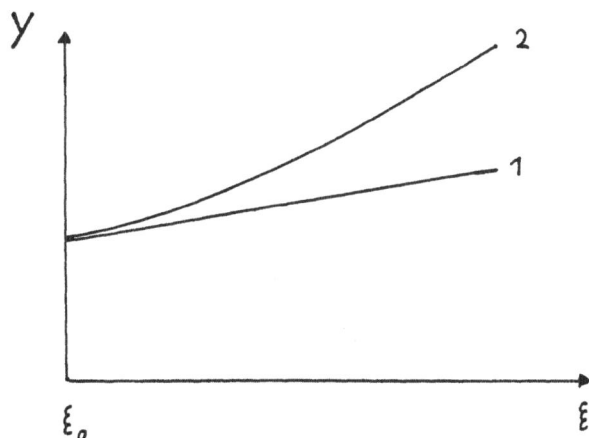

Fig.14. Boarder lines in the $\xi - y$ plane (for explanations see text).

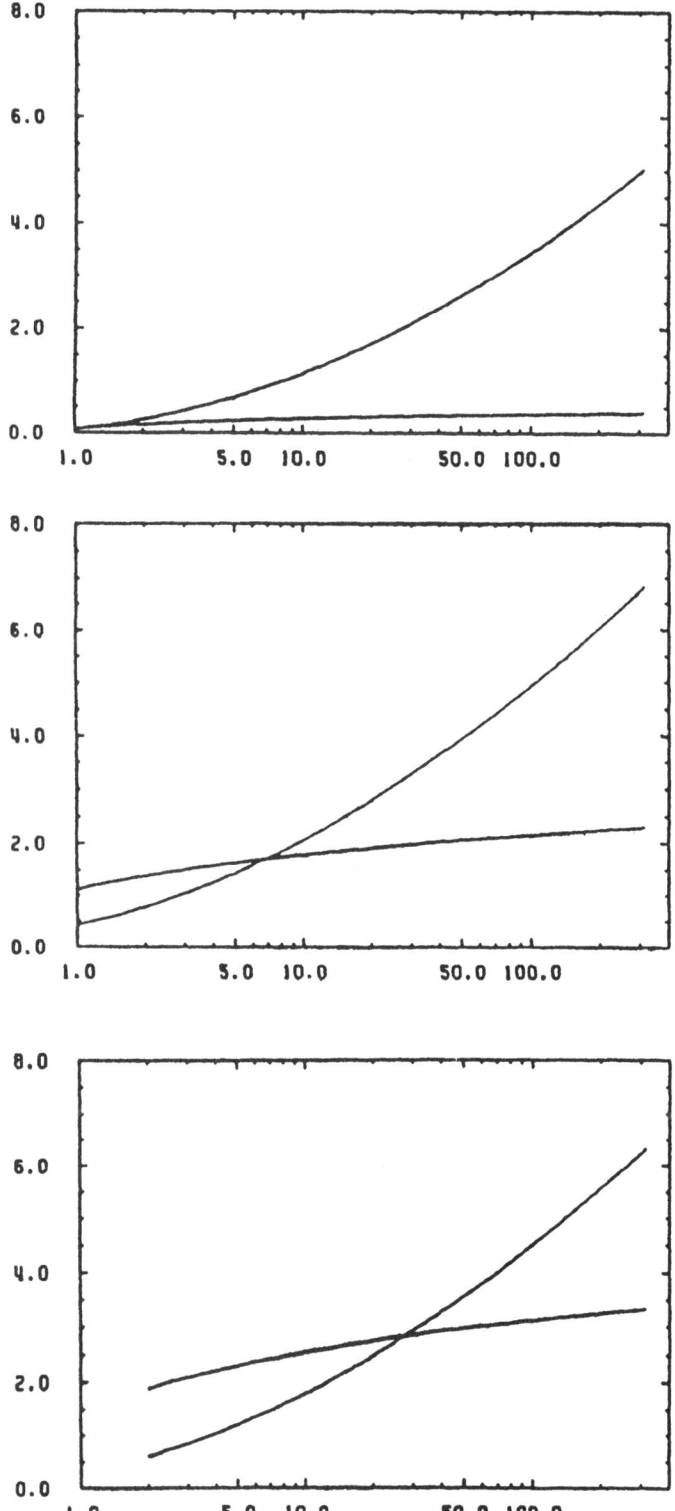

Fig.15. Numerical examples for the curves 1 and 2 of Fig.14 for different values Q_0^2 and ξ_0. (from [35]). We plot $n = -\log_{10} x$ versus $\sqrt{Q^2}$ (in GeV). The values for Q_0 and Λ are: 1 GeV and 440 MeV, 1 GeV and 200 MeV, 2 GeV and 200 MeV, resp.

part of the small-x region where the new ideas can be tested. Theorists should be prepared when HERA starts to work.

In the last part of my talk I would like to say a few word on an issue which goes under the name "Pomeron structure function". Ingelman and Schlein [39] (see also Fritzsch and Streng [40]) proposed to measure the gluon content of the Pomeron in high-mass diffractive $p\bar{p}$-scattering: the antiproton scatters elastically and emits a Pomeron (with low momentum transfer t), which combines with the proton to produce two large-p_T jets inside the high-mass Pomeron-proton system (Fig.16). Viewing the Pomeron as an "incoming particle", the two jets arise from the fusion of a gluon inside the Pomeron and a quark (or gluon) taken from the proton. The Pomeron structure function which enters the calculation of this process is, a priori, a new and unknown function: a first comparison with data [41] seems to favor a rather soft (near x=1) gluon distribution, but there is some uncertainty about the normalization. In fact, it seems to me, that the situation may be more favorable and the Pomeron structure function may be calculable within perturbative QCD (provided we stay away from any extreme kinematical limit). In the above picture it has been assumed [42],[43] that the Pomeron which fuses with the proton and then gives rise to hard processes is really the same as in hadron hadron scattering: factorization holds, and the energy dependence is the same as in the triple-Regge limit of inclusive scattering:

$$\frac{d^2\sigma_{jet\,jet}}{dt\,dM_x^2} = \frac{d^2\sigma_{SD}}{dt\,dM_x^2} \cdot \frac{\sigma_{P\,Pom \to jet\,jet}}{\sigma_{P\,Pom \to X}} \tag{22}$$

$$\frac{d^2\sigma_{SD}}{dt\,dM_x^2} = \frac{1}{16\pi}\beta_{\bar{p}}(t)^2 G_{triple\,Pom}(t)\beta_P(0)$$
$$s^{2\alpha_p(t)-1}(M_x^2)^{\alpha(0)-2\alpha_p(t)} \tag{23}$$

$$\sigma_{P\,Pom \to X} \approx const \approx 1\,mb$$

$$\sigma_{P\,Pom \to jet\,jet} = \int dx_1 dx_2 \sum_i G(x_1, Q^2) f_i(x_2, Q^2)\frac{d\hat{\sigma}_i}{d\hat{t}}$$

$$G = \text{Pomeron structure function}$$

$$f_i = \text{proton structure function}$$

This implies that the Pomeron is still soft when it meets the proton, and only from then on it starts to develop hard constituents. Based upon the QCD picture it seems more probable that the evolution of hard constituents inside the Pomeron starts much earlier, namely as soon as the Pomeron is emitted from the antiproton. To illustrate this in some more detail, it is convenient to consider this "Pomeron structure function" in deep-inelastic ep-scattering (Fig.17). The kinematical configuration analogous to that in $p\bar{p}$-scattering outlined above would be the following: in the subprocess: photon+proton → anything take those final states where a high-mass cluster is produced along the photon direction and, in rapidity, is well-separated from the proton. In other words, there is a "hole" in the final state between this cluster and the proton which goes through elastically. Such a process obviously is just the diffractive cut through the first fan diagram (Fig.17) which I have described above. The two ladders above and below the cutting line then represent the Pomeron. The assumption made in [42] that the Pomeron stays "soft" until it starts to produce the high-mass final state, translated to this QCD picture, then means that the internal momenta inside the Pomeron ladder remain small all the way up to the "triple vertex" and then start to grow in the usual fashion. It seems much more natural that the main contribution should be given by the configuration of internal momenta which I have described in the beginning of this section: the growth of transverse momenta and the decrease of momentum fraction starts right at the end of the diagrams when the Pomeron takes off from the proton. If this expectation is correct, the Pomeron structure function could be calculated (as long as we stay away from

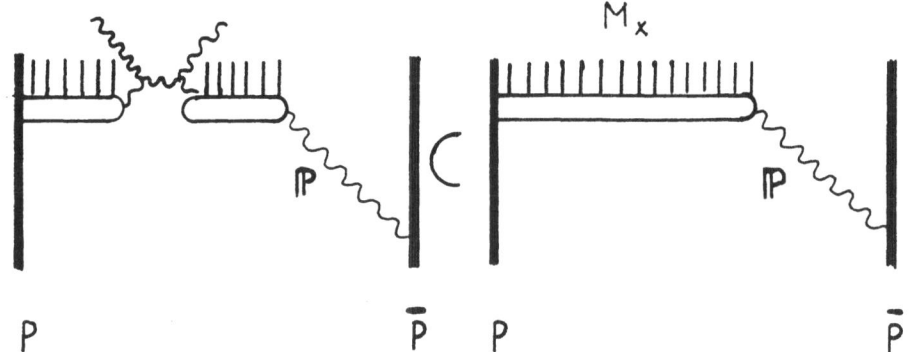

Fig.16. The Ingelman-Schlein final state configuration for extracting the Pomeron structure function.

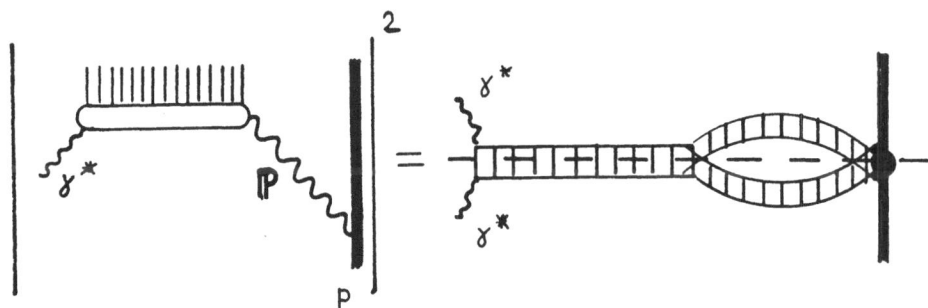

Fig.17. The Pomeron structure function in deep-inelastic scattering.

too small values of x) in QCD. In fact, all this has been discussed already in one of the later sections of [32], and it follows from this that eqs. (22),(23) have to be changed slightly. A numerical study is presently underway [44].

V Conclusion

The aim of this talk has been to give an overview of the present situation of Pomeron-related topics in the framework of QCD. As to the purely hadronic part of this, it seems to me that there the total cross section may very well exhibit some changes when higher energies are reached. The question of the true asymptotic behavior in hadron hadron scattering is, therefore,not settled. The theoretical status of the Pomeron in QCD is, in my opinion, not satisfactory. The Pomeron has its roots both in the perturbative and the nonperturbative parts of QCD, and a clear separation between both seems not possible.It is, therefore, conceivable that we first have to have a better understanding of the confinement dynamics before the Pomeron problem can be solved. In hard scattering the situation is somewhat better: the extreme small-x limit where perturbation theory breaks down is far away. Before experiments will reach this "dangerous" region, we first should test the idea of "improvement" due to the fan diagrams which will give a useful hint whether we are going in the right direction.

Acknowledgements: I would like to thank A. Ali and A. Zichichi for inviting me to this stimulating and very pleasant meeting. I also would like to express my gratitude to the Institute for Nonlinear Science (INLS) at the University of California, San Diego where most of this written version has been prepared.

References

1. M.Bozzo et al.,Phys. Lett. 147B (1984) 392.
2. U. Amaldi et al.,Phys. Lett. 66B(1977) 390.
3. A. Martin in: Proc. Fourth Topical Workshop on pp Collider Physics, Bern, 1984, p.308; C. Bourrely, A. Martin in: Proc.of the ECFA-CERN Workshop on Large Hadron Collider in the LEP Tunnel, held at Lausanne and Geneva, 1984, p.323. M.Haguenauer, G.Matthiae, ibid. p.303.
4. M.M. Block and R.N. Cahn, in "Strong Interactions and Gauge Theories," Proceedings of 21st Recontre de Moriond, Les Ares, France, 1986, edited by Tran Thanh Van (Editions Frontiére, Gif-sur-Yvette, 1986), Vol. 2, p.679.
5. D.Bernard et al.,Phys.Lett. 171B (1986) 142.
6. D.Bernard et al.,Phys.Lett. 198B (1987) 583.
7. P.M.Kluit and J.Timmermans, Phys.Lett. 202B (1988) 458.
8. For general reviews see: H.D.I. Abarbanel, J.B.Bronzan, R.L.Sugar, A.R.White, Phys.Rep. 21C (1975) 119; M.Baker, K.A.Ter-Martirosyan, Phys.Rep. 28C (1976) 3; M.Moshe, Phys. Rep 37C (1978) 255.
9. S.T.Jones, J.W.Dash, Phys.Rev. D37 (1988) 583.
10. C. Bourrely, J. Soffer and T.T.Wu, Nucl.Phys. B247 (1984), 15; Phys.Rev. Lett. 54 (1984) 757; Z. Phys. C37 (1988) 369.
11. P.Gauron, E.Leader, B.Nicolescu, Phys Rev.Lett. 54 (1985), 2656; Phys. Rev. Lett. 55 (1985) 639; preprint IPNO/TH 86-55 (1986).
12. A.Donnachie, P.V. Landshoff, Nucl. Phys. B244 (1984) 322.
13. A.Donnachie, P.V. Landshoff, Phys. Lett. 123B (1983) 345; Nucl. Phys. B231 (1983) 189;P.V.Landshoff, O.Nachtmann, Z. Phys. C 35 (1987) 405.
14. A.Capella, J.Tran Thanh Van, Z. Phys. C38 (1988) 177 and references therein.
15. J.Baumel, M. Feingold, M.Moshe, Nucl. Phys. B198 (1982) 13.
16. M.Baig, J. Bartels, J.W.Dash, Nucl. Phys. B237 (1984) 502.
17. H.Cheng and T.T. Wu, Phys. Rev. Lett. 22 (1969) 666; Phys. Rev. Lett. 23 (1969) 1311; Phys. Rev. Lett. 24 (1970) 759 and 1456; Phys. Rev. 182 (1969) 1852; Phys. Rev. D1 (1970) 2775.

18. Second reference of [10].

19. D. Bernard, P. Gauron and B. Nicolescu, Phys. Lett. 199B (1987) 125.

20. V.M. Schekhter, Sov. Journ. Nucl. Phys. 33 (1981) 426.

21. E.A. Kuraev, L.N. Lipatov, V.S. Fadin, Sov. Phys. JETP 44(1976) 443.

22. E.A. Kuraev, L.N. Lipatov, V.S. Fadin, Sov. Phys. TETP 45(1977) 199.

23. Ya.Ya. Balitzky, L.N. Lipatov, Sov. Journ. Nucl. Phys. 28(1978) 822.

24. Ya.Ya. Balitzky, L.N. Lipatov, JETP Lett.30 (1979) 355.

25. L.N. Lipatov, Sov. Phys. JETP 63 (1986) 904.

26. J. Bartels, Nucl. Phys. B151 (1979) 293; Nucl. Phys. B175(1980) 365.

27. J. Bartels, Acta Phys. Polonica B11(1980) 281; unpublished.

28. R. Kirschner and L.N. Lipatov, Sov. Phys. JETP 56(1982;) Phys. Rev. D26 (1982) 1202.

29. E. Fradkin and S.H. Shenker, Phys. Rev. D26 (1982) 3682.

30. A.R. White, talk given at the IInd International Conference on Elastic and Diffractive Scattering, Rockefeller University, New York, October 1987; preprint ANL-HEP-CP-87-120; in "Hadron Multiparticle Production," edited by P. Carruthers (World Scientific 1988), and references therein.

31. For general reviews see: C.H.Llewellyn Smith, Schladming Lectures 1978, Acta Physica Austr., Suppl. XIX (1978) 331; Yu.Dokshitzer, D.I.Dyakonov, S.I.Troyan, Phys. Rep. 58 (1980) 269;E. Reya, Phys. Rep. 69 (1981) 195; A.H. Mueller, Phys. Rep.73 (1981) 237.

32. L.V. Gribov, E.M. Levin, M.G. Ryskin, Phys. Rep. 100 (1983) 1.

33. J.P. Ralston, D.W. McKay, in 'Physics Simulations at High Energy,' Madison, Wisconsin, 1986,p.30.

34. T. Jaroscewicz, Phys. Lett. 116B (1982) 291.

35. G. Schuler, unpublished.

36. J. Kwiecinski, Z. Phys. C29 (1985) 561.

37. M. Glueck, R.M. Goodbole, E. Reya, Dortmund preprint DO-TH 88/9.

38. E. Eichten, I. Hinchliffe, K. Lane, C. Quigg, Rev. Mod. Phys. 56 (1984) 579.

39. G. Ingelman, P. Schlein, Phys. Lett. 152B (1985) 256.

40. H. Fritzsch, H.H. Streng, Phys. Lett.164B (1985) 391.

41. R. Buino et al., CERN-EP/88-60.

42. E.L. Berger, J.C. Collins, D.E.Soper, G.. Sterman, Nucl.Phvs. B286 (1987) 704.

43. A. Donnachie, P.V. Landshoff, CERN-TH.5020/88.

44. J. Bartels, G. Ingelman, in preparation.

Participants

Ahmed Ali
DESY
Notkestraße 85
D–2000, Hamburg 52, FRG

Guido Altarelli
CERN
TH–Division
CH–1211 Geneve 23, Switzerland

Fernando Barreiro
Universidad Autonoma
Dipartamento Fisica Teorica
Cantoblanco
28029 Madrid, Spain

Wulfrin Bartel
DESY
Notkestraße 85
D–2000 Hamburg 52, FRG

Jochen Bartels
II Institut f. Theoretische Physik
Universität Hamburg
Notkestraße 85
D–2000 Hamburg 52, FRG

Ulrich Baur
CERN
TH–Division
CH–1211 Geneve 23, Switzerland

Ulrich Becker
CERN
EP–Division
CH–1211 Geneve 23, Switzerland

Werner Bernreuther
CERN
TH–Division
CH–1211 Geneve 23, Switzerland

Pierre Binetruy
LAPP
IN2P3
B.P. 909
F–74019 Annecy-le-Vieux Cedex, France

Rudolph Bock
CERN
DD Division
CH–1211 Geneve 23, Switzerland

Stan Brodsky
SLAC
P.O. Box 4349
Stanford, CA 94305, USA

Graziano Bruni
INFN
Via Irnerio, 46
I–40126 Bologna, Italy

Michael Chanowitz
Lawrence Berkeley Laboratory
50 A3115
Berkeley, CA 94720, USA

Luisa Cifarelli
CERN
EP–Division
CH–1211 Geneve 23, Switzerland

David Cline
Physics Department
University of California-LA
Los Angeles, CA 90024, USA

Hubert Cobbaert
IIHE
ELEM-IIHE
Pleinlaan 2
1050 Brussel, Belgium

Daniel Coffman
Cornell University
Wilson Laboratory
Ithaca, NY 14853, USA

Bradley Cox
Fermilab
MS122
P.O. Box 500
Batavia, IL 60510, USA

Giacomo D'Ali
University of Palermo
I-90100 Palermo, Italy

Barbara de Lotto
University of Pavia
Via Bassi, 6
I-27100 Pavia, Italy

Daniel De Negri
CERN
EP–Division
CH–1211 Geneve 23, Switzerland

Salvatore De Pasquale
INFN
Via Irnerio, 46
I-40126 Bologna, Italy

Roger Dixon
Fermilab
P.O. Box 500
Batavia, IL 60510, USA

Yuri Dokshitzer
Leningrad Nuclear Physics Institute
Gatchina
Leningrad 188350, USSR

Antonio Ereditato
INFN
Pad. 20 Mostra D'oltremare
I-80125 Napoli Italy

Roberto Fiore
Department of Physics
University of Calabria
Arcavata di Rende
Rende (CO), Italy

Harald Fritzsch
University of Münich
Theresienstr. 37a
D–8000 München, FRG

Michel Gourdin
Université Pierre et Marie Curie
4, Place Jussieu
F–75230 Paris, France

David Hedin
Northern Illinois University
Physics Department—NIU
DEKALB, IL 60115, USA

Pierre Henrard
LPC
BP 45
F–63170 Aubiere, France

David Hitlin
CALTECH
356–48
Pasadena, CA 91125, USA

Giuseppe Iacobucci
INFN
Via Irnerio, 46
I-40126 Bologna, Italy

Peter Igo-Kemenes
Physics Department
University of Heidelberg
D–6900 Heidelberg 1, FRG

Karl Koller
Institut für Theorethische Physik
Universität München
Theresienstraße 37
D–8000 München 2, FRG

Jurgen Körner
Institut für Physik
Universität Mainz
P.O. Box 3980
D–6500 Mainz, FRG

Alan Krish
Physics Department
University of Michigan
Ann Arbour, MI 48109, USA

Elizabeth Locci
CEN/Saclay
F–91191 Gif-sur-Yvette Cedex, France

David London
DESY
Notkestraße 85
D–2000 Hamburg 52, FRG

Giuseppe Marchesini
Department of Physics
University of Parma
I–43100 Parma, Italy

André Martin
CERN
TH–Division
CH–1211 Geneve 23, Switzerland

Antonio Masiero
INFN
Via Marsolo, 8
35100 Padova, Italy

Bianca Monteleoni
INFN
Largo E. Fermi, 2
I–50125 Firenze, Italy

Istvan Montvay
DESY
Notkestraße 85
D–2000 Hamburg 52, FRG

Aniello Nappi
Department of Physics
University of Pisa
P.zza Torricelli, 2
I–56100 Pisa, Italy

Nello Paver
Dipartamento di Fisica Teorica
Università di Trieste
Strada Costiera, 11
34014 Miramare-Grignano (TS), Italy

Piergiovanni Pelfer
University of Florence
Largo E. Fermi, 2
I–50125 Firenze, Italy

Marcello Piccolo
LNF/INFN
C.P. 13
I–00044 Frascati (Roma), Italy

Luciano Ristori
INFN
Via Vecchia Livornese
I–56010 San Piero a Grado (Pisa), Italy

Reinhold Rückl
DESY
Notkestraße 85
D–2000 Hamburg 52, FRG

Dieter Schildknecht
Fakultät f. Physik
Universität Bielefeld
Postfach 8640
Universitätsstraße
D–4000 Bielefeld 1, FRG

Fridger Schrempp
Theoretische Physik
Universität München
Theresienstr. 37
D–8000 München 2, FRG

Henning Schröder
DESY
Notkestraße 85
D–2000 Hamburg 52, FRG

Gerhard Schuler
DESY
Notkestraße 85
D–2000 Hamburg 52, FRG

James Stirling
Physics Department
Durham University
South Road
Durham DH1 3LE, UK

Giancarlo Susinno
CERN
EP–Division
CH–1211 Geneve 23, Switzerland

Ingrid ten Have
NIKHEF–H
P.O. Box 41882
Amsterdam, The Netherlands

Daniele Treleani
Dipartamento di Fisica Teorica
Università di Trieste
Strada Costiera, 11
I–34014 Miramare-Grignano, Italy

Jochum Van der Bij
CERN
CH–1211 Geneve 23, Switzerland

Bob van Eijk
CERN
EP–Division
1211 Geneva 23, Switzerland

Richard Wigmans
CERN
EF–Division
CH–1211 Geneve 23, Switzerland

Thomas Ypsilantis
College de France
11, Place Marcelin-Berthelot
F–75231 Paris Cedex 05, France

Peter Zerwas
Institut f. Theoretische Physik
RWTH
D–5100, Aachen, FRG

Antonino Zichichi
CERN
EP–Division
1211 Geneva 23, Switzerland

INDEX